Linear Algebra and Matrix Theory

Second Edition

Jimmie Gilbert and Linda Gilbert
University of South Carolina Spartanburg

THOMSON

™

BROOKS/COLE

Australia • Canada • Mexico • Singapore • Spain
United Kingdom • United States

THOMSON

BROOKS/COLE

Acquisitions Editor: *John-Paul Ramin*
Assistant Editor: *Lisa Chow*
Editorial Assistant: *Darlene Amidon-Brent*
Project Manager, Editorial Production: *Janet Hill*
Marketing Manager: *Tom Ziolkowski*
Marketing Assistant: *Jennifer Gee*
Print/Media Buyer: *Karen Hunt*

Production Service: *Hearthside Publishing Services/Anne Seitz*
Text Designer: *John Edeen*
Illustrator: *Hearthside Publishing Services/ Jade Myers*
Cover Designer: *Matt Gilbert*
Cover Image: *Matt Gilbert*
Cover/Interior Printer: *Phoenix Color BTP*

Printed in the United States of America

1 2 3 4 5 6 7 07 06 05 04 03

For more information about our products, contact us at:
Thomson Learning Academic Resource Center
1-800-423-0563

For permission to use material from this text, contact us by:
Phone: 1-800-730-2214
Fax: 1-800-730-2215
Web: http://www.thomsonrights.com

Library of Congress Control Number: 2003117069

ISBN 0-534-40581-9

Brooks/Cole–Thomson Learning
10 Davis Drive
Belmont, CA 94002
USA

Asia
Thomson Learning
5 Shenton Way #01-01
UIC Building
Singapore 068808

Australia/New Zealand
Thomson Learning
102 Dodds Street
South Bank, Victoria 3006
Australia

Canada
Nelson
1120 Birchmount Road
Toronto, Ontario M1K 5G4
Canada

Europe/Middle East/Africa
Thomson Learning
High Holborn House
50/51 Bedford Row
London WC1R 4LR
United Kingdom

Latin America
Thomson Learning
Seneca, 53
Colonia Polanco
11560 Mexico D.F.
Mexico

Spain/Portugal
Paraninfo
Calle/Magallanes, 25
28015 Madrid, Spain

To our children:
Donna, Lisa, Martin, Dan, Beckie, and Matt

And our grandchildren:
*James, Beth, Jill, Mary,
Adrienne, Eric, Malissa, and Cole*

Preface

Intended for a serious first course or a second course, this textbook will carry students beyond eigenvalues and eigenvectors to the classification of bilinear forms, normal matrices, spectral decompositions, the Jordan form, and sequences and series of matrices. We feel that this book presents the straightest, smoothest path to the heart of linear algebra.

As the first edition was, this book is planned for use with one of the following two options:

- The complete text would be suitable for a one-year undergraduate course for mathematics majors and would provide a strong foundation for more abstract courses at higher levels of instruction.
- A one-semester or one-quarter course could be taught from the first five chapters together with selections from the remaining chapters. The selections could be chosen so as to meet the requirements of students in the fields of business, science, economics, or engineering.

Our own experience, together with feedback from users of the first edition, has influenced us to make two major changes in this second edition.

- We have slowed the pace in the early chapters, especially the first, by rewriting the material and expanding the exercise sets and the number of examples. There are four new sections in Chapters 1–10, including one on LU decompositions. Forty-six new examples and 881 new exercises have been added in these chapters.
- We have included a new Chapter 11 on numerical methods that enhances the usefulness of the text for two courses. This new chapter presents the basic material on convergence of sequences and series of vectors and matrices, iterative methods for the solution of systems of linear equations, and iterative methods for determining eigenvalues and eigenvectors.

It is our opinion that linear algebra is well suited to provide the transition from the intuitive developments of courses at a lower level to the more abstract treatments encountered later. Throughout the treatment here, material is presented from a structural point of view: Fundamental algebraic properties of the entities involved are emphasized. This approach is particularly important because the mathematical systems encountered in linear algebra furnish a wealth of examples for the structures studied in more advanced courses.

The unifying concept for the first five chapters is that of elementary operations. This concept provides the pivot for a concise and efficient development of the basic theory of vector spaces, linear transformations, matrix multiplication, and the fundamental equivalence relations on matrices.

A rigorous treatment of determinants from the traditional viewpoint is presented in Chapter 6. For a class already familiar with this material, the chapter can be omitted.

In Chapters 7 through 10, the central theme of the development is the change in the matrix representing a vector function when only certain types of basis changes are admitted. It is from this approach that the classical canonical forms for matrices are derived.

Numerous examples and exercises are provided to illustrate the theory. Exercises are included of both computational and theoretical nature. Those of a theoretical nature amplify the treatment and provide experience in constructing deductive arguments, while those of a computational nature illustrate fundamental techniques. The amount of labor in the computational problems is kept to a minimum. Even so, many of them provide opportunities to utilize current technology, if that is the wish of the instructor. Answers are provided for about half of the computational problems.

The exercises are intended to develop confidence and deepen understanding. It is assumed that students grow in maturity as they progress through the text and the proportion of theoretical problems increases in later chapters.

Acknowledgments

We wish to express our appreciation for the support given us by the University of South Carolina Spartanburg during the writing of this book, since much of the work was done while we were on sabbatical leave. We would especially like to thank James Spencer, Celia Adair, M.B. Ulmer, Judy Prince, and John Stockwell for their approval and encouragement of the project.

We would like to acknowledge with thanks the helpful suggestions made by the following reviewers of the first edition:

- Ed Dixon, Tennessee Technological University
- Jack Garner, University of Arkansas at Little Rock
- Edward Hinson, University of New Hampshire
- Melvyn Jeter, Illinois Wesleyan University
- Bob Jones, Belhaven College

In addition, we thank the following people for their reviews of this edition:

- Melvyn Jeter, Illinois Wesleyan University
- John McMath, Mississippi College
- Rama Rao, University of North Florida
- Wasin So, San Jose State University

We also wish to extend our sincere appreciation to Bob Pirtle for initiating the project at Brooks/Cole, to John-Paul Ramin, Darlene Amidon-Brent, Janet Hill, Anne Seitz, Lisa Chow, Vernon Boes, and Leslie Galen for their supervision of the most efficient production process that we have ever experienced. Finally, our thanks go to Eric T. Howe and Ian Crewe for their accuracy checking of the text.

Jimmie Gilbert
Linda Gilbert

Contents

1 Real Coordinate Spaces   1

1.1 The Vector Spaces $\mathbf{R}^n$ 1
1.2 Linear Independence   8
1.3 Subspaces of $\mathbf{R}^n$ 12
1.4 Spanning Sets   17
1.5 Geometric Interpretations of $\mathbf{R}^2$ and $\mathbf{R}^3$ 23
1.6 Bases and Dimension   33

2 Elementary Operations on Vectors   47

2.1 Elementary Operations and Their Inverses   47
2.2 Elementary Operations and Linear Independence   54
2.3 Standard Bases for Subspaces   57

3 Matrix Multiplication   67

3.1 Matrices of Transition   67
3.2 Properties of Matrix Multiplication   75
3.3 Invertible Matrices   83
3.4 Column Operations and Column-Echelon Forms   94
3.5 Row Operations and Row-Echelon Forms   104
3.6 Row and Column Equivalence   115
3.7 Rank and Equivalence   124
3.8 *LU* Decompositions   133

4 Vector Spaces, Matrices, and Linear Equations   141

4.1 Vector Spaces   141
4.2 Subspaces and Related Concepts   152
4.3 Isomorphisms of Vector Spaces   159
4.4 Standard Bases for Subspaces   165
4.5 Matrices over an Arbitrary Field   171
4.6 Systems of Linear Equations   171
4.7 More on Systems of Linear Equations   179

5   Linear Transformations   189

5.1   Linear Transformations   189
5.2   Linear Transformations and Matrices   202
5.3   Change of Basis   217
5.4   Composition of Linear Transformations   229

6   Determinants   239

6.1   Permutations and Indices   239
6.2   The Definition of a Determinant   243
6.3   Cofactor Expansions   248
6.4   Elementary Operations and Cramer's Rule   257
6.5   Determinants and Matrix Multiplication   264

7   Eigenvalues and Eigenvectors   271

7.1   Eigenvalues and Eigenvectors   271
7.2   Eigenspaces and Similarity   282
7.3   Representation by a Diagonal Matrix   291

8   Functions of Vectors   301

8.1   Linear Functionals   301
8.2   Real Quadratic Forms   310
8.3   Orthogonal Matrices   317
8.4   Reduction of Real Quadratic Forms   323
8.5   Classification of Real Quadratic Forms   331
8.6   Bilinear Forms   339
8.7   Symmetric Bilinear Forms   346
8.8   Hermitian Forms   353

9   Inner Product Spaces   365

9.1   Inner Products   365
9.2   Norms and Distances   370
9.3   Orthonormal Bases   376
9.4   Orthogonal Complements   380
9.5   Isometries   384
9.6   Normal Matrices   389
9.7   Normal Linear Operators   395

10   Spectral Decompositions   403

10.1 Projections and Direct Sums   403
10.2 Spectral Decompositions   409
10.3 Minimal Polynomials and Spectral Decompositions   415
10.4 Nilpotent Transformations   428
10.5 The Jordan Canonical Form   436

11   Numerical Methods   445

11.1 Sequences and Series of Vectors   445
11.2 Sequences and Series of Matrices   452
11.3 The Standard Method of Iteration   463
11.4 Cimmino's Method   466
11.5 An Iterative Method for Determining Eigenvalues   472

Answers to Selected Exercises   481

Index   515

1 Real Coordinate Spaces

There are various approaches to that part of mathematics known as *linear algebra*. Different approaches emphasize different aspects of the subject such as matrices, applications, or computational methods. As presented in this text, linear algebra is in essence a study of vector spaces, and this study of vector spaces is primarily devoted to finite-dimensional vector spaces. The real coordinate spaces, in addition to being important in many applications, furnish excellent intuitive models of abstract finite-dimensional vector spaces. For these reasons, we begin our study of linear algebra with a study of the real coordinate spaces. Later it will be found that many of the results and techniques employed here will easily generalize to more abstract settings.

1.1 The Vector Spaces R^n

Throughout this text the symbol **R** will denote the set of all real numbers. We assume a knowledge of the algebraic properties of **R**, and begin with the following definition of real coordinate spaces.

1.1 DEFINITION

For each positive integer n, $\mathbf{R}^n$ will denote the set of all ordered n-tuples $(u_1, u_2, \ldots, u_n)$ of real numbers u_i. Two n-tuples $(u_1, u_2, \ldots, u_n)$ and $(v_1, v_2, \ldots, v_n)$ are **equal** if and only if $u_i = v_i$ for $i = 1, 2, \ldots, n$. The set $\mathbf{R}^n$ is referred to as an **n-dimensional real coordinate space**. The elements of $\mathbf{R}^n$ are called **n-dimensional real coordinate vectors**, or simply **vectors**. The numbers u_i in a vector $(u_1, u_2, \ldots, u_n)$ will be called the **components** of the vector. The elements of **R** will be referred to as **scalars**.[1]

The real coordinate spaces and the related terminology described in this definition are easily seen to be generalizations and extensions of the two- and three-dimensional vector spaces studied in the calculus.

When we use a single letter to represent a vector, the letter will be printed in boldface lower case Roman, such as **v**, or written with an arrow over it, such as $\vec{v}$. In handwritten work with vectors, the arrow notation $\vec{v}$ is commonly used. Scalars will be represented by letters printed in lower case italics.

---

[1] The terms "vector" and "scalar" are later extended to more general usage, but this will cause no confusion since the context will make the meaning clear.

**1.2
DEFINITION**

Addition in $\mathbf{R}^n$ is defined as follows: for any $\mathbf{u} = (u_1, u_2, \ldots, u_n)$ and $\mathbf{v} = (v_1, v_2, \ldots, v_n)$ in $\mathbf{R}^n$, the sum $\mathbf{u} + \mathbf{v}$ is given by

$$\mathbf{u} + \mathbf{v} = (u_1 + v_1, u_2 + v_2, \ldots, u_n + v_n).$$

For any scalar a and any vector $\mathbf{u} = (u_1, u_2, \ldots, u_n)$ in $\mathbf{R}^n$, the **product** $a\mathbf{u}$ is defined by

$$a\mathbf{u} = (au_1, au_2, \ldots, au_n).$$

The operation that combines the scalar a and the vector $\mathbf{u}$ to yield $a\mathbf{u}$ is referred to as **multiplication of the vector u by the scalar** a, or simply as **scalar multiplication**. Also, the product $a\mathbf{u}$ is called a **scalar multiple** of $\mathbf{u}$.

The following theorem gives the basic properties of the two operations that we have defined.

**1.3
THEOREM**

The following properties are valid for any scalars a and b, and any vectors $\mathbf{u}, \mathbf{v}, \mathbf{w}$ in $\mathbf{R}^n$:

1. $\mathbf{u} + \mathbf{v} \in \mathbf{R}^n$. (Closure under addition)
2. $(\mathbf{u} + \mathbf{v}) + \mathbf{w} = \mathbf{u} + (\mathbf{v} + \mathbf{w})$. (Associative property of addition)
3. There is a vector $\mathbf{0}$ in $\mathbf{R}^n$ such that $\mathbf{u} + \mathbf{0} = \mathbf{u}$ for all $\mathbf{u} \in \mathbf{R}^n$. (Additive identity)
4. For each $\mathbf{u} \in \mathbf{R}^n$, there is a vector $-\mathbf{u}$ in $\mathbf{R}^n$ such that $\mathbf{u} + (-\mathbf{u}) = \mathbf{0}$. (Additive inverses)
5. $\mathbf{u} + \mathbf{v} = \mathbf{v} + \mathbf{u}$. (Commutative property of addition)
6. $a\mathbf{u} \in \mathbf{R}^n$. (Absorption under scalar multiplication)
7. $a(b\mathbf{u}) = (ab)\mathbf{u}$. (Associative property of scalar multiplication)
8. $a(\mathbf{u} + \mathbf{v}) = a\mathbf{u} + a\mathbf{v}$. (Distributive property, vector addition)
9. $(a + b)\mathbf{u} = a\mathbf{u} + b\mathbf{u}$. (Distributive property, scalar addition)
10. $1 \cdot \mathbf{u} = \mathbf{u}$.

The proofs of these properties are easily carried out using the definitions of vector addition and scalar multiplication, as well as the properties of real numbers. As typical examples, properties 3, 4, and 8 will be proved here. The remaining proofs are left as an exercise.

Proof of Property 3   The vector $\mathbf{0} = (0, 0, \ldots, 0)$ is in $\mathbf{R}^n$, and if $\mathbf{u} = (u_1, u_2, \ldots, u_n)$,

$$\mathbf{u} + \mathbf{0} = (u_1 + 0, u_2 + 0, \ldots, u_n + 0) = \mathbf{u}. \qquad \blacksquare\blacksquare\blacksquare$$

Proof of Property 4 If $\mathbf{u} = (u_1, u_2, \ldots, u_n) \in \mathbf{R}^n$, then $\mathbf{v} = (-u_1, -u_2, \ldots, -u_n)$ is in $\mathbf{R}^n$ since all real numbers have additive inverses. And since

$$\mathbf{u} + \mathbf{v} = (u_1 + (-u_1), u_2 + (-u_2), \ldots, u_n + (-u_n)) = \mathbf{0},$$

$\mathbf{v}$ is the vector $-\mathbf{u}$ as required by property (4).   ■ ■ ■

Proof of Property 8 Let $\mathbf{u} = (u_1, u_2, \ldots, u_n)$ and $\mathbf{v} = (v_1, v_2, \ldots, v_n)$. Then

$$
\begin{aligned}
a(\mathbf{u} + \mathbf{v}) &= a(u_1 + v_1, u_2 + v_2, \ldots, u_n + v_n) \\
&= (a(u_1 + v_1), a(u_2 + v_2), \ldots, a(u_n + v_n)) \\
&= (au_1 + av_1, au_2 + av_2, \ldots, au_n + av_n) \\
&= (au_1, au_2, \ldots, au_n) + (av_1, av_2, \ldots, av_n) \\
&= a\mathbf{u} + a\mathbf{v}.
\end{aligned}
$$
   ■ ■ ■

The associative property of addition in $\mathbf{R}^n$ can be generalized so that the terms in a sum such as $a_1\mathbf{u}_1 + a_2\mathbf{u}_2 + \cdots + a_k\mathbf{u}_k$ can be grouped with parentheses in any way without changing the value of the result. Such a sum can be written in the compact form $\sum_{i=1}^{k} a_i\mathbf{u}_i$ or $\sum_i a_i\mathbf{u}_i$ if the number of terms in the sum is not important. It is always understood that only a finite number of nonzero terms is involved in such a sum.

1.4 DEFINITION

Let $\mathcal{A}$ be a nonempty set of vectors in $\mathbf{R}^n$. A vector $\mathbf{v}$ in $\mathbf{R}^n$ is **linearly dependent** on the set $\mathcal{A}$ if there exist vectors $\mathbf{u}_1, \mathbf{u}_2, \ldots, \mathbf{u}_k$ in $\mathcal{A}$ and scalars $a_1, a_2, \ldots, a_k$ such that $\mathbf{v} = \sum_{i=1}^{k} a_i\mathbf{u}_i$. A vector of the form $\sum_{i=1}^{k} a_i\mathbf{u}_i$ is called a **linear combination** of the $\mathbf{u}_i$.

If linear dependence on a finite set $\mathcal{B} = \{\mathbf{v}_1, \mathbf{v}_2, \ldots, \mathbf{v}_r\}$ is under consideration, the statement in Definition 1.4 is equivalent to requiring that *all* of the vectors in $\mathcal{B}$ be involved in the linear combination. That is, a vector $\mathbf{v}$ is linearly dependent on $\mathcal{B} = \{\mathbf{v}_1, \mathbf{v}_2, \ldots, \mathbf{v}_r\}$ if and only if there are scalars $b_1, b_2, \ldots, b_r$ such that $\mathbf{v} = \sum_{i=1}^{r} b_i\mathbf{v}_i$.

EXAMPLE 1 As an illustration, let $\mathcal{B} = \{\mathbf{v}_1, \mathbf{v}_2, \mathbf{v}_3\}$, where

$$\mathbf{v}_1 = (1, 0, 2, 1), \mathbf{v}_2 = (0, -2, 2, 3), \quad \text{and} \quad \mathbf{v}_3 = (1, -2, 4, 4),$$

and let $\mathbf{v} = (3, -4, 10, 9)$. Now $\mathbf{v}$ can be written as

$$\mathbf{v} = 1 \cdot \mathbf{v}_1 + 0 \cdot \mathbf{v}_2 + 2 \cdot \mathbf{v}_3,$$

or

$$\mathbf{v} = 3 \cdot \mathbf{v}_1 + 2 \cdot \mathbf{v}_2 + 0 \cdot \mathbf{v}_3.$$

Either of these combinations shows that $\mathbf{v}$ is linearly dependent on $\mathcal{B}$.

To consider a situation involving an infinite set, let $\mathcal{A}$ be the set of all vectors in $\mathbf{R}^3$ that have integral components, and let $\mathbf{v} = (\sqrt{2}, \frac{2}{3}, 0)$. Now $\mathbf{u}_1 = (1, 0, 1)$, $\mathbf{u}_2 = (0, 1, 0)$, and $\mathbf{u}_3 = (0, 0, 1)$ are in $\mathcal{A}$, and

$$\mathbf{v} = \sqrt{2}\,\mathbf{u}_1 + \tfrac{2}{3}\,\mathbf{u}_2 - \sqrt{2}\,\mathbf{u}_3.$$

Thus $\mathbf{v}$ is linearly dependent on $\mathcal{A}$. It should be noted that other choices of vectors $\mathbf{u}_i$ can be made in order to exhibit this dependence.   ■

Other types of dependence are considered in some situations, but linear dependence is the only type that we will use. For this reason, we will frequently use the term "dependent" in place of "linearly dependent."

In order to decide whether a certain vector is linearly dependent on a given set in $\mathbf{R}^n$, it is usually necessary to solve a system of equations. This is illustrated in the following examples.

EXAMPLE 2   Consider the question as to whether $(6, 0, -1)$ is linearly dependent on the set $\mathcal{A} = \{(2, -1, 1), (0, 1, -1), (-2, 1, 0)\}$. To answer the question, we investigate the conditions on a_1, a_2, and a_3 that are required by the equation

$$a_1(2, -1, 1) + a_2(0, 1, -1) + a_3(-2, 1, 0) = (6, 0, -1).$$

Performing the indicated scalar multiplications and additions in the left member of this equation leads to

$$(2a_1 - 2a_3, -a_1 + a_2 + a_3, a_1 - a_2) = (6, 0, -1).$$

This vector equation is equivalent to the system of equations

$$
\begin{aligned}
2a_1 \qquad\quad - 2a_3 &= 6 \\
-a_1 + a_2 + a_3 &= 0 \\
a_1 - a_2 \qquad\quad &= -1.
\end{aligned}
$$

We decide to work toward the solution of this system by eliminating a_1 from two of the equations in the system. As steps toward this goal, we multiply the first equation by $\frac{1}{2}$ and we add the second equation to the third equation. These steps yield the system

$$
\begin{aligned}
a_1 \qquad\quad - a_3 &= 3 \\
-a_1 + a_2 + a_3 &= 0 \\
a_3 &= -1.
\end{aligned}
$$

Adding the first equation to the second now results in

$$
\begin{aligned}
a_1 \quad - a_3 &= 3 \\
a_2 \qquad\quad &= 3 \\
a_3 &= -1.
\end{aligned}
$$

The solution $a_1 = 2$, $a_2 = 3$, $a_3 = -1$ is now readily obtained. Thus the vector $(6, 0, -1)$ is linearly dependent on the set $\mathcal{A}$. ∎

EXAMPLE 3 In order to show that $(3, 0, 0, 1)$ is dependent on the set

$$\mathcal{A} = \{(1, 0, 1, 0), (0, 2, 2, 0)\},$$

we must exhibit scalars a_1 and a_2 such that

$$a_1(1, 0, 1, 0) + a_2(0, 2, 2, 0) = (3, 0, 0, 1)$$

or

$$(a_1, 0, a_1, 0) + (0, 2a_2, 2a_2, 0) = (3, 0, 0, 1).$$

This vector equation leads to the system of equations

$$
\begin{aligned}
a_1 \quad\quad\quad &= 3 \\
2a_2 &= 0 \\
a_1 + 2a_2 &= 0 \\
0 &= 1
\end{aligned}
$$

which clearly has no solution. Thus the vector $(3, 0, 0, 1)$ is not linearly dependent on $\mathcal{A}$ and we say that $(3, 0, 0, 1)$ is independent of the set $\mathcal{A}$. ∎

In the last section of this chapter, the following definition and theorem are of primary importance. Both are natural extensions of Definition 1.4.

1.5 DEFINITION	A set $\mathcal{A} \subseteq \mathbf{R}^n$ is linearly dependent on the set $\mathcal{B} \subseteq \mathbf{R}^n$ if each $\mathbf{u} \in \mathcal{A}$ is linearly dependent on $\mathcal{B}$.

Thus $\mathcal{A}$ is linearly dependent on $\mathcal{B}$ if and only if each vector in $\mathcal{A}$ is a linear combination of vectors that are contained in $\mathcal{B}$.

EXAMPLE 4 The set $\mathcal{A} = \{(2, 3, 1), (1, 1, 0)\}$ is linearly dependent on $\mathcal{B} = \{(1, 1, 1), (0, 1, 1), (0, 0, 1)\}$ since

$$(2, 3, 1) = 2(1, 1, 1) + 1(0, 1, 1) - 2(0, 0, 1)$$

and

$$(1, 1, 0) = 1(1, 1, 1) + 0(0, 1, 1) - 1(0, 0, 1).$$ ∎

In the proof of the next theorem, it is convenient to have available the notation $\sum_{i=1}^{k}\left(\sum_{j=1}^{m}\mathbf{u}_{ij}\right)$ for the sum of the form $\sum_{j=1}^{m}\mathbf{u}_{1j} + \sum_{j=1}^{m}\mathbf{u}_{2j} + \cdots + \sum_{j=1}^{m}\mathbf{u}_{kj}$. The associative and commutative properties for vector addition [Theorem 1.3, (2) and (5)] imply that

$$\sum_{i=1}^{k}\left(\sum_{j=1}^{m}\mathbf{u}_{ij}\right) = \sum_{j=1}^{m}\left(\sum_{i=1}^{k}\mathbf{u}_{ij}\right).$$

1.6 THEOREM

Let $\mathcal{A}$, $\mathcal{B}$, and $\mathcal{C}$ be subsets of $\mathbf{R}^n$. If $\mathcal{A}$ is dependent on $\mathcal{B}$ and $\mathcal{B}$ is dependent on $\mathcal{C}$, then $\mathcal{A}$ is dependent on $\mathcal{C}$.

Proof Suppose that $\mathcal{A}$ is dependent on $\mathcal{B}$ and $\mathcal{B}$ is dependent on $\mathcal{C}$, and let $\mathbf{u}$ be an arbitrary vector in $\mathcal{A}$. Since $\mathcal{A}$ is dependent on $\mathcal{B}$, there are vectors $\mathbf{v}_1, \mathbf{v}_2, \ldots, \mathbf{v}_m$ in $\mathcal{B}$ and scalars $a_1, a_2, \ldots, a_m$ such that

$$\mathbf{u} = \sum_{j=1}^{m} a_j \mathbf{v}_j.$$

Since $\mathcal{B}$ is dependent on $\mathcal{C}$, each $\mathbf{v}_j$ can be written as a linear combination of certain vectors in $\mathcal{C}$. In general, for different vectors $\mathbf{v}_j$, different sets of vectors from $\mathcal{C}$ would be involved in these linear combinations. But each of these m linear combinations (one for each $\mathbf{v}_j$) would involve only a finite number of vectors. Hence the set of all vectors in $\mathcal{C}$ that appear in a term of at least one of these linear combinations is a finite set, say $\{\mathbf{w}_1, \mathbf{w}_2, \ldots, \mathbf{w}_k\}$, and each $\mathbf{v}_j$ can be written in the form $\mathbf{v}_j = \sum_{i=1}^{k} b_{ij}\mathbf{w}_i$. Replacing the $\mathbf{v}_j$'s in the above expression for $\mathbf{u}$ by these linear combinations, we obtain

$$\mathbf{u} = \sum_{j=1}^{m} a_j \mathbf{v}_j$$

$$= \sum_{j=1}^{m} a_j \left(\sum_{i=1}^{k} b_{ij}\mathbf{w}_i\right)$$

$$= \sum_{j=1}^{m}\left(\sum_{i=1}^{k} a_j b_{ij}\mathbf{w}_i\right)$$

$$= \sum_{i=1}^{k}\left(\sum_{j=1}^{m} a_j b_{ij}\mathbf{w}_i\right)$$

$$= \sum_{i=1}^{k}\left(\sum_{j=1}^{m} a_j b_{ij}\right)\mathbf{w}_i.$$

Letting $c_i = \sum_{j=1}^{m} a_j b_{ij}$ for $i = 1, 2, \ldots, k$, we have $\mathbf{u} = \sum_{i=1}^{k} c_i \mathbf{w}_i$. Thus $\mathbf{u}$ is dependent on $\mathcal{C}$.

Since $\mathbf{u}$ was arbitrarily chosen in $\mathcal{A}$, $\mathcal{A}$ is dependent on $\mathcal{C}$ and the theorem is proved.

■■■

1.1 Exercises

1. Prove the remaining parts of Theorem 1.3.

2. Assuming the properties stated in Theorem 1.3, prove the following statements.

   (a) The vector $\mathbf{0}$ in property (3) is unique.
   (b) $c\mathbf{u} = \mathbf{0}$ if and only if either $c = 0$ or $\mathbf{u} = \mathbf{0}$.
   (c) The vector $-\mathbf{u}$ in $\mathbf{R}^n$ which satisfies $\mathbf{u} + (-\mathbf{u}) = \mathbf{0}$ is unique.
   (d) $-\mathbf{u} = (-1)\mathbf{u}$.
   (e) The vector $\mathbf{u} - \mathbf{v}$ is by definition the vector $\mathbf{w}$ such that $\mathbf{v} + \mathbf{w} = \mathbf{u}$. Prove that $\mathbf{u} - \mathbf{v} = \mathbf{u} + (-1)\mathbf{v}$.

3. In each case, determine whether or not the given vector $\mathbf{v}$ is linearly dependent on the given set $\mathcal{A}$. If $\mathbf{v}$ is linearly dependent on $\mathcal{A}$, write $\mathbf{v}$ as a linear combination of the vectors in $\mathcal{A}$.

   (a) $\mathbf{v} = (-2, 1, 4)$, $\mathcal{A} = \{(1, 1, 1), (0, 1, 1), (0, 0, 1)\}$
   (b) $\mathbf{v} = (-4, 4, 2)$, $\mathcal{A} = \{(2, 1, -3), (1, -1, 3)\}$
   (c) $\mathbf{v} = (1, 2, 1)$, $\mathcal{A} = \{(1, 0, -2), (0, 2, 1), (1, 2, -1), (-1, 2, 3)\}$
   (d) $\mathbf{v} = (2, 13, -5)$, $\mathcal{A} = \{(1, 2, -1), (3, 6, -3), (-1, 1, 0), (0, 6, -2), (2, 4, -2)\}$
   (e) $\mathbf{v} = (0, 1, 2, 0)$, $\mathcal{A} = \{(1, 0, -1, 1), (-2, 0, 2, -2), (1, 1, 1, 1), (2, 1, 0, 2)\}$
   (f) $\mathbf{v} = (2, 3, 5, 5)$, $\mathcal{A} = \{(0, 1, 1, 1), (1, 0, 1, 1), (1, -1, 0, 0), (1, 1, 2, 2)\}$

4. Show that any vector in $\mathbf{R}^3$ is dependent on the set $\{\mathbf{e}_1, \mathbf{e}_2, \mathbf{e}_3\}$ where $\mathbf{e}_1 = (1, 0, 0)$, $\mathbf{e}_2 = (0, 1, 0)$, $\mathbf{e}_3 = (0, 0, 1)$.

5. Show that every vector in $\mathbf{R}^3$ is dependent on the set $\{\mathbf{u}_1, \mathbf{u}_2, \mathbf{u}_3\}$ where $\mathbf{u}_1 = (1, 0, 0)$, $\mathbf{u}_2 = (1, 1, 0)$, and $\mathbf{u}_3 = (1, 1, 1)$.

6. Show that any vector in $\mathbf{R}^4$ is dependent on the set $\{\mathbf{e}_1, \mathbf{e}_2, \mathbf{e}_3, \mathbf{e}_4\}$ where $\mathbf{e}_1 = (1, 0, 0, 0)$, $\mathbf{e}_2 = (0, 1, 0, 0)$, $\mathbf{e}_3 = (0, 0, 1, 0)$, $\mathbf{e}_4 = (0, 0, 0, 1)$.

7. Show that any vector in $\mathbf{R}^4$ is dependent on the set $\{\mathbf{u}_1, \mathbf{u}_2, \mathbf{u}_3, \mathbf{u}_4\}$ where $\mathbf{u}_1 = (1, 1, 0, 0)$, $\mathbf{u}_2 = (0, 1, 1, 0)$, $\mathbf{u}_3 = (0, 0, 1, 1)$, $\mathbf{u}_4 = (0, 0, 0, 1)$.

8. Determine whether or not the given set $\mathcal{A}$ is dependent on the set $\mathcal{B}$.

   (a) $\mathcal{A} = \{(1, 1, 1), (2, 1, 0)\}$, $\mathcal{B} = \{(0, 1, 0), (1, 0, 1), (1, 1, 0)\}$
   (b) $\mathcal{A} = \{(0, 3, 3), (6, 4, 4), (1, 1, 1)\}$, $\mathcal{B} = \{(1, 0, 0), (0, 1, 1), (1, 1, 1)\}$
   (c) $\mathcal{A} = \{(1, 0, 1), (-1, 0, 1)\}$, $\mathcal{B} = \{(2, 2, 0), (2, 0, 0)\}$
   (d) $\mathcal{A} = \{(0, 1, 2), (0, 0, 3)\}$, $\mathcal{B} = \{(-1, 1, 1), (2, -1, 0)\}$

9. Is $\mathcal{A} = \{\mathbf{u}_1 - \mathbf{u}_2, \mathbf{u}_2 - \mathbf{u}_3\}$ linearly dependent on $\mathcal{B} = \{\mathbf{u}_1, \mathbf{u}_2, \mathbf{u}_3\}$?

10. Is $\mathcal{A} = \{\mathbf{u}_1, \mathbf{u}_2, \mathbf{u}_3\}$, linearly dependent on $\mathcal{B} = \{\mathbf{u}_1, \mathbf{u}_1 + \mathbf{u}_2, \mathbf{u}_3 - \mathbf{u}_2\}$?

11. Prove or disprove: The set $\mathcal{A}$ is linearly dependent on the set $\mathcal{B}$ implies that the set $\mathcal{B}$ is linearly dependent on the set $\mathcal{A}$.

12. Prove or disprove: The set $\mathcal{A}$ is linearly dependent on the set $\mathcal{B}$ and the set $\mathcal{A}$ is linearly dependent on the set $\mathcal{C}$ implies that the set $\mathcal{B}$ is linearly dependent on the set $\mathcal{C}$.

13. Prove that the zero vector is linearly dependent on the set $\mathcal{A}$ where $\mathcal{A}$ is any nonempty set of vectors in $\mathbf{R}^n$.

14. For any pair of positive integers i and j, the symbol δ_{ij} is defined by $\delta_{ij} = 0$ if $i \neq j$ and $\delta_{ij} = 1$ if $i = j$. This symbol is known as the **Kronecker delta**.

(a) Find the value of $\sum_{i=1}^{5} \left(\sum_{j=1}^{4} (-1)^{\delta_{ij}} \right)$.

(b) Find the value of $\sum_{i=1}^{n} \left(\sum_{j=1}^{n} (1 - \delta_{ij}) \right)$.

(c) Find the value of $\sum_{i=1}^{n} \left(\sum_{j=1}^{m} \delta_{ij} \right)$.

(d) Find the value of $\sum_{j=1}^{n} \delta_{ij} \delta_{jk}$.

1.2 Linear Independence

Another important type of dependence is given in the definition below. This time, the phrase *linearly dependent* involves only a set instead of involving both a vector and a set.

1.7 **DEFINITION**	A set $\mathcal{A}$ of vectors in $\mathbf{R}^n$ is **linearly dependent** if there is a collection of vectors $\mathbf{u}_1, \mathbf{u}_2, \ldots, \mathbf{u}_k$ in $\mathcal{A}$ and scalars $c_1, c_2, \ldots, c_k$, not all of which are zero, such that $c_1\mathbf{u}_1 + c_2\mathbf{u}_2 + \cdots + c_k\mathbf{u}_k = \mathbf{0}$. If no such collection of vectors exists in $\mathcal{A}$, then $\mathcal{A}$ is called **linearly independent**.

Again, the case involving a finite set of vectors is of special interest. It is readily seen that a finite set $\mathcal{B} = \{\mathbf{v}_1, \mathbf{v}_2, \ldots, \mathbf{v}_r\}$ is linearly dependent if and only if there are scalars $b_1, b_2, \ldots, b_r$, not all zero, such that $\sum_{i=1}^{r} b_i \mathbf{v}_i = \mathbf{0}$.

EXAMPLE 1   The set $\mathcal{B}$ in Example 1 of Section 1.1 is a finite set of vectors that is linearly dependent since

$$\mathbf{v}_1 + \mathbf{v}_2 - \mathbf{v}_3 = (1, 0, 2, 1) + (0, -2, 2, 3) - (1, -2, 4, 4) = \mathbf{0}.$$

Also the set $\mathcal{A}$ of all vectors in $\mathbf{R}^3$ with integral components is an infinite set of vectors that is linearly dependent. For if $\mathbf{u}_1 = (1, 0, 1)$, $\mathbf{u}_2 = (0, 1, 0)$, and $\mathbf{u}_3 = (0, 0, 1)$ as in Example 1 of Section 1.1 and $\mathbf{u}_4 = (-3, -2, -4)$, then

$$3\mathbf{u}_1 + 2\mathbf{u}_2 + \mathbf{u}_3 + \mathbf{u}_4 = \mathbf{0}. \qquad \blacksquare$$

To determine the linear dependence or linear independence of a set $\{\mathbf{u}_1, \mathbf{u}_2, \ldots, \mathbf{u}_k\}$ of vectors in $\mathbf{R}^n$, it is necessary to investigate the conditions on the c_i which are imposed by requiring that $\sum_{j=1}^{k} c_j \mathbf{u}_j = \mathbf{0}$. If $\mathbf{u}_j = (u_{1j}, u_{2j}, \ldots, u_{nj})$, we have

$$\sum_{j=1}^{k} c_j \mathbf{u}_j = \sum_{j=1}^{k} c_j (u_{1j}, u_{2j}, \ldots, u_{nj}) = \left(\sum_{j=1}^{k} c_j u_{1j}, \sum_{j=1}^{k} c_j u_{2j}, \ldots, \sum_{j=1}^{k} c_j u_{nj} \right).$$

Thus $\sum_{j=1}^{k} c_j \mathbf{u}_j = \mathbf{0}$ if and only if $\sum_{j=1}^{k} c_j u_{ij} = 0$ for each $i = 1, 2, \ldots, n$. This shows that the problem of determining the conditions on the c_j is equivalent to investigating the solutions of a system of n equations in k unknowns. If $\sum_{j=1}^{k} c_j \mathbf{u}_j = \mathbf{0}$ implies $c_1 = c_2 = \cdots = c_k = 0$, then $\{\mathbf{u}_1, \mathbf{u}_2, \ldots, \mathbf{u}_k\}$ is linearly independent.

The discussion in the preceding paragraph is illustrated in the next example.

EXAMPLE 2   Consider the set of vectors

$$\{(1, 1, 8, 1), (1, 0, 3, 0), (3, 1, 14, 1)\}$$

in $\mathbf{R}^4$. To determine the linear dependence or linear independence of this set, we investigate the solutions c_1, c_2, c_3 to

$$c_1(1, 1, 8, 1) + c_2(1, 0, 3, 0) + c_3(3, 1, 14, 1) = (0, 0, 0, 0).$$

Performing the indicated vector operations gives the equation

$$(c_1 + c_2 + 3c_3, c_1 + c_3, 8c_1 + 3c_2 + 14c_3, c_1 + c_3) = (0, 0, 0, 0).$$

This vector equation is equivalent to

$$\begin{array}{rrrr} c_1 + & c_2 + & 3c_3 & = 0 \\ c_1 & & + \ \ c_3 & = 0 \\ 8c_1 + & 3c_2 + & 14c_3 & = 0 \\ c_1 & & + \ \ c_3 & = 0. \end{array}$$

To solve this system, we first interchange the first two equations to place the equation $c_1 + c_3 = 0$ at the top.

$$\begin{array}{rrrr} c_1 & & + \ \ c_3 & = 0 \\ c_1 + & c_2 + & 3c_3 & = 0 \\ 8c_1 + & 3c_2 + & 14c_3 & = 0 \\ c_1 & & + \ \ c_3 & = 0 \end{array}$$

By adding suitable multiples of the first equation to each of the other equations, we then eliminate c_1 from all but the first equation. This yields the system

$$c_1 \quad + \quad c_3 = 0$$
$$c_2 + 2c_3 = 0$$
$$3c_2 + 6c_3 = 0$$
$$0 = 0.$$

Eliminating c_2 from the third equation gives

$$c_1 \quad + \quad c_3 = 0$$
$$c_2 + 2c_3 = 0$$
$$0 = 0$$
$$0 = 0.$$

It is now clear that there are many solutions to the system, and they are given by

$$c_1 = -c_3$$
$$c_2 = -2c_3$$
$$c_3 \text{ is arbitrary.}$$

Hence we can write

$$-c_3(1, 1, 8, 1) - 2c_3(1, 0, 3, 0) + c_3(3, 1, 14, 1) = (0, 0, 0, 0)$$

for any choice of c_3. In particular, it is not necessary that c_3 be zero, so the original set of vectors is linearly dependent. ■

The following theorem gives an alternative description of linear dependence for a certain type of set.

1.8 THEOREM

A set $\mathcal{A} = \{\mathbf{u}_1, \mathbf{u}_2, \ldots, \mathbf{u}_k\}$ in $\mathbf{R}^n$ that contains at least two vectors is linearly dependent if and only if some $\mathbf{u}_j$ is a linear combination of the remaining vectors in the set.

Proof If the set $\mathcal{A} = \{\mathbf{u}_1, \mathbf{u}_2, \ldots, \mathbf{u}_k\}$ is linearly dependent, then there are scalars $a_1, a_2, \ldots, a_k$ such that $\sum_{i=1}^{k} a_i \mathbf{u}_i = \mathbf{0}$ with at least one a_i, say a_j, not zero. This implies that

$$a_j \mathbf{u}_j = -a_1 \mathbf{u}_1 - \cdots - a_{j-1} \mathbf{u}_{j-1} - a_{j+1} \mathbf{u}_{j+1} - \cdots - a_k \mathbf{u}_k$$

so that

$$\mathbf{u}_j = \left(-\frac{a_1}{a_j}\right) \mathbf{u}_1 + \cdots + \left(-\frac{a_{j-1}}{a_j}\right) \mathbf{u}_{j-1} + \left(-\frac{a_{j+1}}{a_j}\right) \mathbf{u}_{j+1} + \cdots + \left(-\frac{a_k}{a_j}\right) \mathbf{u}_k.$$

Thus $\mathbf{u}_j$ can be written as a linear combination of the remaining vectors in the set.

Now assume that some $\mathbf{u}_j$ is a linear combination of the remaining vectors in the set, i.e.,

$$\mathbf{u}_j = b_1\mathbf{u}_1 + b_2\mathbf{u}_2 + \cdots + b_{j-1}\mathbf{u}_{j-1} + b_{j+1}\mathbf{u}_{j+1} + \cdots + b_k\mathbf{u}_k.$$

Then

$$b_1\mathbf{u}_1 + \cdots + b_{j-1}\mathbf{u}_{j-1} + (-1)\mathbf{u}_j + b_{j+1}\mathbf{u}_{j+1} + \cdots + b_k\mathbf{u}_k = \mathbf{0},$$

and since the coefficient of $\mathbf{u}_j$ in this linear combination is not zero, the set is linearly dependent. ■ ■ ■

The different meanings of the word "dependent" in Definitions 1.4 and 1.7 should be noted carefully. These meanings, though different, are closely related. The preceding theorem, for example, could be restated as follows: "A set $\{\mathbf{u}_1, \ldots, \mathbf{u}_k\}$ is linearly dependent if and only if some $\mathbf{u}_j$ is linearly dependent on the remaining vectors." This relation is further illustrated in some of the exercises at the end of this section.

1.2 Exercises

1. Determine whether or not the given set $\mathcal{A}$ is linearly dependent.
   (a) $\mathcal{A} = \{(1, 0, -2), (0, 2, 1), (-1, 2, 3)\}$
   (b) $\mathcal{A} = \{(1, 4, 3), (2, 12, 6), (5, 21, 15), (0, 2, -1)\}$
   (c) $\mathcal{A} = \{(1, 2, -1), (-1, 1, 0), (1, 3, -1)\}$
   (d) $\mathcal{A} = \{(1, 0, 1, 2), (2, 1, 0, 0), (4, 5, 6, 0), (1, 1, 1, 0)\}$

2. Show that the given set is linearly dependent and write one of the vectors as a linear combination of the remaining vectors.
   (a) $\{(2, 1, 0), (1, 1, 0), (0, 1, 1), (-1, 1, 1)\}$
   (b) $\{(1, 2, 1, 0), (3, -4, 5, 6), (2, -1, 3, 3), (-2, 6, -4, -6)\}$

3. Show that there is one vector in the set

$$\{(1, 1, 0), (0, 1, 1), (1, 0, -1), (1, 0, 1)\}$$

that cannot be written as a linear combination of the other vectors in the set.

4. Prove that if the set $\{\mathbf{u}_1, \mathbf{u}_2, \ldots, \mathbf{u}_k\}$ of vectors in $\mathbf{R}^n$ contains the zero vector, it is linearly dependent.

5. Prove that a set consisting of exactly one nonzero vector is linearly independent.

6. Prove that a set of two vectors in $\mathbf{R}^n$ is linearly dependent if and only if one of the vectors is a scalar multiple of the other.

7. Prove that a set of nonzero vectors $\{\mathbf{u}_1, \mathbf{u}_2, \ldots, \mathbf{u}_k\}$ in $\mathbf{R}^n$ is linearly dependent if and only if some $\mathbf{u}_r$ is a linear combination of the preceding vectors.

8. Let $\mathcal{A} = \{\mathbf{u}_1, \mathbf{u}_2, \mathbf{u}_3\}$ be a linearly independent set of vectors in $\mathbf{R}^n$.

(a) Prove that the set $\{\mathbf{u}_1 - \mathbf{u}_2, \mathbf{u}_2 - \mathbf{u}_3, \mathbf{u}_1 + \mathbf{u}_3\}$ is linearly independent.

(b) Prove that the set $\{\mathbf{u}_1 - \mathbf{u}_2, \mathbf{u}_2 - \mathbf{u}_3, \mathbf{u}_1 - \mathbf{u}_3\}$ is linearly dependent.

9. Let $\mathcal{A} = \{\mathbf{u}_1, \mathbf{u}_2, \mathbf{u}_3\}$ be a set of vectors in $\mathbf{R}^n$.

(a) Prove that if $\{\mathbf{u}_1, \mathbf{u}_2\}$ is linearly dependent then $\{\mathbf{u}_1, \mathbf{u}_1 + \mathbf{u}_2\}$ is linearly dependent.

(b) Prove that if $\{\mathbf{u}_1, \mathbf{u}_2, \mathbf{u}_3\}$ is linearly dependent then $\{\mathbf{u}_1 - \mathbf{u}_2, \mathbf{u}_2, \mathbf{u}_1 + \mathbf{u}_3\}$ is linearly dependent.

10. Let $\{\mathbf{u}_1, \ldots, \mathbf{u}_{r-1}, \mathbf{u}_r, \mathbf{u}_{r+1}, \ldots, \mathbf{u}_k\}$ be a linearly independent set of k vectors in $\mathbf{R}^n$, and let $\mathbf{u}_r' = \sum_{j=1}^{k} a_j \mathbf{u}_j$ with $a_r \neq 0$. Prove that $\{\mathbf{u}_1, \ldots, \mathbf{u}_{r-1}, \mathbf{u}_r', \mathbf{u}_{r+1}, \ldots, \mathbf{u}_k\}$ is linearly independent.

11. Let $\varnothing$ denote the empty set of vectors in $\mathbf{R}^n$. Determine whether or not $\varnothing$ is linearly dependent, and justify your conclusion.

12. Prove that any subset of a linearly independent set $\mathcal{A} \subseteq \mathbf{R}^n$ is linearly independent.

13. Let $\mathcal{A} \subseteq \mathbf{R}^n$. Prove that if $\mathcal{A}$ contains a linearly dependent subset, then $\mathcal{A}$ is linearly dependent.

1.3 Subspaces of R^n

There are many subsets of $\mathbf{R}^n$ that possess the properties stated in Theorem 1.3. A study of these subsets furnishes a great deal of insight into the structure of the spaces $\mathbf{R}^n$, and is of vital importance in subsequent material.

1.9 DEFINITION

A set $\mathbf{W}$ is a **subspace** of $\mathbf{R}^n$ if $\mathbf{W}$ is contained in $\mathbf{R}^n$ and if the properties of Theorem 1.3 are valid in $\mathbf{W}$. That is,

1. $\mathbf{u} + \mathbf{v} \in \mathbf{W}$ for all $\mathbf{u}, \mathbf{v}$ in $\mathbf{W}$.

2. $(\mathbf{u} + \mathbf{v}) + \mathbf{w} = \mathbf{u} + (\mathbf{v} + \mathbf{w})$ for all $\mathbf{u}, \mathbf{v}, \mathbf{w}$ in $\mathbf{W}$.

3. $\mathbf{0} \in \mathbf{W}$.

4. For each $\mathbf{u} \in \mathbf{W}$, $-\mathbf{u}$ is in $\mathbf{W}$.

5. $\mathbf{u} + \mathbf{v} = \mathbf{v} + \mathbf{u}$ for all $\mathbf{u}, \mathbf{v}$ in $\mathbf{W}$.

6. $a\mathbf{u} \in \mathbf{W}$ for all $a \in \mathbf{R}$ and all $\mathbf{u} \in \mathbf{W}$.

7. $a(b\mathbf{u}) = (ab)\mathbf{u}$ for all $a, b \in \mathbf{R}$ and all $\mathbf{u} \in \mathbf{W}$.

8. $a(\mathbf{u} + \mathbf{v}) = a\mathbf{u} + a\mathbf{v}$ for all $a \in \mathbf{R}$ and all $\mathbf{u}, \mathbf{v} \in \mathbf{W}$.

9. $(a + b)\mathbf{u} = a\mathbf{u} + b\mathbf{u}$ for all $a, b \in \mathbf{R}$ and all $\mathbf{u} \in \mathbf{W}$.

10. $1 \cdot \mathbf{u} = \mathbf{u}$ for all $\mathbf{u} \in \mathbf{W}$.

Before considering some examples of subspaces, we observe that the list of properties in Definition 1.9 can be shortened a great deal. For example, properties (2), (5), (7), (8),

(9), and (10) are valid throughout $\mathbf{R}^n$, and hence are automatically satisfied in any subset of $\mathbf{R}^n$. Thus a subset $\mathbf{W}$ of $\mathbf{R}^n$ is a subspace if and only if properties (1), (3), (4), and (6) hold in $\mathbf{W}$. This reduces the amount of labor necessary in order to determine whether or not a given subset is a subspace, but an even more practical test is given in the following theorem.

1.10 THEOREM

Let $\mathbf{W}$ be a subset of $\mathbf{R}^n$. Then $\mathbf{W}$ is a subspace of $\mathbf{R}^n$ if and only if the following conditions hold:

(i) $\mathbf{W}$ is nonempty;
(ii) for any $a, b \in \mathbf{R}$ and any $\mathbf{u}, \mathbf{v} \in \mathbf{W}$, $a\mathbf{u} + b\mathbf{v} \in \mathbf{W}$.

Proof Suppose that $\mathbf{W}$ is a subspace of $\mathbf{R}^n$. Then $\mathbf{W}$ is nonempty, since $\mathbf{0} \in \mathbf{W}$ by property (3). Let a and b be any elements of $\mathbf{R}$, and let $\mathbf{u}$ and $\mathbf{v}$ be elements of $\mathbf{W}$. By property (6), $a\mathbf{u}$ and $b\mathbf{v}$ are in $\mathbf{W}$. Hence $a\mathbf{u} + b\mathbf{v} \in \mathbf{W}$ by property (1), and condition (ii) is satisfied.

Suppose, on the other hand, that conditions (i) and (ii) are satisfied in $\mathbf{W}$. From our discussion above, it is necessary only to show that properties (1), (3), (4), and (6) are valid in $\mathbf{W}$. Since $\mathbf{W}$ is nonempty, there is at least one vector $\mathbf{u} \in \mathbf{W}$. By condition (ii), $1\mathbf{u} + (-1)\mathbf{u} = \mathbf{0} \in \mathbf{W}$, and property (3) is valid. Again by condition (ii), $(-1)\mathbf{u} = -\mathbf{u} \in \mathbf{W}$, so property (4) is valid. For any $\mathbf{u}, \mathbf{v}$ in $\mathbf{W}$, $1 \cdot \mathbf{u} + 1 \cdot \mathbf{v} = \mathbf{u} + \mathbf{v} \in \mathbf{W}$, and property (1) is valid. For any $a \in \mathbf{R}$ and any $\mathbf{u} \in \mathbf{W}$, $a\mathbf{u} + 0 \cdot \mathbf{u} = a\mathbf{u} \in \mathbf{W}$. Thus property (6) is valid, and $\mathbf{W}$ is a subspace of $\mathbf{R}^n$. ■■■

EXAMPLE 1 The following list gives several examples of subspaces of $\mathbf{R}^n$. We will prove that the third set in the list forms a subspace and leave it as an exercise to verify that the remaining sets are subspaces of $\mathbf{R}^n$.

1. The set $\{\mathbf{0}\}$, called the **zero subspace** of $\mathbf{R}^n$.
2. The set of all scalar multiples of a fixed vector $\mathbf{u} \in \mathbf{R}^n$.
3. The set of all vectors that are dependent on a given set $\{\mathbf{u}_1, \mathbf{u}_2, \ldots, \mathbf{u}_k\}$ of vectors in $\mathbf{R}^n$.
4. The set of all vectors $(x_1, x_2, \ldots, x_n)$ in $\mathbf{R}^n$ that satisfy a fixed equation

$$a_1 x_1 + a_2 x_2 + \cdots + a_n x_n = 0.$$

For $n = 2$ or $n = 3$, this example has a simple geometric interpretation, as we shall see later.

5. The set of all vectors $(x_1, x_2, \ldots, x_n)$ in $\mathbf{R}^n$ that satisfy the system of equations

$$a_{11} x_1 + a_{12} x_2 + \cdots + a_{1n} x_n = 0$$
$$a_{21} x_1 + a_{22} x_2 + \cdots + a_{2n} x_n = 0$$
$$\cdots \cdots$$
$$a_{m1} x_1 + a_{m2} x_2 + \cdots + a_{mn} x_n = 0.$$

■

Proof for the third set Let $\mathbf{W}$ be the set of all vectors that are dependent on the set $\mathcal{A} = \{\mathbf{u}_1, \mathbf{u}_2, \ldots, \mathbf{u}_k\}$ of vectors in $\mathbf{R}^n$. From the discussion in the paragraph following Definition 1.4, we know that $\mathbf{W}$ is the set of all vectors that can be written in the form $\sum_{i=1}^{k} a_i \mathbf{u}_i$. The set $\mathbf{W}$ is nonempty since

$$0 \cdot \mathbf{u}_1 + 0 \cdot \mathbf{u}_2 + \cdots + 0 \cdot \mathbf{u}_k = \mathbf{0} \text{ is in } \mathbf{W}.$$

Let $\mathbf{u}, \mathbf{v}$ be arbitrary vectors in $\mathbf{W}$, and let a, b be arbitrary scalars. From the definition of $\mathbf{W}$, there are scalars c_i and d_i such that

$$\mathbf{u} = \sum_{i=1}^{k} c_i \mathbf{u}_i \quad \text{and} \quad \mathbf{v} = \sum_{i=1}^{k} d_i \mathbf{u}_i.$$

Thus we have

$$a\mathbf{u} + b\mathbf{v} = a \left(\sum_{i=1}^{k} c_i \mathbf{u}_i \right) + b \left(\sum_{i=1}^{k} d_i \mathbf{u}_i \right)$$

$$= \sum_{i=1}^{k} a c_i \mathbf{u}_i + \sum_{i=1}^{k} b d_i \mathbf{u}_i$$

$$= \sum_{i=1}^{k} (a c_i \mathbf{u}_i + b d_i \mathbf{u}_i)$$

$$= \sum_{i=1}^{k} (a c_i + b d_i) \mathbf{u}_i.$$

The last expression is a linear combination of the elements in $\mathcal{A}$, and consequently $a\mathbf{u} + b\mathbf{v}$ is an element of $\mathbf{W}$ since $\mathbf{W}$ contains all vectors that are dependent on $\mathcal{A}$. Both conditions in Theorem 1.10 have been verified, and therefore $\mathbf{W}$ is a subspace of $\mathbf{R}^n$. ■ ■ ■

Our next theorem has a connection with the sets listed as 4 and 5 in Example 1 that should be investigated by the student. In this theorem, we are confronted with a situation which involves a collection that is not necessarily finite. In situations such as this, it is desirable to have available a notational convenience known as *indexing*.

Let $\mathcal{L}$ and $\mathcal{T}$ be nonempty sets. Suppose that with each $\lambda \in \mathcal{L}$ there is associated a unique element t_λ of $\mathcal{T}$, and that each element of $\mathcal{T}$ is associated with at least one $\lambda \in \mathcal{L}$. (That is, suppose that there is given a function with domain $\mathcal{L}$ and range $\mathcal{T}$.) Then we say that the set $\mathcal{T}$ is *indexed* by the set $\mathcal{L}$, and refer to $\mathcal{L}$ as an *index set*. We write $\{t_\lambda \mid \lambda \in \mathcal{L}\}$ to denote that the collection of t_λ's is indexed by $\mathcal{L}$.

If $\{\mathcal{M}_\lambda \mid \lambda \in \mathcal{L}\}$ is a collection of sets $\mathcal{M}_\lambda$ indexed by $\mathcal{L}$, then $\bigcup_{\lambda \in \mathcal{L}} \mathcal{M}_\lambda$ indicates the union of this collection of sets. Thus $\bigcup_{\lambda \in \mathcal{L}} \mathcal{M}_\lambda$ is the set of all elements that are contained in at least one $\mathcal{M}_\lambda$. Similarly, $\bigcap_{\lambda \in \mathcal{L}} \mathcal{M}_\lambda$ denotes the intersection of the sets $\mathcal{M}_\lambda$, and consists of all elements that are in every $\mathcal{M}_\lambda$.

1.11 THEOREM

The intersection of any nonempty collection of subspaces of $\mathbf{R}^n$ is a subspace of $\mathbf{R}^n$.

Proof Let $\{\mathcal{S}_\lambda \mid \lambda \in \mathcal{L}\}$ be any nonempty collection of subspaces $\mathcal{S}_\lambda$ of $\mathbf{R}^n$, and let $\mathbf{W} = \bigcap_{\lambda \in \mathcal{L}} \mathcal{S}_\lambda$. Now $\mathbf{0} \in \mathcal{S}_\lambda$ for each $\lambda \in \mathcal{L}$, so $\mathbf{0} \in \mathbf{W}$ and $\mathbf{W}$ is nonempty. Let $a, b \in \mathbf{R}$, and let $\mathbf{u}, \mathbf{v} \in \mathbf{W}$. Since each of $\mathbf{u}$ and $\mathbf{v}$ is in $\mathcal{S}_\lambda$ for every $\lambda \in \mathcal{L}$, $a\mathbf{u} + b\mathbf{v} \in \mathcal{S}_\lambda$ for every $\lambda \in \mathcal{L}$. Hence $a\mathbf{u} + b\mathbf{v} \in \mathbf{W}$, and $\mathbf{W}$ is a subspace by Theorem 1.10. ■ ■ ■

Thus the operation of intersection can be used to construct new subspaces from given subspaces.

EXAMPLE 2 Consider the sets $\mathbf{W}_1$ and $\mathbf{W}_2$ given by

$$\mathbf{W}_1 = \{(x_1, x_2, x_3) \mid x_1 + x_2 + 2x_3 = 0\}$$

$$\mathbf{W}_2 = \{(x_1, x_2, x_3) \mid x_2 = -x_3\}.$$

Both $\mathbf{W}_1$ and $\mathbf{W}_2$ are examples of the type of subspaces of $\mathbf{R}^3$ as described in part 4 of Example 1. According to the previous theorem, the intersection $\mathbf{W}_1 \cap \mathbf{W}_2$ is a subspace of $\mathbf{R}^3$. To find a vector in both $\mathbf{W}_1$ and $\mathbf{W}_2$, we set $x_2 = -x_3$ in $x_1 + x_2 + 2x_3 = 0$ to obtain $x_1 - x_3 + 2x_3 = 0$ or $x_1 = -x_3$. Thus

$$\mathbf{W}_1 \cap \mathbf{W}_2 = \{(-x_3, -x_3, x_3)\}$$

and $\mathbf{W}_1 \cap \mathbf{W}_2$ consists of all multiples of the vector $(1, 1, -1)$. ■

There is another operation, given in the following definition, that also can be used to form new subspaces from given subspaces.

1.12 DEFINITION

Let $\mathcal{S}_1$ and $\mathcal{S}_2$ be nonempty subsets of $\mathbf{R}^n$. Then the **sum** $\mathcal{S}_1 + \mathcal{S}_2$ is the set of all vectors $\mathbf{u} \in \mathbf{R}^n$ that can be expressed in the form $\mathbf{u} = \mathbf{u}_1 + \mathbf{u}_2$, where $\mathbf{u}_1 \in \mathcal{S}_1$ and $\mathbf{u}_2 \in \mathcal{S}_2$.

Although this definition applies to nonempty subsets of $\mathbf{R}^n$ generally, the more interesting situations are those in which both subsets are subspaces.

1.13 THEOREM

If $\mathbf{W}_1$ and $\mathbf{W}_2$ are subspaces of $\mathbf{R}^n$, then $\mathbf{W}_1 + \mathbf{W}_2$ is a subspace of $\mathbf{R}^n$.

Proof Clearly $\mathbf{0} + \mathbf{0} = \mathbf{0}$ is in $\mathbf{W}_1 + \mathbf{W}_2$, so that $\mathbf{W}_1 + \mathbf{W}_2$ is nonempty.

Let $a, b \in \mathbf{R}$, and let $\mathbf{u}$ and $\mathbf{v}$ be arbitrary elements in $\mathbf{W}_1 + \mathbf{W}_2$. From the definition of $\mathbf{W}_1 + \mathbf{W}_2$, it follows that there are vectors $\mathbf{u}_1, \mathbf{v}_1$ in $\mathbf{W}_1$ and $\mathbf{u}_2, \mathbf{v}_2$ in $\mathbf{W}_2$ such that $\mathbf{u} = \mathbf{u}_1 + \mathbf{u}_2$ and $\mathbf{v} = \mathbf{v}_1 + \mathbf{v}_2$. Now $a\mathbf{u}_1 + b\mathbf{v}_1 \in \mathbf{W}_1$ and $a\mathbf{u}_2 + b\mathbf{v}_2 \in \mathbf{W}_2$, since $\mathbf{W}_1$ and $\mathbf{W}_2$ are subspaces. This gives

$$au + bv = a(\mathbf{u}_1 + \mathbf{u}_2) + b(\mathbf{v}_1 + \mathbf{v}_2)$$
$$= (a\mathbf{u}_1 + b\mathbf{v}_1) + (a\mathbf{u}_2 + b\mathbf{v}_2),$$

which clearly is an element of $\mathbf{W}_1 + \mathbf{W}_2$. Therefore, $\mathbf{W}_1 + \mathbf{W}_2$ is a subspace of $\mathbf{R}^n$ by Theorem 1.10. ■ ■ ■

EXAMPLE 3 Consider the subspaces $\mathbf{W}_1$ and $\mathbf{W}_2$ as follows:

$\mathbf{W}_1$ is the set of all vectors in $\mathbf{R}^4$ dependent on $\{(1, -1, 0, 0), (0, 0, 0, 1)\}$,

$\mathbf{W}_2$ is the set of all vectors in $\mathbf{R}^4$ dependent on $\{(2, -2, 0, 0), (0, 0, 1, 0)\}$.

Then $\mathbf{W}_1$ is the set of all vectors of the form

$$a_1(1, -1, 0, 0) + a_2(0, 0, 0, 1) = (a_1, -a_1, 0, a_2),$$

$\mathbf{W}_2$ is the set of all vectors of the form

$$b_1(2, -2, 0, 0) + b_2(0, 0, 1, 0) = (2b_1, -2b_1, b_2, 0),$$

and $\mathbf{W}_1 + \mathbf{W}_2$ is the set of all vectors of the form

$$a_1(1, -1, 0, 0) + a_2(0, 0, 0, 1) + b_1(2, -2, 0, 0) + b_2(0, 0, 1, 0)$$
$$= (a_1 + 2b_1, -a_1 - 2b_1, b_2, a_2).$$

The last equation describes the vectors in $\mathbf{W}_1 + \mathbf{W}_2$, but it is not the most efficient description possible. Since $a_1 + 2b_1$ can take on any real number c_1 as a value, we see that $\mathbf{W}_1 + \mathbf{W}_2$ is the set of all vectors of the form

$$(c_1, -c_1, c_2, c_3). \qquad ■$$

1.3 Exercises

In Problems 1–13, determine which of the given sets $\mathbf{W}$ is a subspace of the indicated vector space. If a set is not a subspace, state a reason.

1. The set $\mathbf{W}$ of all vectors in $\mathbf{R}^2$ of the form (a, b), where $a \geq 0$ and $b \geq 0$.

2. The set $\mathbf{W}$ of all vectors in $\mathbf{R}^2$ of the form (a, a).

3. The set $\mathbf{W}$ of all vectors in $\mathbf{R}^2$ of the form (a, b), where $ab \geq 0$.

4. The set $\mathbf{W}$ of all vectors in $\mathbf{R}^2$ of the form (a, b), where $ab \leq 0$.

5. The set $\mathbf{W}$ of all vectors in $\mathbf{R}^2$ of the form $(a, 4a)$.

6. The set $\mathbf{W}$ of all vectors in $\mathbf{R}^2$ of the form $(a, a + 1)$.

7. The set **W** of all vectors in $\mathbf{R}^3$ of the form $(a, -a, b)$.

8. The set **W** of all vectors in $\mathbf{R}^3$ of the form $(a, b, 0)$.

9. The set **W** of all vectors in $\mathbf{R}^3$ of the form $(a, b, 0)$, where $b \geq 0$.

10. The set **W** of all vectors in $\mathbf{R}^3$ of the form (a, b, c), where $c^2 = a^2 + b^2$.

11. The set **W** of all vectors in $\mathbf{R}^4$ of the form (a, b, c, d), where $a + b = 0$ and $c - d = 0$.

12. The set **W** of all vectors in $\mathbf{R}^4$ of the form $(a, 0, a, 0)$.

13. The set **W** of all vectors in $\mathbf{R}^4$ of the form (x_1, x_2, x_3, x_4) where $ax_1 + bx_2 + cx_3 + dx_4 = k$, for $k \neq 0$.

14. Prove that each of the sets listed as 1, 2, 4, and 5 in Example 1 of this section is a subspace of $\mathbf{R}^n$.

15. Explain the connection between the sets listed as 4 and 5 in Example 1 and Theorem 1.11.

16. Let $\mathcal{L}$ denote the set of all real numbers λ such that $0 \leq \lambda \leq 1$. For each $\lambda \in \mathcal{L}$, let M_λ be the set of all $x \in \mathbf{R}$ such that $|x| < \lambda$. Find $\bigcup_{\lambda \in \mathcal{L}} M_\lambda$ and $\bigcap_{\lambda \in \mathcal{L}} M_\lambda$.

17. Formulate Definition 1.4 and Definition 1.7 for an indexed set $\mathcal{A} = \{\mathbf{u}_\lambda \mid \lambda \in \mathcal{L}\}$ of vectors in $\mathbf{R}^n$.

18. Let $\mathbf{W}_1$ be the set of vectors in $\mathbf{R}^3$ that are dependent on $\{(2, -1, 5)\}$ and $\mathbf{W}_2$ the set of vectors in $\mathbf{R}^3$ that are dependent on $\{(3, -2, 10)\}$. Determine whether or not the given vector $\mathbf{u}$ is in $\mathbf{W}_1 + \mathbf{W}_2$.

    (a) $\mathbf{u} = (-4, 1, -5)$ **(b)** $\mathbf{u} = (3, 2, -6)$

    (c) $\mathbf{u} = (-5, 3, -2)$ **(d)** $\mathbf{u} = (3, 0, 0)$

19. Let $\mathbf{W}_1 = \{(x_1, x_2) \mid x_2 = 0\}$ and $\mathbf{W}_2 = \{(x_1, x_2) \mid x_1 + x_2 = 0\}$.

    (a) Find $\mathbf{W}_1 \cap \mathbf{W}_2$.   **(b)** Find $\mathbf{W}_1 + \mathbf{W}_2$.

20. Let $\mathbf{W}_1 = \{(x_1, x_2, x_3) \mid x_1 - 2x_2 = 0\}$ and $\mathbf{W}_2 = \{(x_1, x_2, x_3) \mid x_1 + x_3 = 0\}$.

    (a) Find $\mathbf{W}_1 \cap \mathbf{W}_2$.   **(b)** Find $\mathbf{W}_1 + \mathbf{W}_2$.

21. Prove or disprove: If $\mathbf{W}_1$ and $\mathbf{W}_2$ are subspaces of $\mathbf{R}^n$, then $\mathbf{W}_1 \cup \mathbf{W}_2$ is a subspace of $\mathbf{R}^n$.

22. Let **W** be a subspace of $\mathbf{R}^n$. Use condition (ii) of Theorem 1.10 and mathematical induction to show that any linear combination of vectors in **W** is again a vector in **W**.

1.4 Spanning Sets

There are all sorts of subsets in a given subspace **W** of $\mathbf{R}^n$. Some of these have the important property of being *spanning sets* for **W**, or sets that *span* **W**. The following definition describes this property.

1.14
DEFINITION

Let $\mathbf{W}$ be a subspace of $\mathbf{R}^n$. A nonempty set $\mathcal{A}$ of vectors in $\mathbf{R}^n$ **spans** $\mathbf{W}$ if $\mathcal{A} \subseteq \mathbf{W}$ and if every vector in $\mathbf{W}$ is a linear combination of vectors in $\mathcal{A}$. By definition, the empty set $\varnothing$ spans the zero subspace.

Intuitively, the word *span* is a natural choice in Definition 1.14 because a spanning set $\mathcal{A}$ reaches across (hence *spans*) the entire subspace when all linear combinations of $\mathcal{A}$ are formed.

EXAMPLE 1   We shall show that each of the following sets spans $\mathbf{R}^3$:

$$\mathcal{E}_3 = \{(1, 0, 0), (0, 1, 0), (0, 0, 1)\},$$

$$\mathcal{A} = \{(1, 1, 1), (0, 1, 1), (0, 0, 1)\}.$$

The set $\mathcal{E}_3$ spans $\mathbf{R}^3$ since an arbitrary (x, y, z) in $\mathbf{R}^3$ can be written as

$$(x, y, z) = x(1, 0, 0) + y(0, 1, 0) + z(0, 0, 1).$$

In calculus texts, the vectors in $\mathcal{E}_3$ are labeled with the standard notation

$$\mathbf{i} = (1, 0, 0), \quad \mathbf{j} = (0, 1, 0), \quad \mathbf{k} = (0, 0, 1)$$

and this is used to write $(x, y, z) = x\mathbf{i} + y\mathbf{j} + z\mathbf{k}$.

We can take advantage of the way zeros are placed in the vectors of $\mathcal{A}$ and write

$$(x, y, z) = x(1, 1, 1) + (y - x)(0, 1, 1) + (z - y)(0, 0, 1).$$

The coefficients in this equation can be found by inspection if we start with the first component and work from left to right. ∎

We shall see in Theorem 1.17 that the concept of a spanning set is closely related to the set $\langle \mathcal{A} \rangle$ defined as follows.

1.15
DEFINITION

For any nonempty set $\mathcal{A}$ of vectors in $\mathbf{R}^n$, $\langle \mathcal{A} \rangle$ is the set of all vectors in $\mathbf{R}^n$ that are dependent on $\mathcal{A}$. By definition, $\langle \varnothing \rangle$ is the zero subspace $\{\mathbf{0}\}$.

Thus, for $\mathcal{A} \neq \varnothing$, $\langle \mathcal{A} \rangle$ is the set of all vectors $\mathbf{u}$ that can be written as $\mathbf{u} = \sum_{j=1}^{k} a_j \mathbf{u}_j$ with a_j in $\mathbf{R}$ and $\mathbf{u}_j$ in $\mathcal{A}$. Since any $\mathbf{u}$ in $\mathcal{A}$ is dependent on $\mathcal{A}$, the subset relation $\mathcal{A} \subseteq \langle \mathcal{A} \rangle$ always holds.

In Example 1 of the previous section, the third set is $\langle \mathcal{A} \rangle$ where $\mathcal{A} = \{\mathbf{u}_1, \mathbf{u}_2, \ldots, \mathbf{u}_k\}$ in $\mathbf{R}^n$. When the notation $\langle \mathcal{A} \rangle$ is combined with the set notation for this $\mathcal{A}$, the result is a somewhat cumbersome notation:

$$\langle \mathcal{A} \rangle = \langle \{\mathbf{u}_1, \mathbf{u}_2, \ldots, \mathbf{u}_k\} \rangle.$$

We make a notational agreement for situations like this to simply write

$$\langle \mathcal{A} \rangle = \langle \mathbf{u}_1, \mathbf{u}_2, \ldots, \mathbf{u}_k \rangle.$$

For example, if $\mathcal{A} = \{(1, 3, 7), (2, 0, 6)\}$, we would write

$$\langle \mathcal{A} \rangle = \langle (1, 3, 7), (2, 0, 6) \rangle$$

instead of $\langle \mathcal{A} \rangle = \langle \{(1, 3, 7), (2, 0, 6)\} \rangle$ to indicate the set of all vectors that are dependent on $\mathcal{A}$.

We have proved that, for a finite subset $\mathcal{A} = \{\mathbf{u}_1, \mathbf{u}_2, \ldots, \mathbf{u}_k\}$ of $\mathbf{R}^n$, the set $\langle \mathcal{A} \rangle$ is a subspace of $\mathbf{R}^n$. The next theorem generalizes this result to an arbitrary subset $\mathcal{A}$ of $\mathbf{R}^n$.

1.16 THEOREM

For any subset $\mathcal{A}$ of $\mathbf{R}^n$, $\langle \mathcal{A} \rangle$ is a subspace of $\mathbf{R}^n$.

Proof If $\mathcal{A}$ is empty, then $\langle \mathcal{A} \rangle$ is the zero subspace by definition.

Suppose $\mathcal{A}$ is nonempty. Then $\langle \mathcal{A} \rangle$ is nonempty, since $\mathcal{A} \subseteq \langle \mathcal{A} \rangle$. Let $a, b \in \mathbf{R}$, and let $\mathbf{u}, \mathbf{v} \in \langle \mathcal{A} \rangle$. Now $a\mathbf{u} + b\mathbf{v}$ is dependent on $\{\mathbf{u}, \mathbf{v}\}$ and each of $\mathbf{u}$ and $\mathbf{v}$ is dependent on $\mathcal{A}$. Hence $\{a\mathbf{u} + b\mathbf{v}\}$ is dependent on $\{\mathbf{u}, \mathbf{v}\}$ and $\{\mathbf{u}, \mathbf{v}\}$ is dependent on $\mathcal{A}$. Therefore $\{a\mathbf{u} + b\mathbf{v}\}$ is dependent on $\mathcal{A}$ by Theorem 1.6. Thus $a\mathbf{u} + b\mathbf{v} \in \langle \mathcal{A} \rangle$, and $\langle \mathcal{A} \rangle$ is a subspace. ■ ■ ■

We state the relation between Definitions 1.14 and 1.15 as a theorem, even though the proof is almost trivial.

1.17 THEOREM

Let $\mathbf{W}$ be a subspace of $\mathbf{R}^n$, and let $\mathcal{A}$ be a subset of $\mathbf{R}^n$. Then $\mathcal{A}$ spans $\mathbf{W}$ if and only if $\langle \mathcal{A} \rangle = \mathbf{W}$.

Proof The statement is trivial in case $\mathcal{A} = \varnothing$. Suppose, then, that $\mathcal{A}$ is nonempty.

If $\mathcal{A}$ spans $\mathbf{W}$, then every vector in $\mathbf{W}$ is dependent on $\mathcal{A}$, so $\mathbf{W} \subseteq \langle \mathcal{A} \rangle$. Now $\mathcal{A} \subseteq \mathbf{W}$, and repeated application of condition (ii), Theorem 1.10, yields the fact that any linear combination of vectors in $\mathcal{A}$ is again a vector in $\mathbf{W}$. Thus, $\langle \mathcal{A} \rangle \subseteq \mathbf{W}$, and we have $\langle \mathcal{A} \rangle = \mathbf{W}$.

If $\langle \mathcal{A} \rangle = \mathbf{W}$, it follows at once that $\mathcal{A}$ spans $\mathbf{W}$, and this completes the proof. ■ ■ ■

We will refer to $\langle \mathcal{A} \rangle$ as the **subspace spanned by** $\mathcal{A}$. Some of the notations used in various texts for this same subspace are

$$\text{span}(\mathcal{A}), \quad \text{sp}(\mathcal{A}), \quad \text{lin}(\mathcal{A}), \quad \text{Sp}(\mathcal{A}), \quad \text{and} \quad S[\mathcal{A}].$$

We will use span$(\mathcal{A})$ or $\langle \mathcal{A} \rangle$ in this book.

EXAMPLE 2 With $\mathcal{A}$ and $\mathbf{W}$ as follows, we shall determine whether or not $\mathcal{A}$ spans $\mathbf{W}$.

$$\mathcal{A} = \{(1, 0, 1, 0), (0, 1, 0, 1), (2, 3, 2, 3)\}$$

$$\mathbf{W} = \langle (1, 0, 1, 0), (1, 1, 1, 1), (0, 1, 1, 1) \rangle$$

We first check to see whether $\mathcal{A} \subseteq \mathbf{W}$. That is, we check to see if each vector in $\mathcal{A}$ is a linear combination of the vectors listed in the spanning set for $\mathbf{W}$. By inspection, we see that $(1, 0, 1, 0)$ is listed in the spanning set, and

$$(0, 1, 0, 1) = (-1)(1, 0, 1, 0) + (1)(1, 1, 1, 1) + (0)(0, 1, 1, 1).$$

The linear combination is not as apparent with $(2, 3, 2, 3)$, so we place unknown coefficients in the equation

$$a_1(1, 0, 1, 0) + a_2(1, 1, 1, 1) + a_3(0, 1, 1, 1) = (2, 3, 2, 3).$$

Using the same procedure as in Example 2 of Section 1.1, we obtain the system of equations

$$\begin{aligned}
a_1 + a_2 \phantom{{}+ a_3} &= 2 \\
a_2 + a_3 &= 3 \\
a_1 + a_2 + a_3 &= 2 \\
a_2 + a_3 &= 3.
\end{aligned}$$

It is then easy to find the solution $a_1 = -1, a_2 = 3, a_3 = 0$. That is,

$$(-1)(1, 0, 1, 0) + 3(1, 1, 1, 1) + 0(0, 1, 1, 1) = (2, 3, 2, 3).$$

Thus we have $\mathcal{A} \subseteq \mathbf{W}$.

We must now decide if every vector in $\mathbf{W}$ is linearly dependent on $\mathcal{A}$, and Theorem 1.6 is of help here. We have $\mathbf{W}$ dependent on the set

$$\mathcal{B} = \{(1, 0, 1, 0), (1, 1, 1, 1), (0, 1, 1, 1)\}.$$

If $\mathcal{B}$ is dependent on $\mathcal{A}$, then $\mathbf{W}$ is dependent on $\mathcal{A}$, by Theorem 1.6. On the other hand, if $\mathcal{B}$ is not dependent on $\mathcal{A}$, then $\mathbf{W}$ is not dependent on $\mathcal{A}$ since $\mathcal{B} \subseteq \mathbf{W}$. Thus we need only check to see if each vector in $\mathcal{B}$ is a linear combination of vectors in $\mathcal{A}$. We see that $(1, 0, 1, 0)$ is listed in $\mathcal{A}$, and

$$(1, 1, 1, 1) = (1, 0, 1, 0) + (0, 1, 0, 1).$$

Considering $(0, 1, 1, 1)$, we set up the equation

$$b_1(1, 0, 1, 0) + b_2(0, 1, 0, 1) + b_3(2, 3, 2, 3) = (0, 1, 1, 1).$$

This is equivalent to the system

$$\begin{aligned}
b_1 \phantom{{}+ b_2} + 2b_3 &= 0 \\
b_2 + 3b_3 &= 1 \\
b_1 \phantom{{}+ b_2} + 2b_3 &= 1 \\
b_2 + 3b_3 &= 1.
\end{aligned}$$

The first and third equations contradict each other, so there is no solution. Hence **W** is not dependent on $\mathcal{A}$, and the set $\mathcal{A}$ does not span **W**. ■

The next example illustrates how to find a spanning set for a given subspace.

EXAMPLE 3 Consider the subspace $\mathbf{W} = \{(x_1, x_2, x_3) \mid x_1 + 2x_2 = 0\}$ of $\mathbf{R}^3$. To find a spanning set for **W** we note that the components of any vector in **W** must satisfy the condition that $x_1 = -2x_2$ where x_2 and x_3 can be any real numbers. Thus the vectors in **W** have the form $(-2x_2, x_2, x_3)$, which can be written as a linear combination of two vectors in the following way:

$$(-2x_2, x_2, x_3) = x_2(-2, 1, 0) + x_3(0, 0, 1).$$

This shows that $\{(-2, 1, 0), (0, 0, 1)\}$ is a spanning set for **W** consisting of 2 vectors. That is, $\mathbf{W} = \langle(-2, 1, 0), (0, 0, 1)\rangle$. However, this is not the only spanning set for **W** since $\{(-2, 1, 0), (0, 0, 1), (0, 0, 0)\}$ also spans **W**, as would any subset of **W** that contains $(-2, 1, 0)$ and $(0, 0, 1)$. ■

1.4 Exercises

1. Let $\mathbf{V} = \mathbf{R}^3$.

   (a) Exhibit a set of three vectors that spans **V**.

   (b) Exhibit a set of four vectors that spans **V**.

2. Determine whether or not the given vector **v** is in $\langle\mathcal{A}\rangle$ for each $\mathcal{A}$. If so, write **v** as a linear combination of the vectors in $\mathcal{A}$.

   (a) $\mathbf{v} = (2, 3), \mathcal{A} = \{(1, 1), (0, 1)\}$

   (b) $\mathbf{v} = (1, 2), \mathcal{A} = \{(1, -1), (2, 1)\}$

   (c) $\mathbf{v} = (-4, 4), \mathcal{A} = \{(0, -1), (4, 3)\}$

   (d) $\mathbf{v} = (4, -9), \mathcal{A} = \{(1, -1), (0, 2), (1, 1)\}$

   (e) $\mathbf{v} = (-1, 2, 5), \mathcal{A} = \{(0, 1, 3), (1, 0, 1)\}$

   (f) $\mathbf{v} = (5, 3, 1, 0), \mathcal{A} = \{(1, 0, 0, 0), (1, 1, 0, 0), (1, 1, 1, 0)\}$

   (g) $\mathbf{v} = (1, 2, -9, 1), \mathcal{A} = \{(1, 0, 0, 0), (1, 1, 0, 0), (1, 1, 1, 0), (2, 1, 0, 0)\}$

   (h) $\mathbf{v} = (0, 0, 0, 0, 1), \mathcal{A} = \{(0, 1, 0, 0, 0), (1, 1, 0, 0, 0), (1, 1, 1, 0, 1)\}$

3. Find a vector **v** in $\mathbf{R}^3$ that is not in $\langle\mathcal{A}\rangle$ for each $\mathcal{A}$.

   (a) $\mathcal{A} = \{(0, -1, 4)\}$ **(b)** $\mathcal{A} = \{(2, 3, 0), (3, -1, 0)\}$

   (c) $\mathcal{A} = \{(1, 1, 0), (0, 1, 1)\}$

   (d) $\mathcal{A} = \{(0, 0, 1), (1, 1, 0), (3, 3, -2), (2, 2, 7)\}$

4. Determine whether or not the given vectors span $\mathbf{R}^2$.

   (a) $\mathbf{u}_1 = (1, 1), \mathbf{u}_2 = (2, 3)$ **(b)** $\mathbf{u}_1 = (-1, 1), \mathbf{u}_2 = (2, -2)$

(c) $\mathbf{u}_1 = (0, 0)$, $\mathbf{u}_2 = (1, 1)$

(d) $\mathbf{u}_1 = (1, 0)$, $\mathbf{u}_2 = (1, 1)$, $\mathbf{u}_3 = (2, -1)$

5. Determine whether or not the given vectors span $\mathbf{R}^3$.

   (a) $\mathbf{u}_1 = (1, 0, -1)$, $\mathbf{u}_2 = (1, 1, 1)$, $\mathbf{u}_3 = (3, 1, -1)$

   (b) $\mathbf{u}_1 = (2, 3, 0)$, $\mathbf{u}_2 = (0, 1, 1)$

   (c) $\mathbf{u}_1 = (0, 1, 0)$, $\mathbf{u}_2 = (2, 0, 1)$, $\mathbf{u}_3 = (1, 0, 0)$

   (d) $\mathbf{u}_1 = (0, -1, 1)$, $\mathbf{u}_2 = (1, 0, 0)$, $\mathbf{u}_3 = (0, 1, 0)$, $\mathbf{u}_4 = (0, 0, 0)$

   (e) $\mathbf{u}_1 = (-1, -1, 1)$, $\mathbf{u}_2 = (1, 0, 1)$, $\mathbf{u}_3 = (0, 1, 1)$, $\mathbf{u}_4 = (0, -1, 2)$

6. Determine whether or not the given vectors span $\mathbf{R}^4$.

   (a) $\mathbf{u}_1 = (1, 0, 0, 0)$, $\mathbf{u}_2 = (0, 1, 1, 0)$, $\mathbf{u}_3 = (0, 0, 1, 1)$, $\mathbf{u}_4 = (1, 0, 0, 1)$

   (b) $\mathbf{u}_1 = (1, 0, -1, 1)$, $\mathbf{u}_2 = (0, 1, 1, 0)$, $\mathbf{u}_3 = (1, 1, 0, -1)$

7. For each given set $\mathcal{A}$ and subspace $\mathbf{W}$, determine whether or not $\mathcal{A}$ spans $\mathbf{W}$.

   (a) $\mathcal{A} = \{(1, 0, 2), (-1, 1, -3)\}$, $\mathbf{W} = \langle (1, 0, 2) \rangle$

   (b) $\mathcal{A} = \{(1, 0, 2), (-1, 1, -3)\}$, $\mathbf{W} = \langle (1, 1, 1), (2, -1, 5) \rangle$

   (c) $\mathcal{A} = \{(1, -2), (-1, 3)\}$, $\mathbf{W} = \langle (1, 1), (-1, 0), (0, 1) \rangle$

   (d) $\mathcal{A} = \{(2, 3, 0, -1), (2, 1, -1, 2)\}$, $\mathbf{W} = \langle (0, -2, -1, 3), (6, 7, -1, 0) \rangle$

   (e) $\mathcal{A} = \{(3, -1, 2, 1), (4, 0, 1, 0)\}$, $\mathbf{W} = \langle (3, -1, 2, 1), (4, 0, 1, 0), (0, -1, 0, 1) \rangle$

   (f) $\mathcal{A} = \{(3, -1, 2, 1), (4, 0, 1, 0)\}$, $\mathbf{W} = \langle (3, -1, 2, 1), (4, 0, 1, 0), (1, 1, -1, -1) \rangle$

8. Find a spanning set consisting of n vectors for the given subspace $\mathbf{W}$ of the given vector space $\mathbf{V}$.

   (a) $n = 2$, $\mathbf{W} = \{(x_1, x_2) \mid x_2 = 0\}$, $\mathbf{V} = \mathbf{R}^2$

   (b) $n = 3$, $\mathbf{W} = \{(x_1, x_2) \mid x_2 = 0\}$, $\mathbf{V} = \mathbf{R}^2$

   (c) $n = 2$, $\mathbf{W} = \{(x_1, x_2, x_3) \mid x_1 + x_2 = 0\}$, $\mathbf{V} = \mathbf{R}^3$

   (d) $n = 3$, $\mathbf{W} = \{(x_1, x_2, x_3) \mid x_1 + x_2 = 0\}$, $\mathbf{V} = \mathbf{R}^3$

   (e) $n = 4$, $\mathbf{W} = \{(x_1, x_2, x_3) \mid x_1 + x_2 = 0\}$, $\mathbf{V} = \mathbf{R}^3$

9. Find a spanning set for the subspace of $\mathbf{R}^2$ spanned by the set of vectors (x_1, x_2) that satisfy the equation $3x_1 + x_2 = 0$.

10. Find a spanning set for the subspace of $\mathbf{R}^2$ spanned by the set of vectors (x_1, x_2) that satisfy the equation $x_1 + 4x_2 = 0$.

11. Let $\mathcal{A} = \{(1, 2, 0), (1, 1, 1)\}$, and let $\mathcal{B} = \{(0, 1, -1), (0, 2, 2)\}$. By direct verification of the conditions of Theorem 1.10 in each case, show that $\langle \mathcal{A} \rangle$, $\langle \mathcal{B} \rangle$, and $\langle \mathcal{A} \rangle + \langle \mathcal{B} \rangle$ are subspaces of $\mathbf{R}^3$.

12. Let $\mathcal{A}_1 = \{\mathbf{u}, \mathbf{v}\}$ and $\mathcal{A}_2 = \{\mathbf{u}, \mathbf{v}, \mathbf{w}\}$ be subsets of $\mathbf{R}^n$ with $\mathbf{w} = \mathbf{u} + \mathbf{v}$. Show that $\langle \mathcal{A}_1 \rangle = \langle \mathcal{A}_2 \rangle$ by use of Definition 1.15.

13. Prove or disprove: $\langle \mathcal{A} + \mathcal{B} \rangle = \langle \mathcal{A} \rangle + \langle \mathcal{B} \rangle$ for any nonempty subsets $\mathcal{A}$ and $\mathcal{B}$ of $\mathbf{R}^n$.

14. Let $\mathcal{A} = \{\mathbf{u}_1, \mathbf{u}_2, \ldots, \mathbf{u}_k\}$ be a set of vectors in $\mathbf{R}^n$. Prove that $\langle \mathcal{A} \rangle$ is the smallest subspace of $\mathbf{R}^n$ containing $\mathcal{A}$.

15. If $\mathcal{A} \subseteq \mathbf{R}^n$, prove that $\langle \mathcal{A} \rangle$ is the intersection of all the subspaces of $\mathbf{R}^n$ that contain $\mathcal{A}$.

16. Let $\mathbf{W}$ be a subspace of $\mathbf{R}^n$. Prove that $\langle \mathbf{W} \rangle = \mathbf{W}$.

17. Prove that $\langle \mathcal{A} \rangle = \langle \mathcal{B} \rangle$ if and only if every vector in $\mathcal{A}$ is dependent on $\mathcal{B}$ and every vector in $\mathcal{B}$ is dependent on $\mathcal{A}$.

1.5 Geometric Interpretations of R² and R³

For $n = 1$, 2, or 3, the vector space $\mathbf{R}^n$ has a useful geometric interpretation in which a vector is identified with a directed line segment. This procedure is no doubt familiar to the student from the study of the calculus. In this section, we briefly review this interpretation of vectors and relate the geometric concepts to our work. The procedure can be described as follows.

For $n = 1$, the vector $\mathbf{v} = (x)$ is identified with the directed line segment on the real line that has its initial point (tail) at the origin and its terminal point (head) at x. This is shown in Figure 1.1.

Figure 1.1

For $n = 2$ or $n = 3$, the vector $\mathbf{v} = (x, y)$ or $\mathbf{v} = (x, y, z)$ is identified with the directed line segment that has initial point at the origin and terminal point with rectangular coordinates given by the components of $\mathbf{v}$. This is shown in Figure 1.2.

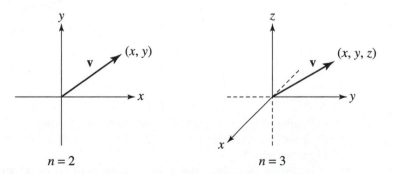

Figure 1.2

In making identifications of vectors with directed line segments, we shall follow the convention that any line segment with the same direction and the same length as the one we have described may be used to represent the same vector $\mathbf{v}$.

In the remainder of this section, we shall concentrate our attention on $\mathbf{R}^3$. However, it should be observed that corresponding results are valid in $\mathbf{R}^2$.

If $\mathbf{u} = (u_1, u_2, u_3)$ and $\mathbf{v} = (v_1, v_2, v_3)$, then $\mathbf{u} + \mathbf{v} = (u_1 + v_1, u_2 + v_2, u_3 + v_3)$. Thus, in the identification above, $\mathbf{u} + \mathbf{v}$ is the diagonal of a parallelogram which has $\mathbf{u}$ and $\mathbf{v}$ as two adjacent sides. This is illustrated in Figure 1.3. The vector $\mathbf{u} + \mathbf{v}$ can be drawn by placing the initial point of $\mathbf{v}$ at the terminal point of $\mathbf{u}$ and then drawing the directed line segment from the initial point of $\mathbf{u}$ to the terminal point of $\mathbf{v}$. The "heads to tails" construction shown in Figure 1.3 is called the *parallelogram rule* for adding vectors.

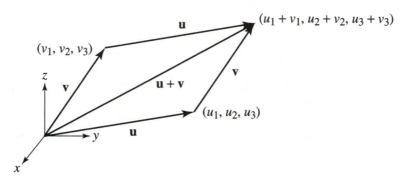

Figure 1.3

Now $\mathbf{u} - \mathbf{v}$ is the vector $\mathbf{w}$ satisfying $\mathbf{v} + \mathbf{w} = \mathbf{u}$, so that $\mathbf{u} - \mathbf{v}$ is the directed line segment from the terminal point of $\mathbf{v}$ to the terminal point of $\mathbf{u}$, as shown in Figure 1.4. Since $\mathbf{u} - \mathbf{v}$ has its head at $\mathbf{u}$ and its tail at $\mathbf{v}$, this construction is sometimes referred to as the "heads minus tails" rule.

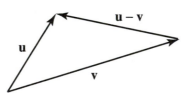

Figure 1.4

In approaching the geometric interpretation of subspaces, it is convenient to consider only those line segments with initial point at the origin. We shall do this in the following four paragraphs.

As mentioned earlier, the set of all scalar multiples of a fixed nonzero vector $\mathbf{v}$ in $\mathbf{R}^3$ is a subspace of $\mathbf{R}^3$. From our interpretation above, it is clear that this subspace $\langle \mathbf{v} \rangle$ consists of all vectors that lie on a line passing through the origin. This is shown in Figure 1.5.

If $\mathcal{A} = \{\mathbf{v}_1, \mathbf{v}_2\}$ is independent, then $\mathbf{v}_1$ and $\mathbf{v}_2$ are not collinear. If P is any point in the plane determined by $\mathbf{v}_1$ and $\mathbf{v}_2$, then the vector $\overrightarrow{OP}$ from the origin to P is the diagonal of a parallelogram with sides parallel to $\mathbf{v}_1$ and $\mathbf{v}_2$, as shown in Figure 1.6.

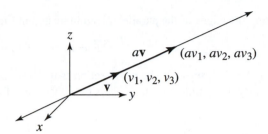

Figure 1.5

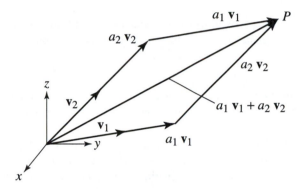

Figure 1.6

In this case, the subspace $\langle \mathcal{A} \rangle$ consists of all vectors in the plane through the origin that contains $\mathbf{v}_1$ and $\mathbf{v}_2$.

If $\mathcal{A} = \{\mathbf{v}_1, \mathbf{v}_2, \mathbf{v}_3\}$ is linearly independent, then $\mathbf{v}_1$ and $\mathbf{v}_2$ are not collinear and $\mathbf{v}_3$ does not lie in the plane of $\mathbf{v}_1$ and $\mathbf{v}_2$. Vectors $\mathbf{v}_1, \mathbf{v}_2,$ and $\mathbf{v}_3$ of this type are shown in Figure 1.7. An arbitrary vector $\overrightarrow{OP}$ in $\mathbf{R}^3$ is the diagonal of a parallelepiped with adjacent edges $a_1\mathbf{v}_1, a_2\mathbf{v}_2,$ and $a_3\mathbf{v}_3$ as shown in Figure 1.7(a). The "heads to tails" construction

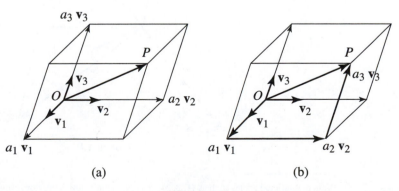

(a)   (b)

Figure 1.7

along the edges of the parallelepiped indicated in Figure 1.7(b) shows that

$$\overrightarrow{OP} = a_1\mathbf{v}_1 + a_2\mathbf{v}_2 + a_3\mathbf{v}_3.$$

We shall prove in the next section that a subset of $\mathbf{R}^3$ cannot contain more than three linearly independent vectors. Thus the subspaces of $\mathbf{R}^3$ fall into one of four categories:

1. the origin;
2. a line through the origin;
3. a plane through the origin;
4. the entire space $\mathbf{R}^3$.

It is shown in calculus courses that a plane in $\mathbf{R}^3$ consists of all points with rectangular coordinates (x, y, z) that satisfy a linear equation

$$ax + by + cz = d$$

in which at least one of a, b, c is not zero. A connection is made in the next example between this fact and our classification of subspaces.

EXAMPLE 1 Consider the problem of finding an equation of the plane $\langle \mathcal{A} \rangle$ if $\mathcal{A} = \{(1, 2, 3), (3, 5, 1)\}$.

Now the line segments[2] extending from the origin $(0, 0, 0)$ to $(1, 2, 3)$ and from the origin to $(3, 5, 1)$ lie in the plane $\langle \mathcal{A} \rangle$, so the three points with coordinates $(0, 0, 0)$, $(1, 2, 3)$, and $(3, 5, 1)$ must all lie in the plane. This is shown in Figure 1.8.

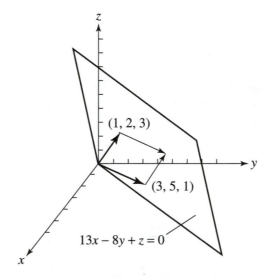

Figure 1.8

---

[2] Note that ordered triples such as $(1, 2, 3)$ are doing double duty here. Sometimes they are coordinates of points, and sometimes they are vectors.

Since the points lie in the plane, their coordinates must satisfy the equation

$$ax + by + cz = d$$

of the plane. Substituting in order for $(0, 0, 0)$, $(1, 2, 3)$, and $(3, 5, 1)$, we obtain

$$0 = d$$
$$a + 2b + 3c = d$$
$$3a + 5b + \ c = d.$$

Using $d = 0$ and subtracting 3 times the second equation from the last one leads to

$$a + 2b + 3c = 0$$
$$-b - 8c = 0.$$

Solving for a and b in terms of c, we obtain the following solutions

$$a = 13c$$
$$b = -8c$$
$$c \text{ is arbitrary.}$$

With $c = 1$, we have

$$13x - 8y + z = 0$$

as the equation of the plane $\langle A \rangle$.   ■

In the remainder of our discussion, we shall need the following definition, which applies to real coordinate spaces in general.

1.18 **DEFINITION**	For any two vectors $\mathbf{u} = (u_1, u_2, \ldots, u_n)$ and $\mathbf{v} = (v_1, v_2, \ldots, v_n)$, the **inner product** (**dot product**, or **scalar product**) of $\mathbf{u}$ and $\mathbf{v}$ is defined by $$\mathbf{u} \cdot \mathbf{v} = u_1 v_1 + u_2 v_2 + \cdots + u_n v_n = \sum_{k=1}^{n} u_k v_k.$$

The inner product defined in this way is a natural extension of the following definitions that are used in the calculus:

$$(x_1, y_1) \cdot (x_2, y_2) = x_1 x_2 + y_1 y_2,$$
$$(x_1, y_1, z_1) \cdot (x_2, y_2, z_2) = x_1 x_2 + y_1 y_2 + z_1 z_2.$$

The distance formulas used in the calculus lead to formulas for the length $\|\mathbf{v}\|$ of a vector $\mathbf{v}$ in $\mathbf{R}^2$ or $\mathbf{R}^3$ as follows:

$$\|(x, y)\| = \sqrt{x^2 + y^2},$$

$$\|(x, y, z)\| = \sqrt{x^2 + y^2 + z^2}.$$

We extend these formulas for length to more general use in the next definition.

1.19
DEFINITION

For any $\mathbf{v} = (v_1, v_2, \ldots, v_n)$ in $\mathbf{R}^n$, the **length** (or **norm**) of $\mathbf{v}$ is denoted by $\|\mathbf{v}\|$ and is defined by

$$\|\mathbf{v}\| = \sqrt{v_1^2 + v_2^2 + \cdots + v_n^2}.$$

The following properties are direct consequences of the definitions involved, and are presented as a theorem for convenient reference.

1.20
THEOREM

For any $\mathbf{u}, \mathbf{v}, \mathbf{w}$ in $\mathbf{R}^n$ and any a in $\mathbf{R}$:

(i) $\mathbf{u} \cdot \mathbf{v} = \mathbf{v} \cdot \mathbf{u}$

(ii) $(a\mathbf{u}) \cdot \mathbf{v} = \mathbf{u} \cdot (a\mathbf{v}) = a(\mathbf{u} \cdot \mathbf{v})$

(iii) $\mathbf{u} \cdot (\mathbf{v} + \mathbf{w}) = \mathbf{u} \cdot \mathbf{v} + \mathbf{u} \cdot \mathbf{w}$

(iv) $\|\mathbf{u}\| = \sqrt{\mathbf{u} \cdot \mathbf{u}}$, or $\mathbf{u} \cdot \mathbf{u} = \|\mathbf{u}\|^2$

(v) $\|a\mathbf{u}\| = |a| \|\mathbf{u}\|$.

Our next theorem gives a geometric interpretation of $\mathbf{u} \cdot \mathbf{v}$ in $\mathbf{R}^2$ or $\mathbf{R}^3$. In the proof, we use the Law of Cosines from trigonometry: *If the sides and angles of an arbitrary triangle are labeled according to the pattern in Figure 1.9, then*

$$\cos C = \frac{a^2 + b^2 - c^2}{2ab}.$$

We state and prove the theorem for $\mathbf{R}^3$, but the same result holds in $\mathbf{R}^2$ with a similar proof.

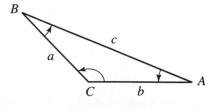

Figure 1.9

1.21
THEOREM

For any two nonzero vectors $\mathbf{u} = (u_1, u_2, u_3)$ and $\mathbf{v} = (v_1, v_2, v_3)$ in $\mathbf{R}^3$,

$$\mathbf{u} \cdot \mathbf{v} = \|\mathbf{u}\|\,\|\mathbf{v}\| \cos\theta,$$

where θ is the angle between the directions of $\mathbf{u}$ and $\mathbf{v}$ and $0° \leq \theta \leq 180°$.

Proof Suppose first that $\theta = 0°$ or $\theta = 180°$. Then $\mathbf{v} = c\mathbf{u}$, where the scalar c is positive if $\theta = 0°$ and negative if $\theta = 180°$. We have

$$\|\mathbf{u}\|\,\|\mathbf{v}\| \cos\theta = \|\mathbf{u}\|\,(|c| \cdot \|\mathbf{u}\|) \cos\theta = |c| \cos\theta \|\mathbf{u}\|^2 = c\|\mathbf{u}\|^2$$

and

$$\mathbf{u} \cdot \mathbf{v} = \mathbf{u} \cdot (c\,\mathbf{u}) = c\,(\mathbf{u} \cdot \mathbf{u}) = c\|\mathbf{u}\|^2.$$

Thus the theorem is true for $\theta = 0°$ or $\theta = 180°$.

Suppose now that $0° < \theta < 180°$. If $\mathbf{u} - \mathbf{v}$ is drawn from the head of $\mathbf{v}$ to the head of $\mathbf{u}$, the vectors $\mathbf{u}$, $\mathbf{v}$ and $\mathbf{u} - \mathbf{v}$ form a triangle with $\mathbf{u} - \mathbf{v}$ as the side opposite θ. (See Figure 1.10.)

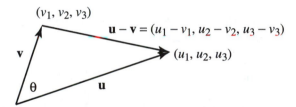

(v_1, v_2, v_3)

$\mathbf{u} - \mathbf{v} = (u_1 - v_1, u_2 - v_2, u_3 - v_3)$

$\mathbf{v}$

(u_1, u_2, u_3)

θ

$\mathbf{u}$

Figure 1.10

From the Law of Cosines, we have

$$\cos\theta = \frac{\|\mathbf{u}\|^2 + \|\mathbf{v}\|^2 - \|\mathbf{u} - \mathbf{v}\|^2}{2\|\mathbf{u}\|\,\|\mathbf{v}\|}.$$

Thus

$$\begin{aligned}
\|\mathbf{u}\|\,\|\mathbf{v}\| \cos\theta &= \tfrac{1}{2}(\|\mathbf{u}\|^2 + \|\mathbf{v}\|^2 - \|\mathbf{u} - \mathbf{v}\|^2) \\
&= \tfrac{1}{2}\{u_1^2 + u_2^2 + u_3^2 + v_1^2 + v_2^2 + v_3^2 \\
&\quad - [(u_1 - v_1)^2 + (u_2 - v_2)^2 + (u_3 - v_3)^2]\} \\
&= \mathbf{u} \cdot \mathbf{v}. \quad\blacksquare\blacksquare\blacksquare
\end{aligned}$$

1.22
COROLLARY

In $\mathbf{R}^2$ or $\mathbf{R}^3$, two nonzero vectors $\mathbf{u}$ and $\mathbf{v}$ are perpendicular (or orthogonal) if and only if $\mathbf{u} \cdot \mathbf{v} = 0$.

Proof This follows at once from the fact that $\mathbf{u} \cdot \mathbf{v} = 0$ if and only if $\cos\theta = 0$.   ■ ■ ■

Suppose that $\mathbf{u}$ and $\mathbf{v}$ are vectors in $\mathbf{R}^2$ or $\mathbf{R}^3$ represented by directed line segments with the same initial point, as shown in Figure 1.11. The vector labeled $\mathbf{Proj_u v}$ in the figure is called the **vector projection of v onto u**. In order to construct $\mathbf{Proj_u v}$, we first draw the straight line that contains $\mathbf{u}$. Next we draw a perpendicular segment joining the head of $\mathbf{v}$ to the line containing $\mathbf{u}$. The vector from the initial point of $\mathbf{u}$ to the foot of the perpendicular segment is $\mathbf{Proj_u v}$. The vector $\mathbf{Proj_u v}$ is also called the **vector component** of $\mathbf{v}$ along $\mathbf{u}$.

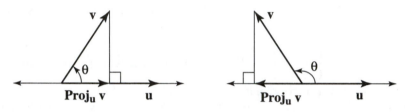

Figure 1.11

Let θ ($0° \leq \theta \leq 180°$) denote the angle between the directions of $\mathbf{u}$ and $\mathbf{v}$ as labeled in Figure 1.11. The number

$$d = \|\mathbf{v}\| \cos\theta = \frac{\mathbf{u} \cdot \mathbf{v}}{\|\mathbf{u}\|}$$

is called the **scalar projection** of $\mathbf{v}$ onto $\mathbf{u}$ or the **scalar component** of $\mathbf{v}$ along $\mathbf{u}$. From Figure 1.11, it is clear that d is the length of $\mathbf{Proj_u v}$ if $0° \leq \theta \leq 90°$ and d is the negative of the length of $\mathbf{Proj_u v}$ if $\theta > 90°$. Thus d can be regarded as the *directed length* of $\mathbf{Proj_u v}$.

The geometry involved in having line segments perpendicular to each other breaks down in $\mathbf{R}^n$ if $n > 3$. Even so, we extend the use of the word *orthogonal* to all $\mathbf{R}^n$. Two vectors $\mathbf{u}, \mathbf{v}$ in $\mathbf{R}^n$ are called **orthogonal** if $\mathbf{u} \cdot \mathbf{v} = 0$. A set $\{\mathbf{u}_\lambda \mid \lambda \in \mathcal{L}\}$ of vectors in $\mathbf{R}^n$ is an orthogonal set if $\mathbf{u}_{\lambda_1} \cdot \mathbf{u}_{\lambda_2} = 0$ whenever $\lambda_1 \neq \lambda_2$.

1.5 Exercises

1. Use Figures 1.2 and 1.3 as patterns and illustrate the parallelogram rule with the vectors $\mathbf{u} = (1, 6)$, $\mathbf{v} = (4, -4)$, and $\mathbf{u} + \mathbf{v}$ in an xy-coordinate system.

2. Use Figures 1.2 and 1.4 as patterns and sketch the vectors $\mathbf{u} = (5, 6)$, $\mathbf{v} = (2, -3)$, and $\mathbf{u} - \mathbf{v}$ in an xy-coordinate system.

3. For each $\lambda \in \mathbf{R}$, let $\mathcal{M}_\lambda$ be the set of all points in the plane with rectangular coordinates (x, y) that satisfy $y = \lambda x$. Find $\bigcap_{\lambda \in \mathcal{L}} \mathcal{M}_\lambda$ and $\bigcup_{\lambda \in \mathcal{L}} \mathcal{M}_\lambda$.

4. Find the equation of the line $\langle \mathcal{A} \rangle$ for the given set $\mathcal{A}$.

    (a) $\mathcal{A} = \{(4, -3)\}$ **(b)** $\mathcal{A} = \{(-2, 1)\}$

5. Find the equation of the plane $\langle \mathcal{A} \rangle$ for the given set $\mathcal{A}$.

    (a) $\mathcal{A} = \{(1, 0, 2), (2, -1, 1)\}$ **(b)** $\mathcal{A} = \{(1, 0, 2), (2, 1, 5)\}$

    (c) $\mathcal{A} = \{(2, 1, 0), (1, 0, -1)\}$ **(d)** $\mathcal{A} = \{(4, 3, -1), (-1, 2, 0)\}$

6. Find the lengths of the following vectors.

    (a) $(3, -4, -12)$ **(b)** $(2, 3, 6)$ **(c)** $(1, -2, 4, 2)$

    (d) $(2, 6, 0, -3)$ **(e)** $(1, -2, -4, 3)$ **(f)** $(3, 0, -5, 8)$

    (g) $3(4, 1, 1)$ **(h)** $-2(1, 1, 2, 5)$ **(i)** $3(4, -3) + 2(-12, 5)$

    (j) $2(2, -2) - 4(-3, 2)$

7. Determine x so that the given vectors are orthogonal.

    (a) $(x, 2)$ and $(-3, 9)$ **(b)** $(1, -2, 4)$ and $(x, 5, 2)$

    (c) $(3, x, 0, 1)$ and $(0, -2, 1, 1)$ **(d)** $(5, 2, -1, x, 1)$ and $(x, -2, 3, -3, 1)$

8. A vector of length 1 is called a **unit vector**.

    (a) Find a unit vector that has the same direction as $(3, -4, 12)$.

    (b) Find a vector in the direction of $\mathbf{u} = (2, -3, 6)$ that has length 4 units.

9. Find each of the following scalar projections.

    (a) The scalar projection of $(2, 3, 1)$ onto $(-1, -2, 4)$.

    (b) The scalar projection of $(-3, 4, 12)$ onto $(2, 3, -6)$.

10. Find the vector projection of $\mathbf{v}$ onto $\mathbf{u}$ for each of the following pairs of vectors.

    (a) $\mathbf{v} = (1, 7), \mathbf{u} = (2, 4)$ **(b)** $\mathbf{v} = (-1, 3), \mathbf{u} = (4, -3)$

    (c) $\mathbf{v} = (0, 1, 2), \mathbf{u} = (-2, 5, 2)$ **(d)** $\mathbf{v} = (1, -1, 5), \mathbf{u} = (0, 2, 4)$

11. Find the length of the projection of the given vector $\mathbf{v}$ onto a vector contained in the given line l.

    (a) $\mathbf{v} = (3, 4), \quad l : x - 2y = 0$ **(b)** $\mathbf{v} = (5, -2), \quad l : 4x + y = 0$

12. Use projections to write the vector $(19, 22)$ as a linear combination of $(3, 4)$ and $(4, -3)$. (Note that $(3, 4)$ and $(4, -3)$ are perpendicular.)

13. Use projections to write the vector $(-26, 39)$ as a linear combination of $(5, 12)$, and $(12, -5)$.

14. Let $\mathcal{A} = \{(1, 0, 2), (2, -1, 1)\}$ and let $\mathcal{B} = \{(1, 1, -1), (2, 1, 1)\}$.

    (a) Find a set of vectors that spans $\langle \mathcal{A} \rangle \cap \langle \mathcal{B} \rangle$.

    (b) Find a set of vectors that spans $\langle \mathcal{A} \rangle + \langle \mathcal{B} \rangle$.

15. Work problem 11 with $\mathcal{A} = \{(3, 1, -2), (-2, 1, 3)\}$ and $\mathcal{B} = \{(0, 1, 0), (1, 1, 1)\}$.

16. Give an example of an orthogonal set of vectors in $\mathbf{R}^n$.

17. Prove that an orthogonal set of nonzero vectors in $\mathbf{R}^n$ is linearly independent.

18. Let **u** and **v** be vectors in $\mathbf{R}^n$. Prove that $\|\mathbf{u}\| = \|\mathbf{v}\|$ if and only if $\mathbf{u} + \mathbf{v}$ and $\mathbf{u} - \mathbf{v}$ are orthogonal.

19. Prove Theorem 1.20.

20. Let **u** and **v** be vectors in $\mathbf{R}^n$. Prove that $\mathbf{u} \cdot \mathbf{v} = \frac{1}{4}\|\mathbf{u} + \mathbf{v}\|^2 - \frac{1}{4}\|\mathbf{u} - \mathbf{v}\|^2$.

21. Let **u** and **v** be vectors in $\mathbf{R}^3$. Prove that $|\mathbf{u} \cdot \mathbf{v}| \le \|\mathbf{u}\|\,\|\mathbf{v}\|$.

22. Let **u** and **v** be orthogonal vectors in $\mathbf{R}^n$. Prove that $\|\mathbf{u} + \mathbf{v}\|^2 = \|\mathbf{u}\|^2 + \|\mathbf{v}\|^2$.

23. The **distance** between two vectors **u** and **v** in $\mathbf{R}^n$, denoted by $d(\mathbf{u}, \mathbf{v})$, is defined by

$$d(\mathbf{u}, \mathbf{v}) = \|\mathbf{u} - \mathbf{v}\|.$$

Find the distance between each of the following pairs of vectors.

(a) $\mathbf{u} = (3, 6)$, $\mathbf{v} = (-1, 9)$ **(b)** $\mathbf{u} = (-1, 4, -2)$, $\mathbf{v} = (5, 6, 5)$
(c) $\mathbf{u} = (0, 3, 4, 1)$, $\mathbf{v} = (1, 2, 0, 0)$
(d) $\mathbf{u} = (-1, 1, 5, 3, 0)$, $\mathbf{v} = (-2, 1, 4, 0, 1)$

24. Prove the following properties concerning the distance where **u**, **v**, and **w** are vectors in $\mathbf{R}^n$.

(a) $d(\mathbf{u}, \mathbf{v}) \ge 0$ **(b)** $d(\mathbf{u}, \mathbf{v}) = 0$ if and only if $\mathbf{u} = \mathbf{v}$.
(c) $d(\mathbf{u}, \mathbf{v}) = d(\mathbf{v}, \mathbf{u})$
(d) $d(\mathbf{u}, \mathbf{v}) \le d(\mathbf{u}, \mathbf{w}) + d(\mathbf{w}, \mathbf{v})$ (The triangle inequality)

25. The **cross product** $\mathbf{u} \times \mathbf{v}$ of two vectors $\mathbf{u} = (u_1, u_2, u_3)$ and $\mathbf{v} = (v_1, v_2, v_3)$ is given by

$$\mathbf{u} \times \mathbf{v} = (u_2 v_3 - u_3 v_2,\ u_3 v_1 - u_1 v_3,\ u_1 v_2 - u_2 v_1)$$

$$= \begin{vmatrix} u_2 & u_3 \\ v_2 & v_3 \end{vmatrix} \mathbf{e}_1 + \begin{vmatrix} u_3 & u_1 \\ v_3 & v_1 \end{vmatrix} \mathbf{e}_2 + \begin{vmatrix} u_1 & u_2 \\ v_1 & v_2 \end{vmatrix} \mathbf{e}_3,$$

where $\mathbf{e}_1 = (1, 0, 0)$, $\mathbf{e}_2 = (0, 1, 0)$, $\mathbf{e}_3 = (0, 0, 1)$. The symbolic determinant below is frequently used as a memory aid, since "expansion" about the first row yields the value of $\mathbf{u} \times \mathbf{v}$.

$$\mathbf{u} \times \mathbf{v} = \begin{vmatrix} \mathbf{e}_1 & \mathbf{e}_2 & \mathbf{e}_3 \\ u_1 & u_2 & u_3 \\ v_1 & v_2 & v_3 \end{vmatrix}.$$

Prove the following facts concerning the cross product.

(a) $\mathbf{u} \times \mathbf{v}$ is perpendicular to each of **u**, **v**. **(b)** $\mathbf{u} \times \mathbf{u} = \mathbf{0}$.
(c) $\mathbf{u} \times \mathbf{v} = -(\mathbf{v} \times \mathbf{u})$.   **(d)** $(a\mathbf{u}) \times \mathbf{v} = a(\mathbf{u} \times \mathbf{v}) = \mathbf{u} \times (a\mathbf{v})$.
(e) $\mathbf{u} \times (\mathbf{v} + \mathbf{w}) = (\mathbf{u} \times \mathbf{v}) + (\mathbf{u} \times \mathbf{w})$.   **(f)** $\|\mathbf{u} \times \mathbf{v}\|^2 = \|\mathbf{u}\|^2\|\mathbf{v}\|^2 - (\mathbf{u} \cdot \mathbf{v})^2$.
(g) $\|\mathbf{u} \times \mathbf{v}\| = \|\mathbf{u}\|\,\|\mathbf{v}\| \sin\theta$, where θ is the angle between the directions of **u** and **v**, and $0° \le \theta \le 180°$.
(h) $\|\mathbf{u} \times \mathbf{v}\|$ is the area of a parallelogram with **u** and **v** as adjacent sides.

1.6 Bases and Dimension

We have seen in Section 1.4 that a subset $\mathcal{A}$ of the subspace $\mathbf{W}$ of $\mathbf{R}^n$ may be a spanning set for $\mathbf{W}$ and also that $\mathcal{A}$ may be a linearly independent set. When both of these conditions are imposed, they form the requirements necessary for the subset to be a basis of $\mathbf{W}$.

1.23
DEFINITION

A set $\mathcal{B}$ of vectors is a **basis** of the subspace $\mathbf{W}$ if (i) $\mathcal{B}$ spans $\mathbf{W}$ and (ii) $\mathcal{B}$ is linearly independent.

The empty set $\varnothing$ is regarded as being linearly independent since the condition for linear dependence in Definition 1.7 cannot be satisfied. Thus $\varnothing$ is a basis of the zero subspace of $\mathbf{R}^n$.

Some of our earlier work helps in providing examples concerning bases.

EXAMPLE 1 We saw in Example 1 of Section 1.4 that each of the sets

$$\mathcal{E}_3 = \{(1, 0, 0), (0, 1, 0), (0, 0, 1)\},$$

$$\mathcal{A} = \{(1, 1, 1), (0, 1, 1), (0, 0, 1)\}$$

spans $\mathbf{R}^3$. To show that the set $\mathcal{A}$ is linearly independent, we set up the equation

$$c_1(1, 1, 1) + c_2(0, 1, 1) + c_3(0, 0, 1) = (0, 0, 0).$$

This equation leads directly to the following system of equations.

$$
\begin{aligned}
c_1 & & & = 0 \\
c_1 + c_2 & & & = 0 \\
c_1 + c_2 + c_3 & & & = 0
\end{aligned}
$$

The only solution to this system is $c_1 = 0, c_2 = 0, c_3 = 0$, and therefore $\mathcal{A}$ is linearly independent. It is even easier to see that $\mathcal{E}_3$ is linearly independent. Thus both $\mathcal{E}_3$ and $\mathcal{A}$ are bases for $\mathbf{R}^3$. ∎

EXAMPLE 2 We saw in Example 2 of Section 1.2 that the set

$$\mathcal{A} = \{(1, 1, 8, 1), (1, 0, 3, 0), (3, 1, 14, 1)\}$$

is linearly dependent. It follows that this set $\mathcal{A}$ is not a basis for the subspace

$$\mathbf{W} = \langle (1, 1, 8, 1), (1, 0, 3, 0), (3, 1, 14, 1) \rangle$$

that it spans. ∎

We are concerned in much of our future work with indexed sets of vectors, and we use a restricted type of equality for this type of set. Two indexed sets $\mathcal{A}$ and $\mathcal{B}$ are **equal** if and only if they are indexed $\mathcal{A} = \{\mathbf{u}_\lambda \mid \lambda \in \mathcal{L}\}$ and $\mathcal{B} = \{\mathbf{v}_\lambda \mid \lambda \in \mathcal{L}\}$ by the same index set $\mathcal{L}$ such that $\mathbf{u}_\lambda = \mathbf{v}_\lambda$ for each $\lambda \in \mathcal{L}$. In particular, two finite sets $\mathcal{A} = \{\mathbf{u}_1, \mathbf{u}_2, \ldots, \mathbf{u}_k\}$ and $\mathcal{B} = \{\mathbf{v}_1, \mathbf{v}_2, \ldots, \mathbf{v}_k\}$ are equal if and only if they consist of the same vectors in the same order.

The equality described in the preceding paragraph is the one we shall use in the remainder of this book. For finite sets $\mathcal{A} = \{\mathbf{u}_1, \mathbf{u}_2, \ldots, \mathbf{u}_k\}$ of vectors, this equality is actually an equality of ordered sets. For example, if $\mathbf{u}_1 \neq \mathbf{u}_2$, then

$$\{\mathbf{u}_1, \mathbf{u}_2, \ldots, \mathbf{u}_k\} \neq \{\mathbf{u}_2, \mathbf{u}_1, \ldots, \mathbf{u}_k\}.$$

When we write

$$\mathcal{A} = \{\mathbf{u}_1, \mathbf{u}_2, \ldots, \mathbf{u}_k\},$$

this notation is meant to imply that $\mathcal{A}$ is an ordered set with $\mathbf{u}_1$ as the first vector, $\mathbf{u}_2$ as the second vector, and so on. Moreover, we make a notational agreement for the remainder of this book that when we list the vectors in a set, this listing from left to right specifies their order. For instance, if we write

$$\mathcal{A} = \{(5, -1, 0, 2), (-4, 0, 3, 7), (1, -1, 3, 9)\}$$

this means that $(5, -1, 0, 2)$ is the first vector in $\mathcal{A}$, $(-4, 0, 3, 7)$ is the second vector in $\mathcal{A}$, and $(1, -1, 3, 9)$ is the third vector in $\mathcal{A}$. That is, the vectors in $\mathcal{A}$ are automatically indexed with positive integers $1, 2, 3, \ldots$ from left to right without this being stated.

1.24 THEOREM

Let $\mathcal{B} = \{\mathbf{v}_1, \mathbf{v}_2, \ldots, \mathbf{v}_k\}$ be a basis of a subspace $\mathbf{W}$ of $\mathbf{R}^n$. Then any $\mathbf{v} \in \mathbf{W}$ can be written uniquely as $\mathbf{v} = \sum_{i=1}^{k} a_i \mathbf{v}_i$.

Proof Since $\mathcal{B}$ is a basis then $\mathcal{B}$ spans $\mathbf{W}$, and any $\mathbf{v} \in \mathbf{W}$ can be written as $\sum_{i=1}^{k} a_i \mathbf{v}_i$. Now if $\mathbf{v} = \sum_{i=1}^{k} b_i \mathbf{v}_i$ as well, then

$$\sum_{i=1}^{k} a_i \mathbf{v}_i = \sum_{i=1}^{k} b_i \mathbf{v}_i$$

and therefore

$$\sum_{i=1}^{k} (a_i - b_i) \mathbf{v}_i = \mathbf{0}.$$

Since $\mathcal{B}$ is linearly independent, this requires that $a_i - b_i = 0$ and $a_i = b_i$ for each i. Thus the representation $\mathbf{v} = \sum_{i=1}^{k} a_i \mathbf{v}_i$ is unique. ∎ ∎ ∎

In connection with this last theorem, we observe that if $\mathbf{u} = \mathbf{0}$, then we write

$$\mathbf{u} = \sum_{i=1}^{k} 0\,\mathbf{v}_i = \mathbf{0}$$

and all a_i are zero. Also this uniqueness of coefficients is not valid for spanning sets in general. The set $\mathcal{B}$ in Example 1 of Section 1.1 furnishes an illustration of this fact.

Although a given subspace usually has many different bases, it happens that the number of vectors in different bases of the same subspace is always the same. The derivation of this result is the principal objective of this section.

1.25 THEOREM

Let $\mathbf{W}$ be a subspace of $\mathbf{R}^n$. Suppose that a finite set $\mathcal{A} = \{\mathbf{u}_1, \mathbf{u}_2, \ldots, \mathbf{u}_r\}$ spans $\mathbf{W}$, and let $\mathcal{B}$ be a linearly independent set of vectors in $\mathbf{W}$. Then $\mathcal{B}$ contains at most r vectors.

Proof Let $\mathbf{W}$, $\mathcal{A}$, and $\mathcal{B}$ be as described in the statement of the theorem. If $\mathcal{B}$ contains less than r vectors, the theorem is true. Suppose then that $\mathcal{B}$ contains at least r vectors, say $\{\mathbf{v}_1, \mathbf{v}_2, \ldots, \mathbf{v}_r\} \subseteq \mathcal{B}$.

Our proof of the theorem follows this plan: We shall show that each of the vectors $\mathbf{v}_i$ in $\mathcal{B}$ can be used in turn to replace a suitably chosen vector in $\mathcal{A}$, with $\mathcal{A}$ dependent on the set obtained after each replacement. The replacement process finally leads to the fact that $\mathcal{A}$ is dependent on the set $\{\mathbf{v}_1, \mathbf{v}_2, \ldots, \mathbf{v}_r\}$. We then prove that this set of r vectors must, in fact, be equal to $\mathcal{B}$.

Since $\mathcal{A}$ spans $\mathbf{W}$, $\mathbf{v}_1 = \sum_{i=1}^{r} a_{i1}\mathbf{u}_i$ with at least one $a_{i1} \neq 0$ because $\mathbf{v}_1 \neq \mathbf{0}$. Without loss of generality, we may assume that $a_{11} \neq 0$ in the equation

$$\mathbf{v}_1 = a_{11}\mathbf{u}_1 + a_{21}\mathbf{u}_2 + \cdots + a_{r1}\mathbf{u}_r.$$

(This assumption is purely for notational convenience. We are assuming that the "suitably chosen" vector in $\mathcal{A}$ is the first vector listed in $\mathcal{A}$.) The equation for $\mathbf{v}_1$ implies that

$$a_{11}\mathbf{u}_1 = \mathbf{v}_1 - a_{21}\mathbf{u}_2 - \cdots - a_{r1}\mathbf{u}_r$$

and therefore

$$\mathbf{u}_1 = \left(\frac{1}{a_{11}}\right)\mathbf{v}_1 + \left(-\frac{a_{21}}{a_{11}}\right)\mathbf{u}_2 + \cdots + \left(-\frac{a_{r1}}{a_{11}}\right)\mathbf{u}_r.$$

Thus $\mathbf{u}_1$ is dependent on $\{\mathbf{v}_1, \mathbf{u}_2, \ldots, \mathbf{u}_r\}$, and this clearly implies that $\mathcal{A}$ is dependent on $\{\mathbf{v}_1, \mathbf{u}_2, \ldots, \mathbf{u}_r\}$.

Assume now that $\mathcal{A}$ is dependent on $\{\mathbf{v}_1, \mathbf{v}_2, \ldots, \mathbf{v}_k, \mathbf{u}_{k+1}, \ldots, \mathbf{u}_r\}$, where $1 \leq k < r$. Since $\mathbf{W}$ is dependent on $\mathcal{A}$, then $\mathbf{W}$ is dependent on $\{\mathbf{v}_1, \mathbf{v}_2, \ldots, \mathbf{v}_k, \mathbf{u}_{k+1}, \ldots, \mathbf{u}_r\}$ by

Theorem 1.6. In particular,

$$\mathbf{v}_{k+1} = \sum_{i=1}^{k} b_{i,k+1}\mathbf{v}_i + \sum_{i=k+1}^{r} a_{i,k+1}\mathbf{u}_i.$$

At least one of the coefficients $a_{i,k+1}$ of the $\mathbf{u}_i$ must be nonzero. For if they were all zero, then $\mathbf{v}_{k+1}$ would be a linear combination of $\mathbf{v}_1, \mathbf{v}_2, \ldots, \mathbf{v}_k$, and this would contradict the linear independence of $\mathcal{B}$. Without loss of generality, we may assume that $a_{k+1,k+1} \neq 0$. Hence we obtain

$$\mathbf{v}_{k+1} = \sum_{i=1}^{k} b_{i,k+1}\mathbf{v}_i + \sum_{i=k+1}^{r} a_{i,k+1}\mathbf{u}_i$$

where $a_{k+1,k+1} \neq 0$. This gives

$$\mathbf{u}_{k+1} = \sum_{i=1}^{k} \left(-\frac{b_{i,k+1}}{a_{k+1,k+1}} \right) \mathbf{v}_i + \left(\frac{1}{a_{k+1,k+1}} \right) \mathbf{v}_{k+1} + \sum_{i=k+2}^{r} \left(-\frac{a_{i,k+1}}{a_{k+1,k+1}} \right) \mathbf{u}_i.$$

Thus

$$\{\mathbf{v}_1, \mathbf{v}_2, \ldots, \mathbf{v}_k, \mathbf{u}_{k+1}, \ldots, \mathbf{u}_r\}$$

is dependent on

$$\{\mathbf{v}_1, \mathbf{v}_2, \ldots, \mathbf{v}_k, \mathbf{v}_{k+1}, \mathbf{u}_{k+2}, \ldots, \mathbf{u}_r\}.$$

Since $\mathcal{A}$ is dependent on $\{\mathbf{v}_1, \mathbf{v}_2, \ldots, \mathbf{v}_k, \mathbf{u}_{k+1}, \ldots, \mathbf{u}_r\}$, Theorem 1.6 implies that $\mathcal{A}$ is dependent on $\{\mathbf{v}_1, \mathbf{v}_2, \ldots, \mathbf{v}_k, \mathbf{v}_{k+1}, \mathbf{u}_{k+2}, \ldots, \mathbf{u}_r\}$.

Letting $k = 1, 2, \ldots, r-1$ in the iterative argument above, we see that each $\mathbf{v}_i$ in $\mathcal{B}$ can be used to replace a suitably chosen vector in $\mathcal{A}$ until we obtain the fact that $\mathcal{A}$ is dependent on $\{\mathbf{v}_1, \mathbf{v}_2, \ldots, \mathbf{v}_r\}$. But $\mathcal{B}$ is dependent on $\mathcal{A}$, so we have $\mathcal{B}$ dependent on $\{\mathbf{v}_1, \mathbf{v}_2, \ldots, \mathbf{v}_r\}$. In particular, if $\mathcal{B}$ had more than r elements, any $\mathbf{v}_{r+1}$ in $\mathcal{B}$ would be dependent on $\{\mathbf{v}_1, \mathbf{v}_2, \ldots, \mathbf{v}_r\}$. But this is impossible since $\mathcal{B}$ is independent. Therefore, $\mathcal{B}$ has r elements, and this completes the proof. ■ ■ ■

1.26 COROLLARY

Any linearly independent set of vectors in $\mathbf{R}^n$ contains at most n vectors.

Proof The set of n vectors $\mathbf{e}_1 = (1, 0, \ldots, 0), \mathbf{e}_2 = (0, 1, \ldots, 0), \ldots, \mathbf{e}_n = (0, 0, \ldots, 1)$ spans $\mathbf{R}^n$ since $\mathbf{v} = (v_1, v_2, \ldots, v_n)$ can be written as $\mathbf{v} = \sum_{i=1}^{n} v_i \mathbf{e}_i$. The corollary follows at once from the theorem. ■ ■ ■

If we think in terms of geometric models as presented in Section 1.5, the next theorem seems intuitively obvious. It certainly seems obvious in $\mathbf{R}^2$ and $\mathbf{R}^3$, and there is no reason to suspect the situation to be different in $\mathbf{R}^n$ for other values of n. On the other hand, there is no *compelling* reason to suspect that the situation would *not* be different in $\mathbf{R}^n$ for other values of n. At any rate, we refuse to accept such an important statement on faith or intuition, and insist that this result be validated by a logical argument based upon our development up to this point. This attitude or frame of mind is precisely what is meant when one refers to the "axiomatic method" of mathematics.

1.27 THEOREM

Every subspace of $\mathbf{R}^n$ has a basis with a finite number of elements.

Proof Let $\mathbf{W}$ be a subspace of $\mathbf{R}^n$. If $\mathbf{W} = \{\mathbf{0}\}$, then $\varnothing$ is a basis of $\mathbf{W}$, and the theorem is true.

Suppose $\mathbf{W} \neq \{\mathbf{0}\}$. Then there is at least one nonzero $\mathbf{v}_1$ in $\mathbf{W}$. The set $\{\mathbf{v}_1\}$ is linearly independent by Problem 5 of Exercises 1.2. Thus, there are nonempty subsets of $\mathbf{W}$ that are linearly independent, and Corollary 1.26 shows that each of these subsets contains at most n elements. Let $\mathcal{T}$ be the set of all positive integers t such that $\mathbf{W}$ contains a set of t linearly independent vectors. Then any t in $\mathcal{T}$ satisfies the inequality $1 \leq t \leq n$. Let r be the largest integer in $\mathcal{T}$, and let $\mathcal{B} = \{\mathbf{v}_1, \mathbf{v}_2, \ldots, \mathbf{v}_r\}$ be a linearly independent set of r vectors in $\mathbf{W}$. We shall show that $\mathcal{B}$ spans $\mathbf{W}$.

Let $\mathbf{v}$ be any vector in $\mathbf{W}$. Then $\{\mathbf{v}_1, \mathbf{v}_2, \ldots, \mathbf{v}_r, \mathbf{v}\}$ is linearly dependent, from the choice of r. Thus, there are scalars $c_1, c_2, \ldots, c_r, c_{r+1}$, not all zero, such that

$$\sum_{i=1}^{r} c_i \mathbf{v}_i + c_{r+1}\mathbf{v} = \mathbf{0}.$$

Now $c_{r+1} \neq 0$ since $\mathcal{B}$ is independent. Therefore, $\mathbf{v} = \sum_{i=1}^{r}(-c_i/c_{r+1})\mathbf{v}_i$. This shows that $\mathcal{B}$ spans $\mathbf{W}$, and hence is a basis of $\mathbf{W}$. ∎

This brings us to the main result of this section.

1.28 THEOREM

Let $\mathbf{W}$ be a subspace of $\mathbf{R}^n$, and let $\mathcal{A}$ and $\mathcal{B}$ be any two bases for $\mathbf{W}$. Then $\mathcal{A}$ and $\mathcal{B}$ have the same number of elements.

Proof If $\mathbf{W} = \{\mathbf{0}\}$, then each of $\mathcal{A}$ and $\mathcal{B}$ must be the empty set $\varnothing$, and the number of elements in both $\mathcal{A}$ and $\mathcal{B}$ is 0.

Suppose $\mathbf{W} \neq \{\mathbf{0}\}$. From Corollary 1.26, $\mathcal{A}$ and $\mathcal{B}$ are both finite. Let $\mathcal{A} = \{\mathbf{u}_1, \mathbf{u}_2, \ldots, \mathbf{u}_r\}$ and $\mathcal{B} = \{\mathbf{v}_1, \mathbf{v}_2, \ldots, \mathbf{v}_t\}$. Since $\mathcal{A}$ spans $\mathbf{W}$ and $\mathcal{B}$ is linearly independent, $t \leq r$ by Theorem 1.25. But $\mathcal{B}$ spans $\mathbf{W}$ and $\mathcal{A}$ is linearly independent, so $r \leq t$ by the same theorem. Thus, $t = r$. ∎

1.29
DEFINITION   If **W** is any subspace of $\mathbf{R}^n$, the number of vectors in a basis of **W** is called the **dimension** of **W** and is abbreviated as dim(**W**).

The following theorem is somewhat trivial, but it serves to confirm that the preceding definition of dimension is consistent with our prior experience.

1.30
THEOREM   The dimension of $\mathbf{R}^n$ is n.

Proof Consider the set $\mathcal{E}_n = \{\mathbf{e}_1, \mathbf{e}_2, \ldots, \mathbf{e}_n\}$, where $\mathbf{e}_1 = (1, 0, \ldots, 0)$, $\mathbf{e}_2 = (0, 1, \ldots, 0)$, $\ldots, \mathbf{e}_n = (0, 0, 0, \ldots, 1)$, are the same as in the proof of Corollary 1.26. It was noted in that proof that an arbitrary vector $\mathbf{v} = (v_1, v_2, \ldots, v_n)$ can be written as

$$\mathbf{v} = \sum_{i=1}^{n} v_i \mathbf{e}_i,$$

and therefore $\mathcal{E}_n$ spans $\mathbf{R}^n$.

The set $\mathcal{E}_n$ is linearly independent since

$$\sum_{i=1}^{n} c_i \mathbf{e}_i = \mathbf{0} \text{ implies } (c_1, c_2, \ldots, c_n) = (0, 0, \ldots, 0)$$

and therefore all $c_i = 0$. Thus $\mathcal{E}_n$ is a basis of $\mathbf{R}^n$ with n elements, and it follows that $\mathbf{R}^n$ has dimension n. ■ ■ ■

1.31
DEFINITION   The set $\mathcal{E}_n = \{\mathbf{e}_1, \mathbf{e}_2, \ldots, \mathbf{e}_n\}$ used in the proof of Theorem 1.30 is called the **standard basis** of $\mathbf{R}^n$.

Theorem 1.24 explains why the coefficients $c_1, c_2, \ldots, c_n$ in $\mathbf{v} = \sum_{i=1}^{n} c_i \mathbf{v}_i$ are unique whenever $\mathcal{B} = \{\mathbf{v}_1, \mathbf{v}_2, \ldots, \mathbf{v}_n\}$ is a basis of $\mathbf{R}^n$. The scalars c_i are called the **coordinates** of **v** relative to $\mathcal{B}$. For the special basis $\mathcal{E}_n = \{\mathbf{e}_1, \mathbf{e}_2, \ldots, \mathbf{e}_n\}$, the components of **v** are the same as the coordinates relative to $\mathcal{E}_n$.

EXAMPLE 3 With the basis

$$\mathcal{E}_n = \{(1, 0, 0), (0, 1, 0), (0, 0, 1)\},$$

the coordinates of $\mathbf{v} = (x, y, z)$ relative to $\mathcal{E}_3$ are the same as the components x, y, z, respectively. But for the basis

$$\mathcal{B} = \{(1, 1, 1), (0, 1, 1), (0, 0, 1)\},$$

the coordinates of $\mathbf{v} = (x, y, z)$ relative to $\mathcal{B}$ are the numbers $x, y - x, z - y$ because

$$(x, y, z) = x(1, 1, 1) + (y - x)(0, 1, 1) + (z - y)(0, 0, 1). \qquad \blacksquare$$

There are several types of problems involving "basis" and "dimension" that occur often in linear algebra. In dealing with a certain subspace $\mathbf{W}$, it may be necessary to find the dimension of $\mathbf{W}$, to find a basis of $\mathbf{W}$, or to determine whether or not a given set is a basis of $\mathbf{W}$. Frequently it is desirable to find a basis of $\mathbf{W}$ that has certain specified properties. The fundamental techniques for attacking problems such as these are developed in the remainder of this section.

1.32 **THEOREM**	Every spanning set of a subspace $\mathbf{W}$ of $\mathbf{R}^n$ contains a basis of $\mathbf{W}$.

Proof Suppose that $\mathbf{W}$ is a subspace of $\mathbf{R}^n$ and that $\mathcal{A}$ is a spanning set for $\mathbf{W}$. If $\mathcal{A}$ is independent, then $\mathcal{A}$ is a basis and the theorem is true. Consider now the possibilities when $\mathcal{A}$ is dependent.

If $\mathcal{A} = \{\mathbf{0}\}$, then $\mathbf{W}$ is the zero subspace. But we have already seen that $\varnothing$ is a basis for the zero subspace and $\varnothing \subseteq \{\mathbf{0}\}$. Thus the theorem is true if $\mathcal{A} = \{\mathbf{0}\}$. If $\mathcal{A} \neq \{\mathbf{0}\}$, then there exists a $\mathbf{v}_1$ in $\mathcal{A}$ that is nonzero. If $\mathcal{A}$ is dependent on $\{\mathbf{v}_1\}$, we have a basis of $\mathbf{W}$. If $\mathcal{A}$ is not dependent on $\{\mathbf{v}_1\}$, then there is a $\mathbf{v}_2$ in $\mathcal{A}$ such that $\{\mathbf{v}_1, \mathbf{v}_2\}$ is linearly independent. This procedure can be repeated until we obtain a set $\mathcal{B} = \{\mathbf{v}_1, \mathbf{v}_2, \ldots, \mathbf{v}_r\}, r \leq n$, that is linearly independent and spans $\mathbf{W}$. For if we did not obtain such a linearly independent set, we could continue until we had a linearly independent set in $\mathbf{W}$ containing more than n vectors, and this would contradict Corollary 1.26 since $\mathbf{W} \subseteq \mathbf{R}^n$. The set $\mathcal{B}$ thus obtained is the required basis of $\mathbf{W}$. $\blacksquare\blacksquare\blacksquare$

Although the details of the work would vary, the procedure given in the proof above provides a method for "refining" a basis from a given spanning set. This refinement procedure is demonstrated in the next example.

EXAMPLE 4 With

$$\mathcal{A} = \{(1, 2, 1, 0), (3, -4, 5, 6), (2, -1, 3, 3), (-2, 6, -4, -6)\},$$

we shall use the procedure in the proof of Theorem 1.32 to find a basis of $\mathbf{W} = \langle \mathcal{A} \rangle$ that is contained in $\mathcal{A}$.

It is natural to start the procedure by choosing $\mathbf{v}_1 = (1, 2, 1, 0)$. We see that $\mathcal{A}$ is not dependent on $\{\mathbf{v}_1\}$ because the second vector in $\mathcal{A}$, $(3, -4, 5, 6)$, is not a multiple of $\mathbf{v}_1$. If we let $\mathbf{v}_2 = (3, -4, 5, 6)$, then $\{\mathbf{v}_1, \mathbf{v}_2\}$ is linearly independent.

We need to check now to see if $\mathcal{A}$ is dependent on $\{\mathbf{v}_1, \mathbf{v}_2\}$. When we set up the equation

$$c_1(1, 2, 1, 0) + c_2(3, -4, 5, 6) = (2, -1, 3, 3),$$

this leads to the system of equations

$$
\begin{aligned}
c_1 + 3c_2 &= 2 \\
2c_1 - 4c_2 &= -1 \\
c_1 + 5c_2 &= 3 \\
6c_2 &= 3.
\end{aligned}
$$

The solution to this system is easily found to be $c_1 = c_2 = \frac{1}{2}$. Thus the third vector in $\mathcal{A}$ is dependent on $\{\mathbf{v}_1, \mathbf{v}_2\}$. In similar fashion, we find that

$$(1)(1, 2, 1, 0) + (-1)(3, -4, 5, 6) = (-2, 6, -4, -6).$$

Thus $\mathcal{A}$ is dependent on $\{\mathbf{v}_1, \mathbf{v}_2\}$, and

$$\{(1, 2, 1, 0), (3, -4, 5, 6)\}$$

is a basis for $\langle \mathcal{A} \rangle$.

The work we have done shows that $\langle \mathcal{A} \rangle$ has dimension 2. After checking to see that no vector in $\mathcal{A}$ is a multiple of another vector in $\mathcal{A}$, we can then see that *any pair of vectors from $\mathcal{A}$ forms a basis of $\langle \mathcal{A} \rangle$.* ■

In the proof of Theorem 1.32, we have seen how a basis of a subspace $\mathbf{W}$ can be refined or extracted from an arbitrary spanning set. The spanning set is not *required* to be a finite set, but it could happen to be finite, of course. If a spanning set is finite, the natural refining procedure demonstrated in Example 4 can be given a simpler description: A basis of $\mathbf{W}$ can be obtained by deleting all vectors in the spanning set that are linear combinations of the preceding vectors. Problem 7 of Exercises 1.2 assures us this will lead to an independent set, and Theorem 1.6 assures us this will lead to a spanning set. Thus a basis will result from the deletion of all vectors in a finite spanning set that are linear combinations of preceding vectors as listed in the spanning set.

Our next theorem looks at a procedure that in a sense is opposite to refining: It considers extending a linearly independent set to a basis.

1.33
THEOREM

Any linearly independent set in a subspace $\mathbf{W}$ of $\mathbf{R}^n$ can be extended to a basis of $\mathbf{W}$.

Proof Let $\mathcal{A} = \{\mathbf{u}_1, \mathbf{u}_2, \ldots, \mathbf{u}_r\}$ be a basis of $\mathbf{W}$, and let $\mathcal{B} = \{\mathbf{v}_1, \mathbf{v}_2, \ldots, \mathbf{v}_t\}$ be a linearly independent set in $\mathbf{W}$. If every $\mathbf{u}_i$ is in $\langle \mathcal{B} \rangle$, then $\mathcal{A}$ is dependent on $\mathcal{B}$. By Theorem 1.6, $\mathbf{W}$ is dependent on $\mathcal{B}$, and $\mathcal{B}$ is a basis of $\mathbf{W}$.

Suppose that some $\mathbf{u}_i \notin \langle \mathcal{B} \rangle$. Let k_1 be the smallest integer such that $\mathbf{u}_{k_1} \notin \langle \mathcal{B} \rangle$. Then $\mathcal{B}_1 = \{\mathbf{v}_1, \mathbf{v}_2, \ldots, \mathbf{v}_t, \mathbf{u}_{k_1}\}$ is linearly independent. If each $\mathbf{u}_i \in \langle \mathcal{B}_1 \rangle$, then $\mathcal{B}_1$ spans $\mathbf{W}$ and forms a basis. If some $\mathbf{u}_i \notin \langle \mathcal{B}_1 \rangle$, we repeat the process. After p steps $(1 \leq p \leq r)$, we arrive at a set $\mathcal{B}_p = \{\mathbf{v}_1, \mathbf{v}_2, \ldots, \mathbf{v}_t, \mathbf{u}_{k_1}, \mathbf{u}_{k_2}, \ldots, \mathbf{u}_{k_p}\}$ such that all vectors of $\mathcal{A}$ are dependent on $\mathcal{B}_p$. Thus $\mathcal{B}_p$ spans $\mathbf{W}$. Since no vector in $\mathcal{B}_p$ is a linear combination of the preceding vectors, $\mathcal{B}_p$ is linearly independent by Problem 7 of Exercises 1.2. Therefore $\mathcal{B}_p$ is a basis of **W**. ∎∎∎

In the next example, we follow the preceding proof to extend a linearly independent set to a basis.

EXAMPLE 5 Given that

$$\mathcal{A} = \{(1, 2, 1, 3), (1, 0, 0, 0), (0, 0, 1, 0), (0, 1, 0, 1)\}$$

is a basis of $\mathbf{R}^4$ and that $\mathcal{B} = \{(1, 0, 1, 0), (0, 2, 0, 3)\}$ is linearly independent, we shall extend $\mathcal{B}$ to a basis of $\mathbf{R}^4$.

In keeping with the notational agreement made earlier in this section about indexing, we assume that $\mathcal{A}$ and $\mathcal{B}$ are indexed with positive integers from left to right so that the notation in the proof of Theorem 1.33 applies with

$$\mathbf{u}_1 = (1, 2, 1, 3), \quad \mathbf{u}_2 = (1, 0, 0, 0), \quad \mathbf{u}_3 = (0, 0, 1, 0), \quad \mathbf{u}_4 = (0, 1, 0, 1)$$

and

$$\mathbf{v}_1 = (1, 0, 1, 0), \quad \mathbf{v}_2 = (0, 2, 0, 3).$$

Following the proof of the theorem, we find that

$$(1, 2, 1, 3) = (1, 0, 1, 0) + (0, 2, 0, 3)$$

and thus $\mathbf{u}_1$ is dependent on $\mathcal{B}$. By inspection, we see that

$$(1, 0, 0, 0) = c_1(1, 0, 1, 0) + c_2(0, 2, 0, 3)$$

has no solution. Thus $\mathbf{u}_2 = (1, 0, 0, 0)$ is not in $\langle \mathcal{B} \rangle$, and $k_1 = 2$ is the smallest integer such that $\mathbf{u}_{k_1} \notin \langle \mathcal{B} \rangle$. Using the notation in the proof of the theorem, the set

$$\mathcal{B}_1 = \{\mathbf{v}_1, \mathbf{v}_2, \mathbf{u}_2\}$$
$$= \{(1, 0, 1, 0), (0, 2, 0, 3), (1, 0, 0, 0)\}$$

is linearly independent. We check now for a vector $\mathbf{u}_{k_2}$ in $\mathcal{A}$ that is not in $\langle \mathcal{B}_1 \rangle$. We find that

$$(0, 0, 1, 0) = (1)(1, 0, 1, 0) + (0)(0, 2, 0, 3) + (-1)(1, 0, 0, 0)$$

and $\mathbf{u}_3 = (0, 0, 1, 0)$ is in $\langle \mathcal{B}_1 \rangle$, but the equation

$$(0, 1, 0, 1) = c_1(1, 0, 1, 0) + c_2(0, 2, 0, 3) + c_3(1, 0, 0, 0)$$

has no solution. Thus $\mathbf{u}_4 = (0, 1, 0, 1)$ is not in $\langle \mathcal{B}_1 \rangle$.

After two steps we have arrived at the set

$$\mathcal{B}_2 = \{\mathbf{v}_1, \mathbf{v}_2, \mathbf{u}_2, \mathbf{u}_4\}$$

such that all vectors in $\mathcal{A}$ are dependent on $\mathcal{B}_2$. According to the proof of Theorem 1.33, this set $\mathcal{B}_2$ is a basis of $\mathbf{R}^4$. ∎

Our last two theorems in this section apply to the very special situations where the number of vectors in a set is the same as the dimension r of the subspace involved. For sets of this special type, only one of the conditions for a basis needs to be checked. This is the substance of the following two theorems.

1.34 THEOREM

Let **W** be a subspace of $\mathbf{R}^n$ of dimension r. Then a set of r vectors in **W** is a basis of **W** if and only if it is linearly independent.

Proof If a set of r vectors in **W** is a basis, then it is linearly independent (and spans **W** as well) according to Definition 1.23.

Let $\mathcal{B} = \{\mathbf{v}_1, \mathbf{v}_2, \ldots, \mathbf{v}_r\}$ be a set of r linearly independent vectors in **W**. Then $\mathcal{B}$ can be extended to a basis of **W**, by Theorem 1.33. Now **W** has a basis of r elements since it is of dimension r, and hence all bases of **W** have r elements. In particular, the basis to which $\mathcal{B}$ is extended has r elements, and therefore is the same as $\mathcal{B}$. ■■■

1.35 THEOREM

Let **W** be a subspace of $\mathbf{R}^n$ of dimension r. Then a set of r vectors in **W** is a basis if and only if it spans **W**.

Proof If a set of r vectors in **W** is a basis, then it spans **W** (and is linearly independent as well) by Definition 1.23.

Suppose $\mathcal{A} = \{\mathbf{v}_1, \mathbf{v}_2, \ldots, \mathbf{v}_r\}$ is a set of r vectors which spans **W**. According to Theorem 1.32, $\mathcal{A}$ contains a basis of **W**. But any basis of **W** contains r vectors. Therefore, the basis contained in $\mathcal{A}$ is not a proper subset of $\mathcal{A}$, and $\mathcal{A}$ is a basis of **W**. ■■■

1.6 Exercises

1. Which of the following sets of vectors in $\mathbf{R}^3$ are linearly dependent?
    (a) $\{(1, 3, 1), (1, 3, 0)\}$
    (b) $\{(1, -1, 0), (0, 1, 1), (1, 1, 1), (0, 0, 1)\}$
    (c) $\{(1, 1, 0), (0, 1, 1), (1, 2, 1), (1, 0, -1)\}$
    (d) $\{(1, 0, 1), (0, 1, 1), (2, 1, 3)\}$
    (e) $\{(1, 0, 0), (1, 1, 0), (1, 1, 1)\}$
    (f) $\{(1, 1, 0), (0, 1, -1), (1, 0, 0)\}$

2. Which of the sets of vectors in Problem 1 span $\mathbf{R}^3$?

3. Which of the following sets are bases for $\mathbf{R}^3$?

(a) $\{(1, 0, 0), (0, 1, 0), (0, 0, 1), (1, 1, 1)\}$ (b) $\{(1, 0, 0), (0, 1, 1)\}$

(c) $\{(1, 0, 0), (1, 0, 1), (1, 1, 1)\}$ (d) $\{(1, 0, 0), (0, 1, 0), (1, 1, 0)\}$

(e) $\{(1, 0, 2), (0, 0, 0), (0, 1, 0)\}$ (f) $\{(-1, -1, 0), (1, 2, 3), (3, 2, 1)\}$

4. Determine whether or not $\mathcal{A}$ is a basis of $\mathbf{R}^4$.

(a) $\mathcal{A} = \{(1, 2, 1, 0), (3, -4, 5, 6), (2, -1, 3, 3), (-2, 6, -4, -6)\}$

(b) $\mathcal{A} = \{(1, -1, 2, -3), (1, 1, 2, 0), (3, -1, 6, -6), (0, 2, 0, 3)\}$

(c) $\mathcal{A} = \{(1, 0, 2, -1), (0, 1, 1, 2), (1, 2, 1, 4), (2, 2, 3, 0)\}$

(d) $\mathcal{A} = \{(1, 2, 1, -1), (0, 1, 2, 3), (1, 4, 5, 5), (2, 7, 0, 2)\}$

5. Show that $\mathcal{A}$ is a basis of $\mathbf{R}^3$ by using Theorem 1.34.

(a) $\mathcal{A} = \{(1, -2, 3), (0, 1, -2), (1, -1, 2)\}$

(b) $\mathcal{A} = \{(1, 0, 0), (1, 1, 0), (1, 1, 1)\}$

(c) $\mathcal{A} = \{(2, 0, 0), (4, 1, 0), (3, 3, 1)\}$

(d) $\mathcal{A} = \{(2, -1, 1), (0, 1, -1), (-2, 1, 0)\}$

6. Show that each of the sets $\mathcal{A}$ in Problem 5 is a basis of $\mathbf{R}^3$ by using Theorem 1.35.

7. By direct use of the definition of a basis, show that each of the sets $\mathcal{A}$ in Problem 5 is a basis of $\mathbf{R}^3$.

8. Write out the coordinates of each of the following vectors $\mathbf{v}$ relative to the given basis $\mathcal{B} = \{(1, -1), (2, 3)\}$ of $\mathbf{R}^2$.

(a) $\mathbf{v} = (1, 0)$ (b) $\mathbf{v} = (0, 5)$ (c) $\mathbf{v} = (8, 2)$ (d) $\mathbf{v} = (0, 0)$

9. Write out the coordinates of each of the following vectors $\mathbf{v}$ relative to the given basis $\mathcal{B} = \{(1, 0, -1), (2, 1, 0), (0, 1, 1)\}$ of $\mathbf{R}^3$.

(a) $\mathbf{v} = (-1, -1, -1)$ (b) $\mathbf{v} = (4, -1, 5)$

(c) $\mathbf{v} = (3, 2, 0)$ (d) $\mathbf{v} = (0, 2, -1)$

10. Write out the coordinates of $\mathbf{v} = (0, 5)$ relative to the given basis $\mathcal{B}$ of $\mathbf{R}^2$.

(a) $\mathcal{B} = \{(-2, 3), (1, 2)\}$ (b) $\mathcal{B} = \{(3, -1), (2, 1)\}$

11. Write out the coordinates of $\mathbf{v} = (2, 0, 0)$ relative to the given basis $\mathcal{B}$ of $\mathbf{R}^3$.

(a) $\mathcal{B} = \{(1, 1, 1), (-1, 2, 0), (0, 1, -1)\}$ (b) $\mathcal{B} = \{(1, 0, 1), (0, 2, -3), (1, 1, 0)\}$

12. Write out the coordinates of $\mathbf{v} = (x, y)$ relative to the given basis $\mathcal{B}$ of $\mathbf{R}^2$.

(a) $\mathcal{B} = \{(0, 1), (1, 2)\}$ (b) $\mathcal{B} = \{(3, -1), (2, -1)\}$

13. Write out the coordinates of $\mathbf{v} = (x, y, z)$ relative to the given basis $\mathcal{B}$ of $\mathbf{R}^3$.

(a) $\mathcal{B} = \{(1, 1, 0), (1, 0, 1), (0, 1, 1)\}$ (b) $\mathcal{B} = \{(-1, 0, 1), (2, 1, 0), (0, 1, -1)\}$

14. Find a basis for the given subspace of $\mathbf{R}^n$.

(a) $\mathbf{W} = \{(a, 2a) \mid a \in \mathbf{R}\}$ (b) $\mathbf{W} = \{(0, a) \mid a \in \mathbf{R}\}$

(c) $W = \{(a, 0, 3a) \mid a \in \mathbf{R}\}$ **(d)** $W = \{(a, b, a + b) \mid a, b \in \mathbf{R}\}$

(e) $W = \{(3a, a - b, 2b) \mid a, b \in \mathbf{R}\}$ **(f)** $W = \{(0, -4a, b, 0) \mid a, b \in \mathbf{R}\}$

(g) $W = \{(a - c, b + c, c, -c) \mid a, b, c \in \mathbf{R}\}$

15. Find a basis for the line in $\mathbf{R}^2$ with the given equation.

    (a) $4x - 3y = 0$ **(b)** $2x + 7y = 0$

16. Find a basis for the plane in $\mathbf{R}^3$ with the given equation.

    (a) $3x + 2y - z = 0$ **(b)** $x - 4y + 3z = 0$

    (c) $7x + z = 0$ **(d)** $y - 2z = 0$

17. Extend each of the following sets to a basis of $\mathbf{R}^3$.

    (a) $\{(1, 2, 0)\}$ **(b)** $\{(3, 0, 1), (-1, 1, 1)\}$

18. Extend each of the following sets to a basis of $\mathbf{R}^4$.

    (a) $\{(0, 0, 4, -5)\}$ **(b)** $\{(1, -1, 0, 1), (0, 3, 0, 1)\}$

    (c) $\{(-2, 1, 0, 0), (0, 1, 1, 1)\}$ **(d)** $\{(0, 2, 3, 1), (1, 1, 0, 0), (1, 0, 1, 1)\}$

19. Given that each set A below spans $\mathbf{R}^3$, find a basis of $\mathbf{R}^3$ that is contained in A. (*Hint*: Follow the proof of Theorem 1.32.)

    (a) $A = \{(2, 6, -3), (5, 15, -8), (3, 9, -5), (1, 3, -2), (5, 3, -2)\}$

    (b) $A = \{(1, 0, 2), (0, 1, 1), (2, 1, 5), (1, 1, 3), (1, 2, 1)\}$

    (c) $A = \{(1, 1, 0), (2, 2, 0), (2, 4, 1), (5, 9, 2), (7, 13, 3), (1, 2, 1)\}$

    (d) $A = \{(1, 1, 2), (2, 2, 4), (1, -1, 1), (2, 0, 3), (3, 1, 5), (1, 1, 1)\}$

20. Find a basis of $\langle A \rangle$ that is contained in A.

    (a) $A = \{(1, 0, 1, -1), (3, -2, 3, 5), (2, -1, 2, 2), (5, -2, 5, 3)\}$

    (b) $A = \{(1, 0, 1, 2), (3, 1, 0, 3), (2, 1, -1, 1), (1, -1, 4, 5)\}$

    (c) $A = \{(2, -1, 0, 1), (1, 2, 1, 0), (3, -4, -1, 2), (5, 3, 2, 1)\}$

    (d) $A = \{(1, 0, 1, 1), (0, 1, -1, 1), (1, -1, 2, 0), (2, 1, 1, 7)\}$

21. Find the dimension of $\langle A \rangle$.

    (a) $A = \{(1, 2, 1, 0), (3, -4, 5, 6), (2, -1, 3, 3), (-2, 6, -4, -6)\}$

    (b) $A = \{(4, 3, 2, -1), (5, 4, 3, -1), (-2, -2, -1, 2), (11, 6, 4, 1)\}$

    (c) $A = \{(1, 0, 1, 2, -1), (0, 1, -2, 1, 3), (2, 1, 0, 5, 1), (1, -1, 3, 1, -4)\}$

    (d) $A = \{(1, 2, 0, 1, 0), (2, 4, 1, 4, 3), (1, 2, 2, 5, -2), (-1, -2, 3, 5, 4)\}$

22. Give an example of a two-dimensional subspace for each of the following.

    (a) $\mathbf{R}^4$

    (b) $\langle (1, 0, 0, 0), (1, 1, 0, 0), (0, 0, 0, 0), (1, 1, 1, 0) \rangle$

    (c) $\langle (-2, 2 - 6, 4), (-1, 1, -3, 2), (2, 1, 1, -1), (-1, 4, -8, 5) \rangle$

23. Give an example of a three-dimensional subspace for each of the following.

(a) $\mathbf{R}^4$

(b) $\langle (1, 2, 0, 0, 1), (1, 1, 0, 0, 1), (1, 1, 1, 0, 1), (1, 1, 1, 1, 1) \rangle$

(c) $\langle (0, 1, 4, 0, 0), (0, 0, 1, 3, 0), (0, 2, 7, -6, 0), (-1, 0, 0, 1, 0), (0, 0, 0, 1, 1) \rangle$

24. Show that the sets $\mathcal{A}$ and $\mathcal{B}$ span the same subspace of $\mathbf{R}^4$.

(a) $\mathcal{A} = \{(1, 1, 3, -1), (1, 0, -2, 0)\}$, $\mathcal{B} = \{(3, 2, 4, -2), (0, 1, 5, -1)\}$

(b) $\mathcal{A} = \{(2, 3, 0, -1), (2, 1, -1, 2)\}$, $\mathcal{B} = \{(0, -2, -1, 3), (6, 7, -1, 0)\}$

(c) $\mathcal{A} = \{(1, 0, 3, 0), (1, 1, 8, 4)\}$, $\mathcal{B} = \{(1, -1, -2, -4), (1, 1, 8, 4), (3, -1, 4, -4)\}$

(d) $\mathcal{A} = \{(1, -1, -1, -2), (1, -5, -1, 0)\}$,
$\mathcal{B} = \{(0, 2, 0, -1), (1, -3, -1, -1), (3, -5, -3, -5)\}$

25. Given that each set $\mathcal{A}$ is a basis of $\mathbf{R}^4$ and that each $\mathcal{B}$ is linearly independent, follow the proof of Theorem 1.33 to extend $\mathcal{B}$ to a basis of $\mathbf{R}^4$.

(a) $\mathcal{A} = \{(1, 1, 0, 0), (0, 1, 1, 0), (0, 0, 0, 1), (0, 1, 0, 1)\}$,
$\mathcal{B} = \{(1, 0, 2, 3), (0, 1, -2, -3)\}$

(b) $\mathcal{A} = \{(1, 0, 0, 0), (0, 0, 1, 0), (5, 1, 11, 0), (-4, 0, -6, 1)\}$,
$\mathcal{B} = \{(1, 0, 1, 0), (0, 2, 0, 3)\}$

(c) $\mathcal{A} = \{(1, 1, 1, 1), (1, 1, -1, -1), (1, 0, 1, 0), (0, 1, 0, -1)\}$,
$\mathcal{B} = \{(1, 1, 0, 0), (0, 0, 1, 1)\}$

(d) $\mathcal{A} = \{(1, 1, 0, 0), (1, 0, 4, 6), (0, 0, 0, 1), (0, 1, 0, 1)\}$,
$\mathcal{B} = \{(1, 0, 2, 3), (0, 1, -2, -3)\}$

26. Let $\mathcal{B} = \{\mathbf{v}_1, \mathbf{v}_2, \ldots, \mathbf{v}_k\}$ be a basis of a subspace $\mathbf{W}$ of $\mathbf{R}^n$. Prove that the set $\mathcal{B}' = \{\mathbf{u}, \mathbf{v}_2, \ldots, \mathbf{v}_k\}$ where $\mathbf{u} = a_1\mathbf{v}_1 + a_2\mathbf{v}_2 + \cdots + a_k\mathbf{v}_k$, $a_1 \neq 0$, is also a basis of $\mathbf{W}$.

27. Let $\mathcal{B} = \{\mathbf{v}_1, \mathbf{v}_2, \ldots, \mathbf{v}_k\}$ be a basis of a subspace $\mathbf{W}$ of $\mathbf{R}^n$. Prove that the set $\mathcal{B}' = \{\mathbf{v}_1, \mathbf{v}_1 + \mathbf{v}_2, \ldots, \mathbf{v}_1 + \mathbf{v}_2 + \cdots + \mathbf{v}_k\}$ is also a basis of $\mathbf{W}$.

28. Let $\mathcal{A} = \{\mathbf{v}_1, \mathbf{v}_2, \ldots, \mathbf{v}_r\}$ be a set of nonzero vectors in $\mathbf{R}^n$, such that every vector $\mathbf{v}$ in $\langle \mathcal{A} \rangle$ can be uniquely written as a linear combination of vectors in $\mathcal{A}$. Show that $\mathcal{A}$ is a basis of $\langle \mathcal{A} \rangle$.

29. Prove that if $\mathbf{W}_1$ and $\mathbf{W}_2$ are subspaces of $\mathbf{R}^n$, then

$$\dim(\mathbf{W}_1 + \mathbf{W}_2) = \dim(\mathbf{W}_1) + \dim(\mathbf{W}_2) - \dim(\mathbf{W}_1 \cap \mathbf{W}_2).$$

(*Hint:* Let $\mathcal{C} = \{\mathbf{w}_1, \ldots, \mathbf{w}_r\}$ be a basis of $\mathbf{W}_1 \cap \mathbf{W}_2$, and extend $\mathcal{C}$ to bases $\mathcal{A} = \{\mathbf{w}_1, \ldots, \mathbf{w}_r, \mathbf{u}_1, \ldots, \mathbf{u}_s\}$ and $\mathcal{B} = \{\mathbf{w}_1, \ldots, \mathbf{w}_r, \mathbf{v}_1, \ldots, \mathbf{v}_t\}$ of $\mathbf{W}_1$ and $\mathbf{W}_2$, respectively. Prove that $\{\mathbf{w}_1, \ldots, \mathbf{w}_r, \mathbf{u}_1, \ldots, \mathbf{u}_s, \mathbf{v}_1, \ldots, \mathbf{v}_t\}$ is a basis of $\mathbf{W}_1 + \mathbf{W}_2$.)

30. The sum $\mathbf{W}_1 + \mathbf{W}_2 + \cdots + \mathbf{W}_k = \sum_{j=1}^{k} \mathbf{W}_j$ of subspaces $\mathbf{W}_j$ of $\mathbf{R}^n$ is defined to be the set of all vectors of the form $\mathbf{v}_1 + \mathbf{v}_2 + \cdots + \mathbf{v}_k$, with $\mathbf{v}_i$ in $\mathbf{W}_i$. The sum $\sum_{j=1}^{k} \mathbf{W}_j$

is called *direct* if

$$\mathbf{W}_i \cap \sum_{\substack{j=1 \\ j \neq i}}^{k} \mathbf{W}_j = \{\mathbf{0}\}$$

for $i = 1, 2, \ldots, k$. A direct sum is written as $\mathbf{W}_1 \oplus \mathbf{W}_2 \oplus \cdots \oplus \mathbf{W}_k$. Prove that

$$\dim(\mathbf{W}_1 \oplus \mathbf{W}_2 \oplus \cdots \oplus \mathbf{W}_k) = \sum_{j=1}^{k} \dim(\mathbf{W}_j).$$

2 Elementary Operations on Vectors

The elementary operations are as fundamental in linear algebra as the operations of differentiation and integration are in the calculus. These elementary operations are indispensable both in the development of the theory of linear algebra and in the applications of this theory.

In many treatments of linear algebra, the elementary operations are introduced after the development of a certain amount of matrix theory, and the matrix theory is used as a tool in establishing the properties of the elementary operations. In the presentation here, this procedure is reversed somewhat. The elementary operations are introduced as operations on sets of vectors and many of the results in matrix theory are developed with the aid of our knowledge of elementary operations. This approach has two main advantages. The material in Chapter 1 can be used to efficiently develop several of the properties of elementary operations, and the statements of many of these properties are simpler when formulated in vector terminology.

2.1 Elementary Operations and Their Inverses

We saw in the proof of Theorem 1.30 that $\mathbf{R}^n$ has the standard basis $\mathcal{E}_n = \{\mathbf{e}_1, \mathbf{e}_2, \ldots, \mathbf{e}_n\}$ in which each vector $\mathbf{e}_i$ has a very simple form. The Kronecker delta that was introduced in Exercises 1.1 can be used to describe the vectors $\mathbf{e}_i$ in a concise way. Using the fact that

$$\delta_{ij} = \begin{cases} 1 \text{ if } i = j \\ 0 \text{ if } i \neq j \end{cases}$$

the vectors $\mathbf{e}_i$ can be written as

$$\mathbf{e}_i = (\delta_{i1}, \delta_{i2}, \ldots, \delta_{in})$$

for $i = 1, 2, \ldots, n$.

EXAMPLE 1 The vectors in the standard basis $\mathcal{E}_4 = \{\mathbf{e}_1, \mathbf{e}_2, \mathbf{e}_3, \mathbf{e}_4\}$ are given by

$$\mathbf{e}_1 = (\delta_{11}, \delta_{12}, \delta_{13}, \delta_{14}) = (1, 0, 0, 0), \qquad \mathbf{e}_2 = (\delta_{21}, \delta_{22}, \delta_{23}, \delta_{24}) = (0, 1, 0, 0),$$

$$\mathbf{e}_3 = (\delta_{31}, \delta_{32}, \delta_{33}, \delta_{34}) = (0, 0, 1, 0), \qquad \mathbf{e}_4 = (\delta_{41}, \delta_{42}, \delta_{43}, \delta_{44}) = (0, 0, 0, 1). \quad \blacksquare$$

We shall see in this chapter that every subspace $\mathbf{W}$ of $\mathbf{R}^n$ has a basis with a certain simple form, and that particular basis is called the *standard basis* of $\mathbf{W}$. In order to develop the concept of this standard basis, we shall need certain operations that change one spanning set of $\mathbf{W}$ into another spanning set of $\mathbf{W}$.

More specifically, we define three types of elementary operations on nonempty ordered finite sets of vectors. These types of elementary operations will be referred to hereafter as types I, II, or III. Let $\mathcal{A} = \{\mathbf{v}_1, \mathbf{v}_2, \ldots, \mathbf{v}_k\}$ be a set of vectors in $\mathbf{R}^n$.

(i) An *elementary operation of type I* multiplies one of the $\mathbf{v}_i$ in $\mathcal{A}$ by a nonzero scalar.

(ii) An *elementary operation of type II* replaces one of the vectors $\mathbf{v}_s$ by the sum of $\mathbf{v}_s$ and a scalar multiple of a $\mathbf{v}_t (s \neq t)$ in $\mathcal{A}$.

(iii) An *elementary operation of type III* interchanges two vectors in $\mathcal{A}$.

If the number 1 is used as the scalar in an elementary operation of type I, the resulting elementary operation is called the *identity operation*. That is, the *identity operation* on a set is that operation that leaves the set unchanged.

The following definition gives a notation that is useful in describing the result obtained when an elementary operation is applied to a set.

2.1 **DEFINITION**	If $\mathcal{A} \subseteq \mathbf{R}^n$ and E denotes an elementary operation that may be applied to $\mathcal{A}$, then $E(\mathcal{A})$ denotes the set that is obtained by applying E to $\mathcal{A}$.

EXAMPLE 2 Consider the set of vectors

$$\mathcal{A} = \{\mathbf{v}_1 = (1, 0, 2), \mathbf{v}_2 = (2, 3, 19), \mathbf{v}_3 = (0, 1, 5)\}.$$

The elementary operation E_1 of type I that multiplies the first vector in $\mathcal{A}$ by 2 results in the set $\mathcal{A}_1$, where

$$\mathcal{A}_1 = E_1(\mathcal{A}) = \{\mathbf{v}_{11} = (2, 0, 4), \mathbf{v}_{12} = (2, 3, 19), \mathbf{v}_{13} = (0, 1, 5)\}.$$

If the vector $\mathbf{v}_{12}$ in $\mathcal{A}_1$ is replaced by $\mathbf{v}_{12} + (-3)\mathbf{v}_{13}$, we have an elementary operation E_2 of type II that yields the set $\mathcal{A}_2$, where

$$\mathcal{A}_2 = E_2(\mathcal{A}_1) = \{\mathbf{v}_{21} = (2, 0, 4), \mathbf{v}_{22} = (2, 0, 4), \mathbf{v}_{23} = (0, 1, 5)\}.$$

An interchange of $\mathbf{v}_{22}$ and $\mathbf{v}_{23}$ is an elementary operation E_3 of type III that produces

$$\mathcal{A}_3 = E_3(\mathcal{A}_2) = \{\mathbf{v}_{31} = (2, 0, 4), \mathbf{v}_{32} = (0, 1, 5), \mathbf{v}_{33} = (2, 0, 4)\}.$$

Application of an elementary operation E_4 of type II that replaces $\mathbf{v}_{33}$ by $\mathbf{v}_{33} + (-1)\mathbf{v}_{31}$ gives

$$\mathcal{A}_4 = E_4(\mathcal{A}_3) = \{\mathbf{v}_{41} = (2, 0, 4), \mathbf{v}_{42} = (0, 1, 5), \mathbf{v}_{43} = (0, 0, 0)\}. \qquad \blacksquare$$

As the example above suggests, the application of a sequence of elementary operations can be used to replace a given set $\mathcal{A}$ by a set $\mathcal{A}'$ which has a simpler appearance. The next definition describes the result obtained when several elementary operations are applied in a sequence. Later in this chapter, we shall turn to an investigation of the properties that $\mathcal{A}$ and $\mathcal{A}'$ have in common. When these properties are known, we shall see that the elementary operations can be chosen so as to make the set $\mathcal{A}'$ display certain important information concerning $\mathcal{A}$.

2.2
DEFINITION

If $A \subseteq \mathbf{R}^n$ and $E_1, E_2, \ldots, E_t$ is a sequence of elementary operations that may be applied to A, then $E_t E_{t-1} \cdots E_2 E_1(A)$ is defined inductively by

$$E_t E_{t-1} \cdots E_2 E_1(A) = E_t \left(E_{t-1} \cdots E_2 E_1(A) \right).$$

EXAMPLE 3 In Example 2, we can thus write

$$E_1(A) = \{2\mathbf{v}_1, \mathbf{v}_2, \mathbf{v}_3\}$$
$$= \{(2, 0, 4), (2, 3, 19), (0, 1, 5)\}$$
$$E_2 E_1(A) = E_2(E_1(A))$$
$$= E_2(\{(2, 0, 4), (2, 3, 19), (0, 1, 5)\})$$
$$= \{(2, 0, 4), (2, 0, 4), (0, 1, 5)\}$$
$$E_3 E_2 E_1(A) = E_3(E_2 E_1(A))$$
$$= E_3(\{(2, 0, 4), (2, 0, 4), (0, 1, 5)\})$$
$$= \{(2, 0, 4), (0, 1, 5), (2, 0, 4)\}$$
$$E_4 E_3 E_2 E_1(A) = E_4 \left(E_3 E_2 E_1(A) \right)$$
$$= E_4\{(2, 0, 4), (0, 1, 5), (2, 0, 4)\}$$
$$= \{(2, 0, 4), (0, 1, 5), (0, 0, 0)\}. \qquad \blacksquare$$

EXAMPLE 4 Let $A = \{\mathbf{v}_1, \mathbf{v}_2, \mathbf{v}_3\}$, where $\mathbf{v}_1 = (1, 0, 2)$, $\mathbf{v}_2 = (2, 1, 6)$, $\mathbf{v}_3 = (0, 3, 8)$, and consider the sequence E_1, E_2, E_3, where the elementary operations E_i are given by:

E_1: Replace the second vector by the sum of the second vector and (-2) times the first vector.

E_2: Replace the third vector by the sum of the third vector and (-3) times the second vector.

E_3: Multiply the third vector by $\frac{1}{2}$.

According to Definition 2.2,

$$E_1(A) = \{\mathbf{v}_1, \mathbf{v}_2 + (-2)\mathbf{v}_1, \mathbf{v}_3\}$$
$$= \{(1, 0, 2), (0, 1, 2), (0, 3, 8)\}$$
$$E_2 E_1(A) = E_2(E_1(A))$$
$$= E_2(\{(1, 0, 2), (0, 1, 2), (0, 3, 8)\})$$
$$= \{(1, 0, 2), (0, 1, 2), (0, 0, 2)\}$$

$$E_3 E_2 E_1(\mathcal{A}) = E_3(E_2 E_1(\mathcal{A}))$$

$$= E_3(\{(1, 0, 2), (0, 1, 2), (0, 0, 2)\})$$

$$= \{(1, 0, 2), (0, 1, 2), (0, 0, 1)\}. \qquad \blacksquare$$

In the investigation of the properties shared by the sets $\mathcal{A}$ and $\mathcal{A}'$, where $\mathcal{A}'$ is obtained from $\mathcal{A}$ by a sequence of elementary operations, it is convenient to have available the concept of an *inverse* of an elementary operation. Suppose that an elementary operation is performed on $\mathcal{A} = \{\mathbf{v}_1, \mathbf{v}_2, \ldots, \mathbf{v}_k\}$ to obtain a new set $\mathcal{A}'$, and consider the operation necessary to obtain the original set $\mathcal{A}$ from the new set $\mathcal{A}'$.

(i) If $\mathcal{A}'$ is obtained by employing a type I elementary operation, then $\mathcal{A}'$ is of the form

$$\mathcal{A}' = \{\mathbf{v}_1, \ldots, \mathbf{v}_{s-1}, a\mathbf{v}_s, \mathbf{v}_{s+1}, \ldots, \mathbf{v}_k\},$$

where $a \neq 0$. It is readily seen that $\mathcal{A}$ is obtained from $\mathcal{A}'$ by replacing $a\mathbf{v}_s$ by $\frac{1}{a}(a\mathbf{v}_s)$. Thus in this case $\mathcal{A}$ is obtained from $\mathcal{A}'$ by an elementary operation of the same type.

(ii) If a type II operation is used to obtain $\mathcal{A}'$ from $\mathcal{A}$, then $\mathcal{A}'$ has the form

$$\mathcal{A}' = \{\mathbf{v}_1, \ldots, \mathbf{v}_{s-1}, \mathbf{v}_s + b\mathbf{v}_t, \mathbf{v}_{s+1}, \ldots, \mathbf{v}_k\},$$

where $s \neq t$. Now $\mathbf{v}_t$ is in $\mathcal{A}'$, and if $\mathbf{v}_s + b\mathbf{v}_t$ in $\mathcal{A}'$ is replaced by $(\mathbf{v}_s + b\mathbf{v}_t) + (-b)\mathbf{v}_t$, the original set $\mathcal{A}$ is obtained. This replacement is an elementary operation of type II, so that, once again, $\mathcal{A}$ is obtained from $\mathcal{A}'$ by an elementary operation of the same type as was used in obtaining $\mathcal{A}'$ from $\mathcal{A}$.

(iii) If $\mathcal{A}'$ is obtained from $\mathcal{A}$ by interchanging the vectors $\mathbf{v}_s$ and $\mathbf{v}_t$ in $\mathcal{A}$, then $\mathcal{A}$ is obtained from $\mathcal{A}'$ by the very same operation of interchanging $\mathbf{v}_s$ and $\mathbf{v}_t$.

We see, then, that once an elementary operation is applied to a set $\mathcal{A}$ to obtain a set $\mathcal{A}'$, we need only apply another elementary operation of the same type to $\mathcal{A}'$ in order to obtain $\mathcal{A}$.

**2.3
DEFINITION**

When an elementary operation E is applied to a set $\mathcal{A}$ to obtain a set $\mathcal{A}'$, the elementary operation that must be applied to $\mathcal{A}'$ in order to obtain $\mathcal{A}$ is called the **inverse elementary operation** of E, and is denoted by E^{-1}.

It is clear from our discussion above that the inverse of an elementary operation E is unique, and is of the same type as E.

**2.4
THEOREM**

If $\mathcal{A}'$ is obtained from $\mathcal{A}$ by a sequence of elementary operations, then $\mathcal{A}$ can be obtained from $\mathcal{A}'$ by applying the inverses of these elementary operations in reverse order.

Proof   Suppose that $\mathcal{A}'$ is obtained from $\mathcal{A}$ by a sequence $E_1, E_2, \ldots, E_t$ of elementary operations. That is, the operations $E_1, E_2, \ldots, E_t$ are applied successively, obtaining a new set $\mathcal{A}_i$ each time an E_i is applied, until we obtain $\mathcal{A}_t = \mathcal{A}'$.

$$\mathcal{A}_1 = E_1(\mathcal{A})$$

$$\mathcal{A}_2 = E_2\,(\mathcal{A}_1) = E_2 E_1(\mathcal{A})$$

$$\mathcal{A}_3 = E_3\,(\mathcal{A}_2) = E_3 E_2 E_1(\mathcal{A})$$

$$\vdots$$

$$\mathcal{A}_{t-1} = E_{t-1}\,(\mathcal{A}_{t-2}) = E_{t-1} \cdots E_3 E_2 E_1(\mathcal{A})$$

$$\mathcal{A}' = \mathcal{A}_t = E_t\,(\mathcal{A}_{t-1}) = E_t E_{t-1} \cdots E_3 E_2 E_1(\mathcal{A})$$

Now consider the sequence $E_t^{-1}, E_{t-1}^{-1}, \ldots, E_2^{-1}, E_1^{-1}$ applied to $\mathcal{A}'$. Applying E_t^{-1} to $\mathcal{A}' = \mathcal{A}_t$ results in $\mathcal{A}_{t-1}$ since E_t yields $\mathcal{A}_t$ when applied to $\mathcal{A}_{t-1}$. Continuing in this manner, we obtain, successively, the sets

$$\mathcal{A}_{t-1} = E_t^{-1}\,(\mathcal{A}_t) = E_t^{-1}\,(\mathcal{A}')$$

$$\mathcal{A}_{t-2} = E_{t-1}^{-1}\,(\mathcal{A}_{t-1}) = E_{t-1}^{-1} E_t^{-1}\,(\mathcal{A}')$$

$$\mathcal{A}_{t-3} = E_{t-2}^{-1}\,(\mathcal{A}_{t-2}) = E_{t-2}^{-1} E_{t-1}^{-1} E_t^{-1}\,(\mathcal{A}')$$

$$\vdots$$

$$\mathcal{A}_1 = E_2^{-1}\,(\mathcal{A}_2) = E_2^{-1} \cdots E_{t-2}^{-1} E_{t-1}^{-1} E_t^{-1}\,(\mathcal{A}')$$

$$\mathcal{A} = E_1^{-1}\,(\mathcal{A}_1) = E_1^{-1} E_2^{-1} \cdots E_{t-2}^{-1} E_{t-1}^{-1} E_t^{-1}\,(\mathcal{A}')\,.$$

Therefore $\mathcal{A}$ results by applying the sequence $E_t^{-1}, E_{t-1}^{-1}, \ldots, E_2^{-1}, E_1^{-1}$ to $\mathcal{A}'$, and the theorem is proved.   ∎ ∎ ∎

An illustration of this theorem and its proof is provided in the next example.

EXAMPLE 5   In Example 2, the set

$$\mathcal{A}' = \mathcal{A}_4 = \{\mathbf{v}_{41} = (2, 0, 4),\ \mathbf{v}_{42} = (0, 1, 5),\ \mathbf{v}_{43} = (0, 0, 0)\}$$

is obtained from the set

$$\mathcal{A} = \{\mathbf{v}_1 = (1, 0, 2),\ \mathbf{v}_2 = (2, 3, 19),\ \mathbf{v}_3 = (0, 1, 5)\}$$

by a sequence E_1, E_2, E_3, E_4 of elementary operations that can be described as follows:

E_1:   Multiply the first vector by 2.

E_2:   Replace the second vector by the sum of the second vector and (-3) times the third vector.

E_3: Interchange the second and third vectors.

E_4: Replace the third vector by the sum of the third vector and (-1) times the first vector.

Utilizing the general discussion preceding Definition 2.3, we formulate the inverse elementary operations as follows.

E_1^{-1}: Multiply the first vector by $\frac{1}{2}$.

E_2^{-1}: Replace the second vector by the sum of the second vector and 3 times the third vector.

E_3^{-1}: Interchange the second and third vectors.

E_4^{-1}: Replace the third vector by the sum of the third vector and 1 times the first vector.

Applying these inverse operations to $\mathcal{A}'$ in reverse order, we find that

$$E_4^{-1}(\mathcal{A}') = E_4^{-1}\{(2, 0, 4), (0, 1, 5), (0, 0, 0)\}$$
$$= \{(2, 0, 4), (0, 1, 5), (2, 0, 4)\} = \mathcal{A}_3$$
$$E_3^{-1}(E_4^{-1}(\mathcal{A})) = E_3^{-1}(\mathcal{A}_3)$$
$$= E_3^{-1}(\{(2, 0, 4), (0, 1, 5), (2, 0, 4)\})$$
$$= \{(2, 0, 4), (2, 0, 4), (0, 1, 5)\} = \mathcal{A}_2$$
$$E_2^{-1}\left(E_3^{-1}E_4^{-1}(\mathcal{A})\right) = E_2^{-1}(\mathcal{A}_2)$$
$$= E_2^{-1}(\{(2, 0, 4), (2, 0, 4), (0, 1, 5)\})$$
$$= \{(2, 0, 4), (2, 3, 19), (0, 1, 5)\} = \mathcal{A}_1$$
$$E_1^{-1}\left(E_2^{-1}E_3^{-1}E_4^{-1}(\mathcal{A})\right) = E_1^{-1}(\mathcal{A}_1)$$
$$= E_1^{-1}\{(2, 0, 4), (2, 3, 19), (0, 1, 5)\}$$
$$= \{(1, 0, 2), (2, 3, 19), (0, 1, 5)\} = \mathcal{A}. \qquad \blacksquare$$

2.1 Exercises

1. Write out the elements of the standard basis of $\mathbf{R}^5$.

2. Find an elementary operation that yields

$$\{(1, 0, 2, 1), (0, 3, 0, 7), (3, 6, 4, 3)\}$$

when applied to $\{(1, 0, 2, 1), (-2, 3, -4, 5), (3, 6, 4, 3)\}$.

3. Find an elementary operation that yields

$$\{(1, 0, 2, 1), (-2, 3, -4, 5), (3, 6, 4, 3)\}$$

when applied to $\{(1, 0, 2, 1), (0, 3, 0, 7), (3, 6, 4, 3)\}$.

4. Show that the set

$$\{(2, 3, 0, -1), (2, 1, -1, 2)\}$$

can be obtained from the set $\{(0, -2, -1, 3), (6, 7, -1, 0)\}$ by a sequence of elementary operations.

5. Show that the set

$$\{(0, -2, -1, 3), (6, 7, -1, 0)\}$$

can be obtained from the set $\{(2, 3, 0, -1), (2, 1, -1, 2)\}$ by a sequence of elementary operations.

6. Show that the standard basis $\mathcal{E}_3$ of $\mathbf{R}^3$ can be obtained from the given set $\mathcal{A}$ by a sequence of elementary operations.

(a) $\mathcal{A} = \{(-3, 5, 3), (1, -2, 0), (-1, -2, 1)\}$
(b) $\mathcal{A} = \{(2, 2, 4), (1, 3, 0), (3, 1, 1)\}$

7. Show that the standard basis $\mathcal{E}_4$ of $\mathbf{R}^4$ can be obtained from the given set $\mathcal{A}$ by a sequence of elementary operations.

(a) $\mathcal{A} = \{(0, 2, 1, 4), (4, 1, 0, 1), (3, 0, 1, 0), (0, 1, 5, 1)\}$
(b) $\mathcal{A} = \{(5, 2, 1, -4), (0, 1, 1, 1), (-5, 0, 1, 4), (6, 1, 0, 0)\}$

8. Can the standard basis $\mathcal{E}_3$ of $\mathbf{R}^3$ be obtained from the given set $\mathcal{A}$ by a sequence of elementary operations?

(a) $\mathcal{A} = \{(1, 1, 1), (2, 2, 2), (3, 3, 3)\}$
(b) $\mathcal{A} = \{(1, 0, 1), (-1, 0, 1), (7, 0, 3)\}$

9. Can the standard basis $\mathcal{E}_4$ of $\mathbf{R}^4$ be obtained from the given set $\mathcal{A}$ by a sequence of elementary operations?

(a) $\mathcal{A} = \{(1, 1, 1, 1), (2, 2, 2, 0), (3, 3, 3, 0)\}$
(b) $\mathcal{A} = \{(1, 0, 1, 0), (-1, 0, 1, 0), (0, 1, 0, 1), (0, 1, 2, 1)\}$

10. Let $\mathcal{A} = \{\mathbf{v}_1, \mathbf{v}_2, \mathbf{v}_3\}$ and $\mathcal{A}' = \{\mathbf{v}'_1, \mathbf{v}'_2, \mathbf{v}'_3\}$ be sets of vectors in $\mathbf{R}^n$ such that

$$\mathbf{v}'_1 = 2\mathbf{v}_1$$
$$\mathbf{v}'_2 = 2\mathbf{v}_2 + 3\mathbf{v}_3$$
$$\mathbf{v}'_3 = \mathbf{v}_3 + \mathbf{v}_1.$$

Write out a sequence of elementary operations that yields $\mathcal{A}'$ when applied to $\mathcal{A}$.

11. Let $\mathcal{A} = \{\mathbf{v}_1, \mathbf{v}_2, \mathbf{v}_3, \mathbf{v}_4\}$ and let $\mathcal{A}' = \{\mathbf{v}'_1, \mathbf{v}'_2, \mathbf{v}'_3, \mathbf{v}'_4\}$ be sets of vectors in $\mathbf{R}^3$ such that

$$\mathbf{v}'_1 = \mathbf{v}_1$$
$$\mathbf{v}'_2 = \mathbf{v}_1 + \mathbf{v}_2$$
$$\mathbf{v}'_3 = \mathbf{v}_2 + \mathbf{v}_3$$
$$\mathbf{v}'_4 = \mathbf{v}_3 + \mathbf{v}_4.$$

Write out a sequence of elementary operations that yields $\mathcal{A}'$ when applied to $\mathcal{A}$.

12. Let $\mathcal{A} = \{\mathbf{v}_1, \mathbf{v}_2, \mathbf{v}_3, \mathbf{v}_4\}$ and let $\mathcal{A}' = \{\mathbf{v}_1', \mathbf{v}_2', \mathbf{v}_3', \mathbf{v}_4'\}$ be sets of vectors in $\mathbf{R}^3$ such that

$$\mathbf{v}_1' = \mathbf{v}_1 - \mathbf{v}_4$$
$$\mathbf{v}_2' = \mathbf{v}_2 + 3\mathbf{v}_1$$
$$\mathbf{v}_3' = \mathbf{v}_3 + 2\mathbf{v}_2$$
$$\mathbf{v}_4' = 2\mathbf{v}_4.$$

Write out a sequence of elementary operations that yields $\mathcal{A}'$ when applied to $\mathcal{A}$.

13. Assume that the set $\mathcal{A}' = \{\mathbf{v}_1', \mathbf{v}_2', \mathbf{v}_3'\}$ is obtained from the set $\mathcal{A} = \{\mathbf{v}_1, \mathbf{v}_2, \mathbf{v}_3\}$ by the sequence E_1, E_2, E_3 defined as follows.

E_1: Multiply the first vector by -3.

E_2: Replace the second vector by the sum of the second vector and 2 times the first vector.

E_3: Replace the third vector by the sum of the third vector and (-2) times the second vector.

Write out a sequence of elementary operations that yields $\mathcal{A}$ when applied to $\mathcal{A}'$.

14. With the sets $\mathcal{A}$ and $\mathcal{A}'$ as given in Problem 10, write out a sequence of elementary operations that yields $\mathcal{A}$ when applied to $\mathcal{A}'$.

15. With the sets $\mathcal{A}$ and $\mathcal{A}'$ as given in Problem 11, write out a sequence of elementary operations that yields $\mathcal{A}$ when applied to $\mathcal{A}'$.

16. With the sets $\mathcal{A}$ and $\mathcal{A}'$ as given in Problem 12, write out a sequence of elementary operations that yields $\mathcal{A}$ when applied to $\mathcal{A}'$.

17. Show that the sequence of elementary operations used to obtain $\mathcal{A}_4$ from $\mathcal{A}$ in Example 2 is *not* unique by exhibiting a different sequence of elementary operations that yields $\mathcal{A}'$ when applied to $\mathcal{A}$.

18. Show that the identity operation on a set with more than one element is an elementary operation of type II.

2.2 Elementary Operations and Linear Independence

One of the properties that is preserved by application of an elementary operation to a set is that of linear independence. This important result is established in the next theorem.

2.5 THEOREM   Suppose that $\mathcal{A}$ and $\mathcal{A}'$ are sets of vectors in $\mathbf{R}^n$ such that $\mathcal{A}'$ is obtained from $\mathcal{A}$ by applying a single elementary operation. Then $\mathcal{A}'$ is linearly independent if and only if $\mathcal{A}$ is linearly independent.

Proof Suppose first that $\mathcal{A} = \{\mathbf{v}_1, \mathbf{v}_2, \ldots, \mathbf{v}_k\}$ is linearly independent.

If $\mathcal{A}'$ is obtained by a type I elementary operation, then

$$\mathcal{A}' = \{\mathbf{v}_1, \ldots, \mathbf{v}_{s-1}, \ a\mathbf{v}_s, \mathbf{v}_{s+1}, \ldots, \mathbf{v}_k\},$$

where $a \neq 0$. Now suppose that

$$\sum_{\substack{i=1 \\ i \neq s}}^{k} c_i \mathbf{v}_i + c_s a \mathbf{v}_s = \mathbf{0}.$$

This implies that $c_1 = \cdots = c_{s-1} = c_s a = c_{s+1} = \cdots = c_k = 0$ since $\mathcal{A}$ is linearly independent. But $c_s a = 0$ implies $c_s = 0$ since $a \neq 0$. Therefore, all c_i are zero, and $\mathcal{A}'$ is linearly independent.

If $\mathcal{A}'$ is obtained from $\mathcal{A}$ by a type II elementary operation, then

$$\mathcal{A}' = \{\mathbf{v}_1, \ldots, \mathbf{v}_{s-1}, \mathbf{v}_s + b\mathbf{v}_t, \mathbf{v}_{s+1}, \ldots, \mathbf{v}_k\},$$

where $s \neq t$. If

$$\sum_{\substack{i=1 \\ i \neq s}}^{k} c_i \mathbf{v}_i + c_s (\mathbf{v}_s + b\mathbf{v}_t) = \mathbf{0},$$

then

$$\sum_{\substack{i=1 \\ i \neq s,t}}^{k} c_i \mathbf{v}_i + c_s \mathbf{v}_s + (c_t + c_s b)\mathbf{v}_t = \mathbf{0},$$

and this implies that $c_1 = \cdots = c_s = \cdots = c_t + c_s b = \cdots = c_k = 0$ since $\mathcal{A}$ is linearly independent. But $c_s = 0$ and $c_t + c_s b = 0$ imply that $c_t = 0$. Hence all c_i are zero, and $\mathcal{A}'$ is linearly independent.

If $\mathcal{A}'$ is obtained by an elementary operation of type III, then $\mathcal{A}'$ consists of exactly the same vectors as does $\mathcal{A}$, except that the order of the vectors is different. It is clear, then, that $\mathcal{A}'$ is linearly independent.

Thus, $\mathcal{A}'$ is linearly independent if $\mathcal{A}$ is linearly independent.

Suppose now that $\mathcal{A}'$ is linearly independent. Since $\mathcal{A}'$ is obtained from $\mathcal{A}$ by a single elementary operation, $\mathcal{A}$ can be obtained from $\mathcal{A}'$ by the inverse elementary operation, which is an elementary operation of the same type. It then follows from the proof of the first part of the theorem that $\mathcal{A}$ is linearly independent. ∎

Frequently, information concerning a set of vectors can be obtained by successive application of a sequence of elementary operations that cannot be obtained by use of a single elementary operation. The corollary below is extremely useful in this respect.

2.6 COROLLARY

If a set $\mathcal{A}'$ is obtained by applying a sequence of elementary operations to a set $\mathcal{A}$ of vectors in $\mathbf{R}^n$, then $\mathcal{A}'$ is linearly independent if and only if $\mathcal{A}$ is linearly independent.

Proof Suppose that $\mathcal{A}'$ is obtained from $\mathcal{A}$ by a sequence $E_1, E_2, \ldots, E_t$ of elementary operations. Put $\mathcal{A}_0 = \mathcal{A}$, and let $\mathcal{A}_i$ be the set obtained by applying E_i to $\mathcal{A}_{i-1}$ for $i = 1, 2, \ldots, t$. Repeated application of Theorem 2.5 yields the following information:

$\mathcal{A}_1$ is independent if and only if $\mathcal{A}_0 = \mathcal{A}$ is independent.

$\mathcal{A}_2$ is independent if and only if $\mathcal{A}_1$ is independent.

$$\vdots$$

$\mathcal{A}' = \mathcal{A}_t$ is independent if and only if $\mathcal{A}_{t-1}$ is independent.

Thus $\mathcal{A}$ is linearly independent if and only if $\mathcal{A}'$ is independent. ■ ■ ■

As an example of the use of this corollary, consider the following.

EXAMPLE 1 In Example 2 of Section 2.1, it is shown that the set

$$\mathcal{A}' = \{(2, 0, 4), (0, 1, 5), (0, 0, 0)\}$$

can be obtained from the set

$$\mathcal{A} = \{(1, 0, 2), (2, 3, 19), (0, 1, 5)\}$$

by a sequence of elementary operations. The set $\mathcal{A}'$ is obviously linearly dependent since it contains the zero vector. It follows that the set $\mathcal{A}$ is linearly dependent. ■

At this point, we have developed a somewhat crude method for investigating the linear dependence of a given set $\mathcal{A}$ of vectors in $\mathbf{R}^n$. If, by application of a sequence of elementary operations to $\mathcal{A}$, it is possible to obtain a set that contains the zero vector (or any set that is clearly dependent), then the given set is linearly dependent. By the same token, if a set can be obtained that is clearly independent, then $\mathcal{A}$ is linearly independent. This method is refined to a systematic procedure later in this chapter.

We conclude this section with a final corollary to Theorem 2.5.

**2.7
COROLLARY**   A set of vectors resulting from applying a sequence of elementary operations to a basis of $\mathbf{R}^n$ is again a basis of $\mathbf{R}^n$.

Proof Let $\mathcal{A}$ be a basis of $\mathbf{R}^n$. According to Corollary 2.6, any set $\mathcal{A}'$ obtained from $\mathcal{A}$ by a sequence of elementary operations is a linearly independent set of n vectors, and hence is a basis by Theorem 1.34. ■ ■ ■

2.2 Exercises

1. Show the set $\mathcal{A} = \{(1, 2, 1, 0), (3, -4, 5, 6), (2, -1, 3, 3), (-2, 6, -4, -6)\}$ is linearly dependent by applying a sequence of elementary operations to $\mathcal{A}$ and obtaining a set $\mathcal{A}'$ that contains the zero vector.

2. Use the method described in Problem 1 to show that the set

$$\{(1, 1, 0), (0, 1, 1), (1, 0, -1), (1, 0, 1)\}$$

is linearly dependent.

3. Show that the set $\mathcal{A} = \{(1, 0, 0), (1, 1, 0), (1, 1, 1)\}$ is linearly independent by obtaining $\mathcal{A}$ from the standard basis of $\mathbf{R}^3$ by a sequence of elementary operations.

4. Show that the set $\mathcal{A} = \{(1, 0, 0), (1, 1, 0), (1, 1, 1)\}$ is linearly independent by obtaining the standard basis of $\mathbf{R}^3$ from $\mathcal{A}$ by a sequence of elementary operations.

5. Show that each of the following sets is linearly independent by obtaining the standard basis of $\mathbf{R}^4$ from $\mathcal{A}$ by a sequence of elementary operations.

   (a) $\mathcal{A} = \{(0, 0, 1, 1), (0, 2, 0, -3), (-3, 0, 2, 2), (0, 2, 0, -2)\}$
   (b) $\mathcal{A} = \{(1, 0, 1, -3), (-2, 1, 0, 0), (0, 0, 0, 4), (-2, 1, 2, -6)\}$

6. Use elementary operations to determine whether or not the given set is linearly independent.

   (a) $\{(1, 0, 2), (2, -1, 1), (1, 1, -1)\}$
   (b) $\{(1, 1, -1), (2, -1, 1), (2, 1, 1)\}$
   (c) $\{(1, 1, 8, -1), (1, 0, 3, 0), (3, 2, 19, -2)\}$
   (d) $\{(1, -1, -2, -4), (1, 1, 8, 4), (3, -1, 4, -4)\}$
   (e) $\{(1, 0, 1, 0), (0, 1, 0, 1), (4, 3, 2, 3), (1, 0, 0, 0)\}$
   (f) $\{(1, 0, 1, 0), (2, 1, 4, 3), (1, 2, 5, -2), (-1, 3, 5, 4)\}$

2.3 Standard Bases for Subspaces

In the last section, we found that linear independence is a property that is preserved by application of elementary operations. As the reader has likely anticipated, we turn next to an investigation of the application of elementary operations to spanning sets of a subspace. We then combine and sharpen our results so as to obtain a systematic method of attacking the types of problems mentioned in Section 1.6. Later, we shall see that our results have even more applications.

   The next theorem in our development is fairly obvious, but it is important enough to be designated as a theorem. A restricted form of the converse is contained in the last theorem of this section, but the proof of that theorem must wait until some intermediate results are established.

2.8 THEOREM   If $\mathcal{A}'$ is obtained by applying a sequence of elementary operations to a set $\mathcal{A}$ of vectors in $\mathbf{R}^n$, then $\langle \mathcal{A}' \rangle = \langle \mathcal{A} \rangle$.

Proof Suppose that $\mathcal{A}'$ is obtained by applying the sequence $E_1, E_2, \ldots, E_t$ of elementary operations E_i to $\mathcal{A}$. Let $\mathcal{A}_0 = \mathcal{A}$ and $\mathcal{A}_i = E_i(\mathcal{A}_{i-1})$ for $i = 1, 2, \ldots, t$, so that $\mathcal{A}' = \mathcal{A}_t$.

Now $\mathcal{A}_i$ is obtained from $\mathcal{A}_{i-1}$ by application of E_i. If we recall the definitions of the elementary operations in Section 2.1, it is evident that each vector of $\mathcal{A}_i$ is a linear combination of vectors in $\mathcal{A}_{i-1}$. Thus $\langle \mathcal{A}_i \rangle \subseteq \langle \mathcal{A}_{i-1} \rangle$. But $\mathcal{A}_{i-1} = E_i^{-1}(\mathcal{A}_i)$ so that each vector of $\mathcal{A}_{i-1}$ is a linear combination of vectors in $\mathcal{A}_i$. Therefore $\langle \mathcal{A}_{i-1} \rangle \subseteq \langle \mathcal{A}_i \rangle$, and $\langle \mathcal{A}_{i-1} \rangle = \langle \mathcal{A}_i \rangle$. Letting $i = 1, 2, \ldots, t$ in succession, we have

$$\langle \mathcal{A} \rangle = \langle \mathcal{A}_0 \rangle = \langle \mathcal{A}_1 \rangle = \cdots = \langle \mathcal{A}_t \rangle = \langle \mathcal{A}' \rangle,$$

and the theorem is proved. ∎

EXAMPLE 1 Consider the set $\mathcal{A} = \{(1, 4, 0), (2, 1, 0), (1, 1, 0)\}$ and the sequence E_1, E_2, E_3, E_4, E_5 of elementary operations given by:

E_1: Replace the second vector by the sum of the second vector and -2 times the first vector.

E_2: Replace the third vector by the sum of the third vector and -1 times the first vector.

E_3: Multiply the second vector by $-\frac{1}{7}$.

E_4: Replace the first vector by the sum of the first vector and -4 times the second vector.

E_5: Replace the third vector by the sum of the third vector and 3 times the second vector.

Applying the first elementary operation E_1 to $\mathcal{A}$, we find that

$$E_1(\mathcal{A}) = E_1 \{(1, 4, 0), (2, 1, 0), (1, 1, 0)\}$$
$$= \{(1, 4, 0), (0, -7, 0), (1, 1, 0)\}.$$

Continuing in this fashion, we have

$$E_2(E_1(\mathcal{A})) = E_2 \{(1, 4, 0), (0, -7, 0), (1, 1, 0)\}$$
$$= \{(1, 4, 0), (0, -7, 0), (0, -3, 0)\}$$
$$E_3(E_2 E_1(\mathcal{A})) = E_3 \{(1, 4, 0), (0, -7, 0), (0, -3, 0)\}$$
$$= \{(1, 4, 0), (0, 1, 0), (0, -3, 0)\}$$
$$E_4(E_3 E_2 E_1(\mathcal{A})) = E_4 \{(1, 4, 0), (0, 1, 0), (0, -3, 0)\}$$
$$= \{(1, 0, 0), (0, 1, 0), (0, -3, 0)\}$$
$$E_5(E_4 E_3 E_2 E_1(\mathcal{A})) = E_5 \{(1, 0, 0), (0, 1, 0), (0, -3, 0)\}$$
$$= \{(1, 0, 0), (0, 1, 0), (0, 0, 0)\} = \mathcal{A}'.$$

Since $\mathcal{A}'$ was obtained from $\mathcal{A}$ by a sequence of elementary operations then $\langle \mathcal{A}' \rangle = \langle \mathcal{A} \rangle$. ∎

Combining Theorem 2.8 and Corollary 2.6, we obtain an important corollary concerning bases of a subspace.

2.9 **COROLLARY**	A set of vectors resulting from the application of a sequence of elementary operations to a basis of a subspace **W** is again a basis of **W**.

The next theorem is our first step in standardizing the bases of subspaces. Example 2 appears just after the end of the proof of this theorem, and the work in that example illustrates the steps described in the proof. If the steps in the proof and the steps in the example are traced together, this should make each of them easier to follow.

2.10 **THEOREM**	Let $\mathcal{A} = \{\mathbf{v}_1, \mathbf{v}_2, \ldots, \mathbf{v}_m\}$ be a set of m vectors in $\mathbf{R}^n$ that spans the subspace $\mathbf{W} = \langle \mathcal{A} \rangle$ of dimension r, where $m \geq r > 0$. Then a set $\mathcal{A}' = \{\mathbf{v}'_1, \mathbf{v}'_2, \ldots, \mathbf{v}'_r, \mathbf{0}, \ldots, \mathbf{0}\}$ of m vectors can be obtained from $\mathcal{A}$ by a finite sequence of elementary operations so that $\{\mathbf{v}'_1, \mathbf{v}'_2, \ldots, \mathbf{v}'_r\}$ has the following properties:

1. The first nonzero component from the left in $\mathbf{v}'_j$ is a 1 in the k_j position for $j = 1, 2, \ldots, r$. (This 1 is called a **leading one**.)
2. $k_1 < k_2 < \cdots < k_r$. (In vectors listed later in $\mathcal{A}'$, the leading ones occur in positions that are farther to the right.)
3. $\mathbf{v}'_j$ is the only vector in $\mathcal{A}'$ with a nonzero k_j component.
4. $\{\mathbf{v}'_1, \mathbf{v}'_2, \ldots, \mathbf{v}'_r\}$ is a basis of **W**.

Proof By Theorem 1.32, $\mathcal{A}$ contains a basis of $\langle \mathcal{A} \rangle$. Thus, there is at least one vector in $\mathcal{A}$ that is not zero. Let k_1 be the smallest positive integer for which some $\mathbf{v}_i$ has nonzero k_1 component. By no more than one interchange of vectors, a spanning set for **W** can be obtained in which the first vector has a nonzero k_1 component. Multiplication of this vector by the reciprocal of its k_1 component yields a spanning set of **W** in which the k_1 component of the first vector is 1. Then each of the other vectors can be replaced by the sum of that vector and a suitable multiple of the new first vector to obtain a spanning set

$$\mathcal{A}_1 = \{\mathbf{v}_1^{(1)}, \mathbf{v}_2^{(1)}, \ldots, \mathbf{v}_m^{(1)}\}$$

of **W** in which

(i) the first nonzero component in $\mathbf{v}_1^{(1)}$ is a 1 in the k_1 component, and
(ii) $\mathbf{v}_1^{(1)}$ is the only vector in $\mathcal{A}_1$ with a nonzero number in any of the first k_1 positions from the left.

If $r = 1$, the theorem follows trivially at this point.
If $r > 1$, let k_2 be the least positive integer for which some $\mathbf{v}_i^{(1)}$, $i \neq 1$, has nonzero k_2 component. At least one such $\mathbf{v}_i^{(1)}$ exists, since all remaining vectors would otherwise be zero. By an interchange of vectors that does not involve the first vector, a spanning set for

W can be obtained in which the second vector has nonzero k_2 component. Multiplication of this vector by the reciprocal of its k_2 component will yield a spanning set of W in which the k_2 component of the second vector is 1. Then each of the other vectors can be replaced by the sum of that vector and a suitable multiple of the new second vector to obtain a spanning set

$$\mathcal{A}_2 = \{\mathbf{v}_1^{(2)}, \mathbf{v}_2^{(2)}, \dots, \mathbf{v}_m^{(2)}\}$$

of W that has the following properties:

(i) The first nonzero component in $\mathbf{v}_1^{(2)}$ is a 1 in the k_1 component, and the first nonzero component in $\mathbf{v}_2^{(2)}$ is a 1 in the k_2 component,

(ii) $k_1 < k_2$,

(iii) $\mathbf{v}_1^{(2)}$ is the only vector in $\mathcal{A}_2$ with a nonzero k_1 component, and $\mathbf{v}_2^{(2)}$ is the only vector in $\mathcal{A}_2$ with a nonzero k_2 component.

That is, the first two vectors in the set $\mathcal{A}_2$ have the first three properties required in the statement of the theorem.

Suppose that a set $\mathcal{A}_i$ that spans W has been obtained in which the first i vectors $(i < r)$ satisfy the first three properties listed in the theorem. Then k_{i+1} is chosen to be the least positive integer for which one of the last $m - i$ vectors in the set has a nonzero k_{i+1} component. Such a vector exists, for $\mathcal{A}_i$ must contain at least r nonzero vectors. The procedure described to obtain $\mathcal{A}_1$ and $\mathcal{A}_2$ can then be repeated to obtain the set $\mathcal{A}_{i+1}$ that spans W in which the first $i + 1$ vectors satisfy the first three conditions.

It is clear, then, that a finite sequence of elementary operations can be applied to $\mathcal{A}$ to obtain a set $\mathcal{A}_r = \{\mathbf{v}_1^{(r)}, \mathbf{v}_2^{(r)}, \dots, \mathbf{v}_m^{(r)}\}$ that spans W and in which the first r vectors satisfy the conditions (1), (2), (3). Now assume that there exists an $\mathbf{v}_j^{(r)}$, $j > r$, with nonzero k_{r+1} component. Then it must be that $k_{r+1} > k_r$, and from this it is easily seen that $\{\mathbf{v}_1^{(r)}, \mathbf{v}_2^{(r)}, \dots, \mathbf{v}_r^{(r)}, \mathbf{v}_j^{(r)}\}$ is linearly independent, contradicting the fact that r is the dimension of W. Thus the remaining $m - r$ vectors in $\mathcal{A}_r$ are zero and $\mathcal{A}_r = \mathcal{A}'$, where $\mathcal{A}'$ satisfies the conditions of the theorem. ∎ ∎ ∎

The proof given for Theorem 2.10 is a constructive one in that it describes a method of obtaining the set $\mathcal{A}'$ from a given set $\mathcal{A}$. This is illustrated in the following example.

EXAMPLE 2 Let $\mathcal{A} = \{\mathbf{v}_1, \mathbf{v}_2, \mathbf{v}_3, \mathbf{v}_4\}$ for $\mathbf{v}_1 = (0, 0, 0, -1, -1)$, $\mathbf{v}_2 = (0, 3, -6, 1, -2)$, $\mathbf{v}_3 = (0, 2, -4, 2, 0)$, $\mathbf{v}_4 = (0, -1, 2, 2, 3)$, and let $W = \langle \mathcal{A} \rangle$.

Following the proof of the theorem, we see that $k_1 = 2$ is the smallest positive integer for which some $\mathbf{v}_i$ has nonzero k_1 component. By an interchange of $\mathbf{v}_1$ and $\mathbf{v}_3$, we obtain the spanning set

$$\{(0, 2, -4, 2, 0), (0, 3, -6, 1, -2), (0, 0, 0, -1, -1), (0, -1, 2, 2, 3)\}$$

in which the first vector has nonzero k_1 component. Multiplication of this vector by $\frac{1}{2}$ yields the spanning set

$$\{(0, 1, -2, 1, 0), (0, 3, -6, 1, -2), (0, 0, 0, -1, -1), (0, -1, 2, 2, 3)\}$$

in which the k_1 component of the first vector is 1. Each of the other vectors is now replaced by the sum of that vector and a suitable multiple of the new first vector as follows.

Replace the second vector by the sum of the second vector and (-3) times the first vector.

No change is needed on the third vector. (A "suitable" multiple to add would be the zero multiple.)

Replace the fourth vector by the sum of the fourth vector and 1 times the first vector.

This yields the spanning set

$$
\begin{aligned}
\mathcal{A}_1 &= \{\mathbf{v}_1^{(1)}, \mathbf{v}_2^{(1)}, \mathbf{v}_3^{(1)}, \mathbf{v}_4^{(1)}\} \\
&= \{(0, 1, -2, 1, 0), (0, 0, 0, -2, -2), (0, 0, 0, -1, -1), (0, 0, 0, 3, 3)\}
\end{aligned}
$$

in which

(i) the first nonzero component in $\mathbf{v}_1^{(1)}$ is a 1 in the second component, and

(ii) $\mathbf{v}_1^{(1)} = (0, 1, -2, 1, 0)$ is the only vector in $\mathcal{A}_1$ with a nonzero component in either of the first two positions.

Now $k_2 = 4$ is the least positive integer for which some $\mathbf{v}_i^{(1)}, i \neq 1$, in $\mathcal{A}_1$ has a nonzero k_2 component, and we have $\mathbf{v}_2^{(1)} = (0, 0, 0, -2, -2)$ with a nonzero k_2 component. Multiplication of $\mathbf{v}_2^{(1)}$ by $-\frac{1}{2}$ yields

$$\{(0, 1, -2, 1, 0), (0, 0, 0, 1, 1), (0, 0, 0, -1, -1), (0, 0, 0, 3, 3)\}.$$

Each of the vectors other than the second is now replaced by the sum of that vector and a suitable multiple of the second vector as follows.

Replace the first vector by the sum of the first vector and (-1) times the second vector.

Replace the third vector by the sum of the third vector and the second vector.

Replace the fourth vector by the sum of the fourth vector and (-3) times the second vector.

This yields

$$
\begin{aligned}
\mathcal{A}_2 &= \{\mathbf{v}_1^{(2)}, \mathbf{v}_2^{(2)}, \mathbf{v}_3^{(2)}, \mathbf{v}_4^{(2)}\} \\
&= \{(0, 1, -2, 0, -1), (0, 0, 0, 1, 1), (0, 0, 0, 0, 0), (0, 0, 0, 0, 0)\}.
\end{aligned}
$$

It is evident at this point that $\mathcal{A}_2 = \mathcal{A}'$. Simultaneously with finding $\mathcal{A}'$, we have found that $\mathbf{W}$ has dimension $r = 2$. ∎

The conditions of Theorem 2.10 are very restrictive, and one might expect that there is only one basis of a given subspace that satisfies these conditions. The next theorem confirms that this is the case.

2.11 THEOREM	There is one and only one basis of a given subspace $\mathbf{W}$ that satisfies the conditions of Theorem 2.10.

Proof Starting in Theorem 2.10 with a *basis* $\mathcal{A}$ of $\mathbf{W}$, that theorem assures us that a basis $\mathcal{A}'$ of $\mathbf{W}$ that satisfies the conditions can be obtained from $\mathcal{A}$ by a sequence of elementary operations. Thus there is at least one basis of the required type.

Suppose now that $\mathcal{A}' = \{\mathbf{v}_1', \ldots, \mathbf{v}_r'\}$ and $\mathcal{A}'' = \{\mathbf{v}_1'', \ldots, \mathbf{v}_r''\}$ are two bases of $\mathbf{W}$ that satisfy the conditions of Theorem 2.10. Let $k_1', \ldots, k_r'$ and $k_1'', \ldots, k_r''$ be the sequences of positive integers described in the conditions for $\mathcal{A}'$ and $\mathcal{A}''$, respectively.

Assume $k_1' < k_1''$. Since $\mathcal{A}''$ spans $\mathbf{W}$, there must exist scalars c_{i1} such that $\mathbf{v}_1' = \sum_{i=1}^{r} c_{i1} \mathbf{v}_i''$. Since each $\mathbf{v}_i''$ has zero jth component for each $j < k_1''$, any linear combination such as $\sum_{i=1}^{r} c_{i1} \mathbf{v}_i''$ must have zero jth component for each $j < k_1''$. But $\mathbf{v}_1'$ has a nonzero k_1' component, and $k_1' < k_1''$. This is a contradiction; hence $k_1' \geq k_1''$. The symmetry of the conditions on $\mathcal{A}'$ and $\mathcal{A}''$ implies that $k_1'' \geq k_1'$, and thus, $k_1' = k_1''$.

Now assume $k_2' < k_2''$. Since $\mathcal{A}''$ spans $\mathbf{W}$, there must exist scalars c_{i2} such that $\mathbf{v}_2' = \sum_{i=1}^{r} c_{i2} \mathbf{v}_i''$. Now $\mathbf{v}_1''$ is the only vector in $\mathcal{A}''$ that has nonzero k_1' component, so a linear combination $\sum_{i=1}^{r} c_{i2} \mathbf{v}_i''$ has zero k_1' component if and only if $c_{i1} = 0$. Since $\mathbf{v}_2'$ has a zero k_1' component, $c_{12} = 0$, and we have $\mathbf{v}_2' = \sum_{i=2}^{r} c_{i2} \mathbf{v}_i''$. For $i \geq 2$, $\mathbf{v}_i''$ has zero jth component for each $j < k_2''$, and thus the linear combination $\sum_{i=2}^{r} c_{i2} \mathbf{v}_i''$ has zero jth component for all $j < k_2''$. But $\mathbf{v}_2'$ has nonzero k_2' component, and $k_2' < k_2''$. As before, we have a contradiction, and therefore, $k_2' \geq k_2''$. From the symmetry of the conditions, $k_2'' \geq k_2'$, and thus, $k_2' = k_2''$.

It is clear that this argument may be repeated to obtain $k_j' = k_j''$ for $j = 1, 2, \ldots, r$.

Now $\mathcal{A}''$ spans $\mathbf{W}$, so for each $\mathbf{v}_j'$ there must exist scalars c_{ij} such that $\mathbf{v}_j' = \sum_{i=1}^{r} c_{ij} \mathbf{v}_i''$. Since $\mathbf{v}_i''$ is the only vector in $\mathcal{A}''$ that has nonzero k_i' component and $\mathbf{v}_i''$ has k_i' component equal to 1, c_{ij} is the k_i' component of $\sum_{i=1}^{r} c_{ij} \mathbf{v}_i''$. But $\mathbf{v}_j'$ has zero k_i' component for $i \neq j$, so $c_{ij} = 0$ for $i \neq j$, and since $\mathbf{v}_j'$ has k_j' component equal to 1, $c_{jj} = 1$. Therefore $\mathbf{v}_j' = \mathbf{v}_j''$, and $\mathcal{A}' = \mathcal{A}''$. ■■■

Theorem 2.11 allows us to make the following definition.

2.12 DEFINITION	Let $\mathbf{W}$ be a subspace of $\mathbf{R}^n$ of dimension r. The **standard basis** of $\mathbf{W}$ is the basis $\{\mathbf{v}_1, \mathbf{v}_2, \ldots, \mathbf{v}_r\}$ of $\mathbf{W}$ that satisfies the conditions of Theorem 2.10.

That is, the standard basis of $\mathbf{W}$ is the unique basis $\{\mathbf{v}_1, \mathbf{v}_2, \ldots, \mathbf{v}_r\}$ that has the following properties.

1. The first nonzero component from the left in the jth vector $\mathbf{v}_j$ is a 1 in the k_j component. (This 1 is called a **leading one**.)
2. $k_1 < k_2 < \cdots < k_r$.
3. The jth vector $\mathbf{v}_j$ is the only vector in the basis with a nonzero k_j component.

Clearly, if $r = m = n$ in Theorem 2.10, the standard basis thus defined is the same as that given in Definition 1.31, and our two definitions are in agreement with each other.

EXAMPLE 3   In Example 2, we found that the standard basis of the subspace

$$\mathbf{W} = \langle (0, 0, 0, -1, -1), (0, 3, -6, 1, -2), (0, 2, -4, 2, 0), (0, -1, 2, 2, 3) \rangle$$

is the set $\{(0, 1, -2, 0, -1), (0, 0, 0, 1, 1)\}$.

If we are only interested in finding the standard basis of $\mathbf{W}$, the amount of work that is necessary is much less than that done in Example 2. As a first step, we might write the vectors of $\mathcal{A}$ in a rectangular array as

$$A = \begin{bmatrix} 0 & 0 & 0 & 0 \\ 0 & 3 & 2 & -1 \\ 0 & -6 & -4 & 2 \\ -1 & 1 & 2 & 2 \\ -1 & -2 & 0 & 3 \end{bmatrix}.$$

In composing this array, we have recorded the components of $\mathbf{v}_i$ from top to bottom in the ith column from the left. (It is admittedly more natural to record these components in rows rather than columns. The reason for the use of columns will become clear in Chapter 3.) Let us use an arrow from the first array to a second to indicate that the set represented by the second array is obtained from the first array by application of one or more elementary operations.

Our work in Example 2 can then be recorded in this manner:

$$A \rightarrow \begin{bmatrix} 0 & 0 & 0 & 0 \\ 2 & 3 & 0 & -1 \\ -4 & -6 & 0 & 2 \\ 2 & 1 & -1 & 2 \\ 0 & -2 & -1 & 3 \end{bmatrix} \rightarrow \begin{bmatrix} 0 & 0 & 0 & 0 \\ 1 & 3 & 0 & -1 \\ -2 & -6 & 0 & 2 \\ 1 & 1 & -1 & 2 \\ 0 & -2 & -1 & 3 \end{bmatrix} \rightarrow \begin{bmatrix} 0 & 0 & 0 & 0 \\ 1 & 0 & 0 & 0 \\ -2 & 0 & 0 & 0 \\ 1 & -2 & -1 & 3 \\ 0 & -2 & -1 & 3 \end{bmatrix}$$

$$\rightarrow \begin{bmatrix} 0 & 0 & 0 & 0 \\ 1 & 0 & 0 & 0 \\ -2 & 0 & 0 & 0 \\ 1 & 1 & -1 & 3 \\ 0 & 1 & -1 & 3 \end{bmatrix} \rightarrow \begin{bmatrix} 0 & 0 & 0 & 0 \\ 1 & 0 & 0 & 0 \\ -2 & 0 & 0 & 0 \\ 0 & 1 & 0 & 0 \\ -1 & 1 & 0 & 0 \end{bmatrix}. \qquad \blacksquare$$

Primarily for future use, we record the following theorems.

2.13 THEOREM   The standard basis of a subspace $\mathbf{W}$ can be obtained from any basis of $\mathbf{W}$ by a sequence of elementary operations.

Proof   Let $\mathbf{W}$ be a subspace of dimension r, and let $\mathcal{A}'$ be the standard basis of $\mathbf{W}$. If $\mathcal{A}$ is any basis of $\mathbf{W}$, then Theorem 2.10 applies with $m = r$, asserting that $\mathcal{A}'$ can be obtained from $\mathcal{A}$ by a finite sequence of elementary operations.   ■■■

> **2.14**
> **THEOREM**
>
> Let $\mathcal{A}$ and $\mathcal{B}$ be two sets of m vectors each in $\mathbf{R}^n$. Then $\langle\mathcal{A}\rangle = \langle\mathcal{B}\rangle$ if and only if $\mathcal{B}$ can be obtained from $\mathcal{A}$ by a sequence of elementary operations.

Proof If $\mathcal{B}$ can be obtained from $\mathcal{A}$ by a sequence of elementary operations, then $\langle\mathcal{A}\rangle = \langle\mathcal{B}\rangle$ by Theorem 2.8.

Suppose now that $\langle\mathcal{A}\rangle = \langle\mathcal{B}\rangle = \mathbf{W}$, where $\mathbf{W}$ has dimension r. Let $\mathcal{A}'$ denote the set of vectors in Theorem 2.10 that can be obtained from $\mathcal{A}$ by a sequence $E_1, E_2, \ldots, E_k$ of elementary operations. Let $\mathcal{B}'$ denote the corresponding set that can be obtained from $\mathcal{B}$ by a sequence $F_1, F_2, \ldots, F_t$ of elementary operations. Then the last $m - r$ vectors in both $\mathcal{A}'$ and $\mathcal{B}'$ are zero, and the first r vectors in both $\mathcal{A}'$ and $\mathcal{B}'$ constitute the standard basis of $\mathbf{W}$, which is unique by Theorem 2.11. Thus $\mathcal{A}' = \mathcal{B}'$.

Now $\mathcal{B}$ can be obtained from $\mathcal{B}' = \mathcal{A}'$ by the sequence $F_t^{-1}, \ldots, F_2^{-1}, F_1^{-1}$. Thus $\mathcal{B}$ can be obtained from $\mathcal{A}$ by application of the sequence

$$E_1, E_2, \ldots, E_k, F_t^{-1}, \ldots, F_2^{-1}, F_1^{-1}$$

of elementary operations. ∎

The value of Theorem 2.14 is mainly theoretical. For arbitrary sets $\mathcal{A}$ and $\mathcal{B}$ of vectors in $\mathbf{R}^n$, a practical way to determine whether or not $\langle\mathcal{A}\rangle = \langle\mathcal{B}\rangle$ is to find and compare the sets $\mathcal{A}'$ and $\mathcal{B}'$ described in the proof. Then $\langle\mathcal{A}\rangle = \langle\mathcal{B}\rangle$ if and only if $\mathcal{A}' = \mathcal{B}'$. If $\mathcal{A}$ and $\mathcal{B}$ happen not to have the same number of vectors, then $\langle\mathcal{A}\rangle = \langle\mathcal{B}\rangle$ if and only if $\mathcal{A}'$ and $\mathcal{B}'$ contain the same nonzero vectors. We illustrate this situation in the next example.

EXAMPLE 4 Consider the problem of determining if $\langle\mathcal{A}\rangle = \langle\mathcal{B}\rangle$ where

$$\mathcal{A} = \{(-2, 1, 4, 8), (-1, 1, 5, 9), (-2, 5, 7, 6)\},$$

$$\mathcal{B} = \{(0, -4, 1, 10)(1, -5, 1, 11), (0, 2, 2, 0), (-1, 1, 0, -1)\}.$$

We first arrange the vectors of $\mathcal{A}$ in a rectangular array A and use elementary operations to determine $\mathcal{A}'$.

$$A = \begin{bmatrix} -2 & -1 & -2 \\ 1 & 1 & 5 \\ 4 & 5 & 7 \\ 8 & 9 & 6 \end{bmatrix} \rightarrow \begin{bmatrix} 0 & -1 & 0 \\ -1 & 1 & 3 \\ -6 & 5 & -3 \\ -10 & 9 & -12 \end{bmatrix} \rightarrow \begin{bmatrix} 1 & 0 & 0 \\ -1 & -1 & 3 \\ -5 & -6 & -3 \\ -9 & -10 & -12 \end{bmatrix}$$

$$\rightarrow \begin{bmatrix} 1 & 0 & 0 \\ 0 & 1 & 0 \\ 1 & 6 & -21 \\ 1 & 10 & -42 \end{bmatrix} \rightarrow \begin{bmatrix} 1 & 0 & 0 \\ 0 & 1 & 0 \\ 1 & 6 & 1 \\ 1 & 10 & 2 \end{bmatrix} \rightarrow \begin{bmatrix} 1 & 0 & 0 \\ 0 & 1 & 0 \\ 0 & 0 & 1 \\ -1 & -2 & 2 \end{bmatrix} = A'.$$

We now read the components of the vectors of $\mathcal{A}'$ from the columns of A'.

$$\mathcal{A}' = \{(1, 0, 0, -1), (0, 1, 0, -2), (0, 0, 1, 2)\}.$$

Similarly, we form the array B using the vectors in $\mathcal{B}$, perform elementary operations on the columns of B, yielding B', whose columns will contain the components of the vectors of $\mathcal{B}'$.

$$B = \begin{bmatrix} 0 & 1 & 0 & -1 \\ -4 & -5 & 2 & 1 \\ 1 & 1 & 2 & 0 \\ 10 & 11 & 0 & -1 \end{bmatrix} \rightarrow \begin{bmatrix} 1 & 0 & 0 & -1 \\ -5 & -4 & 2 & 1 \\ 1 & 1 & 2 & 0 \\ 11 & 10 & 0 & -1 \end{bmatrix} \rightarrow \begin{bmatrix} 1 & 0 & 0 & 0 \\ -5 & -4 & 2 & -4 \\ 1 & 1 & 2 & 1 \\ 11 & 10 & 0 & 10 \end{bmatrix}$$

$$\rightarrow \begin{bmatrix} 1 & 0 & 0 & 0 \\ 0 & 0 & 1 & 0 \\ 6 & 5 & 1 & 5 \\ 11 & 10 & 0 & 10 \end{bmatrix} \rightarrow \begin{bmatrix} 1 & 0 & 0 & 0 \\ 0 & 1 & 0 & 0 \\ 6 & 1 & 1 & 0 \\ 11 & 0 & 2 & 0 \end{bmatrix} \rightarrow \begin{bmatrix} 1 & 0 & 0 & 0 \\ 0 & 1 & 0 & 0 \\ 0 & 0 & 1 & 0 \\ -1 & -2 & 2 & 0 \end{bmatrix} = B'.$$

Thus $\mathcal{B}' = \{(1, 0, 0, -1), (0, 1, 0, -2), (0, 0, 1, 2), (0, 0, 0, 0)\}$. We see that the nonzero vectors of $\mathcal{A}'$ and $\mathcal{B}'$ are equal. Hence $\langle \mathcal{A} \rangle = \langle \mathcal{B} \rangle$. ∎

2.3 Exercises

1. Let $\mathcal{A}$ and $\mathcal{A}'$ be as given in Example 2. Write out a complete sequence of elementary operations that will yield $\mathcal{A}'$ when applied to $\mathcal{A}$.

2. For each of the sets $\mathcal{A}$, find the set $\mathcal{A}'$ of Theorem 2.10 by the method used in the proof of Theorem 2.10 and Example 2.

   (a) $\mathcal{A} = \{(-5, 11, -3, 0, 1), (0, 5, 10, 1, 6), (2, -4, 2, 1, 1), (2, 1, 12, 2, 7)\}$

   (b) $\mathcal{A} = \{(0, 1, 1, 8, -1), (0, 1, 1, 3, 0), (0, 3, 3, 19, -2), (0, 4, 4, 22, -2),$
   $(0, 3, 3, 14, 1)\}$

3. Determine whether or not each of the sets below is linearly independent by finding the dimension of the subspace spanned by the set.

   (a) $\{(1, 2, 0, 1, 0), (2, 4, 1, 4, 3), (1, 2, 2, 5, -2), (-1, -2, 3, 5, 4)\}$

   (b) $\{(1, -1, 2, -3), (3, -1, 6, -6), (1, 0, 1, 0), (1, 1, 2, 0)\}$

4. For each of the sets $\mathcal{A}$ below, find the dimension of $\langle \mathcal{A} \rangle$.

   (a) $\mathcal{A} = \{(1, 1, 3, -1), (1, 0, -2, 0), (3, 2, 4, -2)\}$

   (b) $\mathcal{A} = \{(1, 2, -1, 3), (0, 1, 0, 2), (1, 3, -1, 5), (1, 1, -1, 0)\}$

   (c) $\mathcal{A} = \{(0, 2, 0, 4, -1), (0, 5, -1, 11, 8), (0, 0, 1, -7, 9), (0, 7, 0, 8, 16)\}$

   (d) $\mathcal{A} = \{(0, 3, 0, -1, 3), (0, -2, 0, 1, -1), (0, 5, 0, 1, 13), (0, 4, 0, -2, 2)\}$

5. For each of the sets $\mathcal{A}$ given below, use rectangular arrays as in Example 3 to find the standard basis for $\langle \mathcal{A} \rangle$.

   (a) $\mathcal{A} = \{(1, 1, 0, 0), (0, 1, 1, 0), (0, 0, 1, 1), (1, 2, 2, 1)\}$

   (b) $\mathcal{A} = \{(1, -1, 2, -3), (3, -1, 6, -6), (1, 0, 1, 0), (1, 1, 2, 0)\}$

   (c) $\mathcal{A} = \{(3, 2, 5, 1), (-1, 0, -1, -1), (2, 1, 3, 1), (1, 0, 1, 1)\}$

   (d) $\mathcal{A} = \{(0, 2, 0, 3), (1, 0, 1, 0), (3, -1, 6, -6), (1, 1, 2, 0), (1, -1, 2, -3)\}$

6. Find the standard basis for the given subspace of $\mathbf{R}^n$.

   (a) $\mathbf{W} = \{(2a, -a, 0) \mid a \in \mathbf{R}\}$
   (b) $\mathbf{W} = \{(a - b, b - c, a - c) \mid a, b, c \in \mathbf{R}\}$
   (c) $\mathbf{W} = \{(3a, b, b - a) \mid a, b \in \mathbf{R}\}$
   (d) $\mathbf{W} = \{(0, b - a, 3b - 4a) \mid a, b \in \mathbf{R}\}$
   (e) $\mathbf{W} = \{(0, 0, a + 2b, b) \mid a, b \in \mathbf{R}\}$
   (f) $\mathbf{W} = \{(0, a, 3a - 2b, 0) \mid a, b \in \mathbf{R}\}$
   (g) $\mathbf{W} = \{(-4a, b, 0, c, -a - b + c) \mid a, b, c \in \mathbf{R}\}$
   (h) $\mathbf{W} = \{(0, 0, 0, 4a - b, a - 4b) \mid a, b \in \mathbf{R}\}$

7. Find the standard basis for the line in $\mathbf{R}^2$ with the given equation.

   (a) $2x + 3y = 0$ **(b)** $x - 7y = 0$

8. Find the standard basis for the plane in $\mathbf{R}^3$ that has the given equation.

   (a) $x - y - z = 0$ **(b)** $2x + y - 3z = 0$
   (c) $z - 2y = 0$ **(d)** $3x + 2y = 0$

9. In each case, determine whether or not $\langle \mathcal{A} \rangle = \langle \mathcal{B} \rangle$.

   (a) $\mathcal{A} = \{(1, 1, 1, 1), (0, 0, 2, 2)\}$, $\mathcal{B} = \{(2, 2, 3, 3), (1, 1, 2, 2)\}$
   (b) $\mathcal{A} = \{(1, 1, 0, 0), (1, 0, 1, 1)\}$, $\mathcal{B} = \{(2, -1, 3, 3), (0, 1, -1, -1)\}$
   (c) $\mathcal{A} = \{(2, 2, 0, 0), (1, 2, 1, 0), (1, 1, 1, 1)\}$,
   $\mathcal{B} = \{(1, 1, 0, 0), (0, 1, 1, 0), (0, 0, 1, 2)\}$
   (d) $\mathcal{A} = \{(1, 2, 1, -1), (0, 1, 2, 3), (1, 4, 5, 5)\}$, $\mathcal{B} = \{(2, 4, 2, -2), (0, 3, -2, 4)\}$
   (e) $\mathcal{A} = \{(1, 1, -1), (1, 3, 1), (-1, 1, 2)\}$,
   $\mathcal{B} = \{(2, 1, 2), (1, 0, 1), (3, 2, 4), (2, 6, 3)\}$
   (f) $\mathcal{A} = \{(1, -2, -1), (3, 2, -1), (0, 4, 1), (2, 0 - 1)\}$, $\mathcal{B} = \{(5, -2, -3), (1, 6, 1)\}$

10. For each pair $\mathbf{W}_1, \mathbf{W}_2$, find the dimension of $\mathbf{W}_1 + \mathbf{W}_2$.

    (a) $\mathbf{W}_1 = \langle (1, 1, 3, -1), (1, 0, -2, 0), (3, 2, 4, -2) \rangle$,
    $\mathbf{W}_2 = \langle (1, 0, 0, 1), (1, 1, 7, 1) \rangle$
    (b) $\mathbf{W}_1 = \langle (1, -1, 2, -3), (1, 1, 2, 0), (3, -1, 6, -6) \rangle$,
    $\mathbf{W}_2 = \langle (2, 0, 4, -3), (0, 0, 0, 1) \rangle$

11. Given that the sets $\mathcal{A}$ and $\mathcal{B}$ below span the same subspace $\mathbf{W}$, follow the proof of Theorem 2.14 to find a sequence of elementary operations that can be used to obtain $\mathcal{B}$ from $\mathcal{A}$.

    (a) $\mathcal{A} = \{(1, 1, 0, 0), (1, 0, 1, 1)\}$, $\mathcal{B} = \{(2, -1, 3, 3), (0, 1, -1, -1)\}$
    (b) $\mathcal{A} = \{(1, 1, 3, -1), (1, 0, -2, 0)\}$, $\mathcal{B} = \{(3, 2, 4, -2), (0, 1, 5, -1)\}$
    (c) $\mathcal{A} = \{(4, 5, 2, 2), (1, 1, 1, 0)\}$, $\mathcal{B} = \{(1, -1, 5, -4), (0, 2, -4, 4)\}$
    (d) $\mathcal{A} = \{(1, 1, 8, -1), (1, 0, 3, 0)\}$, $\mathcal{B} = \{(1, -1, -2, 1), (3, -1, 4, 1)\}$

3 Matrix Multiplication

In Example 3 of Section 2.3, it was found that rectangular arrays were a useful notational convenience in recording the results of elementary operations on a set of vectors. These rectangular arrays are specific examples of the more general concept of a *matrix* over a set $\mathcal{M}$, to be defined in the following section. In this chapter, we define the operation of multiplication on matrices with real numbers as elements and establish the basic properties of this operation. As mentioned earlier, the results of Chapter 2 are extremely useful in the development here, particularly in the last two sections.

3.1 Matrices of Transition

The definition of a matrix over a set $\mathcal{M}$ is as follows.

3.1 DEFINITION

An **r by s matrix** over a set $\mathcal{M}$ is a rectangular array of elements of $\mathcal{M}$, arranged in r rows and s columns. Such a matrix will be written in the form

$$A = \begin{bmatrix} a_{11} & a_{12} & \cdots & a_{1s} \\ a_{21} & a_{22} & \cdots & a_{2s} \\ \vdots & \vdots & & \vdots \\ a_{r1} & a_{r2} & \cdots & a_{rs} \end{bmatrix}$$

where a_{ij} denotes the element in the ith row and jth column of the matrix. The numbers r and s are said to be the **dimensions** of A, and r by s is sometimes written as $r \times s$.

The matrix A above may be written more simply as $A = [a_{ij}]_{r \times s}$ or $A = [a_{ij}]$ if the number of rows or columns is not important.

There are several terms that are useful in describing matrices of certain types. An r by r matrix is said to be a **square matrix** or a **matrix of order r**. The elements a_{ii} of $A = [a_{ij}]$ are the **diagonal elements** of A, and a square matrix $A = [a_{ij}]$ with $a_{ij} = 0$ whenever $i \neq j$ is a **diagonal matrix**. The matrix $I_r = [\delta_{ij}]_{r \times r}$ is the **identity matrix** of order r. A matrix $A = [a_{ij}]$ is a **zero matrix** if $a_{ij} = 0$ for all pairs i, j. A matrix that has only one row is a **row matrix**, and a matrix that has only one column is a **column matrix**.

3.2 DEFINITION

Two matrices $A = [a_{ij}]_{r \times s}$ and $B = [b_{ij}]_{p \times q}$ over a set $\mathcal{M}$ are **equal** if and only if $r = p, s = q$, and $a_{ij} = b_{ij}$ for all pairs i, j.

EXAMPLE 1 Consider the following matrices.

$$A = \begin{bmatrix} 1 & 7 & -3 \\ 0 & 5 & -6 \\ -8 & 4 & 9 \end{bmatrix} \quad B = \begin{bmatrix} 0 & 0 & 0 \\ 0 & 0 & 0 \\ 0 & 0 & 0 \end{bmatrix} \quad C = \begin{bmatrix} 2 & 4 & 6 \end{bmatrix} \quad D = \begin{bmatrix} 0 \\ 0 \end{bmatrix}$$

$$E = \begin{bmatrix} 5 & 0 \\ 0 & -3 \end{bmatrix} \quad F = \begin{bmatrix} 1 & 0 & 0 \\ 0 & 1 & 0 \\ 0 & 0 & 1 \end{bmatrix} \quad G = \begin{bmatrix} 1 & 0 & 0 \\ 0 & 1 & 0 \end{bmatrix}$$

The special terms just introduced apply to these matrices in the following ways:

A, B, E, and F are square matrices.
B, E, and F are diagonal matrices.
F is an identity matrix.
Both B and D are zero matrices, but $B \neq D$.
C is a row matrix.
D is a column matrix.
None of the special terms apply to G. ■

At times, we will denote a zero matrix by the same symbol $\mathbf{0}$ that we use for a zero vector. This will not cause confusion, because the context where the symbol is used will make the meaning clear.

Column matrices prove useful in recording the coordinates of a vector $\mathbf{v}$ relative to a basis $\mathcal{B} = \{\mathbf{u}_1, \mathbf{u}_2, \dots, \mathbf{u}_r\}$ of a subspace $\mathbf{W}$ of $\mathbf{R}^n$. Theorem 1.24 guarantees that any vector $\mathbf{v} \in \mathbf{W}$ can be written uniquely as a linear combination of the base vectors

$$\mathbf{v} = \sum_{i=1}^{r} c_i \mathbf{u}_i.$$

The coefficients c_i, $i = 1, \dots, r$, are used to form the coordinate matrix of $\mathbf{v}$ relative to the basis $\mathcal{B}$ as defined next.

3.3 DEFINITION

If $\mathcal{B} = \{\mathbf{u}_1, \mathbf{u}_2, \dots, \mathbf{u}_r\}$ is a basis of the subspace $\mathbf{W}$ and $\mathbf{v} = \sum_{i=1}^{r} c_i \mathbf{u}_i$, the r by 1 matrix C given by

$$C = \begin{bmatrix} c_1 \\ c_2 \\ \vdots \\ c_r \end{bmatrix}$$

is called the **coordinate matrix** of $\mathbf{v}$ relative to $\mathcal{B}$. The coordinate matrix of $\mathbf{v}$ relative to $\mathcal{B}$ is denoted by $[\mathbf{v}]_\mathcal{B}$, and the phrase "with respect to $\mathcal{B}$" is used interchangeably with "relative to $\mathcal{B}$."

For a given basis $\mathcal{B}$, the vector $\mathbf{v}$ and the coordinate matrix $[\mathbf{v}]_{\mathcal{B}}$ uniquely determine each other.

EXAMPLE 2 The coordinate matrix of $\mathbf{v} = (2, 3, 0, -1)$ relative to the basis $\mathcal{B} = \{(1, 0, 1, 0),$ $(0, 1, 1, 0), (0, 0, 1, 1), (1, 1, 1, 1)\}$ is

$$[\mathbf{v}]_{\mathcal{B}} = \begin{bmatrix} 0 \\ 1 \\ -3 \\ 2 \end{bmatrix}$$

since $(2, 3, 0, -1) = 0(1, 0, 1, 0) + 1(0, 1, 1, 0) - 3(0, 0, 1, 1) + 2(1, 1, 1, 1)$. However, the coordinate matrix of $\mathbf{v}$ relative to $\mathcal{E}_4$ is

$$[\mathbf{v}]_{\mathcal{E}_4} = \begin{bmatrix} 2 \\ 3 \\ 0 \\ -1 \end{bmatrix}$$

since $(2, 3, 0, -1) = 2(1, 0, 0, 0) + 3(0, 1, 0, 0) + 0(0, 0, 1, 0) - 1(0, 0, 0, 1)$. In this situation, we have an illustration of how the coordinates of a vector change from basis to basis.

On the other hand, if

$$[\mathbf{v}]_{\mathcal{B}'} = \begin{bmatrix} -5 \\ 2 \\ 1 \end{bmatrix}$$

is the coordinate matrix of $\mathbf{v}$ relative to the basis $\mathcal{B}' = \{(1, 2, 3), (-1, 0, 1), (-4, 1, 0)\}$, then the vector $\mathbf{v}$ is given by

$$\mathbf{v} = -5(1, 2, 3) + 2(-1, 0, 1) + 1(-4, 1, 0)$$
$$= (-11, -9, -13). \qquad \blacksquare$$

Now consider a set of vectors $\mathcal{A} = \{\mathbf{u}_1, \mathbf{u}_2, \ldots, \mathbf{u}_r\}$ in $\mathbf{R}^n$ and a second set $\mathcal{B} = \{\mathbf{v}_1, \mathbf{v}_2, \ldots, \mathbf{v}_s\}$ contained in $\langle \mathcal{A} \rangle$. Since $\mathcal{A}$ spans $\langle \mathcal{A} \rangle$, there are scalars a_{ij} in $\mathbf{R}$ such that $\mathbf{v}_j = \sum_{i=1}^{r} a_{ij} \mathbf{u}_i$ for $j = 1, 2, \ldots, s$. The following definition involves these scalars a_{ij}.

3.4 DEFINITION

If $\mathcal{A} = \{\mathbf{u}_1, \mathbf{u}_2, \ldots, \mathbf{u}_r\}$ is a set of vectors in $\mathbf{R}^n$ and $\mathcal{B} = \{\mathbf{v}_1, \mathbf{v}_2, \ldots, \mathbf{v}_s\}$ is a set of vectors in $\langle \mathcal{A} \rangle$, a **matrix of transition** (or **transition matrix**) from $\mathcal{A}$ to $\mathcal{B}$ is a matrix $A = [a_{ij}]_{r \times s}$ such that

$$\mathbf{v}_j = \sum_{i=1}^{r} a_{ij} \mathbf{u}_i \quad \text{for } j = 1, 2, \ldots, s.$$

The term *matrix of transition* applies only to situations involving nonempty finite sets of vectors, and these sets must be ordered. Whenever a set of vectors is listed without indices, it is understood that the index j is to go with the jth vector from the left. Thus the first vector from the left is to have index 1, the second from the left to have index 2, and so on. This is consistent with the notational agreement made after Example 2 in Section 1.6.

Another point needs to be emphasized in connection with Definition 3.4. The definition of the term *transition matrix* that is found in many elementary linear algebra texts is *not* equivalent to the one given here. The one stated in Definition 3.4 is what we need to present our development of matrix multiplication, and it is the one that leads to simpler proofs of major theorems later in this book.

The following example shows that the matrix of transition A is not always uniquely determined by $\mathcal{A}$ and $\mathcal{B}$.

EXAMPLE 3 Let $\mathcal{A} = \{(1, 2, 0), (2, 4, 0), (0, 0, 1)\}$ and $\mathcal{B} = \{(1, 2, 4), (2, 4, 8)\}$ in $\langle \mathcal{A} \rangle$. Now

$$(1, 2, 4) = (1)(1, 2, 0) + (0)(2, 4, 0) + (4)(0, 0, 1)$$

and

$$(2, 4, 8) = (2)(1, 2, 0) + (0)(2, 4, 0) + (8)(0, 0, 1)$$

so that

$$A_1 = \begin{bmatrix} 1 & 2 \\ 0 & 0 \\ 4 & 8 \end{bmatrix}$$

is a matrix of transition from $\mathcal{A}$ to $\mathcal{B}$. But

$$(1, 2, 4) = (-1)(1, 2, 0) + (1)(2, 4, 0) + (4)(0, 0, 1)$$

and

$$(2, 4, 8) = (0)(1, 2, 0) + (1)(2, 4, 0) + (8)(0, 0, 1)$$

so that

$$A_2 = \begin{bmatrix} -1 & 0 \\ 1 & 1 \\ 4 & 8 \end{bmatrix}$$

is also a matrix of transition from $\mathcal{A}$ to $\mathcal{B}$. Obviously $A_1 \neq A_2$, but we note that $\mathcal{A}$ is not linearly independent and thus is not a basis for $\langle \mathcal{A} \rangle$. ∎

Note that if $\mathcal{A}$ is a basis of a subspace $\mathbf{W}$ and $\mathcal{B} \subseteq \mathbf{W}$, a transition matrix $A = [a_{ij}]_{r \times s}$ from $\mathcal{A}$ to $\mathcal{B}$ always exists. Also, the number of columns in A is the same as the number of vectors in $\mathcal{B}$, and the number of rows in A is the same as the number of vectors in $\mathcal{A}$. The scalars a_{ij} that are the coefficients in the sum

$$\mathbf{v}_j = \sum_{i=1}^{r} a_{ij} \mathbf{u}_i$$

are the *coordinates* of $\mathbf{v}_j$ relative to $\mathcal{A}$. It is evident that the coordinates of $\mathbf{v}_j$ relative to $\mathcal{A}$ are recorded in the jth column of the matrix of transition from $\mathcal{A}$ to $\mathcal{B}$. Consequently, the a_{ij} are unique (since $\mathcal{A}$ is a *basis*), and the matrix $A = [a_{ij}]$ is unique. In particular, any $A = [a_{ij}]_{m \times n}$ may be considered as the matrix of transition from the standard basis $\mathcal{E}_m$ to a set of n vectors in $\mathbf{R}^m$, and the components of these vectors are given by the columns of A. Also, we note that the jth column of A, written as a column matrix

$$\begin{bmatrix} a_{1j} \\ a_{2j} \\ \vdots \\ a_{rj} \end{bmatrix},$$

is $[\mathbf{v}_j]_{\mathcal{A}}$, the coordinate matrix of $\mathbf{v}_j$ relative to $\mathcal{A}$.

EXAMPLE 4 Consider the matrix

$$A = \begin{bmatrix} 1 & 2 & 1 & -3 \\ 2 & 1 & -4 & 5 \\ 3 & 0 & 2 & 1 \end{bmatrix}.$$

This matrix is the transition matrix from $\mathcal{E}_3$ to the set $\mathcal{B} = \{\mathbf{v}_1, \mathbf{v}_2, \mathbf{v}_3, \mathbf{v}_4\}$ where

$$\mathbf{v}_1 = 1 \cdot \mathbf{e}_1 + 2 \cdot \mathbf{e}_2 + 3 \cdot \mathbf{e}_3 = (1, 2, 3),$$

$$\mathbf{v}_2 = 2 \cdot \mathbf{e}_1 + 1 \cdot \mathbf{e}_2 + 0 \cdot \mathbf{e}_3 = (2, 1, 0),$$

$$\mathbf{v}_3 = 1 \cdot \mathbf{e}_1 + (-4) \cdot \mathbf{e}_2 + 2 \cdot \mathbf{e}_3 = (1, -4, 2),$$

$$\mathbf{v}_4 = (-3) \cdot \mathbf{e}_1 + 5 \cdot \mathbf{e}_2 + 1 \cdot \mathbf{e}_3 = (-3, 5, 1).$$

Using the basis $\mathcal{A} = \{\mathbf{u}_1, \mathbf{u}_2, \mathbf{u}_3\}$, where $\mathbf{u}_1 = (0, 1, 1)$, $\mathbf{u}_2 = (1, 0, 1)$, $\mathbf{u}_3 = (1, 1, 0)$, the same matrix A is the matrix of transition from $\mathcal{A}$ to the set $\mathcal{C} = \{\mathbf{w}_1, \mathbf{w}_2, \mathbf{w}_3, \mathbf{w}_4\}$ where

$$\mathbf{w}_1 = 1 \cdot \mathbf{u}_1 + 2 \cdot \mathbf{u}_2 + 3 \cdot \mathbf{u}_3 = (5, 4, 3),$$

$$\mathbf{w}_2 = 2 \cdot \mathbf{u}_1 + 1 \cdot \mathbf{u}_2 + 0 \cdot \mathbf{u}_3 = (1, 2, 3),$$

$$\mathbf{w}_3 = 1 \cdot \mathbf{u}_1 + (-4) \cdot \mathbf{u}_2 + 2 \cdot \mathbf{u}_3 = (-2, 3, -3),$$

$$\mathbf{w}_4 = (-3) \cdot \mathbf{u}_1 + 5 \cdot \mathbf{u}_2 + 1 \cdot \mathbf{u}_3 = (6, -2, 2).$$

Again, we have an illustration of how the coordinates of a vector change from basis to basis. For the vector $\mathbf{v}_1 = \mathbf{w}_2 = (1, 2, 3)$ has the coordinate matrix

$$\begin{bmatrix} 1 \\ 2 \\ 3 \end{bmatrix}$$

relative to $\mathcal{E}_3$ and the coordinate matrix

$$\begin{bmatrix} 2 \\ 1 \\ 0 \end{bmatrix}$$

relative to $\mathcal{A}$. ■

3.1   Exercises

1. Determine x and y so that

$$\begin{bmatrix} 2x+y & x-2y \\ 4x-8y & 3x-y \end{bmatrix}$$

is a diagonal matrix.

2. Each of the sets A below is a basis of $\langle A \rangle$. Find the vector $\mathbf{v}$ that has the given matrix $[\mathbf{v}]_A$ as its coordinate matrix with respect to A.

(a) $A = \{(1, 3), (-2, 1)\}$; $[\mathbf{v}]_A = \begin{bmatrix} 5 \\ -1 \end{bmatrix}$

(b) $A = \{(2, -1, 1), (0, 1, -1), (-2, 1, 0)\}$; $[\mathbf{v}]_A = \begin{bmatrix} 3 \\ 1 \\ -4 \end{bmatrix}$

(c) $A = \{(3, 2, -1, 0), (0, -2, 5, 0), (2, 0, -4, 1)\}$; $[\mathbf{v}]_A = \begin{bmatrix} 0 \\ -2 \\ 1 \end{bmatrix}$

(d) $A = \{(0, 3, 4, 6), (-1, -2, 0, 2), (4, 0, -3, 1)\}$; $[\mathbf{v}]_A = \begin{bmatrix} 2 \\ -3 \\ -1 \end{bmatrix}$

3. With A as given in the corresponding part of Problem 2, the given vector $\mathbf{v}$ is in $\langle A \rangle$. Find $[\mathbf{v}]_A$, the coordinate matrix of $\mathbf{v}$ relative to A.

(a) $\mathbf{v} = (2, 20)$ **(b)** $\mathbf{v} = (4, 0, -1)$

(c) $\mathbf{v} = (4, 8, -8, -1)$ **(d)** $\mathbf{v} = (0, 8, 3, -9)$

4. Assume that the vector $\mathbf{v} = (x_1, x_2, x_3)$ has coordinate matrix

$$[\mathbf{v}]_A = \begin{bmatrix} c_1 \\ c_2 \\ c_3 \end{bmatrix}$$

relative to the basis $A = \{(1, 0, 1), (0, 1, 1), (1, 1, 0)\}$, and express the components x_1, x_2, x_3 of $\mathbf{v}$ in terms of c_1, c_2, and c_3.

5. Suppose that the vector $\mathbf{v} = (x_1, x_2, x_3)$ has coordinate matrix

$$[\mathbf{v}]_A = \begin{bmatrix} d_1 \\ d_2 \\ d_3 \end{bmatrix}$$

with respect to the basis $A = \{(1, 1, 1), (1, 1, 0), (1, 0, 0)\}$. Express each d_i in terms of x_1, x_2, and x_3.

6. Assume that $\mathcal{A} = \{\mathbf{u}, \mathbf{v}, \mathbf{w}\}$ is a basis of $\mathbf{R}^3$, where $\mathbf{u} = (u_1, u_2, u_3)$, $\mathbf{v} = (v_1, v_2, v_3)$, and $\mathbf{w} = (w_1, w_2, w_3)$. If a vector (x_1, x_2, x_3) has coordinates d_1, d_2, d_3 relative to $\mathcal{A}$, express the components x_i in terms of the components of $\mathbf{u}$, $\mathbf{v}$, and $\mathbf{w}$.

7. Let $\mathcal{A} = \{\mathbf{u}_1, \mathbf{u}_2\}$ and $\mathcal{B} = \{\mathbf{u}_1 + \mathbf{u}_2, \mathbf{u}_2\}$ be bases of a subspace $\mathbf{W}$ of $\mathbf{R}^n$. If $\mathbf{v} \in \mathbf{W}$ and

$$[\mathbf{v}]_{\mathcal{A}} = \begin{bmatrix} 3 \\ 2 \end{bmatrix},$$

find $[\mathbf{v}]_{\mathcal{B}}$.

8. Let $\mathcal{A} = \{\mathbf{u}_1, \mathbf{u}_2\}$ and $\mathcal{B} = \{\mathbf{u}_1, 3\mathbf{u}_1 + \mathbf{u}_2\}$ be bases of a subspace $\mathbf{W}$ of $\mathbf{R}^n$. If $\mathbf{v} \in \mathbf{W}$ and

$$[\mathbf{v}]_{\mathcal{A}} = \begin{bmatrix} -1 \\ 5 \end{bmatrix},$$

find $[\mathbf{v}]_{\mathcal{B}}$.

9. Let $\mathcal{A} = \{\mathbf{u}_1, \mathbf{u}_2, \mathbf{u}_3\}$ and $\mathcal{B} = \{\mathbf{u}_1, \mathbf{u}_1 + \mathbf{u}_2, \mathbf{u}_1 + \mathbf{u}_2 + \mathbf{u}_3\}$ be bases of a subspace $\mathbf{W}$ of $\mathbf{R}^n$. If $\mathbf{v} \in \mathbf{W}$ and

$$[\mathbf{v}]_{\mathcal{A}} = \begin{bmatrix} 2 \\ 1 \\ -1 \end{bmatrix},$$

find $[\mathbf{v}]_{\mathcal{B}}$.

10. Let $\mathcal{A} = \{\mathbf{u}_1, \mathbf{u}_2, \mathbf{u}_3\}$ and $\mathcal{B} = \{\mathbf{u}_1, \mathbf{u}_1 - \mathbf{u}_2, \mathbf{u}_1 + 2\mathbf{u}_3\}$ be bases of a subspace $\mathbf{W}$ of $\mathbf{R}^n$. If $\mathbf{v} \in \mathbf{W}$ and

$$[\mathbf{v}]_{\mathcal{A}} = \begin{bmatrix} 1 \\ -3 \\ 4 \end{bmatrix},$$

find $[\mathbf{v}]_{\mathcal{B}}$.

11. Let $\mathcal{A} = \{\mathbf{u}_1, \mathbf{u}_2, \mathbf{u}_3\}$ and $\mathcal{B} = \{\mathbf{u}_1 + \mathbf{u}_2 + \mathbf{u}_3, \mathbf{u}_2 + \mathbf{u}_3, \mathbf{u}_3\}$ be bases of a subspace $\mathbf{W}$ of $\mathbf{R}^n$. If $\mathbf{v} \in \mathbf{W}$ and

$$[\mathbf{v}]_{\mathcal{A}} = \begin{bmatrix} a \\ b \\ c \end{bmatrix},$$

find $[\mathbf{v}]_{\mathcal{B}}$.

12. Let $\mathcal{A} = \{\mathbf{u}_1, \mathbf{u}_2, \mathbf{u}_3\}$ and $\mathcal{B} = \{\mathbf{u}_1 + 2\mathbf{u}_3, \mathbf{u}_2 - \mathbf{u}_3, \mathbf{u}_3\}$ be bases of a subspace $\mathbf{W}$ of $\mathbf{R}^n$. If $\mathbf{v} \in \mathbf{W}$ and

$$[\mathbf{v}]_{\mathcal{A}} = \begin{bmatrix} a \\ b \\ c \end{bmatrix},$$

find $[\mathbf{v}]_{\mathcal{B}}$.

13. Suppose $\mathcal{B}$ is a basis of a subspace $\mathbf{W}$ of $\mathbf{R}^n$ and $\{\mathbf{u}_1, \mathbf{u}_2, \ldots, \mathbf{u}_r\}$ is any subset of $\mathbf{W}$. Prove that the set $\{\mathbf{u}_1, \mathbf{u}_2, \ldots, \mathbf{u}_r\}$ is linearly independent if and only if the set $\{[\mathbf{u}_1]_\mathcal{B}, [\mathbf{u}_2]_\mathcal{B}, \ldots, [\mathbf{u}_r]_\mathcal{B}\}$ is linearly independent.

14. Suppose $\mathcal{B}$ is a basis of a subspace $\mathbf{W}$ of $\mathbf{R}^n$, $\{\mathbf{u}_1, \mathbf{u}_2, \ldots, \mathbf{u}_r\}$ is any subset of $\mathbf{W}$, and $\mathbf{v} \in \mathbf{W}$. Prove that $\mathbf{v}$ is dependent on the set $\{\mathbf{u}_1, \mathbf{u}_2, \ldots, \mathbf{u}_r\}$ if and only if $[\mathbf{v}]_\mathcal{B}$ is dependent on the set $\{[\mathbf{u}_1]_\mathcal{B}, [\mathbf{u}_2]_\mathcal{B}, \ldots, [\mathbf{u}_r]_\mathcal{B}\}$.

15. Find the transition matrix from $\mathcal{A}$ to $\mathcal{B}$.

(a) $\mathcal{A} = \{(1, 1, 0, 0), (0, 0, 1, 1)\}; \quad \mathcal{B} = \{(2, 2, 3, 3), (1, 1, 2, 2)\}$

(b) $\mathcal{A} = \{(2, 2, 3, 3), (1, 1, 2, 2)\}; \quad \mathcal{B} = \{(1, 1, 0, 0), (0, 0, 1, 1)\}$

(c) $\mathcal{A} = \{(1, 3), (1, 1)\}; \quad \mathcal{B} = \{(3, 5), (5, 3)\}$

(d) $\mathcal{A} = \{(3, 3), (1, -1)\}; \quad \mathcal{B} = \{(5, 2), (7, 10)\}$

(e) $\mathcal{A} = \{(2, 2, 0, 0), (1, 2, 1, 0), (1, 1, 1, 1), (0, 0, 0, 1)\};$
$\quad \mathcal{B} = \{(4, 5, 2, 2), (1, 1, -1, 0)\}$

(f) $\mathcal{A} = \{(2, 3), (4, -2)\}; \quad \mathcal{B} = \{(6, 1), (8, -4), (2, -5)\}$

16. For each pair $\mathcal{A}$ and $\mathcal{B}$ in Problem 15, find the transition matrix from $\mathcal{B}$ to $\mathcal{A}$.

17. If C is the matrix of transition from a set of p vectors in $\mathbf{R}^n$ to a set of q vectors in $\mathbf{R}^n$, how many rows are in C?

18. In each part below, a set $\mathcal{A}$ and a matrix A are given. In each case, find $\mathcal{B}$ so that A is the matrix of transition from $\mathcal{A}$ to $\mathcal{B}$.

(a) $\mathcal{A} = \{(1, 2), (0, 1)\}; \quad A = \begin{bmatrix} 4 & -1 \\ -4 & 3 \end{bmatrix}$

(b) $\mathcal{A} = \{(1, 2), (2, 1)\}; \quad A = \begin{bmatrix} 5 & 7 \\ 2 & 10 \end{bmatrix}$

(c) $\mathcal{A} = \{(1, 0, 2, -1), (0, 1, 1, 2), (1, 2, 1, 4), (2, 2, 3, 0)\};$
$$A = \begin{bmatrix} 0 & 2 & 4 & 2 & 0 \\ 0 & 3 & -6 & 1 & -2 \\ 0 & 0 & 0 & -1 & -1 \\ 0 & -1 & 2 & 2 & 3 \end{bmatrix}$$

(d) $\mathcal{A} = \{(3, 1, 2, 4), (1, 0, 1, -1), (0, 2, 0, -4), (0, -3, 1, -3)\};$
$$A = \begin{bmatrix} 0 & 1 & 1 \\ 0 & -3 & 2 \\ 0 & 1 & -2 \\ 0 & 1 & 0 \end{bmatrix}$$

(e) $\mathcal{A} = \{(0, 1, -2), (-1, 1, 2), (1, -1, 0)\}; A = \begin{bmatrix} 1 & 1 & -1 \\ -1 & 3 & -1 \\ -1 & 2 & 0 \end{bmatrix}$

(f) $\mathcal{A} = \{(2, 1, 0, -1, 0), (0, -2, 0, 2, 0), (1, -1, 0, 1, 0)\}; A = \begin{bmatrix} 1 & 2 \\ 1 & 3 \\ 1 & -4 \end{bmatrix}$

19. Determine whether or not there is a matrix of transition from the first set to the second, and find such a matrix if it exists.

(a) $\{(1, 0, 2, 0), (0, 1, 2, 1)\};$ $\{(3, 0, 6, 0), (-8, 4, -8, 4), (-5, 4, -2, 4)\}$

(b) $\{(2, 4, 6, 4), (4, 6, 7, 3), (2, 0, -4, -6)\};$ $\{(2, 2, 1, -1), (0, 2, 5, 5)\}$

(c) $\{(2, 2, 0, 0), (1, 2, 1, 0), (1, 1, 1, 1)\};$ $\{(1, 1, 0, 0), (0, 1, 1, 0), (0, 0, 1, 2)\}$

(d) $\{(2, 4, 2, -2), (0, 3, -2, 4)\};$ $\{(1, 2, 1, -1), (0, 1, 2, 3), (1, 4, 5, 5)\}$

20. If

$$A = \begin{bmatrix} 1 & 2 & 1 & 3 \\ 2 & 1 & 4 & 5 \\ 3 & 0 & 2 & 1 \end{bmatrix}$$

is the matrix of transition from $\mathcal{E}_3$ to $\mathcal{B}$, find the standard basis of $\langle \mathcal{B} \rangle$.

21. If

$$A = \begin{bmatrix} 2 & 4 & 1 & 1 \\ 1 & 2 & 2 & 0 \\ 0 & 0 & 3 & -1 \\ 3 & 6 & 0 & 2 \end{bmatrix}$$

is the matrix of transition from $\mathcal{E}_4$ to $\mathcal{B}$, find the standard basis of $\langle \mathcal{B} \rangle$.

22. Let $\mathcal{A} = \{\mathbf{u}_1, \mathbf{u}_2, \ldots, \mathbf{u}_r\}$ and $\mathcal{B} = \{\mathbf{v}_1, \mathbf{v}_2, \ldots, \mathbf{v}_r\}$ be bases of $\mathbf{W}$, and let $P = [p_{ij}]_{r \times r}$ be the matrix of transition from $\mathcal{A}$ to $\mathcal{B}$. If $\mathbf{v}$ has coordinates $c_1, c_2, \ldots, c_r$ relative to $\mathcal{A}$ and $d_1, d_2, \ldots, d_r$ relative to $\mathcal{B}$, what relation exists between the c_i and the d_i?

3.2 Properties of Matrix Multiplication

Let $A = [a_{ij}]_{r \times s}$ be a given r by s matrix over $\mathbf{R}$. If $\mathbf{W}$ is any subspace of $\mathbf{R}^n$ and $\mathcal{A} = \{\mathbf{u}_1, \mathbf{u}_2, \ldots, \mathbf{u}_r\}$ is any set of r vectors in $\mathbf{W}$, then A is the matrix of transition from $\mathcal{A}$ to a *unique* set $\mathcal{B} = \{\mathbf{v}_1, \mathbf{v}_2, \ldots, \mathbf{v}_s\}$, where $\mathbf{v}_j = \sum_{i=1}^{r} a_{ij} \mathbf{u}_i$. That is, once $\mathbf{W}$ and $\mathcal{A}$ are chosen, the matrix A is the matrix of transition from $\mathcal{A}$ to one and only one subset $\mathcal{B}$. Also, as was pointed out in the last section, there is only one matrix of transition from $\mathcal{A}$ to $\mathcal{B}$ if $\mathcal{A}$ is a basis of $\mathbf{W}$. With these facts in mind, we make the following definition.

3.5 DEFINITION

Let $A = [a_{ij}]_{r \times s}$ and $B = [b_{ij}]_{s \times t}$ be matrices over $\mathbf{R}$, and let $\mathbf{W}$ be a subspace of $\mathbf{R}^n$ of dimension r. Then A is the matrix of transition from a basis $\mathcal{A} = \{\mathbf{u}_1, \mathbf{u}_2, \ldots, \mathbf{u}_r\}$ of $\mathbf{W}$ to a set $\mathcal{B} = \{\mathbf{v}_1, \mathbf{v}_2, \ldots, \mathbf{v}_s\}$ in $\mathbf{W}$, and B is a matrix of transition from $\mathcal{B}$ to a set $\mathcal{C} = \{\mathbf{w}_1, \mathbf{w}_2, \ldots, \mathbf{w}_t\}$ in $\mathbf{W}$. The **product** AB is defined to be the matrix of transition from $\mathcal{A}$ to $\mathcal{C}$. (See Figure 3.1.)

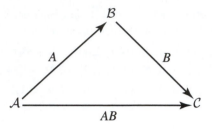

Figure 3.1

We observe that if A is r by s and B is s by t, then AB is r by t. Also, the number of rows in B must be the same as the number of columns in A in order for AB to be defined. When the product AB is defined, we say that B is *conformable* to A. The roles played by the dimensions of the matrices involved are shown in the following diagram.

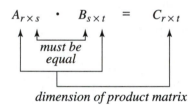

dimension of product matrix

The product AB as given in Definition 3.5 involves not only the matrices A and B, but choices of $\mathcal{A}$ and $\mathbf{W}$ as well. This means that there is a possibility that the product AB may not be unique, but may vary with the choices of $\mathcal{A}$ and $\mathbf{W}$. The next theorem shows that this does not happen, and the product AB actually depends only on A and B.

3.6 THEOREM   If $A = [a_{ij}]_{r \times s}$, $B = [b_{ij}]_{s \times t}$, and $AB = [c_{ij}]_{r \times t}$, then $c_{ij} = \sum_{k=1}^{s} a_{ik} b_{kj}$ for all pairs i, j.

Proof Let $\mathcal{A} = \{\mathbf{u}_1, \mathbf{u}_2, \ldots, \mathbf{u}_r\}$ be a basis of $\mathbf{W}$. Then $A = [a_{ij}]_{r \times s}$ is the matrix of transition from $\mathcal{A}$ to a set $\mathcal{B} = \{\mathbf{v}_1, \mathbf{v}_2, \ldots, \mathbf{v}_s\}$, and $B = [b_{ij}]_{s \times t}$ is a matrix of transition from $\mathcal{B}$ to a set $\mathcal{C} = \{\mathbf{w}_1, \mathbf{w}_2, \ldots, \mathbf{w}_t\}$. That is, $\mathbf{v}_k = \sum_{i=1}^{r} a_{ik} \mathbf{u}_i$ for $k = 1, 2, \ldots, s$ and $\mathbf{w}_j = \sum_{k=1}^{s} b_{kj} \mathbf{v}_k$ for $j = 1, 2, \ldots, t$. Thus

$$\mathbf{w}_j = \sum_{k=1}^{s} b_{kj} \mathbf{v}_k$$

$$= \sum_{k=1}^{s} b_{kj} \left(\sum_{i=1}^{r} a_{ik} \mathbf{u}_i \right)$$

$$= \sum_{k=1}^{s} \left(\sum_{i=1}^{r} a_{ik} b_{kj} \mathbf{u}_i \right)$$

$$= \sum_{i=1}^{r} \left(\sum_{k=1}^{s} a_{ik} b_{kj} \mathbf{u}_i \right)$$

$$= \sum_{i=1}^{r} \left(\sum_{k=1}^{s} a_{ik} b_{kj} \right) \mathbf{u}_i.$$

But $AB = [c_{ij}]_{r \times t}$ is the matrix of transition from $\mathcal{A}$ to $\mathcal{C}$, so that $\mathbf{w}_j = \sum_{i=1}^{r} c_{ij} \mathbf{u}_i$. Since $\mathcal{A}$ is a basis, $\mathbf{w}_j$ can be written as a linear combination of $\mathbf{u}_1, \mathbf{u}_2, \ldots, \mathbf{u}_r$ in only one way. Therefore, $c_{ij} = \sum_{k=1}^{s} a_{ik} b_{kj}$ for all pairs i, j. ∎ ■ ■ ■

If A is a matrix with only one row, $A = [a_1 \quad a_2 \quad \cdots \quad a_s]$, and B is a matrix with a single column,

$$B = \begin{bmatrix} b_1 \\ b_2 \\ \vdots \\ b_s \end{bmatrix},$$

then AB has only one element, given by $a_1 b_1 + a_2 b_2 + \cdots + a_s b_s$. This result can be committed to memory easily by mentally "standing A by B," forming products of pairs of corresponding elements, and then forming the sum of these products:

$$a_1 b_1$$

$$+ a_2 b_2$$

$$+ \cdots$$

$$+ a_s b_s.$$

It is easily seen from the formula in Theorem 3.6 that the element in the ith row and jth column of AB can be found by multiplying the ith row of A by the jth column of B, following the routine given for a single row and a single column. This aid to mental multiplication is known as the *row-by-column-rule.*

The row-by-column rule uses the same pattern as the one for computing the inner product of two vectors (see Definition 1.18). This pattern is illustrated in the following diagram.

$$
\text{row } i \text{ of } A\}
\begin{bmatrix}
\vdots & \vdots & \vdots & & \vdots \\
a_{i1} & a_{i2} & a_{i3} & \cdots & a_{in} \\
\vdots & \vdots & \vdots & & \vdots
\end{bmatrix}
\cdot
\begin{bmatrix}
\cdots & b_{1j} & \cdots \\
\cdots & b_{2j} & \cdots \\
\cdots & b_{3j} & \cdots \\
& \vdots & \\
\cdots & b_{nj} & \cdots
\end{bmatrix}
\overbrace{\hspace{2cm}}^{\text{column } j \text{ of } B}
$$

$$
=
\begin{bmatrix}
& \vdots & \\
\cdots & c_{ij} & \cdots \\
& \vdots &
\end{bmatrix}
\{\text{row } i \text{ of } C
\qquad \overbrace{\hspace{2cm}}^{\text{column } j \text{ of } C}
$$

where

$$
c_{ij} = a_{i1}b_{1j} + a_{i2}b_{2j} + a_{i3}b_{3j} + \cdots + a_{in}b_{nj}.
$$

EXAMPLE 1 Consider the products AB and BA for the matrices

$$
A =
\begin{bmatrix}
2 & -1 \\
3 & 1 \\
6 & -5 \\
0 & 4
\end{bmatrix},
\quad
B =
\begin{bmatrix}
7 & -2 & -3 \\
5 & -4 & 0
\end{bmatrix}.
$$

The number of rows in B is the same as the number of columns in A, so the product AB is defined. Performing the computations, we find that

$$
AB =
\begin{bmatrix}
2 & -1 \\
3 & 1 \\
6 & -5 \\
0 & 4
\end{bmatrix}
\begin{bmatrix}
7 & -2 & -3 \\
5 & -4 & 0
\end{bmatrix}
$$

$$
=
\begin{bmatrix}
(2)(7) + (-1)(5) & (2)(-2) + (-1)(-4) & (2)(-3) + (-1)(0) \\
(3)(7) + (1)(5) & (3)(-2) + (1)(-4) & (3)(-3) + (1)(0) \\
(6)(7) + (-5)(5) & (6)(-2) + (-5)(-4) & (6)(-3) + (-5)(0) \\
(0)(7) + (4)(5) & (0)(-2) + (4)(-4) & (0)(-3) + (4)(0)
\end{bmatrix}
$$

$$
=
\begin{bmatrix}
9 & 0 & -6 \\
26 & -10 & -9 \\
17 & 8 & -18 \\
20 & -16 & 0
\end{bmatrix}.
$$

The product BA is not defined because the number of rows in A is 4 and the number of columns in B is 3. ∎

The major result concerning properties of matrix multiplication is stated next.

3.7 **THEOREM**	Let $A = [a_{ij}]_{r \times s}$, $B = [b_{ij}]_{s \times t}$, and $C = [c_{ij}]_{t \times v}$ be matrices over **R**. Then $A(BC) = (AB)C$.

Proof By Theorem 3.6, $AB = [d_{ij}]_{r \times t}$ where $d_{ij} = \sum_{m=1}^{s} a_{im} b_{mj}$, and $(AB)C = \left[\sum_{k=1}^{t} d_{ik} c_{kj} \right]_{r \times v}$ where

$$\sum_{k=1}^{t} d_{ik} c_{kj} = \sum_{k=1}^{t} \left(\sum_{m=1}^{s} a_{im} b_{mk} \right) c_{kj}$$

$$= \sum_{k=1}^{t} \left(\sum_{m=1}^{s} (a_{im} b_{mk}) c_{kj} \right).$$

Similarly, $BC = [f_{ij}]_{s \times v}$ where $f_{ij} = \sum_{k=1}^{t} b_{ik} c_{kj}$, and $A(BC) = \left[\sum_{m=1}^{s} a_{im} f_{mj} \right]_{r \times v}$ where

$$\sum_{m=1}^{s} a_{im} f_{mj} = \sum_{m=1}^{s} a_{im} \left(\sum_{k=1}^{t} b_{mk} c_{kj} \right)$$

$$= \sum_{m=1}^{s} \left(\sum_{k=1}^{t} a_{im} (b_{mk} c_{kj}) \right)$$

$$= \sum_{k=1}^{t} \left(\sum_{m=1}^{s} a_{im} (b_{mk} c_{kj}) \right).$$

But $a_{im}(b_{mk} c_{kj}) = (a_{im} b_{mk}) c_{kj}$ from the associative property of multiplication of real numbers. Hence $A(BC) = (AB)C$. ■■■

If all necessary conformability is assumed, then matrix multiplication is associative.

Now the matrix product BA may fail to exist, even when AB is defined, so the commutative property of multiplication does not hold in general. In Example 2, matrices A and B are given for which AB and BA are both defined, and yet $AB \neq BA$.

EXAMPLE 2 Let

$$A = \begin{bmatrix} 1 & 2 \\ 4 & 0 \end{bmatrix} \quad \text{and} \quad B = \begin{bmatrix} 3 & -1 \\ 2 & 1 \end{bmatrix}.$$

Then

$$AB = \begin{bmatrix} 7 & 1 \\ 12 & -4 \end{bmatrix} \quad \text{and} \quad BA = \begin{bmatrix} -1 & 6 \\ 6 & 4 \end{bmatrix},$$

so that $AB \neq BA$. ■

There are two other fundamental properties of multiplication of real numbers that are not valid for multiplication of matrices. In general $AB = AC$ and $A \neq \mathbf{0}$ do not imply $B = C$, nor do $BA = CA$ and $A \neq \mathbf{0}$ imply that $B = C$. That is, there is no cancellation property for matrix multiplication. Also, $AB = \mathbf{0}$ does not imply that one of A, B must be zero. Thus, the product of two nonzero matrices may be a zero matrix. Examples of these situations are requested in some of the exercises.

3.2 Exercises

1. Compute the product AB for the given matrices.

(a) $A = \begin{bmatrix} 2 & 0 & 5 \\ 1 & -2 & 4 \\ -3 & 1 & 6 \end{bmatrix}$, $B = \begin{bmatrix} 2 & 0 \\ 8 & 1 \\ 5 & -1 \end{bmatrix}$

(b) $A = \begin{bmatrix} 4 & -5 \\ -2 & 1 \end{bmatrix}$, $B = \begin{bmatrix} 8 & -2 & 7 \\ 3 & 0 & 6 \end{bmatrix}$

(c) $A = \begin{bmatrix} 3 & 2 & -1 \\ 1 & 4 & 2 \end{bmatrix}$, $B = \begin{bmatrix} 1 & -2 \\ 5 & -3 \\ 4 & 6 \end{bmatrix}$

(d) $A = \begin{bmatrix} 5 & -2 & 7 \end{bmatrix}$, $B = \begin{bmatrix} 3 \\ 6 \\ 2 \end{bmatrix}$

(e) $A = \begin{bmatrix} 2 & -5 & 3 & 0 \\ -6 & 8 & 1 & -2 \end{bmatrix}$, $B = \begin{bmatrix} -1 & -1 \\ -8 & -1 \\ -9 & 3 \\ 4 & 4 \end{bmatrix}$

(f) $A = \begin{bmatrix} 2 & 1 & -4 \\ 5 & 0 & 2 \end{bmatrix}$, $B = \begin{bmatrix} 8 \\ 4 \\ -5 \end{bmatrix}$

2. Whenever possible, compute the products BA in Problem 1.

3. For each of the following pairs A, B, find AB.

(a) $A = [a_{ij}]_{3 \times 4}$ where $a_{ij} = (-1)^{i+j}$, $B = [b_{ij}]_{4 \times 2}$ where $b_{ij} = \delta_{ij}$

(b) $A = [a_{ij}]_{3 \times 2}$ where $a_{ij} = i\,(-1)^{i+j}$, $B = [b_{ij}]_{2 \times 4}$ where $b_{ij} = (i + j)\delta_{ij}$

(c) $A = [a_{ij}]_{4 \times 4}$ where $a_{ij} = \begin{cases} 1 & \text{for } i > j \\ 0 & \text{for } i \leq j \end{cases}$, $B = [b_{ij}]_{4 \times 3}$ where $b_{ij} = 2i - j$

(d) $A = [a_{ij}]_{3 \times 3}$ where $a_{ij} = \begin{cases} i & \text{for } i > j \\ 1 & \text{for } i = j \\ 0 & \text{for } i < j \end{cases}$, $B = [b_{ij}]_{3 \times 4}$ where $b_{ij} = (-1)^{\delta_{ij}}$

4. Let $A = [a_{ij}]_{m \times n}$ and $B = [b_{ij}]_{r \times t}$ be matrices over **R**.

(a) What conditions are necessary in order that AB be defined?

(b) What are the dimensions of AB when it exists?

(c) What conditions are necessary in order that both AB and BA be defined?

(d) What conditions are necessary in order that the dimensions of AB be the same as the dimensions of BA?

5. Let $A = [a_{ij}]_{r \times s}$. Under what conditions does A^k exist for every positive integer k? ($A^1 = A$, $A^2 = A \cdot A$, etc.)

6. The Fibonacci sequence is defined by

$$x_0 = 0, \; x_1 = 1, \; x_n = x_{n-1} + x_{n-2}, \quad n = 2, 3, \ldots$$

(a) Show that the matrix

$$A = \begin{bmatrix} 1 & 1 \\ 1 & 0 \end{bmatrix}$$

can be used to generate the Fibonacci sequence by

$$X_n = AX_{n-1} \quad \text{where } X_n = \begin{bmatrix} x_n \\ x_{n-1} \end{bmatrix}, \quad n = 1, 2, \ldots.$$

(b) Show that $X_n = A^{n-1}X_1, \; n = 1, 2, \ldots$.

7. Give an example of nonzero matrices A and B such that $AB = BA$.

8. Give an example of two 2×2 matrices A and B over $\mathbf{R}$ such that $AB = 0$ but $A \neq 0$ and $B \neq 0$.

9. Give an example of 2×2 matrices A, B, and C over $\mathbf{R}$ such that $AB = AC$, but $B \neq C$ and $A \neq 0$.

10. Let

$$B = \begin{bmatrix} 1 & -2 \\ 3 & 2 \end{bmatrix}.$$

Find all 2×2 matrices A such that $AB = BA$.

11. Find all 2×2 matrices A that satisfy the given conditions.

(a) $A^2 = \begin{bmatrix} x & 0 \\ 0 & y \end{bmatrix}$ where $x, y > 0$ **(b)** $A^2 = \begin{bmatrix} x & x \\ x & x \end{bmatrix}$ where $x > 0$

12. Let $A = [a_{ij}]_{2 \times 3}$, $B = [b_{ij}]_{3 \times 2}$, and $C = [c_{ij}]_{2 \times 1}$ over $\mathbf{R}$. For these particular matrices, verify that $A(BC) = (AB)C$ without using Theorem 3.7.

13. For each of the following pairs A, B, let $AB = [c_{ij}]$ and write a formula for c_{ij} in terms of i and j.

(a) $A = [a_{ij}]_{2 \times 3}$ where $a_{ij} = i + j$, $B = [b_{ij}]_{3 \times 4}$ where $b_{ij} = 2i - j$

(b) $A = [a_{ij}]_{3 \times 2}$ where $a_{ij} = i - j$, $B = [b_{ij}]_{2 \times 4}$ where $b_{ij} = i + j$

(c) $A = [a_{ij}]_{3\times2}$ where $a_{ij} = i - j$, $B = [b_{ij}]_{2\times3}$ where $b_{ij} = 2i + j$

(d) $A = [a_{ij}]_{2\times3}$ where $a_{ij} = i + 2j$, $B = [b_{ij}]_{3\times2}$ where $b_{ij} = i - 2j$

14. Show that the matrix equation

$$\begin{bmatrix} 1 & -2 & 1 \\ 2 & 0 & 3 \\ 1 & 4 & -1 \end{bmatrix} \cdot \begin{bmatrix} x_1 \\ x_2 \\ x_3 \end{bmatrix} = \begin{bmatrix} 4 \\ 5 \\ 6 \end{bmatrix}$$

is equivalent to a system of linear equations in x_1, x_2, and x_3. (*Hint:* Use Definition 3.2.)

15. Reversing the procedure in Problem 14, write a matrix equation that is equivalent to the system of equations

$$\begin{aligned} x_1 + 4x_2 + 3x_3 + 2x_4 &= 3 \\ x_2 + x_3 + 7x_4 &= -5 \\ 2x_1 - 3x_2 + x_3 - x_4 &= 0. \end{aligned}$$

16. Suppose

$$A = \begin{bmatrix} 4 & -2 & 1 & 0 \\ 5 & 0 & 2 & 1 \end{bmatrix}$$

is the transition matrix from $\mathcal{A} = \{(1, 1), (0, 1)\}$ to the set $\mathcal{B}$, and

$$B = \begin{bmatrix} 1 & 0 \\ 4 & -3 \\ 0 & -1 \\ -8 & 6 \end{bmatrix}$$

is the transition matrix from $\mathcal{B}$ to $\mathcal{C}$.

(a) Find the matrix of transition from $\mathcal{A}$ to $\mathcal{C}$.

(b) Find $\mathcal{C}$.

17. Suppose

$$A = \begin{bmatrix} 3 & 1 \\ 0 & -1 \end{bmatrix}$$

is the transition matrix from $\mathcal{A} = \{(1, -1), (1, 1)\}$ to $\mathcal{B}$, and B is the transition matrix from $\mathcal{B}$ to $\mathcal{C} = \{(6, -4), (3, 1)\}$.

(a) Find B.

(b) Find the matrix of transition from $\mathcal{A}$ to $\mathcal{C}$.

18. Suppose A is the transition matrix from $\mathcal{A} = \{(1, 0, -1), (1, 1, 0), (0, -1, 1)\}$ to $\mathcal{B} = \{(2, 4, 0), (2, 1, 1)\}$ and

$$B = \begin{bmatrix} 0 & 1 & 1 \\ -2 & 1 & 0 \end{bmatrix}$$

is the transition matrix from $\mathcal{B}$ to $\mathcal{C}$.

(a) Find the transition matrix from $\mathcal{A}$ to $\mathcal{C}$.

(b) Find $\mathcal{C}$.

19. Prove or disprove: The product AB is a diagonal matrix if B is a diagonal matrix.

20. Prove or disprove: If AB is a diagonal matrix then at least one of A or B is a diagonal matrix.

21. Prove or disprove: The products AB and BA are defined if and only if both A and B are square matrices of the same order.

22. The **trace** $t(A)$ of a square matrix $A = [a_{ij}]_n$ is defined as $t(A) = \sum_{i=1}^{n} a_{ii}$.

(a) Prove that if A and B are square matrices of order n then $t(AB) = t(BA)$.

(b) Let A and B be matrices such that AB and BA are defined. Prove or disprove: $t(AB) = t(BA)$.

23. A square matrix $A = [a_{ij}]_n$ with $a_{ij} = 0$ for all $i > j$ is called **upper triangular**. Prove or disprove each of the following statements.

(a) The product AB is upper triangular if both A and B are square upper triangular matrices.

(b) If A and B are square and the product AB is upper triangular then at least one of A or B is upper triangular.

24. Prove or disprove: The columns of AB are linearly independent if the columns of A are linearly independent.

25. Prove that $(AB)^n = A^n B^n$ for all positive integers n, if $AB = BA$.

26. Prove or disprove each of the following statements where the product AB exists.

(a) If the jth column of A contains all zeros, then the jth column of AB contains all zeros.

(b) If the ith row of A contains all zeros, then the ith row of AB contains all zeros.

(c) If the jth column of AB contains all zeros, then at least one row of A contains all zeros.

27. Prove Theorem 3.7 by direct use of Definition 3.5.

3.3 Invertible Matrices

Those matrices that are matrices of transition from one basis to another basis are of particular importance.

3.8 **DEFINITION**	A matrix $A = [a_{ij}]_{n \times n}$ is **nonsingular** if and only if A is a matrix of transition from one basis of $\mathbf{R}^n$ to another basis of $\mathbf{R}^n$. A square matrix that is not nonsingular is called **singular**.

Suppose that $A = [a_{ij}]$ is a square matrix of order n. For each basis $\mathcal{A}$ of $\mathbf{R}^n$, A is the matrix of transition from $\mathcal{A}$ to a set $\mathcal{B}$ of n vectors in $\mathbf{R}^n$. Generally, different choices of $\mathcal{A}$ yield different sets $\mathcal{B}$. In order for the term *nonsingular* to be well-defined, we must show that if one of the sets $\mathcal{B}$ is a basis, then all of them are bases.

Suppose $A = [a_{ij}]_{n \times n}$ is the matrix of transition from the basis $\mathcal{A} = \{\mathbf{u}_1, \mathbf{u}_2, \dots, \mathbf{u}_n\}$ of $\mathbf{R}^n$ to the basis $\mathcal{B} = \{\mathbf{v}_1, \mathbf{v}_2, \dots, \mathbf{v}_n\}$ of $\mathbf{R}^n$. Let $\mathcal{A}' = \{\mathbf{u}_1', \mathbf{u}_2', \dots, \mathbf{u}_n'\}$ be another basis of $\mathbf{R}^n$, and let $\mathcal{B}' = \{\mathbf{v}_1', \mathbf{v}_2', \dots, \mathbf{v}_n'\}$ be the set so that A is the matrix of transition from $\mathcal{A}'$ to $\mathcal{B}'$.

In order to show that $\mathcal{B}'$ is a basis of $\mathbf{R}^n$, it is sufficient to show that $\mathcal{B}'$ is linearly independent (Theorem 1.34). Suppose that

$$\sum_{j=1}^n b_j \mathbf{v}_j' = \mathbf{0}.$$

Since $\mathbf{v}_j' = \sum_{i=1}^n a_{ij} \mathbf{u}_i'$, we have $\sum_{j=1}^n b_j \left(\sum_{i=1}^n a_{ij} \mathbf{u}_i' \right) = \mathbf{0}$ and

$$\sum_{i=1}^n \left(\sum_{j=1}^n b_j a_{ij} \right) \mathbf{u}_i' = \mathbf{0}.$$

But $\mathcal{A}'$ is linearly independent, so this means that $\sum_{j=1}^n b_j a_{ij} = 0$ for $i = 1, 2, \dots, n$. Hence

$$\sum_{j=1}^n b_j \mathbf{v}_j = \sum_{j=1}^n b_j \left(\sum_{i=1}^n a_{ij} \mathbf{u}_i \right)$$

$$= \sum_{i=1}^n \left(\sum_{j=1}^n b_j a_{ij} \right) \mathbf{u}_i$$

$$= \sum_{i=1}^n 0 \cdot \mathbf{u}_i$$

$$= \mathbf{0},$$

and this implies that $b_1 = b_2 = \cdots = b_n = 0$ since $\mathcal{B}$ is linearly independent. Thus, $\mathcal{B}'$ is linearly independent and is a basis of $\mathbf{R}^n$.

An n by n matrix A over $\mathbf{R}$ is a nonsingular matrix if and only if the columns of A record the coordinates of one basis of $\mathbf{R}^n$ with respect to a second (not necessarily different) basis of $\mathbf{R}^n$. That is, A is nonsingular if and only if the jth column of A is the coordinate matrix of the jth vector in a basis of $\mathbf{R}^n$ for $j = 1, 2, \dots, n$.

In the discussion of special types of matrices just before Definition 3.2, an *identity matrix* was defined to be a matrix of the form $I_n = [\delta_{ij}]_{n \times n}$, where δ_{ij} is the Kronecker delta. There are many identity matrices, I_n, but only one for each value of n. As examples,

$$I_2 = \begin{bmatrix} 1 & 0 \\ 0 & 1 \end{bmatrix} \quad \text{and} \quad I_3 = \begin{bmatrix} 1 & 0 & 0 \\ 0 & 1 & 0 \\ 0 & 0 & 1 \end{bmatrix}.$$

From the placing of the 1's and 0's in I_n, it is easy to see that I_n is the unique matrix of transition from a basis $\mathcal{A}$ to the same basis $\mathcal{A}$, and I_n is therefore nonsingular.

Using the fact that I_n is the transition matrix from $\mathcal{A}$ to $\mathcal{A}$, it follows easily from Definition 3.5 that

$$I_m A = A \quad \text{and} \quad A I_n = A$$

for any $m \times n$ matrix A. In particular,

$$I_n A = A = A I_n$$

if A is an $n \times n$ matrix. Thus I_n is a *multiplicative identity* for square matrices of order n, and it is upon this fact that the term *identity matrix* is based.

The existence of a multiplicative identity for square matrices of order n leads naturally to the question as to which square matrices have multiplicative inverses. When working with matrices, it is conventional to use the term *inverse* to mean "multiplicative inverse."

3.9 DEFINITION

Let A be a square matrix of order n. An $n \times n$ matrix B is an **inverse**, of A if $AB = I_n = BA$. Also, a square matrix is called **invertible** if it has an inverse.

We note that inverses of matrices occur in pairs in this sense: If B is an inverse of A, then A is also an inverse of B.

Our next theorem gives an answer to the question as to which square matrices are invertible.

3.10 THEOREM

Let A be a square matrix of order n. Then A is invertible if and only if A is nonsingular.

Proof Suppose that $\mathcal{A}$ is a basis of $\mathbf{R}^n$ and let the $n \times n$ matrix A be the transition matrix from $\mathcal{A}$ to a set $\mathcal{B}$ of n vectors in $\mathbf{R}^n$.

Assume first that A is nonsingular. Then $\mathcal{B}$ is a basis of $\mathbf{R}^n$ and therefore every vector in $\mathcal{A}$ is a linear combination of vectors in $\mathcal{B}$. Hence there exists a transition matrix B from $\mathcal{B}$ to $\mathcal{A}$. By Definition 3.5, AB is the matrix of transition from $\mathcal{A}$ to $\mathcal{A}$. It follows that $AB = I_n$ since I_n is the unique matrix of transition from $\mathcal{A}$ to $\mathcal{A}$. In similar fashion, we can show that BA is the matrix of transition from the basis $\mathcal{B}$ to $\mathcal{B}$, and therefore $BA = I_n$.

To prove the other part of the theorem, assume that A has an inverse B, so that $AB = I_n$ and $BA = I_n$. With the same notation as in the first paragraph of this proof, let B be the transition matrix from the set $\mathcal{B}$ of n vectors to a set $\mathcal{C}$ of n vectors in $\mathbf{R}^n$. Then AB is the transition matrix from $\mathcal{A}$ to $\mathcal{C}$, by Definition 3.5. But $AB = I_n$, so the set $\mathcal{C}$ is exactly the same as $\mathcal{A}$. This means that B is a transition matrix from $\mathcal{B}$ to $\mathcal{A}$. Therefore $\mathcal{A}$ is dependent on $\mathcal{B}$, and this implies that $\mathbf{R}^n$ is dependent on $\mathcal{B}$, by Theorem 1.6. That is, $\mathcal{B}$ spans $\mathbf{R}^n$. It follows then from Theorem 1.35 that $\mathcal{B}$ is a basis of $\mathbf{R}^n$ and hence A is nonsingular.

■ ■ ■

> **3.11**
> **COROLLARY**
>
> Suppose A is the matrix of transition from a basis $\mathcal{A}$ of $\mathbf{R}^n$ to a basis $\mathcal{B}$ of $\mathbf{R}^n$. Then the matrix B is an inverse of A if and only if B is the transition matrix from $\mathcal{B}$ to $\mathcal{A}$.

Proof The corollary follows at once from the proof of the theorem. ■ ■ ■

EXAMPLE 1 The matrix

$$A = \begin{bmatrix} 3 & 15 \\ 4 & 20 \end{bmatrix}$$

is not invertible. It is the transition matrix from the standard basis $\mathcal{E}_2 = \{(1, 0), (0, 1)\}$ to $\mathcal{B} = \{(3, 4), (15, 20)\}$, and $\mathcal{B}$ is clearly dependent because its second vector is equal to 5 times its first vector. As an example of an invertible matrix, consider the matrix

$$A = \begin{bmatrix} 1 & 1 & 1 \\ 0 & 1 & 1 \\ 0 & 0 & 1 \end{bmatrix}.$$

This matrix is the transition matrix from $\mathcal{E}_3 = \{(1, 0, 0), (0, 1, 0), (0, 0, 1)\}$ to $\mathcal{B} = \{1, 0, 0), (1, 1, 0), (1, 1, 1)\}$ and it is easy to see that $\mathcal{B}$ is linearly independent since none of its vectors are linear combinations of the preceding vectors. Of course, it is not always so easy to tell whether a matrix is invertible or not. ■

Up to this point, we have allowed the possibility that a matrix might have two or more distinct inverses. The next theorem shows that this possibility does not actually happen.

> **3.12**
> **THEOREM**
>
> The inverse of an invertible matrix is unique.

Proof Suppose that B and C are both inverses of the $n \times n$ matrix A. Then all of the equations

$$AB = I_n = BA \quad \text{and} \quad AC = I_n = CA$$

are valid. To prove that $B = C$, we use the fact that I_n is a multiplicative identity and that matrix multiplication is associative. Thus

$$B = BI_n = B(AC) = (BA)C = I_nC = C,$$

and the inverse of the invertible matrix A is unique. ■ ■ ■

Theorem 3.12 allows us to make the following definition.

**3.13
DEFINITION**

If A is an invertible matrix, its unique inverse is denoted by A^{-1}.

EXAMPLE 2 We saw in Example 1 of this section that the matrix

$$A = \begin{bmatrix} 1 & 1 & 1 \\ 0 & 1 & 1 \\ 0 & 0 & 1 \end{bmatrix}$$

is invertible and is the transition matrix from the basis $\mathcal{E}_3 = \{(1, 0, 0), (0, 1, 0), (0, 0, 1)\}$ to the basis $\mathcal{B} = \{(1, 0, 0), (1, 1, 0), (1, 1, 1)\}$ of $\mathbf{R}^3$. In order to find A^{-1}, it is sufficient to find the transition matrix from $\mathcal{B}$ to $\mathcal{E}_3$. Since

$$(1, 0, 0) = (1)(1, 0, 0) + (0)(1, 1, 0) + (0)(1, 1, 1)$$

$$(0, 1, 0) = (-1)(1, 0, 0) + (1)(1, 1, 0) + (0)(1, 1, 1)$$

$$(0, 0, 1) = (0)(1, 0, 0) + (-1)(1, 1, 0) + (1)(1, 1, 1),$$

we see that

$$A^{-1} = \begin{bmatrix} 1 & -1 & 0 \\ 0 & 1 & -1 \\ 0 & 0 & 1 \end{bmatrix}.$$

If a given square matrix A is complicated, a more efficient method than the one we used here is needed to find A^{-1}. Such a method is presented in Section 3.5. ■

Some of our earlier results can be rewritten using the exponential notation in Definition 3.13. The equations $AB = I_n = BA$ now read as

$$AA^{-1} = I_n = A^{-1}A.$$

Also, we noted earlier that inverses occur in pairs: When $B = A^{-1}$, then $A = B^{-1}$. Substituting the value $B = A^{-1}$ in the equation $A = B^{-1}$ yields

$$A = (A^{-1})^{-1}.$$

In the definition and all discussion in this section, it has been required that both of the equations $AA^{-1} = I_n$ and $A^{-1}A = I_n$ be satisfied by the inverse matrix. Our next two theorems show that these equations are not independent for square matrices. In fact, *for square matrices*, we shall see that either of them implies the other.

**3.14
THEOREM**

Let A be a square matrix of order n over $\mathbf{R}$. If there is a square matrix B over $\mathbf{R}$ such that $AB = I_n$, then A is invertible and $B = A^{-1}$.

Proof Suppose there is a square matrix B such that $AB = I_n$. Now A is the matrix of transition from $\mathcal{E}_n$ to a set $\mathcal{A}$ of n vectors in $\mathbf{R}^n$, B is a matrix of transition from $\mathcal{A}$ to a set $\mathcal{B}$ of n vectors in $\mathbf{R}^n$, and $AB = I_n$, so $\mathcal{B}$ must be $\mathcal{E}_n$. Thus $\mathcal{B}$ is a matrix of transition from $\mathcal{A}$ to $\mathcal{E}_n$, and this means that each vector in $\mathcal{E}_n$ is a linear combination of vectors in $\mathcal{A}$. That is, $\mathcal{E}_n$ is dependent on $\mathcal{A}$. And since $\mathbf{R}^n$ is dependent on $\mathcal{E}_n$, this means that $\mathbf{R}^n$ is dependent on $\mathcal{A}$. By Theorem 1.35, $\mathcal{A}$ is a basis of $\mathbf{R}^n$, and A is invertible. It follows from Corollary 3.11 that $B = A^{-1}$.   ∎∎∎

In view of Theorem 3.14, we see that a matrix A of order n is invertible if and only if there is a square matrix B such that $AB = I_n$.

The proof of the next theorem is quite similar to that of Theorem 3.14, and is left as an exercise.

3.15 THEOREM

Let A be a square matrix of order n over $\mathbf{R}$. If there is a square matrix B over $\mathbf{R}$ such that $BA = I_n$, then A is invertible and $B = A^{-1}$.

3.16 THEOREM

If A and B are invertible matrices of the same order, then AB is invertible and $(AB)^{-1} = B^{-1}A^{-1}$.

Proof If A and B are invertible matrices of order n, then A^{-1} and B^{-1} exist. Since

$$(AB)(B^{-1}A^{-1}) = A(BB^{-1})A^{-1} = AI_nA^{-1} = I_n,$$

AB is invertible and $(AB)^{-1} = B^{-1}A^{-1}$ by Theorem 3.14.   ∎∎∎

Repeated application of Theorem 3.16 yields the following corollary.

3.17 COROLLARY

If $A_1, A_2, \ldots, A_k$ are square matrices of order n over $\mathbf{R}$ and each A_i is invertible, then $A_1 A_2 \cdots A_k$ is invertible and $(A_1 A_2 \cdots A_k)^{-1} = A_k^{-1} \cdots A_2^{-1} A_1^{-1}$.

3.18 DEFINITION

If $\mathcal{A}$ and $\mathcal{A}'$ are bases of a subspace $\mathbf{W}$ of $\mathbf{R}^n$ such that $\mathcal{A}'$ is obtained from $\mathcal{A}$ by a single elementary operation, then the matrix of transition from $\mathcal{A}$ to $\mathcal{A}'$ is an **elementary matrix**.

That is, a matrix M is an elementary matrix if and only if it is a matrix of transition from a given basis to another basis that is obtained by applying a single elementary operation E to the original basis. We shall say that the elementary matrix M and the elementary operation E are *associated* with each other. The elementary matrix is classified as type I, II, or III according to the type of elementary operation that is applied to the original basis.

3.19 **THEOREM**	A square matrix over **R** is invertible if and only if it is a product of elementary matrices.

Proof Suppose that $A = M_1 M_2 \cdots M_t$, where each M_i is an elementary matrix. Now each M_i is invertible since it is a matrix of transition from one basis to another, and therefore A is invertible by Corollary 3.17.

Suppose now that A is of order n and invertible. Then A is the matrix of transition from $\mathcal{E}_n$ to a basis $\mathcal{A}$ of $\mathbf{R}^n$. By Theorem 2.14, $\mathcal{A}$ can be obtained from $\mathcal{E}_n$ by a sequence $E_1, E_2, \ldots, E_t$ of elementary operations. If M_i is the elementary matrix that is the matrix of transition for the operation E_i, then $M_1 M_2 \cdots M_t$ is a matrix of transition from $\mathcal{E}_n$ to $\mathcal{A}$. But this matrix of transition is unique since $\mathcal{E}_n$ is a basis, and therefore $A = M_1 M_2 \cdots M_t$.

■ ■ ■

If it is desired to write a given invertible $n \times n$ matrix A as a product of elementary matrices, there may be some difficulty in following one of the steps described in the proof of the last theorem. Usually it is not obvious how the basis $\mathcal{A}$ can be obtained from $\mathcal{E}_n$ by a sequence $E_1, E_2, \ldots, E_t$ of elementary operations, and an indirect approach is easier to use. If we start with the set $\mathcal{A}$ and apply a sequence of elementary operations $F_1, F_2, \ldots, F_k$ to find the standard basis of $\langle \mathcal{A} \rangle$, that standard basis will be $\mathcal{E}_n$ since $\langle \mathcal{A} \rangle = \mathbf{R}^n$. We will then have

$$F_k \cdots F_2 F_1(\mathcal{A}) = \mathcal{E}_n$$

and therefore

$$\mathcal{A} = F_1^{-1} F_2^{-1} \cdots F_k^{-1}(\mathcal{E}_n).$$

If M_i is the elementary matrix of transition that is associated with F_i, then this indirect procedure will yield

$$A = M_k^{-1} \cdots M_2^{-1} M_1^{-1}$$

since M_i^{-1} is the elementary matrix associated with the elementary operation F_i^{-1}. Note that we also have

$$A^{-1} = M_1 M_2 \cdots M_k$$

since $M_1 M_2 \cdots M_k$ is the transition matrix from $\mathcal{A}$ to $\mathcal{E}_n$. This is diagrammed in Figure 3.2.

EXAMPLE 3 Consider the problem of expressing the invertible matrix

$$A = \begin{bmatrix} 2 & 3 \\ 6 & 4 \end{bmatrix}$$

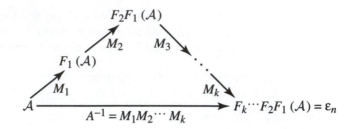

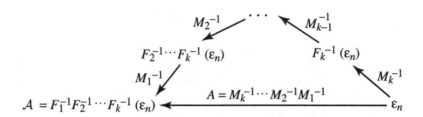

Figure 3.2

as a product of elementary matrices. This matrix A is the transition matrix from $\mathcal{E}_n$ to $\mathcal{A} = \{(2, 6), (3, 4)\}$. We start with $\mathcal{A}$ and apply a single elementary operation at a time until we obtain $\mathcal{E}_2$. Using arrows to indicate that an elementary operation F_i has been applied, this appears as follows.

$$\mathcal{A} = \{(2, 6), (3, 4)\} \quad \xrightarrow{F_1} \quad \{(1, 3), (3, 4)\}$$

$$\xrightarrow{F_2} \quad \{(1, 3), (0, -5)\}$$

$$\xrightarrow{F_3} \quad \{(1, 3), (0, 1)\}$$

$$\xrightarrow{F_4} \quad \{(1, 0), (0, 1)\} = \mathcal{E}_2$$

The following display shows the elementary operations F_i, their associated elementary matrices M_i, and the inverses M_i^{-1}.

Elementary Operation	Associated Matrix	Inverse
F_1 {Multiply the first vector by $\frac{1}{2}$.	$M_1 = \begin{bmatrix} \frac{1}{2} & 0 \\ 0 & 1 \end{bmatrix}$	$M_1^{-1} = \begin{bmatrix} 2 & 0 \\ 0 & 1 \end{bmatrix}$
F_2 { Replace the second vector by the sum of the second vector and (-3) times the first vector.	$M_2 = \begin{bmatrix} 1 & -3 \\ 0 & 1 \end{bmatrix}$	$M_2^{-1} = \begin{bmatrix} 1 & 3 \\ 0 & 1 \end{bmatrix}$

Elementary Operation	Associated Matrix	Inverse
$F_3 \left\{ \text{Multiply the second vector by } \left(-\tfrac{1}{5} \right) \right.$	$M_3 = \begin{bmatrix} 1 & 0 \\ 0 & -\tfrac{1}{5} \end{bmatrix}$	$M_3^{-1} = \begin{bmatrix} 1 & 0 \\ 0 & -5 \end{bmatrix}$
$F_4 \left\{ \begin{array}{l} \text{Replace the first vector by the} \\ \text{sum of the first vector and } (-3) \\ \text{times the second vector.} \end{array} \right.$	$M_4 = \begin{bmatrix} 1 & 0 \\ -3 & 1 \end{bmatrix}$	$M_4^{-1} = \begin{bmatrix} 1 & 0 \\ 3 & 1 \end{bmatrix}$

According to the discussion preceding this example, we can now express A as a product of elementary matrices by using $A = M_4^{-1} M_3^{-1} M_2^{-1} M_1^{-1}$.

$$\begin{bmatrix} 2 & 3 \\ 6 & 4 \end{bmatrix} = \begin{bmatrix} 1 & 0 \\ 3 & 1 \end{bmatrix} \begin{bmatrix} 1 & 0 \\ 0 & -5 \end{bmatrix} \begin{bmatrix} 1 & 3 \\ 0 & 1 \end{bmatrix} \begin{bmatrix} 2 & 0 \\ 0 & 1 \end{bmatrix}$$ ∎

3.3 Exercises

1. For each matrix A, determine $\mathcal{A}$ so that A is the matrix of transition from $\mathcal{E}_3$ to $\mathcal{A}$. Then use Definition 3.8 to decide whether or not A is nonsingular.

(a) $A = \begin{bmatrix} 2 & -6 & 6 \\ -5 & 13 & 1 \\ -2 & 4 & 10 \end{bmatrix}$ **(b)** $A = \begin{bmatrix} 4 & 4 & 4 \\ 3 & 4 & 2 \\ -6 & 1 & 7 \end{bmatrix}$

(c) $A = \begin{bmatrix} 5 & 2 & 7 \\ 2 & 1 & 0 \\ 2 & 9 & 3 \end{bmatrix}$ **(d)** $A = \begin{bmatrix} 2 & 1 & 5 \\ 0 & 3 & 9 \\ 4 & -4 & -8 \end{bmatrix}$

2. Which of the following are elementary matrices?

(a) $\begin{bmatrix} 1 & 4 & 0 \\ 0 & 1 & 0 \\ 0 & 0 & 1 \end{bmatrix}$ **(b)** $\begin{bmatrix} 1 & 0 & 0 \\ 4 & 1 & 0 \\ 0 & 0 & 1 \end{bmatrix}$ **(c)** $\begin{bmatrix} 1 & 4 & 0 \\ 4 & 1 & 0 \\ 0 & 0 & 1 \end{bmatrix}$ **(d)** $\begin{bmatrix} 1 & 1 & 0 \\ 1 & 1 & 0 \\ 0 & 0 & 1 \end{bmatrix}$

(e) $\begin{bmatrix} 0 & 0 & 1 \\ 0 & 1 & 0 \\ 1 & 0 & 0 \end{bmatrix}$ **(f)** $\begin{bmatrix} 1 & 1 & 0 \\ 0 & 2 & 0 \\ 0 & 0 & 1 \end{bmatrix}$ **(g)** $\begin{bmatrix} 1 & 0 & 0 \\ 0 & 1 & 0 \\ 0 & 0 & 2 \end{bmatrix}$ **(h)** $\begin{bmatrix} 0 & 1 & 0 \\ 0 & 0 & 1 \\ 1 & 0 & 0 \end{bmatrix}$

3. Suppose that $\mathcal{A}$ is a set of three vectors in $\mathbf{R}^4$ and that $\mathcal{B}$ is obtained from $\mathcal{A}$ by each of the following elementary operations. In each case, find the matrix of transition from $\mathcal{A}$ to $\mathcal{B}$ and the matrix of transition from $\mathcal{B}$ to $\mathcal{A}$.

(a) Multiply the third vector in $\mathcal{A}$ by -3.

(b) Replace the second vector in $\mathcal{A}$ by the sum of the second vector and 5 times the third vector.

(c) Interchange the second and third vectors in $\mathcal{A}$.

(d) Replace the first vector in $\mathcal{A}$ by the sum of the first vector and 2 times the third vector, then interchange the first and second vectors.

(e) Interchange the first and third vectors in $\mathcal{A}$, then multiply the third vector by -2.

(f) Replace the second vector in $\mathcal{A}$ by the sum of the second vector and -1 times the first vector, then replace the third vector by the sum of the third vector and 3 times the second vector.

4. Write the inverse of each elementary matrix in Problem 2.

5. Given the following factorization of A, write A^{-1} as a product of elementary matrices.

(a) $A = \begin{bmatrix} 1 & 0 \\ 2 & 1 \end{bmatrix}\begin{bmatrix} 1 & 0 \\ 0 & 3 \end{bmatrix}\begin{bmatrix} 0 & 1 \\ 1 & 0 \end{bmatrix}$

(b) $A = \begin{bmatrix} 1 & 3 & 0 \\ 0 & 1 & 0 \\ 0 & 0 & 1 \end{bmatrix}\begin{bmatrix} 0 & 1 & 0 \\ 1 & 0 & 0 \\ 0 & 0 & 1 \end{bmatrix}$

(c) $A = \begin{bmatrix} 4 & 0 & 0 \\ 0 & 1 & 0 \\ 0 & 0 & 1 \end{bmatrix}\begin{bmatrix} 1 & 0 & -2 \\ 0 & 1 & 0 \\ 0 & 0 & 1 \end{bmatrix}\begin{bmatrix} 1 & 0 & 0 \\ 0 & 1 & 3 \\ 0 & 0 & 1 \end{bmatrix}$

(d) $A = \begin{bmatrix} 1 & 0 & 0 & 0 \\ 0 & 0 & 1 & 0 \\ 0 & 1 & 0 & 0 \\ 0 & 0 & 0 & 1 \end{bmatrix}\begin{bmatrix} 1 & 0 & 0 & 0 \\ 0 & 1 & 0 & 0 \\ 0 & 0 & -5 & 0 \\ 0 & 0 & 0 & 1 \end{bmatrix}\begin{bmatrix} 1 & 0 & 0 & 0 \\ 0 & 1 & 0 & 0 \\ -1 & 0 & 1 & 0 \\ 0 & 0 & 0 & 1 \end{bmatrix}\begin{bmatrix} 1 & 0 & 0 & 4 \\ 0 & 1 & 0 & 0 \\ 0 & 0 & 1 & 0 \\ 0 & 0 & 0 & 1 \end{bmatrix}$

6. Each of the matrices A below is nonsingular. Use A as the transition matrix from the basis $\mathcal{A} = \{(1, -2), (2, 1)\}$ of $\mathbf{R}^2$ to the basis $\mathcal{A}'$ of $\mathbf{R}^2$. Determine A^{-1} by finding the matrix of transition from $\mathcal{A}'$ to $\mathcal{A}$.

(a) $A = \begin{bmatrix} 1 & 3 \\ -1 & -2 \end{bmatrix}$

(b) $A = \begin{bmatrix} 2 & -7 \\ -1 & 4 \end{bmatrix}$

(c) $A = \begin{bmatrix} 2 & 1 \\ -1 & 2 \end{bmatrix}$

(d) $A = \begin{bmatrix} 2 & 1 \\ 5 & 3 \end{bmatrix}$

7. Suppose

$$A = \begin{bmatrix} 4 & -2 & 1 & 0 \\ 5 & 0 & 2 & 1 \end{bmatrix}$$

is the transition matrix from $\mathcal{A} = \{(1, 1), (1, 0)\}$ to the set $\mathcal{B}$, and

$$B = \begin{bmatrix} 1 & 0 \\ 4 & -3 \\ 0 & -1 \\ -8 & 6 \end{bmatrix}$$

is the transition matrix from $\mathcal{B}$ to $\mathcal{C}$.

(a) Find the matrix of transition from $\mathcal{A}$ to $\mathcal{C}$.
(b) Find the set $\mathcal{C}$.
(c) Does $(AB)^{-1}$ exist? Explain the reason for your answer.

8. Write each of the following invertible matrices as a product of elementary matrices.

(a) $\begin{bmatrix} 2 & 4 \\ 3 & 4 \end{bmatrix}$
(b) $\begin{bmatrix} -1 & 1 \\ 1 & 0 \end{bmatrix}$
(c) $\begin{bmatrix} 3 & 4 \\ 2 & 1 \end{bmatrix}$

(d) $\begin{bmatrix} 0 & 3 \\ -2 & 6 \end{bmatrix}$
(e) $\begin{bmatrix} 2 & 0 & 0 \\ 0 & -1 & 0 \\ 0 & 0 & 3 \end{bmatrix}$
(f) $\begin{bmatrix} 5 & 0 & 0 & 0 \\ 0 & -6 & 0 & 0 \\ 0 & 0 & 3 & 0 \\ 0 & 0 & 0 & -4 \end{bmatrix}$

9. Find the inverse of each matrix in Problem 8 by use of Corollary 3.17 and the factorization obtained in Problem 8.

10. Given that

$$B^{-1} = \begin{bmatrix} 1 & 3 \\ -1 & -2 \end{bmatrix} \quad \text{and} \quad AB = \begin{bmatrix} -4 & -5 \\ 2 & 3 \end{bmatrix},$$

find the matrix A.

11. Given the matrices

$$B = \begin{bmatrix} 2 & 4 \\ 7 & 8 \end{bmatrix} \quad \text{and} \quad (AB)^{-1} = \begin{bmatrix} 2 & 3 \\ -1 & -2 \end{bmatrix},$$

find A^{-1}.

12. Given that the invertible matrix

$$B = \begin{bmatrix} 4 & 6 & 0 \\ 2 & -5 & 3 \\ 4 & 8 & 2 \end{bmatrix}$$

was obtained from the invertible matrix A by adding 3 times the first column to the second column, find the elementary matrix M such that $BM = A$. (*Hint:* Let B be the matrix of transition from $\mathcal{E}_3$ to $\mathcal{B}$.).

13. Let A be an invertible matrix of order n, X an $n \times p$ matrix, and B an $n \times p$ matrix. Solve the matrix equation $AX = B$ for the matrix X.

14. Use the results of Exercises 13 and 6 to solve the following matrix equations.

(a) $\begin{bmatrix} 1 & 3 \\ -1 & -2 \end{bmatrix} X = \begin{bmatrix} -4 \\ 3 \end{bmatrix}$
(b) $\begin{bmatrix} 2 & -7 \\ -1 & 4 \end{bmatrix} X = \begin{bmatrix} 4 & 7 & -1 \\ -3 & 0 & 1 \end{bmatrix}$

(c) $\begin{bmatrix} 2 & 1 \\ -1 & 2 \end{bmatrix} X = \begin{bmatrix} 5 & -10 \\ 0 & 0 \end{bmatrix}$
(d) $\begin{bmatrix} 2 & 1 \\ 5 & 3 \end{bmatrix} X = \begin{bmatrix} 1 & 0 & -1 & 0 \\ 5 & 0 & 2 & -4 \end{bmatrix}$

15. Prove Theorem 3.15.

16. Let A, B, and C be square matrices of order n.

(a) Prove that if $AB = 0$ where A is invertible, then $B = 0$.

(b) Prove that if A is invertible and $AB = AC$, then $B = C$.

(c) If A is invertible, solve the equation $AB = CA$ for B.

(d) If A and B are invertible, solve the equation $ACB = I$ for C.

(e) Prove that if AB is invertible, then A is invertible and write an expression for A^{-1} in terms of B and AB.

(f) Prove that if AB is invertible, then B is invertible and write an expression for B^{-1} in terms of A and AB.

(g) If A, B and C are invertible matrices such that $ACB = I$, write an expression for C^{-1} in terms of A and B.

17. Prove or disprove: Let A be the matrix of transition from $\mathcal{A}$ to $\mathcal{B}$ with one column of A consisting entirely of zeros. Then A is nonsingular.

18. Use mathematical induction to prove that $(A^n)^{-1} = (A^{-1})^n$, for all positive integers n, where A is an invertible matrix.

19. Prove Corollary 3.17 by mathematical induction.

20. Derive a formula for the inverse of a nonsingular $A = [a_{ij}]_{2\times2}$ by consideration of a system of equations obtained from

$$\begin{bmatrix} a_{11} & a_{12} \\ a_{21} & a_{22} \end{bmatrix} \begin{bmatrix} x_1 & x_2 \\ x_3 & x_4 \end{bmatrix} = \begin{bmatrix} 1 & 0 \\ 0 & 1 \end{bmatrix}.$$

21. Prove that if $A = [a_{ij}]_{n\times n}$ is not invertible, then AB is not invertible for any $n \times p$ matrix B. (*Hint:* Use Theorem 3.14.)

22. Prove that if $A = [a_{ij}]_{n\times n}$ is singular, then BA is singular for any $p \times n$ matrix B. (*Hint:* Use Theorem 3.15.)

3.4 Column Operations and Column-Echelon Forms

The elementary matrices of each type can be conveniently described by comparing them with the identity matrix I_n. (It is clear that the identity operation has I_n as its associated elementary matrix.) The operation of multiplying a matrix by a scalar is helpful in describing these comparisons. In this section, it is especially useful in situations where the matrix is a column matrix. Other uses will be found in later sections.

3.20 DEFINITION   For any $a \in \mathbf{R}$ and any $A = [a_{ij}]_{m\times n}$ over $\mathbf{R}$, the **product**, aA is defined by $aA = [aa_{ij}]_{m\times n}$.

Note that in the product aA, *each element of A is multiplied by the scalar a*. A matrix of the form aI_n is called a **scalar matrix**.

Suppose now that M is an elementary matrix with E as its associated elementary operation.

If M is of type I, then $\mathcal{B} = E(\mathcal{A})$ is obtained from a basis $\mathcal{A}$ of $\mathbf{R}^n$ by multiplying the sth vector of $\mathcal{A}$ by a nonzero scalar a. This means that M differs from I_n only in column s, and the sth column of M is the product of a and the sth column of I_n.

If M is of type II, then E replaces one of the vectors $\mathbf{v}_s$ in a basis $\mathcal{A} = \{\mathbf{v}_1, \mathbf{v}_2, \ldots, \mathbf{v}_n\}$ of $\mathbf{R}^n$ by $\mathbf{v}_s + b\mathbf{v}_t$, where $s \neq t$. This means that M is identical to I_n except in column s. In column s of M, 1 appears in row s and b appears in row t. Thus, M can be obtained from I_n by adding to column s of I_n the product of b and column t of I_n, where $s \neq t$.

Consider now the case where M is of type III. Then E interchanges the sth and tth ($s \neq t$) vectors in a basis of $\mathbf{R}^n$. This means that M can be obtained from I_n by interchanging the sth and tth columns ($s \neq t$).

We have thus found that each elementary matrix can be obtained from I_n by performing a suitable operation on the columns of I_n. This is illustrated in the following example.

EXAMPLE 1 Each of the following illustrates an elementary matrix M, associated with an elementary operation E, on the basis $\mathcal{A} = \{\mathbf{u}_1, \mathbf{u}_2, \mathbf{u}_3\}$ of $\mathbf{R}^3$ resulting in the set $\mathcal{B}$.

(a) E: multiply the first vector by 3;

$$M = \begin{bmatrix} 3 & 0 & 0 \\ 0 & 1 & 0 \\ 0 & 0 & 1 \end{bmatrix}; \quad \mathcal{B} = \{3\mathbf{u}_1, \mathbf{u}_2, \mathbf{u}_3\}.$$

(b) E: replace the second vector by the sum of the second vector and -2 times the third vector;

$$M = \begin{bmatrix} 1 & 0 & 0 \\ 0 & 1 & 0 \\ 0 & -2 & 1 \end{bmatrix}; \quad \mathcal{B} = \{\mathbf{u}_1, \mathbf{u}_2 - 2\mathbf{u}_3, \mathbf{u}_3\}.$$

(c) E: interchange the first and third vectors;

$$M = \begin{bmatrix} 0 & 0 & 1 \\ 0 & 1 & 0 \\ 1 & 0 & 0 \end{bmatrix}; \quad \mathcal{B} = \{\mathbf{u}_3, \mathbf{u}_2, \mathbf{u}_1\}.$$

In each case, we note that $\langle \mathcal{A} \rangle = \langle \mathcal{B} \rangle$. ■

Corresponding to the operations previously performed on I_n, we define three types of elementary column operations on a matrix A.

**3.21
DEFINITION**

(i) An elementary column operation of type I multiplies one of the columns of A by a nonzero scalar.

(ii) An elementary column operation of type II adds to column s of A the product of b and column t of A ($s \neq t$).

(iii) An elementary column operation of type III interchanges two columns of A.

A more unwieldy but more accurate description of an elementary column operation of type II would be this statement: An elementary column operation of type II replaces column s of A by the sum of column s and b times column t ($s \neq t$). This wording more closely matches the descriptions we used with elementary operations on sets of vectors.

Let us consider the effect on an m by n matrix A when A is multiplied on the right by an elementary matrix M of order n. This effect is diagrammed in Figure 3.3.

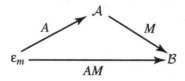

Figure 3.3

Now A is the matrix of transition from the basis $\mathcal{E}_m$ of $\mathbf{R}^m$ to a set $\mathcal{A}$ of n vectors in $\mathbf{R}^m$, M is a matrix of transition from $\mathcal{A}$ to a set $\mathcal{B}$ of n vectors in $\mathbf{R}^m$, and AM is the matrix of transition from $\mathcal{E}_m$ to $\mathcal{B}$.

If M is of type I, then $\mathcal{B}$ is obtained from $\mathcal{A}$ by multiplying the sth vector of $\mathcal{A}$ by a nonzero scalar a. This means that AM can be obtained from A by multiplying the sth column of A by a.

If M is of type II, then $\mathcal{B}$ is obtained from $\mathcal{A} = \{\mathbf{v}_1, \mathbf{v}_2, \ldots, \mathbf{v}_n\}$ by replacing $\mathbf{v}_s$ by $\mathbf{v}_s + b\mathbf{v}_t$ ($s \neq t$). This means that AM is identical to A except in column s, and column s in AM is obtained by adding to column s of A the product of b and column t.

If M is of type III, then $\mathcal{B}$ is obtained from $\mathcal{A}$ by interchanging the sth and tth vectors ($s \neq t$) of $\mathcal{A}$. Thus AM is obtained from A by interchanging the sth and tth columns of A ($s \neq t$).

Thus, multiplication of A on the right by an elementary matrix M of a certain type performs an elementary column operation E of the same type on A. We shall say that the elementary matrix M and the elementary column operation E are *associated* with each other whenever multiplication of A on the right by M performs the elementary column operation E on A.

EXAMPLE 2 Let $\mathcal{A} = \{(1, 1, 0), (0, 1, 0)\}$. Then

$$A = \begin{bmatrix} 1 & 0 \\ 1 & 1 \\ 0 & 0 \end{bmatrix}$$

is the matrix of transition from $\mathcal{E}_3$ to $\mathcal{A}$ and we consider each of the three types of elementary operations and the resulting column operations.

(a) If $\mathcal{B}$ is obtained from $\mathcal{A}$ by multiplying the first vector of $\mathcal{A}$ by 2, then $\mathcal{B} = \{(2, 2, 0), (0, 1, 0)\}$. Now

$$M = \begin{bmatrix} 2 & 0 \\ 0 & 1 \end{bmatrix}$$

is the matrix of transition from $\mathcal{A}$ to $\mathcal{B}$ and we see that

$$AM = \begin{bmatrix} 1 & 0 \\ 1 & 1 \\ 0 & 0 \end{bmatrix} \begin{bmatrix} 2 & 0 \\ 0 & 1 \end{bmatrix} = \begin{bmatrix} 2 & 0 \\ 2 & 1 \\ 0 & 0 \end{bmatrix}$$

can be obtained from A by multiplying the first column by 2.

(b) If $\mathcal{B}$ is obtained from $\mathcal{A}$ by replacing the second vector by the sum of the second vector and 3 times the first vector, then $\mathcal{B} = \{(1, 1, 0), (3, 4, 0)\}$. Now

$$M = \begin{bmatrix} 1 & 3 \\ 0 & 1 \end{bmatrix}$$

is the matrix of transition from $\mathcal{A}$ to $\mathcal{B}$ and

$$AM = \begin{bmatrix} 1 & 0 \\ 1 & 1 \\ 0 & 0 \end{bmatrix} \begin{bmatrix} 1 & 3 \\ 0 & 1 \end{bmatrix} = \begin{bmatrix} 1 & 3 \\ 1 & 4 \\ 0 & 0 \end{bmatrix},$$

which can be obtained from A by replacing the second column by the sum of the second column and 3 times the first column.

(c) Finally, if $\mathcal{B}$ is obtained from $\mathcal{A}$ by interchanging the two vectors, then $\mathcal{B} = \{(0, 1, 0), (1, 1, 0)\}$ and

$$M = \begin{bmatrix} 0 & 1 \\ 1 & 0 \end{bmatrix}.$$

Then

$$AM = \begin{bmatrix} 1 & 0 \\ 1 & 1 \\ 0 & 0 \end{bmatrix} \begin{bmatrix} 0 & 1 \\ 1 & 0 \end{bmatrix} = \begin{bmatrix} 0 & 1 \\ 1 & 1 \\ 0 & 0 \end{bmatrix},$$

which can be obtained from A by interchanging the two columns of A. ■

In each case above, AM is obtained from A by a single column operation. Hence performing an elementary operation on $\mathcal{A}$ corresponds to performing an elementary column operation on A. By Theorem 2.8, $\langle\mathcal{A}\rangle = \langle\mathcal{B}\rangle$, so A and AM are matrices of transition from a given basis to set of vectors that span the same subspace.

In the remainder of this section, we shall be concerned with the reduction of a given matrix A to a certain standard form, known as the *reduced column-echelon* form.

**3.22
DEFINITION**

A matrix $A = [a_{ij}]_{m \times n}$ over **R** that satisfies the following conditions is a matrix in **reduced column-echelon form**, or a **reduced column-echelon matrix**.

1. The first nonzero element in column j is a 1 in row k_j for $j = 1, 2, \ldots, r$. (This 1 is called a **leading one**.)

2. $k_1 < k_2 < \cdots < k_r \leq m$ (That is, for each change in columns from left to right, the leading one appears in a lower row.)

3. For $j = 1, 2, \ldots, r$, the leading one in column j is the only nonzero element in row k_j.

4. Each of the last $n - r$ columns consists entirely of zeros.

For future use, we note that conditions (1) and (3) can be reworded in the following ways.

1. $a_{ij} = 0$ for $i < k_j$, and $a_{k_j j} = 1$ for $j = 1, 2, \ldots, r$.

3. Column j is the only column with a nonzero element in row k_j.

Thus a matrix is in reduced column-echelon form if and only if its nonzero columns record the components of the standard basis of a subspace, or if its nonzero columns form the matrix of transition from $\mathcal{E}_m$ to the standard basis of a subspace.

EXAMPLE 3   Consider the question as to which of the following matrices are in reduced column-echelon form.

$$A = \begin{bmatrix} 1 & 0 & 0 \\ 0 & 2 & 0 \\ 5 & 4 & 3 \\ 0 & 0 & 0 \end{bmatrix}, \quad B = \begin{bmatrix} 1 & 0 & 0 \\ 0 & 0 & 0 \\ 0 & 1 & 0 \\ 3 & 4 & 0 \end{bmatrix}, \quad C = \begin{bmatrix} 1 & 0 & 0 \\ 2 & 0 & 0 \\ 0 & 1 & 0 \\ 3 & 4 & 1 \end{bmatrix}, \quad D = \begin{bmatrix} 1 & 0 & 0 \\ 0 & 0 & 1 \\ 0 & 1 & 0 \\ 0 & 0 & 0 \end{bmatrix}$$

The matrix A is not in reduced column-echelon form. It fails to satisfy condition (1) in the second and third columns because the first nonzero element in each of these columns is not a 1.

The matrix B is in reduced column-echelon form since it satisfies all four conditions. If B is considered as the transition matrix from $\mathcal{E}_4$ to $\mathcal{B}$, then the nonzero columns of B record the components of the standard basis of the subspace $\mathcal{B}$. Thus the standard basis of $\mathcal{B}$ is $\{(1, 0, 0, 3), (0, 0, 1, 4)\}$.

The matrix C fails on condition (3) because the leading 1 in column 3 is not the only nonzero element in row $k_3 = 4$. The matrix D fails on condition (2) because $k_2 = 3$ and $k_3 = 2$ violate $k_2 < k_3$. That is, the leading 1 in column 3 of D fails to be in a lower row than the leading 1 in column 2. ∎

3.23 THEOREM

If A is any $m \times n$ matrix over $\mathbf{R}$, there is an invertible matrix Q over $\mathbf{R}$ such that $AQ = A'$ is a matrix in reduced column-echelon form. The matrix A' is uniquely determined by A.

Proof Let $A = [a_{ij}]$ be an $m \times n$ matrix over $\mathbf{R}$ and consider the set $\mathcal{A} = \{\mathbf{v}_1, \mathbf{v}_2, \ldots, \mathbf{v}_n\}$ of n vectors in $\mathbf{R}^m$ where $\mathbf{v}_j = (a_{1j}, a_{2j}, \ldots, a_{mj})$. (That is, A is the matrix of transition from $\mathcal{E}_m$ to $\mathcal{A}$.) By Theorem 2.10 (with m and n interchanged), there is a sequence $E_1, E_2, \ldots, E_t$ of elementary operations that can be applied to $\mathcal{A}$ to obtain a set $\mathcal{A}' = \{\mathbf{v}_1', \mathbf{v}_2', \ldots, \mathbf{v}_r', \mathbf{0}, \ldots, \mathbf{0}\}$ in which

1. The first nonzero component from the left in $\mathbf{v}_j$ is a 1 in the k_j component for $j = 1, 2, \ldots, r$.
2. $k_1 < k_2 < \cdots < k_r \leq m$.
3. $\mathbf{v}_j$ is the only vector in $\mathcal{A}'$ with a nonzero k_j component.
4. $\{\mathbf{v}_1', \mathbf{v}_2', \ldots, \mathbf{v}_r'\}$ is a basis of $\langle \mathcal{A} \rangle$.

Now each of the elementary operations E_i has an associated elementary matrix Q_i, and when E_i is applied to a set, a matrix of transition from that set to the new set is Q_i. As the diagram in Figure 3.4 shows, this means that a matrix of transition from $\mathcal{A}$ to $\mathcal{A}'$ is $Q_1 Q_2 \cdots Q_t$, and that $A Q_1 Q_2 \cdots Q_t = A'$ is the matrix of transition from $\mathcal{E}_m$ to $\mathcal{A}'$. Thus the ith element in column j of $A' = [a_{ij}']$ is equal to the ith component of $\mathbf{v}_j'$ and therefore:

1. The first nonzero element in column j of A' is a 1 in row k_j, for $j = 1, 2, \ldots, r$.
2. $k_1 < k_2 < \cdots < k_r \leq m$.
3. Column j is the only column with a nonzero element in row k_j.
4. The last $n - r$ columns consist entirely of zeros.

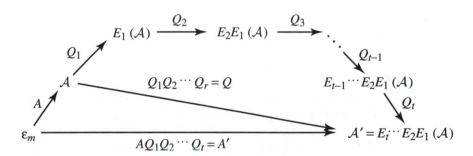

Figure 3.4

The matrix $Q = Q_1 Q_2 \cdots Q_t$ is invertible since each Q_i is invertible, and the matrix A' is uniquely determined by A since $\mathcal{A}'$ is uniquely determined by $\mathcal{A}$. ∎

We saw in Chapter 2 that the set $\mathcal{A}'$ used in the proof of Theorem 3.23 is uniquely determined by the set $\mathcal{A}$, and the nonzero vectors of $\mathcal{A}'$ form the standard basis of $\langle \mathcal{A} \rangle$. It follows from this fact that each matrix A over $\mathbf{R}$ has an associated unique reduced column-echelon form A' as described in the proof. However, we have seen in Chapter 2 that the sequence of elementary operations used to obtain $\mathcal{A}'$ from $\mathcal{A}$ is *not unique*. The invertible matrix Q is similarly *not unique* in spite of the fact that $AQ = A'$ is unique for A. An example that demonstrates this lack of uniqueness is given at the end of this section.

3.24	The matrix A' in Theorem 3.23 is called the **reduced column-echelon form** for A.
DEFINITION	

We shall now show how the proof of Theorem 3.23 can be interpreted so as to give a systematic procedure for finding an invertible Q such that AQ is in reduced column-echelon form. This procedure is closely related to that used in Chapter 2 in finding the standard basis of a subspace.

Suppose that $A = [a_{ij}]$ is a given $m \times n$ matrix over $\mathbf{R}$. Interpreting A as a matrix of transition from $\mathcal{E}_m$ to $\mathcal{A}$ is equivalent to obtaining A by recording the components of vectors in $\mathcal{A}$, as was done in Chapter 2. We have seen that performing an elementary operation on $\mathcal{A}$ corresponds to performing an elementary column operation on A. With these interpretations, the procedure in Example 3 of Section 2.3 can be regarded as a method for obtaining the reduced column-echelon form

$$A' = \begin{bmatrix} 0 & 0 & 0 & 0 \\ 1 & 0 & 0 & 0 \\ -2 & 0 & 0 & 0 \\ 0 & 1 & 0 & 0 \\ -1 & 1 & 0 & 0 \end{bmatrix}$$

for the matrix

$$A = \begin{bmatrix} 0 & 0 & 0 & 0 \\ 0 & 3 & 2 & -1 \\ 0 & -6 & -4 & 2 \\ -1 & 1 & 2 & 2 \\ -1 & -2 & 0 & 3 \end{bmatrix}.$$

This is an efficient method for finding the reduced column-echelon form, but it does not yield the matrix Q of Theorem 3.23.

In order to obtain the matrix Q, one needs a method for recording the products $Q_1 Q_2 \cdots Q_i$ for $i = 1, 2, \ldots, t$. Consider the sequence of products

$$A I_n = A$$

$$A I_n Q_1 = A Q_1$$

$$A I_n Q_1 Q_2 = A Q_1 Q_2$$

$$\vdots$$

$$A I_n Q_1 Q_2 \cdots Q_t = A Q_1 Q_2 \cdots Q_t = A'.$$

Let E_i be the elementary column operation that has Q_i as its associated matrix. Now $A Q_1 = E_1(A)$, and $A Q_1 Q_2 \cdots Q_i = E_i E_{i-1} \cdots E_1(A)$ in general. Thus the right members of the equations above may be found by applying the sequence $E_1, E_2, \ldots, E_t$ to A. The left members have as factors the products $I_n Q_1 Q_2 \cdots Q_i = E_i E_{i-1} \cdots E_1(I_n)$.

Thus Q can be found by applying the same sequence of elementary operations to I_n. What is desired, then, is an efficient method of recording the results $E_i \cdots E_2 E_1(A)$ and $E_i \cdots E_2 E_1(I_n)$. This can be done effectively by recording both A and I_n in a single matrix as

$$\begin{bmatrix} A \\ I_n \end{bmatrix}.$$

Instead of applying the elementary column operations separately to A and I_n,

$$\begin{bmatrix} A \\ I_n \end{bmatrix}, \quad \begin{bmatrix} E_1(A) \\ E_1(I_n) \end{bmatrix}, \quad \begin{bmatrix} E_2 E_1(A) \\ E_2 E_1(I_n) \end{bmatrix}, \quad \text{etc.,}$$

one can simply apply the operation to the entire matrices

$$\begin{bmatrix} A \\ I_n \end{bmatrix}, \quad E_1\left(\begin{bmatrix} A \\ I_n \end{bmatrix}\right), \quad E_2 E_1\left(\begin{bmatrix} A \\ I_n \end{bmatrix}\right), \quad \text{etc.}$$

This procedure is quite valid since the same operations in the same order are to be applied to each of A and I_n.

EXAMPLE 4 Using the matrix A in Example 3 of Section 2.3, we have

$$\begin{bmatrix} A \\ I_n \end{bmatrix} = \left[\begin{array}{cccc} 0 & 0 & 0 & 0 \\ 0 & 3 & 2 & -1 \\ 0 & -6 & -4 & 2 \\ -1 & 1 & 2 & 2 \\ -1 & -2 & 0 & 3 \\ \hline 1 & 0 & 0 & 0 \\ 0 & 1 & 0 & 0 \\ 0 & 0 & 1 & 0 \\ 0 & 0 & 0 & 1 \end{array}\right] \rightarrow \left[\begin{array}{cccc} 0 & 0 & 0 & 0 \\ 2 & 3 & 0 & -1 \\ -4 & -6 & 0 & 2 \\ 2 & 1 & -1 & 2 \\ 0 & -2 & -1 & 3 \\ \hline 0 & 0 & 1 & 0 \\ 0 & 1 & 0 & 0 \\ 1 & 0 & 0 & 0 \\ 0 & 0 & 0 & 1 \end{array}\right] \rightarrow \left[\begin{array}{cccc} 0 & 0 & 0 & 0 \\ 1 & 3 & 0 & -1 \\ -2 & -6 & 0 & 2 \\ 1 & 1 & -1 & 2 \\ 0 & -2 & -1 & 3 \\ \hline 0 & 0 & 1 & 0 \\ 0 & 1 & 0 & 0 \\ \frac{1}{2} & 0 & 0 & 0 \\ 0 & 0 & 0 & 1 \end{array}\right]$$

$$\rightarrow \left[\begin{array}{cccc} 0 & 0 & 0 & 0 \\ 1 & 0 & 0 & 0 \\ -2 & 0 & 0 & 0 \\ 1 & -2 & -1 & 3 \\ 0 & -2 & -1 & 3 \\ \hline 0 & 0 & 1 & 0 \\ 0 & 1 & 0 & 0 \\ \frac{1}{2} & -\frac{3}{2} & 0 & \frac{1}{2} \\ 0 & 0 & 0 & 1 \end{array}\right] \rightarrow \left[\begin{array}{cccc} 0 & 0 & 0 & 0 \\ 1 & 0 & 0 & 0 \\ -2 & 0 & 0 & 0 \\ 0 & 1 & 0 & 0 \\ -1 & 1 & 0 & 0 \\ \hline 0 & 0 & 1 & 0 \\ \frac{1}{2} & -\frac{1}{2} & -\frac{1}{2} & \frac{3}{2} \\ -\frac{1}{4} & \frac{3}{4} & \frac{3}{4} & -\frac{7}{4} \\ 0 & 0 & 0 & 1 \end{array}\right].$$

Thus

$$A' = \begin{bmatrix} 0 & 0 & 0 & 0 \\ 1 & 0 & 0 & 0 \\ -2 & 0 & 0 & 0 \\ 0 & 1 & 0 & 0 \\ -1 & 1 & 0 & 0 \end{bmatrix}$$

and

$$Q = \begin{bmatrix} 0 & 0 & 1 & 0 \\ \frac{1}{2} & -\frac{1}{2} & -\frac{1}{2} & \frac{3}{2} \\ -\frac{1}{4} & \frac{3}{4} & \frac{3}{4} & -\frac{7}{4} \\ 0 & 0 & 0 & 1 \end{bmatrix}$$

is an invertible matrix such that $AQ = A'$. It is easy to show that Q is not unique by performing elementary operations that involve only the last two columns of Q. For instance, adding the third column of Q to its fourth column yields the matrix

$$B = \begin{bmatrix} 0 & 0 & 1 & 1 \\ \frac{1}{2} & -\frac{1}{2} & -\frac{1}{2} & 1 \\ -\frac{1}{4} & \frac{3}{4} & \frac{3}{4} & -1 \\ 0 & 0 & 0 & 1 \end{bmatrix}$$

such that $AB = AQ = A'$. ■

3.4 Exercises

1. Describe the elementary column operation that is associated with the given elementary matrix M.

(a) $M = \begin{bmatrix} 0 & 1 \\ 1 & 0 \end{bmatrix}$

(b) $M = \begin{bmatrix} 0 & 1 & 0 \\ 1 & 0 & 0 \\ 0 & 0 & 1 \end{bmatrix}$

(c) $M = \begin{bmatrix} 1 & 0 & 0 \\ 0 & 1 & 0 \\ 0 & 2 & 1 \end{bmatrix}$

(d) $M = \begin{bmatrix} 1 & 0 & 0 \\ 2 & 1 & 0 \\ 0 & 0 & 1 \end{bmatrix}$

(e) $M = \begin{bmatrix} 1 & 0 & 0 \\ 0 & 1 & 0 \\ 0 & 0 & -3 \end{bmatrix}$

(f) $M = \begin{bmatrix} 1 & 0 & 0 & 0 \\ 0 & 1 & 0 & 0 \\ 0 & -5 & 1 & 0 \\ 0 & 0 & 0 & 1 \end{bmatrix}$

2. Which of the following are in reduced column-echelon form?

(a) $\begin{bmatrix} 0 & 0 & 0 \\ 1 & 0 & 0 \\ 0 & 1 & 0 \\ 2 & 3 & 0 \end{bmatrix}$
(b) $\begin{bmatrix} 0 & 1 & 0 \\ 1 & 0 & 0 \\ 0 & 1 & 0 \\ 1 & 0 & 0 \end{bmatrix}$
(c) $\begin{bmatrix} 0 & 0 & 0 \\ 1 & 0 & 0 \\ 1 & 1 & 0 \\ 1 & 1 & 1 \end{bmatrix}$

(d) $\begin{bmatrix} 0 & 1 & 0 \\ 0 & 0 & 1 \\ 0 & 0 & 0 \\ 0 & 0 & 0 \end{bmatrix}$
(e) $\begin{bmatrix} 1 & 1 & 1 \\ 0 & 1 & 1 \\ 0 & 0 & 1 \\ 0 & 0 & 0 \end{bmatrix}$
(f) $\begin{bmatrix} 1 & 0 & 0 \\ 1 & 1 & 0 \\ 0 & 0 & 1 \end{bmatrix}$

(g) $\begin{bmatrix} 1 & 0 \\ -5 & 0 \end{bmatrix}$
(h) $\begin{bmatrix} 0 & 0 & 0 & 0 \\ 1 & 0 & 0 & 0 \\ 3 & 0 & 0 & 0 \\ 0 & 1 & 0 & 0 \end{bmatrix}$

3. Write the elementary matrix of transition from A to B.

(a) $A = \{(1, 2, -1), (2, 1, 3)\}$, $B = \{(1, 2, -1), (0, -3, 5)\}$

(b) $A = \{(1, 0, 1), (0, 1, 1), (1, 1, 0)\}$, $B = \{(1, 1, 0), (0, 1, 1), (1, 0, 1)\}$

(c) $A = \{(1, 0, 0, 1), (1, 0, 1, 0)\}$, $B = \{(2, 0, 1, 1), (1, 0, 1, 0)\}$

(d) $A = \{(1, 1, 1, 0), (1, 0, 1, 0), (1, 1, 0, 0)\}$,
$B = \{(1, 1, 1, 0), (3, 0, 3, 0), (1, 1, 0, 0)\}$

4. For each of the matrices B and M, suppose that B is obtained from A by multiplying A on the right by M. Find A.

(a) $B = \begin{bmatrix} 1 & 4 & 2 \\ 2 & 0 & 1 \\ -2 & 1 & 3 \end{bmatrix}$, $M = \begin{bmatrix} 1 & 0 & 0 \\ 0 & 1 & 0 \\ 2 & 0 & 1 \end{bmatrix}$

(b) $B = \begin{bmatrix} 4 & 0 \\ 1 & -2 \\ 0 & 6 \end{bmatrix}$, $M = \begin{bmatrix} 1 & -2 \\ 0 & 1 \end{bmatrix}$

(c) $B = \begin{bmatrix} 1 & 0 & 2 \\ 2 & -4 & 7 \end{bmatrix}$, $M = \begin{bmatrix} 1 & 0 & 0 \\ 0 & 0 & 1 \\ 0 & 1 & 0 \end{bmatrix}$

(d) $B = \begin{bmatrix} 2 & 3 & 1 & 0 \\ 1 & 0 & 3 & 1 \\ 1 & 1 & -1 & 7 \end{bmatrix}$, $M = \begin{bmatrix} 1 & 0 & 0 & 0 \\ 0 & 1 & 0 & 0 \\ 0 & 0 & -\frac{1}{2} & 0 \\ 0 & 0 & 0 & 1 \end{bmatrix}$

5. If M is an elementary matrix, are A and MA matrices of transition from a given basis to sets of vectors that span the same subspace? Illustrate with an example.

6. Find the reduced column-echelon form for each matrix.

(a) $\begin{bmatrix} 0 & 0 & 0 \\ 1 & 2 & 1 \\ 2 & 4 & 2 \\ 1 & 1 & 1 \end{bmatrix}$
(b) $\begin{bmatrix} 9 & 6 \\ -4 & -3 \\ -4 & -2 \end{bmatrix}$
(c) $\begin{bmatrix} 1 & 0 & 1 & 1 \\ 2 & 1 & 3 & 1 \\ -1 & 0 & -1 & -1 \\ 3 & 2 & 5 & 1 \end{bmatrix}$

$$\textbf{(d)} \begin{bmatrix} 0 & 0 & 0 & 0 \\ 2 & 3 & 0 & -1 \\ -4 & -6 & 0 & 2 \\ 2 & 1 & -1 & 2 \\ 0 & -2 & -1 & 3 \end{bmatrix} \quad \textbf{(e)} \begin{bmatrix} 1 & 2 & 1 & -1 \\ 2 & 4 & 2 & -2 \\ 0 & 1 & 2 & 3 \\ 1 & 4 & 5 & 5 \\ 0 & 3 & -2 & 4 \\ 1 & 6 & 1 & 6 \end{bmatrix} \quad \textbf{(f)} \begin{bmatrix} 1 & 2 & 1 & -1 \\ 2 & 4 & 2 & -2 \\ 0 & 1 & 2 & 3 \\ 1 & 4 & 5 & 5 \\ 0 & 3 & -2 & 4 \end{bmatrix}$$

$$\textbf{(g)} \begin{bmatrix} 1 & 1 & 3 & 2 & 0 \\ -1 & 1 & -1 & 0 & 0 \\ 2 & 2 & 6 & 4 & 0 \\ -3 & 0 & -6 & -3 & 1 \end{bmatrix} \quad \textbf{(h)} \begin{bmatrix} 1 & 1 & 3 & 1 & 1 \\ 1 & 0 & 2 & 0 & 1 \\ 3 & -2 & 4 & 0 & 7 \\ -1 & 0 & -2 & 1 & 1 \end{bmatrix}$$

7. In each part of Problem 6, let A be the given matrix and find an invertible matrix Q such that AQ is in reduced column-echelon form.

8. In each part of Problem 6, let A be the given matrix where A is the transition matrix from $\mathcal{E}_n$ to a subset $\mathcal{A}$ of $\mathbf{R}^n$. Write out the standard basis of $\langle \mathcal{A} \rangle$.

9. Write the scalar matrix aI_n, $a \neq 0$, as a product of n elementary matrices.

10. Let $A = [a_{ij}]_{r \times s}$ and $B = [b_{ij}]_{s \times t}$ be matrices over $\mathbf{R}$ and $a \in \mathbf{R}$. Prove that $(aA)B = a(AB) = A(aB)$.

11. Let A be an invertible matrix and a any nonzero real number. Prove that aA is invertible.

12. Prove or disprove: If A' is the reduced column-echelon form of A, then aA' is the reduced column-echelon form for aA.

3.5 Row Operations and Row-Echelon Forms

In this section we shall develop elementary row operations that are analogous to the column operations of the preceding section. In later chapters, these row operations will be equally as important as column operations.

The connection that was made between columns and vectors led to a natural and simple correspondence between elementary operations on sets of vectors and elementary column operations. The interpretation of elementary row operations is not as simple, and a difference in the method of investigation is necessary. The fundamental technique is much the same for the different types of operations, with the most complicated situations occurring with operations of type II. For this reason, the derivation is carried out for operations of type II, and those for the other types are left as exercises.

Let us return to the comparisons of elementary matrices with I_n that were made earlier. If we direct our attention to the rows of the elementary matrices rather than the columns, the resulting descriptions are as follows.

If M is the elementary matrix of type I on page 95, then M is obtained from I_n by multiplying row s of I_n by a nonzero scalar a.

If M is the elementary matrix of type II on page 95, then M is obtained from I_n by adding to row t of I_n the product of b and row s of I_n, where $s \neq t$.

If M is the elementary matrix of type III on page 95, then M is obtained from I_n by interchanging row s and t $(s \neq t)$.

Thus, each elementary matrix can be obtained from I_n by performing a suitable operation on the rows of I_n.

EXAMPLE 1   Each of the following illustrates an elementary matrix M obtained by applying an operation to the rows of I_n.

(a) The elementary matrix

$$M = \begin{bmatrix} 1 & 0 \\ 0 & -1 \end{bmatrix}$$

can be obtained by multiplying row 2 of I_2 by -1.

(b) The elementary matrix

$$M = \begin{bmatrix} 1 & 0 & 0 \\ 0 & 1 & 0 \\ 0 & -4 & 1 \end{bmatrix}$$

can be obtained from I_3 by adding -4 times row 2 to row 3.

(c) The elementary matrix

$$M = \begin{bmatrix} 0 & 0 & 1 & 0 \\ 0 & 1 & 0 & 0 \\ 1 & 0 & 0 & 0 \\ 0 & 0 & 0 & 1 \end{bmatrix}$$

can be obtained from I_4, by interchanging the first and third rows of I_4. ■

The next definition defines the three types of elementary row operations on a matrix A.

**3.25
DEFINITION**

(i) An **elementary row operation of type I** multiplies one of the rows of A by a nonzero scalar a.

(ii) An **elementary row operation of type II** adds to row t in A the product of b and row s in A, where $s \neq t$.

(iii) An **elementary row operation of type III** interchanges two rows in A.

The descriptions of the products MA, where M is an elementary matrix, are very much like those obtained for the products AM in Section 3.4, even though the derivations are fundamentally different. As mentioned earlier, we consider a matrix M of type II here and leave the other derivations as exercises.

Suppose that M is an $m \times m$ elementary matrix of type II and $A = [a_{ij}]$ is $m \times n$ over $\mathbf{R}$. The matrix M is the matrix of transition from $\mathcal{E}_m$ to a basis $\mathcal{A} = \{\mathbf{u}_1, \mathbf{u}_2, \ldots, \mathbf{u}_m\}$ of $\mathbf{R}^m$, where $\mathbf{u}_i = \mathbf{e}_i$ for $i \neq s$ and $\mathbf{u}_s = \mathbf{e}_s + b\mathbf{e}_t$ for $s \neq t$. Also, A is the matrix of transition from $\mathcal{A}$ to a set $\mathcal{B} = \{\mathbf{v}_1, \mathbf{v}_2, \ldots, \mathbf{v}_n\}$, and MA is the matrix of transition from $\mathcal{E}_m$ to $\mathcal{B}$. This is shown in Figure 3.5.

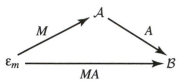

Figure 3.5

We have

$$\mathbf{v}_j = \sum_{i=1}^{m} a_{ij}\mathbf{u}_i$$

$$= \sum_{\substack{i=1 \\ i \neq s}}^{m} a_{ij}\mathbf{u}_i + a_{sj}\mathbf{u}_s$$

$$= \sum_{\substack{i=1 \\ i \neq s}}^{m} a_{ij}\mathbf{e}_i + a_{sj}(\mathbf{e}_s + b\mathbf{e}_t)$$

$$= \sum_{\substack{i=1 \\ i \neq t}}^{m} a_{ij}\mathbf{e}_i + (a_{tj} + ba_{sj})\mathbf{e}_t$$

so that the coordinates of $\mathbf{v}_j$ relative to $\mathcal{E}_m$ are the same as the coordinates of $\mathbf{v}_j$ relative to $\mathcal{A}$ except for the tth coordinate, and the tth coordinate of $\mathbf{v}_j$ relative to $\mathcal{E}_m$ is obtained by adding to the tth coordinate of $\mathbf{v}_j$ relative to $\mathcal{A}$ the product of b and the sth coordinate relative to $\mathcal{A}$. Hence multiplying A on the left by M simply adds the product of b and the sth row to the tth row of A. That is, row t of A is replaced by the sum of row t of A and b times row s of A.

The next example illustrates the effects of multiplying a matrix A on the left by an elementary matrix of type II.

EXAMPLE 2 The matrix

$$M = \begin{bmatrix} 1 & 2 \\ 0 & 1 \end{bmatrix}$$

is an elementary matrix of type II obtained from I_2 by adding 2 times the second row to the first row. This matrix is the matrix of transition from $\mathcal{E}_2 = \{(1, 0), (0, 1)\}$ to the basis

$\mathcal{A} = \{(1, 0), (2, 1)\}$ of $\mathbf{R}^2$. Consider the matrix

$$A = \begin{bmatrix} 1 & 1 & 0 \\ 1 & -1 & 2 \end{bmatrix}.$$

This is a matrix of transition from $\mathcal{A}$ to the set $\mathcal{B} = \{(3, 1), (-1, -1), (4, 2)\}$ and the product

$$MA = \begin{bmatrix} 1 & 2 \\ 0 & 1 \end{bmatrix} \begin{bmatrix} 1 & 1 & 0 \\ 1 & -1 & 2 \end{bmatrix} = \begin{bmatrix} 3 & -1 & 4 \\ 1 & -1 & 2 \end{bmatrix}$$

is the transition matrix from $\mathcal{E}_2$ to the set $\mathcal{B}$. Note that multiplying A on the left by M has the effect of adding 2 times the second row to the first row, and this is the same operation that was performed on I_2 to obtain M. ∎

The descriptions of MA for elementary matrices M of types I and III are as follows.

- If M is obtained from I_m by multiplying row s of I_m by a, then MA is obtained from A by multiplying row s of A by a.
- If M is obtained from I_m by interchanging rows s and t of I_m ($s \neq t$), then MA is obtained from A by interchanging rows s and t of A.

The concept of the transpose of a matrix is extremely useful in obtaining the row analogue of the reduced column-echelon form. This analogous form is called the *reduced row-echelon form* of a matrix.

3.26 DEFINITION If $A = [a_{ij}]$ is any $m \times n$ matrix over $\mathbf{R}$, the **transpose** of A is the $n \times m$ matrix $B = [b_{ij}]$ with $b_{ij} = a_{ji}$ for $i = 1, 2, \ldots, n$; $j = 1, 2, \ldots, m$. The transpose of A is denoted by A^T. If A is a matrix such that $A^T = A$, then A is called **symmetric**. If $A^T = -A$, then A is **skew-symmetric**.

Thus row i of A^T is composed of the elements from column i of A, in order from left to right rather than from top to bottom. We say that A^T is obtained from A "by interchanging rows and columns."

EXAMPLE 3 Consider the matrices

$$B = \begin{bmatrix} 1 & -2 & 3 \\ -2 & -4 & 5 \\ 3 & 5 & -6 \end{bmatrix} \quad \text{and} \quad C = \begin{bmatrix} 0 & 1 & -2 \\ -1 & 0 & 3 \\ 2 & -3 & 0 \end{bmatrix}.$$

It is easy to see that $B^T = B$ and

$$C^T = \begin{bmatrix} 0 & -1 & 2 \\ 1 & 0 & -3 \\ -2 & 3 & 0 \end{bmatrix} = -C.$$

Thus B is a symmetric matrix and C is a skew-symmetric matrix. ■

Our next theorem states that the transpose of a product is equal to the product of the transposes *in reverse order*.

3.27 THEOREM | If $A = [a_{ij}]_{m \times n}$ and $B = [b_{ij}]_{n \times r}$ over **R**, then $(AB)^T = B^T A^T$.

Proof Since $AB = [c_{ij}]_{m \times r}$ with $c_{ij} = \sum_{k=1}^{n} a_{ik} b_{kj}$, then we have

$$(AB)^T = \left[\sum_{k=1}^{n} a_{jk} b_{ki} \right]_{r \times m}.$$

Also, $B^T = [d_{ij}]_{r \times n}$ with $d_{ij} = b_{ji}$ and $A^T = [f_{ij}]_{n \times m}$ with $f_{ij} = a_{ji}$. Thus

$$B^T A^T = \left[\sum_{k=1}^{n} d_{ik} f_{kj} \right]_{r \times m} = \left[\sum_{k=1}^{n} b_{ki} a_{jk} \right]_{r \times m} = (AB)^T. \quad ■■■$$

3.28 DEFINITION | An $m \times n$ matrix $A = [a_{ij}]$ over **R** is in **reduced row-echelon form** if and only if the following conditions are satisfied:

1. The first nonzero element in row i is a 1 in column k_i for $i = 1, 2, \ldots, r$. (This 1 is called a **leading one**.)
2. $k_1 < k_2 < \cdots < k_r \leq n$. (That is, for each change in rows from upper to lower, the leading one appears farther to the right.)
3. For $i = 1, 2, \ldots, r$, the leading one in row i is the only nonzero element in column k_i.
4. Each of the last $m - r$ rows consists entirely of zeros.

EXAMPLE 4 Consider the question as to which of the following matrices are in reduced row-echelon form.

$$A = \begin{bmatrix} 1 & 0 & 0 & 0 \\ 0 & 0 & 0 & 0 \\ 0 & 1 & 0 & 0 \end{bmatrix}, \quad B = \begin{bmatrix} 1 & 0 & 0 & 3 \\ 0 & 1 & 0 & 2 \\ 0 & 0 & 0 & 1 \end{bmatrix},$$

$$C = \begin{bmatrix} 0 & 0 & 0 & 1 \\ 1 & 0 & 0 & 0 \\ 0 & 0 & 0 & 0 \end{bmatrix}, \quad D = \begin{bmatrix} 0 & 0 & 1 & 2 \\ 0 & 0 & 0 & 0 \\ 0 & 0 & 0 & 0 \end{bmatrix}.$$

The matrix A is not in reduced row-echelon form since the row of zeros is placed above a nonzero element, violating condition (4). The matrix B is not in reduced row-echelon form because the leading 1 in row 3 is not the only nonzero element in column $k_3 = 4$. The matrix C fails on condition (2) because the leading 1 in the second row does *not* appear farther to the right than the leading 1 in the first row. That is, $k_1 = 4$ and $k_2 = 1$ do not satisfy $k_1 < k_2$. The matrix D satisfies all four conditions, and D is in reduced row-echelon form. ■

Conditions (1) and (3) in Definition 3.28 can be reworded in the following ways that parallel the rewording we used with the definition of reduced column-echelon form:

1. $a_{ij} = 0$ for $j < k_i$, and $a_{ik_i} = 1$ for $i = 1, 2, \ldots, r$.

3. Row i is the only row with a nonzero element in column k_i for $i = 1, 2, \ldots, r$.

These alternative wordings are more useful in proving our next theorem.

3.29 THEOREM

A matrix A is in reduced column-echelon form if and only if A^T is in reduced row-echelon form.

Proof Let $A = [a_{ij}]_{m \times n}$ and let $B = [b_{ij}]_{n \times m} = A^T$. If we let j denote the row numbers and i denote the column numbers of elements of A in Definition 3.22, then A is in reduced column-echelon form if and only if these conditions hold:

1. $a_{ji} = 0$ for $j < k_i$, and $a_{k_i i} = 1$ for $i = 1, 2, \ldots, r$.

2. $k_1 < k_2 < \cdots < k_r \le m$.

3. Column i is the only column of A with a nonzero element in row k_i.

4. Each of the last $n - r$ columns of A consists entirely of zeros.

The elements in column i of A are the elements in row i of A^T, so the conditions on A are satisfied if and only if:

1. $b_{ij} = a_{ji} = 0$ for $j < k_i$, and $b_{ik_i} = 1$ for $i = 1, 2, \ldots, r$.

2. $k_1 < k_2 < \cdots < k_r \le m$.

3. Row i is the only row of A^T with a nonzero element in column k_i.

4. Each of the last $n - r$ rows of A^T consists entirely of zeros.

Since A^T is an $n \times m$ matrix, these conditions are equivalent to the assertion that A^T is in reduced row-echelon form. ■■■

From this theorem and the fact that $(A^T)^T = A$, it follows that A^T is in reduced column-echelon form if and only if A is in reduced row-echelon form.

3.30
THEOREM

If A is invertible, then A^T is invertible and $(A^T)^{-1} = (A^{-1})^T$.

Proof If A is invertible, then $A^{-1}A = I_n$ and therefore

$$(A^T) \cdot (A^{-1})^T = (A^{-1}A)^T = I_n^T = I_n.$$

According to Theorem 3.14, A^T is invertible and $(A^T)^{-1} = (A^{-1})^T$.   ■ ■ ■

3.31
THEOREM

If A is an $m \times n$ matrix over **R**, there exists an invertible matrix P over **R** such that PA is in reduced row-echelon form.

Proof By Theorem 3.23, there is an invertible matrix Q such that $A^T Q$ is in reduced column-echelon form, and this means that $(A^T Q)^T = Q^T (A^T)^T = Q^T A$ is in reduced row-echelon form. But $P = Q^T$ is an invertible matrix, so the theorem is proved.   ■ ■ ■

In the proof of Theorem 3.31, the reduced column-echelon form for A^T is uniquely determined by A^T, and therefore the reduced row-echelon form PA is uniquely determined by A. We make the following definition.

3.32
DEFINITION

The unique matrix PA in the statement of Theorem 3.31 is called the **reduced row-echelon form** for A.

3.33
THEOREM

A square matrix $A = [a_{ij}]_{n \times n}$ is invertible if and only if the reduced row-echelon form for A is I_n.

Proof Assume that A is invertible, and let P be an invertible matrix such that PA is in reduced row-echelon form. There must be no rows of zeros in PA, for otherwise, A^T would obviously be singular. This means that $k_i = i$ of each i, and $PA = I_n$.

If the reduced row-echelon form for A is I_n, there is an invertible matrix P such that $PA = I_n$. Then A is invertible by Theorem 3.15.   ■ ■ ■

We have seen that a matrix is invertible if and only if it is a product of elementary matrices. Thus the last theorem implies that there are elementary matrices $P_1, P_2, \ldots, P_s$

such that $P_s P_{s-1} \cdots P_2 P_1 A$ is in reduced row-echelon form. But multiplication of a given matrix on the left by P_i performs an elementary row operation on the given matrix. Thus a sequence of elementary row operations may be applied to A in order to obtain a reduced row-echelon matrix. In much the same manner as with column operations, this gives rise to a systematic method for finding the invertible matrix P.

With the notation of the preceding paragraph, consider the sequence of equations

$$I_m A = A$$

$$P_1 I_m A = P_1 A$$

$$\vdots$$

$$P_s \cdots P_2 P_1 I_m A = P_s \cdots P_2 P_1 A = PA.$$

These equations indicate that if the reduced row-echelon form is obtained by application of a certain sequence of elementary row operations to A, the matrix P may be obtained by application of the same sequence of operations, in the same order, to I_m. This can be done efficiently by recording A and I_m in a single matrix as $[A \mid I_m]$ and performing the row operations simultaneously on A and I_m.

EXAMPLE 5 To illustrate the procedure, let

$$A = \begin{bmatrix} 2 & 0 & 2 \\ 0 & 1 & -3 \\ 2 & 1 & 1 \end{bmatrix}$$

and consider the problem of finding an invertible matrix P such that PA is in reduced row-echelon form.

$$[A \mid I_m] = \begin{bmatrix} 2 & 0 & 2 & | & 1 & 0 & 0 \\ 0 & 1 & -3 & | & 0 & 1 & 0 \\ 2 & 1 & 1 & | & 0 & 0 & 1 \end{bmatrix} \rightarrow \begin{bmatrix} 1 & 0 & 1 & | & \frac{1}{2} & 0 & 0 \\ 0 & 1 & -3 & | & 0 & 1 & 0 \\ 0 & 1 & -1 & | & -1 & 0 & 1 \end{bmatrix}$$

$$\rightarrow \begin{bmatrix} 1 & 0 & 1 & | & \frac{1}{2} & 0 & 0 \\ 0 & 1 & -3 & | & 0 & 1 & 0 \\ 0 & 0 & 2 & | & -1 & -1 & 1 \end{bmatrix} \rightarrow \begin{bmatrix} 1 & 0 & 0 & | & 1 & \frac{1}{2} & -\frac{1}{2} \\ 0 & 1 & 0 & | & -\frac{3}{2} & -\frac{1}{2} & \frac{3}{2} \\ 0 & 0 & 1 & | & -\frac{1}{2} & -\frac{1}{2} & \frac{1}{2} \end{bmatrix}$$

$$= [PA \mid PI_3] = [I_3 \mid P]$$

Thus, I_3 is the reduced row-echelon form for A, and

$$P = \begin{bmatrix} 1 & \frac{1}{2} & -\frac{1}{2} \\ -\frac{3}{2} & -\frac{1}{2} & \frac{3}{2} \\ -\frac{1}{2} & -\frac{1}{2} & \frac{1}{2} \end{bmatrix}$$

is an invertible matrix such that $PA = I_3$. ∎

In Example 5, we found P such that $PA = I_3$. According to Theorem 3.33, A is invertible when its reduced row-echelon form is an identity matrix. However, the equation $PA = I_n$ implies more: It implies that $P = A^{-1}$, by Theorem 3.15. Thus the procedure illustrated in Example 5 is an efficient and systematic way of finding A^{-1} when it exists.

Suppose now that P is invertible and PA is in reduced row-echelon form but different from I_n. Then PA must contain at least one row of zeros, and therefore PA is not invertible. This implies that A is not invertible, by Theorem 3.16.

Summarizing our discussion of the procedure in Example 5, we can say that one of the following possibilities must happen:

1. If the procedure leads to $PA = I_n$, then A is invertible and $P = A^{-1}$.

2. If the procedure leads to a row of zeros in PA, then A is not invertible.

EXAMPLE 6 Consider the matrix

$$A = \begin{bmatrix} 1 & -1 & 0 \\ 0 & 1 & -1 \\ 1 & 0 & -1 \end{bmatrix}.$$

Applying elementary row operations to $[A \mid I_3]$ yields the following sequence of matrices.

$$[A \mid I_3] \rightarrow \left[\begin{array}{ccc|ccc} 1 & -1 & 0 & 1 & 0 & 0 \\ 0 & 1 & -1 & 0 & 1 & 0 \\ 1 & 0 & -1 & 0 & 0 & 1 \end{array}\right] \rightarrow \left[\begin{array}{ccc|ccc} 1 & -1 & 0 & 1 & 0 & 0 \\ 0 & 1 & -1 & 0 & 1 & 0 \\ 0 & 1 & -1 & -1 & 0 & 1 \end{array}\right]$$

$$\rightarrow \left[\begin{array}{ccc|ccc} 1 & 0 & -1 & 1 & 1 & 0 \\ 0 & 1 & -1 & 0 & 1 & 0 \\ 0 & 1 & -1 & -1 & 0 & 1 \end{array}\right] \rightarrow \left[\begin{array}{ccc|ccc} 1 & 0 & -1 & 1 & 1 & 0 \\ 0 & 1 & -1 & 0 & 1 & 0 \\ 0 & 0 & 0 & -1 & -1 & 1 \end{array}\right]$$

$$= [PA \mid PI_3]$$

Since the reduced row-echelon form PA of A is not I_3, then A is not invertible. ■

3.5 Exercises

1. Describe the elementary row operation that is associated with the given elementary matrix M.

(a) $M = \begin{bmatrix} 4 & 0 & 0 \\ 0 & 1 & 0 \\ 0 & 0 & 1 \end{bmatrix}$ **(b)** $M = \begin{bmatrix} 1 & 0 & 0 \\ 0 & 1 & 5 \\ 0 & 0 & 1 \end{bmatrix}$ **(c)** $M = \begin{bmatrix} 0 & 0 & 1 \\ 0 & 1 & 0 \\ 1 & 0 & 0 \end{bmatrix}$

(d) $M = \begin{bmatrix} 1 & 3 & 0 \\ 0 & 1 & 0 \\ 0 & 0 & 1 \end{bmatrix}$ **(e)** $M = \begin{bmatrix} 1 & 0 \\ 0 & -3 \end{bmatrix}$ **(f)** $M = \begin{bmatrix} 1 & 0 & 0 & 0 \\ 0 & 1 & 0 & 0 \\ 0 & 0 & 7 & 0 \\ 0 & 0 & 0 & 1 \end{bmatrix}$

2. Let

$$A = \begin{bmatrix} 1 & -2 \\ 0 & 3 \\ 1 & 0 \end{bmatrix}, \quad B = \begin{bmatrix} -1 & 0 \\ 4 & 2 \end{bmatrix}, \quad C = \begin{bmatrix} 0 & 2 & 5 \\ -1 & 4 & 0 \\ 0 & 0 & 3 \end{bmatrix}.$$

Compute each of the following products, if possible.

(a) AA^T **(b)** $A^T A$ **(c)** $(AB)^T$ **(d)** $(BA)^T$

(e) $B^T A^T$ **(f)** $A^T C$ **(g)** $(A^T C)^T$ **(h)** $C^T A$

3. Given

$$A = \begin{bmatrix} 1 & 2 \\ 0 & 1 \end{bmatrix} \begin{bmatrix} 0 & 1 \\ 1 & 0 \end{bmatrix} \begin{bmatrix} 1 & 0 \\ 0 & 3 \end{bmatrix},$$

write A^T as a product of elementary matrices.

4. Which of the following are in reduced row-echelon form?

(a) $\begin{bmatrix} 1 & 0 & 3 & 0 \\ 0 & 2 & 4 & 0 \\ 0 & 0 & 1 & 0 \end{bmatrix}$ **(b)** $\begin{bmatrix} 0 & 1 & 0 & 1 \\ 1 & 0 & 1 & 0 \\ 0 & 0 & 0 & 0 \end{bmatrix}$ **(c)** $\begin{bmatrix} 0 & 1 & 1 & 1 \\ 0 & 0 & 1 & 1 \\ 0 & 0 & 0 & 1 \end{bmatrix}$

(d) $\begin{bmatrix} 0 & 0 & 0 & 0 \\ 1 & 0 & 0 & 0 \\ 0 & 1 & 0 & 0 \end{bmatrix}$ **(e)** $\begin{bmatrix} 1 & 0 & 0 & 0 \\ 1 & 1 & 0 & 0 \\ 1 & 1 & 1 & 0 \end{bmatrix}$ **(f)** $\begin{bmatrix} 1 & 0 & 0 & 0 \\ 1 & 1 & 0 & 0 \\ 0 & 0 & 1 & 0 \end{bmatrix}$

5. Find the reduced row-echelon form for each of the matrices given in Problem 6 of Exercises 3.4.

6. In each part of Problem 6 in Exercises 3.4, let A be the given matrix and find an invertible matrix P such that PA is in reduced row-echelon form.

7. Find the inverse of each of the following invertible matrices.

(a) $\begin{bmatrix} 1 & 2 & 1 \\ 1 & 0 & 1 \\ 0 & 1 & -1 \end{bmatrix}$ **(b)** $\begin{bmatrix} -4 & -5 & 3 \\ 3 & 3 & -2 \\ -1 & -1 & 1 \end{bmatrix}$

(c) $\begin{bmatrix} 1 & 0 & 1 \\ 1 & 1 & 2 \\ 3 & 4 & -2 \end{bmatrix}$ **(d)** $\begin{bmatrix} 1 & 2 & 1 \\ -1 & -1 & 1 \\ 0 & 1 & 3 \end{bmatrix}$

8. Which of the following matrices are invertible? For each invertible matrix, find its inverse.

(a) $\begin{bmatrix} -3 & -4 \\ 1 & 2 \end{bmatrix}$ **(b)** $\begin{bmatrix} 1 & 2 \\ 3 & 6 \end{bmatrix}$ **(c)** $\begin{bmatrix} 1 & -1 & -1 \\ 0 & 2 & -1 \\ 1 & 0 & -1 \end{bmatrix}$

(d) $\begin{bmatrix} 3 & -1 & 0 \\ 1 & 1 & -1 \\ 0 & 1 & -1 \end{bmatrix}$ **(e)** $\begin{bmatrix} 1 & 2 & 2 \\ 0 & -4 & 0 \\ 1 & 0 & 2 \end{bmatrix}$ **(f)** $\begin{bmatrix} -2 & 2 & 1 \\ 2 & 0 & 1 \\ 2 & -1 & 0 \end{bmatrix}$

(g) $\begin{bmatrix} 1 & 1 & -2 & 0 \\ 0 & 1 & 2 & 1 \\ 1 & -3 & -6 & -2 \\ 2 & 3 & 0 & 2 \end{bmatrix}$ (h) $\begin{bmatrix} 0 & -1 & 0 & 1 \\ 1 & 0 & -1 & 0 \\ 0 & 2 & 0 & -1 \\ 1 & 0 & 1 & 0 \end{bmatrix}$

9. Each of the following matrices is a matrix of transition from $\mathcal{E}_n$ to a set $\mathcal{A}$. In each case, find $\mathcal{A}$ and the transition matrix from $\mathcal{A}$ to $\mathcal{E}_n$.

(a) $\begin{bmatrix} 1 & 0 & 2 \\ -1 & 1 & 0 \\ 0 & 1 & -1 \end{bmatrix}$ (b) $\begin{bmatrix} 3 & 0 & -4 \\ -2 & 0 & 3 \\ 0 & 1 & 7 \end{bmatrix}$

(c) $\begin{bmatrix} 0 & 1 & 1 & 0 \\ -1 & 2 & 0 & 0 \\ 0 & 1 & 1 & 1 \\ -1 & 0 & 1 & 1 \end{bmatrix}$ (d) $\begin{bmatrix} 0 & 0 & 4 & 1 \\ 2 & 0 & 1 & 0 \\ -4 & 2 & 0 & 1 \\ 0 & 0 & -3 & -1 \end{bmatrix}$

10. Given that the matrix

$$B = \begin{bmatrix} 4 & 2 & 4 \\ 6 & -5 & 8 \\ 0 & 3 & 2 \end{bmatrix}$$

is obtained from A by the following elementary operations:

i. First, the second and third rows of A are interchanged.

ii. Next, the first column is multiplied by 2 and added to the second column.

Find the matrix A.

11. Prove that if M is obtained from I_m by multiplying row s of I_m by a, then MA is obtained from A by multiplying row s by a.

12. Prove that if M is obtained from I_m by interchanging row s and t of I_m, then MA is obtained from A by interchanging row s and t.

13. Which types of elementary matrices are symmetric?

14. Let M be an elementary matrix of type III. Prove that $M = M^T = M^{-1}$.

15. Prove that $(A^T)^T = A$.

16. Prove that $(A^n)^T = (A^T)^n$ for all square matrices A and all positive integers n.

17. Prove that any two diagonal elements of a skew-symmetric matrix over $\mathbf{R}$ are equal.

18. Prove or disprove.

(a) The set of all symmetric $n \times n$ matrices over $\mathbf{R}$ is closed under multiplication.

(b) The set of all skew-symmetric $n \times n$ matrices over $\mathbf{R}$ is closed under multiplication.

19. Prove or disprove: If $AA^T = 0$, then $A = 0$.

20. Prove that, for any matrix A over $\mathbf{R}$, each of AA^T and A^TA is defined and each is symmetric.

21. Extend Theorem 3.27 to any product with a finite number of factors by proving that

$$(A_1 A_2 \cdots A_k)^T = A_k^T \cdots A_2^T A_1^T.$$

22. Prove that if A is symmetric, then A^k is symmetric for any positive integer k.

23. Prove that if A, B and AB are symmetric matrices, then $AB = BA$.

24. Prove or disprove: If A and B are symmetric and AB exists, then AB is symmetric.

25. Let A be an invertible matrix. Prove or disprove each of the following statements.

    (a) If A is symmetric, then A^{-1} is symmetric.

    (b) If A is skew-symmetric, then A^{-1} is skew-symmetric.

26. Let $A = [a_{ij}]_{n \times n}$ with $a_{ik} = a_{ij} + a_{jk}$ for all i, j, k.

    (a) Prove that A is skew-symmetric.

    (b) Show that A is determined by its first row.

27. Let A be any square matrix of order n over **R**.

    (a) Let $B = [b_{ij}]$ where $b_{ij} = a_{ij} + a_{ji}$ for all i, j. Prove that B is symmetric.

    (b) Let $B = [b_{ij}]$ where $b_{ij} = a_{ij} - a_{ji}$ for all i, j. Prove that B is skew-symmetric.

3.6 Row and Column Equivalence

Sometimes elements of sets are associated with each other in ways that have some of the basic properties of equality. These associations between elements are called *relations*, and certain types of relations are called *equivalence relations*. A knowledge of the various equivalence relations on a set adds depth to our understanding of the structure of the set.

The formal definition of a relation can be made as follows: A **relation** (more precisely, a **binary relation**) on a set $\mathcal{T}$ is a nonempty set "$\sim$", of ordered pairs (a, b) of elements a and b of $\mathcal{T}$. If (a, b) is in "$\sim$", then we say that a has the relation $\sim$ to b. We write $a \sim b$ to indicate that a has the relation $\sim$ to b.

If the relation under consideration is ordinary equality, then the set $\sim$ consists of all ordered pairs (a, a), and we write $a = b$ for $a \sim b$.

3.34 DEFINITION

A relation $\sim$ on a set $\mathcal{T}$ is an **equivalence relation** if and only if the following conditions are satisfied for arbitrary a, b, c in $\mathcal{T}$:

1. $a \sim a$ is true for all $a \in \mathcal{T}$.

2. If $a \sim b$ is true, then $b \sim a$ is true.

3. If $a \sim b$ is true and $b \sim c$ is true, then $a \sim c$ is true.

The properties (1), (2), and (3) are known as the **reflexive**, **symmetric**, and **transitive** properties, respectively.

We are interested here in equivalence relations on matrices. There are two such relations that are intimately connected with the two preceding sections.

**3.35
DEFINITION**

Let A and B be matrices over **R**. The matrix B is **column-equivalent** to A if and only if there exists an invertible matrix Q such that $B = AQ$.

The term "column-equivalent" defines a relation on the set of all matrices over **R**. The relation consists of the set of all ordered pairs (B, A) such that $B = AQ$ for some invertible matrix Q. It is clear that if A is $m \times n$, then any matrix that is column-equivalent to A is $m \times n$.

**3.36
THEOREM**

The relation defined as column-equivalence is an equivalence relation on the set of all matrices over **R**. That is:

1. For any matrix A, A is column-equivalent to A.
2. If A is column-equivalent to B, then B is column-equivalent to A.
3. If A is column-equivalent to B and B is column-equivalent to C, then A is column-equivalent to C.

Proof The statement (1) follows from $A = AI_n$.

If $A = BQ$, where Q is invertible, then $B = AQ^{-1}$ and Q^{-1} is invertible. Thus (2) is valid.

Assume $A = BQ$ and $B = CP$, where Q and P are invertible. Then $A = (CP)Q = C(PQ)$, and PQ is invertible. Thus (3) is valid. ∎ ■ ■ ■

The relation of column-equivalence can be described in several alternate ways. Four of these are stated in the next theorem, and examples illustrating these alternate descriptions follow Theorem 3.38.

**3.37
THEOREM**

Let A and B be $m \times n$ matrices over **R**. For any given basis of $\mathbf{R}^m$, A will be the matrix of transition from the given basis to a set $\mathcal{A}$ of n vectors, and B will be the matrix of transition from the given basis to a set $\mathcal{B}$ of n vectors. Each of the following statements implies the other three:

1. B is column-equivalent to A.
2. B may be obtained from A by a sequence of elementary column operations.
3. $\langle \mathcal{B} \rangle = \langle \mathcal{A} \rangle$.
4. $\mathcal{B}$ may be obtained from $\mathcal{A}$ by a sequence of elementary operations.

Proof Statement (3) is true if and only if (4) is true, according to Theorem 2.14. In view of Theorem 3.19, (1) means the same as the assertion that $B = A Q_1 Q_2 \cdots Q_t$, where each Q_i is elementary. It thus follows from the discussion following Definition 3.21 that (1) and (2) are equivalent. Thus the theorem will be proved if we show that each of (1) and (4) implies the other.

Suppose first that B is column-equivalent to A. Then $B = A Q_1 Q_2 \cdots Q_t$, where each Q_i is elementary. This means that $A Q_1 Q_2 \cdots Q_t$ is the matrix of transition from the given basis to $\mathcal{B}$. And since A is the matrix of transition from the given basis to $\mathcal{A}$, $Q_1 Q_2 \cdots Q_t$ is a matrix of transition from $\mathcal{A}$ to $\mathcal{B}$. Each Q_i has an associated elementary operation E_i, and we have $E_t \cdots E_2 E_1 (\mathcal{A}) = \mathcal{B}$. Thus (1) implies (4).

Assume now that $\mathcal{B}$ may be obtained from $\mathcal{A}$ by a sequence $E_1, E_2, \ldots, E_t$ of elementary operations E_i. If Q_i is the elementary matrix associated with E_i, then $Q_1 Q_2 \cdots Q_t$ is the matrix of transition from $\mathcal{A}$ to $E_t \cdots E_2 E_1 (\mathcal{A}) = \mathcal{B}$. Since A is the matrix of transition from the given basis to $\mathcal{A}$, $A Q_1 Q_2 \cdots Q_t$ is the matrix of transition from the given basis to $\mathcal{B}$. But the matrix of transition from the given basis to $\mathcal{B}$ is B, and this matrix is unique. Hence $B = A Q_1 Q_2 \cdots Q_t$, and B is column-equivalent to A. ∎ ∎ ∎

3.38 **THEOREM**	Any matrix over **R** is column-equivalent to a unique matrix in reduced column-echelon form.

Proof This is a restatement of Theorem 3.23. ∎ ∎ ∎

We consider now some examples that illustrate the matrices and sets of vectors involved in Theorem 3.37 and its proof.

EXAMPLE 1 Let the matrices A and B be given by

$$A = \begin{bmatrix} 2 & 0 & 1 \\ 1 & 1 & 0 \\ 8 & 4 & 2 \\ 6 & 0 & 3 \end{bmatrix}, \qquad B = \begin{bmatrix} 1 & -3 & 0 \\ 0 & 1 & 2 \\ 2 & -2 & 8 \\ 3 & -9 & 0 \end{bmatrix}.$$

Then A is the transition matrix from $\mathcal{E}_4$ to

$$\mathcal{A} = \{(2, 1, 8, 6), (0, 1, 4, 0), (1, 0, 2, 3)\}$$

and B is the transition matrix from $\mathcal{E}_4$ to

$$\mathcal{B} = \{(1, 0, 2, 3), (-3, 1, -2, -9), (0, 2, 8, 0)\}.$$

Theorem 2.14 assures us that $\mathcal{B}$ and $\mathcal{A}$ span the same subspace of $\mathbf{R}^4$ if and only if $\mathcal{B}$ can be obtained from $\mathcal{A}$ by a sequence of elementary operations. By the end of the next example, we can see that $\langle \mathcal{B} \rangle = \langle \mathcal{A} \rangle$. ∎

EXAMPLE 2 For the matrices A and B in Example 1, we shall show that B is column-equivalent to A and find an invertible matrix Q such that $B = AQ$. To show that B is column-equivalent to A we need only confirm that their reduced column-echelon forms are equal. But in order to find an invertible Q such that $B = AQ$, we need first to find invertible matrices M_1 and M_2 such that $AM_1 = A'$ and $BM_2 = B'$ are both in reduced column-echelon form. Using the same procedure as in Example 4 of Section 3.4, we obtain the following results.

$$
\begin{bmatrix} A \\ \hline I_3 \end{bmatrix} =
\left[\begin{array}{ccc}
2 & 0 & 1 \\
1 & 1 & 0 \\
8 & 4 & 2 \\
6 & 0 & 3 \\ \hline
1 & 0 & 0 \\
0 & 1 & 0 \\
0 & 0 & 1
\end{array}\right]
\rightarrow
\left[\begin{array}{ccc}
1 & 0 & 2 \\
0 & 1 & 1 \\
2 & 4 & 8 \\
3 & 0 & 6 \\ \hline
0 & 0 & 1 \\
0 & 1 & 0 \\
1 & 0 & 0
\end{array}\right]
$$

$$
\rightarrow
\left[\begin{array}{ccc}
1 & 0 & 0 \\
0 & 1 & 1 \\
2 & 4 & 4 \\
3 & 0 & 0 \\ \hline
0 & 0 & 1 \\
0 & 1 & 0 \\
1 & 0 & -2
\end{array}\right]
\rightarrow
\left[\begin{array}{ccc}
1 & 0 & 0 \\
0 & 1 & 0 \\
2 & 4 & 0 \\
3 & 0 & 0 \\ \hline
0 & 0 & 1 \\
0 & 1 & -1 \\
1 & 0 & -2
\end{array}\right]
$$

Thus

$$
M_1 = \begin{bmatrix} 0 & 0 & 1 \\ 0 & 1 & -1 \\ 1 & 0 & -2 \end{bmatrix}
$$

is an invertible matrix such that

$$
AM_1 = A' = \begin{bmatrix} 1 & 0 & 0 \\ 0 & 1 & 0 \\ 2 & 4 & 0 \\ 3 & 0 & 0 \end{bmatrix}.
$$

Working in similar fashion with B, we have

$$
\begin{bmatrix} B \\ \hline I_3 \end{bmatrix} =
\left[\begin{array}{ccc}
1 & -3 & 0 \\
0 & 1 & 2 \\
2 & -2 & 8 \\
3 & -9 & 0 \\ \hline
1 & 0 & 0 \\
0 & 1 & 0 \\
0 & 0 & 1
\end{array}\right]
\rightarrow
\left[\begin{array}{ccc}
1 & 0 & 0 \\
0 & 1 & 2 \\
2 & 4 & 8 \\
3 & 0 & 0 \\ \hline
1 & 3 & 0 \\
0 & 1 & 0 \\
0 & 0 & 1
\end{array}\right]
\rightarrow
\left[\begin{array}{ccc}
1 & 0 & 0 \\
0 & 1 & 0 \\
2 & 4 & 0 \\
3 & 0 & 0 \\ \hline
1 & 3 & -6 \\
0 & 1 & -2 \\
0 & 0 & 1
\end{array}\right].
$$

Thus

$$M_2 = \begin{bmatrix} 1 & 3 & -6 \\ 0 & 1 & -2 \\ 0 & 0 & 1 \end{bmatrix}$$

is an invertible matrix such that

$$BM_2 = B' = \begin{bmatrix} 1 & 0 & 0 \\ 0 & 1 & 0 \\ 2 & 4 & 0 \\ 3 & 0 & 0 \end{bmatrix}.$$

As mentioned earlier, the fact that $A' = B'$ shows that B is column-equivalent to A. The equation $BM_2 = AM_1$ implies that $B = AM_1M_2^{-1}$, and thus $Q = M_1M_2^{-1}$ is an invertible matrix such that $B = AQ$. When these computations are performed, we find that

$$M_2^{-1} = \begin{bmatrix} 1 & -3 & 0 \\ 0 & 1 & 2 \\ 0 & 0 & 1 \end{bmatrix}$$

and

$$Q = M_1M_2^{-1} = \begin{bmatrix} 0 & 0 & 1 \\ 0 & 1 & -1 \\ 1 & 0 & -2 \end{bmatrix} \begin{bmatrix} 1 & -3 & 0 \\ 0 & 1 & 2 \\ 0 & 0 & 1 \end{bmatrix} = \begin{bmatrix} 0 & 0 & 1 \\ 0 & 1 & 1 \\ 1 & -3 & -2 \end{bmatrix}$$

is an invertible matrix such that $B = AQ$. According to Theorem 3.37, the sets $\mathcal{B}$ and $\mathcal{A}$ in Example 1 do indeed span the same subspace. Also the column-echelon form $A' = B'$ is the matrix of transition from $\mathcal{E}_4$ to $\mathcal{C}$ where

$$\mathcal{C} = \{(1, 0, 2, 3), (0, 1, 4, 0), (0, 0, 0, 0)\}$$

and $\langle \mathcal{A} \rangle = \langle \mathcal{B} \rangle = \langle \mathcal{C} \rangle$. Hence the standard basis of $\langle \mathcal{A} \rangle$ or of $\langle \mathcal{B} \rangle$ consists of the nonzero vectors in $\mathcal{C}$, that is $\{(1, 0, 2, 3), (0, 1, 4, 0)\}$.   ■

EXAMPLE 3   Suppose now that we wish to find elementary matrices $Q_1, Q_2, \ldots, Q_t$ such that $B = AQ_1Q_2 \cdots Q_t$ as described in the proof of Theorem 3.37. These elementary matrices can be obtained from the work in Example 2 because both A' and B' were found by performing a single elementary column operation in each step. It is easy to see that the work done in Example 2 with A can be represented by

$$\begin{bmatrix} A \\ I_3 \end{bmatrix} \rightarrow \begin{bmatrix} AE_1 \\ I_3E_1 \end{bmatrix} \rightarrow \begin{bmatrix} AE_1E_2 \\ I_3E_1E_2 \end{bmatrix} \rightarrow \begin{bmatrix} AE_1E_2E_3 \\ I_3E_1E_2E_3 \end{bmatrix}$$

where E_1, E_2, and E_3 are elementary matrices given by

$$E_1 = \begin{bmatrix} 0 & 0 & 1 \\ 0 & 1 & 0 \\ 1 & 0 & 0 \end{bmatrix}, \quad E_2 = \begin{bmatrix} 1 & 0 & -2 \\ 0 & 1 & 0 \\ 0 & 0 & 1 \end{bmatrix}, \quad E_3 = \begin{bmatrix} 1 & 0 & 0 \\ 0 & 1 & -1 \\ 0 & 0 & 1 \end{bmatrix}$$

and $M_1 = E_1 E_2 E_3$. Similarly, the work with B can be represented by

$$\begin{bmatrix} B \\ I_3 \end{bmatrix} \rightarrow \begin{bmatrix} BF_1 \\ I_3 F_1 \end{bmatrix} \rightarrow \begin{bmatrix} BF_1 F_2 \\ I_3 F_1 F_2 \end{bmatrix}$$

where F_1 and F_2 are elementary matrices given by

$$F_1 = \begin{bmatrix} 1 & 3 & 0 \\ 0 & 1 & 0 \\ 0 & 0 & 1 \end{bmatrix}, \quad F_2 = \begin{bmatrix} 1 & 0 & 0 \\ 0 & 1 & -2 \\ 0 & 0 & 1 \end{bmatrix}$$

and $M_2 = F_1 F_2$. To find the desired $Q_1, Q_2, \ldots, Q_t$, we write

$$B = A M_1 M_2^{-1}$$
$$= A E_1 E_2 E_3 (F_1 F_2)^{-1}$$
$$= A E_1 E_2 E_3 F_2^{-1} F_1^{-1}.$$

Thus $Q_1 = E_1$, $Q_2 = E_2$, $Q_3 = E_3$, $Q_4 = F_2^{-1}$, $Q_5 = F_1^{-1}$ are elementary matrices such that $B = A Q_1 Q_2 Q_3 Q_4 Q_5$. Writing out all the elementary matrices involved, we have

$$B = A \begin{bmatrix} 0 & 0 & 1 \\ 0 & 1 & 0 \\ 1 & 0 & 0 \end{bmatrix} \begin{bmatrix} 1 & 0 & -2 \\ 0 & 1 & 0 \\ 0 & 0 & 1 \end{bmatrix} \begin{bmatrix} 1 & 0 & 0 \\ 0 & 1 & -1 \\ 0 & 0 & 1 \end{bmatrix} \begin{bmatrix} 1 & 0 & 0 \\ 0 & 1 & 2 \\ 0 & 0 & 1 \end{bmatrix} \begin{bmatrix} 1 & -3 & 0 \\ 0 & 1 & 0 \\ 0 & 0 & 1 \end{bmatrix}. \quad \blacksquare$$

As might be expected, there is an equivalence relation connected with row operations that parallels the one connected with column operations.

3.39
DEFINITION

Let A and B be matrices over **R**. The matrix B is **row-equivalent** to A if and only if there exists an invertible matrix P such that $B = PA$.

Similar to the situation with Definition 3.35 the term "row-equivalent" defines a relation on the set of all matrices over **R**. It is left as an exercise to prove that this relation is an equivalence relation.

We have seen in Section 3.5 that multiplication of a given matrix on the left by a product of elementary matrices yields the same result as the application of a sequence of elementary row operations to the matrix. In combination with Theorem 3.19 this shows that the following conditions are equivalent:

1. B is row-equivalent to A;

2. B may be obtained from A by applying a sequence of elementary row operations.

3.40 THEOREM	Any matrix over $\mathbf{R}$ is row-equivalent to a unique matrix in reduced row-echelon form.

Proof This theorem follows from Theorem 3.31 and the discussion of uniqueness just before Definition 3.32. ∎∎∎

EXAMPLE 4 To determine the reduced row-echelon form B of the matrix

$$A = \begin{bmatrix} 1 & 1 & 0 & 1 \\ 2 & 1 & 1 & 0 \\ 0 & 1 & -1 & 2 \end{bmatrix}$$

we use row operations on the matrix $[A \mid I_4]$ keeping track of the elementary matrices E_i associated with each operation.

$$[A \mid I_4] = \begin{bmatrix} 1 & 1 & 0 & 1 & \mid & 1 & 0 & 0 \\ 2 & 1 & 1 & 0 & \mid & 0 & 1 & 0 \\ 0 & 1 & -1 & 2 & \mid & 0 & 0 & 1 \end{bmatrix} \rightarrow \begin{bmatrix} 1 & 1 & 0 & 1 & \mid & 1 & 0 & 0 \\ 0 & -1 & 1 & -2 & \mid & -2 & 1 & 0 \\ 0 & 1 & -1 & 2 & \mid & 0 & 0 & 1 \end{bmatrix}$$

$$\rightarrow \begin{bmatrix} 1 & 0 & 1 & -1 & \mid & -1 & 1 & 0 \\ 0 & -1 & 1 & -2 & \mid & -2 & 1 & 0 \\ 0 & 1 & -1 & 2 & \mid & 0 & 0 & 1 \end{bmatrix}$$

$$\rightarrow \begin{bmatrix} 1 & 0 & 1 & -1 & \mid & -1 & 1 & 0 \\ 0 & -1 & 1 & -2 & \mid & -2 & 1 & 0 \\ 0 & 0 & 0 & 0 & \mid & -2 & 1 & 1 \end{bmatrix}$$

$$\rightarrow \begin{bmatrix} 1 & 0 & 1 & -1 & \mid & -1 & 1 & 0 \\ 0 & 1 & -1 & 2 & \mid & 2 & -1 & 0 \\ 0 & 0 & 0 & 0 & \mid & -2 & 1 & 1 \end{bmatrix} = [PA \mid P]$$

The elementary matrices associated with each of the above steps are

$$E_1 = \begin{bmatrix} 1 & 0 & 0 \\ -2 & 1 & 0 \\ 0 & 0 & 1 \end{bmatrix}, \quad E_2 = \begin{bmatrix} 1 & 1 & 0 \\ 0 & 1 & 0 \\ 0 & 0 & 1 \end{bmatrix}, \quad E_3 = \begin{bmatrix} 1 & 0 & 0 \\ 0 & 1 & 0 \\ 0 & 1 & 1 \end{bmatrix}, \quad E_4 = \begin{bmatrix} 1 & 0 & 0 \\ 0 & -1 & 0 \\ 0 & 0 & 1 \end{bmatrix}.$$

Now $B = PA$ is the unique reduced row-echelon form of the matrix A where $P = E_4 E_3 E_2 E_1$ and A is row-equivalent to B. However, the matrices E_1, E_2, E_3, and E_4 forming the product P are not unique. ∎

3.6 Exercises

1. Determine whether or not the given matrices are column-equivalent.

(a) $A = \begin{bmatrix} 1 & 3 & 2 \\ -2 & -6 & 4 \end{bmatrix}$, $B = \begin{bmatrix} 3 & -6 & 7 \\ 2 & -4 & 5 \end{bmatrix}$

(b) $A = \begin{bmatrix} 3 & 2 \\ 6 & 4 \\ 13 & 9 \end{bmatrix}$, $B = \begin{bmatrix} 5 & 1 \\ 10 & 2 \\ 9 & 2 \end{bmatrix}$

(c) $A = \begin{bmatrix} 1 & 2 & 3 \\ -2 & -4 & -6 \\ 1 & 0 & -1 \\ -1 & 0 & 1 \end{bmatrix}$, $B = \begin{bmatrix} 1 & 0 & -1 \\ 4 & 1 & -2 \\ 3 & 1 & -1 \\ 2 & 1 & 0 \end{bmatrix}$

(d) $A = \begin{bmatrix} 1 & 0 & 0 & 1 \\ 1 & 1 & 0 & 2 \\ 0 & 1 & 1 & 2 \\ 0 & 0 & 1 & 1 \end{bmatrix}$, $B = \begin{bmatrix} 1 & 0 & 0 & 1 \\ 2 & 1 & 0 & 2 \\ 2 & 2 & 3 & 1 \\ 1 & 1 & 3 & 1 \end{bmatrix}$

2. Which of the pairs of matrices in Problem 1 are row-equivalent?

3. In the following list of matrices, A and B are column-equivalent. Also, Q_1 and Q_2 are invertible such that AQ_1 and BQ_2 are both in reduced column-echelon form. Find an invertible Q such that $B = AQ$.

$$A = \begin{bmatrix} -2 & 0 & 5 \\ -4 & 0 & 10 \\ -2 & 1 & 6 \end{bmatrix}, \qquad B = \begin{bmatrix} 0 & -1 & -1 \\ 0 & -2 & -2 \\ 1 & 1 & -2 \end{bmatrix}$$

$$Q_1 = \begin{bmatrix} -3 & 0 & -5 \\ 0 & 1 & 2 \\ -1 & 0 & -2 \end{bmatrix}, \qquad Q_2 = \begin{bmatrix} -2 & 1 & -3 \\ 0 & 0 & 1 \\ -1 & 0 & -1 \end{bmatrix}$$

4. In the following list of matrices, A and B are column-equivalent and Q_1 is an invertible matrix such that AQ_1 is in reduced column-echelon form. Find an invertible matrix Q such that $AQ = B$.

$$A = \begin{bmatrix} -7 & 21 & -2 \\ 7 & -21 & 2 \\ 1 & -2 & 0 \\ 3 & -9 & 1 \end{bmatrix}, \quad B = \begin{bmatrix} -3 & 1 & 3 \\ 3 & -1 & -3 \\ 4 & -1 & -3 \\ 2 & 0 & -1 \end{bmatrix}, \quad Q_1 = \begin{bmatrix} 2 & 3 & 4 \\ 1 & 1 & 2 \\ 3 & 0 & 7 \end{bmatrix}$$

5. For each pair of matrices in Problem 1 that are column-equivalent, find an invertible matrix Q such that $B = AQ$.

6. For each pair of matrices in Problem 1 that are column-equivalent, find elementary matrices $Q_1, Q_2, \ldots, Q_t$ such that $B = AQ_1Q_2 \cdots Q_t$.

7. In the following list of matrices, A and B are row-equivalent. Also, P_1 and P_2 are invertible such that $P_1 A$ and $P_2 B$ are both in reduced row-echelon form. Find an invertible P such that $B = PA$.

$$A = \begin{bmatrix} 3 & 2 \\ 6 & 4 \\ 13 & 9 \end{bmatrix}, \quad B = \begin{bmatrix} 5 & 1 \\ 10 & 2 \\ 9 & 2 \end{bmatrix},$$

$$P_1 = \begin{bmatrix} 1 & 4 & -2 \\ -1 & -6 & 3 \\ -2 & 1 & 0 \end{bmatrix}, \quad P_2 = \begin{bmatrix} 0 & 1 & -1 \\ 1 & -5 & 5 \\ -2 & 1 & 0 \end{bmatrix}.$$

8. In the following list of matrices, A and B are row-equivalent and P_1 is an invertible matrix such that $P_1 A$ is in reduced row-echelon form. Find an invertible matrix P such that $PA = B$.

$$A = \begin{bmatrix} 1 & 3 & 5 \\ 2 & 7 & 12 \\ 3 & 4 & 5 \end{bmatrix}, \quad B = \begin{bmatrix} 1 & 0 & -1 \\ 2 & 1 & 0 \\ 1 & 2 & 3 \end{bmatrix}, \quad P_1 = \begin{bmatrix} 7 & -3 & 0 \\ -2 & 1 & 0 \\ -13 & 5 & 1 \end{bmatrix}.$$

9. Find PA, the unique reduced row-echelon form of

$$A = \begin{bmatrix} 1 & 2 & -3 \\ 0 & 0 & 1 \\ -2 & -4 & 2 \\ 3 & 6 & 0 \end{bmatrix},$$

and find two different factorizations of P into a product of elementary matrices.

10. Find AQ, the unique reduced column-echelon form of

$$A = \begin{bmatrix} 0 & 0 & 0 & 0 \\ 1 & -2 & -2 & 0 \\ 0 & 0 & 0 & 1 \\ 2 & -4 & -4 & 0 \\ 1 & 2 & -1 & -2 \end{bmatrix},$$

and find two different factorizations of Q into a product of elementary matrices.

11. For each pair of matrices in Problem 1 that are row-equivalent, find an invertible matrix P such that $B = PA$.

12. For each pair of matrices in Problem 1, find the standard basis of $\langle A \rangle$ and that of $\langle B \rangle$, where A is the transition matrix from $\mathcal{E}_n$ to $\mathcal{A}$, and B is the transition matrix from $\mathcal{E}_n$ to $\mathcal{B}$. For each pair, is $\langle A \rangle = \langle B \rangle$?

13. Justify your answer for each of the following questions.

    (a) Which square matrices over **R** are column-equivalent to I_n?

    (b) Which square matrices over **R** are row-equivalent to I_n?

14. Prove that the relation defined by "row-equivalent" is an equivalence relation on the set of all matrices over **R**.

15. Define the relation $\sim$ on the set of all square matrices of order n over **R** as follows: $A \sim B$ if and only if there exists an invertible matrix P such that $B = P^T A P$. Prove that $\sim$ is an equivalence relation.

16. Define the relation $\sim$ on the set of all square matrices of order n over **R** as follows: $A \sim B$ if and only if there exists an invertible matrix P such that $B = P^{-1} A P$. Prove that $\sim$ is an equivalence relation.

17. Prove that B is column-equivalent to A if and only if B^T is row-equivalent to A^T.

18. If B is column-equivalent to A and B is invertible, prove that B^{-1} is row-equivalent to A^{-1}.

19. Prove that A and B are column-equivalent if and only if they have the same reduced column-echelon form.

20. Assuming that A and B are row-equivalent, describe a procedure for finding elementary matrices $P_1, \ldots, P_t$ such that $B = P_1 \cdots P_t A$.

3.7 Rank and Equivalence

There is one more equivalence relation on matrices that we wish to consider at this time. This relation is of fundamental importance in the study of linear transformations, and it has traditionally been referred to by the term "equivalent," although other terms have been used.

3.41 **DEFINITION**	Let A and B be matrices over **R**. The matrix B is **equivalent** to A if and only if there are invertible matrices P and Q such that $B = PAQ$.

Thus, B is equivalent to A if B is row-equivalent to A or if B is column-equivalent to A. We shall see in Problem 10 of the Exercises for this section that B may be equivalent to A, and yet be neither row-equivalent nor column-equivalent to A.

It is readily verified that the relation in Definition 3.41 is a true equivalence relation. From this point on, "equivalence of matrices" will refer to this relation. Equivalence of matrices is intimately connected with the *rank* of a matrix.

3.42 **DEFINITION**	If A is any $m \times n$ matrix over **R**, the **rank** of A is the number r of nonzero columns in the reduced column-echelon form of A. The rank of A will be denoted by rank(A).

3.43
DEFINITION

Let $A = [a_{ij}]_{m \times n}$ over $\mathbf{R}$, and let $\mathbf{v}_j = (a_{1j}, a_{2j}, \ldots, a_{mj})$ for $j = 1, 2, \ldots, n$. The subspace of $\mathbf{R}^m$ spanned by $\{\mathbf{v}_1, \mathbf{v}_2, \ldots, \mathbf{v}_n\}$ is called the **column space** of A, and the vectors $\mathbf{v}_j$ are called the **column vectors** of A.

It is easy to see that A is the matrix of transition from $\mathcal{E}_m$ to a spanning set of the column space of A.

3.44
THEOREM

The rank of A is the dimension of the column space of A.

Proof Let $A = [a_{ij}]_{m \times n}$ have rank r, and let $A' = [a'_{ij}]_{m \times n}$ be the reduced column-echelon form of A. Let $\mathcal{A} = \{\mathbf{v}_1, \mathbf{v}_2, \ldots, \mathbf{v}_n\}$, where $\mathbf{v}_j = (a_{1j}, a_{2j}, \ldots, a_{mj})$, and let $\mathcal{A}' = \{\mathbf{v}'_1, \mathbf{v}'_2, \ldots, \mathbf{v}'_n\}$, where $\mathbf{v}'_j = (a'_{1j}, a'_{2j}, \ldots, a'_{mj})$. Then $\langle \mathcal{A} \rangle$ and $\langle \mathcal{A}' \rangle$ are the column spaces of A and A', respectively. It is clear from conditions (1) and (2) of Definition 3.22 that $\{\mathbf{v}'_1, \mathbf{v}'_2, \ldots, \mathbf{v}'_r\}$ is linearly independent and therefore is a basis of $\langle \mathcal{A}' \rangle$. But $\langle \mathcal{A} \rangle = \langle \mathcal{A}' \rangle$ by Theorem 3.37 since A and A' are column-equivalent. Thus $\langle \mathcal{A} \rangle$ has dimension r.

■ ■ ■

3.45
COROLLARY

If A is $m \times n$ over $\mathbf{R}$ and Q is any invertible $n \times n$ matrix, then $\text{rank}(A) = \text{rank}(AQ)$.

Proof Since A and AQ are column-equivalent, their column spaces are equal. Hence their ranks are equal, by the theorem. ■ ■ ■

We illustrate the preceding definitions and theorem in the next example.

EXAMPLE 1 Let

$$A = \begin{bmatrix} 1 & 0 & 3 & -1 & 0 \\ 0 & 1 & 2 & 0 & -1 \\ -1 & 0 & -3 & 1 & 0 \\ 2 & 0 & 4 & 0 & -2 \end{bmatrix}$$

be the transition matrix from $\mathcal{E}_4$ to $\mathcal{A}$. The subspace

$$\langle \mathcal{A} \rangle = \langle (1, 0, -1, 2), (0, 1, 0, 0), (3, 2, -3, 4), (-1, 0, 1, 0), (0, -1, 0, -2) \rangle$$

is the column space of A. Since the reduced column-echelon form

$$
\begin{bmatrix}
1 & 0 & 0 & 0 & 0 \\
0 & 1 & 0 & 0 & 0 \\
-1 & 0 & 0 & 0 & 0 \\
0 & 0 & 1 & 0 & 0
\end{bmatrix}
$$

of A contains 3 nonzero columns, then rank$(A) = 3$ and the dimension of the column space $\langle A \rangle$ of A is 3. The nonzero columns of the reduced column-echelon form determine the standard basis $\{(1, 0, -1, 0), (0, 1, 0, 0), (0, 0, 0, 1)\}$ of the column space of A. ∎

3.46
THEOREM

> Let A be an $m \times n$ matrix over $\mathbf{R}$, and let r be the rank of A. There exist invertible matrices P and Q such that PAQ has the first r diagonal elements equal to 1, and all other elements zero.

Proof Let Q be an invertible matrix such that $AQ = A' = [a_{ij}]_{m \times n}$ is in reduced column-echelon form. As in the proof of the preceding theorem, the set $\{\mathbf{v}'_1, \mathbf{v}'_2, \ldots, \mathbf{v}'_r\}$ is a basis of the column space $\langle A \rangle = \langle A' \rangle$ of A.

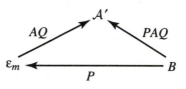

Figure 3.6

According to Theorem 1.33, the linearly independent set $\{\mathbf{v}'_1, \mathbf{v}'_2, \ldots, \mathbf{v}'_r\}$ can be extended to a basis $\mathcal{B} = \{\mathbf{w}_1, \mathbf{w}_2, \ldots, \mathbf{w}_m\}$ of $\mathbf{R}^m$, where $\mathbf{w}_i = \mathbf{v}'_i$ for $i = 1, 2, \ldots, r$. Let P be the invertible $m \times m$ matrix of transition from $\mathcal{B}$ to $\mathcal{E}_m$ (see Figure 3.6). Then PAQ is the matrix of transition from $\mathcal{B}$ to A'. Since the first r vectors of A' are the same as the first r vectors of $\mathcal{B}$, the first r columns of PAQ are the same as the first r columns of I_m. And since the last $n - r$ vectors of A' are zero, the last $n - r$ columns of PAQ are zero. Thus

$$
PAQ = \left[
\begin{array}{ccc|ccc}
1 & \cdots & 0 & 0 & \cdots & 0 \\
\vdots & \ddots & \vdots & \vdots & & \vdots \\
0 & \cdots & 1 & 0 & \cdots & 0 \\
\hline
0 & \cdots & 0 & 0 & \cdots & 0 \\
\vdots & & \vdots & \vdots & & \vdots \\
0 & \cdots & 0 & 0 & \cdots & 0
\end{array}
\right]
=
\left[
\begin{array}{c|c}
I_r & \mathbf{0} \\
\hline
\mathbf{0} & \mathbf{0}
\end{array}
\right],
$$

and the theorem is proved. ∎∎∎

Let

$$D_r = \left[\begin{array}{c|c} I_r & 0 \\ \hline 0 & 0 \end{array}\right]$$

denote the matrix PAQ in the proof of Theorem 3.46 above. The proof of the theorem can be followed literally step by step so as to obtain invertible matrices P and Q such that $PAQ = D_r$. The only difficulty is in finding the matrix of transition P from $\mathcal{B}$ to $\mathcal{E}_m$. This is probably most easily done by writing out the matrix P^{-1} (which is the matrix of transition from $\mathcal{E}_m$ to $\mathcal{B}$), and then finding the inverse of P^{-1}. However, we have seen earlier that multiplication on the right by Q is equivalent to applying a sequence of column operations, and multiplication on the left by P is equivalent to applying a sequence of row operations. Thus one may proceed to use column operations on

$$\begin{bmatrix} A \\ I_n \end{bmatrix} \quad \text{to obtain} \quad \begin{bmatrix} A' \\ Q \end{bmatrix}$$

as we did in Section 3.4, and then use row operations on $[A' \mid I_m]$ to obtain $[D_r \mid P]$, where $D_r = PA' = PAQ$. This is illustrated in the following example.

EXAMPLE 2 For the matrix

$$A = \begin{bmatrix} 1 & 2 & -1 \\ 3 & 6 & -3 \\ -1 & 1 & 0 \\ 2 & 4 & -2 \end{bmatrix},$$

we shall find invertible matrices P and Q such that $PAQ = D_r$.

   We first use column operations to transform

$$\begin{bmatrix} A \\ I_3 \end{bmatrix} \quad \text{into} \quad \begin{bmatrix} A' \\ Q \end{bmatrix},$$

where $A' = AQ$ is in reduced column-echelon form.

$$\begin{bmatrix} A \\ I_3 \end{bmatrix} = \left[\begin{array}{ccc} 1 & 2 & -1 \\ 3 & 6 & -3 \\ -1 & 1 & 0 \\ 2 & 4 & -2 \\ \hline 1 & 0 & 0 \\ 0 & 1 & 0 \\ 0 & 0 & 1 \end{array}\right] \rightarrow \left[\begin{array}{ccc} 1 & 0 & 0 \\ 3 & 0 & 0 \\ -1 & 3 & -1 \\ 2 & 0 & 0 \\ \hline 1 & -2 & 1 \\ 0 & 1 & 0 \\ 0 & 0 & 1 \end{array}\right]$$

$$\rightarrow \begin{bmatrix} 1 & 0 & 0 \\ 3 & 0 & 0 \\ -1 & -1 & 3 \\ 2 & 0 & 0 \\ \hline 1 & 1 & -2 \\ 0 & 0 & 1 \\ 0 & 1 & 0 \end{bmatrix} \rightarrow \begin{bmatrix} 1 & 0 & 0 \\ 3 & 0 & 0 \\ 0 & 1 & 0 \\ 2 & 0 & 0 \\ \hline 0 & -1 & 1 \\ 0 & 0 & 1 \\ -1 & -1 & 3 \end{bmatrix} = \begin{bmatrix} A' \\ Q \end{bmatrix}$$

Next we use row operations to transform $[A' \mid I_4]$ into $[D_r \mid P]$, where $D_r = PA' = PAQ$.

$$[A' \mid I_4] = \begin{bmatrix} 1 & 0 & 0 & | & 1 & 0 & 0 & 0 \\ 3 & 0 & 0 & | & 0 & 1 & 0 & 0 \\ 0 & 1 & 0 & | & 0 & 0 & 1 & 0 \\ 2 & 0 & 0 & | & 0 & 0 & 0 & 1 \end{bmatrix} \rightarrow \begin{bmatrix} 1 & 0 & 0 & | & 1 & 0 & 0 & 0 \\ 0 & 0 & 0 & | & -3 & 1 & 0 & 0 \\ 0 & 1 & 0 & | & 0 & 0 & 1 & 0 \\ 0 & 0 & 0 & | & -2 & 0 & 0 & 1 \end{bmatrix}$$

$$\rightarrow \begin{bmatrix} 1 & 0 & 0 & | & 1 & 0 & 0 & 0 \\ 0 & 1 & 0 & | & 0 & 0 & 1 & 0 \\ 0 & 0 & 0 & | & -3 & 1 & 0 & 0 \\ 0 & 0 & 0 & | & -2 & 0 & 0 & 1 \end{bmatrix} = [D_r \mid P]$$

Thus

$$P = \begin{bmatrix} 1 & 0 & 0 & 0 \\ 0 & 0 & 1 & 0 \\ -3 & 1 & 0 & 0 \\ -2 & 0 & 0 & 1 \end{bmatrix} \quad \text{and} \quad Q = \begin{bmatrix} 0 & -1 & 1 \\ 0 & 0 & 1 \\ -1 & -1 & 3 \end{bmatrix}$$

are invertible matrices such that

$$PAQ = \begin{bmatrix} 1 & 0 & | & 0 \\ 0 & 1 & | & 0 \\ \hline 0 & 0 & | & 0 \\ 0 & 0 & | & 0 \end{bmatrix} = \begin{bmatrix} I_2 & | & \mathbf{0} \\ \hline \mathbf{0} & | & \mathbf{0} \end{bmatrix} = D_2. \qquad \blacksquare$$

3.47 THEOREM Let A and B be $m \times n$ matrices over $\mathbf{R}$. Then B is equivalent to A if and only if B and A have the same rank.

Proof Let r denote the rank of A, and let r' denote the rank of B.

Suppose that $r = r'$. Then there are invertible matrices P, Q, P', and Q' such that $PAQ = D_r = D_{r'} = P'BQ'$. The matrices P and P' are $m \times m$, and the matrices Q and Q' are $n \times n$. Hence the equation $PAQ = P'BQ'$ implies that

$$(P')^{-1}PAQ(Q')^{-1} = B,$$

where $(P')^{-1}P$ and $Q(Q')^{-1}$ are invertible matrices. Therefore B is equivalent to A.

Assume now that B is equivalent to A. Let D_r be the diagonal matrix equivalent to A that has the form described in Theorem 3.46 and let $D_{r'}$ be the corresponding matrix that is equivalent to B. Since A and B are equivalent, D_r and $D_{r'}$ are equivalent. Therefore there are invertible matrices P and Q such that $PD_r Q = D_{r'}$ and

$$PD_r = D_{r'}Q^{-1}.$$

Now PD_r has $n - r$ columns of zeros so that $\text{rank}(PD_r) \leq r$. Thus

$$r' = \text{rank}(D_{r'}) = \text{rank}(D_{r'}Q^{-1}) = \text{rank}(PD_r) \leq r.$$

Similarly,

$$r = \text{rank}(D_r) = \text{rank}(D_r Q) = \text{rank}(P^{-1}D_{r'}) \leq r'$$

so that $r = r'$. ■ ■ ■

The following theorem and corollary are extremely useful in connection with the solution of systems of linear equations in Chapter 4.

3.48 THEOREM

Let $A = [a_{ij}]_{m \times n}$ over **R**. Then A and A^T have the same rank.

Proof Let r denote the rank of A, and let P and Q be invertible matrices such that

$$PAQ = \left[\begin{array}{c|c} I_r & 0 \\ \hline 0 & 0 \end{array} \right] = D_r.$$

Now Q^T and P^T are invertible (Theorem 3.30), and $Q^T A^T P^T = D_r^T$. The dimension of the column space of D_r^T is clearly r, so the D_r^T has rank r. But A^T and D_r^T are equivalent, so A^T must have rank r also. ■ ■ ■

3.49 COROLLARY

If A has rank r, then r is the number of nonzero rows in the reduced row-echelon form of A.

Proof Suppose that A has rank r, and let P be an invertible matrix such that PA is in reduced row-echelon form. By Theorem 3.29, $(PA)^T = A^T P^T$ is in reduced column-echelon form. Since A, A^T, and $A^T P^T$ all have the same rank r, the number of nonzero columns in $(PA)^T = A^T P^T$ is r. But the number of nonzero columns in $(PA)^T$ is the same as the number of nonzero rows in PA. Hence the corollary is proved. ∎ ∎ ∎

This result leads to the next corollary. The proof is requested in Problem 21 of the Exercises.

3.50 COROLLARY

An $n \times n$ matrix A is invertible if and only if A has rank n.

3.7 Exercises

1. Each of the following matrices A are in either reduced row-echelon or reduced column-echelon form. Find the rank(A).

(a) $\begin{bmatrix} 0 & 0 & 0 \\ 1 & 0 & 0 \\ 2 & 0 & 0 \\ 0 & 1 & 0 \end{bmatrix}$

(b) $\begin{bmatrix} 1 & 0 & 0 & 0 \\ 0 & 1 & 0 & 0 \\ 4 & 0 & 0 & 0 \end{bmatrix}$

(c) $\begin{bmatrix} 1 & 0 & 0 & 0 & 0 \\ -1 & 0 & 0 & 0 & 0 \\ 0 & 1 & 0 & 0 & 0 \\ 0 & 0 & 1 & 0 & 0 \end{bmatrix}$

(d) $\begin{bmatrix} 1 & 0 & 0 & 3 & 0 \\ 0 & 1 & 0 & 0 & -2 \\ 0 & 0 & 1 & 1 & 0 \\ 0 & 0 & 0 & 0 & 0 \end{bmatrix}$

(e) $\begin{bmatrix} 1 & -5 & 0 & 0 \\ 0 & 0 & 1 & 0 \\ 0 & 0 & 0 & 0 \\ 0 & 0 & 0 & 0 \end{bmatrix}$

(f) $\begin{bmatrix} 0 & 0 & 1 & 5 & 0 & 2 & 7 \\ 0 & 0 & 0 & 0 & 1 & -1 & 0 \\ 0 & 0 & 0 & 0 & 0 & 0 & 0 \\ 0 & 0 & 0 & 0 & 0 & 0 & 0 \end{bmatrix}$

2. Find the rank of each of the following matrices.

(a) $\begin{bmatrix} 2 & 4 & 1 & 1 \\ 1 & 2 & 2 & 0 \\ 0 & 0 & 3 & -1 \\ 3 & 6 & 0 & 2 \end{bmatrix}$

(b) $\begin{bmatrix} 1 & 3 & 2 & 3 \\ 3 & -2 & 1 & 4 \\ 5 & -7 & 0 & 5 \\ -2 & 5 & 1 & -1 \end{bmatrix}$

(c) $\begin{bmatrix} 1 & 0 & 2 \\ 4 & 1 & 3 \\ 3 & 1 & 1 \\ 2 & 1 & -1 \end{bmatrix}$

(d) $\begin{bmatrix} 1 & 2 & 3 \\ -2 & -4 & -6 \\ 0 & 2 & 4 \\ 0 & 0 & 0 \end{bmatrix}$

3. Find the standard basis of the column space of each matrix A in Problem 2.

4. For each matrix A in Problem 2, let C be the column space of A and find a basis of C that contains only column vectors of A. (*Hint:* Use the procedure in the proof of Theorem 1.33.)

5. Find a matrix A whose column space is **W**. Find the standard basis of **W** and rank(A).

   (a) $W = \langle (1, 2, 0, 3), (2, 0, 1, 4), (4, 1, 0, 0), (1, 1, 1, 1) \rangle$

   (b) $W = \{(a, -a, b) \mid a, b \in \mathbf{R}\}$

   (c) $W = \{(a, 0, b, 0) \mid a, b \in \mathbf{R}\}$

   (d) $W = \{(a, b, c, d) \mid a + b = 0, c - d = 0 \in \mathbf{R}\}$

   (e) $W = \{(x_1, x_2, x_3) \mid x_1 + x_3 = 0, x_1 + x_2 = 0\}$

   (f) $W = \{(x_1, x_2, x_3) \mid x_1 + x_3 = 0, x_1 + x_2 - x_3 = 0, x_1 - x_2 + x_3 = 0\}$

   (g) **W** is the line $2x - y = 0$ in $\mathbf{R}^2$.

   (h) **W** is the plane $x - 2z = 0$ in $\mathbf{R}^3$.

   (i) **W** is the plane $x + y - z = 0$ in $\mathbf{R}^3$.

6. Suppose the matrix A has the following dimensions. Determine the largest possible value for rank(A).

   (a) 3×3 **(b)** 3×4 **(c)** 4×3 **(d)** 5×5 **(e)** 5×7 **(f)** 7×6

7. Prove or disprove each of the following statements.

   (a) If A is a square matrix of order n, then the columns of A are linearly independent.

   (b) If A is a matrix of dimension $m \times n$ with $m < n$, then the columns of A are linearly dependent.

8. Prove or disprove: rank(A) = rank(A^2) for all square matrices A over $\mathbf{R}$.

9. Prove or disprove each of the following statements for all matrices A over $\mathbf{R}$.

   (a) rank(A) = rank($A^T A$)

   (b) rank(A) = rank($A A^T$)

10. Which of the pairs A, B of matrices in Problem 1 of Exercises 3.6 are equivalent? (Compare this with the results of Problems 1 and 2 of Exercises 3.6.)

11. Which of the matrices in Problem 2 above are equivalent?

12. Which of the following matrices are equivalent?

$$A = \begin{bmatrix} 1 & 3 & 2 \\ -2 & -6 & 4 \end{bmatrix}, \quad B = \begin{bmatrix} 3 & -6 & 7 \\ 2 & -4 & 5 \end{bmatrix}, \quad C = \begin{bmatrix} 3 & 2 \\ 6 & 4 \\ 13 & 9 \end{bmatrix}, \quad D = \begin{bmatrix} 1 & 0 & -1 \\ 4 & 1 & -2 \\ 3 & 1 & -1 \\ 2 & 1 & 0 \end{bmatrix}$$

13. For each of the following pairs of matrices answer each question:

   Are A and B column-equivalent?

   Are A and B row-equivalent?

   Are A and B equivalent?

   (a) $A = \begin{bmatrix} 2 & 1 & -3 \\ 1 & 2 & 0 \\ 1 & 1 & -1 \end{bmatrix}$ $B = \begin{bmatrix} 1 & 1 & -1 \\ 2 & 3 & -1 \\ 3 & 3 & -3 \end{bmatrix}$

(b) $A = \begin{bmatrix} 1 & 1 & 0 \\ 0 & -1 & 2 \\ 1 & 2 & -2 \end{bmatrix}$ $B = \begin{bmatrix} 2 & 1 & -2 \\ -1 & 5 & 7 \\ 3 & -4 & -9 \end{bmatrix}$

(c) $A = \begin{bmatrix} 3 & 2 & 1 \\ 0 & 0 & -2 \\ 0 & 0 & 0 \\ 1 & 1 & 0 \end{bmatrix}$ $B = \begin{bmatrix} 1 & 2 & 1 \\ 0 & 0 & 1 \\ -1 & 1 & 0 \\ 0 & 0 & 0 \end{bmatrix}$

(d) $A = \begin{bmatrix} 0 & 2 & 1 & 0 & -1 \\ 1 & 0 & 1 & 0 & 0 \\ 0 & 0 & 0 & 0 & 0 \\ 2 & 0 & 0 & 0 & 0 \end{bmatrix}$ $B = \begin{bmatrix} 0 & 2 & -1 & -3 & 0 \\ 4 & 0 & 2 & 0 & 0 \\ 1 & 0 & 0 & 0 & 0 \\ 0 & 0 & 0 & 0 & 0 \end{bmatrix}$

14. For

$$A = \begin{bmatrix} 1 & 2 & 1 \\ 2 & 4 & 2 \\ 0 & 1 & 2 \end{bmatrix},$$

it is given that

$$Q = \begin{bmatrix} 1 & -2 & 3 \\ 0 & 1 & -2 \\ 0 & 0 & 1 \end{bmatrix}$$

is an invertible matrix such that

$$AQ = A' = \begin{bmatrix} 1 & 0 & 0 \\ 2 & 0 & 0 \\ 0 & 1 & 0 \end{bmatrix}.$$

Let $\mathcal{A}' = \{(1, 2, 0), (0, 0, 1), (0, 0, 0)\}$. Find a basis $\mathcal{B}$ of $\mathbf{R}^3$ such that the matrix of transition from $\mathcal{B}$ to $\mathcal{A}'$ is

$$D_2 = \begin{bmatrix} 1 & 0 & 0 \\ 0 & 1 & 0 \\ 0 & 0 & 0 \end{bmatrix}$$

and an invertible P such that $PAQ = D_2$. (*Hint:* See the proof of Theorem 3.46.)

15. For each matrix A below, follow the proof of Theorem 3.46 step by step to find invertible matrices P and Q such that $PAQ = D_r$. In your development, write out the sets $\mathcal{A}$, $\mathcal{A}'$, and $\mathcal{B}$.

(a) $A = \begin{bmatrix} 1 & 0 & 0 & 1 \\ 1 & 1 & 0 & 2 \\ 0 & 1 & 1 & 2 \\ 0 & 0 & 1 & 1 \end{bmatrix}$

(b) $A = \begin{bmatrix} 3 & 2 & 1 & 0 \\ -4 & -3 & -2 & 0 \\ 1 & 0 & 3 & 1 \\ -3 & -3 & 1 & 1 \end{bmatrix}$

16. For each matrix A, find invertible matrices P and Q such that PAQ has the first r elements of the main diagonal equal to 1, and all other elements 0.

(a) $A = \begin{bmatrix} 0 & 2 & 4 & 1 \\ 0 & 1 & 2 & 0 \\ 0 & 3 & 6 & 1 \end{bmatrix}$
$\qquad\qquad$
(b) $A = \begin{bmatrix} -1 & 2 & 2 \\ 2 & 2 & 2 \\ -3 & -6 & -6 \end{bmatrix}$

17. In Problem 2 above, let A be the matrix in part (c), and let B be the matrix in part (d). Given that A and B are equivalent, find invertible matrices P and Q such that $B = PAQ$. (*Hint:* See the proof of Theorem 3.47.)

18. In Problem 2 above, let A be the matrix in part (a), and let B be the matrix in part (b). Given that A and B are equivalent, find invertible matrices P and Q such that $B = PAQ$.

19. Prove that the relation "equivalence of matrices" is an equivalence relation on the set of all matrices over **R**.

20. Assume B is conformable to A and prove each statement.

(a) $\text{rank}(AB) \leq \min\{\text{rank}(A), \text{rank}(B)\}$

(b) The columns of A span the column space of (AB).

21. Prove Corollary 3.50.

3.8 *LU* Decompositions

Previously, some special types of products of matrices have been presented. For example, any invertible matrix A can be written as a product of certain elementary matrices $M_1, M_2, \ldots, M_k$:

$$A = M_1 M_2 \cdots M_k.$$

In this section we consider another type of factorization of an $m \times n$ matrix A. Such a factorization might exist even if A is not invertible or even square. The factorization, known as the LU decomposition, is expressed in terms of triangular matrices. A square matrix, where any element is zero if its row number is larger than its column number, was labeled an *upper triangular* matrix in Exercises 3.2. More generally, we state the following definition.

3.51
DEFINITION
$\quad$ A matrix $U = [u_{ij}]$ is **upper triangular** if $u_{ij} = 0$ for all $i > j$. A matrix $L = [l_{ij}]$ is **lower triangular** if $l_{ij} = 0$ for all $i < j$. The lower triangular matrix L is called **unit lower triangular** if $l_{ii} = 1$ for all i. Any upper triangular or lower triangular matrix is called a **triangular matrix**.

The following example illustrates this terminology.

EXAMPLE 1 Each of the following matrices is a triangular matrix.

$$A = \begin{bmatrix} 1 & 0 & 0 & 0 \\ 2 & 0 & 0 & 0 \\ 0 & -1 & 1 & 0 \end{bmatrix} \quad B = \begin{bmatrix} 2 & 1 & 0 \\ 0 & 1 & 3 \end{bmatrix} \quad C = \begin{bmatrix} 1 & 0 & 0 \\ -2 & 1 & 0 \\ 0 & 3 & 1 \end{bmatrix}$$

$$I = \begin{bmatrix} 1 & 0 & 0 & 0 \\ 0 & 1 & 0 & 0 \\ 0 & 0 & 1 & 0 \\ 0 & 0 & 0 & 1 \end{bmatrix} \quad D = \begin{bmatrix} 3 & 0 \\ 0 & 3 \\ 0 & 0 \end{bmatrix}$$

Additionally, the matrices A, C, I, and D are all lower triangular with C and I being unit lower triangular. Also, the matrices B, I, and D are all upper triangular. ∎

3.52
THEOREM

If the matrix A can be transformed into an upper triangular matrix U using only type II elementary row operations, then $A = LU$ where L is a unit lower triangular matrix.

Proof Suppose that A can be transformed into an upper triangular matrix U using only k type II elementary row operations with associated elementary matrices $E_1, E_2, \ldots, E_k$. Then

$$U = E_k \cdots E_2 E_1 A.$$

Since elementary matrices are invertible, A can be expressed as

$$A = E_1^{-1} E_2^{-1} \cdots E_k^{-1} U.$$

Each of $E_1, E_2, \ldots, E_k$ is unit lower triangular and hence (see Problems 24 and 25 of the Exercises) each of $E_1^{-1}, E_2^{-1}, \ldots, E_k^{-1}$ and the product $E_1^{-1} E_2^{-1} \cdots E_k^{-1}$ is a unit lower triangular matrix. Thus A can be written as $A = LU$ where $L = E_1^{-1} E_2^{-1} \cdots E_k^{-1}$ is unit lower triangular and U is upper triangular. ∎∎∎

The factorization of A described in Theorem 3.52 is known as the **LU decomposition** of A. The proof of the theorem reveals a systematic procedure for decomposing A into the product LU as illustrated in the next examples.

EXAMPLE 2 The matrix

$$A = \begin{bmatrix} -1 & 2 & 0 \\ 1 & -3 & 1 \\ 0 & 2 & 1 \end{bmatrix}$$

can be reduced to an upper triangular matrix U using only type II elementary operations.

- First add row 1 of A to row 2 yielding

$$\begin{bmatrix} -1 & 2 & 0 \\ 0 & -1 & 1 \\ 0 & 2 & 1 \end{bmatrix} \quad \text{with associated elementary matrix} \quad E_1 = \begin{bmatrix} 1 & 0 & 0 \\ 1 & 1 & 0 \\ 0 & 0 & 1 \end{bmatrix}.$$

- Next add 2 times row 2 to row 3 yielding

$$\begin{bmatrix} -1 & 2 & 0 \\ 0 & -1 & 1 \\ 0 & 0 & 3 \end{bmatrix} \quad \text{with associated elementary matrix} \quad E_2 = \begin{bmatrix} 1 & 0 & 0 \\ 0 & 1 & 0 \\ 0 & 2 & 1 \end{bmatrix}.$$

Each of E_1 and E_2 is a unit lower triangular matrix and the product

$$E_2 E_1 A = \begin{bmatrix} 1 & 0 & 0 \\ 0 & 1 & 0 \\ 0 & 2 & 1 \end{bmatrix} \begin{bmatrix} 1 & 0 & 0 \\ 1 & 1 & 0 \\ 0 & 0 & 1 \end{bmatrix} \begin{bmatrix} -1 & 2 & 0 \\ 1 & -3 & 1 \\ 0 & 2 & 1 \end{bmatrix} = \begin{bmatrix} -1 & 2 & 0 \\ 0 & -1 & 1 \\ 0 & 0 & 3 \end{bmatrix} = U$$

is an upper triangular matrix. Also both E_2 and E_1 are invertible and A can be expressed as

$$E_1^{-1} E_2^{-1} U = \begin{bmatrix} 1 & 0 & 0 \\ -1 & 1 & 0 \\ 0 & 0 & 1 \end{bmatrix} \begin{bmatrix} 1 & 0 & 0 \\ 0 & 1 & 0 \\ 0 & -2 & 1 \end{bmatrix} \begin{bmatrix} -1 & 2 & 0 \\ 0 & -1 & 1 \\ 0 & 0 & 3 \end{bmatrix} = \begin{bmatrix} -1 & 2 & 0 \\ 1 & -3 & 1 \\ 0 & 2 & 1 \end{bmatrix} = A.$$

Also note that each of E_1^{-1}, E_2^{-1} and their product

$$E_1^{-1} E_2^{-1} = \begin{bmatrix} 1 & 0 & 0 \\ -1 & 1 & 0 \\ 0 & 0 & 1 \end{bmatrix} \begin{bmatrix} 1 & 0 & 0 \\ 0 & 1 & 0 \\ 0 & -2 & 1 \end{bmatrix} = \begin{bmatrix} 1 & 0 & 0 \\ -1 & 1 & 0 \\ 0 & -2 & 1 \end{bmatrix}$$

is unit lower triangular. Hence A can be factored as a product of a unit lower triangular matrix and an upper triangular matrix

$$A = (E_1^{-1} E_2^{-1}) U = \begin{bmatrix} 1 & 0 & 0 \\ -1 & 1 & 0 \\ 0 & -2 & 1 \end{bmatrix} \begin{bmatrix} -1 & 2 & 0 \\ 0 & -1 & 1 \\ 0 & 0 & 3 \end{bmatrix} = \begin{bmatrix} -1 & 2 & 0 \\ 1 & -3 & 1 \\ 0 & 2 & 1 \end{bmatrix}. \qquad \blacksquare$$

The next example illustrates an LU decomposition of a nonsquare matrix.

EXAMPLE 3 The matrix

$$A = \begin{bmatrix} 0 & 1 & 2 & -1 & 0 \\ 0 & 3 & 2 & 3 & -1 \\ 0 & -2 & 4 & 1 & 0 \end{bmatrix}$$

can be reduced to upper triangular form U using the same procedure as in Example 2.

$$A = \begin{bmatrix} 0 & 1 & 2 & -1 & 0 \\ 0 & 3 & 2 & 3 & -1 \\ 0 & -2 & 4 & 1 & 0 \end{bmatrix}$$

$$\rightarrow \begin{bmatrix} 0 & 1 & 2 & -1 & 0 \\ 0 & 0 & -4 & 6 & -1 \\ 0 & -2 & 4 & 1 & 0 \end{bmatrix} \quad \text{with} \quad E_1 = \begin{bmatrix} 1 & 0 & 0 \\ -3 & 1 & 0 \\ 0 & 0 & 1 \end{bmatrix}$$

$$\rightarrow \begin{bmatrix} 0 & 1 & 2 & -1 & 0 \\ 0 & 0 & -4 & 6 & -1 \\ 0 & 0 & 8 & -1 & 0 \end{bmatrix} \quad \text{with} \quad E_2 = \begin{bmatrix} 1 & 0 & 0 \\ 0 & 1 & 0 \\ 2 & 0 & 1 \end{bmatrix}$$

$$\rightarrow \begin{bmatrix} 0 & 1 & 2 & -1 & 0 \\ 0 & 0 & -4 & 6 & -1 \\ 0 & 0 & 0 & 11 & -2 \end{bmatrix} \quad \text{with} \quad E_3 = \begin{bmatrix} 1 & 0 & 0 \\ 0 & 1 & 0 \\ 0 & 2 & 1 \end{bmatrix}$$

$$= U$$

So we now have

$$E_3 E_2 E_1 A = U \quad \text{and} \quad A = E_1^{-1} E_2^{-1} E_3^{-1} U.$$

The product

$$E_1^{-1} E_2^{-1} E_3^{-1} = \begin{bmatrix} 1 & 0 & 0 \\ 3 & 1 & 0 \\ 0 & 0 & 1 \end{bmatrix} \begin{bmatrix} 1 & 0 & 0 \\ 0 & 1 & 0 \\ -2 & 0 & 1 \end{bmatrix} \begin{bmatrix} 1 & 0 & 0 \\ 0 & 1 & 0 \\ 0 & -2 & 1 \end{bmatrix} = \begin{bmatrix} 1 & 0 & 0 \\ 3 & 1 & 0 \\ -2 & -2 & 1 \end{bmatrix}$$

is a unit lower triangular matrix L such that

$$LU = \begin{bmatrix} 1 & 0 & 0 \\ 3 & 1 & 0 \\ -2 & -2 & 1 \end{bmatrix} \begin{bmatrix} 0 & 1 & 2 & -1 & 0 \\ 0 & 0 & -4 & 6 & -1 \\ 0 & 0 & 0 & 11 & -2 \end{bmatrix} = \begin{bmatrix} 0 & 1 & 2 & -1 & 0 \\ 0 & 3 & 2 & 3 & -1 \\ 0 & -2 & 4 & 1 & 0 \end{bmatrix} = A.$$
∎

Not all matrices A have an LU decomposition. Consider the next example.

EXAMPLE 4 The matrix

$$A = \begin{bmatrix} 0 & 2 & 1 \\ 2 & -1 & 0 \\ 1 & 0 & -1 \end{bmatrix}$$

cannot be reduced to an upper triangular matrix with only type II row operations. An interchange of two rows (a type III elementary operation) would have to be performed. However, the matrix

$$PA = \begin{bmatrix} 0 & 0 & 1 \\ 0 & 1 & 0 \\ 1 & 0 & 0 \end{bmatrix} \begin{bmatrix} 0 & 2 & 1 \\ 2 & -1 & 0 \\ 1 & 0 & -1 \end{bmatrix} = \begin{bmatrix} 1 & 0 & -1 \\ 2 & -1 & 0 \\ 0 & 2 & 1 \end{bmatrix}$$

can be reduced to an upper triangular matrix and thus factored as a product of a unit lower triangular matrix and an upper triangular matrix. Following the same procedure as in the previous example we have

$$PA = \begin{bmatrix} 1 & 0 & -1 \\ 2 & -1 & 0 \\ 0 & 2 & 1 \end{bmatrix} \rightarrow \begin{bmatrix} 1 & 0 & -1 \\ 0 & -1 & 2 \\ 0 & 2 & 1 \end{bmatrix} \quad \text{with} \quad E_1 = \begin{bmatrix} 1 & 0 & 0 \\ -2 & 1 & 0 \\ 0 & 0 & 1 \end{bmatrix}$$

$$\rightarrow \begin{bmatrix} 1 & 0 & -1 \\ 0 & -1 & 2 \\ 0 & 0 & 5 \end{bmatrix} \quad \text{with} \quad E_2 = \begin{bmatrix} 1 & 0 & 0 \\ 0 & 1 & 0 \\ 0 & 2 & 1 \end{bmatrix}$$

Thus

$$E_2 E_1 (PA) = U$$

and

$$PA = LU$$

where

$$L = E_1^{-1} E_2^{-1} = \begin{bmatrix} 1 & 0 & 0 \\ 2 & 1 & 0 \\ 0 & 0 & 1 \end{bmatrix} \begin{bmatrix} 1 & 0 & 0 \\ 0 & 1 & 0 \\ 0 & -2 & 1 \end{bmatrix} = \begin{bmatrix} 1 & 0 & 0 \\ 2 & 1 & 0 \\ 0 & -2 & 1 \end{bmatrix}. \quad \blacksquare$$

The usefulness of *LU* decompositions will become evident in Section 4.7.

3.8 Exercises

Find the LU decomposition of each of the following square matrices.

1. $\begin{bmatrix} 1 & 3 \\ 4 & -2 \end{bmatrix}$
2. $\begin{bmatrix} 3 & 0 \\ -1 & 2 \end{bmatrix}$
3. $\begin{bmatrix} 1 & -3 & -2 \\ 3 & 2 & 0 \\ -1 & 4 & 2 \end{bmatrix}$

4. $\begin{bmatrix} 2 & 0 & 4 \\ 0 & 4 & -8 \\ 3 & -1 & 2 \end{bmatrix}$
5. $\begin{bmatrix} -1 & 2 & -3 & 0 \\ 0 & -2 & 1 & -1 \\ 2 & 4 & 0 & 3 \\ 0 & 0 & 0 & 5 \end{bmatrix}$
6. $\begin{bmatrix} 2 & 0 & 1 & 1 \\ -8 & 1 & 0 & 3 \\ 6 & 3 & 1 & 10 \\ 4 & -2 & -2 & 0 \end{bmatrix}$

Find the LU decomposition of each of the following nonsquare matrices.

7. $\begin{bmatrix} 1 & -1 & -2 & 0 \\ 2 & 0 & 3 & -2 \\ 0 & 1 & 1 & -1 \end{bmatrix}$
8. $\begin{bmatrix} 0 & -1 & 3 & 0 & 3 \\ 0 & 1 & 1 & 2 & 0 \\ 0 & 2 & -1 & 0 & -4 \end{bmatrix}$

9. $\begin{bmatrix} 1 & 0 & 0 & 4 & 2 \\ 1 & 0 & -2 & 2 & -1 \\ -2 & 0 & 4 & 2 & 0 \\ 0 & 0 & 6 & 0 & 11 \end{bmatrix}$

10. $\begin{bmatrix} 1 & 1 & -1 & 4 & 0 & 2 \\ 0 & 1 & -3 & 2 & 1 & -1 \\ 2 & -2 & 4 & 0 & 1 & 1 \\ 1 & 1 & -2 & 1 & 0 & 1 \end{bmatrix}$

11. $\begin{bmatrix} 1 & 0 & 1 \\ 2 & 2 & 0 \\ 3 & 4 & -2 \\ 1 & -4 & 1 \end{bmatrix}$

12. $\begin{bmatrix} 0 & 1 & 3 & -1 \\ 0 & -1 & 1 & 2 \\ 0 & 0 & 1 & 5 \\ 0 & 2 & 5 & 1 \\ 0 & 1 & -5 & -2 \end{bmatrix}$

For each of the following matrices A, find matrices P, L and U such that PA has an LU decomposition.

13. $\begin{bmatrix} 0 & 2 \\ 2 & -3 \end{bmatrix}$

14. $\begin{bmatrix} 0 & 2 & 1 \\ -1 & -3 & 2 \\ 1 & 1 & -1 \end{bmatrix}$

15. $\begin{bmatrix} 1 & -2 & -3 \\ -1 & 2 & 5 \\ 2 & 0 & 1 \end{bmatrix}$

16. $\begin{bmatrix} 0 & 1 & 3 & 4 \\ 0 & 2 & 1 & -1 \\ -1 & 4 & 1 & 5 \end{bmatrix}$

17. $\begin{bmatrix} 0 & 0 & 1 & 1 \\ 1 & 2 & -1 & 0 \\ 1 & -1 & 3 & 2 \\ 0 & 3 & 1 & -2 \end{bmatrix}$

18. $\begin{bmatrix} 0 & 1 & 4 & -3 & 1 \\ 0 & 0 & 2 & -3 & 1 \\ 1 & 4 & -2 & 5 & 1 \\ 2 & -3 & 0 & 1 & 1 \end{bmatrix}$

Find a factorization of A in the form $A = L'U'$ where L' is a lower triangular matrix and U' is an upper triangular matrix with diagonal elements 1. (That is, U' is a unit upper triangular matrix.) This type of factorization allows type I elementary row operations.

19. $\begin{bmatrix} 3 & 1 \\ -6 & 3 \end{bmatrix}$

20. $\begin{bmatrix} -1 & 0 & 2 \\ 1 & 3 & -1 \\ 0 & -1 & 4 \end{bmatrix}$

21. $\begin{bmatrix} -2 & 4 & 6 \\ 1 & -4 & 2 \\ 1 & 0 & -1 \end{bmatrix}$

22. $\begin{bmatrix} 2 & 0 & 1 & -1 \\ 2 & 1 & -4 & 1 \\ 0 & 3 & -11 & 2 \\ 1 & 1 & -3 & 4 \end{bmatrix}$

23. Let L be unit lower triangular and U upper triangular. If $L = U$ then prove that $L = U = I$.

24. Prove that the product of two unit lower triangular matrices is a unit lower triangular matrix.

25. Prove that a square unit lower triangular matrix L is invertible and L^{-1} is also unit lower triangular.

26. Show that a lower triangular matrix L can be factored as DL_1, where D is a diagonal matrix and L_1 is a unit lower triangular matrix.

27. Let L be a square lower triangular matrix with nonzero diagonal entries. Prove that L is invertible and L^{-1} is also lower triangular.

4 Vector Spaces, Matrices, and Linear Equations

As promised earlier, the preceding results will now be extended to more general situations. This extension is followed by an application of these results to the solution of systems of linear equations.

4.1 Vector Spaces

The theory developed thus far depended basically on the fact that the set of all real numbers forms a field. Our first objective is to extend this theory to other fields and other vector spaces.

4.1 DEFINITION

> Suppose that $\mathcal{F}$ is a set of elements in which a relation of equality and operations of addition and multiplication, denoted by $+$ and $\cdot$, respectively, are defined. Then $\mathcal{F}$ is a **field** with respect to these operations if the conditions below are satisfied for all a, b, c in $\mathcal{F}$:
>
> 1. $a + b$ is in $\mathcal{F}$. (Closure property for addition)
> 2. $(a + b) + c = a + (b + c)$. (Associative property of addition)
> 3. There is an element 0 in $\mathcal{F}$ such that $a + 0 = a$ for every a in $\mathcal{F}$. (Additive identity)
> 4. For each a in $\mathcal{F}$, there is an element $-a$ in $\mathcal{F}$ such that $a + (-a) = 0$. (Additive inverses)
> 5. $a + b = b + a$. (Commutative property of addition)
> 6. $a \cdot b$ is in $\mathcal{F}$. (Closure property for multiplication)
> 7. $(a \cdot b) \cdot c = a \cdot (b \cdot c)$. (Associative property of multiplication)
> 8. There is an element 1 in $\mathcal{F}$ such that $a \cdot 1 = a$ for every a in $\mathcal{F}$. (Multiplicative identity)
> 9. For each $a \neq 0$ in $\mathcal{F}$, there is an element a^{-1} in $\mathcal{F}$ such that $a \cdot (a^{-1}) = 1$. (Multiplicative inverses)
> 10. $a \cdot b = b \cdot a$. (Commutative property of multiplication)
> 11. $(a + b) \cdot c = a \cdot c + b \cdot c$. (Distributive property)

The notation ab will be used interchangeably with $a \cdot b$ to indicate multiplication. Elements of a field will be referred to as **scalars**.

There are many fields other than the real numbers. Two familiar examples are provided by the set of all rational numbers or the set of all complex numbers. Fields that are subsets

141

of the complex numbers are called **number fields**. There are many fields other than the number fields, and some of them may be known to the student. In the material from here on, the results obtained may be interpreted in as much generality as the student's background permits. If that background includes no knowledge of fields other than number fields, $\mathcal{F}$ may be regarded as being a number field throughout the development.

The following definition should be compared with Definition 1.2 and Theorem 1.3 of Chapter 1.

4.2
DEFINITION

Let $\mathcal{F}$ be a field, and suppose that $\mathbf{V}$ is a set of elements such that for each a in $\mathcal{F}$ and each $\mathbf{v}$ in $\mathbf{V}$, there is a product $a\mathbf{v}$ defined. The operation that yields this product is called **scalar multiplication**. Moreover, suppose that an operation of addition is defined in $\mathbf{V}$. Then $\mathbf{V}$ is a **vector space over** $\mathcal{F}$ with respect to these operations if the conditions below are satisfied for any a, b in $\mathcal{F}$ and any $\mathbf{u}$, $\mathbf{v}$, $\mathbf{w}$ in $\mathbf{V}$:

1. $\mathbf{u} + \mathbf{v}$ is in $\mathbf{V}$. (Closure of $\mathbf{V}$ under addition)
2. $(\mathbf{u} + \mathbf{v}) + \mathbf{w} = \mathbf{u} + (\mathbf{v} + \mathbf{w})$. (Associative property of addition in $\mathbf{V}$)
3. There is an element $\mathbf{0}$ in $\mathbf{V}$ such that $\mathbf{v} + \mathbf{0} = \mathbf{v}$ for all $\mathbf{v}$ in $\mathbf{V}$. (Additive identity)
4. For each $\mathbf{v}$ in $\mathbf{V}$, there is an element $-\mathbf{v}$ in $\mathbf{V}$ such that $\mathbf{v} + (-\mathbf{v}) = \mathbf{0}$. (Additive inverses in $\mathbf{V}$)
5. $\mathbf{u} + \mathbf{v} = \mathbf{v} + \mathbf{u}$. (Commutative property of addition in $\mathbf{V}$)
6. $a\mathbf{v}$ is in $\mathbf{V}$. (Closure of $\mathbf{V}$ under scalar multiplication)
7. $a(b\mathbf{v}) = (ab)\mathbf{v}$. (Associative property of scalar multiplication)
8. $a(\mathbf{u} + \mathbf{v}) = a\mathbf{u} + a\mathbf{v}$. (Distributive property, addition in $\mathbf{V}$)
9. $(a + b)\mathbf{v} = a\mathbf{v} + b\mathbf{v}$. (Distributive property, addition in $\mathcal{F}$)
10. $1 \cdot \mathbf{v} = \mathbf{v}$.

The elements of $\mathbf{V}$ are called **vectors** if $\mathbf{V}$ is a vector space over $\mathcal{F}$.

With the exception of Section 1.5, all of the definitions, theorems, and proofs in Chapter 1 through Definition 1.23 apply to arbitrary vector spaces $\mathbf{V}$. Each corresponding statement may be obtained by replacing $\mathbf{R}$ by an arbitrary field $\mathcal{F}$ and $\mathbf{R}^n$ by an arbitrary vector space $\mathbf{V}$ over $\mathcal{F}$. Some restrictions are necessary in the remainder of Chapter 1.

4.3
DEFINITION

Let $\mathbf{V}$ be a vector space over $\mathcal{F}$. Then $\mathbf{V}$ is a **finite-dimensional vector space** (or a **vector space of finite dimension over** $\mathcal{F}$) if there exists a basis of $\mathbf{V}$ with a finite number of elements. If no such basis exists, $\mathbf{V}$ is **infinite-dimensional**, or of **infinite dimension over** $\mathcal{F}$.

The results in Chapter 1 after Definition 1.23 apply only to finite-dimensional vector spaces. If $\mathbf{R}$ is replaced by $\mathcal{F}$ and $\mathbf{R}^n$ is replaced by a finite-dimensional vector space $\mathbf{V}$

over $\mathcal{F}$, Theorems 1.24, 1.25, 1.27, and 1.28 and Definition 1.29 remain valid *with the proofs unchanged*, except for notation. In particular, any two bases of a finite-dimensional vector space have the same number of elements, and the number of elements in a basis is the **dimension** of the vector space. In Theorems 1.32 through 1.35, $\mathbf{R}^n$ is replaced by an n-dimensional vector space $\mathbf{V}$ and $\mathbf{W}$ is a subspace of $\mathbf{V}$.

Let us consider now some examples of vector spaces. In each case, $\mathcal{F}$ denotes a field.

EXAMPLE 1   For a fixed positive integer n, let $\mathcal{F}^n$ denote the set of all n-tuples $(u_1, u_2, \ldots, u_n)$ with u_i in $\mathcal{F}$. Two elements $\mathbf{u} = (u_1, u_2, \ldots, u_n)$ and $\mathbf{v} = (v_1, v_2, \ldots, v_n)$ are **equal** if and only if $u_i = v_i$ for $i = 1, 2, \ldots, n$. With addition defined in $\mathcal{F}^n$ by

$$(u_1, u_2, \ldots, u_n) + (v_1, v_2, \ldots, v_n) = (u_1 + v_1, u_2 + v_2, \ldots, u_n + v_n)$$

and scalar multiplication by

$$a(u_1, u_2, \ldots, u_n) = (au_1, au_2, \ldots, au_n)$$

the same techniques used to prove Theorem 1.3 in Section 1.1 can be used to prove that $\mathcal{F}^n$ is a vector space over $\mathcal{F}$. Denoting the multiplicative identity in $\mathcal{F}$ by 1, the vectors

$$(1, 0, 0, \ldots, 0), (0, 1, 0, \ldots, 0), \ldots, (0, 0, 0, \ldots, 1)$$

can be shown to form a basis of $\mathcal{F}^n$, and therefore $\mathcal{F}^n$ has dimension n over $\mathcal{F}$. For $n = 1$, we can identify (a_1) with a_1. That is, $\mathcal{F}$ is a vector space of dimension one over $\mathcal{F}$.   ■

EXAMPLE 2   For a fixed nonnegative integer n, let $\mathcal{P}_n$ denote the set of all polynomials of the form

$$a_0 + a_1 x + a_2 x^2 + \cdots + a_n x^n$$

with each a_i in $\mathcal{F}$. We shall refer to this set as "$\mathcal{P}_n$ over $\mathcal{F}$." That is, $\mathcal{P}_n$ over $\mathcal{F}$ is the set of all polynomials $\sum_{i=0}^{n} a_i x^i$ in the variable x with coefficients in $\mathcal{F}$ and degree less than or equal to n.[1] Let

$$p(x) = a_0 + a_1 x + \cdots + a_n x^n$$

and

$$q(x) = b_0 + b_1 x + \cdots + b_n x^n$$

denote two elements of $\mathcal{P}_n$. Then $p(x)$ and $q(x)$ are **equal** if and only if $a_i = b_i$ for $i = 0, 1, 2, \ldots, n$. With addition and scalar multiplication defined in the usual ways by

$$p(x) + q(x) = (a_0 + b_0) + (a_1 + b_1)x + \cdots + (a_n + b_n)x^n$$

and

$$cp(x) = ca_0 + ca_1 x + \cdots + ca_n x^n,$$

---

[1] Note that we are using a_0 interchangeably with $a_0 x^0$ in the sigma notation.

it is easy to verify that $\mathcal{P}_n$ is a vector space over $\mathcal{F}$ with the zero polynomial

$$\mathbf{0} = 0 + 0x + 0x^2 + \cdots + 0x^n$$

as its additive identity. With $p(x)$ as given above, its additive inverse is the polynomial

$$-p(x) = (-a_0) + (-a_1)x + \cdots + (-a_n)x^n$$
$$= (-1)p(x).$$

The set $\mathcal{B}$ of $n + 1$ polynomials given by

$$\mathcal{B} = \{1, x, x^2, \ldots, x^n\}$$

spans $\mathcal{P}_n$ since an arbitrary $p(x) = \sum_{i=0}^{n} a_i x^i$ is automatically a linear combination of vectors in $\mathcal{B}$:

$$p(x) = a_0(1) + a_1(x) + a_2(x^2) + \cdots + a_n(x^n).$$

Also, a linear combination of vectors in $\mathcal{B}$ yields the zero polynomial $\mathbf{0}$ if and only if all coefficients are zero. Thus $\mathcal{B}$ is linearly independent and forms a basis of $\mathcal{P}_n$. It follows from this fact that all bases of $\mathcal{P}_n$ have $n + 1$ elements, and $\mathcal{P}_n$ is of dimension $n + 1$. However, the basis $\mathcal{B} = \{1, x, x^2, \ldots, x^n\}$ of $\mathcal{P}_n$ is referred to as the **standard basis** of $\mathcal{P}_n$. ∎

For our next two examples, we draw on topics from the calculus.

EXAMPLE 3 Let $\mathbf{V}$ denote the set of all infinite sequences

$$\{a_n\} = a_1, a_2, \ldots, a_n, \ldots$$

of real numbers a_n. Two sequences $\{a_n\}$ and $\{b_n\}$ in $\mathbf{V}$ are **equal** if and only if $a_n = b_n$ for all positive integers n. This set $\mathbf{V}$ is a vector space over $\mathbf{R}$ with respect to the operation of addition

$$\{a_n\} + \{b_n\} = \{a_n + b_n\}$$

and scalar multiplication

$$c\{a_n\} = \{ca_n\}$$

for c in $\mathbf{R}$. The zero vector in $\mathbf{V}$ is the sequence $\{c_n\}$ with $c_n = 0$ for all n, and the additive inverse of $\{a_n\}$ in $\mathbf{V}$ is

$$-\{a_n\} = \{-a_n\}$$
$$= (-1)\{a_n\}.$$

We are not equipped in this text to prove it, but this vector space $\mathbf{V}$ is of infinite dimension over $\mathbf{R}$. ∎

EXAMPLE 4 Let **V** be the set of all real-valued functions of the real variable t with domain the set **R** of all real numbers. Two functions f and g in **V** are **equal** if and only if $f(t) = g(t)$ for all real numbers t. With respect to the ordinary operations of addition

$$(f + g)(t) = f(t) + g(t)$$

and scalar multiplication

$$(cf)(t) = c(f(t))$$

used in the calculus, the set **V** is a vector space over **R**. The zero vector in **V** is the constant function that is identically zero for all values of the variable t. The additive inverse of f is the function $-f$ given by

$$(-f)(t) = -(f(t))$$
$$= (-1)(f(t)).$$

We are not equipped to prove it here, but this **V** is an infinite-dimensional vector space over **R**. ∎

EXAMPLE 5 Let $\mathcal{F}_{m \times n}$ denote the set of all m by n matrices $A = [a_{ij}]_{m \times n}$ with elements a_{ij} in $\mathcal{F}$. With addition defined by

$$A + B = [a_{ij}]_{m \times n} + [b_{ij}]_{m \times n} = [a_{ij} + b_{ij}]_{m \times n}$$

and scalar multiplication defined by

$$cA = c[a_{ij}]_{m \times n} = [ca_{ij}]_{m \times n},$$

$\mathcal{F}_{m \times n}$ is a finite-dimensional vector space over $\mathcal{F}$. The additive identity in $\mathcal{F}_{m \times n}$ is the **zero matrix**

$$O_{m \times n} = \begin{bmatrix} 0 & 0 & \cdots & 0 \\ 0 & 0 & \cdots & 0 \\ \vdots & \vdots & & \vdots \\ 0 & 0 & \cdots & 0 \end{bmatrix} = [0]_{m \times n},$$

and the additive inverse of $A = [a_{ij}]_{m \times n}$ is the matrix

$$-A = [-a_{ij}]_{m \times n}$$
$$= (-1)A.$$

In many instances, we shall also use the zero vector symbol **0** to indicate a zero matrix. ∎

EXAMPLE 6 As illustrations of the operations defined in Example 5, we have

$$\begin{bmatrix} 1 & -2 & 0 \\ 7 & 4 & -3 \end{bmatrix} + \begin{bmatrix} 6 & 5 & -9 \\ -5 & -1 & 8 \end{bmatrix} = \begin{bmatrix} 7 & 3 & -9 \\ 2 & 3 & 5 \end{bmatrix}$$

and

$$3 \begin{bmatrix} 5 & -4 & 7 \\ 1 & 0 & 2 \end{bmatrix} = \begin{bmatrix} 15 & -12 & 21 \\ 3 & 0 & 6 \end{bmatrix}$$

in the vector space $\mathbf{R}_{2\times3}$ of all 2×3 matrices over $\mathbf{R}$. The six 2×3 matrices given by

$$M_{11} = \begin{bmatrix} 1 & 0 & 0 \\ 0 & 0 & 0 \end{bmatrix}, \quad M_{12} = \begin{bmatrix} 0 & 1 & 0 \\ 0 & 0 & 0 \end{bmatrix}, \quad M_{13} = \begin{bmatrix} 0 & 0 & 1 \\ 0 & 0 & 0 \end{bmatrix}$$

$$M_{21} = \begin{bmatrix} 0 & 0 & 0 \\ 1 & 0 & 0 \end{bmatrix}, \quad M_{22} = \begin{bmatrix} 0 & 0 & 0 \\ 0 & 1 & 0 \end{bmatrix}, \quad M_{23} = \begin{bmatrix} 0 & 0 & 0 \\ 0 & 0 & 1 \end{bmatrix}$$

spans $\mathbf{R}_{2\times3}$ since an arbitrary matrix

$$A = \begin{bmatrix} a_{11} & a_{12} & a_{13} \\ a_{21} & a_{22} & a_{23} \end{bmatrix}$$

can be expressed as a linear combination of vectors in $\mathcal{E}_{2\times3} = \{M_{11}, M_{12}, M_{13}, M_{21}, M_{22}, M_{23}\}$:

$$A = a_{11}M_{11} + a_{12}M_{12} + a_{13}M_{13} + a_{21}M_{21} + a_{22}M_{22} + a_{23}M_{23}.$$

Clearly the set $\mathcal{E}_{2\times3}$ is linearly independent and hence is a basis of $\mathbf{R}_{2\times3}$ and we see that $\mathbf{R}_{2\times3}$ is of dimension 6. This particular basis is called the **standard basis** of $\mathbf{R}_{2\times3}$.   ■

In connection with the previous example, the standard basis $\mathcal{E}_{m\times n}$ of $\mathbf{R}_{m\times n}$ consists of mn matrices $M_{ij}, i = 1, \ldots, m, j = 1, \ldots, n$. Each matrix M_{ij} contains 0's in all entries except one. The only nonzero element is a 1 located in row i column j.

Our last example in this section demonstrates that the operations used in a set to form a vector space are not unique and do not have to be defined in a certain way. Some operations that can be used as addition and scalar multiplication may look somewhat strange when first encountered.

EXAMPLE 7 Let $\mathbf{V}$ be the set of all ordered pairs of real numbers with the usual equality:

$$(x_1, x_2) = (y_1, y_2) \quad \text{if and only if} \quad x_1 = y_1 \text{ and } x_2 = y_2.$$

Addition and scalar multiplication are defined in $\mathbf{V}$ as follows:

$$(x_1, x_2) + (y_1, y_2) = (x_1 + y_1 + 1, x_2 + y_2 + 1),$$

$$c(x_1, x_2) = (c + cx_1 - 1, c + cx_2 - 1).$$

We shall systematically check the ten conditions required by Definition 4.2 in order that $\mathbf{V}$ be a vector space over $\mathbf{R}$. To this end, let $\mathbf{u} = (u_1, u_2)$, $\mathbf{v} = (v_1, v_2)$, and $\mathbf{w} = (w_1, w_2)$ be arbitrary elements in $\mathbf{V}$, and let a and b represent arbitrary real numbers.

1. The sum

$$\mathbf{u} + \mathbf{v} = (u_1 + v_1 + 1, u_2 + v_2 + 1)$$

is in **V** since both $u_1 + v_1 + 1$ and $u_2 + v_2 + 1$ are real numbers.

2. We have

$$(\mathbf{u} + \mathbf{v}) + \mathbf{w} = (u_1 + v_1 + 1, u_2 + v_2 + 1) + (w_1, w_2)$$
$$= ((u_1 + v_1 + 1) + w_1 + 1, (u_2 + v_2 + 1) + w_2 + 1)$$
$$= (u_1 + v_1 + w_1 + 2, u_2 + v_2 + w_2 + 2)$$

and

$$\mathbf{u} + (\mathbf{v} + \mathbf{w}) = (u_1, u_2) + (v_1 + w_1 + 1, v_2 + w_2 + 1)$$
$$= (u_1 + (v_1 + w_1 + 1) + 1, u_2 + (v_2 + w_2 + 1) + 1)$$
$$= (u_1 + v_1 + w_1 + 2, u_2 + v_2 + w_2 + 2).$$

Thus addition in **V** is associative.

3. The element $(-1, -1)$ in **V** is an additive identity since

$$(v_1, v_2) + (-1, -1) = (v_1 - 1 + 1, v_2 - 1 + 1)$$
$$= (v_1, v_2)$$

for all $\mathbf{v} = (v_1, v_2)$ in **V**. Thus we write $\mathbf{0} = (-1, -1)$.

4. The additive inverse of $\mathbf{v} = (v_1, v_2)$ is $(-v_1 - 2, -v_2 - 2)$ in **V** since

$$(v_1, v_2) + (-v_1 - 2, -v_2 - 2) = (v_1 - v_1 - 2 + 1, v_2 - v_2 - 2 + 1)$$
$$= (-1, -1)$$
$$= \mathbf{0}.$$

5. Addition in **V** is commutative since

$$\mathbf{u} + \mathbf{v} = (u_1 + v_1 + 1, u_2 + v_2 + 1)$$
$$= (v_1 + u_1 + 1, v_2 + u_2 + 1)$$
$$= \mathbf{v} + \mathbf{u}.$$

6. The product

$$a(v_1, v_2) = (a + av_1 - 1, a + av_2 - 1)$$

is always in **V** since both $a + av_1 - 1$ and $a + av_2 - 1$ are real numbers.

7. We have

$$a(b\mathbf{v}) = a(b + bv_1 - 1, b + bv_2 - 1)$$
$$= (a + a(b + bv_1 - 1) - 1, a + a(b + bv_2 - 1) - 1)$$
$$= (a + ab + a(bv_1) - a - 1, a + ab + a(bv_2) - a - 1)$$
$$= ((ab) + (ab)v_1 - 1, ab + (ab)v_2 - 1)$$
$$= (ab)\mathbf{v},$$

so scalar multiplication has the associative property.

8. Verifying the distributive property with addition in **V**, we find that

$$a(\mathbf{u} + \mathbf{v}) = a(u_1 + v_1 + 1, u_2 + v_2 + 1)$$
$$= (a + au_1 + av_1 + a - 1, a + au_2 + av_2 + a - 1)$$
$$= (a + au_1 - 1, a + au_2 - 1) + (a + av_1 - 1, a + av_2 - 1)$$
$$= a\mathbf{u} + a\mathbf{v}.$$

9. The distributive property with addition in **R** is valid since

$$(a + b)\mathbf{v} = ((a + b) + (a + b)v_1 - 1, (a + b) + (a + b)v_2 - 1)$$
$$= (a + b + av_1 + bv_1 - 1, a + b + av_2 + bv_2 - 1)$$
$$= (a + av_1 - 1, a + av_2 - 1) + (b + bv_1 - 1, b + bv_2 - 1)$$
$$= a\mathbf{v} + b\mathbf{v}.$$

10. We have

$$1 \cdot \mathbf{v} = (1 + 1(v_1) - 1, 1 + 1(v_2) - 1)$$
$$= (v_1, v_2).$$

We have verified all the conditions in Definition 4.2, so this set **V** is a vector space over **R** with respect to these operations, even though these operations are dramatically different from the standard operations in the familiar vector space $\mathbf{R}^2$. ■

4.1 Exercises

1. Verify that the set in the indicated example from this section is actually a vector space.
   (a) Example 1   **(b)** Example 2   **(c)** Example 3   **(d)** Example 4
   (e) Example 5

2. Write out a basis for the vector space $\mathbf{R}_{3 \times 2}$.

3. What is the dimension of $\mathcal{F}_{m \times n}$?

4. Find a set that forms a basis for the vector space **V** in Example 7 of this section, and prove that your set is a basis for **V**.

In Problems 5–22, assume that equality in the given set is the same as in the example of this section that involves the same elements, and determine if the given set is a vector space over **R** *with respect to the operations defined in the problem. If it is not, list all conditions in Definition 4.2 that fail to hold.*

5. The set **V** of all ordered pairs of real numbers with operations defined by

$$(x_1, x_2) + (y_1, y_2) = (x_1 + y_1, 0),$$
$$c(x_1, x_2) = (cx_1, cx_2).$$

6. The set **V** of all ordered pairs of real numbers with operations defined by

$$(x_1, x_2) + (y_1, y_2) = (x_1 + y_1, x_2 + y_2),$$
$$c(x_1, x_2) = (cx_1, x_2).$$

7. The set **V** of all ordered pairs of *positive* real numbers with operations defined by

$$(x_1, x_2) + (y_1, y_2) = (x_1 y_1, x_2 y_2),$$
$$c(x_1, x_2) = (x_1^c, x_2^c).$$

8. The set **V** of all ordered pairs of real numbers with operations defined by

$$(x_1, x_2) + (y_1, y_2) = (x_1 y_1, x_2 y_2),$$
$$c(x_1, x_2) = (cx_1, cx_2).$$

9. The set **V** of all ordered pairs of real numbers with operations defined by

$$(x_1, x_2) + (y_1, y_2) = (x_1 + y_1 + 1, x_2 + y_2 + 1),$$
$$c(x_1, x_2) = (cx_1, cx_2).$$

10. The set **V** of all ordered pairs of real numbers with operations defined by

$$(x_1, x_2) + (y_1, y_2) = (x_1 + y_1, x_2 + y_2),$$
$$c(x_1, x_2) = (x_1, x_2).$$

11. The set **V** of all ordered triples of real numbers with operations defined by

$$(x_1, x_2, x_3) + (y_1, y_2, y_3) = (x_1 + y_1, x_2 + y_2, x_3 + y_3),$$
$$c(x_1, x_2, x_3) = (cx_1, 0, 0).$$

12. The set **V** of all ordered triples of real numbers with operations defined by

$$(x_1, x_2, x_3) + (y_1, y_2, y_3) = (x_1 + y_1, x_2 + y_2, x_3 + y_3),$$
$$c(x_1, x_2, x_3) = (x_1^c, x_2^c, x_3^c).$$

13. The set **V** of all ordered triples of real numbers with operations defined by

$$(x_1, x_2, x_3) + (y_1, y_2, y_3) = (x_1 + y_1, x_2 + y_2, x_3 + y_3),$$
$$c(x_1, x_2, x_3) = (0, 0, 0).$$

14. The set **V** of all ordered triples of real numbers with operations defined by

$$(x_1, x_2, x_3) + (y_1, y_2, y_3) = (x_1 + y_1, x_2 + y_2, x_3 + y_3),$$
$$c(x_1, x_2, x_3) = (3cx_1, 3cx_2, 3cx_3).$$

15. The set **V** of all ordered triples of the form $(1, x_1, x_2)$ where $x_1, x_2 \in \mathbf{R}$, with operations defined by

$$(1, x_1, x_2) + (1, y_1, y_2) = (1, x_1 + y_1, x_2 + y_2),$$
$$c(1, x_1, x_2) = (1, cx_1, cx_2).$$

16. The set **V** of all polynomials of degree 3 or less over **R** with operations defined by

$$(a_0 + a_1 x + a_2 x^2 + a_3 x^3) + (b_0 + b_1 x + b_2 x^2 + b_3 x^3) = a_0 + b_0,$$
$$c(a_0 + a_1 x + a_2 x^2 + a_3 x^3) = ca_0.$$

17. The set **V** of all polynomials of degree 3 or less over **R** with operations defined by

$$(a_0 + \cdots + a_3 x^3) + (b_0 + \cdots + b_3 x^3) = (a_0 + b_0) + \cdots + (a_3 + b_3)x^3,$$
$$c(a_0 + a_1 x + a_2 x^2 + a_3 x^3) = a_0^c + a_1^c x + a_2^c x^2 + a_3^c x^3.$$

18. The set **V** of all 2×2 matrices over **R** with operations defined by

$$\begin{bmatrix} a_{11} & a_{12} \\ a_{21} & a_{22} \end{bmatrix} + \begin{bmatrix} b_{11} & b_{12} \\ b_{21} & b_{22} \end{bmatrix} = \begin{bmatrix} 0 & a_{12} + b_{12} \\ a_{21} + b_{21} & 0 \end{bmatrix},$$
$$c\begin{bmatrix} a_{11} & a_{12} \\ a_{21} & a_{22} \end{bmatrix} = \begin{bmatrix} ca_{11} & ca_{12} \\ ca_{21} & ca_{22} \end{bmatrix}.$$

19. The set **V** of all 2×2 matrices over **R** with operations defined by

$$\begin{bmatrix} a_{11} & a_{12} \\ a_{21} & a_{22} \end{bmatrix} + \begin{bmatrix} b_{11} & b_{12} \\ b_{21} & b_{22} \end{bmatrix} = \begin{bmatrix} a_{11} + b_{11} + 1 & a_{12} + b_{12} \\ a_{21} + b_{21} & a_{22} + b_{22} - 1 \end{bmatrix},$$
$$c\begin{bmatrix} a_{11} & a_{12} \\ a_{21} & a_{22} \end{bmatrix} = \begin{bmatrix} ca_{11} & ca_{12} \\ ca_{21} & ca_{22} \end{bmatrix}.$$

20. The set **V** of all real-valued functions of the real variable t with domain the set **R** with operations defined by

$$(f + g)(t) = 2f(t) + g(t),$$
$$(cf)(t) = c(f(t)).$$

21. The set $\mathbf{V}$ of invertible $n \times n$ matrices over $\mathbf{R}$ with operations defined by

$$A + B = AB,$$

$$cA = [ca_{ij}], \text{ where } A = [a_{ij}].$$

22. The set $\mathbf{V}$ of $m \times n$ matrices over $\mathbf{R}$ with operations defined by

$$A + B = [a_{ij} + b_{ij}], \text{ where } A = [a_{ij}] \text{ and } B = [b_{ij}],$$

$$cA = \mathbf{0}_{m \times n}.$$

23. Prove that the additive identity in a vector space is unique.

24. Prove that the additive inverse $-\mathbf{v}$ of an element $\mathbf{v}$ in a vector space is unique.

25. Let $\mathbf{V}$ be a vector space over $\mathcal{F}$.

(a) Prove that $0\mathbf{v} = \mathbf{0}$ for an arbitrary vector $\mathbf{v} \in \mathbf{V}$.

(b) If $\mathbf{0}$ denotes the additive identity in $\mathbf{V}$, prove that $c\mathbf{0} = \mathbf{0}$ for all c in $\mathcal{F}$.

26. Suppose $\mathbf{V}$ is a vector space over $\mathcal{F}$. Prove that if $c\mathbf{v} = \mathbf{0}$ for c in $\mathcal{F}$ and $\mathbf{v}$ in $\mathbf{V}$, then either $c = 0$ or $\mathbf{v} = \mathbf{0}$.

27. Prove that $-\mathbf{v} = (-1)\mathbf{v}$ for an arbitrary vector $\mathbf{v}$ in a vector space.

28. The vector $\mathbf{u} - \mathbf{v}$ is defined as the vector $\mathbf{w}$ that satisfies the equation $\mathbf{v} + \mathbf{w} = \mathbf{u}$. Prove that $\mathbf{u} - \mathbf{v} = \mathbf{u} + (-1)\mathbf{v}$.

29. Prove that $-(-\mathbf{v}) = \mathbf{v}$ for an arbitrary vector $\mathbf{v}$ in a vector space.

30. Let $\mathbf{u}, \mathbf{v}$ and $\mathbf{w}$ be arbitrary vectors in a vector space $\mathbf{V}$. Prove that if $\mathbf{u} + \mathbf{v} = \mathbf{u} + \mathbf{w}$, then $\mathbf{v} = \mathbf{w}$.

31. Let A and B be elements in the vector space $\mathcal{F}_{m \times n}$. Prove each of the following statements.

(a) $(A + B)^T = A^T + B^T$

(b) $(kA)^T = kA^T$

(c) $A + A^T$ is symmetric, for all matrices $A \in \mathcal{F}_{n \times n}$

(d) $A - A^T$ is skew-symmetric, for all matrices $A \in \mathcal{F}_{n \times n}$

32. As mentioned just after Definition 4.1, the set $\mathcal{C}$ of all complex numbers is a field. Assuming this fact, prove that each of the sets below is a field.

(a) The set of all complex numbers c that can be written in the form $c = a + b\sqrt{2}$ with a and b rational numbers.

(b) The set of all complex numbers c that can be written as $c = a + bi$ with a and b rational numbers. (The symbol i denotes the complex number such that $i^2 = -1$.)

33. Let $\mathbf{R}_{2 \times 2}$ be as defined in Example 5, and let $\mathcal{F}$ be the subset of $\mathbf{R}_{2 \times 2}$ that consists of all invertible matrices in $\mathbf{R}_{2 \times 2}$, together with the zero matrix of order 2. With multiplication as usual and addition as given in Example 5, determine whether or not $\mathcal{F}$ is a field, and justify your answer.

4.2 Subspaces and Related Concepts

Some of the discussion in the last section outlined the way that the theory in Chapter 1 extends to general vector spaces. In order to add depth and meaning to this outline, we shall illustrate here the concepts of subspace, spanning set, basis, and transition matrix in vector spaces other than $\mathbf{R}^n$. We begin with the concept of a subspace.

If $\mathbf{W}$ is a subset of the vector space $\mathbf{V}$ over $\mathcal{F}$, then $\mathbf{W}$ may also be a vector space over $\mathcal{F}$. The terminology used in $\mathbf{R}^n$ generalizes to arbitrary vector spaces, and such a subset is called a **subspace** of $\mathbf{V}$.

Theorem 1.10 also extends to an arbitrary vector space $\mathbf{V}$ over $\mathcal{F}$. That is, a subset $\mathbf{W}$ of the vector space $\mathbf{V}$ is a subspace of $\mathbf{V}$ if and only if the following conditions hold:

(i) $\mathbf{W}$ is nonempty;

(ii) for any $a, b \in \mathcal{F}$ and any $\mathbf{u}, \mathbf{v} \in \mathbf{W}$, $a\mathbf{u} + b\mathbf{v} \in \mathbf{W}$.

EXAMPLE 1 Let $\mathbf{W}$ be the set of all symmetric matrices in the vector space $\mathbf{R}_{n \times n}$ of all square matrices of order n over $\mathbf{R}$. That is, for an $n \times n$ matrix A over $\mathbf{R}$,

$$A \in \mathbf{W} \quad \text{if and only if} \quad A^T = A.$$

We shall show that $\mathbf{W}$ is a subspace of $\mathbf{R}_{n \times n}$. The set $\mathbf{W}$ is nonempty since the zero matrix $\mathbf{0}_{n \times n} = [0]_{n \times n}$ is in $\mathbf{W}$. Let r and s be arbitrary real numbers, and let $A = [a_{ij}]_{n \times n}$ and $B = [b_{ij}]_{n \times n}$ be arbitrary elements in $\mathbf{W}$. Then A and B are symmetric, so that $a_{ji} = a_{ij}$ and $b_{ji} = b_{ij}$ for $i = 1, 2, \ldots, n$; $j = 1, 2, \ldots, n$. Evaluating $rA + sB$, we have

$$
\begin{aligned}
rA + sB &= [ra_{ij}]_{n \times n} + [sb_{ij}]_{n \times n} \\
&= [c_{ij}]_{n \times n} \\
&= C,
\end{aligned}
$$

where $c_{ij} = ra_{ij} + sb_{ij}$ for $i = 1, 2, \ldots, n$; $j = 1, 2, \ldots, n$. Thus

$$
\begin{aligned}
c_{ji} &= ra_{ji} + sb_{ji} \\
&= ra_{ij} + sb_{ij} \quad \text{since } a_{ji} = a_{ij} \text{ and } b_{ji} = b_{ij} \\
&= c_{ij}
\end{aligned}
$$

for all pairs i, j. This means that $rA + sB$ is symmetric and hence a member of $\mathbf{W}$. Therefore $\mathbf{W}$ is a subspace of $\mathbf{R}_{n \times n}$. ∎

For a nonempty subset $\mathcal{A}$ of a general vector space $\mathbf{V}$, the set $\langle \mathcal{A} \rangle$ is the set of all linear combinations of vectors in $\mathcal{A}$. The same development used in Section 1.4 applies in $\mathbf{V}$, and $\langle \mathcal{A} \rangle$ is the **subspace spanned** by $\mathcal{A}$. As in Section 1.4, $\langle \varnothing \rangle$ is the zero subspace $\{\mathbf{0}\}$ of $\mathbf{V}$.

EXAMPLE 2 In the vector space $\mathbf{R}_{2\times2}$, consider the sets of vectors $\mathcal{A} = \{A_1, A_2\}$ and $\mathcal{B} = \{B_1, B_2\}$, where

$$A_1 = \begin{bmatrix} 2 & -1 \\ 0 & 1 \end{bmatrix}, \quad A_2 = \begin{bmatrix} 1 & 2 \\ 1 & 0 \end{bmatrix}, \quad B_1 = \begin{bmatrix} 3 & -4 \\ -1 & 2 \end{bmatrix}, \quad B_2 = \begin{bmatrix} 4 & 3 \\ 2 & 1 \end{bmatrix}.$$

We shall show that $\langle \mathcal{A} \rangle = \langle \mathcal{B} \rangle$.

By inspection, we see that A_2 is not a multiple of A_1. Hence $\mathcal{A} = \{A_1, A_2\}$ is linearly independent and $\langle \mathcal{A} \rangle$ has dimension 2. Similarly, B_2 is not a multiple of B_1 and therefore $\langle \mathcal{B} \rangle$ has dimension 2. Knowing that $\langle \mathcal{A} \rangle$ and $\langle \mathcal{B} \rangle$ have the same dimension, it is sufficient to show that $\langle \mathcal{B} \rangle \subseteq \langle \mathcal{A} \rangle$, in order to prove that $\langle \mathcal{B} \rangle = \langle \mathcal{A} \rangle$. But $\langle \mathcal{B} \rangle$ is dependent on $\mathcal{B}$, so we need only show that $\mathcal{B} \subseteq \langle \mathcal{A} \rangle$, by Theorem 1.6. That is, we need only demonstrate that each of B_1 and B_2 is a linear combination of A_1 and A_2.

Setting up the equation $c_1 A_1 + c_2 A_2 = B_1$, we have

$$c_1 \begin{bmatrix} 2 & -1 \\ 0 & 1 \end{bmatrix} + c_2 \begin{bmatrix} 1 & 2 \\ 1 & 0 \end{bmatrix} = \begin{bmatrix} 3 & -4 \\ -1 & 2 \end{bmatrix}.$$

This leads to the system of equations

$$\begin{aligned} 2c_1 + c_2 &= 3 \\ -c_1 + 2c_2 &= -4 \\ c_2 &= -1 \\ c_1 &= 2, \end{aligned}$$

so $B_1 = 2A_1 - A_2$. Similarly, we find that $B_2 = A_1 + 2A_2$. Thus we have proved that $\langle \mathcal{A} \rangle = \langle \mathcal{B} \rangle$. ■

Just after Example 4 in Section 1.6, we noted that a basis of a subspace can be refined from a finite spanning set for the subspace by deleting all vectors in the spanning set that are linear combinations of preceding vectors. This procedure is illustrated in the following example.

EXAMPLE 3 Let

$$p_1(x) = 1 + x + 2x^2, \quad p_2(x) = 6 + 6x + 12x^2, \quad p_3(x) = 1 + x^2 + x^3,$$
$$p_4(x) = 2 + x + 3x^2 + x^3, \quad p_5(x) = 1 + x^2$$

in the vector space $\mathcal{P}_3$ over $\mathbf{R}$, and let

$$\mathcal{A} = \{p_1(x), p_2(x), p_3(x), p_4(x), p_5(x)\}.$$

We shall find a subset of $\mathcal{A}$ that forms a basis of $\langle \mathcal{A} \rangle$.

Employing the procedure described just before this example, we see that $p_2(x) = 6p_1(x)$, so $p_2(x)$ can be deleted from $\mathcal{A}$ to obtain the spanning set

$$\{p_1(x), p_3(x), p_4(x), p_5(x)\}$$

of $\langle \mathcal{A} \rangle$. We see that $p_3(x)$ is not a multiple of $p_1(x)$, so we then check to see if $p_4(x)$ is a linear combination of $p_1(x)$ and $p_3(x)$. It is easy to discover that

$$p_4(x) = p_1(x) + p_3(x),$$

so $p_4(x)$ can be deleted from the last spanning set of $\langle \mathcal{A} \rangle$ we obtained, leaving

$$\{p_1(x), p_3(x), p_5(x)\}$$

as a spanning set of $\langle \mathcal{A} \rangle$. Setting up the equation

$$p_5(x) = c_1 p_1(x) + c_2 p_3(x)$$

leads to

$$1 + x^2 = c_1(1 + x + 2x^2) + c_2(1 + x^2 + x^3)$$
$$= (c_1 + c_2) + c_1 x + (2c_1 + c_2)x^2 + c_2 x^3$$

and the resulting system of equations

$$\begin{aligned} c_1 + c_2 &= 1 \\ c_1 \quad\ &= 0 \\ 2c_1 + c_2 &= 1 \\ c_2 &= 0, \end{aligned}$$

which clearly has no solution. Thus $\{p_1(x), p_3(x), p_5(x)\}$ is linearly independent and therefore forms a basis of $\langle \mathcal{A} \rangle$ that is contained in $\mathcal{A}$. ∎

The concept of a transition matrix from one finite set of vectors to another in an arbitrary vector space applies in every vector space. We consider an example now in the vector space $\mathcal{P}_2$ of all polynomials in x with degree ≤ 2 and coefficients in $\mathbf{R}$.

EXAMPLE 4 Consider the bases $\mathcal{A} = \{p_1(x), p_2(x), p_3(x)\}$ and $\mathcal{B} = \{q_1(x), q_2(x), q_3(x)\}$ of $\mathcal{P}_2$ over $\mathbf{R}$, where

$$\begin{aligned} p_1(x) &= 1 + x, & p_2(x) &= x, & p_3(x) &= x + x^2, \\ q_1(x) &= 1, & q_2(x) &= 1 + x, & q_3(x) &= 1 + x + x^2. \end{aligned}$$

We shall find the matrix of transition A from $\mathcal{A}$ to $\mathcal{B}$. In the matrix $A = [a_{ij}]_{3\times 3}$, the jth column is the coordinate matrix of $q_j(x)$ with respect to $\mathcal{A}$. Thus we must write each $q_j(x)$ as a linear combination of the polynomials in $\mathcal{A}$. The required linear combinations are given by

$$1 = (1)(1 + x) + (-1)(x) + (0)(x + x^2),$$
$$1 + x = (1)(1 + x) + (0)(x) + (0)(x + x^2),$$
$$1 + x + x^2 = (1)(1 + x) + (-1)(x) + (1)(x + x^2).$$

That is,

$$q_1(x) = (1)p_1(x) + (-1)p_2(x) + (0)p_3(x),$$

$$q_2(x) = (1)p_1(x) + (0)p_2(x) + (0)p_3(x),$$

$$q_3(x) = (1)p_1(x) + (-1)p_2(x) + (1)p_3(x).$$

Thus the transition matrix from $\mathcal{A}$ to $\mathcal{B}$ is

$$A = \begin{bmatrix} 1 & 1 & 1 \\ -1 & 0 & -1 \\ 0 & 0 & 1 \end{bmatrix}.$$

We note that the coordinates of the $q_j(x)$ are entered as *columns* of A, not as *rows* of A.

∎

4.2 Exercises

In Problems 1–30, determine which of the given sets **W** *are subspaces of the indicated vector space. If a set is not a subspace, state a reason.*

1. The set **W** of all vectors in $\mathbf{R}^3$ of the form (a, b, c), where $a + b = 1$.

2. The set **W** of all vectors in $\mathbf{R}^3$ of the form (a, b, c), where $b = a^2$.

3. The set **W** of all vectors in $\mathbf{R}^3$ of the form (a, b, c), where $c = 2a + b$.

4. The set **W** of all vectors in $\mathbf{R}^3$ of the form (a, b, c), where $c = ab$.

5. The set **W** of all polynomials $a_0 + a_1 x + a_2 x^2$ in $\mathcal{P}_2$ over **R** that have $a_1 = 0$.

6. The set **W** of all polynomials $a_0 + a_1 x + a_2 x^2$ in $\mathcal{P}_2$ over **R** that have $a_0 + a_1 = 0$.

7. The set **W** of all polynomials $p(x)$ of degree n over **R** that have zero constant term.

8. The set **W** of all matrices

$$\begin{bmatrix} a & b \\ c & d \end{bmatrix}$$

in $\mathbf{R}_{2\times 2}$ that have $a + b = 0$.

9. The set **W** of all matrices of the form

$$\begin{bmatrix} 1 & a \\ b & c \end{bmatrix}$$

in $\mathbf{R}_{2\times 2}$.

10. The set **W** of all matrices in $\mathbf{R}_{2\times 2}$ that have the form

$$\begin{bmatrix} a & b \\ c & d \end{bmatrix}$$

with a, b, c, and d integers.

11. The set **W** that consists of the zero matrix

$$\begin{bmatrix} 0 & 0 \\ 0 & 0 \end{bmatrix}$$

together with all invertible matrices in $\mathbf{R}_{2\times 2}$.

12. The set **W** of all matrices of the form

$$\begin{bmatrix} 0 & a \\ b & 0 \end{bmatrix}$$

in $\mathbf{R}_{2\times 2}$.

13. The set **W** of all matrices in $\mathbf{R}_{2\times 2}$ that have the form

$$\begin{bmatrix} a & b \\ c & d \end{bmatrix}$$

where $a^2 = d^2$.

14. The set **W** of all matrices of the form

$$\begin{bmatrix} a & a+b \\ a+b & b \end{bmatrix}$$

in $\mathbf{R}_{2\times 2}$.

15. The set **W** of all matrices of the form

$$\begin{bmatrix} a & b \\ c & d \end{bmatrix}$$

in $\mathbf{R}_{2\times 2}$ where $ad - bc = 0$.

16. The set **W** of all upper triangular matrices in $\mathbf{R}_{3\times 3}$.

17. The set **W** of all unit lower triangular matrices in $\mathbf{R}_{3\times 3}$.

18. The set **W** of all diagonal matrices in $\mathbf{R}_{n\times n}$.

19. The set **W** of all skew-symmetric matrices in $\mathbf{R}_{n\times n}$.

20. The set $\mathbf{W} = \{A \in \mathbf{R}_{2\times 2} \mid AB = BA\}$ where B is a fixed matrix in $\mathbf{R}_{2\times 2}$.

21. The set $\mathbf{W} = \{A \in \mathbf{R}_{2\times 2} \mid BA = \mathbf{0}\}$ where B is a fixed matrix in $\mathbf{R}_{2\times 2}$.

22. The set **W** of all square matrices over **R** of order n whose trace is 0.

23. The set **W** of all ordered triples that lie on the plane $ax + by + cz = 0$ through the origin.

24. The set **W** of all ordered triples that lie on the plane $ax + by + cz = d$ where $d \neq 0$ (that is, a plane that does not go through the origin).

25. The set **W** of all real-valued functions f with domain **R** such that $f(x) = f(-x)$.

26. The set **W** of all real-valued functions f with domain **R** such that f is differentiable.

27. The set **W** of all real-valued functions f with domain **R** such that $f(0) = 0$.

28. The set **W** of all real-valued functions f with domain **R** such that $f(0) = 1$.

29. The set **W** of all real-valued functions f with domain **R** that are continuous on the interval $[0, 1]$.

30. The set **W** of all real-valued functions f with domain **R** that are continuous on $[0, 1]$ and $f(0) = f(1)$.

31. Determine whether the given set of vectors is linearly dependent or linearly independent in the vector space $\mathcal{P}_2$ over **R**.

(a) $\{1 - 2x, x + x^2, 1 - x + x^2\}$ (b) $\{1 + x, x + x^2, 1 + 2x + x^2\}$

(c) $\{1 - x^2, x + 2x^2, 1 + 3x^2\}$ (d) $\{1 + x, 1 + x^2, 1 + x + x^2\}$

(e) $\{1 - 2x, 2 - x^2, 4 + x^2, 1 + x + x^2\}$ (f) $\{1 + x, 1 - x\}$

32. Determine whether the given set of vectors in $\mathbf{R}_{2 \times 2}$ is linearly dependent or linearly independent.

(a) $\left\{ \begin{bmatrix} 1 & 1 \\ 0 & 0 \end{bmatrix}, \begin{bmatrix} 0 & 1 \\ 1 & 0 \end{bmatrix}, \begin{bmatrix} 1 & 0 \\ 1 & 0 \end{bmatrix}, \begin{bmatrix} 1 & -1 \\ 1 & 0 \end{bmatrix} \right\}$

(b) $\left\{ \begin{bmatrix} 1 & 0 \\ 0 & 0 \end{bmatrix}, \begin{bmatrix} 0 & 1 \\ 0 & 0 \end{bmatrix}, \begin{bmatrix} 0 & 0 \\ 1 & 0 \end{bmatrix}, \begin{bmatrix} 1 & 1 \\ 1 & 0 \end{bmatrix} \right\}$

(c) $\left\{ \begin{bmatrix} 1 & 1 \\ 0 & 0 \end{bmatrix}, \begin{bmatrix} 1 & 1 \\ 1 & 0 \end{bmatrix}, \begin{bmatrix} 0 & 0 \\ 1 & 0 \end{bmatrix} \right\}$

(d) $\left\{ \begin{bmatrix} 1 & 1 \\ 0 & 0 \end{bmatrix}, \begin{bmatrix} 0 & 1 \\ 1 & 0 \end{bmatrix}, \begin{bmatrix} 0 & 0 \\ 1 & 0 \end{bmatrix} \right\}$

33. Determine whether the given set of polynomials spans $\mathcal{P}_2$ over **R**.

(a) $\{1 - x, 1 + x, 1 - x^2\}$ (b) $\{1 - x, x + x^2, 1 + x^2\}$

(c) $\{1 + x, x + x^2, 1 + 2x + x^2, 1 - x^2\}$ (d) $\{1 - x, x + x^2, 1 + x + x^2, 1 + x^2\}$

34. Which of the following sets of vectors are bases for the vector space $\mathcal{P}_2$ over **R**?

(a) $\{1 - x, 1 - x + x^2, 1 + x^2, 1 - x - x^2\}$

(b) $\{1 - x, 1 - x^2, x - x^2\}$

(c) $\{1 - x, x - x^2, 1 + x^2\}$ (d) $\{1 + x, 1 + x^2\}$

35. Find a subset of the given set of vectors that forms a basis for the subspace spanned by the given set.

(a) $p_1(x) = x + x^2$, $p_2(x) = 1 + x + x^2$, $p_3(x) = 1$, $p_4(x) = 1 + 2x + 2x^2$, in $\mathcal{P}_2$ over **R**

(b) $p_1(x) = 1 + x^2$, $p_2(x) = x + x^2$, $p_3(x) = 1 - x$, $p_4(x) = 1 + x + 2x^2$, in $\mathcal{P}_2$ over **R**

(c) $\left\{ \begin{bmatrix} 1 & 0 \\ 1 & 1 \end{bmatrix}, \begin{bmatrix} 0 & 1 \\ -1 & 1 \end{bmatrix}, \begin{bmatrix} 1 & -1 \\ 2 & 0 \end{bmatrix}, \begin{bmatrix} 2 & 1 \\ 1 & 3 \end{bmatrix} \right\}$ in $\mathbf{R}_{2 \times 2}$

(d) $\left\{ \begin{bmatrix} 1 & 1 \\ 2 & 1 \end{bmatrix}, \begin{bmatrix} 1 & 0 \\ -3 & 1 \end{bmatrix}, \begin{bmatrix} 0 & 1 \\ 5 & 0 \end{bmatrix}, \begin{bmatrix} 0 & 0 \\ 1 & 1 \end{bmatrix}, \begin{bmatrix} 1 & 1 \\ 3 & 2 \end{bmatrix} \right\}$ in $\mathbf{R}_{2 \times 2}$

(e) $p_1(x) = x^2 + x + 1$, $p_2(x) = x^2 - x - 2$, $p_3(x) = x^3 + x - 1$, $p_4(x) = x - 1$ in $\mathcal{P}_3$ over **R**

(f) $p_1(x) = x^2 - 3x + 2$, $p_2(x) = x^2 + x - 2$, $p_3(x) = x^3 - 1$, $p_4(x) = x - 1$ in $\mathcal{P}_3$ over **R**

36. Verify that the given set **W** is a subspace of $\mathbf{R}^4$ and find a basis for **W**.

(a) **W** is the set of all vectors in $\mathbf{R}^4$ of the form $(a, 0, b, 0)$.

(b) **W** is the set of all vectors in $\mathbf{R}^4$ of the form (a, b, c, d) with $c = a$ and $d = a + b$.

37. Find the coordinate matrix $[\mathbf{v}]_\mathcal{B}$ of **v** relative to the given basis $\mathcal{B}$ of $\mathcal{P}_2$ over **R** for each vector **v**.

(a) $\mathbf{v} = 2 - x + x^2$, $\quad \mathcal{B} = \{1 - x, 1 + x^2, 2x + x^2\}$

(b) $\mathbf{v} = 2 + 3x - 2x^2$, $\quad \mathcal{B} = \{1 + x + x^2, 3 + x^2, 2 + x\}$

38. Find the coordinate matrix $[\mathbf{v}]_\mathcal{B}$ of **v** relative to the given basis $\mathcal{B}$ of $\mathbf{R}_{2 \times 2}$ for each vector **v**.

(a) $\mathbf{v} = \begin{bmatrix} 1 & 1 \\ 2 & 0 \end{bmatrix}$, $\quad \mathcal{B} = \left\{ \begin{bmatrix} 1 & 1 \\ 0 & 0 \end{bmatrix}, \begin{bmatrix} 1 & 0 \\ 1 & 0 \end{bmatrix}, \begin{bmatrix} 1 & 0 \\ 0 & 1 \end{bmatrix}, \begin{bmatrix} 0 & 0 \\ 1 & 1 \end{bmatrix} \right\}$

(b) $\mathbf{v} = \begin{bmatrix} 0 & 3 \\ 0 & 1 \end{bmatrix}$, $\quad \mathcal{B} = \left\{ \begin{bmatrix} 0 & 1 \\ 0 & 1 \end{bmatrix}, \begin{bmatrix} -1 & 0 \\ 1 & 0 \end{bmatrix}, \begin{bmatrix} 0 & 1 \\ 1 & 0 \end{bmatrix}, \begin{bmatrix} 0 & 0 \\ 1 & 1 \end{bmatrix} \right\}$

39. Find the coordinate matrix $[\mathbf{v}]_\mathcal{B}$ of **v** relative to the given basis $\mathcal{B}$ of $\mathbf{R}_{1 \times 3}$ for each vector **v**.

(a) $\mathbf{v} = [1 \quad 3 \quad 4]$, $\quad \mathcal{B} = \{[1 \quad 2 \quad 3], [1 \quad 0 \quad -1], [0 \quad 1 \quad 1]\}$

(b) $\mathbf{v} = [-2 \quad 1 \quad 4]$, $\quad \mathcal{B} = \{[-1 \quad 0 \quad -1], [0 \quad 2 \quad 1], [0 \quad -1 \quad 1]\}$

40. Let **W** be the subspace of $\mathcal{P}_2$ over **R** containing polynomials of the form $a + bx^2$. Find the coordinate matrix $[\mathbf{v}]_\mathcal{B}$ of **v** relative to the given basis $\mathcal{B}$ of **W** for each vector **v**.

(a) $\mathbf{v} = 4 + 5x^2$, $\quad \mathcal{B} = \{2 + x^2, 1 - x^2\}$

(b) $\mathbf{v} = -1 - 3x^2$, $\quad \mathcal{B} = \{1 - 2x^2, 3 - x^2\}$

41. Find the transition matrix from the basis $\mathcal{A}$ to the basis $\mathcal{B}$ in $\mathcal{P}_2$ over **R**.

(a) $\mathcal{A} = \{p_1(x), p_2(x), p_3(x)\}$, $\quad \mathcal{B} = \{q_1(x), q_2(x), q_3(x)\}$, where
$p_1(x) = 1 - x$, $p_2(x) = x + x^2$, $p_3(x) = x$,
$q_1(x) = 1$, $q_2(x) = x + x^2$, $q_3(x) = 1 + x^2$

(b) $\mathcal{A} = \{p_1(x), p_2(x), p_3(x)\}$, $\mathcal{B} = \{q_1(x), q_2(x), q_3(x)\}$, where
$p_1(x) = x + x^2$, $p_2(x) = x$, $p_3(x) = 1 + x^2$,
$q_1(x) = x$, $q_2(x) = x - x^2$, $q_3(x) = 1 + x - x^2$

42. Find the transition matrix from the basis $\mathcal{B}$ to the basis $\mathcal{A}$ in each part of Problem 41.

43. Find the transition matrix from the basis $\mathcal{A}$ to the basis $\mathcal{B}$ in $\mathbf{R}_{m \times n}$.

(a) $\mathcal{A} = \left\{ \begin{bmatrix} 1 & 0 \\ 0 & 0 \end{bmatrix}, \begin{bmatrix} 0 & 1 \\ 0 & 0 \end{bmatrix}, \begin{bmatrix} 0 & 0 \\ 1 & 0 \end{bmatrix}, \begin{bmatrix} 0 & 0 \\ 0 & 1 \end{bmatrix} \right\}$

$\mathcal{B} = \left\{ \begin{bmatrix} 1 & 1 \\ 0 & 0 \end{bmatrix}, \begin{bmatrix} 0 & 1 \\ 1 & 0 \end{bmatrix}, \begin{bmatrix} 0 & 0 \\ 1 & 1 \end{bmatrix}, \begin{bmatrix} 0 & 1 \\ 0 & 1 \end{bmatrix} \right\}$

(b) $\mathcal{A} = \left\{ \begin{bmatrix} 1 \\ 0 \\ 0 \end{bmatrix}, \begin{bmatrix} 1 \\ 1 \\ 0 \end{bmatrix}, \begin{bmatrix} 1 \\ 1 \\ 1 \end{bmatrix} \right\}$, $\mathcal{B} = \left\{ \begin{bmatrix} 1 \\ 0 \\ 1 \end{bmatrix}, \begin{bmatrix} 0 \\ -1 \\ 1 \end{bmatrix}, \begin{bmatrix} 0 \\ 1 \\ 1 \end{bmatrix} \right\}$

44. Find the transition matrix from the basis $\mathcal{B}$ to the basis $\mathcal{A}$ in each part of Problem 43.

45. Let $\mathcal{B}$ be a basis of a subspace $\mathbf{W}$ of the vector space $\mathbf{V}$ over $\mathcal{F}$. Prove each of the following statements, where $[\mathbf{u}]_\mathcal{B}$ represents the coordinate matrix of $\mathbf{u}$ relative to the basis $\mathcal{B}$ of $\mathbf{W}$.

(a) $[\mathbf{u}]_\mathcal{B} + [\mathbf{v}]_\mathcal{B} = [\mathbf{u} + \mathbf{v}]_\mathcal{B}$ for all $\mathbf{u}, \mathbf{v} \in \mathbf{W}$.

(b) $k[\mathbf{u}]_\mathcal{B} = [k\mathbf{u}]_\mathcal{B}$ for all $\mathbf{u} \in \mathbf{W}$ and $k \in \mathcal{F}$.

4.3 Isomorphisms of Vector Spaces

The objective of this section is to derive the most important single result concerning finite-dimensional vector spaces. The concept of isomorphism between two vector spaces is essential to an understanding of this fundamental result, which is contained in Theorem 4.7. For this reason it is necessary to first consider the concept of an isomorphism. An isomorphism is a certain type of mapping, so we begin with a discussion of mappings in general.

4.4 DEFINITION

A **mapping** (or **function**, or **transformation**) from the set S into the set T is a set f of ordered pairs (s, t) of elements $s \in S, t \in T$ that has the following property: For each s in S, there is exactly one element t in T such that $(s, t) \in f$.

The notation $t = f(s)$ indicates that t is the unique element of T that is associated with s by the rule that $(s, t) \in f$. We say that $f(s)$ is the **image** of s, and that s is an **inverse image** of $f(s)$. Two mappings f and g of S into T are **equal** if and only if $f(s) = g(s)$ for every s in S.

If f is a mapping of S into T, then it may happen that $f(s_1) = f(s_2)$, even though $s_1 \neq s_2$. Another point of interest is that it is not required that every element of T be an image of an element in S.

> **4.5**   If f is a mapping of S into T, such that $f(s_1) = f(s_2)$ always implies $s_1 = s_2$, then
> **DEFINITION** f is called **injective** or **one-to-one**. If every t in T is the image of at least one s in
> S under f, we say that f is a **surjective** mapping of S into T, or that f maps S
>   **onto** T. If f is both injective and surjective, then f is called **bijective**.

The one-to-one and onto properties depend on the sets S and T as well as the rule that defines the mapping as illustrated in the next example.

EXAMPLE 1 The rule $f(x) = \sin x$ defines a mapping of the set $\mathbf{R}$ of real numbers into $\mathbf{R}$. This mapping f is clearly not one-to-one (for example, $\sin \frac{\pi}{3} = \sin \frac{2\pi}{3} = \frac{\sqrt{3}}{2}$). Also, f is not onto since there is no x in $\mathbf{R}$ such that $f(x) = 2$.

If $S = \{x \in \mathbf{R} \mid 0 \le x \le \pi\}$ and $T = \{t \in \mathbf{R} \mid 0 \le t \le 1\}$, then the rule $f(x) = \sin x$ defines a mapping of S into T that is onto but is not one-to-one.

If $S = \{x \in \mathbf{R} \mid 0 \le x \le \frac{\pi}{2}\}$ and $T = \{t \in \mathbf{R} \mid 0 \le t \le 1\}$, the rule $f(x) = \sin x$ defines a mapping of S into T that is both onto and one-to-one.   ∎

The set S in the next example is $\mathbf{R}_{2\times 2}$, and the mapping f is considered a function of matrices.

EXAMPLE 2 The rule

$$f\left(\begin{bmatrix} a & b \\ c & d \end{bmatrix}\right) = ad - bc$$

defines a mapping of the set $\mathbf{R}_{2\times 2}$ of matrices into $\mathbf{R}$. This mapping is clearly onto since for any real number r, there exists a 2×2 matrix

$$\begin{bmatrix} r & 0 \\ 0 & 1 \end{bmatrix}$$

over $\mathbf{R}$ such that

$$f\left(\begin{bmatrix} r & 0 \\ 0 & 1 \end{bmatrix}\right) = r(1) - 0(0) = r.$$

However, f is not one-to-one since

$$f\left(\begin{bmatrix} 2 & 0 \\ 0 & 0 \end{bmatrix}\right) = f\left(\begin{bmatrix} 3 & 0 \\ 0 & 0 \end{bmatrix}\right) = 0 \quad \text{and} \quad \begin{bmatrix} 2 & 0 \\ 0 & 0 \end{bmatrix} \ne \begin{bmatrix} 3 & 0 \\ 0 & 0 \end{bmatrix}.$$ ∎

The mapping f in the next example maps 3×1 matrices into $\mathcal{P}_2$ over $\mathbf{R}$, the set of all second degree polynomials with real coefficients.

EXAMPLE 3 The rule

$$f\left(\begin{bmatrix} a \\ b \\ c \end{bmatrix}\right) = a + (b + c)x^2$$

defines a mapping from $\mathbf{R}_{3 \times 1}$ into $\mathcal{P}_2$ over $\mathbf{R}$. This mapping is neither onto nor one-to-one. It is not onto since there is no matrix

$$\begin{bmatrix} a \\ b \\ c \end{bmatrix}$$

that is an inverse image of the polynomial $1 + x + x^2$. Also it is not one-to-one since

$$f\left(\begin{bmatrix} 1 \\ 0 \\ 2 \end{bmatrix}\right) = f\left(\begin{bmatrix} 1 \\ 1 \\ 1 \end{bmatrix}\right) = 1 + 2x^2 \quad \text{but} \quad \begin{bmatrix} 1 \\ 0 \\ 2 \end{bmatrix} \neq \begin{bmatrix} 1 \\ 1 \\ 1 \end{bmatrix}. \qquad \blacksquare$$

Of special interest are bijective mappings that satisfy a particular condition as described in the next definition.

4.6 DEFINITION

Let $\mathbf{U}$ and $\mathbf{V}$ be vector spaces over the same field $\mathcal{F}$. An **isomorphism** from $\mathbf{U}$ to $\mathbf{V}$ is a bijective mapping f of $\mathbf{U}$ into $\mathbf{V}$ that has the property that

$$f(a\mathbf{u} + b\mathbf{v}) = af(\mathbf{u}) + bf(\mathbf{v})$$

for all a, b in $\mathcal{F}$ and all $\mathbf{u}, \mathbf{v}$ in $\mathbf{U}$. If an isomorphism from $\mathbf{U}$ to $\mathbf{V}$ exists, $\mathbf{U}$ and $\mathbf{V}$ are said to be **isomorphic** vector spaces.

Since an isomorphism f from $\mathbf{U}$ to $\mathbf{V}$ is a bijective mapping, it pairs the elements of $\mathbf{U}$ and $\mathbf{V}$ in a one-to-one fashion. In view of the property $f(a\mathbf{u} + b\mathbf{v}) = af(\mathbf{u}) + bf(\mathbf{v})$, f is said to "preserve linear combinations." As particular instances of this property, $f(\mathbf{u} + \mathbf{v}) = f(\mathbf{u}) + f(\mathbf{v})$ and $f(a\mathbf{u}) = af(\mathbf{u})$. Thus f preserves sums and scalar products, and the vector spaces $\mathbf{U}$ and $\mathbf{V}$ are structurally the same.

EXAMPLE 4 The mapping f from $\mathcal{P}_3$ over $\mathbf{R}$ into $\mathbf{R}_{2 \times 2}$ defined by

$$f(a_0 + a_1x + a_2x^2 + a_3x^3) = \begin{bmatrix} a_0 & a_1 \\ a_2 & a_3 \end{bmatrix}$$

is clearly both onto and one-to-one and hence a bijective mapping. It also preserves linear combinations. To demonstrate this property let $p(x)$ and $q(x)$ be elements in $\mathcal{P}_3$, say $p(x) = a_0 + a_1x + a_2x^2 + a_3x^3$ and $q(x) = b_0 + b_1x + b_2x^2 + b_3x^3$. Then for any a, b in $\mathbf{R}$,

$$f(ap(x) + bq(x)) = f(a(a_0 + a_1x + a_2x^2 + a_3x^3) + b(b_0 + b_1x + b_2x^2 + b_3x^3))$$

$$= f((aa_0 + bb_0) + (aa_1 + bb_1)x + (aa_2 + bb_2)x^2 + (aa_3 + bb_3)x^3)$$

$$= \begin{bmatrix} aa_0 + bb_0 & aa_1 + bb_1 \\ aa_2 + bb_2 & aa_3 + bb_3 \end{bmatrix}$$

$$= \begin{bmatrix} aa_0 & aa_1 \\ aa_2 & aa_3 \end{bmatrix} + \begin{bmatrix} bb_0 & bb_1 \\ bb_2 & bb_3 \end{bmatrix}$$

$$= a \begin{bmatrix} a_0 & a_1 \\ a_2 & a_3 \end{bmatrix} + b \begin{bmatrix} b_0 & b_1 \\ b_2 & b_3 \end{bmatrix}$$

$$= af(a_0 + a_1x + a_2x^2 + a_3x^3) + bf(b_0 + b_1x + b_2x^2 + b_3x^3)$$

$$= af(p(x)) + bf(q(x)),$$

and f is an isomorphism from $\mathcal{P}_3$ to $\mathbf{R}_{2\times 2}$. ■

We are now in a position to prove the principal result of this section. This result shows that the first example in Section 4.1 furnishes a completely typical pattern for n-dimensional vector spaces over a field $\mathcal{F}$.

**4.7
THEOREM**

Any n-dimensional vector space over the field $\mathcal{F}$ is isomorphic to $\mathcal{F}^n$.

Proof Let $\mathbf{V}$ be an n-dimensional vector space over the field $\mathcal{F}$, and let

$$\mathcal{B} = \{\mathbf{v}_1, \mathbf{v}_2, \ldots, \mathbf{v}_n\}$$

be a basis of $\mathbf{V}$. Each $\mathbf{u} \in \mathbf{V}$ can be written *uniquely* as $\mathbf{u} = \sum_{i=1}^n a_i \mathbf{v}_i$, so the rule

$$f(\mathbf{u}) = (a_1, a_2, \ldots, a_n)$$

defines a one-to-one mapping of $\mathbf{V}$ into $\mathcal{F}^n$. It is clear that f is onto.

Let $\mathbf{u} = \sum_{i=1}^n a_i \mathbf{v}_i$ and $\mathbf{v} = \sum_{i=1}^n b_i \mathbf{v}_i$ be arbitrary vectors in $\mathbf{V}$, and let a and b be arbitrary scalars in $\mathcal{F}$. Then

$$f(a\mathbf{u} + b\mathbf{v}) = f\left(\sum_{i=1}^n (aa_i + bb_i)\mathbf{v}_i \right)$$

$$= (aa_1 + bb_1, aa_2 + bb_2, \ldots, aa_n + bb_n)$$

$$= a(a_1, a_2, \ldots, a_n) + b(b_1, b_2, \ldots, b_n)$$

$$= af(\mathbf{u}) + bf(\mathbf{v}),$$

so that f is an isomorphism from $\mathbf{V}$ to $\mathcal{F}^n$. ■■■

Theorem 4.7 not only proves that any n-dimensional vector space $\mathbf{V}$ over $\mathcal{F}$ is isomorphic to $\mathcal{F}^n$ but it also furnishes an example of an isomorphism f from $\mathbf{V}$ onto $\mathcal{F}^n$.

EXAMPLE 5 The set $\mathbf{W}$ of all diagonal matrices of order n is a subspace of $\mathbf{R}_{n \times n}$ (see Problem 18 of Exercises 4.2). This subspace has dimension n since

$$
\left\{
\begin{bmatrix}
1 & 0 & \cdots & 0 \\
0 & 0 & \cdots & 0 \\
\vdots & \vdots & \ddots & \vdots \\
0 & 0 & \cdots & 0
\end{bmatrix},
\begin{bmatrix}
0 & 0 & \cdots & 0 \\
0 & 1 & \cdots & 0 \\
\vdots & \vdots & \ddots & \vdots \\
0 & 0 & \cdots & 0
\end{bmatrix},
\ldots,
\begin{bmatrix}
0 & 0 & \cdots & 0 \\
0 & 0 & \cdots & 0 \\
\vdots & \vdots & \ddots & \vdots \\
0 & 0 & \cdots & 1
\end{bmatrix}
\right\}
$$

is a basis for $\mathbf{W}$. Now Theorem 4.7 guarantees the existence of an isomorphism f from $\mathbf{W}$ to $\mathbf{R}^n$, and the mapping defined by

$$
f\left(
\begin{bmatrix}
a_{11} & 0 & \cdots & 0 \\
0 & a_{22} & \cdots & 0 \\
\vdots & \vdots & \ddots & \vdots \\
0 & 0 & \cdots & a_{nn}
\end{bmatrix}
\right) = (a_{11}, a_{22}, \ldots, a_{nn})
$$

is one such isomorphism. ∎

4.3 Exercises

1. Determine whether each of the following mappings f is onto or one-to-one. Is f an isomorphism?

   (a) f maps $\mathbf{R}^2$ into $\mathbf{R}^2$ and is defined by $f(x, y) = (x - 2y, x + y)$.

   (b) f maps $\mathbf{R}^2$ into $\mathbf{R}^3$ and is defined by $f(x, y) = (x, y, x + y)$.

   (c) f maps $\mathbf{R}^2$ into $\mathbf{R}_{1 \times 2}$ and is defined by $f(x, y) = [x \quad 0]$.

   (d) f maps $\mathbf{R}^3$ into $\mathbf{R}_{1 \times 3}$ and is defined by $f(x, y, z) = [x^2 \quad y^2 \quad z^2]$.

   (e) f maps $\mathbf{R}^4$ into $\mathcal{P}_2$ over $\mathbf{R}$ and is defined by $f(a, b, c, d) = a + (b - c)x + dx^2$.

   (f) f maps $\mathbf{R}^3$ into $\mathcal{P}_2$ over $\mathbf{R}$ and is defined by $f(a, b, c) = c + (b - a)x + cx^2$.

   (g) f maps $\mathbf{R}_{2 \times 2}$ into $\mathbf{R}$ and is defined by

   $$
   f\left(\begin{bmatrix} a & b \\ c & d \end{bmatrix}\right) = a.
   $$

   (h) f maps $\mathbf{R}^2$ into $\mathbf{R}_{2 \times 2}$ and is defined by

   $$
   f(a, b) = \begin{bmatrix} a & b \\ b & a \end{bmatrix}.
   $$

   (i) f maps $\mathbf{R}^3$ into $\mathcal{P}_2$ over $\mathbf{R}$, defined by $f(a_1, a_2, a_3) = a_2 - a_3 x + (1 - a_1)x^2$.

   (j) f maps $\mathbf{R}_{m \times n}$ into $\mathbf{R}_{n \times m}$ and is defined by $f(A) = A^T$ for all A in $\mathbf{R}_{m \times n}$.

(k) f maps $\mathcal{P}_2$ over $\mathbf{R}$ into $\mathcal{P}_3$ over $\mathbf{R}$ and is defined by $f(p(x)) = xp(x)$ for all $p(x)$ in $\mathcal{P}_2$.

2. Let $\mathbf{W}$ be the subspace in Example 5. Exhibit another isomorphism (different from the one given in the example) from $\mathbf{W}$ to $\mathbf{R}^n$.

3. Let $\mathbf{V}$ be the vector space in the second example in Section 4.1. Exhibit an isomorphism from $\mathbf{V}$ to $\mathcal{F}^{n+1}$.

4. Define a mapping f of $\mathbf{R}_{3\times 2}$ into $\mathbf{R}^6$ that is an isomorphism from $\mathbf{R}_{3\times 2}$ to $\mathbf{R}^6$, and prove that your mapping is an isomorphism.

5. Let $\mathbf{W}$ be the set of all polynomials $a_0 + a_1 x + a_2 x^2$ in $\mathcal{P}_2$ over $\mathbf{R}$ such that $a_0 + a_1 + a_2 = 0$. Show that $\mathbf{W}$ is isomorphic to $\mathbf{R}^2$ and exhibit an isomorphism from $\mathbf{W}$ to $\mathbf{R}^2$.

6. Let $\mathbf{W}$ be the set of all 2×2 matrices over $\mathbf{R}$ of the form

$$\begin{bmatrix} 0 & a \\ b & a+b \end{bmatrix}.$$

Show that $\mathbf{W}$ is isomorphic to $\mathbf{R}^2$ and exhibit an isomorphism from $\mathbf{W}$ to $\mathbf{R}^2$.

7. Let $\mathbf{W}$ be the set of all 2×2 symmetric matrices over $\mathbf{R}$. Show that $\mathbf{W}$ is isomorphic to $\mathbf{R}^3$ and exhibit an isomorphism from $\mathbf{W}$ to $\mathbf{R}^3$.

8. Let $\mathbf{W}$ be the line through the origin $x - 5y = 0$. Show that $\mathbf{W}$ is isomorphic to $\mathbf{R}^1$ and exhibit an isomorphism from $\mathbf{W}$ to $\mathbf{R}^1$.

9. Let $\mathbf{W}$ be the plane through the origin $3x - 2y + z = 0$. Show that $\mathbf{W}$ is isomorphic to $\mathbf{R}^2$ and exhibit an isomorphism from $\mathbf{W}$ to $\mathbf{R}^2$.

10. Let $\mathbf{W}_1$ be the subspace of vectors of the form (a, a, b, c) over $\mathbf{R}$ and $\mathbf{W}_2$ the subspace of polynomials over $\mathbf{R}$ of the form $a_1 x + a_2 x^2 + a_3 x^3$. Show that $\mathbf{W}_1$ is isomorphic to $\mathbf{W}_2$ and exhibit an isomorphism from $\mathbf{W}_1$ to $\mathbf{W}_2$.

11. Let $\mathbf{W}_1$ be the subspace of 3×3 symmetric matrices over $\mathbf{R}$ and $\mathbf{W}_2$ the set of 2×3 matrices over $\mathbf{R}$. Show that $\mathbf{W}_1$ is isomorphic to $\mathbf{W}_2$ and exhibit an isomorphism from $\mathbf{W}_1$ to $\mathbf{W}_2$.

12. For each subspace $\mathbf{W}$ below, determine the dimension r of $\mathbf{W}$ and find an isomorphism from $\mathbf{W}$ to $\mathbf{R}^r$.

(a) $\mathbf{W} = \langle (1, -1, 1), (2, -2, 2), (0, 1, 0), (1, 0, 1) \rangle$ in $\mathbf{R}^3$

(b) $\mathbf{W} = \langle (4, -2, 5), (0, -4, 0), (12, -18, 15), (4, 0, 5) \rangle$ in $\mathbf{R}^3$

(c) $\mathbf{W} = \langle (2, 0, 4, -1), (5, -1, 11, 8), (0, 1, -7, 9), (7, 0, 8, 16) \rangle$ in $\mathbf{R}^4$

(d) $\mathbf{W} = \langle (4, -2, 5, 5), (0, -4, 0, 4), (12, -18, 15, 27), (4, 0, 5, 3) \rangle$ in $\mathbf{R}^4$

(e) $\mathbf{W} = \langle p_1(x), p_2(x), p_3(x), p_4(x) \rangle$ in $\mathcal{P}_3$ over $\mathbf{R}$, where

$$p_1(x) = x^3 + x^2 + 1, \qquad\qquad p_2(x) = x^2 + x + 1,$$
$$p_3(x) = 2x^3 + 3x^2 + x + 3, \qquad p_4(x) = x^3 - x.$$

(f) $W = \langle p_1(x), p_2(x), p_3(x), p_4(x) \rangle$ in $\mathcal{P}_3$ over $\mathbf{R}$, where

$$p_1(x) = x^3 + x + 1, \qquad\qquad p_2(x) = x^2,$$
$$p_3(x) = 2x^3 + 3x^2 + 2x + 2, \qquad p_4(x) = x + 1.$$

(g) $W = \left\langle \begin{bmatrix} 1 & 0 \\ -1 & 1 \end{bmatrix}, \begin{bmatrix} -2 & 0 \\ 2 & -2 \end{bmatrix}, \begin{bmatrix} 1 & 1 \\ 1 & 1 \end{bmatrix}, \begin{bmatrix} 2 & 1 \\ 0 & 2 \end{bmatrix}, \begin{bmatrix} 0 & 1 \\ 2 & 0 \end{bmatrix} \right\rangle$ in $\mathbf{R}_{2 \times 2}$

(h) $W = \left\langle \begin{bmatrix} 1 & 1 \\ 2 & 0 \end{bmatrix}, \begin{bmatrix} 3 & 3 \\ 6 & 0 \end{bmatrix}, \begin{bmatrix} 1 & 2 \\ 3 & 1 \end{bmatrix}, \begin{bmatrix} 2 & 1 \\ 3 & -1 \end{bmatrix} \right\rangle$ in $\mathbf{R}_{2 \times 2}$

13. (a) Let $\mathbf{V}$ be the subspace spanned by the set $\{p_1(x), p_2(x), p_3(x), p_4(x)\}$ of polynomials in Problem 35(e) of Exercises 4.2. Find an isomorphism from $\mathbf{V}$ to a subspace of $\mathbf{R}^4$.

(b) Let $\mathbf{V}$ be the subspace spanned by the set $\{p_1(x), p_2(x), p_3(x), p_4(x)\}$ of polynomials in Problem 35(f) of Exercises 4.2. Find an isomorphism from $\mathbf{V}$ to a subspace of $\mathbf{R}^4$.

14. Suppose f and g are isomorphisms from $\mathbf{U}$ to $\mathbf{V}$. Prove or disprove each of the following statements:

(a) The mapping $f + g$ is an isomorphism from $\mathbf{U}$ to $\mathbf{V}$.

(b) The mapping kf for $k \neq 0$ is an isomorphism from $\mathbf{U}$ to $\mathbf{V}$.

15. Let f be an isomorphism from the vector space $\mathbf{U}$ to the vector space $\mathbf{V}$. Prove that if $f(\mathbf{u}) = \mathbf{0} \in \mathbf{V}$, then $\mathbf{u} = \mathbf{0}$ in $\mathbf{U}$.

16. Let f be an isomorphism from the vector space $\mathbf{U}$ to the vector space $\mathbf{V}$ and suppose that $\{\mathbf{u}_1, \mathbf{u}_2, \ldots, \mathbf{u}_n\}$ is a linearly independent set of vectors in $\mathbf{U}$. Prove that $\{f(\mathbf{u}_1), f(\mathbf{u}_2), \ldots, f(\mathbf{u}_n)\}$ is linearly independent.

17. Let f be an isomorphism from the vector space $\mathbf{U}$ to the vector space $\mathbf{V}$. Prove that if $\{\mathbf{u}_1, \mathbf{u}_2, \ldots, \mathbf{u}_n\}$ spans $\mathbf{U}$ then $\{f(\mathbf{u}_1), f(\mathbf{u}_2), \ldots, f(\mathbf{u}_n)\}$ spans $\mathbf{V}$.

18. Prove that the relation of being isomorphic is an equivalence relation on the set of all vector spaces over $\mathcal{F}$.

4.4 Standard Bases for Subspaces

The definitions of the elementary operations on sets of vectors as given in Chapter 2 apply unchanged to arbitrary vector spaces. The following statements and proofs also apply unchanged, and may be used in arbitrary vector spaces: Definition 2.1, Definition 2.2, Definition 2.3, Theorem 2.4, Theorem 2.5, Corollary 2.6, Theorem 2.8, and Corollary 2.9. These results may be summarized in the following statement: If $\mathcal{A}$ and $\mathcal{A}'$ are sets of vectors in a vector space $\mathbf{V}$ such that $\mathcal{A}'$ is obtained from $\mathcal{A}$ by a sequence of elementary operations, then (i) $\mathcal{A}$ and $\mathcal{A}'$ span the same subspace of $\mathbf{V}$ and (ii) $\mathcal{A}'$ is linearly independent if and only if $\mathcal{A}$ is linearly independent. The significance of this statement lies in the fact that elementary operations may be used to obtain simpler forms of spanning sets for a subspace and also to investigate linear independence.

From Corollary 2.9 to the end of Chapter 2, the statements and proofs of results are intimately connected with $\mathbf{R}^n$. Thus more changes are necessary in formulating the corresponding development for subspaces of arbitrary vector spaces. However, the proofs of the theorems are basically unchanged. For this reason these theorems are stated here with only indications as to the changes in notation necessary in order to obtain the proofs.

4.8 THEOREM

Let $\mathcal{B} = \{\mathbf{u}_1, \mathbf{u}_2, \ldots, \mathbf{u}_n\}$ be a fixed basis of the vector space $\mathbf{V}$ over $\mathcal{F}$, and let $\mathcal{A} = \{\mathbf{v}_1, \mathbf{v}_2, \ldots, \mathbf{v}_m\}$ be a set of m vectors in $\mathbf{V}$ that spans the subspace $\mathbf{W} = \langle \mathcal{A} \rangle$ of dimension $r > 0$. Then a set $\mathcal{A}' = \{\mathbf{v}'_1, \mathbf{v}'_2, \ldots, \mathbf{v}'_r, \mathbf{0}, \ldots, \mathbf{0}\}$ of m vectors can be obtained from $\mathcal{A}$ by a finite sequence of elementary operations so that $\{\mathbf{v}'_1, \mathbf{v}'_2, \ldots, \mathbf{v}'_r\}$ has the following properties:

1. The first nonzero coordinate of $\mathbf{v}'_j$ with respect to $\mathcal{B}$ is a 1 for the k_j coordinate for $j = 1, 2, \ldots, r$. That is, $\mathbf{v}'_j = \sum_{i=k_j}^{n} a'_{ij} \mathbf{u}_i$ with $a'_{k_j j} = 1$.

2. $k_1 < k_2 < \cdots < k_r$.

3. $\mathbf{v}'_j$ is the only vector in $\mathcal{A}'$ with a nonzero k_j coordinate relative to $\mathcal{B}$.

4. $\{\mathbf{v}'_1, \mathbf{v}'_2, \ldots, \mathbf{v}'_r\}$ is a basis of $\mathbf{W}$.

Proof The proof can be obtained from the proof of Theorem 2.10 by replacing $\mathbf{R}^n$ by $\mathbf{V}$, $\mathcal{E}_n$ by $\mathcal{B}$, and $\mathbf{e}_i$ by $\mathbf{u}_i$ so the a'_{ij} represents the ith coordinate of $\mathbf{v}'_j$ relative to $\mathcal{B}$ instead of the ith component of $\mathbf{v}'_j$. ∎

4.9 THEOREM

For a fixed basis $\mathcal{B}$, there is one and only one basis of a given subspace $\mathbf{W}$ that satisfies the conditions of Theorem 4.8.

Proof It follows from Theorem 4.8 that there is at least one such basis of $\mathbf{W}$. Let $\mathcal{A}' = \{\mathbf{v}'_1, \mathbf{v}'_2, \ldots, \mathbf{v}'_r\}$ and $\mathcal{A}'' = \{\mathbf{v}''_1, \mathbf{v}''_2, \ldots, \mathbf{v}''_r\}$ be two bases of $\mathbf{W}$ that satisfy the conditions. Then the same replacements used in the proof of Theorem 4.8 can be used in the proof of Theorem 2.11 to obtain a proof that $\mathcal{A}' = \mathcal{A}''$. ∎

4.10 DEFINITION

Let $\mathcal{B}$ be a fixed basis of the n-dimensional vector space $\mathbf{V}$ over $\mathcal{F}$, and let $\mathbf{W}$ be a subspace of $\mathbf{V}$ of dimension r. The basis of $\mathbf{W}$ that satisfies the conditions of Theorem 4.8 is called the **standard basis** of $\mathbf{W}$ relative to $\mathcal{B}$.

That is, the standard basis of **W** relative to $\mathcal{B} = \{\mathbf{u}_1, \mathbf{u}_2, \ldots, \mathbf{u}_n\}$ is the unique basis $\mathcal{A} = \{\mathbf{v}_1, \mathbf{v}_2, \ldots, \mathbf{v}_r\}$ that has the following properties:

1. $\mathbf{v}_j = \sum_{i=k_j}^{n} a_{ij}\mathbf{u}_i$ with $a_{k_j j} = 1$

2. $k_1 < k_2 < \cdots < k_r$

3. $\mathbf{v}_j$ is the only vector in $\mathcal{A}$ that has a nonzero k_j coordinate relative to $\mathcal{B}$.

When referring to standard bases, we use the phrase "with respect to $\mathcal{B}$" interchangeably with "relative to $\mathcal{B}$."

EXAMPLE 1   The set $\mathcal{B} = \{\mathbf{u}_1 = (1, 0, 2), \mathbf{u}_2 = (2, 1, 6), \mathbf{u}_3 = (0, 3, 8)\}$ is a basis of $\mathbf{R}^3$. Let $\mathcal{A} = \{(4, -2, 2), (6, 26, 78), (5, 12, 40), (14, 22, 82)\}$ and consider the problem of finding the standard basis of $\langle \mathcal{A} \rangle$ relative to $\mathcal{B}$.

In order to obtain the coordinates a'_{ij} as in Theorem 4.8, we determine the coordinates of the vectors in $\mathcal{A}$ with respect to $\mathcal{B}$ and then record these coordinates in the columns of a matrix. (That is, we follow the same procedure as in Chapter 2, using coordinates rather than components.) The vectors of $\mathcal{A}$ are found to be given by

$$
\begin{aligned}
(4, -2, 2) &= 2\mathbf{u}_1 + \mathbf{u}_2 - \mathbf{u}_3, \\
(6, 26, 78) &= -4\mathbf{u}_1 + 5\mathbf{u}_2 + 7\mathbf{u}_3, \\
(5, 12, 40) &= -\mathbf{u}_1 + 3\mathbf{u}_2 + 3\mathbf{u}_3, \\
(14, 22, 82) &= \phantom{-4\mathbf{u}_1 +} 7\mathbf{u}_2 + 5\mathbf{u}_3,
\end{aligned}
$$

so that the matrix of coordinates is

$$
A = \begin{bmatrix} 2 & -4 & -1 & 0 \\ 1 & 5 & 3 & 7 \\ -1 & 7 & 3 & 5 \end{bmatrix}.
$$

By using elementary column operations on A (which corresponds to using elementary operations on $\mathcal{A}$), we find the reduced column-echelon form for A is

$$
A' = \begin{bmatrix} 1 & 0 & 0 & 0 \\ 0 & 1 & 0 & 0 \\ -\frac{6}{7} & \frac{5}{7} & 0 & 0 \end{bmatrix}.
$$

Thus the standard basis for $\langle \mathcal{A} \rangle$ with respect to $\mathcal{B}$ is $\{\mathbf{v}'_1, \mathbf{v}'_2\}$, where

$$
\mathbf{v}'_1 = \mathbf{u}_1 - \tfrac{6}{7}\mathbf{u}_3 = \left(1, -\tfrac{18}{7}, -\tfrac{34}{7}\right),
$$

$$
\mathbf{v}'_2 = \mathbf{u}_2 + \tfrac{5}{7}\mathbf{u}_3 = \left(2, \tfrac{22}{7}, \tfrac{82}{7}\right). \qquad \blacksquare
$$

The proofs of the following two theorems can be obtained from those of Theorems 2.13 and 2.14 by making the same changes as were indicated in Theorems 4.8 and 4.9.

4.11	Let $\mathcal{B}$ be a fixed basis of the finite-dimensional vector space **V** over $\mathcal{F}$. For any
THEOREM	subspace **W** of **V**, the standard basis of **W** relative to $\mathcal{B}$ can be obtained from any basis of **W** by a sequence of elementary operations.

4.12	Let $\mathcal{A}$ and $\mathcal{B}$ be two sets of m vectors each in the finite-dimensional vector space **V**
THEOREM	over $\mathcal{F}$. Then $\langle \mathcal{A} \rangle = \langle \mathcal{B} \rangle$ if and only if $\mathcal{B}$ can be obtained from $\mathcal{A}$ by a sequence of elementary operations.

EXAMPLE 2 To decide whether or not $\langle \mathcal{A} \rangle = \langle \mathcal{B} \rangle$, where

$$\mathcal{A} = \left\{ \begin{bmatrix} 1 & 2 \\ -1 & 3 \end{bmatrix}, \begin{bmatrix} 2 & 4 \\ -2 & -6 \end{bmatrix}, \begin{bmatrix} 1 & 4 \\ -3 & 13 \end{bmatrix} \right\}$$

and

$$\mathcal{B} = \left\{ \begin{bmatrix} 1 & -2 \\ 3 & 0 \end{bmatrix}, \begin{bmatrix} 0 & 4 \\ -4 & 1 \end{bmatrix}, \begin{bmatrix} 1 & 1 \\ 0 & 1 \end{bmatrix}, \begin{bmatrix} 2 & -1 \\ 3 & 1 \end{bmatrix} \right\},$$

we compare the standard bases of $\langle \mathcal{A} \rangle$ and $\langle \mathcal{B} \rangle$ relative to the basis

$$\mathcal{E}_{2\times 2} = \left\{ \begin{bmatrix} 1 & 0 \\ 0 & 0 \end{bmatrix}, \begin{bmatrix} 0 & 1 \\ 0 & 0 \end{bmatrix}, \begin{bmatrix} 0 & 0 \\ 1 & 0 \end{bmatrix}, \begin{bmatrix} 0 & 0 \\ 0 & 1 \end{bmatrix} \right\}$$

of $\mathbf{R}_{2\times 2}$. The matrix of coordinates of the vectors in $\mathcal{A}$ relative to the basis $\mathcal{E}_{2\times 2}$ is

$$A = \begin{bmatrix} 1 & 2 & 1 \\ 2 & 4 & 4 \\ -1 & -2 & -3 \\ 3 & -6 & 13 \end{bmatrix},$$

whose reduced column-echelon form is

$$\begin{bmatrix} 1 & 0 & 0 \\ 0 & 1 & 0 \\ 1 & -1 & 0 \\ 0 & 0 & 1 \end{bmatrix}.$$

Similarly, the matrix of coordinates for the vectors in $\mathcal{B}$ relative to $\mathcal{E}_{2\times 2}$ is

$$B = \begin{bmatrix} 1 & 0 & 1 & 2 \\ -2 & 4 & 1 & -1 \\ 3 & -4 & 0 & 3 \\ 0 & 1 & 1 & 1 \end{bmatrix},$$

whose reduced column-echelon form is

$$\begin{bmatrix} 1 & 0 & 0 & 0 \\ 0 & 1 & 0 & 0 \\ 1 & -1 & 0 & 0 \\ 0 & 0 & 1 & 0 \end{bmatrix}.$$

Hence the standard basis of both $\langle \mathcal{A} \rangle$ and $\langle \mathcal{B} \rangle$ relative to $\mathcal{E}_{2 \times 2}$ is

$$\left\{ \begin{bmatrix} 1 & 0 \\ 1 & 0 \end{bmatrix}, \begin{bmatrix} 0 & 1 \\ -1 & 0 \end{bmatrix}, \begin{bmatrix} 0 & 0 \\ 0 & 1 \end{bmatrix} \right\}$$

and $\langle \mathcal{A} \rangle = \langle \mathcal{B} \rangle$. ∎

4.4 Exercises

1. Let $\mathcal{B} = \{1, x, x^2, \ldots, x^n\}$ be the standard basis of $\mathcal{P}_n$ over $\mathbf{R}$. Find the standard basis of $\langle \mathcal{A} \rangle$ relative to $\mathcal{B}$ for each of the following sets $\mathcal{A}$.

   (a) $\mathcal{A} = \{1 + 2x^2 + 6x^3, x + 3x^2 - 2x^3, 1 + x + 5x^2 + 4x^3, 3 + x + 9x^2 + 10x^3\}$

   (b) $\mathcal{A} = \{-x + 4x^2, 1 + x, 2 + 3x - 4x^2, 2 + x + 4x^2, 3 + 2x + 4x^2\}$

2. Let $\mathcal{E}_{m \times n}$ be the standard basis of $\mathbf{R}_{m \times n}$. Find the standard basis of $\langle \mathcal{A} \rangle$ relative to $\mathcal{E}_{m \times n}$ for each of the following sets $\mathcal{A}$.

   (a) $\mathcal{A} = \left\{ \begin{bmatrix} 1 & 0 \\ -1 & 1 \end{bmatrix}, \begin{bmatrix} -1 & 3 \\ 0 & -2 \end{bmatrix}, \begin{bmatrix} 1 & 3 \\ -2 & 0 \end{bmatrix}, \begin{bmatrix} 3 & 6 \\ -5 & 1 \end{bmatrix} \right\}$

   (b) $\mathcal{A} = \left\{ \begin{bmatrix} 0 & -1 & 1 \\ 1 & 0 & 2 \end{bmatrix}, \begin{bmatrix} 0 & 2 & 0 \\ 0 & 1 & -4 \end{bmatrix}, \begin{bmatrix} 0 & 1 & 1 \\ 1 & 1 & -2 \end{bmatrix}, \begin{bmatrix} 0 & 0 & 2 \\ 2 & 1 & 0 \end{bmatrix}, \begin{bmatrix} 0 & 1 & -1 \\ -1 & 0 & -2 \end{bmatrix} \right\}$

3. Let $p_1(x) = 2x^2 + 2$, $p_2(x) = x + 1$, $p_3(x) = 2x^2 - 3x + 1$ in the vector space $\mathcal{P}_2$ over $\mathbf{R}$. Given that $\mathcal{B} = \{p_1(x), p_2(x), p_3(x)\}$ is a basis of $\mathcal{P}_2$ over $\mathbf{R}$, find the standard basis of $\langle \mathcal{A} \rangle$ with respect to $\mathcal{B}$ if $\mathcal{A}$ is given by

   (a) $\mathcal{A} = \{q_1(x), q_2(x), q_3(x)\}$, where $q_1(x) = 2p_1(x) - 3p_2(x) - p_3(x)$, $q_2(x) = p_1(x) - p_2(x) - p_3(x)$, and $q_3(x) = 2p_1(x) - p_2(x) - p_3(x)$.

   (b) $\mathcal{A} = \{q_1(x), q_2(x), q_3(x), q_4(x)\}$, where $q_1(x) = -p_3(x)$, $q_2(x) = 3p_1(x) - 6p_2(x) + p_3(x)$, $q_3(x) = 2p_1(x) - 4p_2(x) + 2p_3(x)$, and $q_4(x) = -p_1(x) + 2p_2(x) + 2p_3(x)$.

4. Using $\mathcal{B} = \{(1, 0, 0, 0), (1, 1, 0, 0), (1, 1, 1, 0), (1, 1, 1, 1)\}$ as the fixed basis of $\mathbf{R}^4$, find the standard basis of $\langle \mathcal{A} \rangle$ relative to $\mathcal{B}$ for each set $\mathcal{A}$ below. Write your answers in component form.

   (a) $\mathcal{A} = \{(4, 4, 3, 1), (7, 7, 5, 1), (4, 4, 3, 1)\}$

   (b) $\mathcal{A} = \{(5, 4, 2, 3), (3, 3, 2, 2), (8, 7, 4, 5), (2, 1, 0, 1)\}$

   (c) $\mathcal{A} = \{(-1, -2, -1, -3), (4, 3, 2, 0), (2, -1, 0, -6), (3, 1, 1, -3), (1, 1, 1, 1)\}$

   (d) $\mathcal{A} = \{(4, 3, 2, -1), (-1, -2, -2, 0), (7, 4, 2, -2), (2, 1, 1, 1), (10, 9, 8, 1)\}$

5. Let

$$\mathbf{u}_1 = \begin{bmatrix} 1 & 0 \\ 0 & 0 \end{bmatrix}, \quad \mathbf{u}_2 = \begin{bmatrix} 1 & 1 \\ 0 & 0 \end{bmatrix}, \quad \mathbf{u}_3 = \begin{bmatrix} 1 & 1 \\ 1 & 0 \end{bmatrix}, \quad \mathbf{u}_4 = \begin{bmatrix} 1 & 1 \\ 1 & 1 \end{bmatrix}$$

in the vector space $\mathbf{R}_{2\times 2}$. It is given that $\mathcal{B} = \{\mathbf{u}_1, \mathbf{u}_2, \mathbf{u}_3, \mathbf{u}_4\}$ is a basis of $\mathbf{R}_{2\times 2}$. For each set $\mathcal{A}$, find the standard basis of $\langle \mathcal{A} \rangle$ relative to $\mathcal{B}$.

(a) $\mathcal{A} = \{A_1, A_2, A_3\}$, where

$$A_1 = \begin{bmatrix} 2 & 1 \\ 0 & 1 \end{bmatrix}, \quad A_2 = \begin{bmatrix} 1 & 2 \\ 2 & 1 \end{bmatrix}, \quad A_3 = \begin{bmatrix} 4 & 5 \\ 4 & 3 \end{bmatrix}.$$

(b) $\mathcal{A} = \{A_1, A_2, A_3, A_4\}$, where

$$A_1 = \begin{bmatrix} -2 & -2 \\ -2 & -1 \end{bmatrix}, \quad A_2 = \begin{bmatrix} -4 & -7 \\ -1 & -2 \end{bmatrix}, \quad A_3 = \begin{bmatrix} 0 & -2 \\ 2 & 0 \end{bmatrix}, \quad A_4 = \begin{bmatrix} 6 & 7 \\ 5 & 3 \end{bmatrix}.$$

6. Let $p_1(x) = 1 + x + x^2$, $p_2(x) = x + x^2$, $p_3(x) = x^2$ in the vector space $\mathcal{P}_2$ over $\mathbf{R}$. Given that $\mathcal{B} = \{p_1(x), p_2(x), p_3(x)\}$ is a basis of $\mathcal{P}_2$ over $\mathbf{R}$, find the standard basis of $\langle \mathcal{A} \rangle$ with respect to $\mathcal{B}$ for each of the following sets $\mathcal{A}$.

(a) $\mathcal{A} = \{1 + x + 4x^2, x, 2 + 3x + 8x^2, 3 + 2x + 12x^2\}$

(b) $\mathcal{A} = \{2 + 3x + 6x^2, 1 + 3x^2, 1 + 3x + 3x^2, 1 + 2x + 3x^2, -x\}$

7. Determine in each case whether or not the given sets span the same subspace of $\mathcal{P}_2$ over $\mathbf{R}$.

(a) $\mathcal{A} = \{2x^2 + 2, 2x^2 - 2x + 2\}, \quad \mathcal{B} = \{4x^2 - 6x + 4, 2x, x^2 + 1\}.$

(b) $\mathcal{A} = \{2x^2 + 4x + 4, 4x^2 - 7x + 7, 18x^2 - 10x + 18\},$
$\mathcal{B} = \{4x^2 + 8x + 8, 14x^2 - 14x + 8\}.$

8. Determine in each case whether or not the given sets span the same subspace of $\mathbf{R}_{m\times n}$ over $\mathbf{R}$.

(a) $\mathcal{A} = \left\{ \begin{bmatrix} 1 & 0 \\ 0 & 3 \\ -1 & 2 \end{bmatrix}, \begin{bmatrix} 0 & -1 \\ 1 & 0 \\ 2 & 1 \end{bmatrix}, \begin{bmatrix} 2 & -1 \\ 1 & 6 \\ 0 & 5 \end{bmatrix}, \begin{bmatrix} 1 & 1 \\ -1 & 3 \\ -3 & 1 \end{bmatrix} \right\}$

$\mathcal{B} = \left\{ \begin{bmatrix} 1 & -2 \\ 2 & 3 \\ 3 & 4 \end{bmatrix}, \begin{bmatrix} -2 & 0 \\ 0 & -6 \\ 2 & -4 \end{bmatrix}, \begin{bmatrix} 0 & -3 \\ 3 & 0 \\ 6 & 3 \end{bmatrix}, \begin{bmatrix} 2 & -1 \\ 3 & 0 \\ 6 & 3 \end{bmatrix} \right\}$

(b) $\mathcal{A} = \left\{ \begin{bmatrix} 0 & -1 \\ 1 & 1 \end{bmatrix}, \begin{bmatrix} 2 & 1 \\ -1 & 1 \end{bmatrix}, \begin{bmatrix} 1 & 0 \\ 0 & 2 \end{bmatrix} \right\},$

$\mathcal{B} = \left\{ \begin{bmatrix} -1 & -1 \\ 1 & -1 \end{bmatrix}, \begin{bmatrix} 0 & 1 \\ -1 & -1 \end{bmatrix}, \begin{bmatrix} 4 & 1 \\ -1 & 5 \end{bmatrix}, \begin{bmatrix} 1 & 1 \\ -1 & -1 \end{bmatrix} \right\}$

9. Let $\mathbf{V}$ be the vector space $\mathcal{P}_2$ over $\mathbf{R}$.

(a) Find the matrix of transition from the basis $\{1, x, x^2\}$ of $\mathbf{V}$ to the basis $\mathcal{B}$ in Problem 3.

(b) Find the matrix of transition from the basis $\mathcal{B}$ in Problem 3 to the basis $\{1, x, x^2\}$ of $\mathbf{V}$.

4.5 Matrices over an Arbitrary Field

At this point the generalization of Chapters 1 and 2 to arbitrary vector spaces is complete. This makes available the basis for the theory of multiplication of matrices over an arbitrary field $\mathcal{F}$. An examination of the theory in Chapter 3 reveals that the entire development rests only on the facts that $\mathbf{R}$ is a field and $\mathbf{R}^n$ is an n-dimensional vector space over $\mathbf{R}$. With a single exception, in order to obtain the general form of a result, it is necessary only to replace $\mathbf{R}$ by $\mathcal{F}$ and $\mathbf{R}^n$ by an n-dimensional vector space over $\mathcal{F}$. The single exception is in the definition of the column space of a matrix in Definition 3.43. In this definition, $\mathbf{R}^m$ is replaced by $\mathcal{F}^m$.

From this point on, we shall make free use of the general forms of all the properties of matrix multiplication derived in Chapter 3.

4.6 Systems of Linear Equations

Let $\mathbf{Q}_n$ denote the set of all linear equations in n unknowns $x_1, x_2, \ldots, x_n$ with coefficients in a field $\mathcal{F}$. That is,

$$\mathbf{Q}_n = \{a_1 x_1 + a_2 x_2 + \cdots + a_n x_n = b \mid a_i \in \mathcal{F} \text{ and } b \in \mathcal{F}\},$$

and an element of $\mathbf{Q}_n$ is an equation $a_1 x_1 + a_2 x_2 + \cdots + a_n x_n = b$. Two elements of $\mathbf{Q}_n$ are **equal** if their corresponding coefficients are equal and their constant terms are equal. **Addition**, in $\mathbf{Q}_n$ is defined by

$$[a_1 x_1 + a_2 x_2 + \cdots + a_n x_x = b] + [a_1' x_1 + a_2' x_2 + \cdots + a_n' x_n = b']$$
$$= [(a_1 + a_1') x_1 + (a_2 + a_2') x_2 + \cdots + (a_n + a_n') x_n = (b + b')].$$

Scalar multiplication is given by

$$a \cdot [a_1 x_1 + a_2 x_2 + \cdots + a_n x_n = b] = [a a_1 x_1 + a a_2 x_2 + \cdots + a a_n x_n = ab].$$

It is readily verified that $\mathbf{Q}_n$ is a vector space over $\mathcal{F}$.

A system of linear equations in $x_1, x_2, \ldots, x_n$ with coefficients in $\mathcal{F}$ such as

$$a_{11} x_1 + a_{12} x_2 + \cdots + a_{1n} x_n = b_1$$
$$a_{21} x_1 + a_{22} x_2 + \cdots + a_{2n} x_n = b_2$$
$$\vdots$$
$$a_{m1} x_1 + a_{m2} x_2 + \cdots + a_{mn} x_n = b_m$$

can be regarded as a set of vectors $\mathcal{A} = \{\mathbf{v}_1, \mathbf{v}_2, \ldots, \mathbf{v}_m\}$ in $\mathbf{Q}_n$, where $\mathbf{v}_i$ is the ith equation in the system. A *solution* of the system $\mathcal{A}$ is a set of values for $x_1, x_2, \ldots, x_n$ that satisfies each equation in $\mathcal{A}$. The set of all solutions is the *solution set* of $\mathcal{A}$. In some cases, it is more convenient to write these solutions in vector form as $\mathbf{v} = (x_1, x_2, \ldots, x_n)$, while in others, the matrix form

$$X = \begin{bmatrix} x_1 \\ x_2 \\ \vdots \\ x_n \end{bmatrix}$$

is more convenient to use.

4.13 THEOREM

With the notation of the preceding paragraph, let $\langle \mathcal{A} \rangle$ denote the subspace of $\mathbf{Q}_n$ that is spanned by $\mathcal{A}$. Then each solution of the system $\mathcal{A}$ is a solution of every equation in the subspace $\langle \mathcal{A} \rangle$.

Proof Any equation $\mathbf{v}$ in $\langle \mathcal{A} \rangle$ is a linear combination of the equations in $\mathcal{A}$, $\mathbf{v} = \sum_{i=1}^{m} c_i \mathbf{v}_i$. Thus, $\mathbf{v}$ has the form

$$c_1[a_{11}x_1 + \cdots + a_{1n}x_n = b_1]$$
$$+ c_2[a_{21}x_1 + \cdots + a_{2n}x_n = b_2]$$
$$+ \cdots + c_m[a_{m1}x_1 + \cdots + a_{mn}x_n = b_m]$$

or

$$\left(\sum_{j=1}^{m} c_j a_{j1} \right) x_1 + \left(\sum_{j=1}^{m} c_j a_{j2} \right) x_2 + \cdots + \left(\sum_{j=1}^{m} c_j a_{jn} \right) x_n = \left(\sum_{j=1}^{m} c_j b_j \right).$$

If $x_1 = d_1, x_2 = d_2, \ldots, x_n = d_n$ is a solution of the system $\mathcal{A}$, then

$$a_{i1}d_1 + a_{i2}d_2 + \cdots + a_{in}d_n = b_i \text{ for } i = 1, 2, \ldots, m.$$

Therefore

$$\left(\sum_{j=1}^{m} c_j a_{j1} \right) d_1 + \left(\sum_{j=1}^{m} c_j a_{j2} \right) d_2 + \cdots + \left(\sum_{j=1}^{m} c_j a_{jn} \right) d_n$$
$$= c_1(a_{11}d_1 + \cdots + a_{1n}d_n) + c_2(a_{21}d_1 + \cdots + a_{2n}d_n) + \cdots$$
$$+ c_m(a_{m1}d_1 + \cdots + a_{mn}d_n)$$
$$= c_1 b_1 + c_2 b_2 + \cdots + c_m b_m,$$

and $x_1 = d_1, x_2 = d_2, \ldots, x_n = d_n$ is a solution of $\mathbf{v}$. ■■■

4.14 DEFINITION

Two systems $\mathcal{A}$ and $\mathcal{B}$ contained in $\mathbf{Q}_n$ are **equivalent** if and only if they have the same solutions.

The concept of elementary operations on sets of vectors applies in $\mathbf{Q}_n$ and is a useful tool in solving systems of equations.

4.15 THEOREM

Let $\mathcal{A}$ be a system of m equations contained in $\mathbf{Q}_n$. If the set $\mathcal{B}$ in $\mathbf{Q}_n$ is obtained from $\mathcal{A}$ by a sequence of elementary operations, then $\mathcal{B}$ and $\mathcal{A}$ are equivalent systems of equations.

Proof Suppose that $\mathcal{B}$ is obtained from $\mathcal{A}$ by a sequence of elementary operations. Then $\mathcal{B} \subseteq \langle \mathcal{A} \rangle$, so that every solution of $\mathcal{A}$ is a solution of $\mathcal{B}$ by Theorem 4.13.

But since $\mathcal{B}$ is obtained from $\mathcal{A}$ by a sequence of elementary operations, $\mathcal{A}$ can be obtained from $\mathcal{B}$ by a sequence of elementary operations (Theorem 2.4). Therefore, $\mathcal{A} \subseteq \langle \mathcal{B} \rangle$, and every solution of $\mathcal{B}$ is a solution of $\mathcal{A}$. This completes the proof. ∎ ∎ ∎

EXAMPLE 1 Consider the system of equations $\mathcal{A} = \{\mathbf{v}_1, \mathbf{v}_2, \mathbf{v}_3\}$ where:

$$\mathbf{v}_1 = [x_1 + 2x_2 - x_3 = 1]$$

$$\mathbf{v}_2 = [2x_1 - x_2 + x_3 = 10]$$

$$\mathbf{v}_3 = [5x_2 + 2x_3 = -3]$$

Applying the following elementary operations on $\mathcal{A}$:

$$\text{Replace } \mathbf{v}_2 \text{ with } -2\mathbf{v}_1 + \mathbf{v}_2: \quad \mathbf{v}_2' = [-5x_2 + 3x_3 = 8]$$

$$\text{Replace } \mathbf{v}_3 \text{ with } \mathbf{v}_2' + \mathbf{v}_3: \quad \mathbf{v}_3' = [5x_3 = 5]$$

yields $\mathcal{B} = \{\mathbf{v}_1, \mathbf{v}_2', \mathbf{v}_3'\}$. Now $\mathcal{A}$ and $\mathcal{B}$ are equivalent systems since the solution set of each in vector form is $\mathbf{v} = (4, -1, 1)$. ∎

Matrices are valuable tools in the solution of systems of linear equations. Consider the system $\mathcal{A}$ given by

$$a_{11}x_1 + a_{12}x_2 + \cdots + a_{1n}x_n = b_1$$

$$a_{21}x_1 + a_{22}x_2 + \cdots + a_{2n}x_n = b_2$$

$$\vdots$$

$$a_{m1}x_1 + a_{m2}x_2 + \cdots + a_{mn}x_n = b_m.$$

This system can be written compactly as a single matrix equation $AX = B$, where

$$A = [a_{ij}]_{m \times n}, \quad X = \begin{bmatrix} x_1 \\ x_2 \\ \vdots \\ x_n \end{bmatrix}, \quad \text{and} \quad B = \begin{bmatrix} b_1 \\ b_2 \\ \vdots \\ b_m \end{bmatrix}.$$

EXAMPLE 2   The system $\mathcal{A}$ of linear equations from Example 1 written in the form

$$\begin{aligned} x_1 + 2x_2 - x_3 &= 1 \\ 2x_1 - x_2 + x_3 &= 10 \\ 5x_2 + 2x_3 &= -3 \end{aligned}$$

is equivalent to the single matrix equation

$$\begin{bmatrix} x_1 + 2x_2 - x_3 \\ 2x_1 - x_2 + x_3 \\ 5x_2 + 2x_3 \end{bmatrix} = \begin{bmatrix} 1 \\ 10 \\ -3 \end{bmatrix}.$$

This matrix equation can be factored as

$$\begin{bmatrix} 1 & 2 & -1 \\ 2 & -1 & 1 \\ 0 & 5 & 2 \end{bmatrix} \begin{bmatrix} x_1 \\ x_2 \\ x_3 \end{bmatrix} = \begin{bmatrix} 1 \\ 10 \\ -3 \end{bmatrix}.$$

Thus we have a matrix equation that is equivalent to the original system and has the form $AX = B$, with

$$A = \begin{bmatrix} 1 & 2 & -1 \\ 2 & -1 & 1 \\ 0 & 5 & 2 \end{bmatrix}, \quad X = \begin{bmatrix} x_1 \\ x_2 \\ x_3 \end{bmatrix}, \quad B = \begin{bmatrix} 1 \\ 10 \\ -3 \end{bmatrix}. \qquad \blacksquare$$

4.16
DEFINITION

In the system $AX = B$, the matrix A is called the **coefficient matrix**, X is the **matrix of unknowns**, and B is the **matrix of constants**. The matrix

$$[A \mid B] = \begin{bmatrix} a_{11} & a_{12} & \cdots & a_{1n} & b_1 \\ a_{21} & a_{22} & \cdots & a_{2n} & b_2 \\ \vdots & \vdots & & \vdots & \vdots \\ a_{m1} & a_{m2} & \cdots & a_{mn} & b_m \end{bmatrix}$$

is called the **augmented matrix** of the system.

Each system in a certain set of variables such as $\mathcal{A}$ above has a unique augmented matrix. That is, each system has an augmented matrix, and different systems have different augmented matrices. Any elementary operation performed on $\mathcal{A}$ is reflected in the augmented matrix as an elementary row operation, and any elementary row operation on the augmented matrix produces a corresponding elementary operation on $\mathcal{A}$.

4.17
THEOREM

Two systems $AX = B$ and $A'X = B'$ of m linear equations in n unknowns $x_1, x_2, \ldots, x_n$ are equivalent systems if their augmented matrices $[A \mid B]$ and $[A' \mid B']$ are row-equivalent.

Proof Let $\mathcal{A}$ and $\mathcal{A}'$ be the sets of vectors in $\mathbf{Q}_n$ that consist of the equations in the systems $AX = B$ and $A'X = B'$, respectively.

If the augmented matrices $[A \mid B]$ and $[A' \mid B']$ are row-equivalent, then $[A' \mid B']$ can be obtained from $[A \mid B]$ by a sequence of elementary row operations. Hence $\mathcal{A}'$ can be obtained from $\mathcal{A}$ by a sequence of elementary operations and $\mathcal{A}$ and $\mathcal{A}'$ have the same solutions, by Theorem 4.15. ■■■

4.18 COROLLARY

Let A be $m \times n$ in the system $AX = B$, and let P be any invertible $m \times m$ matrix. Then $AX = B$ and $PAX = PB$ have the same solutions.

Proof This follows from the fact that the augmented matrices $[A \mid B]$ and $[PA \mid PB] = P[A \mid B]$ are row-equivalent. ■■■

EXAMPLE 3 The augmented matrix $[A \mid B]$ for the system $\mathcal{A}$ of linear equations in Example 1 is

$$\begin{bmatrix} 1 & 2 & -1 & 1 \\ 2 & -1 & 1 & 10 \\ 0 & 5 & 2 & -3 \end{bmatrix}.$$

The augmented matrix $[A' \mid B']$ for the system $\mathcal{B}$ of linear equations in that same example can be obtained from $[A \mid B]$ by applying two elementary row operations corresponding to the two elementary operations applied to the set of vectors (equations) in $\mathcal{A}$.

$$[A \mid B] = \begin{bmatrix} 1 & 2 & -1 & 1 \\ 2 & -1 & 1 & 10 \\ 0 & 5 & 2 & -3 \end{bmatrix} \rightarrow \begin{bmatrix} 1 & 2 & -1 & 1 \\ 0 & -5 & 3 & 8 \\ 0 & 5 & 2 & -3 \end{bmatrix} \rightarrow \begin{bmatrix} 1 & 2 & -1 & 1 \\ 0 & -5 & 3 & 8 \\ 0 & 0 & 5 & 5 \end{bmatrix}$$

$$= [A' \mid B']$$

The systems $AX = B$ and $A'X = B'$ are equivalent since their augmented matrices $[A \mid B]$ and $[A' \mid B']$ are row-equivalent. ■

4.19 THEOREM

The system $AX = B$ has a solution

1. if and only if $\mathbf{b} = (b_1, b_2, \ldots, b_m)$ is in the column space of A or equivalently

2. if and only if $\text{rank}([A \mid B]) = \text{rank}(A)$.

Proof The system $AX = B$ can be rewritten in the form

$$\begin{bmatrix} a_{11} \\ a_{21} \\ \vdots \\ a_{m1} \end{bmatrix} x_1 + \begin{bmatrix} a_{12} \\ a_{22} \\ \vdots \\ a_{m2} \end{bmatrix} x_2 + \cdots + \begin{bmatrix} a_{1n} \\ a_{2n} \\ \vdots \\ a_{mn} \end{bmatrix} x_n = \begin{bmatrix} b_1 \\ b_2 \\ \vdots \\ b_m \end{bmatrix}.$$

Thus the system has a solution if and only if there are scalars $x_1, x_2, \ldots, x_n$ in $\mathcal{F}$ such that

$$x_1(a_{11}, a_{21}, \ldots, a_{m1}) + x_2(a_{12}, a_{22}, \ldots, a_{m2}) + \cdots$$
$$+ x_n(a_{1n}, a_{2n}, \ldots, a_{mn}) = (b_1, b_2, \ldots, b_m).$$

Let $\mathbf{S}$ denote the column space of A, let $\mathbf{S}^*$ denote the column space of $[A \mid B]$, and let $\mathbf{b} = (b_1, b_2, \ldots, b_m) \in \mathcal{F}^m$. According to the last statement of the preceding paragraph, $AX = B$ has a solution if and only if $\mathbf{b}$ is in $\mathbf{S}$. But $\mathbf{b}$ is in $\mathbf{S}$ if and only if $\mathbf{S}$ and $\mathbf{S}^*$ have the same dimension, i.e., if and only if A and $[A \mid B]$ have the same rank. ■ ■ ■

**4.20
THEOREM**

If A is an $m \times n$ matrix over $\mathcal{F}$ and rank($[A \mid B]$) = rank(A) = r, then the solutions to $AX = B$ can be expressed in terms of $n - r$ parameters.

Proof Let P be an invertible matrix such that $PA = A'$ is in reduced row-echelon form. The original system is equivalent to $A'X = B'$ where $PB = B'$ (Theorem 4.17). By Corollary 3.49, A' has r nonzero rows. Thus the system $A'X = B'$ has the form

$$x_{k_1} + a'_{1,k_1+1}x_{k_1+1} + \cdots + a'_{1,k_2-1}x_{k_2-1} + \cdots$$
$$+ a'_{1,k_2+1}x_{k_2+1} + \cdots + a'_{1,n}x_n = b'_1$$
$$x_{k_2} + a'_{2,k_2+1}x_{k_2+1} + \cdots + a'_{2,n}x_n = b'_2$$
$$\vdots$$
$$x_{k_r} + \cdots + a'_{r,n}x_n = b'_r$$
$$0 = b'_{r+1}$$
$$\vdots$$
$$0 = b'_m.$$

In this system each variable x_{k_i}, $i = 1, 2, \ldots, r$ occurs just once in the ith equation with a coefficient 1. Hence each of $x_{k_1}, \ldots, x_{k_r}$ can be expressed in terms of the remaining $n - r$ variables. ■ ■ ■

The variables $x_{k_1}, x_{k_2}, \ldots, x_{k_r}$ in the last paragraph are called the **leading variables**, and the remaining $n - r$ variables are called the **parameters** in the solution of the system.

The proof of Theorem 4.20 furnishes a method for determining the existence of solutions and also a method for obtaining them. To solve the system $AX = B$, we can use elementary row operations to transform the augmented matrix $[A \mid B]$ into reduced row-echelon form $[A' \mid B']$. The condition that rank($[A \mid B]$) = rank(A) is reflected in the conditions $0 = b'_{r+1}, \ldots, 0 = b'_m$ since

$$\text{rank}(A) = \text{rank}(PA) = \text{rank}(A')$$

and

$$\text{rank}([A \mid B]) = \text{rank}(P[A \mid B]) = \text{rank}([A' \mid B']).$$

If $\text{rank}([A \mid B]) > \text{rank}(A)$, then $\text{rank}([A' \mid B']) > r$ and at least one of the equations $0 = b'_{r+1}, \ldots, 0 = b'_n$ will be contradictory. The system thus has no solution and we say that the system in **inconsistent**. If $\text{rank}([A \mid B]) = \text{rank}(A)$, then there are solutions (the system is **consistent**), and the solutions can be obtained by solving for the leading variables in terms of the parameters. This method of solution is called **Gauss-Jordan elimination**. A system is solved by Gauss-Jordan elimination in Example 4.

EXAMPLE 4 Consider the system of equations

$$\begin{aligned}
x_1 + 2x_2 + x_3 - 4x_4 + x_5 &= 1 \\
x_1 + 2x_2 - x_3 + 2x_4 + x_5 &= 5 \\
2x_1 + 4x_2 + x_3 - 5x_4 &= 2 \\
x_1 + 2x_2 + 3x_3 - 10x_4 + x_5 &= -3.
\end{aligned}$$

The augmented matrix $[A \mid B]$ can be transformed to reduced row-echelon form as follows.

$$[A \mid B] = \begin{bmatrix} 1 & 2 & 1 & -4 & 1 & 1 \\ 1 & 2 & -1 & 2 & 1 & 5 \\ 2 & 4 & 1 & -5 & 0 & 2 \\ 1 & 2 & 3 & -10 & 1 & -3 \end{bmatrix} \rightarrow \begin{bmatrix} 1 & 2 & 1 & -4 & 1 & 1 \\ 0 & 0 & -2 & 6 & 0 & 4 \\ 0 & 0 & -1 & 3 & -2 & 0 \\ 0 & 0 & 2 & -6 & 0 & -4 \end{bmatrix}$$

$$\rightarrow \begin{bmatrix} 1 & 2 & 0 & -1 & 1 & 3 \\ 0 & 0 & 1 & -3 & 0 & -2 \\ 0 & 0 & 0 & 0 & -2 & -2 \\ 0 & 0 & 0 & 0 & 0 & 0 \end{bmatrix} \rightarrow \begin{bmatrix} 1 & 2 & 0 & -1 & 0 & 2 \\ 0 & 0 & 1 & -3 & 0 & -2 \\ 0 & 0 & 0 & 0 & 1 & 1 \\ 0 & 0 & 0 & 0 & 0 & 0 \end{bmatrix} = [A' \mid B']$$

The reduced row-echelon form $[A' \mid B']$ corresponds to the system

$$\begin{aligned}
x_1 + 2x_2 - x_4 &= 2 \\
x_3 - 3x_4 &= -2 \\
x_5 &= 1.
\end{aligned}$$

When we solve for the leading variables x_1, x_3, x_5 in terms of the parameters x_2, x_4, we obtain the solutions to the system:

$$\begin{aligned}
x_1 &= 2 - 2x_2 + x_4 \\
x_3 &= -2 + 3x_4 \\
x_5 &= 1,
\end{aligned}$$

where x_2 and x_4 are arbitrary. ∎

4.6 Exercises

*In Problems 1–12, (**a**) find all solutions of the given system of equations, and (**b**) write the column of constants as a linear combination of the columns of the coefficient matrix.*

1.
$$x_1 + 2x_2 + x_3 = 5$$
$$-x_1 - x_2 + x_3 = 2$$
$$x_2 + 3x_3 = 1$$

2.
$$x_1 + x_2 - x_3 = 4$$
$$3x_1 + 3x_2 - 2x_3 = 11$$
$$4x_1 + 5x_2 - 3x_3 = 17$$

3.
$$2x_1 \qquad + x_3 = 1$$
$$x_1 + x_2 \qquad = -2$$
$$2x_1 + 4x_2 - x_3 = -3$$

4.
$$x_1 - x_2 - x_3 = 0$$
$$2x_1 \qquad - x_3 = 1$$
$$3x_1 + x_2 - x_3 = 2$$

5.
$$2x_1 + x_2 = -4$$
$$x_1 - x_2 = 4$$
$$-3x_1 + 3x_2 = 2$$

6.
$$9x_1 - 6x_2 = 15$$
$$15x_1 - 10x_2 = 25$$
$$6x_1 - 4x_2 = 10$$

7.
$$x_1 - 2x_2 - 2x_3 - 3x_4 = 1$$
$$2x_1 - 4x_2 + 2x_3 \qquad = 2$$
$$3x_1 - 6x_2 + x_3 - 2x_4 = 3$$
$$x_3 + x_4 = 0$$

8.
$$4x_1 + 3x_2 + 2x_3 - x_4 = 4$$
$$5x_1 + 4x_2 + 3x_3 - x_4 = 4$$
$$-2x_1 - 2x_2 - x_3 + 2x_4 = -3$$
$$11x_1 + 6x_2 + 4x_3 + x_4 = 11$$

9.
$$x_1 + 3x_2 - x_3 \qquad + 2x_5 = 2$$
$$2x_1 + 6x_2 + x_3 + 6x_4 + 4x_5 = 13$$
$$-x_1 - 3x_2 \qquad - 2x_4 - 2x_5 = -5$$

10.
$$x_1 + x_2 + x_3 + x_4 + x_5 = 2$$
$$x_1 + x_2 + 2x_3 + 3x_4 + 4x_5 = 4$$
$$2x_1 + 2x_2 + 3x_3 + 4x_4 + 5x_5 = 6$$

11.
$$x_1 + 2x_2 \qquad + x_4 \qquad = 1$$
$$2x_1 + 4x_2 + x_3 + 4x_4 + 3x_5 = 6$$
$$x_1 + 2x_2 + 2x_3 + 4x_4 - 2x_5 = 1$$
$$-x_1 - 2x_2 + 3x_3 + 5x_4 + 4x_5 = 6$$

12.
$$x_1 + x_2 + 3x_3 + 2x_4 \qquad = 0$$
$$x_1 - x_2 + x_3 \qquad = 2$$
$$2x_1 + 2x_2 + 6x_3 + 4x_4 \qquad = 0$$
$$3x_1 \qquad + 6x_3 + 3x_4 - x_5 = 3$$

*In Problems 13–18, (**a**) find the rank of the coefficient matrix, (**b**) find the rank of the augmented matrix, and (**c**) determine if the system is consistent by comparing these ranks. It is not necessary to solve the systems.*

13.
$$x_1 + 2x_2 - x_3 = 3$$
$$x_2 + x_3 = 1$$
$$x_1 \qquad - 3x_3 = 0$$

14.
$$x_1 + 2x_2 + 3x_3 + x_4 = 0$$
$$x_2 + x_3 + x_4 = 0$$
$$x_1 \qquad + x_3 - x_4 = 1$$

15.
$$\begin{aligned} x_1 + 2x_2 - x_3 &= 2 \\ -3x_1 - x_2 + x_3 &= -3 \\ -x_1 + 3x_2 - x_3 &= 1 \end{aligned}$$

16.
$$\begin{aligned} x_1 \quad\quad + x_3 - x_4 &= 3 \\ 2x_2 - x_3 + 2x_4 &= -2 \\ 3x_1 - 4x_2 + 5x_3 - 7x_4 &= 13 \end{aligned}$$

17.
$$\begin{aligned} 2x_1 - x_2 \quad\quad + x_4 &= 5 \\ 3x_2 + 5x_3 - x_4 &= -7 \\ x_1 \quad\quad - x_3 + x_4 &= 4 \\ x_1 - 2x_2 - 6x_3 + x_4 &= 6 \end{aligned}$$

18.
$$\begin{aligned} x_2 + x_3 - 4x_4 &= 6 \\ -x_1 \quad\quad x_3 + x_4 &= 0 \\ 2x_1 \quad - 2x_3 - x_4 &= -1 \\ x_2 + x_3 + 3x_4 &= 5 \end{aligned}$$

19. Find all real numbers a and b for which the system of equations below does not have a solution.

$$\begin{aligned} x_1 \quad\quad + x_3 &= 1 \\ ax_1 + x_2 + 2x_3 &= 0 \\ 3x_1 + 4x_2 + bx_3 &= 2 \end{aligned}$$

In Problems 20 and 21, find the values of the real number a for which the given system (**a**) *has no solution,* (**b**) *has exactly one solution,* (**c**) *has infinitely many solutions.*

20.
$$\begin{aligned} x_1 + 2x_2 - x_3 &= 2 \\ 2x_1 + 6x_2 + 3x_3 &= 4 \\ 3x_1 + 8x_2 + (a^2 - 2)x_3 &= a + 8 \end{aligned}$$

21.
$$\begin{aligned} x_1 + x_2 + x_3 &= 3 \\ 2x_1 + 3x_2 + 3x_3 &= 8 \\ 3x_1 + 3x_2 + (a^2 - 6)x_3 &= a + 6 \end{aligned}$$

22. Let $B = \{[x_1 = 0], [x_2 = 0], \ldots, [x_n = 0], [x_1 = 1]\}$.

(**a**) Prove that B is a basis of Q_n.

(**b**) What is the dimension of Q_n over $\mathcal{F}$?

23. Prove or disprove each of the following statements:

(**a**) If v is a solution of the system of equations $\mathcal{A} = \{v_1, v_2, \ldots, v_m\}$ in Q_n over $\mathcal{F}$, then cv is also a solution for any c in $\mathcal{F}$.

(**b**) If v and v' are solutions to the system of equations $\mathcal{A} = \{v_1, v_2, \ldots, v_m\}$ in Q_n over $\mathcal{F}$, then $v + v'$ is also a solution.

4.7 More on Systems of Linear Equations

Of particular interest are the systems of equations $AX = B$ with $B = 0$. A system of the form $AX = 0$ is called a **homogeneous system** and always has at least one solution $X = 0$, called the **trivial solution**. However, the trivial solution may not be the only solution to the homogeneous system and the proof of Theorem 4.20 furnishes a method of finding these solutions.

There is another point of view to Theorem 4.20 and its proof that is useful in Chapter 5. The fifth set in Example 1 of Section 1.3 generalizes immediately from $\mathbf{R}^n$ to $\mathcal{F}^n$. That is, the set of all vectors $v = (x_1, x_2, \ldots, x_n)$ in $\mathcal{F}^n$ with components x_i that satisfy a given system of equations $AX = 0$ is a subspace W of $\mathcal{F}^n$. This subspace is called the **nullspace** of the matrix A and its dimension is called the **nullity** of A. The $n - r$ parameters in the proof of Theorem 4.20 represent the components of $\mathbf{v}$ that can be assigned values arbitrarily. We can solve for the leading variables $x_{k_1}, x_{k_2}, \ldots, x_{k_r}$ in the equation $A'X = \mathbf{0}$ and express them in terms of the $n-r$ parameters. If we then replace these leading variables by their values in terms of the parameters, the vector $(x_1, x_2, \ldots, x_n)$ that represents the general solution of the system can be obtained as a linear combination of the vectors in a basis of the nullspace $\mathbf{W}$ of A. There will be $n - r$ vectors in this basis since there are $n - r$ parameters present and the nullity of A is $n - r$. This is illustrated in the following example.

EXAMPLE 1   Consider the system of equations

$$
\begin{aligned}
x_1 + 2x_2 + x_3 - 4x_4 + x_5 &= 0 \\
x_1 + 2x_2 - x_3 + 2x_4 + x_5 &= 0 \\
2x_1 + 4x_2 + x_3 - 5x_4 &= 0 \\
x_1 + 2x_2 + 3x_3 - 10x_4 + x_5 &= 0.
\end{aligned}
$$

The coefficient matrix in this system is the same as the one in Example 4 of Section 4.6, and the reduced row-echelon form for this system can be obtained simply by replacing the constants B' in $[A' \mid B']$ by a column of zeros to obtain $[A' \mid \mathbf{0}]$. This gives

$$
[A' \mid \mathbf{0}] = \begin{bmatrix}
1 & 2 & 0 & -1 & 0 & 0 \\
0 & 0 & 1 & -3 & 0 & 0 \\
0 & 0 & 0 & 0 & 1 & 0 \\
0 & 0 & 0 & 0 & 0 & 0
\end{bmatrix},
$$

which corresponds to the system

$$
\begin{aligned}
x_1 + 2x_2 - x_4 &= 0 \\
x_3 - 3x_4 &= 0 \\
x_5 &= 0.
\end{aligned}
$$

Solving for the leading variables, we get

$$
\begin{aligned}
x_1 &= -2x_2 + x_4 \\
x_3 &= 3x_4 \\
x_5 &= 0.
\end{aligned}
$$

Replacing the leading variables by their values in terms of the parameters yields

$$
\begin{aligned}
(x_1, x_2, x_3, x_4, x_5) &= (-2x_2 + x_4, x_2, 3x_4, x_4, 0) \\
&= (-2x_2, x_2, 0, 0, 0) + (x_4, 0, 3x_4, x_4, 0) \\
&= x_2(-2, 1, 0, 0, 0) + x_4(1, 0, 3, 1, 0).
\end{aligned}
$$

Thus the subspace $\mathbf{W}$ of all solutions to $AX = \mathbf{0}$ is given by

$$
\mathbf{W} = \langle (-2, 1, 0, 0, 0), (1, 0, 3, 1, 0) \rangle,
$$

and $\{(-2, 1, 0, 0, 0), (1, 0, 3, 1, 0)\}$ is a basis for the nullspace **W**. As a final remark concerning these solutions, we note that the matrix form

$$
X = x_2 \begin{bmatrix} -2 \\ 1 \\ 0 \\ 0 \\ 0 \end{bmatrix} + x_4 \begin{bmatrix} 1 \\ 0 \\ 3 \\ 1 \\ 0 \end{bmatrix}
$$

can be easily predicted by listing *all* the variables in a column with the *leading* variables expressed in terms of the parameters:

$$
\begin{aligned}
x_1 &= -2x_2 + x_4 \\
x_2 &= x_2 \\
x_3 &= 3x_4 \\
x_4 &= x_4 \\
x_5 &= 0.
\end{aligned}
$$
■

If **W** is the nullspace of A and $\mathbf{c} = (c_1, c_2, \ldots, c_n)$, where $x_1 = c_1, x_2 = c_2, \ldots, x_n = c_n$ is a particular solution to the system $AX = B$, then it can be shown that any vector in $\mathbf{c} + \mathbf{W}$ is a solution to $AX = B$ and that any solution to the system is in $\mathbf{c} + \mathbf{W}$.

EXAMPLE 2 In Example 4 of Section 4.6, the solution to

$$
\begin{aligned}
x_1 + 2x_2 + x_3 - 4x_4 + x_5 &= 1 \\
x_1 + 2x_2 - x_3 + 2x_4 + x_5 &= 5 \\
2x_1 + 4x_2 + x_3 - 5x_4 &= 2 \\
x_1 + 2x_2 + 3x_3 - 10x_4 + x_5 &= -3
\end{aligned}
$$

can be written in matrix form

$$
X = \begin{bmatrix} 2 \\ 0 \\ -2 \\ 0 \\ 1 \end{bmatrix} + x_2 \begin{bmatrix} -2 \\ 1 \\ 0 \\ 0 \\ 0 \end{bmatrix} + x_4 \begin{bmatrix} 1 \\ 0 \\ 3 \\ 1 \\ 0 \end{bmatrix}.
$$

Thus any solution $\mathbf{v}$ to $AX = B$ can be expressed as $\mathbf{v} = \mathbf{c} + \mathbf{w}$ where $\mathbf{c} = (2, 0, -2, 0, 1)$ is a solution of $AX = B$ and $\mathbf{w} = x_2(-2, 1, 0, 0, 0) + x_4(1, 0, 3, 1, 0)$ is in the nullspace **W** of A.
■

The LU decomposition of the coefficient matrix A in the system $AX = B$ can be useful in solving the system especially when this system must be solved for several different matrices B. If A has an LU decomposition then the system $AX = B$ can be written as

$$
LUX = B
$$

and if we set $Y = UX$, then the system becomes

$$
LY = B.
$$

Since L is a unit lower triangular matrix, this system is easily solved for Y. Once Y is known then the coefficient matrix U in the system

$$Y = UX$$

is upper triangular and this system is easily solved for X.

EXAMPLE 3   The coefficient matrix A for the system of equations

$$\begin{aligned} x_1 + x_2 - x_3 + 4x_4 &= 7 \\ x_1 + 3x_2 - 2x_3 + 6x_4 &= 13 \\ 2x_1 \phantom{{}+3x_2} - x_3 + 7x_2 &= 9 \end{aligned}$$

can be factored as

$$A = LU = \begin{bmatrix} 1 & 0 & 0 \\ 1 & 1 & 0 \\ 2 & -1 & 1 \end{bmatrix} \begin{bmatrix} 1 & 1 & -1 & 4 \\ 0 & 2 & -1 & 2 \\ 0 & 0 & 0 & 1 \end{bmatrix}.$$

Setting $UX = Y$ in the matrix equation $AX = B$ gives

$$LY = \begin{bmatrix} 1 & 0 & 0 \\ 1 & 1 & 0 \\ 2 & -1 & 1 \end{bmatrix} \begin{bmatrix} y_1 \\ y_2 \\ y_3 \end{bmatrix} = \begin{bmatrix} 7 \\ 13 \\ 9 \end{bmatrix} = B.$$

Solving for the unknowns in the system of equations

$$\begin{aligned} y_1 \phantom{{}+ y_2 + y_3} &= 7 \\ y_1 + y_2 \phantom{{}+ y_3} &= 13 \\ 2y_1 - y_2 + y_3 &= 9 \end{aligned}$$

gives $y_1 = 7$, $y_2 = 6$, $y_3 = 1$. Using these values in $UX = Y$ we have the matrix equation

$$\begin{bmatrix} 1 & 1 & -1 & 4 \\ 0 & 2 & -1 & 2 \\ 0 & 0 & 0 & 1 \end{bmatrix} \begin{bmatrix} x_1 \\ x_2 \\ x_3 \\ x_4 \end{bmatrix} = \begin{bmatrix} 7 \\ 6 \\ 1 \end{bmatrix}.$$

The system of equations corresponding to this matrix equation is

$$\begin{aligned} x_1 + x_2 - x_3 + 4x_4 &= 7 \\ 2x_2 - x_3 + 2x_4 &= 6 \\ x_4 &= 1 \end{aligned}$$

whose solution written in matrix form is

$$\begin{bmatrix} x_1 \\ x_2 \\ x_3 \\ x_4 \end{bmatrix} = \begin{bmatrix} 1 \\ 2 \\ 0 \\ 1 \end{bmatrix} + x_3 \begin{bmatrix} \frac{1}{2} \\ \frac{1}{2} \\ 1 \\ 0 \end{bmatrix}.$$

For matrices that cannot be reduced to an upper triangular form without using an interchange of rows it is possible to find an elementary matrix P of type III such that the product PA can be reduced to upper triangular form. This type of situation is illustrated in the next example.

EXAMPLE 4 The system of equations

$$
\begin{array}{rcr}
2x_2 + x_3 = & 0 \\
2x_1 - x_2 \quad\quad = & 8 \\
x_1 \quad\quad - x_3 = & -1
\end{array}
$$

has matrix form $AX = B$ with

$$
A = \begin{bmatrix} 0 & 2 & 1 \\ 2 & -1 & 0 \\ 1 & 0 & -1 \end{bmatrix} \quad \text{and} \quad B = \begin{bmatrix} 0 \\ 8 \\ -1 \end{bmatrix}.
$$

The coefficient matrix A does not have an LU decomposition. However the product

$$
PA = \begin{bmatrix} 0 & 0 & 1 \\ 0 & 1 & 0 \\ 1 & 0 & 0 \end{bmatrix} \begin{bmatrix} 0 & 2 & 1 \\ 2 & -1 & 0 \\ 1 & 0 & -1 \end{bmatrix} = \begin{bmatrix} 1 & 0 & -1 \\ 2 & -1 & 0 \\ 0 & 2 & 1 \end{bmatrix}
$$

can be factored as

$$
PA = \begin{bmatrix} 1 & 0 & -1 \\ 2 & -1 & 0 \\ 0 & 2 & 1 \end{bmatrix} = \begin{bmatrix} 1 & 0 & 0 \\ 2 & 1 & 0 \\ 0 & -2 & 1 \end{bmatrix} \begin{bmatrix} 1 & 0 & -1 \\ 0 & -1 & 2 \\ 0 & 0 & 5 \end{bmatrix} = LU.
$$

Multiplying both sides of $AX = B$ by P on the left gives

$$
PAX = PB \quad \text{or} \quad LUX = PB
$$

where

$$
PB = \begin{bmatrix} 0 & 0 & 1 \\ 0 & 1 & 0 \\ 1 & 0 & 0 \end{bmatrix} \begin{bmatrix} 0 \\ 8 \\ -1 \end{bmatrix} = \begin{bmatrix} -1 \\ 8 \\ 0 \end{bmatrix}.
$$

The system $LUX = PB$ can be solved using the same techniques as the previous example. Again we replace UX with Y yielding the matrix equation $LY = PB$ or

$$
\begin{bmatrix} 1 & 0 & 0 \\ 2 & 1 & 0 \\ 0 & -2 & 1 \end{bmatrix} \begin{bmatrix} y_1 \\ y_2 \\ y_3 \end{bmatrix} = \begin{bmatrix} -1 \\ 8 \\ 0 \end{bmatrix}.
$$

Solving the resulting system

$$
\begin{aligned}
y_1 &= -1 \\
2y_1 + y_2 &= 8 \\
-2y_2 + y_3 &= 0
\end{aligned}
$$

gives $y_1 = -1$, $y_2 = 10$, and $y_3 = 20$. Now using these values in $UX = Y$, we have the matrix equation

$$
\begin{bmatrix} 1 & 0 & -1 \\ 0 & -1 & 2 \\ 0 & 0 & 5 \end{bmatrix}
\begin{bmatrix} x_1 \\ x_2 \\ x_3 \end{bmatrix}
=
\begin{bmatrix} -1 \\ 10 \\ 20 \end{bmatrix}.
$$

This leads to the system of equations

$$
\begin{aligned}
x_1 - x_3 &= -1 \\
-x_2 + 2x_3 &= 10 \\
5x_3 &= 20
\end{aligned}
$$

with solution $x_1 = 3$, $x_2 = -2$, $x_3 = 4$.   ∎

Although each of Examples 3 and 4 could have been solved more efficiently with Gauss-Jordan elimination, the LU decomposition is more popular for use on large (more than 30 equations) systems. As mentioned earlier, it is also more efficient if many systems with the same coefficient matrix and different constant matrices must be solved.

4.7   Exercises

*In Problems 1–12, (**a**) find all solutions of the system of equations, (**b**) find a basis of the nullspace of the coefficient matrix of the system, and (**c**) find the nullity of the coefficient matrix.*

1.
$$
\begin{aligned}
x_1 + x_2 - 2x_3 &= 0 \\
3x_1 - x_2 &= 0 \\
-x_1 + 3x_2 - 4x_3 &= 0
\end{aligned}
$$

2.
$$
\begin{aligned}
2x_1 - x_2 - x_3 &= 0 \\
-x_1 + x_2 - x_3 &= 0 \\
x_1 + x_2 - 5x_3 &= 0
\end{aligned}
$$

3.
$$
\begin{aligned}
x_1 - 2x_2 - x_3 + 2x_4 &= 0 \\
-x_1 - 2x_3 + x_4 &= 0 \\
x_1 + x_2 - x_3 - x_4 &= 0 \\
-2x_2 - 3x_3 + 3x_4 &= 0
\end{aligned}
$$

4.
$$
\begin{aligned}
x_1 + x_2 + 3x_4 &= 0 \\
x_1 - x_2 + 2x_3 - x_4 &= 0 \\
-x_1 + 5x_2 - 6x_3 + 9x_4 &= 0 \\
x_1 - x_2 + 6x_3 + 3x_4 &= 0
\end{aligned}
$$

5.
$$
\begin{aligned}
2x_1 - x_2 - x_3 &= 0 \\
-x_1 - 2x_2 + x_4 &= 0 \\
-5x_2 - x_3 + 2x_4 &= 0 \\
x_1 + 2x_2 - x_4 &= 0
\end{aligned}
$$

6.
$$
\begin{aligned}
x_1 + x_3 + x_5 &= 0 \\
-x_1 - x_2 + x_3 + 2x_4 - x_5 &= 0 \\
x_1 + x_2 - x_3 + x_4 + x_5 &= 0 \\
-x_2 + 2x_3 + 5x_4 &= 0 \\
x_1 - x_2 + 3x_3 + 5x_4 + x_5 &= 0
\end{aligned}
$$

7. $\begin{aligned} x_1 + 2x_2 + 5x_3 \qquad\quad &= 0 \\ 4x_1 + 12x_2 + 21x_3 + 2x_4 &= 0 \\ 3x_1 + 6x_2 + 15x_3 - 3x_4 &= 0 \end{aligned}$

8. $\begin{aligned} 4x_1 + 4x_2 - 7x_3 + 3x_4 &= 0 \\ 3x_1 + 3x_2 - 5x_3 + 2x_4 &= 0 \end{aligned}$

9. $\begin{aligned} x_1 - x_2 + 2x_3 - x_4 + x_5 &= 0 \\ -x_3 + x_4 + x_5 &= 0 \\ -x_1 + x_2 + x_3 + x_4 + 2x_5 &= 0 \end{aligned}$

10. $\begin{aligned} x_2 + 3x_3 - x_4 + 2x_5 &= 0 \\ x_1 + x_2 - x_3 \qquad\quad - x_5 &= 0 \\ -3x_1 - x_2 + 9x_3 - 2x_4 + 7x_5 &= 0 \end{aligned}$

11. $\begin{aligned} -2x_2 + 3x_3 &= 0 \\ x_1 + 2x_2 - 2x_3 &= 0 \\ 2x_1 + x_2 - x_3 &= 0 \end{aligned}$

12. $\begin{aligned} x_1 - x_2 + x_3 + 2x_4 &= 0 \\ -x_1 + 2x_2 - 2x_3 - x_4 &= 0 \\ -x_2 - 2x_3 - x_4 &= 0 \\ x_1 - 3x_2 \qquad\quad + 3x_4 &= 0 \end{aligned}$

In Problems 13 and 14, find a basis of the nullspace of $A - xI$ for the given matrix A and each value of x.

13. $A = \begin{bmatrix} -4 & 1 & 3 \\ 4 & 1 & -4 \\ -3 & 1 & 2 \end{bmatrix}; x = 1, -1$

14. $A = \begin{bmatrix} 0 & -4 & -2 & 2 \\ 1 & 5 & -1 & 1 \\ -3 & -3 & 1 & 3 \\ 0 & 0 & 0 & 4 \end{bmatrix}; x = -2, 4$

In Problems 15–18, express the solution to the system in the form $\mathbf{v} = \mathbf{c} + \mathbf{w}$, where $\mathbf{c}$ is a particular solution to the system and $\mathbf{w}$ is in the nullspace of the coefficient matrix.

15. $\begin{aligned} x_1 - 4x_2 - 4x_3 + 5x_4 &= 2 \\ -x_1 \qquad\quad - 2x_3 + x_4 &= -4 \\ -2x_2 - 3x_3 + 3x_4 &= -1 \end{aligned}$

16. $\begin{aligned} 2x_1 - x_2 - 2x_3 \qquad\quad &= -2 \\ -x_1 \qquad\quad - 2x_3 + x_4 &= -2 \end{aligned}$

17. $\begin{aligned} x_1 + x_2 - 3x_3 \qquad\quad + x_5 &= 1 \\ x_1 - x_2 + 3x_3 + x_4 + x_5 &= 2 \\ -x_2 + 3x_3 + 5x_4 \qquad\quad &= -4 \\ x_1 - x_2 + 3x_3 + 5x_4 + 2x_5 &= -3 \end{aligned}$

18. $\begin{aligned} 2x_1 + x_2 \qquad\quad + 2x_4 &= 14 \\ -x_2 + 2x_3 - x_4 &= -8 \\ 2x_2 + x_3 + x_4 &= 2 \\ x_1 - x_2 + x_3 + x_4 &= 5 \end{aligned}$

In Problems 19–22, solve the system of equations using the fact that the coefficient matrix A can be decomposed into the product LU where L and U are given.

19. $L = \begin{bmatrix} 1 & 0 & 0 \\ -1 & 1 & 0 \\ 0 & 2 & 1 \end{bmatrix}, \quad U = \begin{bmatrix} 2 & 1 & 0 \\ 0 & 1 & -3 \\ 0 & 0 & -1 \end{bmatrix},$
$\begin{aligned} 2x_1 + x_2 \qquad\quad &= -2 \\ -2x_1 \qquad\quad - 3x_3 &= 3 \\ 2x_2 - 7x_3 &= 3 \end{aligned}$

20. $L = \begin{bmatrix} 1 & 0 & 0 \\ 0 & 1 & 0 \\ -2 & 4 & 1 \end{bmatrix}, \quad U = \begin{bmatrix} 1 & -1 & 1 \\ 0 & 3 & 2 \\ 0 & 0 & -1 \end{bmatrix},$
$\begin{aligned} x_1 - x_2 + x_3 &= 12 \\ 3x_2 + 2x_3 &= 5 \\ -2x_1 + 14x_2 + 5x_3 &= -11 \end{aligned}$

21. $L = \begin{bmatrix} 1 & 0 & 0 \\ 3 & 1 & 0 \\ -2 & 0 & 1 \end{bmatrix}$, $U = \begin{bmatrix} -1 & 0 & 2 & 1 \\ 0 & 2 & 0 & -1 \\ 0 & 0 & 0 & 2 \end{bmatrix}$,

$$\begin{aligned} - x_1 \qquad + 2x_3 + x_4 &= -2 \\ -3x_1 + 2x_2 + 6x_3 + 2x_4 &= -5 \\ 2x_1 \qquad - 4x_3 \qquad &= 6 \end{aligned}$$

22. $L = \begin{bmatrix} 1 & 0 & 0 \\ 1 & 1 & 0 \\ 1 & 4 & 1 \end{bmatrix}$, $U = \begin{bmatrix} 2 & 1 & 0 & -1 \\ 0 & 0 & 2 & 1 \\ 0 & 0 & 0 & 3 \end{bmatrix}$, $\begin{aligned} 2x_1 + x_2 \qquad - x_4 &= -4 \\ 2x_1 + x_2 + 2x_3 \qquad &= 0 \\ 2x_1 + x_2 + 8x_3 + 6x_4 &= 18 \end{aligned}$

In Problems 23–28, use an LU decomposition to solve the system $AX = B_i$ for the coefficient matrix A and each of the constant matrices B_i, $i = 1, 2, 3$.

23. $A = \begin{bmatrix} 2 & 1 & 1 \\ 6 & 4 & 3 \\ 2 & -1 & 1 \end{bmatrix}$; $B_1 = \begin{bmatrix} 5 \\ 14 \\ 7 \end{bmatrix}$, $B_2 = \begin{bmatrix} 2 \\ 11 \\ -8 \end{bmatrix}$, $B_3 = \begin{bmatrix} -1 \\ 1 \\ -9 \end{bmatrix}$

24. $A = \begin{bmatrix} 1 & 0 & 1 & 0 \\ 0 & 1 & -1 & 1 \\ 1 & -2 & 5 & 1 \\ -1 & 0 & 3 & 5 \end{bmatrix}$; $B_1 = \begin{bmatrix} 5 \\ -8 \\ 9 \\ -23 \end{bmatrix}$, $B_2 = \begin{bmatrix} 3 \\ -5 \\ 19 \\ 9 \end{bmatrix}$, $B_3 = \begin{bmatrix} -3 \\ 11 \\ 1 \\ 47 \end{bmatrix}$

25. $A = \begin{bmatrix} 1 & 1 & 1 & 0 \\ 4 & 6 & 4 & 1 \\ 0 & -4 & -1 & 1 \end{bmatrix}$, $B_1 = \begin{bmatrix} 5 \\ 18 \\ -6 \end{bmatrix}$, $B_2 = \begin{bmatrix} 2 \\ 13 \\ -14 \end{bmatrix}$, $B_3 = \begin{bmatrix} -2 \\ -7 \\ -4 \end{bmatrix}$

26. $A = \begin{bmatrix} 1 - 1 & 1 - 2 & 0 \\ -3 & 3 - 1 & 7 & 3 \\ 0 & 0 & 4 & 110 \end{bmatrix}$, $B_1 = \begin{bmatrix} -3 \\ 16 \\ 9 \end{bmatrix}$, $B_2 = \begin{bmatrix} -2 \\ 15 \\ 5 \end{bmatrix}$,

$B_3 = \begin{bmatrix} -11 \\ 37 \\ 11 \end{bmatrix}$

27. $A = \begin{bmatrix} 0 & -1 & 1 \\ 4 & 1 & 0 \\ 20 & 8 & -3 \end{bmatrix}$, $B_1 = \begin{bmatrix} -2 \\ 5 \\ 31 \end{bmatrix}$, $B_2 = \begin{bmatrix} -4 \\ -12 \\ -48 \end{bmatrix}$, $B_3 = \begin{bmatrix} 12 \\ -4 \\ -56 \end{bmatrix}$

28. $A = \begin{bmatrix} -1 & 1 & 0 & 3 \\ -1 & 1 & 4 & -4 \\ 2 & -1 & 0 & -2 \\ 0 & 1 & 2 & 0 \end{bmatrix}$, $B_1 = \begin{bmatrix} 27 \\ -7 \\ -25 \\ 9 \end{bmatrix}$, $B_2 = \begin{bmatrix} 4 \\ 13 \\ -4 \\ 8 \end{bmatrix}$, $B_3 = \begin{bmatrix} 10 \\ -26 \\ -1 \\ -1 \end{bmatrix}$

29. Prove or disprove each of the following statements.

(a) If X_1 and X_2 are solutions to the matrix equation $AX = B$, then $X_1 - X_2$ is a solution to the homogeneous system $AX = 0$.

(b) If X_1 and X_2 are solutions to the matrix equation $AX = 0$, then $aX_1 + bX_2$ is a solution to $AX = 0$ for all a, b in $\mathcal{F}$.

30. Let **W** be the nullspace of the matrix A; that is, **W** is the subspace of all vectors $(x_1, x_2, \ldots, x_n)$ in $\mathcal{F}^n$ that satisfy the system $AX = 0$. Let $\mathbf{c} = (c_1, c_2, \ldots, c_n)$, where $x_1 = c_1, x_2 = c_2, \ldots, x_n = c_n$ is a particular solution to the system $AX = B$. Prove that $\mathbf{c} + \mathbf{W}$ is the complete set of solutions to $AX = B$.

5 Linear Transformations

In this chapter, the important concept of a linear transformation of a vector space is introduced. Matrices prove to be a powerful tool in the study of linear transformations of finite-dimensional vector spaces. They can be used to classify linear transformations according to certain equivalence relations that are based on fundamental properties common to different linear transformations.

5.1 Linear Transformations

Suppose that f is a mapping of $\mathcal{S}$ into $\mathcal{T}$, and let $\mathcal{A}$ be an arbitrary subset of $\mathcal{S}$. The set

$$f(\mathcal{A}) = \{t \mid t = f(s) \quad \text{for some } s \in \mathcal{A}\}$$

is called the **image** of $\mathcal{A}$ under f. Thus, f is an onto mapping of $\mathcal{S}$ into $\mathcal{T}$ if and only if $f(\mathcal{S}) = \mathcal{T}$. For any subset $\mathcal{B}$ of $\mathcal{T}$, the set

$$f^{-1}(\mathcal{B}) = \{s \in \mathcal{S} \mid f(s) \in \mathcal{B}\}$$

is called the **inverse image** of $\mathcal{B}$. In particular, if $\mathcal{B}$ consists of a single element t,

$$f^{-1}(t) = \{s \in \mathcal{S} \mid f(s) = t\}.$$

Thus, f is one-to-one if, for every $t \in f(\mathcal{S})$, $f^{-1}(t)$ consists of exactly one element.

We write $f : \mathcal{S} \to \mathcal{T}$ or $\mathcal{S} \xrightarrow{f} \mathcal{T}$ to indicate that f is a mapping of $\mathcal{S}$ into $\mathcal{T}$. Our interest in this chapter is with those mappings of one vector space into another. Throughout the chapter, **U** and **V** will denote vector spaces over the *same* field $\mathcal{F}$.

**5.1
DEFINITION**

A **linear transformation** T is a mapping $T: \mathbf{U} \to \mathbf{V}$ which has the property that

$$T(a\mathbf{u} + b\mathbf{w}) = aT(\mathbf{u}) + bT(\mathbf{w})$$

for all $\mathbf{u}, \mathbf{w}$ in **U** and all a, b in $\mathcal{F}$. If $\mathbf{U} = \mathbf{V}$, then the linear transformation $T: \mathbf{U} \to \mathbf{U}$ is called a **linear operator on U**.

We recall from Section 4.3 that an isomorphism is a bijective mapping f that has the property $f(a\mathbf{u} + b\mathbf{w}) = af(\mathbf{u}) + bf(\mathbf{w})$ required in Definition 5.1. Hence every isomorphism is a linear transformation. However, a linear transformation of **U** into **V** may be neither one-to-one nor onto, even though it preserves linear combinations just as an isomorphism does.

In addition to the isomorphisms, another family of examples of linear transformations is provided by the *zero transformations*. For a given pair of vector spaces **U** and **V**, the **zero linear transformation** is the mapping $Z : \mathbf{U} \to \mathbf{V}$ defined by $Z(\mathbf{u}) = \mathbf{0}$ for all **u** in U.

The following examples provide some more detailed illustrations concerning linear transformations.

EXAMPLE 1   Consider the mapping $T : \mathbf{R}^2 \to \mathbf{R}^2$ defined by[1]

$$T(x, y) = (4x + 5y, 6x - y).$$

For arbitrary $\mathbf{u} = (u_1, u_2)$, $\mathbf{w} = (w_1, w_2)$ in $\mathbf{R}^2$ and arbitrary a, b in **R**, we have

$$T(a\mathbf{u} + b\mathbf{w}) = T(au_1 + bw_1, au_2 + bw_2)$$
$$= (4(au_1 + bw_1) + 5(au_2 + bw_2), 6(au_1 + bw_1) - (au_2 + bw_2))$$
$$= (4au_1 + 5au_2, 6au_1 - au_2) + (4bw_1 + 5bw_2, 6bw_1 - bw_2)$$
$$= aT(\mathbf{u}) + bT(\mathbf{w}),$$

and therefore T is a linear operator on $\mathbf{R}^2$.   ∎

Recall that $\mathcal{P}_n$ denotes the vector space of all polynomials in x with degree n or less and coefficients in **R**. In the next example a mapping is defined on $\mathcal{P}_1$.

EXAMPLE 2   Let $\mathcal{P}_n$ denote the vector space that consists of all polynomials in x with degree n or less and coefficients in **R**, and consider the mapping $T : \mathcal{P}_1 \to \mathcal{P}_2$ defined by

$$T(p(x)) = (1 + x)p(x)$$

for all $p(x)$ in $\mathcal{P}_1$. A specific computation of a value of T is provided by

$$T(3 + 2x) = (1 + x)(3 + 2x)$$
$$= 3 + 5x + 2x^2.$$

For arbitrary $p(x), q(x)$ in $\mathcal{P}_1$ and arbitrary a, b in **R**, the polynomials $T(p(x))$ and $T(q(x))$ are in $\mathcal{P}_2$ and

$$T(ap(x) + bq(x)) = (1 + x)(ap(x) + bq(x))$$
$$= a(1 + x)p(x) + b(1 + x)q(x)$$
$$= aT(p(x)) + bT(q(x)).$$

Thus T is a linear transformation of $\mathcal{P}_1$ into $\mathcal{P}_2$.   ∎

---

[1] For notational convenience, the outer parentheses in $T((x_1, x_2, \ldots, x_n))$ are dropped for mappings defined on $\mathbf{R}^n$.

EXAMPLE 3 Let the mapping $T : \mathbf{R}^2 \to \mathbf{R}^2$ be defined by

$$T(x, y) = (x + 1, 2x + y).$$

For any $\mathbf{u} = (u_1, u_2)$, $\mathbf{w} = (w_1, w_2)$ in $\mathbf{R}^2$ and scalars a, b in $\mathbf{R}$,

$$\begin{aligned}
T(a\mathbf{u} + b\mathbf{w}) &= T(au_1 + bw_1, au_2 + bw_2) \\
&= (au_1 + bw_1 + 1, 2au_1 + 2bw_1 + au_2 + bw_2)
\end{aligned}$$

and

$$\begin{aligned}
aT(\mathbf{u}) + bT(\mathbf{w}) &= a(u_1 + 1, 2u_1 + u_2) + b(w_1 + 1, 2w_1 + w_2) \\
&= (au_1 + bw_1 + a + b, 2au_1 + au_2 + 2bw_1 + bw_2).
\end{aligned}$$

We see that the equality

$$T(a\mathbf{u} + b\mathbf{w}) = aT(\mathbf{u}) + bT(\mathbf{w})$$

holds if and only if $a + b = 1$. Since this equation is not always true for a and b in $\mathbf{R}$, we conclude that T is *not* a linear transformation. ∎

In order to indicate some of the variety in linear transformations, we consider some more examples.

EXAMPLE 4 Let $A = [a_{ij}]_{m \times n}$ be a fixed (that is, constant) matrix over $\mathbf{R}$ and define $T : \mathbf{R}_{n \times 1} \to \mathbf{R}_{m \times 1}$ by

$$T(X) = AX$$

for all

$$X = \begin{bmatrix} x_1 \\ x_2 \\ \vdots \\ x_n \end{bmatrix}$$

in $\mathbf{R}_{n \times 1}$. Let X, Y be arbitrary vectors in $\mathbf{R}_{n \times 1}$, and let a, b be arbitrary real numbers. We have both $T(X) = AX$ and $T(Y) = AY$ in $\mathbf{R}_{m \times 1}$, and

$$\begin{aligned}
T(aX + bY) &= A(aX + bY) \\
&= A(aX) + A(bY) \\
&= aAX + bAY \\
&= aT(X) + bT(Y).
\end{aligned}$$

Thus T is a linear transformation of $\mathbf{R}_{n \times 1}$ into $\mathbf{R}_{m \times 1}$. This type of linear transformation is called a **matrix transformation**. ∎

EXAMPLE 5   We saw in Example 4 of Section 4.1 and Problem 26 of Section 4.2 that the set **V** of all real-valued functions of t with domain **R** and the set **W** of all differentiable functions of t with domain **R** form vector spaces over **R** with respect to the usual operations of addition and scalar multiplication. Consider the mapping $T : \mathbf{W} \to \mathbf{V}$ defined by

$$T(f(t)) = \frac{df}{dt} = f'(t).$$

That is, T maps each differentiable function onto its derivative. Using familiar facts from the calculus, we get

$$
\begin{aligned}
T(af(t) + bg(t)) &= \frac{d}{dt}(af(t) + bg(t)) \\
&= \frac{d}{dt}(af(t)) + \frac{d}{dt}(bg(t)) \\
&= a\frac{df}{dt} + b\frac{dg}{dt} \\
&= aT(f(t)) + bT(g(t)).
\end{aligned}
$$

Thus the process of differentiating functions in **W** is a linear transformation from **W** to **V**. ∎

In this chapter, we study linear transformations as mathematical quantities themselves. In our study, we need to define three operations on them. Addition and multiplication by a scalar are considered first.

5.2 DEFINITION   If S and T are linear transformations of **U** into **V**, the **sum** $S + T$ is defined by

$$(S + T)(\mathbf{u}) = S(\mathbf{u}) + T(\mathbf{u})$$

for all **u** in **U**. Also, for each linear transformation T of **U** into **V** and each scalar a in $\mathcal{F}$, we define the **product** of a and T to be the mapping aT of **U** into **V** given by

$$(aT)(\mathbf{u}) = a(T(\mathbf{u})).$$

EXAMPLE 6   Let $\mathbf{U} = \mathbf{R}^3$ and $\mathbf{V} = \mathbf{R}^2$, and let S and T be defined by

$$S(x_1, x_2, x_3) = (2x_1 + x_3, 3x_1 - x_2),$$

$$T(x_1, x_2, x_3) = (-x_1 + 3x_2, 2x_1 - x_2 + x_3).$$

It is easy to verify that S and T are linear transformations. For example, if $\mathbf{u} = (u_1, u_2, u_3)$, $\mathbf{w} = (w_1, w_2, w_3)$ and a, b are in **R**, then

$$S(a\mathbf{u} + b\mathbf{w}) = S(au_1 + bw_1, au_2 + bw_2, au_3 + bw_3)$$

$$= (2(au_1 + bw_1) + au_3 + bw_3, 3(au_1 + bw_1) - (au_2 + bw_2))$$
$$= (2au_1 + au_3, 3au_1 - au_2) + (2bw_1 + bw_3, 3bw_1 - bw_2)$$
$$= a(2u_1 + u_3, 3u_1 - u_2) + b(2w_1 + w_3, 3w_1 - w_2)$$
$$= aS(\mathbf{u}) + bS(\mathbf{w}).$$

Thus S is a linear transformation. Similarly, it can be shown that T is also a linear transformation. The sum $S + T$ is given by

$$(S + T)(x_1, x_2, x_3) = (2x_1 + x_3, 3x_1 - x_2) + (-x_1 + 3x_2, 2x_1 - x_2 + x_3)$$
$$= (x_1 + 3x_2 + x_3, 5x_1 - 2x_2 + x_3),$$

and the product aT is given by

$$(aT)(x_1, x_2, x_3) = a(-x_1 + 3x_2, 2x_1 - x_2 + x_3)$$
$$= (-ax_1 + 3ax_2, 2ax_1 - ax_2 + ax_3). \qquad \blacksquare$$

In Example 6, we have exhibited the mappings $S + T$ and aT where S and T are linear transformations and a is a scalar. However, it is not clear at this point that the mappings $S + T$ and aT are themselves linear transformations. With the definitions of addition and scalar multiplication given in Definition 5.2, the linear transformations of $\mathbf{U}$ into $\mathbf{V}$ can be regarded as possible vectors. The next theorem shows that they are indeed vectors and hence guarantees that the sum $S + T$ and product aT are always linear transformations.

5.3
THEOREM

Let $\mathbf{U}$ and $\mathbf{V}$ be vector spaces over the same field $\mathcal{F}$. Then the set of all linear transformations of $\mathbf{U}$ into $\mathbf{V}$ is a vector space[2] over $\mathcal{F}$.

Proof For a complete proof, each of the ten conditions of Definition 4.2 must be verified. We verify the first six here, leaving the others as exercises. Let T_1, T_2, and T_3 denote arbitrary linear transformations of $\mathbf{U}$ into $\mathbf{V}$, let $\mathbf{u}$ and $\mathbf{w}$ be arbitrary vectors in $\mathbf{U}$, and let a, b, and c be scalars.

Since

$$(T_1 + T_2)(a\mathbf{u} + b\mathbf{w}) = T_1(a\mathbf{u} + b\mathbf{w}) + T_2(a\mathbf{u} + b\mathbf{w})$$
$$= aT_1(\mathbf{u}) + bT_1(\mathbf{w}) + aT_2(\mathbf{u}) + bT_2(\mathbf{w})$$
$$= a[T_1(\mathbf{u}) + T_2(\mathbf{u})] + b[T_1(\mathbf{w}) + T_2(\mathbf{w})]$$
$$= a(T_1 + T_2)(\mathbf{u}) + b(T_1 + T_2)(\mathbf{w}),$$

$T_1 + T_2$ is a linear transformation of $\mathbf{U}$ into $\mathbf{V}$.

---

[2] The notation $L(\mathbf{U}, \mathbf{V})$ is often used to denote this vector space.

Addition is associative, since

$$(T_1 + (T_2 + T_3))(\mathbf{u}) = T_1(\mathbf{u}) + (T_2 + T_3)(\mathbf{u})$$
$$= T_1(\mathbf{u}) + [T_2(\mathbf{u}) + T_3(\mathbf{u})]$$
$$= [T_1(\mathbf{u}) + T_2(\mathbf{u})] + T_3(\mathbf{u})$$
$$= (T_1 + T_2)(\mathbf{u}) + T_3(\mathbf{u})$$
$$= ((T_1 + T_2) + T_3)(\mathbf{u}).$$

The zero linear transformation Z is an additive identity, since

$$(T_1 + Z)(\mathbf{u}) = T_1(\mathbf{u}) + Z(\mathbf{u}) = T_1(\mathbf{u}) + \mathbf{0} = T_1(\mathbf{u})$$

for all $\mathbf{u}$ in $\mathbf{U}$.

The additive inverse of T_1 is the linear transformation $-T_1$ of $\mathbf{U}$ into $\mathbf{V}$ defined by $(-T_1)(\mathbf{u}) = -T_1(\mathbf{u})$ since

$$(T_1 + (-T_1))(\mathbf{u}) = T_1(\mathbf{u}) + (-T_1(\mathbf{u})) = \mathbf{0}$$

for all $\mathbf{u}$ in $\mathbf{U}$.

For any $\mathbf{u}$ in $\mathbf{U}$,

$$(T_1 + T_2)(\mathbf{u}) = T_1(\mathbf{u}) + T_2(\mathbf{u})$$
$$= T_2(\mathbf{u}) + T_1(\mathbf{u})$$
$$= (T_2 + T_1)(\mathbf{u}),$$

so $T_1 + T_2 = T_2 + T_1$.

Since

$$(cT_1)(a\mathbf{u} + b\mathbf{w}) = c(T_1(a\mathbf{u} + b\mathbf{w}))$$
$$= c(aT_1(\mathbf{u}) + bT_1(\mathbf{w}))$$
$$= a(cT_1(\mathbf{u})) + b(cT_1(\mathbf{w}))$$

cT_1 is a linear transformation of $\mathbf{U}$ into $\mathbf{V}$. The verification of remaining conditions of Definition 4.2 are left for the exercises. ■ ■ ■

There is a third operation involving linear transformations, that of composition, that will be considered in Section 5.4. At this point, we examine the image of a subspace under a linear transformation.

5.4 THEOREM

Let T be a linear transformation of $\mathbf{U}$ into $\mathbf{V}$. If $\mathbf{U}_1$ is any subspace of $\mathbf{U}$, then $T(\mathbf{U}_1)$ is a subspace of $\mathbf{V}$.

Proof Let $\mathbf{U}_1$ be a subspace of $\mathbf{U}$. Then $\mathbf{0} \in \mathbf{U}_1$ and since

$$T(\mathbf{0}) = T(0 \cdot \mathbf{0}) = 0 \, T(\mathbf{0}) = \mathbf{0},$$

then $\mathbf{0} \in T(\mathbf{U}_1)$. Therefore $T(\mathbf{U}_1)$ is nonempty.

Let $\mathbf{v}_1, \mathbf{v}_2$ be in $T(\mathbf{U}_1)$, and a_1, a_2 be in $\mathcal{F}$. There exist vectors $\mathbf{u}_1, \mathbf{u}_2$ in $\mathbf{U}_1$ such that $T(\mathbf{u}_1) = \mathbf{v}_1$ and $T(\mathbf{u}_2) = \mathbf{v}_2$. Since $\mathbf{U}_1$ is a subspace of $\mathbf{U}$, $a_1\mathbf{u}_1 + a_2\mathbf{u}_2$ is in $\mathbf{U}_1$ and

$$T(a_1\mathbf{u}_1 + a_2\mathbf{u}_2) = a_1 T(\mathbf{u}_1) + a_2 T(\mathbf{u}_2)$$

$$= a_1\mathbf{v}_1 + a_2\mathbf{v}_2.$$

Thus, $a_1\mathbf{v}_1 + a_2\mathbf{v}_2$ is in $T(\mathbf{U}_1)$, and $T(\mathbf{U}_1)$ is a subspace of $\mathbf{V}$. ■■■

5.5 DEFINITION

The subspace $T(\mathbf{U})$ of $\mathbf{V}$ is called the **range** of T. The dimension of $T(\mathbf{U})$ is called the **rank** of T. The rank of T will be denoted by $\text{rank}(T)$.

The essence of a method for finding the rank of a linear transformation T is contained in the proof of Theorem 5.6 for the case where $\mathbf{U}$ is finite dimensional.

5.6 THEOREM

If T is a linear transformation of $\mathbf{U}$ into $\mathbf{V}$ and $\mathcal{A} = \{\mathbf{u}_1, \mathbf{u}_2, \ldots, \mathbf{u}_n\}$ is a basis of $\mathbf{U}$, then $T(\mathcal{A})$ spans $T(\mathbf{U})$.

Proof Suppose that $\mathcal{A} = \{\mathbf{u}_1, \mathbf{u}_2, \ldots, \mathbf{u}_n\}$ is a basis of $\mathbf{U}$, and consider the set $T(\mathcal{A}) = \{T(\mathbf{u}_1), T(\mathbf{u}_2), \ldots, T(\mathbf{u}_n)\}$. For any vector $\mathbf{v}$ in $T(\mathbf{U})$, there is a vector $\mathbf{u}$ in $\mathbf{U}$ such that $T(\mathbf{u}) = \mathbf{v}$. The vector $\mathbf{u}$ can be written as $\mathbf{u} = \sum_{i=1}^{n} a_i\mathbf{u}_i$ since $\mathcal{A}$ is a basis of $\mathbf{U}$. This gives $\mathbf{v} = T\left(\sum_{i=1}^{n} a_i\mathbf{u}_i\right) = \sum_{i=1}^{n} a_i T(\mathbf{u}_i)$, and $T(\mathcal{A})$ spans $T(\mathbf{U})$. ■■■

The proof of Theorem 5.6 shows that the set $T(\mathcal{A})$ contains a basis of $T(\mathbf{U})$, and the number of elements in the basis is $\text{rank}(T)$. Our next example demonstrates the use of this method to find a basis for the range of a linear transformation T. A more efficient and systematic method for finding a basis is developed in the next section.

EXAMPLE 7 It is given that the mapping $T : \mathbf{R}_{2\times 3} \to \mathbf{R}^5$ defined by

$$T\left(\begin{bmatrix} a_{11} & a_{12} & a_{13} \\ a_{21} & a_{22} & a_{23} \end{bmatrix}\right) = (a_{11} + a_{21}, 2a_{11} + a_{21} - a_{12}, a_{22} + a_{13}, 0, a_{23})$$

is a linear transformation of $\mathbf{R}_{2\times 3}$ into $\mathbf{R}^5$. In order to find a basis for $T(\mathbf{R}_{2\times 3})$, we first obtain a spanning set $T(\mathcal{A})$ as described in the proof of Theorem 5.6. The set $\mathcal{A} = \{M_{11}, M_{12}, M_{13}, M_{21}, M_{22}, M_{23}\}$ is the standard basis of $\mathbf{R}_{2\times 3}$, where

$$M_{11} = \begin{bmatrix} 1 & 0 & 0 \\ 0 & 0 & 0 \end{bmatrix}, \quad M_{12} = \begin{bmatrix} 0 & 1 & 0 \\ 0 & 0 & 0 \end{bmatrix}, \quad M_{13} = \begin{bmatrix} 0 & 0 & 1 \\ 0 & 0 & 0 \end{bmatrix},$$

$$M_{21} = \begin{bmatrix} 0 & 0 & 0 \\ 1 & 0 & 0 \end{bmatrix}, \quad M_{22} = \begin{bmatrix} 0 & 0 & 0 \\ 0 & 1 & 0 \end{bmatrix}, \quad M_{23} = \begin{bmatrix} 0 & 0 & 0 \\ 0 & 0 & 1 \end{bmatrix}.$$

The set $T(\mathcal{A})$ is given by

$$T(\mathcal{A}) = \{(1, 2, 0, 0, 0), (0, -1, 0, 0, 0), (0, 0, 1, 0, 0), (1, 1, 0, 0, 0),$$
$$(0, 0, 1, 0, 0), (0, 0, 0, 0, 1)\},$$

and we know that $T(\mathcal{A})$ contains a basis of $T(\mathbf{R}_{2 \times 3})$. Using the refinement process from Section 1.6 and Example 3 in Section 4.2, we find that the first three vectors in $T(\mathcal{A})$ are linearly independent and the fourth vector can be written as

$$(1, 1, 0, 0, 0) = (1)(1, 2, 0, 0, 0) + (1)(0, -1, 0, 0, 0) + (0)(0, 0, 1, 0, 0).$$

Thus the fourth vector can be deleted from the spanning set $T(\mathcal{A})$. The fifth vector in $T(\mathcal{A})$ is a repetition of the third vector, so it can also be deleted. The last vector is clearly not a linear combination of the preceding vectors since it is the only one with a nonzero fifth component. Thus

$$\{(1, 2, 0, 0, 0), (0, -1, 0, 0, 0), (0, 0, 1, 0, 0), (0, 0, 0, 0, 1)\}$$

is a basis for the range of T. From the number of vectors in our basis, we see that

$$\mathrm{rank}(T) = \dim(T(\mathbf{R}_{2 \times 3})) = 4. \qquad \blacksquare$$

We now turn our attention to the inverse image of a subspace.

5.7
THEOREM   Let T be a linear transformation of **U** into **V**. If **W** is any subspace of **V**, the inverse image $T^{-1}(\mathbf{W})$ is a subspace of **U**.

Proof   Let **W** be a subspace of **V**. Then $\mathbf{0} \in \mathbf{W}$ and hence $\mathbf{0} \in T^{-1}(\mathbf{W})$ since $T(\mathbf{0}) = \mathbf{0}$. Thus $T^{-1}(\mathbf{W})$ is nonempty. Let $\mathbf{u}_1, \mathbf{u}_2$ be in $T^{-1}(\mathbf{W})$, and let a_1, a_2 be in $\mathcal{F}$. Then $\mathbf{v}_1 = T(\mathbf{u}_1)$ and $\mathbf{v}_2 = T(\mathbf{u}_2)$ are in **W**, and this means that

$$a_1 \mathbf{v}_1 + a_2 \mathbf{v}_2 = a_1 T(\mathbf{u}_1) + a_2 T(\mathbf{u}_2) = T(a_1 \mathbf{u}_1 + a_2 \mathbf{u}_2)$$

is in **W**. Therefore, $a_1 \mathbf{u}_1 + a_2 \mathbf{u}_2$ is in $T^{-1}(\mathbf{W})$, and $T^{-1}(\mathbf{W})$ is a subspace of **U**. $\blacksquare\blacksquare\blacksquare$

5.8 DEFINITION

The subspace $T^{-1}(\mathbf{0})$ is called the **kernel** of the linear transformation T. The dimension of $T^{-1}(\mathbf{0})$ is the **nullity** of T, denoted by nullity(T).

We return to the mapping T of Example 7 and find a basis of its kernel.

EXAMPLE 8 The mapping $T : \mathbf{R}_{2\times 3} \rightarrow \mathbf{R}^5$ was defined by

$$T\left(\begin{bmatrix} a_{11} & a_{12} & a_{13} \\ a_{21} & a_{22} & a_{23} \end{bmatrix}\right) = (a_{11} + a_{21}, 2a_{11} + a_{21} - a_{12}, a_{22} + a_{13}, 0, a_{23}).$$

To find a basis for $T^{-1}(\mathbf{0})$, we set $T(\mathbf{v}) = \mathbf{0}$ and obtain the system of equations

$$\begin{aligned} a_{11} + a_{21} & = 0 \\ 2a_{11} + a_{21} - a_{12} & = 0 \\ a_{22} + a_{13} & = 0 \\ 0 & = 0 \\ a_{23} & = 0. \end{aligned}$$

Using Gauss-Jordan elimination, we find that the reduced row-echelon form for the augmented matrix is

$$\begin{bmatrix} 1 & 0 & -1 & 0 & 0 & 0 & 0 \\ 0 & 1 & 1 & 0 & 0 & 0 & 0 \\ 0 & 0 & 0 & 1 & 1 & 0 & 0 \\ 0 & 0 & 0 & 0 & 0 & 1 & 0 \\ 0 & 0 & 0 & 0 & 0 & 0 & 0 \end{bmatrix}.$$

Solving for the leading variables in terms of the parameters, we get

$$\begin{aligned} a_{11} & = & a_{12} \\ a_{21} & = & -a_{12} \\ a_{12} & = & a_{12} \\ a_{22} & = & -a_{13} \\ a_{13} & = & a_{13} \\ a_{23} & = & 0. \end{aligned}$$

Substituting for the leading variables leads to

$$\begin{bmatrix} a_{11} & a_{12} & a_{13} \\ a_{21} & a_{22} & a_{23} \end{bmatrix} = \begin{bmatrix} a_{12} & a_{12} & a_{13} \\ -a_{12} & -a_{13} & 0 \end{bmatrix}$$

$$= a_{12} \begin{bmatrix} 1 & 1 & 0 \\ -1 & 0 & 0 \end{bmatrix} + a_{13} \begin{bmatrix} 0 & 0 & 1 \\ 0 & -1 & 0 \end{bmatrix}.$$

Thus the set

$$\left\{ \begin{bmatrix} 1 & 1 & 0 \\ -1 & 0 & 0 \end{bmatrix}, \begin{bmatrix} 0 & 0 & 1 \\ 0 & -1 & 0 \end{bmatrix} \right\}$$

forms a basis for the kernel $T^{-1}(\mathbf{0})$ and

$$\text{nullity}(T) = \dim(T^{-1}(\mathbf{0})) = 2. \qquad \blacksquare$$

We note that the sum of the rank and nullity of T is equal to the dimension of the domain of T in the example above. Our next theorem states that this equality always holds.

5.9 THEOREM

Let T be a linear transformation of $\mathbf{U}$ into $\mathbf{V}$. If $\mathbf{U}$ has finite dimension, then

$$\text{rank}(T) + \text{nullity}(T) = \dim(\mathbf{U}).$$

Proof Suppose that $\mathbf{U}$ has dimension n, and let k be the nullity of T. Choose $\{\mathbf{u}_1, \mathbf{u}_2, \dots, \mathbf{u}_k\}$ to be a basis of the kernel $T^{-1}(\mathbf{0})$. This linearly independent set can be extended to a basis

$$\mathcal{A} = \{\mathbf{u}_1, \mathbf{u}_2, \dots, \mathbf{u}_k, \mathbf{u}_{k+1}, \dots, \mathbf{u}_n\}$$

of $\mathbf{U}$. According to Theorem 5.6, the set $T(\mathcal{A})$ spans $T(\mathbf{U})$. But $T(\mathbf{u}_1) = T(\mathbf{u}_2) = \cdots = T(\mathbf{u}_k) = \mathbf{0}$, so this means that the set of $n - k$ vectors $\{T(\mathbf{u}_{k+1}), T(\mathbf{u}_{k+2}), \dots, T(\mathbf{u}_n)\}$ spans $T(\mathbf{U})$. To show that this set is linearly independent, suppose that

$$c_{k+1}T(\mathbf{u}_{k+1}) + c_{k+2}T(\mathbf{u}_{k+2}) + \cdots + c_n T(\mathbf{u}_n) = \mathbf{0}.$$

Then

$$T(c_{k+1}\mathbf{u}_{k+1} + c_{k+2}\mathbf{u}_{k+2} + \cdots + c_n\mathbf{u}_n) = \mathbf{0},$$

and $\sum_{i=k+1}^{n} c_i \mathbf{u}_i$ is in $T^{-1}(\mathbf{0})$. Thus there are scalars $d_1, d_2, \dots, d_k$ such that

$$\sum_{i=1}^{k} d_i \mathbf{u}_i = \sum_{i=k+1}^{n} c_i \mathbf{u}_i$$

and

$$\sum_{i=1}^{k} d_i \mathbf{u}_i - \sum_{i=k+1}^{n} c_i \mathbf{u}_i = \mathbf{0}.$$

Since $\mathcal{A}$ is a basis, each c_i and each d_i must be zero. Hence $\{T(\mathbf{u}_{k+1}), T(\mathbf{u}_{k+2}), \dots, T(\mathbf{u}_n)\}$ is a basis of $T(\mathbf{U})$. Since $\text{rank}(T)$ is the dimension of $T(\mathbf{U})$, $n - k = \text{rank}(T)$ and

$$\text{rank}(T) + \text{nullity}(T) = n = \dim(\mathbf{U}). \qquad \blacksquare\blacksquare\blacksquare$$

In the next section, a systematic method for finding a basis of $T^{-1}(\mathbf{0})$ is described.

5.1 Exercises

1. Let S and T be the mappings of $\mathbf{R}^2$ into $\mathbf{R}^2$ defined by $S(x_1, x_2) = (x_1 + x_2, x_1 - x_2)$ and $T(x_1, x_2) = (-x_2, -x_1)$.
   (a) Prove that each of S and T is a linear operator.
   (b) Find $S + T$ and $2S - 3T$.

2. Determine whether the given mapping $T : \mathbf{R}^2 \to \mathbf{R}^2$ is a linear operator.
   (a) $T(x, y) = (x - y, 0)$ (b) $T(x, y) = (xy, x)$
   (c) $T(x, y) = (x + 1, y - 1)$ (d) $T(x, y) = (x + y, x - y)$
   (e) $T(x, y) = x(2, 1)$ (f) $T(x, y) = (x - y)(x + y, 0)$

3. Determine whether the given mapping $T : \mathbf{R}^3 \to \mathbf{R}^2$ is a linear transformation.
   (a) $T(x, y, z) = (2x + y, x + z)$ (b) $T(x, y, z) = (x - y, x^2 - y^2)$
   (c) $T(x, y, z) = (x + y + 1, x + y - 1)$ (d) $T(x, y, z) = (x + y + z, 0)$

4. Determine which of the following mappings $T : \mathcal{P}_1 \to \mathcal{P}_1$ over $\mathbf{R}$ are linear operators.
   (a) $T(a_0 + a_1 x) = a_0 x$ (b) $T(a_0 + a_1 x) = a_0 + a_1(x + 1)$
   (c) $T(a_0 + a_1 x) = a_0 a_1 + a_0 x$ (d) $T(a_0 + a_1 x) = a_0 + a_1 + a_0 x$

5. Determine which of the following mappings $T : \mathbf{R}_{2 \times 2} \to \mathbf{R}_{2 \times 2}$ are linear operators. In parts (a)–(c), A denotes a constant nonzero 2×2 matrix over $\mathbf{R}$.
   (a) $T(X) = A + AX$ (b) $T(X) = AX - XA$
   (c) $T(X) = AXA$ (d) $T(X) = 2X$

6. Determine whether the given mapping is a linear transformation.
   (a) $T : \mathbf{R}_{m \times n} \to \mathbf{R}_{n \times m}, T(X) = X^T$, where X^T is the transpose of X
   (b) $T : \mathbf{R}_{m \times n} \to \mathbf{R}_{n \times n}, T(X) = X^T X$, where X^T is the transpose of X
   (c) $T : \mathcal{P}_2 \to \mathcal{P}_1$ over $\mathbf{R}, T(a_0 + a_1 x + a_2 x^2) = (a_0 + a_1) + (a_1 + a_2)x$
   (d) $T : \mathcal{P}_2 \to \mathcal{P}_2$ over $\mathbf{R}, T(p(x)) = p(x - 1)$
   (e) $T : \mathcal{P}_2 \to \mathcal{P}_3$ over $\mathbf{R}, T(p(x)) = xp(x) + p(1)$
   (f) $T : \mathbf{R}_{2 \times 2} \to \mathbf{R}, T\left(\begin{bmatrix} a & b \\ c & d \end{bmatrix}\right) = ad - bc$
   (g) $T : \mathbf{R}^2 \to \mathbf{R}, T(\mathbf{v}) = \|\mathbf{v}\|$
   (h) $T : \mathbf{R}^n \to \mathbf{R}, T(\mathbf{v}) = \mathbf{v} \cdot \mathbf{v}$

7. Let $\mathbf{U}$ be a finite dimensional vector space and $\mathcal{A} = \{\mathbf{u}_1, \mathbf{u}_2, \ldots, \mathbf{u}_n\}$ be a basis of $\mathbf{U}$. Define $T : \mathbf{U} \to \mathbf{R}_{n \times 1}$ by $T(\mathbf{v}) = [\mathbf{v}]_{\mathcal{A}}$.
   (a) Show that T is a linear transformation.
   (b) Find the kernel of T.

8. Let $\mathcal{A} = \{\mathbf{u}_1, \mathbf{u}_2, \ldots, \mathbf{u}_n\}$ be a basis of $\mathbf{R}^n$. Define $T : \mathbf{R}^n \to \mathbf{R}^n$ by

$$T(\mathbf{v}) = (\mathbf{v} \cdot \mathbf{u}_1)\mathbf{u}_1 + (\mathbf{v} \cdot \mathbf{u}_2)\mathbf{u}_2 + \cdots + (\mathbf{v} \cdot \mathbf{u}_n)\mathbf{u}_n.$$

Show that T is a linear operator.

9. Let T be the linear transformation of $\mathbf{R}^4$ into $\mathbf{R}^3$ given by

$$T(x_1, x_2, x_3, x_4)$$
$$= (3x_1 - 2x_2 - x_3 - 4x_4, x_1 + x_2 - 2x_3 - 3x_4, 2x_1 - 3x_2 + x_3 - x_4).$$

(a) Find a basis of $T(\mathbf{R}^4)$.

(b) Find two linearly independent vectors in the kernel of T.

10. Let T be the mapping of $\mathbf{R}_{2\times2}$ into $\mathbf{R}_{2\times2}$ defined by

$$T\left(\begin{bmatrix} a_{11} & a_{12} \\ a_{21} & a_{22} \end{bmatrix}\right) = \begin{bmatrix} 2a_{11} + a_{12} & 2a_{21} - a_{22} \\ a_{11} + 3a_{22} & a_{21} - 3a_{12} \end{bmatrix}.$$

(a) Prove that T is a linear transformation.

(b) Find a basis for the range of T.

11. Let T be the mapping of $\mathbf{R}_{2\times3}$ into $\mathcal{P}_2$ over $\mathbf{R}$ defined by

$$T\left(\begin{bmatrix} a_{11} & a_{12} & a_{13} \\ a_{21} & a_{22} & a_{23} \end{bmatrix}\right) = (a_{11} - a_{12}) + a_{23}x.$$

(a) Prove that T is a linear transformation.

(b) Find a basis of the range of T.

(c) State the rank and nullity of T.

12. Find the rank and nullity of the given linear transformation of $\mathbf{U}$ into $\mathbf{V}$.

(a) $\mathbf{U} = \mathbf{V} = \mathbf{R}^4$,

$$T(x_1, x_2, x_3, x_4) = (x_1 + x_2 + x_3 - x_4, x_1 + 2x_2 + x_3 - x_4,$$
$$x_1 - x_2 + 3x_3, -x_1 + 5x_2 - 5x_3 - x_4)$$

(b) $\mathbf{U} = \mathbf{R}^4, \mathbf{V} = \mathbf{R}^3$,

$$T(x_1, x_2, x_3, x_4) = (x_1 + 2x_2 + x_3 + 3x_4, 2x_3 - 4x_4, x_1 + 2x_2 + 3x_3 - x_4)$$

(c) $\mathbf{U} = \mathcal{P}_2$ over $\mathbf{R}, \mathbf{V} = \mathbf{R}^3$,

$$T(a + bx + cx^2) = (a + b, b + c, a - c)$$

(d) $\mathbf{U} = \mathbf{R}^4, \mathbf{V} = \mathbf{R}_{2\times2}$,

$$T(x_1, x_2, x_3, x_4) = \begin{bmatrix} x_1 - x_2 + x_3 & x_1 - x_4 \\ x_2 - x_3 - x_4 & x_1 - 2x_2 + 2x_3 + x_4 \end{bmatrix}$$

13. In each part of Problem 12, find the standard basis of the kernel of T by solving the system of equations that results from setting $T(\mathbf{u}) = \mathbf{0}$.

14. Let T be a linear transformation of $\mathbf{U}$ into $\mathbf{V}$. Prove the following statements.

(a) $T(-\mathbf{u}) = -T(\mathbf{u})$ for all $\mathbf{u} \in \mathbf{U}$

(b) $T(\mathbf{u} - \mathbf{w}) = T(\mathbf{u}) - T(\mathbf{w})$ for all $\mathbf{u}, \mathbf{w}$ in $\mathbf{U}$

(c) $T\left(\sum_{i=1}^{n} a_i \mathbf{u}_i\right) = \sum_{i=1}^{n} a_i T(\mathbf{u}_i)$ for all scalars a_i in $\mathcal{F}$ and vectors $\mathbf{u}_i$ in $\mathbf{U}$

15. **(a)** Give an example of a linear transformation that is not one-to-one.

    (b) Give an example of a linear transformation that is not onto.

16. **(a)** Verify that the zero mapping $Z : \mathbf{U} \to \mathbf{V}$, defined by $Z(\mathbf{u}) = \mathbf{0}$ for all $\mathbf{u} \in \mathbf{U}$, is a linear transformation.

    (b) Let $T : \mathbf{U} \to \mathbf{V}$ be a linear transformation on a finite dimensional vector space $\mathbf{U}$. If $\mathcal{A} = \{\mathbf{u}_1, \mathbf{u}_2, \ldots, \mathbf{u}_n\}$ is a basis of $\mathbf{U}$, and $T(\mathbf{u}_i) = \mathbf{0}$ for $i = 1, 2, \ldots, n$, then prove that T is the zero transformation.

17. **(a)** The mapping $I : \mathbf{U} \to \mathbf{U}$ defined by $I(\mathbf{u}) = \mathbf{u}$ for all $\mathbf{u} \in \mathbf{U}$ is called the identity mapping. Show that I is a linear operator.

    (b) Let $T : \mathbf{U} \to \mathbf{U}$ be a linear operator on a finite dimensional vector space $\mathbf{U}$. If $\mathcal{A} = \{\mathbf{u}_1, \mathbf{u}_2, \ldots, \mathbf{u}_n\}$ is a basis of $\mathbf{U}$, and $T(\mathbf{u}_i) = \mathbf{u}_i$ for $i = 1, 2, \ldots, n$, then prove that T is the identity operator.

18. Let T be a linear transformation of $\mathbf{U}$ into $\mathbf{V}$, and let $\mathbf{U}_1$, be a subspace of $\mathbf{U}$. Prove that if $\mathcal{A}$ spans $\mathbf{U}_1$, then $T(\mathcal{A})$ spans $T(\mathbf{U}_1)$. (*Note:* $\mathcal{A}$ is not necessarily finite.)

19. In each part below, T is a linear transformation of $\mathbf{R}_{2\times2}$ into $\mathbf{R}^3$. For the given subspace $\mathbf{U}_1$ of $\mathbf{R}_{2\times2}$, find the standard basis of $T(\mathbf{U}_1)$.

    (a) $T\left(\begin{bmatrix} a_{11} & a_{12} \\ a_{21} & a_{22} \end{bmatrix}\right) = (a_{11} - 2a_{21} + a_{22}, a_{12} + a_{21} - a_{22}, a_{11} + 2a_{12} + a_{22})$,

    $\mathbf{U}_1 = \left\langle \begin{bmatrix} 1 & 1 \\ 0 & 0 \end{bmatrix}, \begin{bmatrix} 1 & 0 \\ 1 & 0 \end{bmatrix}, \begin{bmatrix} 0 & 0 \\ -1 & 2 \end{bmatrix} \right\rangle$

    (b) $T\left(\begin{bmatrix} a_{11} & a_{12} \\ a_{21} & a_{22} \end{bmatrix}\right) = (a_{11} + a_{12} - a_{21} - a_{22}, 0, a_{21} - 3a_{22})$,

    $\mathbf{U}_1 = \left\langle \begin{bmatrix} 2 & 1 \\ -1 & 0 \end{bmatrix}, \begin{bmatrix} 0 & 1 \\ 1 & 0 \end{bmatrix}, \begin{bmatrix} 0 & 0 \\ -1 & -2 \end{bmatrix} \right\rangle$

20. In each part below, T is a linear transformation of $\mathbf{R}^4$ into $\mathbf{R}^3$. For the given subspace $\mathbf{U}_1$ and the basis $\mathcal{B}$ of $\mathbf{R}^3$, find the standard basis of $T(\mathbf{U}_1)$ relative to $\mathcal{B}$.

    (a) $T(x_1, x_2, x_3, x_4) = (x_1 + 3x_2 + x_4, 3x_1 + 5x_2 - x_3 + 2x_4, 5x_1 + 2x_2 - 2x_3 - 2x_4)$,
    $\mathbf{U}_1 = \langle (1, 0, 1, 0), (1, 0, 2, 0), (0, 1, 0, -1) \rangle$, $\quad \mathcal{B} = \{(1, 0, 1), (0, 1, 1), (1, 1, 1)\}$

    (b) $T(x_1, x_2, x_3, x_4) = (x_1 - x_3 + 3x_4, x_2 + 2x_4, 2x_1 - x_2 - 2x_3 + 4x_4)$,
    $\mathbf{U}_1 = \langle (2, 0, 1, 1), (-1, 1, 0, 0), (4, 0, -2, 0) \rangle$, $\mathcal{B} = \{(1, 2, 0), (1, 0, 2), (0, 0, 1)\}$

21. Complete the proof of Theorem 5.3.

22. Let T be a linear transformation of $\mathbf{U}$ into $\mathbf{V}$ with nullity 0. Prove that if $T(\mathbf{u}_4)$ is dependent on $\{T(\mathbf{u}_1), T(\mathbf{u}_2), T(\mathbf{u}_3)\}$, then $\mathbf{u}_4$ is dependent on $\{\mathbf{u}_1, \mathbf{u}_2, \mathbf{u}_3\}$.

23. Let T be a linear transformation of $\mathbf{U}$ into $\mathbf{V}$. Prove that T is one-to-one if and only if $T^{-1}(\mathbf{0}) = \{\mathbf{0}\}$.

24. If T is a linear transformation of $\mathbf{U}$ into $\mathbf{V}$ and $\mathcal{A} = \{\mathbf{u}_1, \mathbf{u}_2, \ldots, \mathbf{u}_n\}$ is a basis of $\mathbf{U}$, prove that nullity(T) is zero if and only if $\{T(\mathbf{u}_1), T(\mathbf{u}_2), \ldots, T(\mathbf{u}_n)\}$ is linearly independent.

25. Let T be a linear transformation of $\mathbf{U}$ into $\mathbf{V}$, and let $\mathbf{U}_1$ and $\mathbf{U}_2$ be subspaces of $\mathbf{U}$. Prove that $T(\mathbf{U}_1 + \mathbf{U}_2) = T(\mathbf{U}_1) + T(\mathbf{U}_2)$.

26. Let T be a linear transformation of $\mathbf{U}$ into $\mathbf{V}$, and let $\mathbf{U}_1$ and $\mathbf{W}$ denote subspaces of $\mathbf{U}$ and $\mathbf{V}$, respectively. Prove or disprove the statements below.

(a) $T(T^{-1}(\mathbf{W})) = \mathbf{W}$ **(b)** $T^{-1}(T(\mathbf{U}_1)) = \mathbf{U}_1$

5.2 Linear Transformations and Matrices

In this and the following section, $\mathbf{U}$ and $\mathbf{V}$ will denote vector spaces of dimension n and m, respectively, over the same field $\mathcal{F}$, and T will denote a linear transformation of $\mathbf{U}$ into $\mathbf{V}$.

Suppose that $\mathcal{A} = \{\mathbf{u}_1, \mathbf{u}_2, \ldots, \mathbf{u}_n\}$ is a basis of $\mathbf{U}$. Any $\mathbf{u}$ in $\mathbf{U}$ can be written uniquely in the form $\mathbf{u} = \sum_{j=1}^{n} x_j \mathbf{u}_j$, and

$$T(\mathbf{u}) = \sum_{j=1}^{n} x_j T(\mathbf{u}_j).$$

This shows that the value of T at every $\mathbf{u}$ in $\mathbf{U}$ is determined by the values of T at the basis vectors $\mathbf{u}_1, \mathbf{u}_2, \ldots, \mathbf{u}_n$.

EXAMPLE 1 Consider the basis $\mathcal{A} = \{1, 1 + x, 1 + x + x^2\}$ of $\mathcal{P}_2$ over $\mathbf{R}$ and the linear transformation $T : \mathcal{P}_2 \rightarrow \mathbf{R}^2$ where

$$T(1) = (2, 0)$$

$$T(1 + x) = (-3, 4)$$

$$T(1 + x + x^2) = (1, 2).$$

The linear transformation T is completely determined by these images of the base vectors. Since any $p(x) = a + bx + cx^2$ in $\mathcal{P}_2$ can be written as

$$a + bx + cx^2 = (a - b)(1) + (b - c)(1 + x) + c(1 + x + x^2),$$

then the image of $p(x)$ under T can be evaluated as

$$
\begin{aligned}
T(p(x)) &= T(a + bx + cx^2) \\
&= T((a - b)(1) + (b - c)(1 + x) + c(1 + x + x^2)) \\
&= (a - b)T(1) + (b - c)T(1 + x) + cT(1 + x + x^2) \\
&= (a - b)(2, 0) + (b - c)(-3, 4) + c(1, 2) \\
&= (2a - 5b + 4c, 4b - 2c).
\end{aligned}
$$

■

With $\mathcal{A} = \{\mathbf{u}_1, \mathbf{u}_2, \ldots, \mathbf{u}_n\}$ a basis of $\mathbf{U}$, and if $\mathcal{B} = \{\mathbf{v}_1, \mathbf{v}_2, \ldots, \mathbf{v}_m\}$ is a basis of $\mathbf{V}$, then each $T(\mathbf{u}_j)$ can be written uniquely as

$$T(\mathbf{u}_j) = \sum_{i=1}^{m} a_{ij} \mathbf{v}_i.$$

Thus, with each choice of bases $\mathcal{A}$ and $\mathcal{B}$, a linear transformation T of $\mathbf{U}$ into $\mathbf{V}$ determines a unique indexed set $\{a_{ij}\}$ of mn elements of $\mathcal{F}$. These elements make up the matrix of T relative to the bases $\mathcal{A}$ and $\mathcal{B}$.

**5.10
DEFINITION**

Suppose that $\mathcal{A} = \{\mathbf{u}_1, \mathbf{u}_2, \ldots, \mathbf{u}_n\}$ and $\mathcal{B} = \{\mathbf{v}_1, \mathbf{v}_2, \ldots, \mathbf{v}_m\}$ are bases of $\mathbf{U}$ and $\mathbf{V}$, respectively. Let T be a linear transformation of $\mathbf{U}$ into $\mathbf{V}$. The **matrix of T relative to the bases $\mathcal{A}$ and $\mathcal{B}$** is the matrix

$$A = [a_{ij}]_{m \times n} = [T]_{\mathcal{B}, \mathcal{A}}$$

where the a_{ij} are determined by the conditions

$$T(\mathbf{u}_j) = \sum_{i=1}^{m} a_{ij} \mathbf{v}_i$$
$$= a_{1j} \mathbf{v}_1 + a_{2j} \mathbf{v}_2 + \cdots + a_{mj} \mathbf{v}_m$$

for $j = 1, 2, \ldots, n$.

The symbols $A = [a_{ij}]_{m \times n}$ and $[T]_{\mathcal{B}, \mathcal{A}}$ in Definition 5.10 denote the same matrix, but the first one places notational emphasis on the elements of the matrix, while the second one places emphasis on T and the bases $\mathcal{A}$ and $\mathcal{B}$. This matrix A is also referred to as the **matrix of T with respect to $\mathcal{A}$ and $\mathcal{B}$**, and we say that T is **represented by the matrix A**. If T is a linear operator on $\mathbf{U}$ and $\mathcal{A} = \mathcal{B}$, then the matrix of the linear operator T relative to $\mathcal{A}$ is represented by the symbol $[T]_{\mathcal{A}}$.

As mentioned earlier, the elements a_{ij} are uniquely determined by T for given bases $\mathcal{A}$ and $\mathcal{B}$. Another way to describe A is to observe that the jth column of A is the coordinate matrix of $T(\mathbf{u}_j)$ with respect to $\mathcal{B}$. That is,

$$\begin{bmatrix} a_{1j} \\ a_{2j} \\ \vdots \\ a_{mj} \end{bmatrix} = [T(\mathbf{u}_j)]_{\mathcal{B}}$$

and

$$A = [T]_{\mathcal{B}, \mathcal{A}} = \begin{bmatrix} [T(\mathbf{u}_1)]_{\mathcal{B}} & [T(\mathbf{u}_2)]_{\mathcal{B}} & \cdots & [T(\mathbf{u}_n)]_{\mathcal{B}} \end{bmatrix}.$$

EXAMPLE 2 With **R** as the field of scalars, let $T : \mathcal{P}_2 \to \mathcal{P}_3$ be the linear transformation defined by

$$T(a_0 + a_1 x + a_2 x^2) = (2a_0 + 2a_2) + (a_0 + a_1 + 3a_2)x + (a_1 + 2a_2)x^2 + (a_0 + a_2)x^3.$$

We shall find the matrix A of T relative to the bases

$$\mathcal{A} = \{1, 1 - x, x^2\} \text{ of } \mathcal{P}_2$$

and

$$\mathcal{B} = \{1, x, 1 - x^2, 1 + x^3\} \text{ of } \mathcal{P}_3.$$

To find the first column of A, we compute $T(1)$ and write it as a linear combination of the vectors in $\mathcal{B}$.

$$T(1) = 2 + x + x^3$$
$$= (1)(1) + (1)(x) + (0)(1 - x^2) + (1)(1 + x^3).$$

Thus the first column of A is

$$[T(1)]_{\mathcal{B}} = \begin{bmatrix} 1 \\ 1 \\ 0 \\ 1 \end{bmatrix}.$$

For the remaining columns of A, we follow the same procedure with the second and third base vectors in $\mathcal{A}$.

$$T(1 - x) = 2 - x^2 + x^3$$
$$= (0)(1) + (0)(x) + (1)(1 - x^2) + (1)(1 + x^3)$$
$$T(x^2) = 2 + 3x + 2x^2 + x^3$$
$$= (3)(1) + (3)(x) + (-2)(1 - x^2) + (1)(1 + x^3).$$

Thus the matrix of T with respect to $\mathcal{A}$ and $\mathcal{B}$ is given by

$$A = [T]_{\mathcal{B},\mathcal{A}} = \begin{bmatrix} [T(1)]_{\mathcal{B}} & [T(1-x)]_{\mathcal{B}} & [T(x^2)]_{\mathcal{B}} \end{bmatrix} = \begin{bmatrix} 1 & 0 & 3 \\ 1 & 0 & 3 \\ 0 & 1 & -2 \\ 1 & 1 & 1 \end{bmatrix}. \qquad \blacksquare$$

If T is a linear transformation of $\mathbf{R}^n$ into $\mathbf{R}^m$, it is usually easier to find the matrix of T relative to $\mathcal{E}_n$ and $\mathcal{E}_m$ than with any other choice of bases, because the coordinates of a vector are the same as the components when working with the standard bases. However, we shall see in the next section that other choices of bases may give a much simpler matrix for T.

EXAMPLE 3 Consider the linear transformation $T : \mathbf{R}^4 \to \mathbf{R}^3$ given by

$$T(x_1, x_2, x_3, x_4) = (x_1 - x_3 + x_4, 2x_1 + x_2 + 3x_4, x_1 + 2x_2 + 3x_3 + 3x_4).$$

Using the standard bases $\mathcal{E}_4$ and $\mathcal{E}_3$, we compute

$$T(1, 0, 0, 0) = (1, 2, 1) \qquad T(0, 1, 0, 0) = (0, 1, 2)$$
$$T(0, 0, 1, 0) = (-1, 0, 3) \qquad T(0, 0, 0, 1) = (1, 3, 3).$$

Thus the matrix of T relative to $\mathcal{E}_4$ and $\mathcal{E}_3$ is

$$[T]_{\mathcal{E}_3, \mathcal{E}_4} = \begin{bmatrix} 1 & 0 & -1 & 1 \\ 2 & 1 & 0 & 3 \\ 1 & 2 & 3 & 3 \end{bmatrix}. \qquad \blacksquare$$

The mapping $f : \mathbf{R}^n \to \mathbf{R}_{n \times 1}$ defined by

$$f(x_1, x_2, \ldots, x_n) = \begin{bmatrix} x_1 \\ x_2 \\ \vdots \\ x_n \end{bmatrix}$$

is a natural isomorphism. It is so natural, in fact, that some texts make no distinction between

$$(x_1, x_2, \ldots, x_n) \quad \text{and} \quad \begin{bmatrix} x_1 \\ x_2 \\ \vdots \\ x_n \end{bmatrix},$$

and consider them as being the same entity. This isomorphism leads to a natural connection between the preceding example and Example 4 of Section 5.1. From the example in Section 5.1, we know that the matrix transformation $S : \mathbf{R}_{4 \times 1} \to \mathbf{R}_{3 \times 1}$ defined by

$$S\left(\begin{bmatrix} x_1 \\ x_2 \\ x_3 \\ x_4 \end{bmatrix} \right) = \begin{bmatrix} 1 & 0 & -1 & 1 \\ 2 & 1 & 0 & 3 \\ 1 & 2 & 3 & 3 \end{bmatrix} \begin{bmatrix} x_1 \\ x_2 \\ x_3 \\ x_4 \end{bmatrix} = \begin{bmatrix} x_1 - x_3 + x_4 \\ 2x_1 + x_2 + 3x_4 \\ x_1 + 2x_2 + 3x_3 + 3x_4 \end{bmatrix}$$

is a linear transformation. It is clear at a glance that the matrix transformation S and the mapping T in Example 3 are the same except for notation. That is, the two mappings differ only by an isomorphism.

The work in Examples 2 and 3 of this section illustrates that, for given bases $\mathcal{A}$ and $\mathcal{B}$, the matrix $[T]_{\mathcal{B}, \mathcal{A}}$ is uniquely determined by T. On the other hand, for a given matrix $A = [a_{ij}]_{m \times n}$ and fixed bases $\mathcal{A}$ and $\mathcal{B}$, there is only one linear transformation that has A as its matrix relative to $\mathcal{A}$ and $\mathcal{B}$. For if A is the matrix of both S and T, then we have

$S(\mathbf{u}_j) = \sum_{i=1}^{m} a_{ij}\mathbf{v}_i = T(\mathbf{u}_j)$ and, for any $\mathbf{u} = \sum_{j=1}^{n} x_j\mathbf{u}_j$ in **U**,

$$S(\mathbf{u}) = S\left(\sum_{j=1}^{n} x_j\mathbf{u}_j\right) = \sum_{j=1}^{n} x_j S(\mathbf{u}_j) = \sum_{j=1}^{n} x_j T(\mathbf{u}_j) = T\left(\sum_{j=1}^{n} x_j\mathbf{u}_j\right) = T(\mathbf{u}).$$

But $S(\mathbf{u}) = T(\mathbf{u})$ for all $\mathbf{u}$ in **U** means $S = T$. Thus, for fixed bases $\mathcal{A}$ and $\mathcal{B}$, T and $A = [T]_{\mathcal{B},\mathcal{A}}$ determine each other uniquely by the rule

$$T(\mathbf{u}_j) = \sum_{i=1}^{m} a_{ij}\mathbf{v}_i, \quad \text{or} \quad \begin{bmatrix} a_{1j} \\ a_{2j} \\ \vdots \\ a_{mj} \end{bmatrix} = [T(\mathbf{u}_j)]_{\mathcal{B}}.$$

Generally speaking, a change in either or both of the bases $\mathcal{A}$ and $\mathcal{B}$ is reflected in a change in the matrix of the linear transformation. With each particular choice of $\mathcal{A}$ and $\mathcal{B}$, we say that the matrix A of T relative to $\mathcal{A}$ and $\mathcal{B}$ *represents* T relative to these bases. Thus *different* matrices may represent the *same* linear transformation with different choices of bases. These matrices, though different, have certain properties in common. The exact relationship between these matrices is revealed in Section 5.3, but the next theorem gives some useful information on this subject.

5.11 THEOREM

If A is any matrix that represents the linear transformation T of **U** into **V**, then rank$(A) = $ rank(T). That is, rank$([T]_{\mathcal{B},\mathcal{A}}) = $ rank(T).

Proof Suppose $\mathcal{A} = \{\mathbf{u}_1, \mathbf{u}_2, \ldots, \mathbf{u}_n\}$ and $\mathcal{B} = \{\mathbf{v}_1, \mathbf{v}_2, \ldots, \mathbf{v}_m\}$ are bases of **U** and **V**, respectively, and let $A = [a_{ij}]_{m \times n}$ represent T relative to these bases. Then $T(\mathbf{u}_j) = \sum_{i=1}^{m} a_{ij}\mathbf{v}_i$, so that A is the matrix of transition from $\mathcal{B}$ to $T(\mathcal{A})$. According to Theorem 5.6, $T(\mathcal{A})$ spans $T(\mathbf{U})$. This means that the columns of A record the coordinates relative to $\mathcal{B}$ of a spanning set for $T(\mathbf{U})$. By Theorem 3.37, the reduced column-echelon form A' of A also records the coordinates relative to $\mathcal{B}$ of a spanning set for $T(\mathbf{U})$. Hence the number of nonzero columns in A' is the dimension of $T(\mathbf{U})$, and we have rank$(A) = $ rank(T).

∎

Thus, a convenient method for finding the rank of a linear transformation T is to find the rank of a matrix that represents T relative to a pair of bases $\mathcal{A}$ and $\mathcal{B}$. As a matter of fact, the reduced column-echelon form of such a matrix will disclose not only the rank of T, but the standard basis of $T(\mathbf{U})$ relative to $\mathcal{B}$.

EXAMPLE 4 In Example 2 of this section, we found that $T : \mathcal{P}_2 \rightarrow \mathcal{P}_3$ defined by

$$T(a_0 + a_1x + a_2x^2) = (2a_0 + 2a_2) + (a_0 + a_1 + 3a_2)x + (a_1 + 2a_2)x^2 + (a_0 + a_2)x^3$$

has the matrix

$$A = \begin{bmatrix} 1 & 0 & 3 \\ 1 & 0 & 3 \\ 0 & 1 & -2 \\ 1 & 1 & 1 \end{bmatrix}$$

relative to $\mathcal{A} = \{1, 1 - x, x^2\}$ and $\mathcal{B} = \{1, x, 1 - x^2, 1 + x^3\}$. The matrix A can be transformed to reduced column-echelon form as follows.

$$A = \begin{bmatrix} 1 & 0 & 3 \\ 1 & 0 & 3 \\ 0 & 1 & -2 \\ 1 & 1 & 1 \end{bmatrix} \rightarrow \begin{bmatrix} 1 & 0 & 0 \\ 1 & 0 & 0 \\ 0 & 1 & -2 \\ 1 & 1 & -2 \end{bmatrix} \rightarrow \begin{bmatrix} 1 & 0 & 0 \\ 1 & 0 & 0 \\ 0 & 1 & 0 \\ 1 & 1 & 0 \end{bmatrix} = A'$$

Thus $\mathrm{rank}(T) = \mathrm{rank}(A) = 2$, and the computations

$$(1)(1) + (1)(x) + (0)(1 - x^2) + (1)(1 + x^3) = 2 + x + x^3$$

$$(0)(1) + (0)(x) + (1)(1 - x^2) + (1)(1 + x^3) = 2 - x^2 + x^3$$

show that the standard basis of $T(\mathcal{P}_2)$ relative to $\mathcal{B}$ is

$$\{2 + x + x^3, 2 - x^2 + x^3\}. \qquad \blacksquare$$

5.12 THEOREM

Let $\mathcal{A} = \{\mathbf{u}_1, \mathbf{u}_2, \ldots, \mathbf{u}_n\}$ and $\mathcal{B} = \{\mathbf{v}_1, \mathbf{v}_2, \ldots, \mathbf{v}_m\}$ be bases of $\mathbf{U}$ and $\mathbf{V}$, respectively, and let $A = [a_{ij}]_{m \times n} = [T]_{\mathcal{B},\mathcal{A}}$ be the matrix of T relative to the bases $\mathcal{A}$ and $\mathcal{B}$. If $\mathbf{u}$ is an arbitrary vector in $\mathbf{U}$, then

$$[T(\mathbf{u})]_\mathcal{B} = [T]_{\mathcal{B},\mathcal{A}}[\mathbf{u}]_\mathcal{A}.$$

Proof Let

$$[\mathbf{u}]_\mathcal{A} = X = \begin{bmatrix} x_1 \\ x_2 \\ \vdots \\ x_n \end{bmatrix} \quad \text{and} \quad [T(\mathbf{u})]_\mathcal{B} = Y = \begin{bmatrix} y_1 \\ y_2 \\ \vdots \\ y_m \end{bmatrix}$$

so that $\mathbf{u} = \sum_{j=1}^{n} x_j \mathbf{u}_j$ and $T(\mathbf{u}) = \sum_{i=1}^{m} y_i \mathbf{v}_i$. Since $A = [a_{ij}]_{m \times n}$ is the matrix of T relative to $\mathcal{A}$ and $\mathcal{B}$, we have

$$\sum_{i=1}^{m} y_i \mathbf{v}_i = T(\mathbf{u})$$

$$= T\left(\sum_{j=1}^{n} x_j \mathbf{u}_j \right)$$

$$= \sum_{j=1}^{n} x_j T(\mathbf{u}_j)$$

$$= \sum_{j=1}^{n} x_j \left(\sum_{i=1}^{m} a_{ij} \mathbf{v}_i \right)$$

$$= \sum_{i=1}^{m} \left(\sum_{j=1}^{n} a_{ij} x_j \right) \mathbf{v}_i.$$

But the coordinates of $T(\mathbf{u})$ relative to $\mathcal{B}$ are unique, so this means that $y_i = \sum_{j=1}^{n} a_{ij} x_j$, and

$$Y = \begin{bmatrix} y_1 \\ y_2 \\ \vdots \\ y_m \end{bmatrix} = \begin{bmatrix} a_{11}x_1 + a_{12}x_2 + \cdots + a_{1n}x_n \\ a_{21}x_1 + a_{22}x_2 + \cdots + a_{2n}x_n \\ \vdots \\ a_{m1}x_1 + a_{m2}x_2 + \cdots + a_{mn}x_n \end{bmatrix}$$

$$= \begin{bmatrix} a_{11} & a_{12} & \cdots & a_{1n} \\ a_{21} & a_{22} & \cdots & a_{2n} \\ \vdots & \vdots & & \vdots \\ a_{m1} & a_{m2} & \cdots & a_{mn} \end{bmatrix} \begin{bmatrix} x_1 \\ x_2 \\ \vdots \\ x_n \end{bmatrix} = AX.$$

That is,

$$[T(\mathbf{u})]_\mathcal{B} = [T]_{\mathcal{B},\mathcal{A}}[\mathbf{u}]_\mathcal{A}. \qquad \blacksquare\blacksquare\blacksquare$$

Theorem 5.12 shows that the matrix transformation from $\mathbf{R}_{n \times 1}$ to $\mathbf{R}_{m \times 1}$ defined in Section 5.1 by $T(X) = AX$ generalizes to coordinates with arbitrary linear transformations of finite-dimensional vector spaces. However, it should be kept in mind that X, A, and Y in the equation $Y = AX$ are taken relative to the bases $\mathcal{A}$ and $\mathcal{B}$, and consequently change when $\mathcal{A}$ and $\mathcal{B}$ are changed.

EXAMPLE 5 Let T be the linear transformation of $\mathbf{R}^4$ into $\mathbf{R}^3$ that has the matrix

$$A = \begin{bmatrix} 1 & 0 & -1 & 1 \\ 2 & 1 & 0 & 3 \\ 1 & 2 & 3 & 3 \end{bmatrix}$$

with respect to the bases $\mathcal{A} = \{(1, 1, 1, 1), (1, 1, 1, 0), (1, 1, 0, 0), (1, 0, 0, 0)\}$ of $\mathbf{R}^4$ and $\mathcal{B} = \{(0, 1, 1), (1, 0, 0), (0, 0, 1)\}$ of $\mathbf{R}^3$. Suppose we wish to use this matrix A to find the value of $T(3, 4, -1, -4)$.

To find the coordinates of $\mathbf{u} = (3, 4, -1, -4)$ relative to $\mathcal{A}$, we write $\mathbf{u}$ as a linear combination of the vectors in $\mathcal{A}$. The strategic placement of zeros in the vectors of $\mathcal{A}$ allows us to obtain the coefficients of these vectors by inspection: The first vector in $\mathcal{A}$ is the only one with a nonzero fourth component, the first and second vectors in $\mathcal{A}$ are the only ones with nonzero third components, and so on. We obtain

$$(3, 4, -1, -4) = (-4)(1, 1, 1, 1) + (3)(1, 1, 1, 0) + 5(1, 1, 0, 0) + (-1)(1, 0, 0, 0).$$

Thus

$$[\mathbf{u}]_A = \begin{bmatrix} -4 \\ 3 \\ 5 \\ -1 \end{bmatrix}$$

and therefore

$$[T(\mathbf{u})]_B = \begin{bmatrix} 1 & 0 & -1 & 1 \\ 2 & 1 & 0 & 3 \\ 1 & 2 & 3 & 3 \end{bmatrix} \begin{bmatrix} -4 \\ 3 \\ 5 \\ -1 \end{bmatrix} = \begin{bmatrix} -10 \\ -8 \\ 14 \end{bmatrix}.$$

Using these coordinates with the vectors in the basis $\mathcal{B}$, we find that

$$T(3, 4, -1, -4) = (-10)(0, 1, 1) + (-8)(1, 0, 0) + 14(0, 0, 1)$$

$$= (-8, -10, 4).$$ ■

In Example 5 the matrix A is the matrix of the linear transformation T relative to the bases $\mathcal{A}$ and $\mathcal{B}$; that is,

$$A = [T]_{\mathcal{B},\mathcal{A}} = \begin{bmatrix} 1 & 0 & -1 & 1 \\ 2 & 1 & 0 & 3 \\ 1 & 2 & 3 & 3 \end{bmatrix}.$$

The image of $\mathbf{u} = (3, 4, -1, -4)$ under T was found to be

$$T(\mathbf{u}) = T(3, 4, -1, -4) = (-8, -10, 4).$$

However, in Example 3 this same matrix A represents a different linear transformation T' relative to the standard bases $\mathcal{E}_4$ and $\mathcal{E}_3$; that is,

$$A = [T']_{\mathcal{E}_3, \mathcal{E}_4} = \begin{bmatrix} 1 & 0 & -1 & 1 \\ 2 & 1 & 0 & 3 \\ 1 & 2 & 3 & 3 \end{bmatrix}.$$

The image of the same $\mathbf{u} = (3, 4, -1, -4)$ under T' is computed as

$$T'(\mathbf{u}) = T'(3, 4, -1, -4) = (0, -2, -4)$$

or using matrices as

$$[T'(\mathbf{u})]_{\mathcal{E}_3} = [T']_{\mathcal{E}_3,\mathcal{E}_4}[\mathbf{u}]_{\mathcal{E}_4} = \begin{bmatrix} 1 & 0 & -1 & 1 \\ 2 & 1 & 0 & 3 \\ 1 & 2 & 3 & 3 \end{bmatrix} \begin{bmatrix} 3 \\ 4 \\ -1 \\ -4 \end{bmatrix} = \begin{bmatrix} 0 \\ -2 \\ -4 \end{bmatrix}.$$

Since the coordinates of the image $T'(\mathbf{u})$ are the same as its components we have, using either method of computation, that $T'(\mathbf{u}) = (0, -2, -4)$. Hence, we see that the same matrix A may represent two different linear transformations.

Theorem 5.12 reveals a practical approach to the problem of finding a basis for the kernel $T^{-1}(\mathbf{0})$. For $\mathbf{u} \in T^{-1}(\mathbf{0})$ if and only if the coordinates X of $\mathbf{u}$ satisfy $AX = \mathbf{0}$. Thus, the solutions to the system of equations $AX = \mathbf{0}$ furnish the coordinates of the vectors in $T^{-1}(\mathbf{0})$.

EXAMPLE 6 With $\mathbf{R}$ as the field of scalars for $\mathcal{P}_4$, let $T : \mathcal{P}_4 \to \mathbf{R}^4$ be the linear transformation defined by

$$T(a_0 + a_1 x + a_2 x^2 + a_3 x^3 + a_4 x^4)$$
$$= (a_0 + a_1 + a_2 + 2a_4, a_1 + 2a_2 + a_3 - a_4, a_3 - a_4, a_0 + a_1 + a_2 + 2a_4).$$

A natural choice of bases are the standard bases $\mathcal{A} = \{1, x, x^2, x^3, x^4\}$ for $\mathcal{P}_4$ and $\mathcal{E}_4 = \{\mathbf{e}_1, \mathbf{e}_2, \mathbf{e}_3, \mathbf{e}_4\}$ for $\mathbf{R}^4$. Straightforward computations yield

$$T(1) = (1, 0, 0, 1), \qquad T(x) = (1, 1, 0, 1)$$
$$T(x^2) = (1, 2, 0, 1), \qquad T(x^3) = (0, 1, 1, 0),$$
$$T(x^4) = (2, -1, -1, 2).$$

Thus the matrix of T relative to $\mathcal{A}$ and $\mathcal{B}$ is

$$A = \begin{bmatrix} 1 & 1 & 1 & 0 & 2 \\ 0 & 1 & 2 & 1 & -1 \\ 0 & 0 & 0 & 1 & -1 \\ 1 & 1 & 1 & 0 & 2 \end{bmatrix},$$

and we need to solve the linear system $AX = \mathbf{0}$. Using the augmented matrix $[A \mid \mathbf{0}]$, we have

$$[A \mid \mathbf{0}] \to \begin{bmatrix} 1 & 1 & 1 & 0 & 2 & 0 \\ 0 & 1 & 2 & 1 & -1 & 0 \\ 0 & 0 & 0 & 1 & -1 & 0 \\ 1 & 1 & 1 & 0 & 2 & 0 \end{bmatrix} \to \begin{bmatrix} 1 & 1 & 1 & 0 & 2 & 0 \\ 0 & 1 & 2 & 1 & -1 & 0 \\ 0 & 0 & 0 & 1 & -1 & 0 \\ 0 & 0 & 0 & 0 & 0 & 0 \end{bmatrix}$$

$$\to \begin{bmatrix} 1 & 1 & 1 & 0 & 2 & 0 \\ 0 & 1 & 2 & 0 & 0 & 0 \\ 0 & 0 & 0 & 1 & -1 & 0 \\ 0 & 0 & 0 & 0 & 0 & 0 \end{bmatrix} \to \begin{bmatrix} 1 & 0 & -1 & 0 & 2 & 0 \\ 0 & 1 & 2 & 0 & 0 & 0 \\ 0 & 0 & 0 & 1 & -1 & 0 \\ 0 & 0 & 0 & 0 & 0 & 0 \end{bmatrix}.$$

The solutions to $AX = \mathbf{0}$ are given by

$$\begin{aligned} x_1 &= & x_3 - 2x_5 \\ x_2 &= & -2x_3 \\ x_3 &= & x_3 \\ x_4 &= & & x_5 \\ x_5 &= & & x_5 \end{aligned}$$

and in matrix form by

$$X = x_3 \begin{bmatrix} 1 \\ -2 \\ 1 \\ 0 \\ 0 \end{bmatrix} + x_5 \begin{bmatrix} -2 \\ 0 \\ 0 \\ 1 \\ 1 \end{bmatrix}.$$

Using the coordinate matrices $[1 \quad -2 \quad 1 \quad 0 \quad 0]^T$ and $[-2 \quad 0 \quad 0 \quad 1 \quad 1]^T$ with the standard basis $\mathcal{A}$ of $\mathcal{P}_4$, we obtain the basis

$$\{1 - 2x + x^2, -2 + x^3 + x^4\}$$

for the kernel of T.   ■

**5.13
THEOREM**

Let $\mathcal{A} = \{\mathbf{u}_1, \mathbf{u}_2, \ldots, \mathbf{u}_n\}$ and $\mathcal{B} = \{\mathbf{v}_1, \mathbf{v}_2, \ldots, \mathbf{v}_m\}$ be bases of $\mathbf{U}$ and $\mathbf{V}$, respectively. If $A = [a_{ij}]_{m \times n}$ is a matrix such that the equation

$$[T(\mathbf{u})]_{\mathcal{B}} = A[\mathbf{u}]_{\mathcal{A}}$$

is satisfied for all $\mathbf{u}$ in $\mathbf{U}$, then A is the matrix of the linear transformation T relative to $\mathcal{A}$ and $\mathcal{B}$.

Proof For $\mathbf{u} = \mathbf{u}_j$, we have

$$[\mathbf{u}_j] = \begin{bmatrix} \delta_{1j} \\ \delta_{2j} \\ \vdots \\ \delta_{nj} \end{bmatrix}$$

and

$$T[(\mathbf{u}_j)]_{\mathcal{B}} = A[\mathbf{u}_j]_{\mathcal{A}} = \begin{bmatrix} \sum_{k=1}^{n} a_{1k}\delta_{kj} \\ \sum_{k=1}^{n} a_{2k}\delta_{kj} \\ \vdots \\ \sum_{k=1}^{n} a_{mk}\delta_{kj} \end{bmatrix} = \begin{bmatrix} a_{1j} \\ a_{2j} \\ \vdots \\ a_{mj} \end{bmatrix}.$$

Thus $T(\mathbf{u}_j) = \sum_{i=1}^{m} a_{ij}\mathbf{v}_i$, and A is the matrix of T relative to $\mathcal{A}$ and $\mathcal{B}$.   ■ ■ ■

5.2 Exercises

1. Let $v_1 = (2, -1)$ and $v_2 = (1, 0)$ in $\mathbf{R}^2$. Find a formula for $T(x, y)$ if T is the linear transformation of $\mathbf{R}^2$ into $\mathbf{R}^3$ for which $T(v_1) = (1, 0, 1)$ and $T(v_2) = (0, 1, 1)$.

2. Suppose $T : \mathbf{R}_{2 \times 1} \to \mathbf{R}_{3 \times 1}$ is a linear transformation such that

$$T\left(\begin{bmatrix} 1 \\ 0 \end{bmatrix}\right) = \begin{bmatrix} 1 \\ 2 \\ 3 \end{bmatrix} \quad \text{and} \quad T\left(\begin{bmatrix} 0 \\ 1 \end{bmatrix}\right) = \begin{bmatrix} 0 \\ -1 \\ 1 \end{bmatrix}.$$

Find $T\left(\begin{bmatrix} x \\ y \end{bmatrix}\right)$.

3. Consider the basis $\{p_1(x), p_2(x), p_3(x)\}$ for $\mathcal{P}_2$ over $\mathbf{R}$, where $p_1(x) = 1 + x + x^2$, $p_2(x) = x + x^2$, $p_3(x) = x^2$. Find a formula for $T(a_0 + a_1 x + a_2 x^2)$ if $T : \mathcal{P}_2 \to \mathbf{R}^2$ is the linear transformation such that $T(p_1(x)) = (1, 0)$, $T(p_2(x)) = (1, 0)$, and $T(p_3(x)) = (0, 1)$.

4. Let $T : \mathbf{R}_{2 \times 2} \to \mathbf{R}_{3 \times 1}$ be a linear transformation for which it is known that

$$T\left(\begin{bmatrix} 1 & 0 \\ 0 & 0 \end{bmatrix}\right) = \begin{bmatrix} 2 \\ 3 \\ 0 \end{bmatrix}, \quad T\left(\begin{bmatrix} 0 & 1 \\ 0 & 0 \end{bmatrix}\right) = \begin{bmatrix} 2 \\ 3 \\ 1 \end{bmatrix},$$

$$T\left(\begin{bmatrix} 0 & 0 \\ 1 & 0 \end{bmatrix}\right) = \begin{bmatrix} 2 \\ 3 \\ 2 \end{bmatrix}, \quad T\left(\begin{bmatrix} 0 & 0 \\ 0 & 1 \end{bmatrix}\right) = \begin{bmatrix} 3 \\ 3 \\ 0 \end{bmatrix}.$$

Find $T\left(\begin{bmatrix} a & b \\ c & d \end{bmatrix}\right)$.

In Problems 5–8, find the matrix of the given linear transformation T relative to the bases $\mathcal{A}$ and $\mathcal{B}$.

5. $T : \mathbf{R}^3 \to \mathbf{R}^4, \mathcal{A} = \mathcal{E}_3, \mathcal{B} = \mathcal{E}_4$
$T(x_1, x_2, x_3) = (x_1 + x_2 - x_3, 5x_2 - 2x_3, 4x_1 + x_3, 2x_1 + 3x_2 + x_3)$

6. $T : \mathbf{R}^4 \to \mathbf{R}^3, \mathcal{A} = \mathcal{E}_4, \mathcal{B} = \mathcal{E}_3$
$T(x_1, x_2, x_3, x_4) = (x_1 + x_2 + x_3 - x_4, x_1 + 2x_2 + x_3 - x_4, x_1 - x_2 + x_3)$

7. $T : \mathcal{P}_2 \to \mathcal{P}_3$ over $\mathbf{R}, \mathcal{A} = \{1, 1 - x, x - x^2\}, \mathcal{B} = \{1, 1 + x, 1 - x^2, x + x^3\}$
$T(p(x)) = (1 - x)p(x)$

8. $T : \mathcal{P}_2 \to \mathcal{P}_1$ over $\mathbf{R}, \mathcal{A} = \{1 + x^2, 1 + x, 1\}, \mathcal{B} = \{1 - x, x\}$
$T(a_0 + a_1 x + a_2 x^2) = (a_0 + a_1 + a_2) + (a_1 - 2a_2)x$

9. If T is the linear transformation that has the matrix

$$\begin{bmatrix} 1 & 2 \\ -1 & 1 \\ 1 & 1 \end{bmatrix}$$

relative to the bases $\{(1, 1), (-3, 1)\}$ of $\mathbf{R}^2$ and $\mathcal{E}_3$ of $\mathbf{R}^3$, find $T(-2, 2)$.

10. The linear operator T on $\mathbf{R}^2$ has the matrix

$$\begin{bmatrix} 4 & -1 \\ -4 & 3 \end{bmatrix}$$

relative to the basis $\mathcal{A} = \mathcal{B} = \{(1, 2), (0, 1)\}$. A vector $\mathbf{u}$ has coordinates

$$\begin{bmatrix} 1 \\ 1 \end{bmatrix}$$

relative to this basis. Find $T(\mathbf{u})$ in component form (x, y).

11. Let T be the linear transformation of $\mathbf{R}^3$ into $\mathbf{R}^2$ that has the matrix

$$\begin{bmatrix} 2 & 3 & 1 \\ 1 & 2 & 1 \end{bmatrix}$$

relative to the bases $\{(1, -1, 1), (0, 1, 0), (1, 0, 0)\}$ of $\mathbf{R}^3$ and $\{(3, 2), (2, 1)\}$ of $\mathbf{R}^2$. Find $T(2, 0, 1)$.

12. Let

$$A = \begin{bmatrix} 1 & 0 & 0 \\ 2 & 1 & 1 \\ 3 & 2 & 1 \\ 4 & 3 & 1 \end{bmatrix}$$

be the matrix of the linear transformation $T : \mathcal{P}_2 \rightarrow \mathcal{P}_3$ over $\mathbf{R}$ with respect to the bases $\{4, 1 + x, 1 + x^2\}$ and $\{1, x, x^2, x^3\}$. Find $T(2 - 2x + x^2)$.

13. Let

$$A = \begin{bmatrix} 1 & 2 & 3 \\ 0 & 1 & 2 \end{bmatrix}$$

be the matrix of the linear transformation $T : \mathcal{P}_2 \rightarrow \mathcal{P}_1$ over $\mathbf{R}$ with respect to the bases $\{1 + x + x^2, x + x^2, x^2\}$ and $\{1, 2 + 3x\}$. Find $T(2 + 5x + x^2)$.

14. Suppose the bases $\mathcal{A}$ of $\mathbf{R}_{2 \times 2}$ and $\mathcal{B}$ of $\mathcal{P}_1$ over $\mathbf{R}$ are given by

$$\mathcal{A} = \left\{ \begin{bmatrix} 1 & 0 \\ 0 & 1 \end{bmatrix}, \begin{bmatrix} 0 & 0 \\ 1 & 1 \end{bmatrix}, \begin{bmatrix} 0 & 0 \\ 1 & 0 \end{bmatrix}, \begin{bmatrix} 0 & 1 \\ 0 & 0 \end{bmatrix} \right\},$$

$$\mathcal{B} = \{1 + x, 1 - x\}.$$

Let T be the linear transformation $T : \mathbf{R}_{2 \times 2} \to \mathcal{P}_1$ with matrix

$$\begin{bmatrix} 2 & 0 & 1 & 0 \\ 1 & 1 & -1 & 0 \end{bmatrix}$$

relative to $\mathcal{A}$ and $\mathcal{B}$. Find

$$T\left(\begin{bmatrix} 0 & 0 \\ 3 & 2 \end{bmatrix} \right).$$

15. A linear operator T on $\mathbf{R}^4$ has the matrix

$$A = \begin{bmatrix} 1 & 0 & 1 & 1 \\ 2 & 1 & 3 & 1 \\ -1 & 0 & -1 & -1 \\ 3 & 2 & 5 & 1 \end{bmatrix}$$

with respect to the bases $\mathcal{A} = \mathcal{B} = \mathcal{E}_4$.

(a) Find a basis for $T(\mathbf{R}^4)$.

(b) Find a basis of the kernel of T.

(c) Find the standard basis of the kernel of T.

(d) Give the rank and nullity of T.

16. Suppose T is the linear transformation of $\mathbf{R}^n$ into $\mathbf{R}^m$ that has the given matrix A relative to the standard bases $\mathcal{E}_n$ and $\mathcal{E}_m$. Find a basis for the range of T.

(a) $A = \begin{bmatrix} 1 & 1 & 0 & -1 \\ 2 & 3 & 1 & 0 \\ 1 & 3 & 2 & 3 \end{bmatrix}$

(b) $A = \begin{bmatrix} 3 & -2 & -1 & -4 \\ 1 & 1 & -2 & -3 \\ -2 & 3 & -1 & 1 \end{bmatrix}$

(c) $A = \begin{bmatrix} 1 & 2 & -1 & 2 & 1 \\ 1 & 4 & 4 & -3 & -1 \\ 2 & 6 & 3 & -1 & 0 \\ 3 & 8 & 2 & 1 & 2 \end{bmatrix}$

(d) $A = \begin{bmatrix} 1 & 2 & 0 & 1 & 0 \\ 2 & 4 & 1 & 4 & 3 \\ 1 & 2 & 2 & 5 & -2 \\ -1 & -2 & 3 & 5 & 4 \end{bmatrix}$

17. In each part of Problem 16,

(a) find a basis of the kernel of T,

(b) find the standard basis of the kernel of T,

(c) give the rank and nullity of T.

18. Let T be the linear operator on $\mathbf{R}^3$ that has the matrix

$$\begin{bmatrix} 1 & 1 & 0 \\ 2 & 2 & 0 \\ 3 & 3 & 0 \end{bmatrix}$$

relative to the basis $\mathcal{A} = \mathcal{B} = \{(1, 0, 1), (1, 1, 0), (0, 1, 1)\}$ of $\mathbf{R}^3$. Find a basis for the kernel of T.

19. Let T be the linear transformation of $\mathbf{R}^5$ into $\mathbf{R}^3$ that has the matrix A relative to the bases $\{(1, 1, 1, 1, 1), (1, 1, 1, 1, 0), (1, 1, 0, 0, 0), (1, 0, 0, 0, 0), (0, 0, 0, 0, 1)\}$ of $\mathbf{R}^5$ and $\{(1, 1, 1), (0, 1, 0), (1, 0, 0)\}$ of $\mathbf{R}^3$. Find a basis for the range of T.

(a) $A = \begin{bmatrix} 1 & 3 & 2 & 0 & -1 \\ 2 & 6 & 4 & 6 & 4 \\ 1 & 3 & 2 & 2 & 1 \end{bmatrix}$
(b) $A = \begin{bmatrix} 1 & -2 & 0 & 1 & -4 \\ 2 & -4 & 1 & 3 & -5 \\ 1 & -2 & 0 & 0 & -2 \end{bmatrix}$

20. Find a basis for the kernel of T in each part of Problem 19 and give the rank and nullity of T.

21. Let T be a linear transformation from P_2 into P_3 over $\mathbf{R}$ defined by $T(p(x)) = xp(x)$.

    (a) Find $[T]_{\mathcal{B},\mathcal{A}}$ the matrix of T relative to the bases $\mathcal{A} = \{1 - x, 1 - x^2, x\}$ and $\mathcal{B} = \{1, 1 + x, 1 + x + x^2, 1 - x^3\}$.
    (b) Use $[T]_{\mathcal{B},\mathcal{A}}$ to find a basis for the range of T.
    (c) Use $[T]_{\mathcal{B},\mathcal{A}}$ to find a basis for the kernel of T.
    (d) State the rank and nullity of T.

22. Let T be a linear transformation from $\mathbf{R}_{2\times2}$ into P_3 over $\mathbf{R}$ defined by

$$T\left(\begin{bmatrix} a_{11} & a_{12} \\ a_{21} & a_{22} \end{bmatrix}\right) = (a_{11} - a_{12}) + (a_{11} - a_{22})x + (a_{12} - a_{21})x^2 + (a_{21} - a_{22})x^3.$$

    (a) Find $[T]_{\mathcal{B},\mathcal{A}}$ the matrix of T relative to the bases $\mathcal{A}$ and $\mathcal{B}$ where

$$\mathcal{A} = \left\{ \begin{bmatrix} 1 & 0 \\ 1 & 0 \end{bmatrix}, \begin{bmatrix} 0 & 1 \\ 0 & 1 \end{bmatrix}, \begin{bmatrix} 1 & 0 \\ 0 & 1 \end{bmatrix}, \begin{bmatrix} 0 & 0 \\ 1 & 1 \end{bmatrix} \right\}$$

and

$$\mathcal{B} = \{x, x - x^2, x - x^3, x - 1\}.$$

    (b) Use $[T]_{\mathcal{B},\mathcal{A}}$ to find a basis for the range of T.
    (c) Use $[T]_{\mathcal{B},\mathcal{A}}$ to find a basis for the kernel of T.
    (d) State the rank and nullity of T.

23. Let T be a linear transformation from $\mathbf{R}_{1\times3}$ into $\mathbf{R}_{2\times2}$ defined by

$$T\left([a_{11} \quad a_{12} \quad a_{13}]\right) = \begin{bmatrix} a_{11} & 5a_{11} \\ a_{13} & 3a_{13} \end{bmatrix}.$$

    (a) Find $[T]_{\mathcal{B},\mathcal{A}}$ the matrix of T relative to the following bases.

$$\mathcal{A} = \{[0 \quad 1 \quad 0], [0 \quad 1 \quad 1], [1 \quad 1 \quad 0]\}$$

$$\mathcal{B} = \left\{ \begin{bmatrix} 1 & -1 \\ 0 & 0 \end{bmatrix}, \begin{bmatrix} 0 & 1 \\ -1 & 0 \end{bmatrix}, \begin{bmatrix} 0 & 0 \\ 1 & 1 \end{bmatrix}, \begin{bmatrix} 0 & 1 \\ 0 & -1 \end{bmatrix} \right\}$$

    (b) Use $[T]_{\mathcal{B},\mathcal{A}}$ to find a basis for the range of T.

(c) Use $[T]_{B,A}$ to find a basis for the kernel of T.

(d) State the rank and nullity of T.

24. Let T be a linear transformation from P_2 over $\mathbf{R}$ into $\mathbf{R}_{2\times2}$ defined by

$$T(a_0 + a_1x + a_2x^2) = \begin{bmatrix} a_2 & -a_2 \\ a_0 - a_2 & a_0 + a_2 \end{bmatrix}.$$

(a) Find $[T]_{B,A}$ the matrix of T relative to the bases

$$\mathcal{A} = \{x^2 - x, x - 1, x^2 + 1\}$$

$$\mathcal{B} = \left\{ \begin{bmatrix} 1 & 0 \\ 0 & 1 \end{bmatrix}, \begin{bmatrix} 0 & 1 \\ 1 & 0 \end{bmatrix}, \begin{bmatrix} 0 & 0 \\ 1 & 1 \end{bmatrix}, \begin{bmatrix} -1 & 1 \\ 0 & 0 \end{bmatrix} \right\}.$$

(b) Use $[T]_{B,A}$ to find a basis for the range of T.

(c) Use $[T]_{B,A}$ to find a basis for the kernel of T.

(d) State the rank and nullity of T.

25. Find the matrix of T in Problem 9 relative to the bases $\{(-1, 3), (-1, 1)\}$ of $\mathbf{R}^2$ and $\mathcal{E}_3$ of $\mathbf{R}^3$.

26. Let T be the linear transformation of $\mathbf{R}^3$ into $\mathbf{R}^2$ whose matrix is

$$A = \begin{bmatrix} 9 & -4 & -4 \\ 6 & -3 & -2 \end{bmatrix}$$

relative to the standard bases $\mathcal{E}_3$ and $\mathcal{E}_2$. Find the matrix of T relative to the bases $\{(1, 2, 0), (1, 1, 1), (1, 1, 0)\}$ of $\mathbf{R}^3$ and $\{(1, 0), (1, 1)\}$ of $\mathbf{R}^2$.

27. Let T be a linear operator on $\mathbf{R}^2$ that maps $(2, 1)$ onto $(5, 2)$ and $(1, 2)$ onto $(7, 10)$. Determine the matrix of T with respect to the bases $\mathcal{A} = \mathcal{B} = \{(3, 3), (1, -1)\}$.

28. Give an example where two different matrices represent the same linear transformation.

29. Let T be the identity linear operator on an n-dimensional vector space $\mathbf{U}$. Show that the matrix of T relative to any basis $\mathcal{A}$ of $\mathbf{U}$ is I_n, the identity matrix.

30. Let T be the zero linear operator on an n-dimensional vector space $\mathbf{U}$. Show that the matrix of T relative to any basis $\mathcal{A}$ of $\mathbf{U}$ is $\mathbf{0}$, the $n \times n$ zero matrix.

31. Let T be a linear transformation of $\mathbf{U}$ into $\mathbf{V}$ and A the matrix of T relative to the bases $\mathcal{A}$ of $\mathbf{U}$ and $\mathcal{B}$ of $\mathbf{V}$.

(a) Prove that $\mathbf{u}$ is in the kernel of T if and only if $[\mathbf{u}]_A$ is in the nullspace of A.

(b) Prove that $\mathbf{v}$ is in the range of T if and only if $[\mathbf{v}]_B$ is in the column space of A.

32. Let T be a linear transformation of $\mathbf{U}$ into $\mathbf{V}$, and let $\mathbf{U}_1$ and $\mathbf{U}_2$ be subspaces of $\mathbf{U}$. Prove or disprove that $T(\mathbf{U}_1 \cap \mathbf{U}_2) = T(\mathbf{U}_1) \cap T(\mathbf{U}_2)$.

33. Let T be a linear transformation of $\mathbf{U}$ into $\mathbf{V}$, and let $\mathbf{W}_1$ and $\mathbf{W}_2$ be subspaces of $\mathbf{V}$. Prove or disprove the statements below.

(a) $T^{-1}(\mathbf{W}_1 + \mathbf{W}_2) = T^{-1}(\mathbf{W}_1) + T^{-1}(\mathbf{W}_2)$.

(b) $T^{-1}(\mathbf{W}_1 \cap \mathbf{W}_2) = T^{-1}(\mathbf{W}_1) \cap T^{-1}(\mathbf{W}_2)$.

5.3 Change of Basis

It is the purpose of this section to give a complete description of the relation between those matrices that represent the same linear transformation. This description is found by examining the effect that a change in the bases of $\mathbf{U}$ and $\mathbf{V}$ has on the matrix of T.

5.14 THEOREM

Let $\mathcal{C} = \{\mathbf{w}_1, \mathbf{w}_2, \ldots, \mathbf{w}_k\}$ and $\mathcal{C}' = \{\mathbf{w}_1', \mathbf{w}_2', \ldots, \mathbf{w}_k'\}$ be two bases of the vector space $\mathbf{W}$ over $\mathcal{F}$. For an arbitrary vector $\mathbf{w}$ in $\mathbf{W}$, let

$$[\mathbf{w}]_{\mathcal{C}} = C = \begin{bmatrix} c_1 \\ c_2 \\ \vdots \\ c_k \end{bmatrix} \quad \text{and} \quad [\mathbf{w}]_{\mathcal{C}'} = C' = \begin{bmatrix} c_1' \\ c_2' \\ \vdots \\ c_k' \end{bmatrix}$$

denote the coordinate matrices of $\mathbf{w}$ relative to $\mathcal{C}$ and $\mathcal{C}'$, respectively. If P is the matrix of transition from $\mathcal{C}$ to $\mathcal{C}'$, then $C = PC'$. That is,

$$[\mathbf{w}]_{\mathcal{C}} = P[\mathbf{w}]_{\mathcal{C}'}.$$

Proof Let $P = [p_{ij}]_{k \times k}$, and assume that the hypotheses of the theorem are satisfied. Then

$$\mathbf{w} = \sum_{i=1}^{k} c_i \mathbf{w}_i = \sum_{j=1}^{k} c_j' \mathbf{w}_j'$$

and

$$\mathbf{w}_j' = \sum_{i=1}^{k} p_{ij} \mathbf{w}_i.$$

Combining these equalities, we have

$$\mathbf{w} = \sum_{i=1}^{k} c_i \mathbf{w}_i$$

$$= \sum_{j=1}^{k} c_j' \left(\sum_{i=1}^{k} p_{ij} \mathbf{w}_i \right) = \sum_{i=1}^{k} \left(\sum_{j=1}^{k} p_{ij} c_j' \right) \mathbf{w}_i.$$

Therefore $c_i = \sum_{j=1}^{k} p_{ij}c_j'$ and

$$\begin{bmatrix} c_1 \\ c_2 \\ \vdots \\ c_k \end{bmatrix} = \begin{bmatrix} p_{11}c_1' + p_{12}c_2' + \cdots + p_{1k}c_k' \\ p_{21}c_1' + p_{22}c_2' + \cdots + p_{2k}c_k' \\ \vdots \\ p_{k1}c_1' + p_{k2}c_2' + \cdots + p_{kk}c_k' \end{bmatrix} = PC'. \qquad ■■■$$

The result of this theorem can easily be remembered if we utilize a notational convenience by writing $P_{C \to C'}$ to indicate the matrix of transition from C to C'. Then the result is written as

$$[\mathbf{w}]_C = P_{C \to C'}[\mathbf{w}]_{C'}.$$

EXAMPLE 1 Consider the bases $C = \{x, 2 + x\}$ and $C' = \{4 + x, 4 - x\}$ of the vector space $\mathbf{W} = \mathcal{P}_1$ over $\mathbf{R}$. Since

$$4 + x = (-1)(x) + (2)(2 + x),$$
$$4 - x = (-3)(x) + (2)(2 + x)$$

the matrix of transition from C to C' is given by

$$P_{C \to C'} = \begin{bmatrix} -1 & -3 \\ 2 & 2 \end{bmatrix}.$$

The vector $\mathbf{w} = 4 + 3x$ can be written as

$$4 + 3x = 2(4 + x) + (-1)(4 - x)$$

so

$$[\mathbf{w}]_{C'} = \begin{bmatrix} 2 \\ -1 \end{bmatrix}.$$

According to Theorem 5.14, the coordinate matrix $[\mathbf{w}]_C$ may be found from

$$[\mathbf{w}]_C = P_{C \to C'}[\mathbf{w}]_{C'} = \begin{bmatrix} -1 & -3 \\ 2 & 2 \end{bmatrix} \begin{bmatrix} 2 \\ -1 \end{bmatrix} = \begin{bmatrix} 1 \\ 2 \end{bmatrix}.$$

This result can be checked by using the base vectors in C. We get

$$(1)(x) + 2(2 + x) = 4 + 3x = \mathbf{w},$$

so the value for $[\mathbf{w}]_C$ is correct. ■

The following theorem gives a full description of the effect of a change in the bases $\mathcal{A}$ and $\mathcal{B}$.

5.15 **THEOREM**	Suppose that T has matrix A relative to the bases $\mathcal{A}$ of **U** and $\mathcal{B}$ of **V**. If Q is the matrix of transition from $\mathcal{A}$ to the basis $\mathcal{A}'$ of **U** and P is the matrix of transition from $\mathcal{B}$ to the basis $\mathcal{B}'$ of **V**, then the matrix of T relative to $\mathcal{A}'$ and $\mathcal{B}'$ is $P^{-1}AQ$. (See Figure 5.1.)

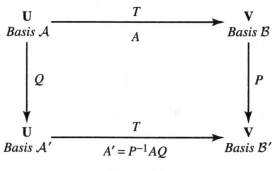

Figure 5.1

Proof Assume that the hypotheses of the theorem are satisfied. Let **u** be an arbitrary vector in **U**, with $[\mathbf{u}]_\mathcal{A}$ and $[\mathbf{u}]_{\mathcal{A}'}$ the coordinate matrices of **u** relative to $\mathcal{A}$ and $\mathcal{A}'$, respectively, and $[T(\mathbf{u})]_\mathcal{B}$ and $[T(\mathbf{u})]_{\mathcal{B}'}$ the coordinate matrices of $T(\mathbf{u})$ relative to $\mathcal{B}$ and $\mathcal{B}'$, respectively. If we write $Q_{\mathcal{A}\to\mathcal{A}'}$ as the matrix of transition Q from $\mathcal{A}$ to $\mathcal{A}'$ and similarly $P_{\mathcal{B}\to\mathcal{B}'}$ for P, by Theorem 5.14, we have

$$[\mathbf{u}]_\mathcal{A} = Q_{\mathcal{A}\to\mathcal{A}'}[\mathbf{u}]_{\mathcal{A}'} \quad \text{and} \quad [T(\mathbf{u})]_\mathcal{B} = P_{\mathcal{B}\to\mathcal{B}'}[T(\mathbf{u})]_{\mathcal{B}'}.$$

But since $\mathcal{B}$ and $\mathcal{B}'$ are bases, then P is invertible and P^{-1} is the matrix of transition from $\mathcal{B}'$ to $\mathcal{B}$. If we write P^{-1} as $P_{\mathcal{B}\to\mathcal{B}'}^{-1}$ then

$$[T(\mathbf{u})]_{\mathcal{B}'} = P_{\mathcal{B}\to\mathcal{B}'}^{-1}[T(\mathbf{u})]_\mathcal{B}.$$

With $A = [T]_{\mathcal{B},\mathcal{A}}$, by Theorem 5.12 we have

$$\begin{aligned}
[T(\mathbf{u})]_{\mathcal{B}'} &= P_{\mathcal{B}\to\mathcal{B}'}^{-1}[T(\mathbf{u})]_\mathcal{B} \\
&= P_{\mathcal{B}\to\mathcal{B}'}^{-1}([T]_{\mathcal{B},\mathcal{A}}[\mathbf{u}]_\mathcal{A}) \\
&= P_{\mathcal{B}\to\mathcal{B}'}^{-1}[T]_{\mathcal{B},\mathcal{A}}(Q_{\mathcal{A}\to\mathcal{A}'}[\mathbf{u}]_{\mathcal{A}'}) \\
&= (P_{\mathcal{B}\to\mathcal{B}'}^{-1}[T]_{\mathcal{B},\mathcal{A}}Q_{\mathcal{A}\to\mathcal{A}'})[\mathbf{u}]_{\mathcal{A}'}.
\end{aligned}$$

By Theorem 5.13 $P_{\mathcal{B}\to\mathcal{B}'}^{-1}[T]_{\mathcal{B},\mathcal{A}}Q_{\mathcal{A}\to\mathcal{A}'}$ or $P^{-1}AQ$ is the matrix of T relative to $\mathcal{A}'$ and $\mathcal{B}'$. ∎

EXAMPLE 2 Suppose that T is the linear transformation of $\mathbf{R}_{2\times2}$ into $\mathcal{P}_2$ over $\mathbf{R}$ defined by

$$T\begin{bmatrix} a_{11} & a_{12} \\ a_{21} & a_{22} \end{bmatrix} = a_{11} + (a_{12} - a_{21})x + a_{22}x^2.$$

The matrix of T relative to the standard bases $\mathcal{A}$ of $\mathbf{R}_{2\times2}$ and $\mathcal{B}$ of $\mathcal{P}_2$ over $\mathbf{R}$ is

$$[T]_{\mathcal{B},\mathcal{A}} = \begin{bmatrix} 1 & 0 & 0 & 0 \\ 0 & 1 & -1 & 0 \\ 0 & 0 & 0 & 1 \end{bmatrix}$$

and the matrix of T relative to the bases

$$\mathcal{A}' = \left\{ \begin{bmatrix} 1 & 1 \\ 0 & 0 \end{bmatrix}, \begin{bmatrix} 0 & 1 \\ 0 & 1 \end{bmatrix}, \begin{bmatrix} 1 & 0 \\ 1 & 0 \end{bmatrix}, \begin{bmatrix} 1 & 0 \\ 0 & 1 \end{bmatrix} \right\}$$

of $\mathbf{R}_{2\times2}$ and $\mathcal{B}' = \{1 + x, x + x^2, 1 + x + x^2\}$ of $\mathcal{P}_2$ over $\mathbf{R}$ is

$$[T]_{\mathcal{B}',\mathcal{A}'} = \begin{bmatrix} 1 & 0 & -1 & -1 \\ 0 & 1 & -2 & -1 \\ 0 & 0 & 2 & 2 \end{bmatrix}.$$

Now the matrix of transition from $\mathcal{A}$ to $\mathcal{A}'$ is

$$Q_{\mathcal{A}\to\mathcal{A}'} = \begin{bmatrix} 1 & 0 & 1 & 1 \\ 1 & 1 & 0 & 0 \\ 0 & 0 & 1 & 0 \\ 0 & 1 & 0 & 1 \end{bmatrix}$$

and from $\mathcal{B}$ to $\mathcal{B}'$ is

$$P_{\mathcal{B}\to\mathcal{B}'} = \begin{bmatrix} 1 & 0 & 1 \\ 1 & 1 & 1 \\ 0 & 1 & 1 \end{bmatrix}.$$

According to Theorem 5.15, the matrix of T relative to $\mathcal{A}'$ and $\mathcal{B}'$ is

$$P_{\mathcal{B}\to\mathcal{B}'}^{-1}[T]_{\mathcal{B},\mathcal{A}}Q_{\mathcal{A}\to\mathcal{A}'} = \begin{bmatrix} 1 & 0 & 1 \\ 1 & 1 & 1 \\ 0 & 1 & 1 \end{bmatrix}^{-1} \begin{bmatrix} 1 & 0 & 0 & 0 \\ 0 & 1 & -1 & 0 \\ 0 & 0 & 0 & 1 \end{bmatrix} \begin{bmatrix} 1 & 0 & 1 & 1 \\ 1 & 1 & 0 & 0 \\ 0 & 0 & 1 & 0 \\ 0 & 1 & 0 & 1 \end{bmatrix}$$

$$= \begin{bmatrix} 0 & 1 & -1 \\ -1 & 1 & 0 \\ 1 & -1 & 1 \end{bmatrix} \begin{bmatrix} 1 & 0 & 0 & 0 \\ 0 & 1 & -1 & 0 \\ 0 & 0 & 0 & 1 \end{bmatrix} \begin{bmatrix} 1 & 0 & 1 & 1 \\ 1 & 1 & 0 & 0 \\ 0 & 0 & 1 & 0 \\ 0 & 1 & 0 & 1 \end{bmatrix}$$

$$= \begin{bmatrix} 1 & 0 & -1 & -1 \\ 0 & 1 & -2 & -1 \\ 0 & 0 & 2 & 2 \end{bmatrix} = [T]_{\mathcal{B}',\mathcal{A}'}. \qquad \blacksquare$$

5.16 **THEOREM**	Two $m \times n$ matrices A and B represent the same linear transformation T of $\mathbf{U}$ into $\mathbf{V}$ if and only if A and B are equivalent.

Proof If A and B represent T relative to set of bases $\mathcal{A}$, $\mathcal{B}$ and $\mathcal{A}'$, $\mathcal{B}'$, respectively, then $B = P^{-1}AQ$, where Q is the matrix of transition from $\mathcal{A}$ to $\mathcal{A}'$ and P is the matrix of transition from $\mathcal{B}$ to $\mathcal{B}'$. Hence A and B are equivalent.

If B is equivalent to A, then $B = P^{-1}AQ$ for invertible P and Q. If A represents T relative to $\mathcal{A}$ and $\mathcal{B}$, then B represents T relative to $\mathcal{A}'$ and $\mathcal{B}'$, where Q is the matrix of transition from $\mathcal{A}$ to $\mathcal{A}'$ and P is the matrix of transition from $\mathcal{B}$ to $\mathcal{B}'$. ■ ■ ■

The proof of Theorem 5.16 can be modified so as to obtain two similar results concerning row equivalence and column equivalence. For requiring $\mathcal{B}' = \mathcal{B}$ is the same as requiring $P = I_m$, and requiring $\mathcal{A}' = \mathcal{A}$ is the same as requiring $Q = I_n$. Thus we have the following theorems.

5.17 **THEOREM**	Two $m \times n$ matrices A and B represent the same linear transformation T of $\mathbf{U}$ into $\mathbf{V}$ relative to the same basis of $\mathbf{U}$ if and only if they are row-equivalent.

5.18 **THEOREM**	Two $m \times n$ matrices A and B represent the same linear transformation T of $\mathbf{U}$ into $\mathbf{V}$ relative to the same basis of $\mathbf{V}$ if and only if they are column-equivalent.

EXAMPLE 3   Suppose the linear transformation $T : \mathcal{P}_2 \to \mathcal{P}_3$ over $\mathbf{R}$ has the matrix

$$A = \begin{bmatrix} 1 & 2 & 0 \\ 1 & 1 & 1 \\ 0 & -1 & 1 \\ 1 & 0 & 1 \end{bmatrix}$$

relative to $\mathcal{A} = \{1, x, x^2\}$ and $\mathcal{B} = \{1, x, x^2, x^3\}$, the standard bases of $\mathcal{P}_2$ and $\mathcal{P}_3$, respectively. We can find bases $\mathcal{A}'$ of $\mathcal{P}_2$ and $\mathcal{B}'$ of $\mathcal{P}_3$ such that the matrix B of T relative to $\mathcal{A}'$ and $\mathcal{B}'$ is the reduced row-echelon form of A. Also we can find the invertible matrix P such that $B = P^{-1}A$ by following the procedure of Section 3.6, that is, we use row operations to transform $[A \mid I_4]$ into $[B \mid P^{-1}]$.

$$[A \mid I_4] = \left[\begin{array}{ccc|cccc} 1 & 2 & 0 & 1 & 0 & 0 & 0 \\ 1 & 1 & 1 & 0 & 1 & 0 & 0 \\ 0 & -1 & 1 & 0 & 0 & 1 & 0 \\ 1 & 0 & 1 & 0 & 0 & 0 & 1 \end{array}\right] \to \left[\begin{array}{ccc|cccc} 1 & 2 & 0 & 1 & 0 & 0 & 0 \\ 0 & -1 & 1 & -1 & 1 & 0 & 0 \\ 0 & -1 & 1 & 0 & 0 & 1 & 0 \\ 0 & -2 & 1 & -1 & 0 & 0 & 1 \end{array}\right]$$

$$\rightarrow \begin{bmatrix} 1 & 0 & 2 & | & -1 & 2 & 0 & 0 \\ 0 & 1 & -1 & | & 1 & -1 & 0 & 0 \\ 0 & 0 & 0 & | & 1 & -1 & 1 & 0 \\ 0 & 0 & -1 & | & 1 & -2 & 0 & 1 \end{bmatrix}$$

$$\rightarrow \begin{bmatrix} 1 & 0 & 0 & | & 1 & -2 & 0 & 2 \\ 0 & 1 & 0 & | & 0 & 1 & 0 & -1 \\ 0 & 0 & 1 & | & -1 & 2 & 0 & -1 \\ 0 & 0 & 0 & | & 1 & -1 & 1 & 0 \end{bmatrix} = [B \mid P^{-1}]$$

Now $B = P^{-1}A$ where P^{-1} is the matrix of transition from $\mathcal{B}'$ to $\mathcal{E}_4$. So P is the matrix of transition from $\mathcal{E}_4$ to $\mathcal{B}'$. Using row operations on $[P^{-1} \mid I_4]$ to transform it to $[I_4 \mid P]$, we find

$$P = \begin{bmatrix} 1 & 2 & 0 & 0 \\ 1 & 1 & 1 & 0 \\ 0 & -1 & 1 & 1 \\ 1 & 0 & 1 & 0 \end{bmatrix}.$$

Thus B is the matrix of the linear transformation T relative to the basis $\mathcal{A}' = \mathcal{A}$ of $\mathcal{P}_2$ and $\mathcal{B}' = \{1 + x + x^3, 2 + x - x^2, x + x^2 + x^3, x^2\}$ of $\mathcal{P}_3$. ∎

The three preceding theorems give a full exposition of the connection between linear transformations and the equivalence relations on matrices that were studied in Chapter 3. However, there is one more major application of matrix theory to the study of linear transformations. This application is contained in the following theorem.

5.19
THEOREM

Let T be an arbitrary linear transformation of $\mathbf{U}$ into $\mathbf{V}$, and let r be the rank of T. Then there exist bases $\mathcal{A}'$ of $\mathbf{U}$ and $\mathcal{B}'$ of $\mathbf{V}$ such that the matrix of T relative to $\mathcal{A}'$ and $\mathcal{B}'$ has the first r diagonal elements equal to 1, and all other elements zero.

Proof With the stated hypotheses, suppose that $\mathcal{A}$ and $\mathcal{B}$ are bases of $\mathbf{U}$ and $\mathbf{V}$, respectively, and that T has matrix A relative to $\mathcal{A}$ and $\mathcal{B}$. By Theorem 3.46, there exist invertible matrices P^{-1} and Q such that $P^{-1}AQ$ has the first r diagonal elements equal to 1, and all other elements zero. Let $\mathcal{A}'$ and $\mathcal{B}'$ be bases such that Q is the matrix of transition from $\mathcal{A}$ to $\mathcal{A}'$ and P is the matrix of transition from $\mathcal{B}$ to $\mathcal{B}'$. Then $P^{-1}AQ$ is the matrix of T relative to $\mathcal{A}'$ and $\mathcal{B}'$, and the theorem is proved. ∎∎∎

Thus, with suitable choice of bases, each linear transformation T of $\mathbf{U}$ into $\mathbf{V}$ can be represented by a matrix of the form

$$D_r = \left[\begin{array}{c|c} I_r & \mathbf{0} \\ \hline \mathbf{0} & \mathbf{0} \end{array} \right],$$

where r is the rank of T. From a different point of view, this means that two linear transformations of $\mathbf{U}$ into $\mathbf{V}$ can be represented by the same matrix if and only if they have the same rank. It is easy to see that the relation of having the same rank is an equivalence relation on the set of all linear transformations of $\mathbf{U}$ into $\mathbf{V}$.

EXAMPLE 4 Let T be the linear transformation from $\mathbf{R}^4$ to $\mathcal{P}_2$ over $\mathbf{R}$ defined by

$$T(a_1, a_2, a_3, a_4) = (a_1 - a_2 + 2a_4) + (a_1 - a_2 + a_3 + a_4)x + (2a_1 - 2a_2 + a_3 + 3a_4)x^2.$$

We shall find bases $\mathcal{A}'$ of $\mathbf{R}^4$ and $\mathcal{B}'$ of $\mathcal{P}_2$ such that the matrix of T relative to $\mathcal{A}'$ and $\mathcal{B}'$ has the form

$$D_r = \left[\begin{array}{c|c} I_r & 0 \\ \hline 0 & 0 \end{array}\right].$$

We first find the matrix A of T with respect to $\mathcal{E}_4$ and $\mathcal{B} = \{1, x, x^2\}$. Since

$$T(1, 0, 0, 0) = (1)(1) + (1)(x) + (2)(x^2),$$
$$T(0, 1, 0, 0) = (-1)(1) + (-1)(x) + (-2)(x^2),$$
$$T(0, 0, 1, 0) = (0)(1) + (1)(x) + (1)(x^2),$$
$$T(0, 0, 0, 1) = (2)(1) + (1)(x) + (3)(x^2),$$

the matrix A is given by

$$A = \begin{bmatrix} 1 & -1 & 0 & 2 \\ 1 & -1 & 1 & 1 \\ 2 & -2 & 1 & 3 \end{bmatrix}.$$

Following the proof of Theorem 5.19 and using the same procedure as in Section 3.7, we first find an invertible matrix Q such that $AQ = A'$ is in reduced column-echelon form.

$$\left[\begin{array}{c} A \\ \hline I_4 \end{array}\right] = \left[\begin{array}{cccc} 1 & -1 & 0 & 2 \\ 1 & -1 & 1 & 1 \\ 2 & -2 & 1 & 3 \\ \hline 1 & 0 & 0 & 0 \\ 0 & 1 & 0 & 0 \\ 0 & 0 & 1 & 0 \\ 0 & 0 & 0 & 1 \end{array}\right] \rightarrow \left[\begin{array}{cccc} 1 & 0 & 0 & 0 \\ 1 & 0 & 1 & -1 \\ 2 & 0 & 1 & -1 \\ \hline 1 & 1 & 0 & -2 \\ 0 & 1 & 0 & 0 \\ 0 & 0 & 1 & 0 \\ 0 & 0 & 0 & 1 \end{array}\right] \rightarrow \left[\begin{array}{cccc} 1 & 0 & 0 & 0 \\ 0 & 1 & 0 & 0 \\ 1 & 1 & 0 & 0 \\ \hline 1 & 0 & 1 & -2 \\ 0 & 0 & 1 & 0 \\ -1 & 1 & 0 & 1 \\ 0 & 0 & 0 & 1 \end{array}\right]$$

$$= \left[\begin{array}{c} AQ \\ \hline Q \end{array}\right]$$

Next we use row operations to transform $[A' \mid I_3]$ into $[D_r \mid P^{-1}]$, where $D_r = P^{-1}A' = P^{-1}AQ$.

$$[A' \mid I_3] = \begin{bmatrix} 1 & 0 & 0 & 0 & | & 1 & 0 & 0 \\ 0 & 1 & 0 & 0 & | & 0 & 1 & 0 \\ 1 & 1 & 0 & 0 & | & 0 & 0 & 1 \end{bmatrix} \rightarrow \begin{bmatrix} 1 & 0 & 0 & 0 & | & 1 & 0 & 0 \\ 0 & 1 & 0 & 0 & | & 0 & 1 & 0 \\ 0 & 0 & 0 & 0 & | & -1 & -1 & 1 \end{bmatrix}$$

$$= [D_r \mid P^{-1}].$$

Thus

$$P^{-1} = \begin{bmatrix} 1 & 0 & 0 \\ 0 & 1 & 0 \\ -1 & -1 & 1 \end{bmatrix} \quad \text{and} \quad Q = \begin{bmatrix} 1 & 0 & 1 & -2 \\ 0 & 0 & 1 & 0 \\ -1 & 1 & 0 & 1 \\ 0 & 0 & 0 & 1 \end{bmatrix}$$

are invertible matrices such that

$$P^{-1}AQ = \begin{bmatrix} 1 & 0 & 0 & 0 \\ 0 & 1 & 0 & 0 \\ 0 & 0 & 0 & 0 \end{bmatrix} = \begin{bmatrix} I_2 & | & \mathbf{0} \\ - & + & - \\ \mathbf{0} & | & \mathbf{0} \end{bmatrix} = D_2.$$

According to the proof of Theorem 5.19, the desired basis $\mathcal{A}'$ and $\mathcal{B}'$ can be found by using Q as the transition matrix from $\mathcal{A}$ to $\mathcal{A}'$ and P as the transition matrix from $\mathcal{B}$ to $\mathcal{B}'$. Using Q to find $\mathcal{A}'$, we get

$$\mathcal{A}' = \{(1, 0, -1, 0), (0, 0, 1, 0), (1, 1, 0, 0), (-2, 0, 1, 1)\}.$$

To find $\mathcal{B}'$, we first obtain P by taking the inverse of P^{-1} and then use P as the transition matrix from $\mathcal{B}$ to $\mathcal{B}'$. We find

$$P = \begin{bmatrix} 1 & 0 & 0 \\ 0 & 1 & 0 \\ -1 & -1 & 1 \end{bmatrix}^{-1} = \begin{bmatrix} 1 & 0 & 0 \\ 0 & 1 & 0 \\ 1 & 1 & 1 \end{bmatrix}$$

and $\mathcal{B}' = \{1 + x^2, x + x^2, x^2\}$. The original defining equation for $T(a_1, a_2, a_3, a_4)$ can be used to check that D_2 is in fact the matrix of T with respect to $\mathcal{A}'$ and $\mathcal{B}'$. ∎

5.3 Exercises

1. Let $\mathcal{A} = \{x + x^2, 1 + x^2, x\}$ and $\mathcal{B} = \{1, 1 + x, 1 + x + x^2\}$ be bases of $\mathcal{P}_2$ over **R**.

   (a) Find P the matrix of transition from $\mathcal{B}$ to $\mathcal{A}$.

   (b) If

$$[\mathbf{v}]_{\mathcal{A}} = \begin{bmatrix} 2 \\ 1 \\ 0 \end{bmatrix},$$

   find $[\mathbf{v}]_{\mathcal{B}}$ and $\mathbf{v}$.

2. Let $\mathcal{A} = \{(2, -1, 1), (0, 1, -1), (-2, 1, 0)\}$ and $\mathcal{B} = \{(-2, 1, 0), (-2, 0, 1), (2, 0, 0)\}$.

(a) Find P the matrix of transition from $\mathcal{B}$ to $\mathcal{A}$.

(b) If

$$[\mathbf{v}]_\mathcal{A} = \begin{bmatrix} -1 \\ 1 \\ 2 \end{bmatrix},$$

find $[\mathbf{v}]_\mathcal{B}$ and $\mathbf{v}$.

3. Let

$$\mathcal{A} = \left\{ \begin{bmatrix} 0 & 0 \\ 1 & -1 \end{bmatrix}, \begin{bmatrix} 0 & -1 \\ 1 & 0 \end{bmatrix}, \begin{bmatrix} 0 & 1 \\ 0 & -1 \end{bmatrix} \right\}, \quad \mathcal{B} = \left\{ \begin{bmatrix} 1 & 0 \\ 0 & 1 \end{bmatrix}, \begin{bmatrix} 1 & 0 \\ 1 & 0 \end{bmatrix}, \begin{bmatrix} 1 & 1 \\ 0 & 0 \end{bmatrix} \right\}.$$

(a) Find P the matrix of transition from $\mathcal{B}$ to $\mathcal{A}$.

(b) If

$$[\mathbf{v}]_\mathcal{A} = \begin{bmatrix} 2 \\ 2 \\ -1 \end{bmatrix},$$

find $[\mathbf{v}]_\mathcal{B}$ and $\mathbf{v}$.

4. Let $\mathcal{A} = \{(1, 1), (2, 0)\}$ and $\mathcal{B} = \{(0, 2), (2, 1)\}$ in $\mathbf{R}^2$.

(a) Find $[\mathbf{u}]_\mathcal{A}$ if $[\mathbf{u}]_\mathcal{B} = \begin{bmatrix} 3 \\ -2 \end{bmatrix}$.   **(b)** Find $[\mathbf{v}]_\mathcal{A}$ if $[\mathbf{v}]_\mathcal{B} = \begin{bmatrix} 1 \\ -1 \end{bmatrix}$.

5. Let $\mathcal{A}$ be the basis of $\mathcal{P}_2$ over $\mathbf{R}$ given by $\mathcal{A} = \{x^2, 1 + x, x + x^2\}$. Given that

$$P = \begin{bmatrix} 1 & 0 & 0 \\ 0 & 1 & 1 \\ -1 & 0 & 1 \end{bmatrix}$$

is the transition matrix from $\mathcal{B}$ to $\mathcal{A}$ and that

$$[\mathbf{u}]_\mathcal{B} = \begin{bmatrix} 1 \\ 1 \\ 1 \end{bmatrix},$$

find $\mathbf{u}$.

6. Let $\mathcal{A} = \{(1, 1, 1, 1), (0, 1, 1, 1), (0, 0, 1, 1), (0, 0, 0, 1)\}$ be a basis of $\mathbf{R}^4$ and

$$P = \begin{bmatrix} 1 & 0 & 0 & 0 \\ 0 & 1 & -2 & 0 \\ 1 & 0 & 1 & 0 \\ 0 & -1 & 0 & 1 \end{bmatrix}$$

be the matrix of transition from $\mathcal{B}$ to $\mathcal{A}$. If

$$[\mathbf{v}]_\mathcal{B} = \begin{bmatrix} -1 \\ 1 \\ 2 \\ 0 \end{bmatrix},$$

find $\mathbf{v}$.

7. Suppose the linear operator T on $\mathcal{P}_1$ over $\mathbf{R}$ has the matrix

$$\begin{bmatrix} 3 & -2 \\ 1 & 0 \end{bmatrix}$$

with respect to $\mathcal{A} = \mathcal{B} = \{1 - x, x\}$. Find the matrix of T with respect to $\mathcal{A}' = \mathcal{B}' = \{2 - x, -1\}$.

8. The linear operator $S : \mathbf{R}^2 \to \mathbf{R}^2$ has the matrix

$$\begin{bmatrix} 1 & 4 \\ 2 & 3 \end{bmatrix}$$

with respect to the bases $\mathcal{A} = \mathcal{B} = \{(1, 1), (0, 1)\}$. Find the matrix of S with respect to $\mathcal{A}' = \mathcal{B}' = \{(2, 1), (1, 2)\}$.

9. Let T be the linear operator on $\mathcal{P}_2$ over $\mathbf{R}$ that has the matrix

$$\begin{bmatrix} 2 & 1 & 0 \\ 0 & 2 & 0 \\ 2 & 3 & 1 \end{bmatrix}$$

relative to the bases $\mathcal{A} = \mathcal{B} = \{x + x^2, -1 + x, x\}$. Find the matrix of T relative to the basis $\mathcal{A}' = \mathcal{B}' = \{2 - x + x^2, -6x - 2x^2, x\}$.

10. Let S be the linear transformation from $\mathbf{R}^3$ to $\mathbf{R}^2$ that has the matrix

$$\begin{bmatrix} 2 & 3 & 1 \\ 1 & 2 & 1 \end{bmatrix}$$

relative to $\mathcal{E}_3$ and $\mathcal{E}_2$. Given that

$$\begin{bmatrix} 1 & 1 & 1 \\ 0 & 0 & -1 \\ 0 & 1 & 1 \end{bmatrix}$$

is the transition matrix from $\mathcal{E}_3$ to $\mathcal{A}'$ and

$$\begin{bmatrix} 2 & 3 \\ 1 & 2 \end{bmatrix}$$

is the transition matrix from $\mathcal{E}_2$ to $\mathcal{B}'$, find the matrix of S relative to $\mathcal{A}'$ and $\mathcal{B}'$.

11. Suppose $S : \mathbf{R}^3 \to \mathbf{R}^2$ is the linear transformation with matrix

$$\begin{bmatrix} 1 & -3 & 1 \\ 2 & -6 & 2 \end{bmatrix}$$

relative to the bases $\mathcal{E}_3$ and $\mathcal{E}_2$. Find the matrix of S with respect to the bases $\{(1, 0, 1), (1, 0, 0), (1, 1, 0)\}$ and $\{(1, -1), (2, 0)\}$.

12. Let $\mathcal{A} = \mathcal{E}_4$ in $\mathbf{R}^4$ and $\mathcal{B} = \{x^2, x, 1\}$ in $\mathcal{P}_2$ over $\mathbf{R}$. If T is the linear transformation that is represented by

$$\begin{bmatrix} 1 & 1 & 0 & 1 \\ 0 & 0 & 1 & -1 \\ 1 & 1 & 0 & 1 \end{bmatrix}$$

relative to $\mathcal{A}$ and $\mathcal{B}$, find the matrix that represents T with respect to $\mathcal{A}'$ and $\mathcal{B}'$ where

$$\mathcal{A}' = \{(1, 0, 0, 0), (0, 0, 1, 0), (1, -1, 0, 0), (0, -1, 1, 1)\},$$
$$\mathcal{B}' = \{x^2 + 1, x, 1\}.$$

13. Let T be the linear transformation of $\mathbf{R}^3$ into $\mathbf{R}^2$ that has the matrix

$$\begin{bmatrix} 1 & -1 & 1 \\ 2 & 1 & 1 \end{bmatrix}$$

relative to the bases $\{(1, 2, 0), (1, 1, 1), (1, 1, 0)\}$ of $\mathbf{R}^3$ and $\{(1, 1), (1, -1)\}$ of $\mathbf{R}^2$. Find the matrix of T relative to the bases $\{(2, 3, 0), (1, 1, 1), (2, 3, 1)\}$ of $\mathbf{R}^3$ and $\{(3, -1), (1, -1)\}$ of $\mathbf{R}^2$.

14. Let T be the linear operator in Problem 10 of Exercises 5.2. Use matrices of transition to find the matrix of T relative to $\mathcal{A} = \mathcal{B} = \mathcal{E}_2$. Then use the new matrix to compute $T(\mathbf{u})$, thus making a check on the answer previously obtained.

15. Work Problem 25 of Exercises 5.2 using matrices of transition.

16. Suppose the linear transformation $S : \mathbf{R}^3 \to \mathbf{R}^2$ has the matrix

$$A = \begin{bmatrix} 1 & -1 & 1 \\ 2 & 1 & 0 \end{bmatrix}$$

relative to $\mathcal{E}_3$ and $\mathcal{E}_2$.

(a) Find bases $\mathcal{A}'$ of $\mathbf{R}^3$ and $\mathcal{B}'$ of $\mathbf{R}^2$ such that the matrix A' of S relative to $\mathcal{A}'$ and $\mathcal{B}'$ is the reduced column-echelon form for A.

(b) Find bases $\mathcal{A}'$ of $\mathbf{R}^3$ and $\mathcal{B}'$ of $\mathbf{R}^2$ such that the matrix A' of S relative to $\mathcal{A}'$ and $\mathcal{B}'$ is the reduced row-echelon form for A.

17. Suppose the linear transformation $T : \mathcal{P}_3 \to \mathcal{P}_2$, over $\mathbf{R}$ has the matrix

$$A = \begin{bmatrix} 1 & 2 & 0 & 0 \\ 0 & 1 & 2 & 1 \\ 1 & 1 & 1 & 1 \end{bmatrix}$$

relative to the standard bases of $\mathcal{P}_3$ and $\mathcal{P}_2$.

(a) Find bases $\mathcal{A}'$ of $\mathcal{P}_3$ and $\mathcal{B}'$ of $\mathcal{P}_2$ such that the matrix A' of T relative to $\mathcal{A}'$ and $\mathcal{B}'$ is the reduced column-echelon form for A.

(b) Find bases $\mathcal{A}'$ of $\mathcal{P}_3$ and $\mathcal{B}'$ of $\mathcal{P}_2$ such that the matrix A' of T relative to $\mathcal{A}'$ and $\mathcal{B}'$ is the reduced row-echelon form for A.

18. Suppose the linear transformation $T : \mathbf{R}^3 \to \mathbf{R}^2$ has the matrix

$$A = \begin{bmatrix} 2 & 3 & 1 \\ 1 & 2 & 1 \end{bmatrix}$$

relative to $\mathcal{E}_3$ and $\mathcal{E}_2$. Given that

$$B_1 = \begin{bmatrix} 2 & -3 \\ -1 & 2 \end{bmatrix} \quad \text{and} \quad B_2 = \begin{bmatrix} 1 & 0 & 1 \\ 0 & 1 & -1 \\ 0 & 0 & 1 \end{bmatrix}$$

are invertible matrices such that

$$B_1 A B_2 = D_2 = \begin{bmatrix} 1 & 0 & 0 \\ 0 & 1 & 0 \end{bmatrix},$$

find bases $\mathcal{A}'$ of $\mathbf{R}^3$ and $\mathcal{B}'$ of $\mathbf{R}^2$ such that T has matrix D_2 relative to $\mathcal{A}'$ and $\mathcal{B}'$.

19. Suppose T is the linear transformation of $\mathbf{R}^n$ into $\mathbf{R}^m$ that has the given matrix A relative to $\mathcal{E}_n$ and $\mathcal{E}_m$. Find bases $\mathcal{A}'$ of $\mathbf{R}^n$ and $\mathcal{B}'$ of $\mathbf{R}^m$ that satisfy the conditions given in Theorem 5.19.

(a) $A = \begin{bmatrix} 1 & 3 & 2 & 0 & -1 \\ 2 & 6 & 5 & 6 & 1 \\ 1 & 3 & 2 & 2 & 0 \end{bmatrix}$ **(b)** $A = \begin{bmatrix} 3 & -2 & -1 & -4 \\ 1 & 1 & -2 & -3 \\ -2 & 3 & -1 & 1 \end{bmatrix}$

20. Suppose the linear transformation $T : \mathbf{R}^3 \to \mathbf{R}^2$ is defined by

$$T(x_1, x_2, x_3) = (x_1 - x_2, x_2 + 2x_3).$$

Find bases $\mathcal{A}'$ of $\mathbf{R}^3$ and $\mathcal{B}'$ of $\mathbf{R}^2$ that satisfy the conditions given in Theorem 5.19.

21. Suppose the linear transformation $T : \mathcal{P}_2 \to \mathcal{P}_3$ over $\mathbf{R}$ is defined by

$$T(p(x)) = xp(x).$$

Find bases $\mathcal{A}'$ of $\mathcal{P}_2$ and $\mathcal{B}'$ of $\mathcal{P}_3$ that satisfy the conditions given in Theorem 5.19.

22. Let T be the linear transformation from $\mathcal{P}_3$ over $\mathbf{R}$ to $\mathbf{R}_{2\times 2}$ defined by

$$T(a_0 + a_1 x + a_2 x^2 + a_3 x^3) = \begin{bmatrix} a_0 - a_1 & a_1 - a_2 \\ a_2 - a_3 & a_0 + a_3 \end{bmatrix}$$

Find bases $\mathcal{A}'$ of $\mathcal{P}_3$ and $\mathcal{B}'$ of $\mathbf{R}_{2\times 2}$ that satisfy the conditions given in Theorem 5.19.

23. Let T be the linear transformation from $\mathbf{R}_{2\times 2}$ to $\mathcal{P}_2$ over $\mathbf{R}$ defined by

$$T\left(\begin{bmatrix} a_{11} & a_{12} \\ a_{21} & a_{22} \end{bmatrix}\right) = a_{22} + (a_{11} - a_{22})x + (a_{12} - a_{21})x^2$$

Find bases $\mathcal{A}'$ of $\mathbf{R}_{2\times 2}$ and $\mathcal{B}'$ of $\mathcal{P}_2$ that satisfy the conditions given in Theorem 5.19.

24. Show that the relation of having the same rank is an equivalence relation on the set of all linear transformations of $\mathbf{U}$ into $\mathbf{V}$.

5.4 Composition of Linear Transformations

In this section we conclude our study of the relation between a linear transformation and its associated matrix. We shall see that there are simple and direct connections between performing binary operations on linear transformations and performing binary operations on their associated matrices.

At various points, we have considered operations of addition and scalar multiplication on both matrices and linear transformations (see Example 5 of Section 4.1 and Definition 5.2). The intimate relation between these operations is described in the following theorem.

**5.20
THEOREM**

Let $\mathbf{U}$ and $\mathbf{V}$ be vector spaces over $\mathcal{F}$ of dimensions n and m, respectively, and let S and T be linear transformations of $\mathbf{U}$ into $\mathbf{V}$. If S has matrix A and T has matrix B relative to certain bases of $\mathbf{U}$ and $\mathbf{V}$, then $S + T$ has matrix $A + B$ relative to these same bases. Also, for any a in $\mathcal{F}$, aT has matrix aB relative to these bases.

Proof Let $\mathcal{A} = \{\mathbf{u}_1, \mathbf{u}_2, \ldots, \mathbf{u}_n\}$ and $\mathcal{B} = \{\mathbf{v}_1, \mathbf{v}_2, \ldots, \mathbf{v}_m\}$ be bases of $\mathbf{U}$ and $\mathbf{V}$, respectively. If S has matrix $A = [a_{ij}]_{m\times n}$ and T has matrix $B = [b_{ij}]_{m\times n}$ relative to $\mathcal{A}$ and $\mathcal{B}$, then $S(\mathbf{u}_j) = \sum_{i=1}^m a_{ij}\mathbf{v}_i$ and $T(\mathbf{u}_j) = \sum_{i=1}^m b_{ij}\mathbf{v}_i$. Hence

$$(S + T)(\mathbf{u}_j) = \sum_{i=1}^m (a_{ij} + b_{ij})\mathbf{v}_i,$$

and $S + T$ has matrix $A + B$ relative to $\mathcal{A}$ and $\mathcal{B}$. Also,

$$(aT)(\mathbf{u}_j) = \sum_{i=1}^m (ab_{ij})\mathbf{v}_i,$$

and aT has matrix aB relative to $\mathcal{A}$ and $\mathcal{B}$. ∎

We turn our attention now to the "third operation" on linear transformations that was mentioned on page 194. This third operation is the composition of linear transformations, defined by the same kind of rule as used for composite functions in the calculus.

In the calculus, two given functions f and g are combined to produce the composite function $f \circ g$ by the rule

$$(f \circ g)(x) = f(g(x)).$$

The domain of $f \circ g$ is the set of all x in the domain of g such that f is defined at $g(x)$.

In linear algebra, it is common to refer to the composite of two linear transformations as their *product*. We adopt this usage here, stated formally in the following definition.

5.21
DEFINITION

Let $\mathbf{U}$, $\mathbf{V}$, and $\mathbf{W}$ be vector spaces over the same field $\mathcal{F}$, and suppose that S is a linear transformation of $\mathbf{U}$ into $\mathbf{V}$ and that T is a linear transformation of $\mathbf{V}$ into $\mathbf{W}$. Then the **product** TS is the mapping of $\mathbf{U}$ into $\mathbf{W}$ defined by

$$TS(\mathbf{u}) = (T \circ S)(\mathbf{u}) = T(S(\mathbf{u}))$$

for each $\mathbf{u}$ in $\mathbf{U}$.

EXAMPLE 1 Consider the linear transformations $S : \mathbf{R}_{2 \times 2} \to \mathbf{R}^3$ and $T : \mathbf{R}^3 \to \mathcal{P}_1$ over $\mathbf{R}$ defined by

$$S \left(\begin{bmatrix} a & b \\ c & d \end{bmatrix} \right) = (a + 2b, b - 3c, c + d)$$

and

$$T(a_1, a_2, a_3) = (a_1 - 2a_2 - 6a_3) + (a_2 + 3a_3)x.$$

Computing the product TS, we have

$$TS \left(\begin{bmatrix} a & b \\ c & d \end{bmatrix} \right) = T \left(S \left(\begin{bmatrix} a & b \\ c & d \end{bmatrix} \right) \right)$$

$$= T(a + 2b, b - 3c, c + d)$$

$$= (a + 2b - 2(b - 3c) - 6(c + d)) + (b - 3c + 3(c + d))x$$

$$= (a - 6d) + (b + 3d)x. \qquad \blacksquare$$

Exercises for this section ask for verification that the product in Definition 5.21 is associative but not commutative. The most important property, as far as our study is concerned, is stated in Theorem 5.22.

For the remainder of this section, $\mathbf{U}, \mathbf{V}$, and $\mathbf{W}$ will denote vector spaces over the same field $\mathcal{F}$. Also, S and T will denote linear transformations of $\mathbf{U}$ into $\mathbf{V}$ and $\mathbf{V}$ into $\mathbf{W}$, respectively.

5.22 THEOREM	The product of two linear transformations is a linear transformation.

Proof It is clear from Definition 5.21 that TS is a mapping of $\mathbf{U}$ into $\mathbf{W}$. To show that TS is a linear transformation, let $\mathbf{u}_1, \mathbf{u}_2 \in \mathbf{U}$ and $a, b \in \mathcal{F}$. Then

$$
\begin{aligned}
TS(a\mathbf{u}_1 + b\mathbf{u}_2) &= T(S(a\mathbf{u}_1 + b\mathbf{u}_2)) \\
&= T(aS(\mathbf{u}_1) + bS(\mathbf{u}_2)) \\
&= aT(S(\mathbf{u}_1)) + bT(S(\mathbf{u}_2)) \\
&= aTS(\mathbf{u}_1) + bTS(\mathbf{u}_2),
\end{aligned}
$$

and TS is indeed a linear transformation of $\mathbf{U}$ into $\mathbf{W}$. ∎

The relation between multiplication of linear transformation and multiplication of matrices is given in the next theorem.

5.23 THEOREM	Suppose that $\mathbf{U}$, $\mathbf{V}$, and $\mathbf{W}$ are finite-dimensional vector spaces with bases $\mathcal{A}$, $\mathcal{B}$, and $\mathcal{C}$, respectively. If S has matrix A relative to $\mathcal{A}$ and $\mathcal{B}$ and T has matrix B relative to $\mathcal{B}$ and $\mathcal{C}$, then TS has matrix BA relative to $\mathcal{A}$ and $\mathcal{C}$.

Proof Assume that the hypotheses are satisfied with $\mathcal{A} = \{\mathbf{u}_1, \mathbf{u}_2, \ldots, \mathbf{u}_n\}$, $\mathcal{B} = \{\mathbf{v}_1, \mathbf{v}_2, \ldots, \mathbf{v}_m\}$, $\mathcal{C} = \{\mathbf{w}_1, \mathbf{w}_2, \ldots, \mathbf{w}_p\}$, $A = [a_{ij}]_{m \times n}$, and $B = [b_{ij}]_{p \times m}$. Let $C = [c_{ij}]_{p \times n}$ be the matrix of TS relative to $\mathcal{A}$ and $\mathcal{C}$. For $j = 1, 2, \ldots, n$ we have

$$
\begin{aligned}
\sum_{i=1}^{p} c_{ij}\mathbf{w}_i &= TS(\mathbf{u}_j) \\
&= T\left(\sum_{k=1}^{m} a_{kj}\mathbf{v}_k\right) \\
&= \sum_{k=1}^{m} a_{kj} T(\mathbf{v}_k) \\
&= \sum_{k=1}^{m} a_{kj} \left(\sum_{i=1}^{p} b_{ik}\mathbf{w}_i\right) \\
&= \sum_{k=1}^{m} \sum_{i=1}^{p} (a_{kj}b_{ik}\mathbf{w}_i) \\
&= \sum_{i=1}^{p} \sum_{k=1}^{m} (b_{ik}a_{kj}\mathbf{w}_i)
\end{aligned}
$$

$$= \sum_{i=1}^{p} \left(\sum_{k=1}^{m} b_{ik} a_{kj} \right) \mathbf{w}_i,$$

and consequently $c_{ij} = \sum_{k=1}^{m} b_{ik} a_{kj}$ for all values of i and j. Therefore $C = BA$ and the theorem is proved. ∎ ∎ ∎

EXAMPLE 2 Let $\mathcal{A}$, $\mathcal{B}$ and $\mathcal{C}$ be the standard bases of $\mathbf{R}_{2\times 2}$, $\mathbf{R}^3$, and $\mathcal{P}_1$ over $\mathbf{R}$, respectively; that is,

$$\mathcal{A} = \left\{ \begin{bmatrix} 1 & 0 \\ 0 & 0 \end{bmatrix}, \begin{bmatrix} 0 & 1 \\ 0 & 0 \end{bmatrix}, \begin{bmatrix} 0 & 0 \\ 1 & 0 \end{bmatrix}, \begin{bmatrix} 0 & 0 \\ 0 & 1 \end{bmatrix} \right\},$$

$\mathcal{B} = \mathcal{E}_3$, and $\mathcal{C} = \{1, x\}$. With S and T as in Example 1, we shall find the matrices of S, T, and TS relative to these bases and verify that the matrix of TS is the product of the matrix of T times the matrix of S.

Since

$$S\left(\begin{bmatrix} 1 & 0 \\ 0 & 0 \end{bmatrix} \right) = (1, 0, 0), \qquad S\left(\begin{bmatrix} 0 & 1 \\ 0 & 0 \end{bmatrix} \right) = (2, 1, 0),$$

$$S\left(\begin{bmatrix} 0 & 0 \\ 1 & 0 \end{bmatrix} \right) = (0, -3, 1), \qquad S\left(\begin{bmatrix} 0 & 0 \\ 0 & 1 \end{bmatrix} \right) = (0, 0, 1),$$

S has matrix

$$A = [S]_{\mathcal{B},\mathcal{A}} = \begin{bmatrix} 1 & 2 & 0 & 0 \\ 0 & 1 & -3 & 0 \\ 0 & 0 & 1 & 1 \end{bmatrix}$$

relative to $\mathcal{A}$ and $\mathcal{B}$. Since

$$T(1, 0, 0) = (1)(1) + (0)(x)$$
$$T(0, 1, 0) = (-2)(1) + (1)(x)$$
$$T(0, 0, 1) = (-6)(1) + (3)(x),$$

the matrix B of T relative to $\mathcal{B}$ and $\mathcal{C}$ is

$$B = [T]_{\mathcal{C},\mathcal{B}} = \begin{bmatrix} 1 & -2 & -6 \\ 0 & 1 & 3 \end{bmatrix}.$$

For the product TS, we have

$$TS\left(\begin{bmatrix} 1 & 0 \\ 0 & 0 \end{bmatrix} \right) = (1)(1) + (0)(x), \qquad TS\left(\begin{bmatrix} 0 & 1 \\ 0 & 0 \end{bmatrix} \right) = (0)(1) + (1)(x),$$

$$TS\left(\begin{bmatrix} 0 & 0 \\ 1 & 0 \end{bmatrix} \right) = (0)(1) + (0)(x), \qquad TS\left(\begin{bmatrix} 0 & 0 \\ 0 & 1 \end{bmatrix} \right) = (-6)(1) + (3)(x),$$

so the matrix of TS relative to $\mathcal{A}$ and $\mathcal{C}$ is

$$[TS]_{\mathcal{C},\mathcal{A}} = \begin{bmatrix} 1 & 0 & 0 & -6 \\ 0 & 1 & 0 & 3 \end{bmatrix}.$$

Routine arithmetic verifies that

$$BA = [T]_{\mathcal{C},\mathcal{B}}[S]_{\mathcal{B},\mathcal{A}} = \begin{bmatrix} 1 & -2 & -6 \\ 0 & 1 & 3 \end{bmatrix} \begin{bmatrix} 1 & 2 & 0 & 0 \\ 0 & 1 & -3 & 0 \\ 0 & 0 & 1 & 1 \end{bmatrix} = \begin{bmatrix} 1 & 0 & 0 & -6 \\ 0 & 1 & 0 & 3 \end{bmatrix} = [TS]_{\mathcal{C},\mathcal{A}}.$$

We note that the product AB is not defined, and this is consistent with the fact that ST is not defined.　■

The operations of addition and multiplication of matrices are connected by the distributive property. The statement of this fact is in Theorem 5.24. Its proof is requested in the exercises.

5.24 THEOREM

Let $A = [a_{ij}]_{m \times n}$, $B = [b_{ij}]_{n \times p}$, and $C = [c_{ij}]_{n \times p}$ over $\mathcal{F}$. Then

$$A(B + C) = AB + AC.$$

There is the possibility in Definition 5.10 that the vector spaces $\mathbf{U}$ and $\mathbf{V}$ may be identical, and yet the bases $\mathcal{A}$ and $\mathcal{B}$ may be different. In many instances, there is no condition present that requires that $\mathcal{A}$ and $\mathcal{B}$ be different. In these instances, it is convenient and conventional to choose $\mathcal{A} = \mathcal{B}$. If it is our intention to use just one base $\mathcal{A}$ of $\mathbf{V}$, we simply use the phrase "matrix of T relative to $\mathcal{A}$" rather than "matrix of T relative to $\mathcal{A}$ and $\mathcal{B}$, where $\mathcal{A}$ and $\mathcal{B}$ are equal." Similarly, "A represents T relative to $\mathcal{A}$" means that A represents T relative to $\mathcal{A}$ and $\mathcal{B}$, where $\mathcal{A}$ and $\mathcal{B}$ are equal.

Let us consider the case where $\mathbf{U} = \mathbf{V} = \mathbf{W}$ in Definition 5.21. This allows us to define positive integral powers of T inductively by $T^{k+1} = T^k \circ T$ for each positive integer k. We define T^0 to be the identity transformation of $\mathbf{V}$. In combination with Definition 5.2, this determines the value of each polynomial $a_r T^r + a_{r-1} T^{r-1} + \cdots + a_1 T + a_0 T^0$ in T with coefficients in $\mathcal{F}$, and such a polynomial is always a linear transformation of $\mathbf{V}$ into $\mathbf{V}$. In such a polynomial we shall write a_0 in place of $a_0 T^0$.

If T has matrix A relative to the basis $\mathcal{A}$, then Theorems 5.20 and 5.23 show that $a_r T^r + a_{r-1} T^{r-1} + \cdots + a_1 T + a_0$ has matrix $a_r A^r + a_{r-1} A^{r-1} + \cdots + a_1 A + a_0 I$ relative to $\mathcal{A}$. Consequently, $\sum_{i=0}^{r} a_i T^i$ is the zero linear transformation if and only if $\sum_{i=0}^{r} a_i A^i$ is the zero matrix. This means that T and A satisfy the same polynomial equations.

A linear transformation T of $\mathbf{U}$ into $\mathbf{V}$ is called **invertible** or **nonsingular** if there is a mapping S of $\mathbf{V}$ into $\mathbf{U}$ such that $ST(\mathbf{u}) = \mathbf{u}$ for all $\mathbf{u} \in \mathbf{U}$ and $TS(\mathbf{v}) = \mathbf{v}$ for all $\mathbf{v} \in \mathbf{V}$. Whenever such a mapping S exists, it is denoted by $S = T^{-1}$ and is called the **inverse** of T. It is left as an exercise to prove that the inverse of T is a linear transformation of $\mathbf{V}$ into $\mathbf{U}$. It follows from Theorem 5.23 that if T is invertible and has matrix A relative to the bases $\mathcal{A}$ of $\mathbf{U}$ and $\mathcal{B}$ of $\mathbf{V}$, then T^{-1} has matrix A^{-1} relative to $\mathcal{B}$ and $\mathcal{A}$.

The next example shows how polynomials in matrices and linear transformations can sometimes give interesting and surprising results.

EXAMPLE 3 Let T be the linear operator on $\mathbf{R}^3$ defined by

$$T(x_1, x_2, x_3) = (2x_1 + 2x_3, x_1 + 2x_2 + 3x_3, x_3).$$

Then T has the matrix

$$A = \begin{bmatrix} 2 & 0 & 2 \\ 1 & 2 & 3 \\ 0 & 0 & 1 \end{bmatrix}$$

relative to the standard basis $\mathcal{E}_3$. Straightforward computations show that A is a zero of the polynomial $x^3 - 5x^2 + 8x - 4$. That is,

$$A^3 - 5A^2 + 8A - 4I = \begin{bmatrix} 8 & 0 & 14 \\ 12 & 8 & 31 \\ 0 & 0 & 1 \end{bmatrix} - 5 \begin{bmatrix} 4 & 0 & 6 \\ 4 & 4 & 11 \\ 0 & 0 & 1 \end{bmatrix} + 8 \begin{bmatrix} 2 & 0 & 2 \\ 1 & 2 & 3 \\ 0 & 0 & 1 \end{bmatrix} - 4 \begin{bmatrix} 1 & 0 & 0 \\ 0 & 1 & 0 \\ 0 & 0 & 1 \end{bmatrix}$$

$$= \begin{bmatrix} 0 & 0 & 0 \\ 0 & 0 & 0 \\ 0 & 0 & 0 \end{bmatrix}.$$

This equation implies that

$$A^3 - 5A^2 + 8A = 4I$$

and therefore

$$A\left(\tfrac{1}{4}A^2 - \tfrac{5}{4}A + 2I\right) = I.$$

By Theorem 3.14, A is invertible and

$$A^{-1} = \tfrac{1}{4}A^2 - \tfrac{5}{4}A + 2I.$$

It follows from this that T^{-1} exists and T^{-1} can be expressed as a polynomial in T:

$$T^{-1} = \tfrac{1}{4}T^2 - \tfrac{5}{4}T + 2.$$

The equation $A^3 - 5A^2 + 8A - 4I = \mathbf{0}$ also has some implications concerning positive integral powers of A. For instance,

$$A^3 = 5A^2 - 8A + 4I$$

and this implies that

$$A^4 = 5A^3 - 8A^2 + 4A.$$

Substituting for A^3, we have

$$A^4 = 5(5A^2 - 8A + 4I) - 8A^2 + 4A = 17A^2 - 36A + 20I.$$

This substitution procedure can be repeated so as to express any higher integral power of A as a quadratic polynomial in A, and the corresponding powers of T can be expressed as quadratic polynomials in T. ∎

5.4 Exercises

1. Let S and T be the linear transformations of $\mathbf{R}^4$ into $\mathbf{R}^3$ given by

$$S(x_1, x_2, x_3, x_4) = (5x_1 - 3x_3, x_2 + 6x_3 - x_4, 2x_1 - 9x_2 + 5x_3 + 2x_4)$$

and

$$T(x_1, x_2, x_3, x_4) = (4x_1 + 10x_2, 6x_1 - x_2 + 7x_4, -3x_1 + 8x_2 - 5x_4).$$

   (a) Find the matrices of S, T, and $S + T$ relative to $\mathcal{E}_4$ and $\mathcal{E}_3$.
   (b) Find the matrix of $2S - 3T$ relative to $\mathcal{E}_4$ and $\mathcal{E}_3$.

2. Let S and T be as given in Problem 1. Consider the bases $\mathcal{A}$ and $\mathcal{B}$ of $\mathbf{R}^4$ and $\mathbf{R}^3$, respectively, where $\mathcal{A} = \{(1, 1, 1, 1), (0, 1, 1, 1), (0, 0, 1, 1), (0, 0, 0, 1)\}$, and $\mathcal{B} = \{(1, 1, 1), (1, 1, 0), (1, 0, 0)\}$.

   (a) Find the matrices of S, T, and $S + T$ relative to $\mathcal{A}$ and $\mathcal{B}$.
   (b) Find the matrix of $2S - 3T$ relative to $\mathcal{A}$ and $\mathcal{B}$.

3. Let $T : \mathbf{R}^2 \to \mathbf{R}^2$ and $S : \mathbf{R}^2 \to \mathbf{R}^2$ be given by

$$T(x, y) = (x + y, 2x), \qquad S(x, y) = (3y, x - y).$$

   (a) Find formulas for $ST(x, y)$ and $TS(x, y)$.
   (b) Find matrix representations for each of T, S, ST, and TS.

4. Suppose that S and T are linear operators on $\mathbf{R}^2$ such that S has matrix

$$\begin{bmatrix} 1 & -1 \\ 0 & 2 \end{bmatrix}$$

and T has matrix

$$\begin{bmatrix} 3 & 0 \\ -2 & 1 \end{bmatrix}$$

relative to the standard basis $\mathcal{E}_2$.

   (a) Find the matrix that represents TS relative to $\mathcal{E}_2$.
   (b) Write out a formula for $TS(x, y)$.
   (c) Find a basis for the kernel of TS.

5. Suppose the linear operator $S : \mathbf{R}^2 \to \mathbf{R}^2$ has the matrix

$$\begin{bmatrix} 1 & -2 \\ -1 & -3 \end{bmatrix}$$

relative to the basis $\{(1, 2), (0, 1)\}$ and the linear transformation $T : \mathbf{R}^2 \to \mathbf{R}^3$ has the matrix

$$\begin{bmatrix} 2 & 1 \\ 1 & -1 \\ 1 & 1 \end{bmatrix}$$

relative to the bases $\{(1, 2), (0, 1)\}$ of $\mathbf{R}^2$ and $\mathcal{E}_3$ of $\mathbf{R}^3$.

(a) Find a matrix representation for TS.

(b) Find a basis for the range of TS.

(c) Find a basis for the kernel of TS.

6. Find the matrix representation of S^{-1} in Problem 5 with respect to the basis $\{(1, 2), (0, 1)\}$.

7. Assume that $\mathcal{A} = \{\mathbf{u}_1, \mathbf{u}_2, \mathbf{u}_3\}$ is a basis for $\mathbf{R}^3$, and let

$$\mathbf{v}_1 = \mathbf{u}_2 + 2\mathbf{u}_3$$

$$\mathbf{v}_2 = \mathbf{u}_3$$

$$\mathbf{v}_3 = \mathbf{u}_1 + 2\mathbf{u}_2 + 5\mathbf{u}_3.$$

(a) Prove that $\mathcal{B} = \{\mathbf{v}_1, \mathbf{v}_2, \mathbf{v}_3\}$ is a basis, and express each $\mathbf{u}_i$ as a linear combination of $\mathbf{v}_1$, $\mathbf{v}_2$, and $\mathbf{v}_3$.

(b) Let $T : \mathbf{R}^3 \to \mathbf{R}^3$ be defined by $T(\mathbf{u}_i) = \mathbf{v}_i$ for $i = 1, 2, 3$. Find the matrix that represents T relative to $\mathcal{A}$.

8. Let $\mathcal{A}$, $\mathcal{B}$ and T be as defined in Problem 7. Define $S : \mathbf{R}^3 \to \mathbf{R}^3$ by $S(\mathbf{v}_i) = \mathbf{u}_i$ for $i = 1, 2, 3$.

(a) Find the matrix that represents S relative to $\mathcal{B}$.

(b) Find the matrix that represents TS relative to $\mathcal{A}$.

9. Let S be the linear transformation on $\mathcal{P}_2$ into $\mathcal{P}_3$ over $\mathbf{R}$ defined by $S(p(x)) = xp(x)$, and T the linear transformation on $\mathcal{P}_3$ over $\mathbf{R}$ into $\mathbf{R}_{2 \times 2}$ defined by

$$T(a_0 + a_1 x + a_2 x^2 + a_3 x^3) = \begin{bmatrix} a_0 & a_1 \\ a_2 & a_3 \end{bmatrix}.$$

(a) Find a formula for $TS(p(x))$.

(b) Find a basis for the range of TS.

(c) Find a basis for the kernel of TS.

(d) Find the matrix of TS relative to the bases $\mathcal{A}$ of $\mathcal{P}_2$ and $\mathcal{B}$ of $\mathbf{R}_{2 \times 2}$ where $\mathcal{A} = \{1, 1 + x, 1 + x + x^2\}$ and

$$\mathcal{B} = \left\{ \begin{bmatrix} 1 & 0 \\ 1 & 0 \end{bmatrix}, \begin{bmatrix} 1 & 1 \\ 0 & 0 \end{bmatrix}, \begin{bmatrix} 1 & 0 \\ 0 & 1 \end{bmatrix}, \begin{bmatrix} 1 & 1 \\ 1 & 1 \end{bmatrix} \right\}$$

10. Let S be the linear operator on $\mathcal{P}_2$ over $\mathbf{R}$ and T the linear transformation from $\mathcal{P}_2$ to $\mathcal{P}_3$ over $\mathbf{R}$ defined by $S(p(x)) = p(x + 1)$ and $T(p(x)) = (x - 1)p(x)$.

(a) Find a formula for $TS(p(x))$.

(b) Find a basis for the range of TS.

(c) Find a basis for the kernel of TS.

(d) Find the matrix of TS relative to the bases $\mathcal{A} = \{1, 1 - x, 1 + x - x^2\}$ of $\mathcal{P}_2$ and $\mathcal{B} = \{1, x, 1 + x^2, x - x^3\}$ of $\mathcal{P}_3$.

11. Let S be the linear operator on $\mathbf{R}_{2 \times 2}$ and T the linear transformation of $\mathbf{R}_{2 \times 2}$ into $\mathbf{R}$ defined by $S(A) = -A^T$ and

$$T\left(\begin{bmatrix} a_{11} & a_{12} \\ a_{21} & a_{22} \end{bmatrix}\right) = a_{11} + a_{22}.$$

(a) Find a formula for $TS(A)$.

(b) Find a basis for the range of TS.

(c) Find a basis for the kernel of TS.

(d) Find the matrix of TS relative to the basis $\mathcal{A}$ of $\mathbf{R}_{2 \times 2}$ and the standard basis $\mathcal{E}_1$ of $\mathbf{R}$ where

$$\mathcal{A} = \left\{ \begin{bmatrix} 1 & 1 \\ 0 & 0 \end{bmatrix}, \begin{bmatrix} 0 & 1 \\ -1 & 0 \end{bmatrix}, \begin{bmatrix} 0 & 0 \\ 1 & 1 \end{bmatrix}, \begin{bmatrix} 1 & 1 \\ 1 & 0 \end{bmatrix} \right\}.$$

12. Given that T is a linear operator on $\mathbf{R}^2$ with matrix

$$\begin{bmatrix} 3 & 2 \\ -4 & -2 \end{bmatrix}$$

relative to $\mathcal{E}_2$, find the matrix of $2T^3 + T^2 - 3T + 7$ relative to $\mathcal{E}_2$.

13. Let T be the linear operator on $\mathbf{R}^2$ that has the matrix

$$\begin{bmatrix} 1 & 2 \\ 2 & -2 \end{bmatrix}$$

relative to $\mathcal{E}_2$.

(a) Find the matrix of $T^2 + 3T - 6$ relative to $\mathcal{E}_2$.

(b) Given that $T^2 + T - 6 = \mathbf{0}$, write T^{-1} as a polynomial in T.

(c) Write T^3 as a first-degree polynomial in T.

14. Suppose T is a linear operator that satisfies the polynomial equation

$$T^3 - 3T^2 + 4T + 6 = \mathbf{0}.$$

(a) Write each of T^4 and T^5 as polynomials in T with degree less than 3.

(b) Write T^{-1} as a polynomial in T.

15. Suppose A is any matrix such that $A^2 - 5A + 12I = \mathbf{0}$.

(a) Prove that A is invertible and that A^{-1} is a polynomial in A.

(b) Given that

$$A = \begin{bmatrix} 1 & 2 \\ -4 & 4 \end{bmatrix}$$

is such a matrix, find A^{-1} by writing A^{-1} as a polynomial in A.

16. Prove the associative property for multiplication of linear transformations.

17. Give an example which shows that it may happen that $ST \neq TS$, even when both ST and TS are defined.

18. Let T_1 and T_2 be linear transformations of **U** into **V**, and let S be a linear transformation of **V** into **W**. Suppose that S, T_1, and T_2 have matrices A, B, and C, respectively, relative to certain bases of **U**, **V**, and **W**. Prove that $S(T_1 + T_2)$ has matrix $AB + AC$ relative to these same bases.

19. Use Problem 18 to prove Theorem 5.24.

20. Use Theorem 3.6 and the definition of addition of matrices to prove Theorem 5.24.

21. Prove that if T is an invertible linear transformation of **U** into **V**, then $T^{-1}(\mathbf{u}) = \mathbf{v}$ if and only if $T(\mathbf{v}) = \mathbf{u}$.

22. Let T be an invertible linear transformation of **U** into **V**. Prove that T^{-1} is a linear transformation of **V** into **U**.

23. Let T be a linear operator on **V** with matrix A relative to the basis $\mathcal{A}$ of **V**. Prove that T is invertible if and only if A is invertible.

24. Let S and T be invertible linear operators on **V**. Prove that ST is invertible, and that $(ST)^{-1} = T^{-1}S^{-1}$.

25. Let $T_1, T_2, \ldots, T_n$ be invertible linear operators on a **V**. Prove that $T_1 T_2 \cdots T_n$ is invertible and

$$(T_1 T_2 \cdots T_n)^{-1} = T_n^{-1} T_{n-1}^{-1} \cdots T_1^{-1}$$

26. Let S and T be linear operators on **V**. Prove or disprove: $(S + T)^2 = S^2 + 2ST + T^2$.

27. Prove or disprove for linear transformations $S : \mathbf{U} \to \mathbf{V}$ and $T : \mathbf{V} \to \mathbf{W}$.

(a) $S^{-1}(\mathbf{0}) \subseteq (TS)^{-1}(\mathbf{0})$

(b) $(TS)(\mathbf{U}) \subseteq T(\mathbf{V})$

6 Determinants

In this chapter, the fundamentals of the theory of determinants are developed. A knowledge of this material is necessary in the study of eigenvalues and eigenvectors of linear transformations, and many of the applications of linear algebra involve a use of eigenvalues and eigenvectors. These topics will be studied in Chapter 7.

6.1 Permutations and Indices

The definition of a determinant presented in Section 6.2 depends on the concept of a permutation of a set of integers and on certain properties of these permutations. A **permutation** of a set $\{x_1, x_2, \ldots, x_n\}$ is simply an arrangement of the elements of the set into a particular sequence or order. For example, the arrangements 1, 4, 5, 3, 2, 6 and 6, 3, 2, 1, 4, 5 are permutations of the set $\{1, 2, 3, 4, 5, 6\}$. If an element x_i appears to the left of an element x_j in a permutation of $\{x_1, x_2, \ldots, x_n\}$, then we say that x_i **precedes** x_j.

Our interest in permutations is limited to the permutations of the first n positive integers. The permutation $1, 2, 3, \ldots, n$ is referred to as the **natural ordering**.

6.1 DEFINITION	The **index** of an integer j_k in a permutation $j_1, j_2, \ldots, j_n$ of $\{1, 2, \ldots, n\}$ is the number of integers greater than j_k that precede j_k in the permutation $j_1, j_2, \ldots, j_n$. The index of j_k is denoted by $\mathcal{I}(j_k)$.

EXAMPLE 1 In the permutation 1, 4, 5, 3, 2, 6, the index of 3 is given by $\mathcal{I}(3) = 2$ since 4 and 5 are greater than 3 and precede 3 in the permutation. No integer greater than 4 precedes 4 in this permutation, so $\mathcal{I}(4) = 0$. Since the three integers 4, 5, and 3 are greater than 2 and precede 2 in the permutation, $\mathcal{I}(2) = 3$. ∎

6.2 DEFINITION	The **index** of a permutation $j_1, j_2, \ldots, j_n$ of $\{1, 2, \ldots, n\}$ is the integer $\mathcal{I}$ given by $$\mathcal{I} = \sum_{k=1}^{n} \mathcal{I}(j_k),$$ where $\mathcal{I}(j_k)$ is the index of j_k in the permutation $j_1, j_2, \ldots, j_n$.

That is, the index of a given permutation is the sum of the indices of all of the elements in that permutation.

EXAMPLE 2 The index of the permutation 1, 4, 5, 3, 2, 6 in Example 1 is

$$\mathcal{I} = \mathcal{I}(1) + \mathcal{I}(4) + \mathcal{I}(5) + \mathcal{I}(3) + \mathcal{I}(2) + \mathcal{I}(6)$$

$$= 0 + 0 + 0 + 2 + 3 + 0$$

$$= 5.$$ ■

6.3 THEOREM

Any single interchange of adjacent elements in a permutation $j_1, j_2, \ldots, j_n$ of $\{1, 2, \ldots, n\}$ changes the index of the permutation by 1.

Proof Let $\mathcal{I} = \sum_{k=1}^{n} \mathcal{I}(j_k)$ be the index of the given permutation

$$j_1, j_2, \ldots, j_m, j_{m+1}, \ldots, j_n$$

and consider the index $\mathcal{I}'$ of the permutation

$$j_1, j_2, \ldots, j_{m+1}, j_m, \ldots, j_n$$

that results from interchanging j_m and j_{m+1} in the original permutation. Now

$$\mathcal{I}' = \sum_{k=1}^{m-1} \mathcal{I}'(j_k) + \mathcal{I}'(j_{m+1}) + \mathcal{I}'(j_m) + \sum_{k=m+2}^{n} \mathcal{I}'(j_k),$$

where $\mathcal{I}'(j_k)$ denotes the index of j_k in the new permutation. It is clear that $\mathcal{I}'(j_k) = \mathcal{I}(j_k)$ if k is different from m and $m + 1$. If $j_m > j_{m+1}$, then $\mathcal{I}'(j_m) = \mathcal{I}(j_m)$, $\mathcal{I}'(j_{m+1}) = \mathcal{I}(j_{m+1}) - 1$, and consequently $\mathcal{I}' = \mathcal{I} - 1$. On the other hand, if $j_m < j_{m+1}$, then $\mathcal{I}'(j_m) = \mathcal{I}(j_m) + 1$ and $\mathcal{I}'(j_{m+1}) = \mathcal{I}(j_{m+1})$, so that $\mathcal{I}' = \mathcal{I} + 1$. In either case, the index is changed by 1, and the theorem is proven. ■ ■ ■

In light of Theorem 6.3, the index $\mathcal{I}(j_k)$ can be thought of as the number of interchanges of adjacent elements necessary to make the index $\mathcal{I}'(j_k)$ of j_k in the new permutation equal to zero.

EXAMPLE 3 The index $\mathcal{I}(3) = 2$ in the permutation 1, 4, 5, 3, 2, 6. In order to make the index $\mathcal{I}'(3) = 0$, two interchanges of adjacent elements are required.

$$1, 4, 5, 3, 2, 6 \rightarrow 1, 4, 3, 5, 2, 6 \rightarrow 1, 3, 4, 5, 2, 6$$
$$\mathcal{I}(3) = 2 \qquad\qquad\qquad\qquad \mathcal{I}'(3) = 0$$ ■

Theorem 6.3 paves the way for obtaining the corresponding result concerning the interchange of any two (not necessarily adjacent) elements in a permutation.

6.4 THEOREM

Any interchange of two elements in a permutation $j_1, j_2, \ldots, j_n$ of the set $\{1, 2, \ldots, n\}$ changes the index by an odd integer.

Proof Let

$$j_1, j_2, \ldots, j_r, \ldots, j_s, \ldots, j_n$$

be the given permutation, and consider the interchange of j_r and j_s. Let m be the number of elements between j_r and j_s.

Now the permutation that results from the interchange of j_r and j_s can be accomplished by using only interchanges of adjacent elements. The element j_r can be moved to the position initially occupied by j_s by $m + 1$ interchanges with the adjacent element on the right. Then j_s can be moved to the position that j_r initially occupied by m interchanges with the adjacent element on the left. Thus, the interchange of j_r and j_s can be accomplished by $2m + 1$ interchanges of adjacent elements. These $2m + 1$ interchanges cause $2m + 1$ changes of 1 in the index of the ordering, and consequently the index has changed by an odd number. ■ ■ ■

EXAMPLE 4 As an illustration of the proof of Theorem 6.4, we shall accomplish the interchange of $j_r = 4$ and $j_s = 2$ in the permutation 1, 4, 5, 3, 2, 6 by using only interchanges of adjacent elements. We note that there are $m = 2$ elements between 4 and 2, and that the index $\mathcal{I}$ of this permutation is 5.

$$1, 4, 5, 3, 2, 6 \rightarrow 1, 5, 4, 3, 2, 6$$
$$\rightarrow 1, 5, 3, 4, 2, 6$$
$$\rightarrow 1, 5, 3, 2, 4, 6$$
$$\rightarrow 1, 5, 2, 3, 4, 6$$
$$\rightarrow 1, 2, 5, 3, 4, 6$$

We first moved 4 to the position initially occupied by 2 using 3 interchanges with the adjacent element on the right, and then we moved 2 to the position that 4 initially occupied by using 2 interchanges with the adjacent element on the left. The total number of interchanges of adjacent elements was $3 + 2 = 5$, an odd integer, and the index of the resulting permutation 1, 2, 5, 3, 4, 6 is

$$\mathcal{I} = \mathcal{I}(1) + \mathcal{I}(2) + \mathcal{I}(5) + \mathcal{I}(3) + \mathcal{I}(4) + \mathcal{I}(6)$$
$$= 0 + 0 + 0 + 1 + 1 + 0$$
$$= 2.$$ ■

The main objective of this section is to establish Theorem 6.7 for use in Section 6.2. The following two lemmas are basic to our proof of Theorem 6.7.

**6.5
LEMMA**   If a given permutation of $\{1, 2, \ldots, n\}$ is carried into another permutation by an odd number of interchanges of elements, then the index of the given permutation differs from the index of the final permutation by an odd number.

Proof Suppose that a given permutation of $\{1, 2, \ldots, n\}$ is carried into another permutation by an odd number of interchanges of elements. According to Theorem 6.4, each of the interchanges of elements changes the index by an odd number. Thus, the index of the original permutation differs from the index of the final permutation by an odd number, since the sum of an odd number of odd integers is an odd integer. ■ ■ ■

6.6
LEMMA

If a given permutation of $\{1, 2, \ldots, n\}$ is carried into another permutation by an even number of interchanges of elements, then the index of the given permutation differs from the index of the final permutation by an even number.

Proof The proof is an exact parallel to that of Lemma 6.5, except that the sum of an even number of odd integers is an even integer. ■ ■ ■

6.7
THEOREM

The number of interchanges used to carry a permutation $j_1, j_2, \ldots, j_n$ of $\{1, 2, \ldots, n\}$ into the natural ordering is either always odd or always even.

Proof Since the index of the natural ordering is zero, the difference in the index of the ordering $j_1, j_2, \ldots, j_n$ and the index of the natural ordering is the same as the index $\mathcal{I}$ of $j_1, j_2, \ldots, j_n$. If $j_1, j_2, \ldots, j_n$ can be carried into $1, 2, \ldots, n$ by an odd number of interchanges, then $\mathcal{I}$ must be odd by Lemma 6.5. If $j_1, j_2, \ldots, j_n$ can be carried into $1, 2, \ldots, n$ by an even number of interchanges, then $\mathcal{I}$ must be even by Lemma 6.6. Thus the number of interchanges used to carry $j_1, j_2, \ldots, j_n$ into $1, 2, \ldots, n$ must always be odd if $\mathcal{I}$ is odd and must always be even if $\mathcal{I}$ is even. ■ ■ ■

EXAMPLE 5 We shall determine whether the index of the permutation 6, 4, 1, 7, 5, 2, 3 is odd or even by counting the number of interchanges used to carry this permutation into the natural ordering.

$$6, 4, 1, 7, 5, 2, 3 \rightarrow 1, 4, 6, 7, 5, 2, 3$$
$$\rightarrow 1, 2, 6, 7, 5, 4, 3$$
$$\rightarrow 1, 2, 3, 7, 5, 4, 6$$
$$\rightarrow 1, 2, 3, 4, 5, 7, 6$$
$$\rightarrow 1, 2, 3, 4, 5, 6, 7$$

Because we have used an odd number of interchanges to carry the original permutation into the natural ordering, the index $\mathcal{I}$ of

$$6, 4, 1, 7, 5, 2, 3$$

must be odd by Lemma 6.5, and the number of interchanges used to carry the original permutation into the natural ordering would always be odd by Theorem 6.7. Although always odd, this number of interchanges may be different from the index of the permutation. In this example, we used 5 interchanges of (not necessarily adjacent) elements, and the index of

$$6, 4, 1, 7, 5, 2, 3$$

is

$$\mathcal{I} = \mathcal{I}(1) + \mathcal{I}(2) + \mathcal{I}(3) + \mathcal{I}(4) + \mathcal{I}(5) + \mathcal{I}(6) + \mathcal{I}(7)$$
$$= 2 + 4 + 4 + 1 + 2 + 0 + 0$$
$$= 13.$$

6.1 Exercises

1. Find the index of the following permutations.

    (a) 5, 3, 1, 4, 2   **(b)** 2, 4, 5, 3, 1   **(c)** 6, 2, 4, 5, 3, 1

    (d) 5, 3, 4, 2, 1, 6   **(e)** 3, 4, 6, 1, 2, 7, 5   **(f)** 2, 5, 1, 4, 3, 7, 6

2. Determine whether the index $\mathcal{I}$ of the given permutation is odd or even by counting the number t of interchanges used to carry the given permutation into the natural ordering. Then compute $\mathcal{I}$.

    (a) 3, 4, 2, 5, 1   **(b)** 5, 1, 3, 2, 4   **(c)** 6, 2, 4, 5, 3, 1

    (d) 5, 2, 4, 6, 1, 3   **(e)** 4, 1, 7, 5, 2, 6, 3   **(f)** 3, 7, 5, 4, 6, 2, 1

3. Write out a sequence of interchanges of *adjacent* elements that will accomplish the interchange of the given pair of elements in the permutation 6, 2, 4, 5, 3, 1.

    (a) 2 and 3   **(b)** 6 and 3   **(c)** 2 and 1   **(d)** 4 and 1

4. Prove that the number of interchanges used to carry a permutation $j_1, j_2, \ldots, j_n$ of $\{1, 2, \ldots, n\}$ into itself (i.e., into the same permutation) must always be even.

5. If $n \geq 2$, what is the index of the permutation $n, n-1, \ldots, 3, 2, 1$? Justify your answer.

6. **(a)** If $n \geq 2$, find the number of permutations of $\{1, 2, \ldots, n\}$ whose index is even.

    (b) If $n \geq 2$, find the number of permutations of $\{1, 2, \ldots, n\}$ whose index is odd.

6.2 The Definition of a Determinant

With the results of Section 6.1 at our disposal, we can proceed to our definition of a determinant. Only square matrices are considered in this chapter since the determinant is defined only for this type of matrix. We shall simplify our notation from $A = [a_{ij}]_{n \times n}$ to $A = [a_{ij}]_n$ or even $A = [a_{ij}]$ if the order of A is not significant.

6.8
DEFINITION

The **determinant** of the square matrix $A = [a_{ij}]_n$ over $\mathcal{F}$ is the scalar $\det(A)$ defined by

$$\det(A) = \sum_{(j)} (-1)^t a_{1j_1} a_{2j_2} \cdots a_{nj_n},$$

where $\sum_{(j)}$ denotes the sum of all terms of the form $(-1)^t a_{1j_1} a_{2j_2} \cdots a_{nj_n}$ as $j_1, j_2, \ldots, j_n$ assumes all possible permutations of the numbers $1, 2, \ldots, n$, and the exponent t is the number of interchanges used to carry $j_1, j_2, \ldots, j_n$ into the natural ordering $1, 2, \ldots, n$.

The notations $\det A$ and $|A|$ are used interchangeably with $\det(A)$. When the elements of $A = [a_{ij}]_n$ are written as a rectangular array, we would have

$$\det(A) = |A| = \begin{vmatrix} a_{11} & a_{12} & \cdots & a_{1n} \\ a_{21} & a_{22} & \cdots & a_{2n} \\ \vdots & \vdots & \ddots & \vdots \\ a_{n1} & a_{n2} & \cdots & a_{nn} \end{vmatrix}.$$

Although the number of interchanges used to carry $j_1, j_2, \ldots, j_n$ into $1, 2, \ldots, n$ is not always the same, Theorem 6.7 assures us that this number is either always even or always odd. Hence the sign $(-1)^t$ of each term is well-defined, and $\det(A)$ is uniquely determined by A.

We observe that there are $n!$ terms in the sum $\det(A)$ since there are $n!$ possible orderings of $1, 2, \ldots, n$. The determinant of an $n \times n$ matrix is referred to as an $n \times n$ **determinant,** or a **determinant of order** n.

EXAMPLE 1 Consider a 3×3 matrix

$$A = \begin{bmatrix} a_{11} & a_{12} & a_{13} \\ a_{21} & a_{22} & a_{23} \\ a_{31} & a_{32} & a_{33} \end{bmatrix}.$$

By the definition,

$$\det(A) = (-1)^{t_1} a_{11} a_{22} a_{33} + (-1)^{t_2} a_{11} a_{23} a_{32} + (-1)^{t_3} a_{12} a_{23} a_{31}$$

$$+ (-1)^{t_4} a_{12} a_{21} a_{33} + (-1)^{t_5} a_{13} a_{22} a_{31} + (-1)^{t_6} a_{13} a_{21} a_{32}.$$

The exponents $t_1, t_2, \ldots, t_6$ are determined by either computing the index of each permutation of the column subscripts or by counting interchanges necessary to carry each permutation of the column subscripts to natural order. If we count interchanges we have:

Term	Interchanges of column subscripts	Exponent
$a_{11}\, a_{22}\, a_{33}$:	$1, 2, 3$	$t_1 = 0$ (even)
$a_{11}\, a_{23}\, a_{32}$:	$1, 3, 2 \rightarrow 1, 2, 3$	$t_2 = 1$ (odd)
$a_{12}\, a_{23}\, a_{31}$:	$2, 3, 1 \rightarrow 1, 3, 2 \rightarrow 1, 2, 3$	$t_3 = 2$ (even)
$a_{12}\, a_{21}\, a_{33}$:	$2, 1, 3 \rightarrow 1, 2, 3$	$t_4 = 1$ (odd)
$a_{13}\, a_{22}\, a_{31}$:	$3, 2, 1 \rightarrow 1, 2, 3$	$t_5 = 1$ (odd)
$a_{13}\, a_{21}\, a_{32}$:	$3, 1, 2 \rightarrow 1, 3, 2 \rightarrow 1, 2, 3$	$t_6 = 2$ (even)

Hence

$$\det(A) = a_{11}\, a_{22}\, a_{33} - a_{11}\, a_{23}\, a_{32} + a_{12}\, a_{23}\, a_{31}$$

$$- a_{12}\, a_{21}\, a_{33} - a_{13}\, a_{22}\, a_{31} + a_{13}\, a_{21}\, a_{32}.$$ ∎

It is worth noting that the value of $\det(A)$ obtained in Example 1 agrees with the value yielded by evaluation routines that are taught in high school algebra. One of the most popular of these routines evaluates a 3×3 determinant $|A|$ by reproducing the first two columns and forming signed products according to the following diagram.

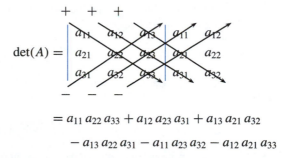

$$= a_{11}\, a_{22}\, a_{33} + a_{12}\, a_{23}\, a_{31} + a_{13}\, a_{21}\, a_{32}$$

$$- a_{13}\, a_{22}\, a_{31} - a_{11}\, a_{23}\, a_{32} - a_{12}\, a_{21}\, a_{33}$$

A similar diagram for a 2×2 determinant given by

$$\det(A) = \begin{vmatrix} a_{11} & a_{12} \\ a_{21} & a_{22} \end{vmatrix} = a_{11}\, a_{22} - a_{12}\, a_{21}$$

gives a correct value, but it is important to know that **there is no similar scheme that works for determinants of order 4 or any order greater than 3.**

Definition 6.8 is frequently referred to as the "row" definition of a determinant, since the row subscripts on the factors $a_{i_j j_i}$ are held fixed in the natural ordering.

The next theorem presents an alternate formulation (the "column" definition) in which the column subscripts are held fixed in the natural ordering.

6.9
THEOREM

For any matrix $A = [a_{ij}]_n$, $\det(A)$ is given by

$$\det(A) = \sum_{(i)} (-1)^s a_{i_1 1} \, a_{i_2 2} \cdots a_{i_n n},$$

where $\sum_{(i)}$ denotes the sum over all possible permutations $i_1, i_2, \ldots, i_n$ of $1, 2, \ldots, n$, and s is the number of interchanges used to carry $i_1, i_2, \ldots, i_n$ into the natural ordering.

Proof Let $S = \sum_{(i)} (-1)^s a_{i_1 1} \, a_{i_2 2} \cdots a_{i_n n}$. Now both S and $\det(A)$ have $n!$ terms. Except possibly for sign, each term of S is a term of $\det(A)$, and each term of $\det(A)$ is a term of S. Thus, S and $\det(A)$ consist of the same terms, with a possible difference in sign.

Consider a certain term $(-1)^s a_{i_1 1} \, a_{i_2 2} \cdots a_{i_n n}$ and let $(-1)^t a_{1 j_1} \, a_{2 j_2} \cdots a_{n j_n}$ be the corresponding term in $\det(A)$. Then $a_{i_1 1} \, a_{i_2 2} \cdots a_{i_n n}$ can be carried into $a_{1 j_1} \, a_{2 j_2} \cdots a_{n j_n}$ by s interchanges of factors since the permutation $i_1, i_2, \ldots, i_n$ can be changed into the natural ordering $1, 2, \ldots, n$ by s interchanges of elements.

$$a_{i_1 1} \, a_{i_2 2} \cdots a_{i_n n} \xrightarrow{\;s \text{ interchanges}\;} a_{1 j_1} a_{2 j_2} \cdots a_{n j_n}$$

This means that the natural ordering $1, 2, \ldots, n$ can be changed into the permutation $j_i, j_2, \ldots, j_n$ by s interchanges since the column subscripts have been interchanged each time the factors were interchanged.

$$1, 2, \ldots, n \xrightarrow{\;s \text{ interchanges}\;} j_1, j_2, \ldots, j_n$$

But $j_1, j_2, \ldots, j_n$ can be carried into $1, 2, \ldots, n$ by t interchanges, by the definition of $\det(A)$.

$$j_1, j_2, \ldots, j_n \xrightarrow{\;t \text{ interchanges}\;} 1, 2, \ldots, n$$

Thus $1, 2, \ldots, n$ can be carried into $j_1, j_2, \ldots, j_n$ and then back into itself by $s + t$ interchanges.

$$1, 2, \ldots, n \xrightarrow[\;s \text{ interchanges}\;]{} j_1, j_2, \ldots, j_n \xrightarrow[\;t \text{ interchanges}\;]{\overset{s+t \text{ interchanges}}{}} 1, 2, \ldots, n$$

Since $1, 2, \ldots, n$ can be carried into itself by an even number (zero) of interchanges,

$$1, 2, \ldots, n \xrightarrow{\;0 \text{ interchanges}\;} 1, 2, \ldots, n$$

then $s + t$ is even by Theorem 6.7. Therefore $(-1)^{s+t} = 1$ and $(-1)^s = (-1)^t$. Now we have the corresponding terms in $\det(A)$ and S with the same sign, and therefore $\det(A) = S$. ∎ ■ ■ ■

We recall that the transpose of an $m \times n$ matrix $A = [a_{ij}]$ is the $n \times m$ matrix A^T with a_{ji} as the element in the ith row and jth column.

6.10 THEOREM	If $A = [a_{ij}]_n$, then $\det(A^T) = \det(A)$.

Proof Let $B = A^T$, so that $b_{ij} = a_{ji}$ for all pairs i, j. Thus

$$\det(B) = \sum_{(j)} (-1)^t b_{1j_1} b_{2j_2} \cdots b_{nj_n}$$

by the definition of $\det(B)$, and

$$\det(B) = \sum_{(j)} (-1)^t a_{j_1 1} a_{j_2 2} \cdots a_{j_n n}.$$

Therefore $\det(B) = \det(A)$, by Theorem 6.9.   ∎

6.2   Exercises

1. Determine whether t is even or odd in the given term of $\det(A)$, where $A = [a_{ij}]_n$.

(a) $(-1)^t a_{13} a_{21} a_{34} a_{42}$ (b) $(-1)^t a_{14} a_{21} a_{33} a_{42}$ (c) $(-1)^t a_{14} a_{23} a_{32} a_{41}$

(d) $(-1)^t a_{12} a_{24} a_{31} a_{43}$ (e) $(-1)^t a_{21} a_{42} a_{13} a_{34}$ (f) $(-1)^t a_{41} a_{22} a_{33} a_{14}$

(g) $(-1)^t a_{24} a_{43} a_{12} a_{31}$ (h) $(-1)^t a_{21} a_{33} a_{14} a_{42}$ (i) $(-1)^t a_{34} a_{21} a_{42} a_{13}$

(j) $(-1)^t a_{34} a_{41} a_{12} a_{23}$ (k) $(-1)^t a_{13} a_{25} a_{31} a_{42} a_{54}$ (l) $(-1)^t a_{31} a_{12} a_{53} a_{24} a_{45}$

2. Write out the complete expression for $\det(A)$ if $A = [a_{ij}]_4$.

3. If $A = [a_{ij}]_n$ is a diagonal matrix, what is the value of $\det(A)$?

4. Evaluate $\det(A)$ if $A = [\delta_{ij}]_n$, where δ_{ij} is the Kronecker delta.

5. If $A = [a_{ij}]_n$ is an upper triangular matrix, what is the value of $\det(A)$?

6. If $A = [a_{ij}]_n$ is a unit lower triangular matrix, what is the value of $\det(A)$?

7. If A is a 3×3 matrix, find the value of $\det(A + A)$ in terms of $\det(A)$.

8. Work Problem 7 for an $n \times n$ matrix A.

9. If $A = [a_{ij}]_n$ and c is any scalar, express the value of $\det(cA)$ in terms of $\det(A)$.

10. Find the value of the given determinant by use of Definition 6.8.

(a) $\begin{vmatrix} 4 & 0 & 0 \\ 0 & -3 & 0 \\ 0 & 0 & 6 \end{vmatrix}$ (b) $\begin{vmatrix} 4 & -5 & 7 \\ 0 & -3 & -8 \\ 0 & 0 & 6 \end{vmatrix}$ (c) $\begin{vmatrix} 3 & -1 & 0 \\ 2 & 5 & 1 \\ 0 & 0 & 0 \end{vmatrix}$

(d) $\begin{vmatrix} 0 & 1 & 1 \\ 2 & 0 & 1 \\ 2 & 2 & 0 \end{vmatrix}$ (e) $\begin{vmatrix} a_{11} & 0 & 0 \\ a_{21} & a_{22} & 0 \\ a_{31} & a_{32} & a_{33} \end{vmatrix}$ (f) $\begin{vmatrix} a_{11} & 0 & 0 \\ 0 & 0 & a_{23} \\ 0 & a_{32} & 0 \end{vmatrix}$

(g) $\begin{vmatrix} 1 & 0 & 0 & 0 \\ 5 & 1 & 0 & 0 \\ 7 & 3 & 1 & 0 \\ -3 & 4 & -5 & 1 \end{vmatrix}$

(h) $\begin{vmatrix} -1 & 0 & 1 & 0 \\ 1 & 1 & 2 & 2 \\ 1 & 1 & 2 & 2 \\ 0 & 1 & 1 & -2 \end{vmatrix}$

11. Find the value of the given determinant by use of Theorem 6.9.

(a) $\begin{vmatrix} 2 & 0 & 1 \\ 0 & 1 & 3 \\ 0 & 0 & -3 \end{vmatrix}$

(b) $\begin{vmatrix} -1 & 0 & 1 \\ 1 & 3 & 2 \\ 2 & 0 & 0 \end{vmatrix}$

(c) $\begin{vmatrix} 1 & -1 & 2 \\ 1 & -1 & 2 \\ 3 & 2 & 1 \end{vmatrix}$

(d) $\begin{vmatrix} 2 & 2 & 0 \\ 1 & 1 & 3 \\ 0 & 0 & 2 \end{vmatrix}$

(e) $\begin{vmatrix} 0 & 0 & a_{13} \\ 0 & a_{22} & a_{23} \\ a_{31} & a_{32} & a_{33} \end{vmatrix}$

(f) $\begin{vmatrix} a_{11} & 0 & a_{13} \\ a_{21} & 0 & a_{23} \\ a_{31} & 0 & a_{33} \end{vmatrix}$

(g) $\begin{vmatrix} 0 & 1 & 0 & -1 \\ 1 & 2 & -1 & 0 \\ 2 & 4 & -1 & 0 \\ 0 & 0 & 1 & -1 \end{vmatrix}$

(h) $\begin{vmatrix} 1 & -1 & 0 & 0 \\ 3 & 2 & 1 & 0 \\ 0 & 1 & 3 & 2 \\ 0 & 0 & 1 & 4 \end{vmatrix}$

12. For each matrix A its LU decomposition is given. Evaluate $|A|$ and $|L||U|$ and compare the values.

(a) $A = \begin{bmatrix} -1 & -3 & 2 \\ 0 & 2 & 1 \\ 0 & -2 & 1 \end{bmatrix} = \begin{bmatrix} 1 & 0 & 0 \\ 0 & 1 & 0 \\ 0 & -1 & 1 \end{bmatrix} \begin{bmatrix} -1 & -3 & 2 \\ 0 & 2 & 1 \\ 0 & 0 & 2 \end{bmatrix} = LU$

(b) $A = \begin{bmatrix} 1 & -2 & -3 \\ 2 & 0 & 1 \\ -1 & 2 & 5 \end{bmatrix} = \begin{bmatrix} 1 & 0 & 0 \\ 2 & 1 & 0 \\ -1 & 0 & 1 \end{bmatrix} \begin{bmatrix} 1 & -2 & -3 \\ 0 & 4 & 7 \\ 0 & 0 & 2 \end{bmatrix} = LU$

13. Prove or disprove each of the given statements.

(a) $|A + B| = |A| + |B|$

(b) $|A + A^T| = |A| + |A^T|$

6.3 Cofactor Expansions

A comparison of Definition 6.8 and Theorem 6.9 shows that there is a certain duality in the determinant of a matrix, since emphasis can be placed on either the row subscripts or the column subscripts. This duality is complete in that the entire theory can be developed with either point of emphasis. For each property formulated in terms of rows, there is a dual formulation in terms of columns, and vice versa. With a few adjustments, the derivation of a property from one point of view can be changed into a derivation from the other point of view. It is for this reason that we include derivations from only one point of view. The approach that is adopted here is to use a formulation in terms of rows. Each result stated in terms of rows has a dual result stated in terms of columns. In some instances it is necessary to state the dual result as a separate theorem. In others, the dual result is indicated by the insertion of the word "column" in parentheses at appropriate places. In these instances, the dual statement is obtained simply by replacing the word "row" with the word "column."

In the course of our development, it will become apparent that the elementary operations play a fundamental role in the theory. In addition, they furnish a valuable tool for use in the evaluation of determinants.

In our first example, we examine the effect on the determinant of interchanging two rows in a matrix.

EXAMPLE 1 The matrix

$$B = \begin{bmatrix} 2 & 0 & 1 \\ 0 & 1 & -1 \\ 0 & 0 & 3 \end{bmatrix}$$

can be obtained from

$$A = \begin{bmatrix} 0 & 0 & 3 \\ 0 & 1 & -1 \\ 2 & 0 & 1 \end{bmatrix}$$

by interchanging the first and third rows. By Problems 5 and 11(e) in Exercises 6.2, we see that $|B| = 6$, $|A| = -6$ and we note that $|B| = -|A|$. ∎

The following theorem describes the effect on the determinant if an elementary row or column operation of type III is applied to a matrix. This result is invaluable in the derivation of the cofactor expansion in Theorem 6.14.

6.11 THEOREM

If the square matrix B is obtained from the matrix A by an elementary row (column) operation of type III, then $\det(B) = -\det(A)$.

Proof Let B be obtained from A by interchanging rows u and v of A, so that $b_{uj} = a_{vj}$ and $b_{vj} = a_{uj}$ for all j. Then

$$\det(B) = \sum_{(j)} (-1)^{t_1} b_{1j_1} \cdots b_{uj_u} \cdots b_{vj_v} \cdots b_{nj_n}$$

$$= \sum_{(j)} (-1)^{t_1} a_{1j_1} \cdots a_{vj_u} \cdots a_{uj_v} \cdots a_{nj_n}$$

$$= \sum_{(j)} (-1)^{t_1} a_{1j_1} \cdots a_{uj_v} \cdots a_{vj_u} \cdots a_{nj_n},$$

where t_1 is the number of interchanges used to carry $j_1, \ldots, j_u, \ldots, j_v, \ldots, j_n$ into $1, 2, \ldots, n$.

$$j_1, \ldots, j_u, \ldots, j_v, \ldots, j_n \xrightarrow{t_1 \text{ interchanges}} 1, 2, \ldots, n$$

Now

$$\det(A) = \sum_{(j)} (-1)^{t_2} a_{1j_1} \cdots a_{uj_v} \cdots a_{vj_u} \cdots a_{nj_n},$$

where t_2 is the number of interchanges used to carry $j_1, \ldots, j_v, \ldots, j_u, \ldots, j_n$ into $1, 2, \ldots, n$.

$$j_1, \ldots, j_v, \ldots, j_u, \ldots, j_n \xrightarrow{t_2 \text{ interchanges}} 1, 2, \ldots, n$$

Except for the exponents t_1 and t_2, $\det(B)$ would be equal to $\det(A)$. Since only one interchange is necessary to obtain $j_1, \ldots, j_u, \ldots, j_v, \ldots, j_n$ from $j_1, \ldots, j_v, \ldots, j_u, \ldots, j_n$,

$$j_1, \ldots, j_v, \ldots, j_u, \ldots, j_n \xrightarrow{1 \text{ interchange}} j_1, \ldots, j_u, \ldots, j_v, \ldots, j_n$$

t_1 and t_2 must differ by an odd number, and $(-1)^{t_1} = (-1)(-1)^{t_2}$. Therefore $\det(B) = -\det(A)$, and the proof is complete.   ■ ■ ■

The main purpose of this section is to establish an expression for the value of a determinant that is known as "an expansion by cofactors." Some new notation and terminology are needed to state this expansion by cofactors.

**6.12
DEFINITION**   If $n > 1$, the **minor** of the element a_{ij} in $A = [a_{ij}]_n$ is the determinant M_{ij} of the $(n-1) \times (n-1)$ submatrix of A obtained by deleting row i and column j of A.

EXAMPLE 2   To obtain the minor of a_{12} in $A = [a_{ij}]_3$, we first delete row 1 and column 2 of A as indicated below.

$$\begin{bmatrix} a_{11} & a_{12} & a_{13} \\ a_{21} & a_{22} & a_{23} \\ a_{31} & a_{32} & a_{33} \end{bmatrix}$$

We then evaluate the determinant of the submatrix that remains. The minor M_{12} of a_{12} is given by

$$M_{12} = \begin{vmatrix} a_{21} & a_{23} \\ a_{31} & a_{33} \end{vmatrix} = a_{21} a_{33} - a_{23} a_{31}.$$ ■

**6.13
DEFINITION**   If $n > 1$, the **cofactor** of a_{ij} in $A = [a_{ij}]_n$ is the product of $(-1)^{i+j}$ and the minor of a_{ij}. The cofactor of a_{ij} is denoted by A_{ij}.

EXAMPLE 3 The minor of $a_{32} = 2$ in

$$A = [a_{ij}]_3 = \begin{bmatrix} 7 & 8 & 9 \\ 4 & 5 & 6 \\ 1 & 2 & 3 \end{bmatrix}$$

is

$$M_{32} = \begin{vmatrix} 7 & 9 \\ 4 & 6 \end{vmatrix} = 42 - 36 = 6,$$

and the cofactor of 2 is

$$A_{32} = (-1)^{3+2} M_{32} = -M_{32} = -6.$$ ■

Our next theorem shows that the evaluation of an $n \times n$ determinant can be reduced to the evaluation of n determinants of order $n - 1$. This is of little practical use except when used in combination with elementary operations. Aside from this fact, however, the theorem has substantial theoretical value. The expression given in the theorem is referred to as "an expansion by cofactors" or more precisely as "the expansion about the ith row." This expansion is the main result of this section.

**6.14
THEOREM**

If $A = [a_{ij}]_n$, then

$$\det(A) = a_{i1} A_{i1} + a_{i2} A_{i2} + \cdots + a_{in} A_{in}.$$

Proof For a fixed integer i, we collect all of the terms in the sum

$$\det(A) = \sum_{(j)} (-1)^t a_{1j_1} a_{2j_2} \cdots a_{nj_n}$$

that contain a_{i1} as a factor in one group, all of the terms that contain a_{i2} as a factor in another group, and so on for each column number. This separates the terms in $\det(A)$ into n groups with no overlapping since each term contains exactly one factor from row i. In each of the terms containing a_{i1}, we factor out a_{i1} and let F_{i1} denote the remaining factor. Repeating this process for each of $a_{i2}, a_{i3}, \ldots, a_{in}$ in turn, we obtain

$$\det(A) = a_{i1} F_{i1} + a_{i2} F_{i2} + \cdots + a_{in} F_{in}.$$

To finish the proof, we need only show that $F_{ij} = A_{ij} = (-1)^{i+j} M_{ij}$, where M_{ij} is the minor of a_{ij}.

Consider first the case where $i = 1$ and $j = 1$. We shall show that $a_{11} F_{11} = a_{11} M_{11}$. Each term in F_{11} was obtained by factoring a_{11} from a term $(-1)^{t_1} a_{11} a_{2j_2} \cdots a_{nj_n}$ in the expansion of $\det(A)$. Thus each term in F_{11} has the form $(-1)^{t_2} a_{2j_2} a_{3j_3} \cdots a_{nj_n}$ where t_2 is the number of interchanges used to carry $j_2, j_3, \ldots, j_n$ into $2, 3, \ldots, n$.

$$j_2, j_3, \ldots, j_n \xrightarrow{t_2 \text{ interchanges}} 2, 3, \ldots, n$$

Letting $j_2, j_3, \ldots, j_n$ range over all permutations of $2, 3, \ldots, n$, we see that each of F_{11} and M_{11} has $(n - 1)!$ terms. Now $1, j_2, \ldots, j_n$ can be carried into the natural ordering by the same interchanges used to carry $j_2, \ldots, j_n$ into $2, \ldots, n$. That is, we may take $t_1 = t_2$.

$$1, j_2, j_3, \ldots, j_n \xrightarrow{t_1 = t_2 \text{ interchanges}} 1, 2, 3, \ldots, n$$

This means that F_{11} and M_{11} have exactly the same terms, yielding $F_{11} = M_{11}$ and $a_{11} F_{11} = a_{11} M_{11}$.

Consider now an arbitrary a_{ij}. By $i - 1$ interchanges of the original row i with the adjacent row above and then $j - 1$ interchanges of column j with the adjacent column on the left, we obtain a matrix B that has a_{ij} in the first row, first column position. Since the order of the remaining rows and columns of A was not changed, the minor of a_{ij} in B is the same M_{ij} as it is in A. By Theorem 6.11,

$$\det(B) = (-1)^{i-1+j-1} \det(A) = (-1)^{i+j} \det(A).$$

This gives $\det(A) = (-1)^{i+j} \det(B)$.

The sum of all the terms in $\det(B)$ that contain a_{ij} as a factor is $a_{ij} M_{ij}$, from our first case. Since $\det(A) = (-1)^{i+j} \det(B)$, the sum of all the terms in $\det(A)$ that contain a_{ij} as a factor is $(-1)^{i+j} a_{ij} M_{ij}$. Thus $a_{ij} F_{ij} = (-1)^{i+j} a_{ij} M_{ij} = a_{ij} A_{ij}$, and the theorem is proved. ∎ ■ ■ ■

EXAMPLE 4 If $A = [a_{ij}]_3$, the expansion of $\det(A)$ about the 2nd row is given by

$$\begin{vmatrix} a_{11} & a_{12} & a_{13} \\ a_{21} & a_{22} & a_{23} \\ a_{31} & a_{32} & a_{33} \end{vmatrix} = a_{21} A_{21} + a_{22} A_{22} + a_{23} A_{23}$$

$$= a_{21}(-1)^{2+1} \begin{vmatrix} a_{12} & a_{13} \\ a_{32} & a_{33} \end{vmatrix} + a_{22}(-1)^{2+2} \begin{vmatrix} a_{11} & a_{13} \\ a_{31} & a_{33} \end{vmatrix}$$

$$+ a_{23}(-1)^{2+3} \begin{vmatrix} a_{11} & a_{12} \\ a_{31} & a_{32} \end{vmatrix}$$

$$= -a_{21}(a_{12} a_{33} - a_{13} a_{32}) + a_{22}(a_{11} a_{33} - a_{13} a_{31})$$

$$- a_{23}(a_{11} a_{32} - a_{12} a_{31}).$$

This value agrees with that given in Example 1 of Section 6.2. ■

The dual statement for Theorem 6.14 describes the expansion by cofactors about the jth column.

**6.15
THEOREM**

If $A = [a_{ij}]_n$, then

$$\det(A) = a_{1j} A_{1j} + a_{2j} A_{2j} + \cdots + a_{nj} A_{nj}.$$

Theorems 6.14 and 6.15 can be used together to evaluate a determinant as illustrated in the next example.

EXAMPLE 5 Expanding

$$|A| = \begin{vmatrix} 1 & 0 & -1 & 4 \\ 1 & 2 & 5 & 1 \\ 2 & -1 & 2 & 7 \\ 0 & 0 & 3 & 0 \end{vmatrix}$$

about the fourth row gives

$$\begin{vmatrix} 1 & 0 & -1 & 4 \\ 1 & 2 & 5 & 1 \\ 2 & -1 & 2 & 7 \\ 0 & 0 & 3 & 0 \end{vmatrix} = 0 \cdot A_{41} + 0 \cdot A_{42} + 3 \cdot A_{43} + 0 \cdot A_{44}$$

$$= 3(-1)^{4+3} M_{43}$$

$$= 3(-1)^{4+3} \begin{vmatrix} 1 & 0 & 4 \\ 1 & 2 & 1 \\ 2 & -1 & 7 \end{vmatrix}.$$

We continue by expanding the 3×3 minor M_{43} about its second column.

$$3(-1)^{4+3} \begin{vmatrix} 1 & 0 & 4 \\ 1 & 2 & 1 \\ 2 & -1 & 7 \end{vmatrix} = -3 \left(2 \cdot (-1)^{2+2} \begin{vmatrix} 1 & 4 \\ 2 & 7 \end{vmatrix} + (-1)(-1)^{3+2} \begin{vmatrix} 1 & 4 \\ 1 & 1 \end{vmatrix} \right)$$

$$= -3(2(7 - 8) + (-1)(-1)(1 - 4))$$

$$= -3(-2 + (-3))$$

$$= 15$$

Thus $|A| = 15$. ∎

Theorem 6.14 proves to be extremely useful. For instance, it provides the key step in establishing our next theorem.

6.16 THEOREM   The expression $c_1 A_{i1} + c_2 A_{i2} + \cdots + c_n A_{in}$ is equal to the determinant of a matrix which is the same as $A = [a_{ij}]_n$ except that the elements a_{ij} of the ith row have been replaced by the scalars c_j.

Proof By Theorem 6.14,

$$\det(A) = a_{i1} A_{i1} + a_{i2} A_{i2} + \cdots + a_{in} A_{in}.$$

Let B be obtained from A by replacing a_{i1} by c_1, a_{i2} by c_2, a_{in} by c_n. Then $\det(B)$ can be found from the above expansion by replacing each a_{ik} by c_k for $k = 1, 2, \ldots, n$. In evaluating A_{ik}, the ith row and kth column are deleted from A. Thus, the values of the A_{ik} do not depend on any of the elements $a_{i1}, a_{i2}, \ldots, a_{in}$. Therefore, the values of the A_{ik} do not change when each a_{ik} is replaced by c_k, and

$$\det(B) = c_1 A_{i1} + c_2 A_{i2} + \cdots + c_n A_{in}.$$ ■■■

The result parallel to Theorem 6.16 has the following formulation in terms of columns.

6.17 THEOREM The expression $c_1 A_{1j} + c_2 A_{2j} + \cdots + c_n A_{nj}$ is equal to the determinant of a matrix which is the same as $A = [a_{ij}]_n$ except that the elements a_{ij} of the jth column have been replaced by the scalars c_i.

6.18 THEOREM The determinant of a matrix $A = [a_{ij}]_n$ that has two identical rows (columns) is zero.

Proof In contrast to the development thus far in this chapter, we must take into account the field $\mathcal{F}$ that contains the elements a_{ij} of A.

Suppose first that $1 + 1 \neq 0$ in $\mathcal{F}$.[1] If the uth and vth rows of A are identical, let B be the matrix formed from A by the interchange of the uth and vth rows. Then $B = A$, but $\det(B) = -\det(A)$ by Theorem 6.11. Thus we have $\det(A) = -\det(A)$, and $(1 + 1) \det(A) = 0$. Since $1 + 1 \neq 0$, $\det(A) = 0$.

Consider now the case where $1 + 1 = 0$ in $\mathcal{F}$. According to Theorem 6.11, an interchange of two rows in A produces a matrix whose determinant has the value $-\det(A)$. But $\det(A) = -\det(A)$ since $\det(A)$ is in $\mathcal{F}$ and $c + c = (1 + 1)c = 0$ for all c in $\mathcal{F}$. Thus, an interchange of rows does not change the value of the determinant, and there is no loss of generality if we assume that the first two rows are equal. That is, $a_{1j} = a_{2j}$ for all j. Since $-1 = 1$ in $\mathcal{F}$, $\det(A) = \sum_{(j)} a_{1j_1} a_{2j_2} \cdots a_{nj_n}$. For each term $a_{1j_1} a_{2j_2} a_{3j_3} \cdots a_{nj_n}$ in $\det(A)$, there is a corresponding term $a_{1j_2} a_{2j_1} a_{3j_3} \cdots a_{nj_n}$. If we group these terms in pairs and sum over only those orderings with $j_1 < j_2$, we have

$$\det(A) = \sum_{j_1 < j_2} (a_{1j_1} a_{2j_2} a_{3j_3} \cdots a_{nj_n} + a_{1j_2} a_{2j_1} a_{3j_3} \cdots a_{nj_n})$$

$$= \sum_{j_1 < j_2} (a_{1j_1} a_{2j_2} a_{3j_3} \cdots a_{nj_n} + a_{2j_2} a_{1j_1} a_{3j_3} \cdots a_{nj_n})$$

$$= \sum_{j_1 < j_2} (1 + 1) a_{1j_1} a_{2j_2} a_{3j_3} \cdots a_{nj_n}$$

$$= 0.$$

This completes the proof.   ■■■

[1] A field in which $1 + 1 = 0$ is a field of characteristic 2. For a discussion of characteristics of a field, the reader may consult any standard text in abstract algebra.

6.19
THEOREM

The sum of the products of the elements of a row of $A = [a_{ij}]_n$ by the cofactors of the corresponding elements of a different row of A is zero. Hence

$$a_{i1} A_{k1} + a_{i2} A_{k2} + \cdots + a_{in} A_{kn} = \delta_{ik} \det(A),$$

where δ_{ik} is the Kronecker delta.

Proof If $i = k$, the equality follows from Theorem 6.14.
Suppose that $i \neq k$. According to Theorem 6.16,

$$a_{i1} A_{k1} + a_{i2} A_{k2} + \cdots + a_{in} A_{kn}$$

is the determinant of a matrix B that is the same as A except that the elements of row k have been replaced by $a_{i1}, a_{i2}, \ldots, a_{in}$. Thus the matrix B has the ith and kth rows identical, and $\det(B) = 0$ by Theorem 6.18. ∎ ■ ■ ■

The dual statement of Theorem 6.19 is given next.

6.20
THEOREM

The sum of the products of the elements of a column of $A = [a_{ij}]_n$ by the cofactors of the corresponding elements of a different column is zero. Hence

$$a_{1j} A_{1k} + a_{2j} A_{2k} + \cdots + a_{nj} A_{nk} = \delta_{jk} \det(A).$$

6.3 Exercises

1. Compute the cofactor of the indicated element a_{ij} in $A = [a_{ij}]$, where

$$A = \begin{bmatrix} 2 & -2 & 2 & 1 \\ 2 & -1 & -2 & -1 \\ 0 & 2 & -4 & -6 \\ 2 & -3 & 10 & 4 \end{bmatrix}.$$

(a) a_{23} **(b)** a_{34} **(c)** a_{42} **(d)** a_{31} **(e)** a_{14} **(f)** a_{32}

2. Evaluate $\det(A)$ with A as in Problem 1.

3. Evaluate the given determinant by a cofactor expansion.

(a) $\begin{vmatrix} 2 & -1 & 2 & 1 \\ 0 & 1 & -4 & -2 \\ 0 & 0 & 4 & -2 \\ 0 & 0 & 4 & 1 \end{vmatrix}$ **(b)** $\begin{vmatrix} 2 & 2 & 0 \\ 2 & 1 & 1 \\ -7 & 2 & -3 \end{vmatrix}$

$$\text{(c)} \quad \begin{vmatrix} 2 & 3 & 0 & -1 \\ -4 & -6 & 0 & 3 \\ 2 & 1 & -1 & 2 \\ 0 & -2 & -1 & 3 \end{vmatrix} \qquad \text{(d)} \quad \begin{vmatrix} 1 & 2 & 1 & 2 \\ 3 & 4 & -1 & 5 \\ -2 & 2 & -1 & -1 \\ 1 & -3 & -2 & -1 \end{vmatrix}$$

4. For each of the following matrices C, evaluate $\det(C - xI)$.

$$\text{(a)} \quad \begin{bmatrix} 0 & 1 & 0 \\ 0 & 0 & 1 \\ -3 & 5 & -1 \end{bmatrix} \qquad \text{(b)} \quad \begin{bmatrix} 0 & 1 & 0 \\ 0 & 0 & 1 \\ 0 & 2 & 1 \end{bmatrix}$$

$$\text{(c)} \quad \begin{bmatrix} 0 & 1 & 0 & 0 \\ 0 & 0 & 1 & 0 \\ 0 & 0 & 0 & 1 \\ 0 & 4 & 7 & 2 \end{bmatrix} \qquad \text{(d)} \quad \begin{bmatrix} 0 & 1 & 0 & 0 \\ 0 & 0 & 1 & 0 \\ 0 & 0 & 0 & 1 \\ -4 & 4 & 3 & -2 \end{bmatrix}$$

5. For each of the following matrices A, find the values of x for which $|A - xI| = 0$.

$$\text{(a)} \quad \begin{bmatrix} -1 & 2 \\ 7 & 4 \end{bmatrix} \qquad \text{(b)} \quad \begin{bmatrix} 5 & 3 \\ 2 & 4 \end{bmatrix} \qquad \text{(c)} \quad \begin{bmatrix} 1 & -1 & -1 \\ -1 & 1 & -1 \\ -1 & -1 & 1 \end{bmatrix}$$

$$\text{(d)} \quad \begin{bmatrix} 4 & -2 & -4 \\ -1 & 5 & 4 \\ 1 & -7 & -6 \end{bmatrix} \qquad \text{(e)} \quad \begin{bmatrix} 3 & 0 & 0 & 0 \\ -2 & 4 & 0 & 0 \\ 1 & 0 & 2 & 0 \\ 0 & 3 & 3 & 5 \end{bmatrix} \qquad \text{(f)} \quad \begin{bmatrix} 5 & 7 & 9 & 11 \\ 0 & 7 & 9 & 11 \\ 0 & 0 & 9 & 11 \\ 0 & 0 & 0 & 11 \end{bmatrix}$$

6. Prove that if $A = [a_{ij}]_n$ has a row with all elements zero, then $\det(A) = 0$.

7. Let $A = [a_{ij}]$ be given by

$$A = \begin{bmatrix} -4 & -5 & 3 \\ 3 & 3 & -2 \\ -1 & -1 & 1 \end{bmatrix}.$$

(a) Find the matrix $B = [A_{ij}]^T$, where A_{ij} is the cofactor of a_{ij}.

(b) Compute AB, where A and B are as in part (a).

(c) Use the results of parts (a) and (b) to find A^{-1}.

8. Follow the instructions for Problem 7 for

$$A = \begin{bmatrix} 2 & 1 & 0 & 0 \\ 1 & 0 & -1 & 1 \\ 0 & 1 & -1 & 1 \\ 2 & -1 & -1 & 0 \end{bmatrix}.$$

9. (a) The matrix

$$\begin{bmatrix} -x & 1 & 0 \\ 0 & -x & 1 \\ -c_0 & -c_1 & -c_2 \end{bmatrix}$$

is called the **companion matrix** of the polynomial $(-1)(c_2 x^2 + c_1 x + c_0)$. Show that

$$\begin{vmatrix} -x & 1 & 0 \\ 0 & -x & 1 \\ -c_0 & -c_1 & -c_2 \end{vmatrix} = (-1)(c_2 x^2 + c_1 x + c_0).$$

(b) Show that

$$\begin{vmatrix} 1 & 1 & 1 \\ a & b & c \\ a^2 & b^2 & c^2 \end{vmatrix} = (a-b)(b-c)(c-a).$$

This is called a **Vandermonde** determinant.

10. For an arbitrary matrix $A = [a_{ij}]_3$, let $B = [A_{ij}]^T$.

   (a) Evaluate AB.

   (b) From the product AB in part (a), deduce a formula for A^{-1} whenever $\det(A) \neq 0$ and $A = [a_{ij}]_3$.

11. Prove Theorem 6.11 for an elementary column operation of type III.

12. Prove Theorem 6.15.

13. Prove Theorem 6.17.

14. Prove the dual statement of Theorem 6.18.

15. Prove Theorem 6.20.

6.4 Elementary Operations and Cramer's Rule

In Theorem 6.11 of the preceding section, we have seen that the application of a single operation of type III to a square matrix has the effect of changing the sign of the determinant. We propose now to investigate the effects of the other types of elementary operations, and to indicate the usefulness of these operations in combination with the cofactor expansions.

For an elementary operation of type I, we have the following result.

6.21 THEOREM

If the square matrix B is obtained from the matrix A by an elementary row (column) operation of type I that multiplies every element of a row (column) by $c \neq 0$, then $\det(B) = c \det(A)$.

Proof Suppose that B is obtained from $A = [a_{ij}]_n$ by multiplying each entry of the kth row by $c \neq 0$.

By Definition 6.8,

$$\det(B) = \sum_{(j)} (-1)^t b_{1j_1} b_{2j_2} \cdots b_{kj_k} \cdots b_{nj_n},$$

and since $b_{ij} = a_{ij}$ if $i \neq k$ and $b_{kj_k} = ca_{kj_k}$, we have

$$\det(B) = \sum_{(j)} (-1)^t a_{1j_1} a_{2j_2} \cdots (ca_{kj_k}) \cdots a_{nj_n}$$

$$= c \sum_{(j)} (-1)^t a_{1j_1} a_{2j_2} \cdots a_{kj_k} \cdots a_{nj_n}$$

$$= c \det(A).$$

■■■

For elementary operations of type II, we have the following description.

6.22 THEOREM

If B is obtained from the square matrix A by an elementary row (column) operation of type II, then $\det(B) = \det(A)$.

Proof Let $A = [a_{ij}]_n$ and suppose that $B = [b_{ij}]_n$ is formed by adding to each element a_{uj} of the uth row of A the product of the scalar c and the corresponding element a_{vj} of the vth row of A ($u \neq v$). Then B and A are the same except in the uth rows, and the cofactor of b_{uj} in the uth row of B is the same as the cofactor A_{uj} of the corresponding element a_{uj} in A. When $\det(B)$ is expanded about the uth row, we find

$$\det(B) = b_{u1} A_{u1} + b_{u2} A_{u2} + \cdots + b_{un} A_{un}$$

$$= (a_{u1} + ca_{v1})A_{u1} + (a_{u2} + ca_{v2})A_{u2} + \cdots + (a_{un} + ca_{vn})A_{un}$$

$$= a_{u1} A_{u1} + a_{u2} A_{u2} + \cdots + a_{un} A_{un} + c(a_{v1} A_{u1} + a_{v2} A_{u2} + \cdots + a_{vn} A_{un}).$$

By Theorems 6.14 and 6.19,

$$\det(B) = \det(A) + c(0) = \det(A).$$

■■■

EXAMPLE 1 As an illustration of the usefulness of the elementary operations, consider the evaluation of the determinant

$$\det(A) = \begin{vmatrix} 5 & 2 & 2 & 15 \\ 2 & 2 & -4 & 6 \\ 2 & -4 & 2 & 6 \\ 0 & 5 & 7 & 1 \end{vmatrix}.$$

By Theorem 6.21,

$$\det(A) = 2 \begin{vmatrix} 5 & 2 & 2 & 15 \\ 1 & 1 & -2 & 3 \\ 2 & -4 & 2 & 6 \\ 0 & 5 & 7 & 1 \end{vmatrix}.$$

According to Theorem 6.22, the value of the determinant is unchanged if we (a) add to row 1 the product of -5 and row 2 and (b) add to row 3 the product of -2 and row 2. Thus

$$\det(A) = 2 \begin{vmatrix} 0 & -3 & 12 & 0 \\ 1 & 1 & -2 & 3 \\ 0 & -6 & 6 & 0 \\ 0 & 5 & 7 & 1 \end{vmatrix}.$$

Expanding about the first column, we have

$$\det(A) = 2(-1)^{2+1} \begin{vmatrix} -3 & 12 & 0 \\ -6 & 6 & 0 \\ 5 & 7 & 1 \end{vmatrix} = (-2)(-3)(6) \begin{vmatrix} 1 & -4 & 0 \\ -1 & 1 & 0 \\ 5 & 7 & 1 \end{vmatrix}.$$

Expanding now about the third column, we have

$$\det(A) = (-2)(-3)(6)(-1)^{3+3} \begin{vmatrix} 1 & -4 \\ -1 & 1 \end{vmatrix} = -108. \qquad \blacksquare$$

Our final result of this section makes an important connection between determinants and the solution of certain types of systems of linear equations. This theorem presents a formula for the unknowns in terms of certain determinants. This formula is commonly known as **Cramer's rule**.

6.23 THEOREM

Consider a system of linear equations $AX = B$, in which $A = [a_{ij}]_{n \times n}$, $X = [x_1 \quad x_2 \quad \cdots \quad x_n]^T$, and $B = [b_1 \quad b_2 \quad \cdots \quad b_n]^T$. If $\det(A) \neq 0$, the unique solution of the system is given by

$$x_j = \frac{\sum_{k=1}^{n} b_k A_{kj}}{\det(A)}, \quad j = 1, 2, \dots, n.$$

Proof We first show that the given values are a solution of the system. Substitution of these values for x_j into the left member of the ith equation of the system yields

$$a_{i1} x_1 + \cdots + a_{in} x_n = \frac{1}{\det(A)} \left(a_{i1} \left(\sum_{k=1}^{n} b_k A_{k1} \right) + \cdots + a_{in} \left(\sum_{k=1}^{n} b_k A_{kn} \right) \right)$$

$$= \frac{1}{\det(A)} \sum_{j=1}^{n} \left(\sum_{k=1}^{n} a_{ij} b_k A_{kj} \right)$$

$$= \frac{1}{\det(A)} \sum_{k=1}^{n} \left(b_k \sum_{j=1}^{n} a_{ij} A_{kj} \right)$$

$$= \frac{1}{\det(A)} \sum_{k=1}^{n} b_k (\delta_{ik} \det(A))$$

$$= \frac{1}{\det(A)} b_i (\delta_{ii} \det(A))$$

$$= b_i.$$

Thus, the values

$$x_j = \frac{\sum_{k=1}^n b_k A_{kj}}{\det(A)}$$

furnish a solution of the system.

To prove uniqueness, suppose that $x_j = y_j$, $j = 1, 2, \ldots, n$, represents any solution to the system. Then the ith equation $\sum_{k=1}^n a_{ik} y_k = b_i$ is satisfied for $i = 1, 2, \ldots, n$. If we multiply both members of the ith equation by A_{ij} (j fixed) and form the sum of these equations, we find that

$$\sum_{i=1}^n \left(\sum_{k=1}^n a_{ik} A_{ij} y_k \right) = \sum_{i=1}^n b_i A_{ij}$$

or

$$\sum_{k=1}^n \left(\sum_{i=1}^n a_{ik} A_{ij} \right) y_k = \sum_{i=1}^n b_i A_{ij}.$$

But, for each k, $\sum_{i=1}^n a_{ik} A_{ij} = \delta_{kj} \det(A)$. Thus

$$\sum_{k=1}^n \delta_{kj} \det(A) y_k = \sum_{i=1}^n b_i A_{ij},$$

and

$$y_j = \frac{\sum_{i=1}^n b_i A_{ij}}{\det(A)}.$$

Hence these y_j's are the same as the solution given in the statement of the theorem. ∎ ■ ■ ■

We note that the sum $\sum_{k=1}^n b_k A_{kj}$ is the determinant of the matrix obtained by replacing the jth column of A by the column of constants $B = [\, b_1 \quad b_2 \quad \cdots \quad b_n \,]^T$. For a system in three unknowns with $\det(A) \neq 0$,

$$a_{11}x_1 + a_{12}x_2 + a_{13}x_3 = b_1$$

$$a_{21}x_1 + a_{22}x_2 + a_{23}x_3 = b_2$$

$$a_{31}x_1 + a_{32}x_2 + a_{33}x_3 = b_3,$$

the solutions stated in Theorem 6.23 can be written as

$$x_1 = \frac{\begin{vmatrix} b_1 & a_{12} & a_{13} \\ b_2 & a_{22} & a_{23} \\ b_3 & a_{32} & a_{33} \end{vmatrix}}{|A|}, \quad x_2 = \frac{\begin{vmatrix} a_{11} & b_1 & a_{13} \\ a_{21} & b_2 & a_{23} \\ a_{31} & b_3 & a_{33} \end{vmatrix}}{|A|}, \quad x_3 = \frac{\begin{vmatrix} a_{11} & a_{12} & b_1 \\ a_{21} & a_{22} & b_2 \\ a_{31} & a_{32} & b_3 \end{vmatrix}}{|A|}.$$

EXAMPLE 2 Cramer's Rule can be used to solve the system

$$
\begin{aligned}
x_1 + x_2 + x_3 &= 0 \\
2x_1 - x_2 &= 11 \\
x_2 + 4x_3 &= 3
\end{aligned}
$$

since

$$
\det(A) = \begin{vmatrix} 1 & 1 & 1 \\ 2 & -1 & 0 \\ 0 & 1 & 4 \end{vmatrix} = -10 \neq 0.
$$

We have

$$
x_1 = \frac{\begin{vmatrix} 0 & 1 & 1 \\ 11 & -1 & 0 \\ 3 & 1 & 4 \end{vmatrix}}{-10}, \quad
x_2 = \frac{\begin{vmatrix} 1 & 0 & 1 \\ 2 & 11 & 0 \\ 0 & 3 & 4 \end{vmatrix}}{-10}, \quad
x_3 = \frac{\begin{vmatrix} 1 & 1 & 0 \\ 2 & -1 & 11 \\ 0 & 1 & 3 \end{vmatrix}}{-10},
$$

and

$$
x_1 = \frac{-30}{-10} = 3, \quad x_2 = \frac{50}{-10} = -5, \quad x_3 = \frac{-20}{-10} = 2. \qquad \blacksquare
$$

6.4 Exercises

1. Evaluate the given determinant by an appropriate combination of elementary operations and expansion by cofactors.

   (a) $\begin{vmatrix} 1 & 0 & 1 \\ 1 & 1 & 2 \\ 3 & 4 & -2 \end{vmatrix}$
   (b) $\begin{vmatrix} -4 & -5 & 3 \\ 3 & 3 & -2 \\ -1 & -1 & 1 \end{vmatrix}$
   (c) $\begin{vmatrix} a & -b \\ b & a \end{vmatrix}$

   (d) $\begin{vmatrix} a-b & 2a-b \\ b-a & 2b-a \end{vmatrix}$
   (e) $\begin{vmatrix} 4 & -2 & -9 \\ 7 & 2 & 10 \\ 4 & 1 & -3 \end{vmatrix}$
   (f) $\begin{vmatrix} 2 & 4 & -3 \\ 4 & 8 & 9 \\ 4 & 4 & -9 \end{vmatrix}$

   (g) $\begin{vmatrix} 1 & 1 & 1 & 1 \\ 2 & 3 & 1 & -2 \\ 4 & 6 & 1 & 4 \\ 8 & 9 & 1 & -8 \end{vmatrix}$
   (h) $\begin{vmatrix} -1 & 1 & 0 & 2 \\ 10 & 2 & 3 & -1 \\ 4 & 0 & 1 & -1 \\ 1 & 1 & 1 & 1 \end{vmatrix}$

2. Justify each of the following statements using elementary operations.

   (a) If $\begin{vmatrix} 2 & 5 & -6 \\ -1 & 2 & 4 \\ 3 & 1 & 5 \end{vmatrix} = 139$, then $\begin{vmatrix} 4 & 15 & -6 \\ -2 & 6 & 4 \\ 6 & 3 & 5 \end{vmatrix} = 834.$

   (b) If $\begin{vmatrix} 4 & 5 & -6 \\ 0 & 1 & 5 \\ 0 & 1 & -3 \end{vmatrix} = -32$, then $\begin{vmatrix} 5 & 4 & -6 \\ 1 & 0 & 5 \\ 1 & 0 & -3 \end{vmatrix} = 32.$

(c) If $\begin{vmatrix} 8 & 4 & -6 \\ 7 & 14 & -49 \\ 0 & -1 & 1 \end{vmatrix} = -266,$ then $\begin{vmatrix} 4 & 2 & -3 \\ 1 & 2 & -7 \\ 0 & 1 & -1 \end{vmatrix} = 19.$

(d) If $\begin{vmatrix} 9 & -3 & 5 \\ 7 & -2 & 3 \\ 1 & 4 & 7 \end{vmatrix} = 54,$ then $\begin{vmatrix} -9 & 3 & -5 \\ -7 & 2 & -3 \\ -1 & -4 & -7 \end{vmatrix} = -54.$

(e) If $\begin{vmatrix} 5 & -1 & 3 & 6 \\ 2 & -3 & 9 & 0 \\ 1 & -3 & 5 & 6 \\ 0 & 2 & 5 & -4 \end{vmatrix} = 752,$ then $\begin{vmatrix} 5 & -1 & 3 & 6 \\ 2 & -3 & 9 & 0 \\ -1 & 0 & -4 & 6 \\ 0 & 2 & 5 & -4 \end{vmatrix} = 752.$

(f) If $\begin{vmatrix} 6 & 1 & -2 & 3 \\ 9 & -3 & 4 & 5 \\ 1 & -2 & 5 & -5 \\ 1 & 4 & 8 & 2 \end{vmatrix} = -2318,$ then $\begin{vmatrix} 6 & 1 & -2 & 0 \\ 9 & -3 & 4 & 14 \\ 1 & -2 & 5 & 1 \\ 1 & 4 & 8 & -10 \end{vmatrix} = -2318.$

3. Solve for x:

$$\begin{vmatrix} x & 0 & 0 & -8 \\ -1 & x & 0 & -10 \\ 0 & -1 & x & -1 \\ 0 & 0 & -1 & 1 \end{vmatrix} = 0.$$

4. Show that

$$\begin{vmatrix} a & a & a \\ a & b & b \\ a & b & c \end{vmatrix} = a(a-b)(b-c).$$

5. Evaluate the following determinant.

$$\begin{vmatrix} (a+b)^2 & c^2 & c^2 \\ a^2 & (b+c)^2 & a^2 \\ b^2 & b^2 & (a+c)^2 \end{vmatrix}$$

6. Show that

$$\begin{vmatrix} -x & 1 & 0 & 0 \\ 0 & -x & 1 & 0 \\ 0 & 0 & -x & 1 \\ -c_0 & -c_1 & -c_2 & -c_3-x \end{vmatrix} = c_0 + c_1 x + c_2 x^2 + c_3 x^3 + x^4.$$

7. Show that

$$\begin{vmatrix} a & b & c \\ d & e & f \\ h & i & j \end{vmatrix} + \begin{vmatrix} a & b & c \\ d & e & f \\ k & m & n \end{vmatrix} = \begin{vmatrix} a & b & c \\ d & e & f \\ h+k & i+m & j+n \end{vmatrix}.$$

8. Use elementary operations to show that the value of the 4×4 Vandermonde determinant

$$
\begin{vmatrix}
1 & 1 & 1 & 1 \\
a & b & c & d \\
a^2 & b^2 & c^2 & d^2 \\
a^3 & b^3 & c^3 & d^3
\end{vmatrix}
$$

is $(a - b)(a - c)(a - d)(b - c)(b - d)(c - d)$.

9. Prove that an equation of the straight line through two distinct points with rectangular coordinates (x_1, y_1) and (x_2, y_2) is given by

$$
\begin{vmatrix}
x & y & 1 \\
x_1 & y_1 & 1 \\
x_2 & y_2 & 1
\end{vmatrix} = 0.
$$

10. Prove that the area of the triangle ABC whose vertices are (x_1, y_1), (x_2, y_2), and (x_3, y_3) is given by

$$
\pm \frac{1}{2}
\begin{vmatrix}
1 & x_1 & y_1 \\
1 & x_2 & y_2 \\
1 & x_3 & y_3
\end{vmatrix}.
$$

11. Suppose that f_{11}, f_{12}, f_{21}, and f_{22} are differentiable functions of x and that g is defined by

$$
g(x) =
\begin{vmatrix}
f_{11}(x) & f_{12}(x) \\
f_{21}(x) & f_{22}(x)
\end{vmatrix}.
$$

Use calculus formulas and show that

$$
g' =
\begin{vmatrix}
f'_{11} & f'_{12} \\
f_{21} & f_{22}
\end{vmatrix} +
\begin{vmatrix}
f_{11} & f_{12} \\
f'_{21} & f'_{22}
\end{vmatrix}.
$$

12. Solve only for the indicated unknown in the following systems using Cramer's Rule.

(a) x_3 in:
$$
\begin{aligned}
2x_1 - x_2 + x_3 &= -7 \\
4x_1 + 2x_2 - x_3 &= -1 \\
x_1 + x_2 - x_3 &= 1
\end{aligned}
$$

(b) x_2 in:
$$
\begin{aligned}
5x_1 - x_2 - 5x_3 &= -2 \\
7x_1 + x_2 + 3x_3 &= 4 \\
x_2 + x_3 &= 6
\end{aligned}
$$

(c) x_1 in:
$$
\begin{aligned}
5x_1 - x_2 - 5x_3 &= -2 \\
7x_1 + x_2 + 3x_3 &= 4 \\
x_2 + x_3 &= 6
\end{aligned}
$$

(d) x_3 in:
$$
\begin{aligned}
2x_1 + x_2 - x_3 + x_4 &= 5 \\
-x_1 + 2x_2 + x_3 &= -6 \\
-3x_2 + x_3 - x_4 &= 1 \\
x_1 + 7x_2 + 3x_4 &= -1
\end{aligned}
$$

(e) x_4 in:
$$\begin{aligned} x_1 - x_2 \qquad\; + 2x_4 + x_5 &= 0 \\ 2x_1 - x_2 - 7x_3 \qquad\; - x_5 &= 1 \\ 2x_2 \qquad + 3x_4 - x_5 &= 3 \\ x_1 \qquad\qquad + 5x_4 \qquad &= 0 \\ x_1 \qquad + 3x_3 \qquad + x_5 &= 0 \end{aligned}$$

13. Use Cramer's Rule to evaluate two of the unknowns, and then find the remaining unknown by substitution of these values into one of the equations.

(a)
$$\begin{aligned} 2x_1 \qquad + 2x_3 &= 2 \\ x_2 - 3x_3 &= -4 \\ 2x_1 + x_2 + x_3 &= 6 \end{aligned}$$

(b)
$$\begin{aligned} 4x_1 - 2x_2 + 9x_3 &= 1 \\ 7x_1 + 2x_2 - 10x_3 &= 0 \\ 4x_1 + x_2 + 3x_3 &= 2 \end{aligned}$$

(c)
$$\begin{aligned} 4x_1 + 2x_2 + 4x_3 &= 2 \\ 6x_1 - 5x_2 + 8x_3 &= -3 \\ 3x_2 + 2x_3 &= 5 \end{aligned}$$

(d)
$$\begin{aligned} 3x_1 - 4x_2 + 6x_3 &= 1 \\ 9x_1 + 8x_2 - 12x_3 &= 3 \\ 9x_1 - 4x_2 + 12x_3 &= 4 \end{aligned}$$

(e)
$$\begin{aligned} 2x_1 - x_2 + 3x_3 &= 17 \\ 5x_1 - 2x_2 + 4x_3 &= 28 \\ 3x_1 + 3x_2 - x_3 &= -1 \end{aligned}$$

(f)
$$\begin{aligned} 2x_1 + 4x_2 + x_3 &= 17 \\ 5x_1 + 3x_2 - x_3 &= 4 \\ 2x_1 - 7x_2 + 3x_3 &= -36 \end{aligned}$$

14. Use Theorem 6.21 to express $|cA|$ in terms of $|A|$ for $A = [a_{ij}]_n$.

15. Prove the dual statement of Theorem 6.21.

16. Prove the dual statement of Theorem 6.22.

6.5 Determinants and Matrix Multiplication

We turn now to an investigation of the determinant of a product of two matrices. Our principal goal is to obtain the result that $\det(AB) = \det(A)\det(B)$. Although the statement of this result is simple enough, the derivation is a bit involved. We adopt an approach that utilizes elementary matrices in this derivation.

6.24 THEOREM

The determinant of any elementary matrix is not zero.

Proof Let M be an elementary matrix. If M is of type I, then $|M| = c|I_n| = c$, where $c \neq 0$, by Theorem 6.21. If M is of type II, then $|M| = |I_n| = 1$ by Theorem 6.22. If M is of type III, then $|M| = -|I_n| = -1$ by Theorem 6.11. ■ ■ ■

6.25 THEOREM

If A and M are $n \times n$ matrices and M is an elementary matrix, then

$$\det(MA) = \det(M)\det(A).$$

Proof We recall that multiplying A on the left by an elementary matrix M performs the same row operation on A as that used to obtain M from the identity matrix I_n (see page 107).

If M is of type I as in Theorem 6.21, then

$$\det(MA) = c \det(A) = \det(M) \det(A),$$

where $c \neq 0$. If M is of type II, then

$$\det(MA) = \det(A) = \det(M) \det(A),$$

by Theorem 6.22. If M is of type III as in Theorem 6.11, then

$$\det(MA) = -\det(A) = (-1) \det(A) = \det(M) \det(A).$$

Thus the theorem is valid in all cases.   ■ ■ ■

The result of the theorem extends readily to the following corollary, the proof of which is left as an exercise (Problem 7).

6.26 COROLLARY

If $M_1, M_2, \ldots, M_k$ are elementary $n \times n$ matrices and A is $n \times n$, then

$$\det(M_1 M_2 \cdots M_k A) = \det(M_1) \det(M_2) \cdots \det(M_k) \det(A).$$

6.27 COROLLARY

If $M_1, M_2, \ldots, M_k$ are elementary $n \times n$ matrices, then

$$\det(M_1 M_2 \cdots M_k) = \det(M_1) \det(M_2) \cdots \det(M_k).$$

Proof The asserted equality follows at once from Corollary 6.26 with $A = I_n$.   ■ ■ ■

Corollary 6.27 enables us to extend the result of Theorem 6.24 to arbitrary invertible matrices, and we are also able to establish the converse.

6.28 THEOREM

An $n \times n$ matrix A is invertible if and only if $\det(A) \neq 0$.

Proof Assume first that A is invertible. Then A is a product of elementary matrices (Theorem 3.19), say $A = M_1 M_2 \cdots M_k$. From Corollary 6.27,

$$\det(A) = \det(M_1) \det(M_2) \cdots \det(M_k),$$

and this product is not zero by Theorem 6.24.

Assume now that $\det(A) \neq 0$. Let P be an invertible matrix such that PA is in reduced row-echelon form. Now P is a product of elementary matrices, $P = M_1 M_2 \cdots M_k$, so that

$$\det(PA) = \det(M_1 M_2 \cdots M_k A) = \det(M_1) \det(M_2) \cdots \det(M_k) \det(A).$$

Since $\det(A) \neq 0$ and each $\det(M_i) \neq 0$, $\det(PA) \neq 0$. This implies that PA does not have a row of zeros, and consequently is of rank n by Corollary 3.49. But A and PA have the same rank since they are equivalent, so this means that A has rank n. It follows from Corollary 3.50 that A is invertible. ■ ■ ■

When the theory of matrices and determinants was in its early stages of development, the word *nonsingular* was normally used instead of invertible, and the word *singular* was used to indicate that a square matrix did not have an inverse. The last theorem gives the basis for the early terminology: Singular corresponds to zero determinant, and nonsingular corresponds to nonzero determinant.

We are now in a position to prove the main theorem in this section.

**6.29
THEOREM**

If A and B are $n \times n$ matrices, then

$$\det(AB) = \det(A)\det(B).$$

Proof If A is not invertible, then AB is not invertible (for AB invertible implies $AB(AB)^{-1} = I_n$, which in turn implies A is invertible by Theorem 3.14). By Theorem 6.28, $\det(A) = 0$ and $\det(AB) = 0$ so that

$$\det(AB) = \det(A)\det(B)$$

in this case.

Assume that A is invertible. Then A is a product of elementary matrices, $A = M_1 M_2 \cdots M_k$, and Corollary 6.26 implies that

$$\det(AB) = \det(M_1 M_2 \cdots M_k B)$$
$$= \det(M_1)\det(M_2)\cdots\det(M_k)\det(B).$$

But

$$\det(A) = \det(M_1)\det(M_2)\cdots\det(M_k)$$

by Corollary 6.27, and we have

$$\det(AB) = \det(A)\det(B). \qquad ■ ■ ■$$

The next theorem is of great interest since it provides a formula for the computation of the inverse of a matrix. Unfortunately the use of this formula is not very practical for matrices of higher orders, due to the large number of computations that are necessary.

The formula is conveniently expressed in terms of the adjoint of a matrix, defined as follows.

6.30 DEFINITION

If $A = [a_{ij}]_n$, then the **adjoint** of A, denoted by adj(A), is the matrix given by adj(A) = $[A_{ij}]_n^T$, where A_{ij} is the cofactor of a_{ij} in A.

EXAMPLE 1 Let

$$A = [a_{ij}]_3 = \begin{bmatrix} 11 & -6 & 2 \\ 3 & -2 & 1 \\ 2 & -2 & 2 \end{bmatrix}.$$

The cofactor of 11 in A is

$$A_{11} = (-1)^{1+1} \begin{vmatrix} -2 & 1 \\ -2 & 2 \end{vmatrix} = -4 + 2 = -2,$$

and the cofactor of -6 is

$$A_{12} = (-1)^{1+2} \begin{vmatrix} 3 & 1 \\ 2 & 2 \end{vmatrix} = (-1)(6 - 2) = -4.$$

Continuing in this fashion, we find that

$$\text{adj}(A) = \begin{bmatrix} -2 & -4 & -2 \\ 8 & 18 & 10 \\ -2 & -5 & -4 \end{bmatrix}^T = \begin{bmatrix} -2 & 8 & -2 \\ -4 & 18 & -5 \\ -2 & 10 & -4 \end{bmatrix}. \qquad ∎$$

6.31 THEOREM

If $A = [a_{ij}]_n$ is invertible, then

$$A^{-1} = \frac{1}{\det(A)} \text{adj}(A).$$

Proof Let $A \text{ adj}(A) = [c_{ij}]_n$. Now

$$A \text{ adj}(A) = \begin{bmatrix} a_{11} & a_{12} & \cdots & a_{1n} \\ a_{21} & a_{22} & \cdots & a_{2n} \\ \vdots & \vdots & \ddots & \vdots \\ a_{n1} & a_{n2} & \cdots & a_{nn} \end{bmatrix} \begin{bmatrix} A_{11} & A_{21} & \cdots & A_{n1} \\ A_{12} & A_{22} & \cdots & A_{n2} \\ \vdots & \vdots & \ddots & \vdots \\ A_{1n} & A_{2n} & \cdots & A_{nn} \end{bmatrix},$$

and hence $c_{ij} = \sum_{k=1}^n a_{ik} A_{jk}$. By Theorem 6.19, $\sum_{k=1}^n a_{ik} A_{jk} = \delta_{ij} \det(A)$. Thus

$$A \text{ adj}(A) = \det(A) I_n.$$

Since $\det(A) \neq 0$, this yields

$$A\left(\frac{1}{\det(A)} \operatorname{adj}(A)\right) = I_n,$$

and it follows that

$$A^{-1} = \frac{1}{\det(A)} \operatorname{adj}(A). \qquad \blacksquare\blacksquare\blacksquare$$

6.32
COROLLARY   For any $A = [a_{ij}]_n$, $A \operatorname{adj}(A) = \det(A)I_n$.

EXAMPLE 2   Using the matrices A and $\operatorname{adj}(A)$ from Example 1, we have

$$\det(A) = \begin{vmatrix} 11 & -6 & 2 \\ 3 & -2 & 1 \\ 2 & -2 & 2 \end{vmatrix} = -2$$

and

$$A^{-1} = \frac{1}{-2}\begin{bmatrix} -2 & -4 & -2 \\ 8 & 18 & 10 \\ -2 & -5 & -4 \end{bmatrix}^T = \begin{bmatrix} 1 & -4 & 1 \\ 2 & -9 & \frac{5}{2} \\ 1 & -5 & 2 \end{bmatrix}. \qquad \blacksquare$$

EXAMPLE 3   The formula in Theorem 6.31 has its greatest usefulness with 2×2 matrices. If

$$A = \begin{bmatrix} a & b \\ c & d \end{bmatrix}$$

and $ad - bc \neq 0$, then

$$A^{-1} = \frac{1}{ad - bc}\begin{bmatrix} d & -b \\ -c & a \end{bmatrix}. \qquad \blacksquare$$

6.5 Exercises

1. Find the inverse of the given matrix, if it exists.

(a) $\begin{bmatrix} 2 & 4 \\ -3 & -2 \end{bmatrix}$ (b) $\begin{bmatrix} -3 & 7 \\ 6 & -14 \end{bmatrix}$ (c) $\begin{bmatrix} 1 & 0 \\ 5 & 6 \end{bmatrix}$ (d) $\begin{bmatrix} -1 & -2 \\ 3 & -4 \end{bmatrix}$

2. Use the formula of Theorem 6.31 to find the inverse of the given matrix.

(a) $\begin{bmatrix} 1 & 2 & 1 \\ 1 & 0 & 1 \\ 0 & 1 & -1 \end{bmatrix}$ (b) $\begin{bmatrix} -4 & -5 & 3 \\ 3 & 3 & -2 \\ -1 & -1 & 1 \end{bmatrix}$ (c) $\begin{bmatrix} 1 & 0 & 1 \\ 1 & 1 & 2 \\ 3 & 4 & -2 \end{bmatrix}$

(d) $\begin{bmatrix} 1 & 2 & 1 \\ -1 & -1 & 1 \\ 0 & 1 & 3 \end{bmatrix}$
(e) $\begin{bmatrix} 4 & 2 & 4 \\ 6 & -5 & 8 \\ 3 & 3 & 2 \end{bmatrix}$
(f) $\begin{bmatrix} 3 & 3 & 2 \\ 2 & 4 & 2 \\ 7 & 10 & 6 \end{bmatrix}$

(g) $\begin{bmatrix} 0 & 1 & 0 & -1 \\ 1 & 3 & 0 & 0 \\ 0 & -1 & 3 & -1 \\ -2 & 1 & 0 & 1 \end{bmatrix}$
(h) $\begin{bmatrix} -4 & 2 & 0 & 1 \\ 0 & 0 & 2 & 4 \\ 1 & -2 & -1 & 0 \\ 0 & 1 & 0 & -1 \end{bmatrix}$

3. We have seen that it may happen that $AB \neq BA$, even when both products AB and BA exist. Prove or disprove that $\det(AB) = \det(BA)$, for square matrices A and B of order n.

4. Use Theorem 6.29 to express $\det(A^m)$ in terms of $\det(A)$ for an arbitrary integer m ($A^0 = I$). In particular, obtain the result $\det(A^{-1}) = \frac{1}{\det(A)}$.

5. Given that A and B are square matrices of the same order, $\det(A) = 2$, and $\det(B) = 6$. Find the value of $\det(A^{-1}B)$.

6. If A and B are invertible and of the same order, express $\det((AB)^{-1})$ in terms of $\det(A)$ and $\det(B)$.

7. Prove Corollary 6.26.

8. Prove that if $\det(A^k) = 0$, for any $k > 0$, then A is not invertible.

9. A square matrix A such that $A^T A = I$ is said to be **orthogonal**. Prove that if A is orthogonal, then $\det(A) = \pm 1$ and hence A is invertible.

10. An $n \times n$ matrix A is said to be **nilpotent** if $A^k = \mathbf{0}$, for some positive integer k. Show that if A is nilpotent then $\det(A) = 0$.

11. A square matrix A is said to be **idempotent** if $A^2 = A$. Prove that if A is idempotent then either $\det(A) = 1$ or $\det(A) = 0$.

12. A matrix A is called **skew-symmetric** if $A^T = -A$. Prove that any skew-symmetric matrix of odd order is singular if $1 + 1 \neq 0$ in $\mathcal{F}$.

13. Prove that $\mathrm{adj}(\mathrm{adj}(A)) = A$ if $\det(A) = 1$.

14. Prove that $\det(\mathrm{adj}(A)) = (\det(A))^{n-1}$, where A is a square matrix of order $n > 1$.

15. Suppose that the matrices $A = [a_{ij}]_n$, $B = [b_{ij}]_n$, and $C = [c_{ij}]_n$ are such that $a_{ij} = b_{ij} = c_{ij}$ whenever $i \neq k$, and $c_{kj} = a_{kj} + b_{kj}$ for each j. Prove that $\det(C) = \det(A) + \det(B)$.

16. For the matrix

$$A = \begin{bmatrix} -1 & -2 & 4 \\ -3 & 0 & 1 \\ 0 & -1 & 1 \end{bmatrix},$$

evaluate $\det(A)$ and use this value to find each of the following.

(a) $\det(2A)$
(b) $\det(A^3)$
(c) $\det(-A^{-1})$

(d) $\det(\operatorname{adj}(A))$ **(e)** $\det(A^T A)$

(f) $\det(P^{-1}AP)$, where P is any invertible 3×3 matrix

(g) $\det((P^{-1}AP)^4)$, where P is any invertible 3×3 matrix

(h) A^{-1}

17. For the matrix

$$A = \begin{bmatrix} 0 & 1 & 2 & -1 \\ -1 & 1 & 3 & 0 \\ 0 & 0 & 1 & -5 \\ 1 & -1 & -2 & 0 \end{bmatrix},$$

evaluate $\det(A)$ and use this value to find each of the following.

(a) $\det(-A)$ **(b)** $\det(A^4)$ **(c)** $\det(3A^{-1})$ **(d)** $\det(\operatorname{adj}(A))$ **(e)** $\det(AA^T)$

(f) $\det(P^{-1}AP)$, where P is any invertible 4×4 matrix

(g) $\det(P^{-1}A^{-2}P)$, where P is any invertible 4×4 matrix

(h) A^{-1}

18. A **submatrix** of an $m \times n$ matrix A is a matrix obtained by deleting certain rows and/or columns of A. For an arbitrary matrix A (not necessarily square), the number $\rho(A)$ is defined as follows:

(i) if $A = \mathbf{0}$, then $\rho(A) = 0$,

(ii) if $A \neq \mathbf{0}$, then $\rho(A)$ is the largest possible order for a square submatrix of A that has a nonzero determinant.

(a) Prove that if the matrix B is obtained from A by an elementary row operation, then $\rho(B) \leq \rho(A)$. Hence conclude that $\rho(B) = \rho(A)$.

(b) Prove that if the matrix B is row-equivalent to A, then $\rho(B) = \rho(A)$.

(c) Prove that $\rho(A)$ is the rank of A. (*Hint*: See Corollary 3.49.)

19. Suppose that A is a nonzero square matrix of order 3 that is singular. What is the most precise statement that can be made about the rank of A?

20. Find all values of x and y in the field of complex numbers for which the matrix

$$\begin{bmatrix} x & y & 8 \\ 1 & x & y \end{bmatrix}$$

has rank 1.

7 Eigenvalues and Eigenvectors

In Chapter 5, we studied linear transformations of a vector space $\mathbf{U}$ into a vector space $\mathbf{V}$, where $\mathbf{U}$ and $\mathbf{V}$ were vector spaces over the same field $\mathcal{F}$. We turn our attention now to the special case in which $\mathbf{U} = \mathbf{V}$. More precisely, we shall be concerned here with linear operators on a finite-dimensional vector space $\mathbf{V}$. Throughout this chapter, $\mathcal{F}$ will denote a field, $\mathbf{V}$ will denote a finite-dimensional vector space over $\mathcal{F}$, and T will denote a linear operator on $\mathbf{V}$.

7.1 Eigenvalues and Eigenvectors

In many applications of linear algebra, particularly in engineering, those vectors in a vector space that are mapped onto multiples of themselves by a certain linear operator are of key importance. This type of vector also arises with the use of quadratic forms in statistics and other areas of mathematics.

7.1 DEFINITION

An **eigenvector** of T is a nonzero vector $\mathbf{v}$ such that $T(\mathbf{v}) = \lambda\mathbf{v}$ for some scalar λ. The scalar λ is called an **eigenvalue** of T.

EXAMPLE 1 Consider the linear operator T on $\mathbf{R}^2$, defined by

$$T(x, y) = (4x + y, 4x + y).$$

Since

$$T(1, 1) = (5, 5) = 5(1, 1)$$

and

$$T(1, -4) = (0, 0) = 0(1, -4),$$

then we see that 5 and 0 are eigenvalues of T and that $(1, 1)$ and $(1, -4)$ are eigenvectors of T. ∎

Note that in the above example for this particular linear operator T, the choice of the eigenvector $(1, 1)$ depends on the eigenvalue 5 just as the eigenvector $(1, -4)$ depends on the eigenvalue 0. For this reason, we say that $\mathbf{v}$ and λ are **associated** with each other, or that they **correspond** to each other. Thus a scalar λ is an eigenvalue of T if and only if there exists a nonzero vector $\mathbf{v}$ such that $T(\mathbf{v}) = \lambda\mathbf{v}$. The set of all eigenvalues of T is called the **spectrum** of T.

The term "eigenvalue" in the definition is not completely standardized. Other terms used interchangeably are **characteristic value, characteristic root, proper value**, and **proper number**. The German word "eigen" translates into English as "characteristic," but the hybrid word eigenvalue seems to be more widely used than any of the other terms. Similarly, other terms for eigenvector are **characteristic vector** and **proper vector**.

We have already studied the intimate relations between a linear transformation and the various matrices that represent it relative to different bases. We shall see presently that the matrices that represent a linear operator are also useful tools in investigating the eigenvalues of the operator. The principal connection between a linear operator T and a matrix A that represents it is provided here by the characteristic matrix of A.

7.2 DEFINITION   Let A be an $n \times n$ matrix over $\mathcal{F}$, and let x represent an indeterminate scalar. Then the matrix $A - xI$ is called the **characteristic matrix** of A.

EXAMPLE 2   The characteristic matrix of

$$A = \begin{bmatrix} 8 & 5 & 6 & 0 \\ 0 & -2 & 0 & 0 \\ -10 & -5 & -8 & 0 \\ 2 & 1 & 1 & 2 \end{bmatrix}$$

is the matrix given by

$$A - xI = \begin{bmatrix} 8-x & 5 & 6 & 0 \\ 0 & -2-x & 0 & 0 \\ -10 & -5 & -8-x & 0 \\ 2 & 1 & 1 & 2-x \end{bmatrix}.$$ ■

If we concentrate on the *degrees* of the terms in the definition of

$$\det(A - xI) = \begin{vmatrix} a_{11}-x & a_{12} & \cdots & a_{1n} \\ a_{21} & a_{22}-x & \cdots & a_{2n} \\ \vdots & \vdots & \ddots & \vdots \\ a_{n1} & a_{n2} & \cdots & a_{nn}-x \end{vmatrix}$$

it is clear that the term with highest degree is the product of the diagonal elements

$$(a_{11} - x)(a_{22} - x) \cdots (a_{nn} - x).$$

Hence $\det(A - xI)$ is a polynomial in x of degree n with lead coefficient $(-1)^n$, say

$$\det(A - xI) = (-1)^n x^n + c_{n-1}x^{n-1} + \cdots + c_1 x + c_0.$$

Upon setting $x = 0$, we find that $c_0 = \det(A)$.

7.3 DEFINITION	For any square matrix A over $\mathcal{F}$, the polynomial $\det(A - xI)$ is the **characteristic polynomial** of A. The equation $\det(A - xI) = 0$ is the **characteristic equation** of A, and the solutions of $\det(A - xI) = 0$ are called the **eigenvalues** of A. The set of all eigenvalues of A is called the **spectrum** of A.

EXAMPLE 3 With A as in Example 2,

$$\det(A - xI) = \begin{vmatrix} 8 - x & 5 & 6 & 0 \\ 0 & -2 - x & 0 & 0 \\ -10 & -5 & -8 - x & 0 \\ 2 & 1 & 1 & 2 - x \end{vmatrix}$$

$$= x^4 - 8x^2 + 16$$

is the *characteristic polynomial* of A. The equation

$$x^4 - 8x^2 + 16 = 0$$

is the *characteristic equation* of A. Since

$$x^4 - 8x^2 + 16 = (x + 2)^2 (x - 2)^2,$$

the *eigenvalues* of A are given by $\lambda_1 = \lambda_2 = -2, \lambda_3 = \lambda_4 = 2$. The *spectrum* of A is $\{-2, 2\}$. ■

The important connection between the eigenvalues of linear operators and those of matrices is given in our next theorem.

7.4 THEOREM	Let A be any matrix that represents the linear operator T. Then T and A have the same spectrum.

Proof Suppose that T is represented by the matrix A relative to the basis $\mathcal{A}$ of $\mathbf{V}$.

Let λ be an eigenvalue of T, and let $X = [x_1 \ \ x_2 \ \ \cdots \ \ x_n]^T$ be the coordinate matrix relative to $\mathcal{A}$ of a corresponding eigenvector $\mathbf{v}$. Then $T(\mathbf{v}) = \lambda\mathbf{v}$ and therefore $AX = \lambda X$ since A represents T. This gives $AX - \lambda X = \mathbf{0}$, and $(A - \lambda I)X = \mathbf{0}$. Since X is a nonzero matrix, the system of equations $(A - \lambda I)X = \mathbf{0}$ has a nontrivial solution given by the coordinates recorded in X. By Theorems 6.28 and 4.20, this implies that $\det(A - \lambda I) = 0$, and λ is an eigenvalue of A.

On the other hand, if $\det(A - \lambda I) = 0$, then the system $(A - \lambda I)X = \mathbf{0}$ has a nontrivial solution by Theorems 6.28 and 4.20. This solution provides the coordinate matrix X relative to $\mathcal{A}$ of a nonzero vector $\mathbf{v}$. Since $AX = \lambda X, T(\mathbf{v}) = \lambda\mathbf{v}$, and λ is an eigenvalue of T. ■■■

The proof of the preceding theorem rests primarily on the equivalence of the equations

$$T(\mathbf{v}) = \lambda \mathbf{v} \quad \text{and} \quad AX = \lambda X.$$

We see that λ is an eigenvalue of A if and only if there exists a nonzero $n \times 1$ matrix X such that $AX = \lambda X$. This motivates the following definition of eigenvectors of matrices.

7.5 DEFINITION	Let λ be an eigenvalue of the $n \times n$ matrix A. Then an **eigenvector** of A associated with λ is a nonzero $n \times 1$ matrix X such that $AX = \lambda X$.

Thus the eigenvectors of matrices and linear operators are related in this way: If the $n \times n$ matrix A represents T relative to the basis $\mathcal{A}$ of $\mathbf{V}$, then X is an eigenvector of A corresponding to λ if and only if X is the coordinate matrix relative to $\mathcal{A}$ of an eigenvector of T corresponding to the same eigenvalue.

The method of proof of Theorem 7.4 provides a systematic method for determining the eigenvalues of a given linear operator. Any convenient choice of a basis $\mathcal{A}$ of $\mathbf{V}$ will determine the matrix A that represents T relative to $\mathcal{A}$, and the eigenvalues of T are precisely the solutions of the characteristic equation $\det(A - xI) = 0$. The eigenvectors $\mathbf{v}$ corresponding to a particular eigenvalue λ are just those nonzero vectors $\mathbf{v}$ in $\mathbf{V}$ with coordinates X relative to $\mathcal{A}$ that satisfy $(A - \lambda I)X = \mathbf{0}$. An illustration of this procedure is given in the next example.

EXAMPLE 4 Consider the linear operator on $\mathbf{R}^4$ defined by

$$T(x_1, x_2, x_3, x_4) = (8x_1 + 5x_2 + 6x_3, -2x_2, -10x_1 - 5x_2 - 8x_3, 2x_1 + x_2 + x_3 + 2x_4).$$

The matrix of T relative to the standard basis of $\mathbf{R}^4$ is the matrix

$$A = \begin{bmatrix} 8 & 5 & 6 & 0 \\ 0 & -2 & 0 & 0 \\ -10 & -5 & -8 & 0 \\ 2 & 1 & 1 & 2 \end{bmatrix}$$

considered in Examples 2 and 3. From Example 3, we know that the characteristic polynomial of A is

$$\det(A - xI) = x^4 - 8x^2 + 16 = (x + 2)^2(x - 2)^2,$$

and the eigenvalues of T are $\lambda_1 = \lambda_2 = -2$ and $\lambda_3 = \lambda_4 = 2$.

To determine the eigenvectors corresponding to $\lambda_1 = \lambda_2 = -2$, we consider the system of equations $(A + 2I)X = \mathbf{0}$:

$$\begin{bmatrix} 10 & 5 & 6 & 0 \\ 0 & 0 & 0 & 0 \\ -10 & -5 & -6 & 0 \\ 2 & 1 & 1 & 4 \end{bmatrix} \begin{bmatrix} x_1 \\ x_2 \\ x_3 \\ x_4 \end{bmatrix} = \begin{bmatrix} 0 \\ 0 \\ 0 \\ 0 \end{bmatrix}.$$

Solving this system, we have

$$\begin{bmatrix} 10 & 5 & 6 & 0 & 0 \\ 0 & 0 & 0 & 0 & 0 \\ -10 & -5 & -6 & 0 & 0 \\ 2 & 1 & 1 & 4 & 0 \end{bmatrix} \rightarrow \begin{bmatrix} 2 & 1 & 1 & 4 & 0 \\ 0 & 0 & 1 & -20 & 0 \\ 0 & 0 & 0 & 0 & 0 \\ 0 & 0 & 0 & 0 & 0 \end{bmatrix} \rightarrow \begin{bmatrix} 1 & \frac{1}{2} & 0 & 12 & 0 \\ 0 & 0 & 1 & -20 & 0 \\ 0 & 0 & 0 & 0 & 0 \\ 0 & 0 & 0 & 0 & 0 \end{bmatrix}.$$

The solutions to this system are given by

$$\begin{aligned} x_1 &= -\tfrac{1}{2}x_2 - 12x_4 \\ x_2 &= x_2 \\ x_3 &= 20x_4 \\ x_4 &= x_4, \end{aligned}$$

where x_2 and x_4 are arbitrary. The eigenvectors of T corresponding to the eigenvalue -2 are those vectors of the form

$$\begin{aligned} (x_1, x_2, x_3, x_4) &= \left(-\tfrac{1}{2}x_2 - 12x_4, x_2, 20x_4, x_4\right) \\ &= x_2\left(-\tfrac{1}{2}, 1, 0, 0\right) + x_4(-12, 0, 20, 1). \end{aligned}$$

For the eigenvalue $\lambda_3 = \lambda_4 = 2$, the system of equations $(A - 2I)X = \mathbf{0}$ is given by

$$\begin{bmatrix} 6 & 5 & 6 & 0 \\ 0 & -4 & 0 & 0 \\ -10 & -5 & -10 & 0 \\ 2 & 1 & 1 & 0 \end{bmatrix} \begin{bmatrix} x_1 \\ x_2 \\ x_3 \\ x_4 \end{bmatrix} = \begin{bmatrix} 0 \\ 0 \\ 0 \\ 0 \end{bmatrix}.$$

The reduced row-echelon form for the augmented matrix here is

$$\begin{bmatrix} 1 & 0 & 0 & 0 & 0 \\ 0 & 1 & 0 & 0 & 0 \\ 0 & 0 & 1 & 0 & 0 \\ 0 & 0 & 0 & 0 & 0 \end{bmatrix},$$

so the solutions are given by $x_1 = 0$, $x_2 = 0$, $x_3 = 0$, x_4 arbitrary. The eigenvectors of T corresponding to the eigenvalue 2 are those vectors of the form

$$(x_1, x_2, x_3, x_4) = x_4(0, 0, 0, 1). \qquad \blacksquare$$

The procedure described just before Example 4 is not always as simple and effective as it might seem, for there are several complications that may arise. There is usually no difficulty in determining the matrix A. (In most cases, the matrix A is already known.) But one may encounter trouble in the solution of the resulting characteristic equation $\det(A - xI) = 0$.

First of all, some or all of the solutions to $\det(A - xI) = 0$ may not lie in the field $\mathcal{F}$. If an eigenvalue λ is not in $\mathcal{F}$, then the nonzero coordinates x_i in a solution of $AX = \lambda X$ are not in $\mathcal{F}$. For the nonzero elements of AX are in $\mathcal{F}$ whenever those of X are, whereas those of λX are not. Thus there are no eigenvectors of T in **V** corresponding to λ whenever $\lambda \notin \mathcal{F}$.

Even when the eigenvalues are all in $\mathcal{F}$, difficulties may be encountered in the solution of the polynomial equation $\det(A - xI) = 0$. This is the typical situation in applications, and a large number of numerical methods have been devised to obtain approximate solutions to such problems.

Problems that call for the determination of eigenvalues or eigenvectors of a certain linear transformation or matrix are called **eigenvalue problems**. They are one of the most common types of problems encountered in the applications of linear algebra. Unfortunately, most of them are quite complicated in their formulation, and their solution frequently involves a knowledge of several areas of mathematics. One of the simplest types that occurs in physical situations is illustrated in our next example.

EXAMPLE 5   Before stating the physical problem to be solved, we note that the force required to stretch or compress a spring by an amount x is directly proportional to x, and that the constant of proportionality is called the *spring constant*. Thus $F = cx$, where F is the force on the spring and x is the change in length caused by F.

Consider now the mechanical system shown in Figure 7.1.

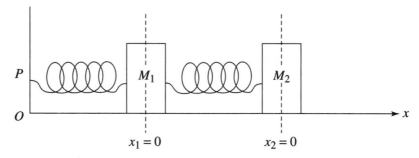

Objects in Equilibrium Positions

Figure 7.1

On the horizontal plane containing the line OX, an object M_1 of mass 1 unit is connected to the fixed point P by a first spring with spring constant $c_1 = 3$. A second object M_2 of mass 1 unit is then connected to M_1 by a second spring with spring constant $c_2 = 2$. The centers of gravity of M_1 and M_2 lie on a horizontal line through P. The object M_1 is displaced 1 unit toward P from its equilibrium position, and the object M_2 is displaced from its equilibrium position 2 units away from P. The two objects are released at time $t = 0$, and it is desired to find the positions of the objects at any subsequent time t. The masses of the springs and frictional forces are to be neglected, and no external forces act on the system.

Let x_1 and x_2 denote the displacement from the equilibrium positions of M_1 and M_2, respectively. Each displacement x_i is measured with the positive direction to the right as shown in Figure 7.2.

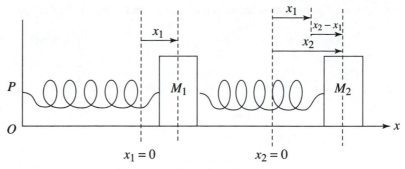

Objects in Motion

Figure 7.2

There are two forces acting on M_1 at any time $t > 0$, one from each spring. The first spring exerts a force F_1 given by $F_1 = -3x_1$ since F_1 acts in the direction opposite to the displacement. The second spring exerts a force F_2 given by $F_2 = 2(x_2 - x_1)$ since $x_2 - x_1$ is the (directed) change in the distance from the center of gravity in M_1 to that in M_2. According to Newton's second law, the sum of the forces acting on M_1 is equal to the product of its mass and its acceleration. Since M_1 has unit mass, this requires that

$$\frac{d^2 x_1}{dt^2} = -3x_1 + 2(x_2 - x_1) = -5x_1 + 2x_2.$$

The only force acting on the object M_2 is the force $-2(x_2 - x_1)$ due to the second spring, and Newton's second law yields

$$\frac{d^2 x_2}{dt^2} = 2x_1 - 2x_2.$$

Thus the original problem has been reduced to that of solving the system of differential equations:

$$\frac{d^2 x_1}{dt^2} = -5x_1 - 2x_2$$

$$\frac{d^2 x_2}{dt^2} = 2x_1 - 2x_2.$$

We assume the existence of a solution of the form

$$x_1 = a_1 \cos \omega_1 t + a_2 \cos \omega_2 t,$$

$$x_2 = b_1 \cos \omega_1 t + b_2 \cos \omega_2 t.$$

(A justification for this assumption would be quite a digression, and we shall see momentarily that it is a valid one, at any rate.)

Substitution for $x_1, x_2, d^2x_1/dt^2$, and d^2x_2/dt^2 in the system yields

$$-a_1\omega_1^2 \cos \omega_1 t - a_2\omega_2^2 \cos \omega_2 t = -5a_1 \cos \omega_1 t - 5a_2 \cos \omega_2 t$$
$$+ 2b_1 \cos \omega_1 t + 2b_2 \cos \omega_2 t,$$
$$-b_1\omega_1^2 \cos \omega_1 t - b_2\omega_2^2 \cos \omega_2 t = 2a_1 \cos \omega_1 t + 2a_2 \cos \omega_2 t$$
$$- 2b_1 \cos \omega_1 t - 2b_2 \cos \omega_2 t.$$

In matrix form, we have

$$\omega_1^2 \cos \omega_1 t \begin{bmatrix} a_1 \\ b_1 \end{bmatrix} + \omega_2^2 \cos \omega_2 t \begin{bmatrix} a_2 \\ b_2 \end{bmatrix} = \cos \omega_1 t \begin{bmatrix} 5 & -2 \\ -2 & 2 \end{bmatrix} \begin{bmatrix} a_1 \\ b_1 \end{bmatrix}$$
$$+ \cos \omega_2 t \begin{bmatrix} 5 & -2 \\ -2 & 2 \end{bmatrix} \begin{bmatrix} a_2 \\ b_2 \end{bmatrix}.$$

Since this equation is an identity in t, we must have

$$\omega_1^2 \begin{bmatrix} a_1 \\ b_1 \end{bmatrix} = \begin{bmatrix} 5 & -2 \\ -2 & 2 \end{bmatrix} \begin{bmatrix} a_1 \\ b_1 \end{bmatrix}$$

and

$$\omega_2^2 \begin{bmatrix} a_2 \\ b_2 \end{bmatrix} = \begin{bmatrix} 5 & -2 \\ -2 & 2 \end{bmatrix} \begin{bmatrix} a_2 \\ b_2 \end{bmatrix}.$$

Thus, ω_1^2 and ω_2^2 must be eigenvalues of the matrix

$$A = \begin{bmatrix} 5 & -2 \\ -2 & 2 \end{bmatrix},$$

and

$$\begin{bmatrix} a_i \\ b_i \end{bmatrix}$$

must be an eigenvector corresponding to ω_i^2.

The eigenvalues of A are found to be $\lambda_1 = 1, \lambda_2 = 6$. With $\omega_1 = 1$ and $\omega_2 = \sqrt{6}$, the eigenvectors are given by

$$\begin{bmatrix} a_1 \\ b_1 \end{bmatrix} = a_1 \begin{bmatrix} 1 \\ 2 \end{bmatrix} \quad \text{and} \quad \begin{bmatrix} a_2 \\ b_2 \end{bmatrix} = b_2 \begin{bmatrix} -2 \\ 1 \end{bmatrix}.$$

When the initial conditions $x_1 = -1, dx_1/dt = 0, x_2 = 2, dx_2/dt = 0$ are imposed, we find that $a_1 = \frac{3}{5}, b_2 = \frac{4}{5}$. The desired solution is given by

$$x_1 = \tfrac{3}{5} \cos t - \tfrac{8}{5} \cos \sqrt{6}t,$$
$$x_2 = \tfrac{6}{5} \cos t + \tfrac{4}{5} \cos \sqrt{6}t.$$

∎

7.1 Exercises

Find the eigenvalues of the given matrix A.

1. $A = \begin{bmatrix} -1 & 2 \\ -2 & 3 \end{bmatrix}$

2. $A = \begin{bmatrix} 1 & 2 \\ 2 & -2 \end{bmatrix}$

3. $A = \begin{bmatrix} 1 & 2 \\ 3 & 4 \end{bmatrix}$

4. $A = \begin{bmatrix} 1 & 2 \\ 5 & 4 \end{bmatrix}$

5. $A = \begin{bmatrix} 2 & -2 & 3 \\ 1 & 1 & 1 \\ 1 & 3 & -1 \end{bmatrix}$

6. $A = \begin{bmatrix} -1 & 2 & 2 \\ 2 & 2 & 2 \\ -3 & -6 & -6 \end{bmatrix}$

7. $A = \begin{bmatrix} 1 & -1 & -1 \\ 1 & 3 & 2 \\ -1 & -1 & 0 \end{bmatrix}$

8. $A = \begin{bmatrix} 1 & 1 & -1 \\ -1 & 3 & -1 \\ -1 & 2 & 0 \end{bmatrix}$

9. $A = \begin{bmatrix} 3 & 1 & -1 \\ -1 & 1 & 1 \\ 1 & 1 & 1 \end{bmatrix}$

10. $A = \begin{bmatrix} -1 & 2 & 2 \\ -2 & 3 & 2 \\ -1 & 1 & 2 \end{bmatrix}$

11. $A = \begin{bmatrix} 3 & 0 & 4 & 4 \\ 0 & -1 & 0 & 0 \\ 0 & -4 & -1 & -4 \\ 0 & 4 & 0 & 3 \end{bmatrix}$

12. $A = \begin{bmatrix} 2 & 0 & 0 & 0 \\ 0 & 1 & -1 & 1 \\ 1 & 0 & 1 & 0 \\ 1 & 0 & -1 & 2 \end{bmatrix}$

In Problems 13–16, find the spectrum of the given linear operator on $\mathbf{R}^n$.

13. $T(x_1, x_2) = (x_1 + 2x_2, 2x_1 - 2x_2)$

14. $T(x_1, x_2) = (2x_1 + x_2, x_1 - x_2)$

15. $T(x_1, x_2, x_3) = (x_1 + x_2 - x_3, -x_1 + 3x_2 - x_3, -x_1 + 2x_2)$

16. $T(x_1, x_2, x_3, x_4) = (x_1 + x_3 + x_4, 2x_1 + x_2 + 3x_3 + x_4, -x_1 - x_3 - x_4, 3x_1 + 2x_2 + 5x_3 + x_4)$

17. The linear operator T on $\mathbf{R}^2$ has the matrix

$$\begin{bmatrix} 4 & -5 \\ -4 & 3 \end{bmatrix}$$

relative to the basis $\{(1, 2), (0, 1)\}$. Find the eigenvalues of T, and obtain an eigenvector corresponding to each eigenvalue.

18. Let T be the linear operator on $\mathbf{R}^2$ with matrix

$$\begin{bmatrix} -1 & 3 \\ 4 & -2 \end{bmatrix}$$

relative to the basis $\{(1, 1), (2, -1)\}$. Find the eigenvalues of T, and obtain an eigenvector corresponding to each eigenvalue.

19. Let T be the linear operator on P_2 over $\mathbf{R}$ that has the matrix

$$A = \begin{bmatrix} 2 & 1 & 0 \\ 0 & 2 & 0 \\ 2 & 3 & 1 \end{bmatrix}$$

with respect to the basis $\{x^2, x - 2, x + 1\}$. Find the eigenvalues of T, and find an eigenvector of T corresponding to each eigenvalue.

20. Let T be the linear operator on $\mathbf{R}_{2 \times 2}$ with matrix

$$\begin{bmatrix} 1 & 0 & 0 & 1 \\ 1 & -1 & 0 & 0 \\ 0 & 0 & 1 & 1 \\ 1 & 0 & -1 & 0 \end{bmatrix},$$

relative to the basis

$$\left\{ \begin{bmatrix} 1 & -1 \\ 0 & 0 \end{bmatrix}, \begin{bmatrix} 0 & 1 \\ -1 & 0 \end{bmatrix}, \begin{bmatrix} 0 & 0 \\ 1 & 1 \end{bmatrix}, \begin{bmatrix} 0 & 1 \\ 0 & -1 \end{bmatrix} \right\}.$$

Find the eigenvalues of T, and obtain an eigenvector corresponding to each eigenvalue.

21. Find the eigenvalues and their corresponding eigenvectors of the matrix

$$A = \begin{bmatrix} 1 & 1 \\ 1 & 0 \end{bmatrix},$$

used to generate the Fibonacci sequence. (See Problem 6 in Exercises 3.2.)

In Problems 22–25, find the spectrum of the given linear operator T on $\mathbf{V}$ and find an eigenvector of T corresponding to each eigenvalue.

22. $\mathbf{V} = P_2$ over $\mathbf{R}$, $T(a_0 + a_1 x + a_2 x^2) = (2a_0 + a_1) + a_1 x + (2a_0 + 2a_1 + a_2)x^2$

23. $\mathbf{V} = P_2$ over $\mathbf{R}$, $T(a_0 + a_1 x + a_2 x^2) = (3a_0 - 2a_1 + a_2) + 2a_1 x + (a_0 - 2a_1 + 3a_2)x^2$

24. $\mathbf{V} = \mathbf{R}_{2 \times 2}$, $T\left(\begin{bmatrix} a_{11} & a_{12} \\ a_{21} & a_{22} \end{bmatrix} \right) = \begin{bmatrix} a_{11} & a_{11} + a_{12} \\ a_{12} + a_{21} & a_{22} \end{bmatrix}$

25. $\mathbf{V} = \mathbf{R}_{2 \times 2}$, $T\left(\begin{bmatrix} a_{11} & a_{12} \\ a_{21} & a_{22} \end{bmatrix} \right) = \begin{bmatrix} -2a_{11} - a_{12} & a_{11} \\ a_{21} & 2a_{22} \end{bmatrix}$

26. Given that T is a linear operator on $\mathbf{R}^2$ with $T(2, 1) = (5, 2)$ and $T(1, 2) = (7, 10)$, determine the eigenvalues of T and a corresponding eigenvector for each eigenvalue.

27. Given that T is the linear operator on $\mathbf{R}^3$ with $T(1, 1, 1) = (0, 0, 1)$, $T(1, 1, 0) = (1, 2, 1)$ and $T(1, 0, 0) = (0, -1, 0)$, determine the eigenvalues of T and a corresponding eigenvector for each eigenvalue.

28. Prove that a square matrix A is not invertible if and only if 0 is an eigenvalue of A.

29. Let A be an invertible matrix. Prove that if λ is an eigenvalue of A, then λ^{-1} is an eigenvalue of A^{-1}.

30. Assume that T is a linear operator on $\mathbf{R}^3$ that has eigenvalues 1, 2, and 3, with associated eigenvectors $(2, 1, 3)$, $(1, 4, 0)$, and $(1, 0, 0)$, respectively. Find the eigenvalues of T^{-1}, and give an eigenvector associated with each eigenvalue.

31. Prove that A and A^T have the same eigenvalues.

32. Prove or disprove: Equivalent matrices have the same eigenvalues.

33. (a) Let $A = [a_{ij}]_n$ such that $\sum_{j=1}^{n} a_{ij} = \lambda$ for all $i = 1, 2, \ldots, n$. Prove that λ is an eigenvalue of A.

(b) Let $A = \left[a_{ij}\right]_n$ such that $\sum_{i=1}^{n} a_{ij} = \lambda$ for all $j = 1, 2, \ldots, n$. Prove that λ is an eigenvalue of A. (A **probability matrix** is a square matrix of order n such that $a_{ij} \geq 0$ for all i, j and $\sum_{i=1}^{n} a_{ij} = 1$ for all $j = 1, 2, \ldots, n$. It follows that $\lambda = 1$ is an eigenvalue for any probability matrix.)

34. Prove or disprove each of the following statements.

(a) If λ_1 and λ_2 are eigenvalues of A, then $c_1\lambda_1 + c_2\lambda_2$ is an eigenvalue of A for any $c_1, c_2 \in \mathcal{F}$.

(b) If X_1 and X_2 are eigenvectors of A, then $c_1 X_1 + c_2 X_2$ is an eigenvector of A for any $c_1, c_2 \in \mathcal{F}$.

35. Suppose λ_1 is an eigenvalue of A and λ_2 is an eigenvalue of B. Prove or disprove each of the following statements.

(a) $\lambda_1 + \lambda_2$ is an eigenvalue of $A + B$.

(b) $\lambda_1\lambda_2$ is an eigenvalue of AB.

36. Prove that the only eigenvalue of a nilpotent matrix A is 0.

37. Show that the only eigenvalues of an idempotent matrix are 0 and 1.

38. Let λ be an eigenvalue of the linear operator T on $\mathbf{V}$ with corresponding eigenvector $\mathbf{v}$. Prove that if $T^k(\mathbf{v}) \neq 0$, then $T^k(\mathbf{v})$ is also an eigenvector of T corresponding to the eigenvalue λ for $k = 1, 2, \ldots$.

39. If 2 is an eigenvalue of T and $\mathbf{v} = (2, -1, 3)$ is a corresponding eigenvector, find an eigenvalue of the linear transformation $S = T^2 - 2T + 1$ ($1 = T^0$ as in Section 5.4.)

40. Let T be a linear operator on $\mathbf{V}$. Prove each of the following statements.

(a) If λ is an eigenvalue of T with associated eigenvector $\mathbf{v}$, then λ^k is an eigenvalue of T^k with associated eigenvector $\mathbf{v}$, where k is a positive integer.

(b) If λ is an eigenvalue of T with associated eigenvector $\mathbf{v}$, then $\lambda - c$ is an eigenvalue of $T - c$ with associated eigenvector $\mathbf{v}$ ($1 = T^0$ as in Section 5.4).

(c) Prove that if λ is an eigenvalue of T and if S is a polynomial in T given by $S = p(T) = \sum_{i=0}^{r} a_i T^i$, then $p(\lambda) = \sum_{i=0}^{r} a_i \lambda^i$ is an eigenvalue of S. Find an eigenvector of S corresponding to the eigenvalue $p(\lambda)$.

41. Translate the results of Problem 40 into statements concerning the eigenvalues and eigenvectors of a polynomial $p(A)$ in a square matrix A.

42. Let λ be a eigenvalue of A. Prove that if $\text{adj}(A - \lambda I) \neq \mathbf{0}$, then a nonzero column of $\text{adj}(A - \lambda I)$ is an eigenvector of A corresponding to the eigenvalue λ.

7.2 Eigenspaces and Similarity

If $\mathbf{v}$ is an eigenvector associated with the eigenvalue λ of T, then $T(c\mathbf{v}) = cT(\mathbf{v}) = c\lambda\mathbf{v} = \lambda(c\mathbf{v})$ for every choice of the scalar c. Thus, there are many eigenvectors associated with the same eigenvalue. The next theorem gives a great deal of insight into this collection of eigenvectors.

7.6 **THEOREM**	For each eigenvalue λ of T in $\mathcal{F}$, let $\mathbf{V}_\lambda$ be the set consisting of the zero vector together with all eigenvectors of T in $\mathbf{V}$ that are associated with λ. Then each $\mathbf{V}_\lambda$ is a subspace of $\mathbf{V}$.

Proof The zero vector is in $\mathbf{V}_\lambda$, so $\mathbf{V}_\lambda$ is nonempty. Let $\mathbf{v}_1, \mathbf{v}_2 \in \mathbf{V}_\lambda$ and $a, b \in \mathcal{F}$. Then

$$T(a\mathbf{v}_1 + b\mathbf{v}_2) = aT(\mathbf{v}_1) + bT(\mathbf{v}_2) = a\lambda\mathbf{v}_1 + b\lambda\mathbf{v}_2 = \lambda(a\mathbf{v}_1 + b\mathbf{v}_2),$$

so $a\mathbf{v}_1 + b\mathbf{v}_2 \in \mathbf{V}_\lambda$, and $\mathbf{V}_\lambda$ is a subspace of $\mathbf{V}$. ∎

7.7 **DEFINITION**	The subspace $\mathbf{V}_\lambda$ of Theorem 7.6 is called the **eigenspace** of T that is associated with the eigenvalue λ. The dimension of $\mathbf{V}_\lambda$ is called the **geometric multiplicity** of λ.

There is another approach that we could have taken in proving the last theorem. For $T(\mathbf{v}) = \lambda\mathbf{v}$ if and only if $(T - \lambda)\mathbf{v} = \mathbf{0}$, i.e., if and only if $\mathbf{v}$ is in the kernel of $T - \lambda$. Thus $\mathbf{V}_\lambda$ is nothing more than the kernel of $T - \lambda$, and the geometric multiplicity of λ is precisely the dimension of the kernel of $T - \lambda$. This translates the problem of finding an eigenspace into the familiar problem of finding a kernel. In Example 4 of Section 7.1, the geometric multiplicity of $\lambda = -2$ is 2 since

$$\mathbf{V}_{-2} = (T + 2)^{-1}(\mathbf{0}) = \langle(-\tfrac{1}{2}, 1, 0, 0), (-12, 0, 20, 1)\rangle$$

and the geometric multiplicity of $\lambda = 2$ is 1 since

$$\mathbf{V}_2 = (T - 2)^{-1}(\mathbf{0}) = \langle(0, 0, 0, 1)\rangle.$$

Concerning those eigenvectors of T that are associated with different eigenvalues, we have the following result.

7.8 **THEOREM**	Let $\{\lambda_1, \lambda_2, \ldots, \lambda_r\}$ be a set of distinct eigenvalues of the linear operator T. For each $i, 1 \leq i \leq r$, let $\mathbf{v}_i$ be an eigenvector of T corresponding to λ_i. Then $\{\mathbf{v}_1, \mathbf{v}_2, \ldots, \mathbf{v}_r\}$ is a linearly independent set.

Proof The proof is by induction on r. The theorem is true for $r = 1$ since any eigenvector is nonzero.

Assume that the theorem is true for any set of k distinct eigenvalues. Let the set $\{\lambda_1, \ldots, \lambda_k, \lambda_{k+1}\}$ be a set of $k + 1$ distinct eigenvalues of T, with $\mathbf{v}_i$ an eigenvector corresponding to λ_i. Suppose that $c_1, \ldots, c_k, c_{k+1}$ are scalars such that

$$c_1\mathbf{v}_1 + \cdots + c_k\mathbf{v}_k + c_{k+1}\mathbf{v}_{k+1} = \mathbf{0}. \tag{7.1}$$

Applying T to each member and using $T(\mathbf{v}_i) = \lambda_i\mathbf{v}_i$, we have

$$c_1\lambda_1\mathbf{v}_1 + \cdots + c_k\lambda_k\mathbf{v}_k + c_{k+1}\lambda_{k+1}\mathbf{v}_{k+1} = \mathbf{0}. \tag{7.2}$$

Multiplying (7.1) by λ_{k+1}, we have

$$c_1\lambda_{k+1}\mathbf{v}_1 + \cdots + c_k\lambda_{k+1}\mathbf{v}_k + c_{k+1}\lambda_{k+1}\mathbf{v}_{k+1} = \mathbf{0}. \tag{7.3}$$

Subtracting (7.3) from (7.2), we obtain

$$c_1(\lambda_1 - \lambda_{k+1})\mathbf{v}_1 + \cdots + c_k(\lambda_k - \lambda_{k+1})\mathbf{v}_k = \mathbf{0}.$$

From the induction hypothesis, $\{\mathbf{v}_1, \ldots, \mathbf{v}_k\}$ is linearly independent so that

$$c_i(\lambda_i - \lambda_{k+1}) = 0$$

for $i = 1, 2, \ldots, k$. Since $\lambda_i - \lambda_{k+1} \neq 0$ for $i = 1, 2, \ldots, k$, we conclude that

$$c_1 = c_2 = \cdots = c_k = 0.$$

Thus in (7.1) we have $c_{k+1}\mathbf{v}_{k+1} = \mathbf{0}$, and hence $c_{k+1} = 0$. This shows that the set of $k + 1$ eigenvectors is linearly independent, and it follows that the theorem is true for all positive integers r. ∎ ■■■

7.9 COROLLARY

If $\mathbf{V}_{\lambda_1}, \mathbf{V}_{\lambda_2}, \ldots, \mathbf{V}_{\lambda_r}$ are distinct eigenspaces of T and $\{\mathbf{v}_1, \mathbf{v}_2, \ldots, \mathbf{v}_r\}$ is a set of eigenvectors such that $\mathbf{v}_i \in \mathbf{V}_{\lambda_i}$ for $i = 1, 2, \ldots, r$, then $\{\mathbf{v}_1, \mathbf{v}_2, \ldots, \mathbf{v}_r\}$ is linearly independent.

Proof The eigenvalues $\lambda_1, \lambda_2, \ldots, \lambda_r$ must be distinct in order for the eigenspaces $\mathbf{V}_{\lambda_1}, \mathbf{V}_{\lambda_2}, \ldots, \mathbf{V}_{\lambda_r}$ to be distinct. Since $\mathbf{v}_i$ is an eigenvector associated with λ_i, then the set $\{\mathbf{v}_1, \mathbf{v}_2, \ldots, \mathbf{v}_r\}$ is linearly independent by the theorem. ■■■

We recall that a sum $\mathbf{W}_1 + \mathbf{W}_2 + \cdots + \mathbf{W}_r$ of subspaces $\mathbf{W}_i$ of $\mathbf{V}$ is called *direct* if

$$\mathbf{W}_i \cap \sum_{\substack{j=1 \\ j \neq i}}^{r} \mathbf{W}_j = \{\mathbf{0}\}$$

for $i = 1, 2, \ldots, r$.

7.10
COROLLARY

If $\mathbf{V}_{\lambda_1}, \mathbf{V}_{\lambda_2}, \ldots, \mathbf{V}_{\lambda_r}$ are distinct eigenspaces of T, the sum $\mathbf{V}_{\lambda_1} + \mathbf{V}_{\lambda_2} + \cdots + \mathbf{V}_{\lambda_r}$ is direct.

Proof Suppose that the sum is not direct, and let $\mathbf{v}_k$ be a nonzero vector in $\mathbf{V}_{\lambda_k}$ that is also contained in $\sum_{\substack{j=1 \\ j \neq k}}^{r} \mathbf{V}_{\lambda_j}$. Then there are vectors $\mathbf{v}_j$ in $\mathbf{V}_{\lambda_j}$ and scalars a_j such that

$$\mathbf{v}_k = \sum_{\substack{j=1 \\ j \neq k}}^{r} a_j \mathbf{v}_j.$$

That is, $\mathbf{v}_k$ is linearly dependent on the set

$$\mathcal{A} = \{\mathbf{v}_1, \ldots, \mathbf{v}_{k-1}, \mathbf{v}_{k+1}, \ldots, \mathbf{v}_r\}.$$

Since $\mathbf{v}_k \neq \mathbf{0}$, there are nonzero vectors in $\mathcal{A}$. Let $\mathcal{A}' = \{\mathbf{v}_1', \mathbf{v}_2', \ldots, \mathbf{v}_t'\}$ be the nonempty set obtained by deleting all zero vectors from $\mathcal{A}$. Then $\mathbf{v}_k$ is dependent on $\mathcal{A}'$ so that the set $\{\mathbf{v}_1', \mathbf{v}_2', \ldots, \mathbf{v}_t', \mathbf{v}_k\}$ is linearly dependent. But $\{\mathbf{v}_1', \mathbf{v}_2', \ldots, \mathbf{v}_t', \mathbf{v}_k\}$ is a set of eigenvectors that satisfies the hypothesis of Corollary 7.9. Thus we have a contradiction, and it follows that the sum $\sum_{i=1}^{r} \mathbf{V}_{\lambda_i}$ is direct. ■ ■ ■

Suppose that T has matrix A relative to the basis $\mathcal{A}$ of $\mathbf{V}$, and that T has matrix B relative to the basis $\mathcal{A}'$ of $\mathbf{V}$. If P is the (invertible) matrix of transition from $\mathcal{A}$ to $\mathcal{A}'$, then Theorem 5.15 asserts that $B = P^{-1}AP$. This leads to the following definition.

7.11
DEFINITION

Let A and B be $n \times n$ matrices over $\mathcal{F}$. Then B is **similar** to A over $\mathcal{F}$ if there is an invertible matrix P with elements in $\mathcal{F}$ such that $B = P^{-1}AP$.

It is left as an exercise (Problem 19) to show that this relation of similarity is a true equivalence relation on the set of $n \times n$ matrices over $\mathcal{F}$. This relation proves to be a useful tool in the investigation of the eigenvalues of matrices.

The remarks just before Definition 7.11 show that two $n \times n$ matrices over $\mathcal{F}$ are similar over $\mathcal{F}$ if and only if they represent the same linear operator on an n-dimensional vector space $\mathbf{V}$ over $\mathcal{F}$.

The strong connection between the relation of similarity and the eigenvalues of matrices becomes apparent in our next theorem.

7.12
THEOREM

Similar matrices have the same characteristic polynomial.

Proof If B is similar to A over $\mathcal{F}$, then there is an invertible matrix P such that $B = P^{-1}AP$. Thus

$$
\begin{aligned}
\det(B - xI) &= \det(P^{-1}AP - xI) \\
&= \det(P^{-1}AP - xP^{-1}P) \\
&= \det\left(P^{-1}(A - xI)P\right) \\
&= \det(P^{-1}) \cdot \det(A - xI) \cdot \det(P) \\
&= \det(P^{-1}P) \cdot \det(A - xI) \\
&= \det(A - xI),
\end{aligned}
$$

so that B and A have the same characteristic polynomial. ■ ■ ■

7.13 COROLLARY

Similar matrices have the same spectrum.

Near the beginning of this section, we defined the geometric multiplicity of an eigenvalue of a linear transformation. There is a second type of multiplicity for eigenvalues, the *algebraic multiplicity*.

7.14 DEFINITION

Let λ be an eigenvalue of T, and let A be any matrix that represents T. The **algebraic multiplicity** of λ is the multiplicity of λ as a root of $\det(A - xI) = 0$.

From Theorem 7.12 and our discussion concerning Definition 7.11, it is clear that the algebraic multiplicity of an eigenvalue is well-defined. That is, the algebraic multiplicity is independent of the choice of the matrix A.

EXAMPLE 1 In Example 4 of Section 7.1, the linear operator T had the two distinct eigenvalues $\lambda_1 = -2$ and $\lambda_3 = 2$. Examining the characteristic polynomial of A, we see that the algebraic multiplicity of each eigenvalue is 2. Upon comparing the algebraic and geometric multiplicities, we find that the two are equal for λ_1, but that the algebraic multiplicity of λ_3 exceeds the geometric multiplicity, which is 1. Our next theorem shows that the situation in this example illustrates the only possibilities. ■

7.15 THEOREM

The geometric multiplicity of an eigenvalue does not exceed its algebraic multiplicity.

Proof Suppose that the geometric multiplicity of an eigenvalue λ of T is r, and let $\{v_1, v_2, \ldots, v_r\}$ be a basis of $\mathbf{V}_\lambda$. This basis of $\mathbf{V}_\lambda$ can be extended to a basis $\mathcal{A} = \{v_1, v_2, \ldots, v_r, \ldots, v_n\}$ of $\mathbf{V}$. The matrix of T relative to this basis is

$$A = \begin{bmatrix} \lambda & 0 & \cdots & 0 & a_{1,r+1} & \cdots & a_{1n} \\ 0 & \lambda & \cdots & 0 & a_{2,r+1} & \cdots & a_{2n} \\ \vdots & \vdots & \ddots & \vdots & \vdots & & \vdots \\ 0 & 0 & \cdots & \lambda & a_{r,r+1} & \cdots & a_{rn} \\ \vdots & \vdots & & \vdots & \vdots & & \vdots \\ 0 & 0 & \cdots & 0 & a_{n,r+1} & \cdots & a_{nn} \end{bmatrix},$$

and

$$\det(A - xI) = (\lambda - x)^r \begin{vmatrix} a_{r+1,r+1} - x & \cdots & a_{r+1,n} \\ \vdots & \ddots & \vdots \\ a_{n,r+1} & \cdots & a_{nn} - x \end{vmatrix}.$$

Thus the algebraic multiplicity of λ is at least r. That is, the geometric multiplicity does not exceed the algebraic multiplicity. ∎

EXAMPLE 2 Let T be the linear operator on $\mathcal{P}_2$ over $\mathbf{R}$ defined by

$$T(a_0 + a_1 x + a_2 x^2) = (-4a_0 - a_1 - a_2) + (4a_0 + 2a_2)x + (2a_0 + a_1 - a_2)x^2.$$

We shall find the algebraic and geometric multiplicity of each eigenvalue of T and obtain a basis for each eigenspace. The set $\{1, x, x^2\}$ is a more convenient choice of basis, but we shall use the basis

$$\mathcal{A} = \{1, 1 + x, 1 + x + x^2\}$$

to emphasize the difference between the eigenvectors of T and the eigenvectors of the matrix A that represents T relative to $\mathcal{A}$.

Since

$$T(1) = -4 + 4x + 2x^2$$
$$= (-8)(1) + 2(1 + x) + 2(1 + x + x^2),$$
$$T(1 + x) = -5 + 4x + 3x^2$$
$$= (-9)(1) + 1(1 + x) + 3(1 + x + x^2),$$
$$T(1 + x + x^2) = -6 + 6x + 2x^2$$
$$= (-12)(1) + 4(1 + x) + 2(1 + x + x^2),$$

the matrix of T relative to $\mathcal{A}$ is

$$A = [T]_{\mathcal{A}} = \begin{bmatrix} -8 & -9 & -12 \\ 2 & 1 & 4 \\ 2 & 3 & 2 \end{bmatrix}.$$

We find that the characteristic polynomial of A is

$$\det(A - xI) = \begin{vmatrix} -8-x & -9 & -12 \\ 2 & 1-x & 4 \\ 2 & 3 & 2-x \end{vmatrix}$$

$$= -(x^3 + 5x^2 + 8x + 4)$$

$$= -(x+1)(x^2 + 4x + 4)$$

$$= -(x+1)(x+2)^2.$$

Thus the eigenvalues of T and A are -2 with algebraic multiplicity 2 and -1 with algebraic multiplicity 1.

For the eigenvalue $\lambda = -2$, the system $(A + 2I)X = \mathbf{0}$ appears as

$$\begin{bmatrix} -6 & -9 & -12 \\ 2 & 3 & 4 \\ 2 & 3 & 4 \end{bmatrix} \begin{bmatrix} x_1 \\ x_2 \\ x_3 \end{bmatrix} = \begin{bmatrix} 0 \\ 0 \\ 0 \end{bmatrix}.$$

The reduced row-echelon form for the corresponding augmented matrix is

$$\begin{bmatrix} 1 & \frac{3}{2} & 2 & 0 \\ 0 & 0 & 0 & 0 \\ 0 & 0 & 0 & 0 \end{bmatrix}.$$

From this, we see that the solutions to $(A+2I)X = \mathbf{0}$ are given by $x_1 = -\frac{3}{2}x_2 - 2x_3$, with x_2 and x_3 arbitrary. That is, the coordinates of vectors in the eigenspace $\mathbf{V}_{-2}$ are given by the eigenvectors

$$X = \begin{bmatrix} x_1 \\ x_2 \\ x_3 \end{bmatrix} = \begin{bmatrix} -\frac{3}{2}x_2 - 2x_3 \\ x_2 \\ x_3 \end{bmatrix} = x_2 \begin{bmatrix} -\frac{3}{2} \\ 1 \\ 0 \end{bmatrix} + x_3 \begin{bmatrix} -2 \\ 0 \\ 1 \end{bmatrix}$$

of A. Thus $\mathbf{V}_{-2}$ has dimension 2. We can find coordinates for a basis of $\mathbf{V}_{-2}$ by first setting $x_2 = -2$ and $x_3 = 0$ to obtain

$$X = \begin{bmatrix} 3 \\ -2 \\ 0 \end{bmatrix}$$

and then setting $x_2 = 0$ and $x_3 = -1$ to obtain

$$X = \begin{bmatrix} 2 \\ 0 \\ -1 \end{bmatrix}.$$

(Any other linearly independent pair of coordinate matrices X would serve as well, of course.) Corresponding to these coordinates, we have the vectors $3(1) + (-2)(1 + x) = 1 - 2x$ and $2(1) + (-1)(1 + x + x^2) = 1 - x - x^2$ that form a basis of $\mathbf{V}_{-2}$.

For the eigenvalue $\lambda = -1$, $(A + I)X = \mathbf{0}$ is given by

$$\begin{bmatrix} -7 & -9 & -12 \\ 2 & 2 & 4 \\ 2 & 3 & 3 \end{bmatrix} \begin{bmatrix} x_1 \\ x_2 \\ x_3 \end{bmatrix} = \begin{bmatrix} 0 \\ 0 \\ 0 \end{bmatrix}.$$

The reduced row-echelon form this time is

$$\begin{bmatrix} 1 & 0 & 3 & 0 \\ 0 & 1 & -1 & 0 \\ 0 & 0 & 0 & 0 \end{bmatrix}$$

and the coordinates of vectors in $\mathbf{V}_{-1}$ are given by

$$X = \begin{bmatrix} x_1 \\ x_2 \\ x_3 \end{bmatrix} = x_3 \begin{bmatrix} -3 \\ 1 \\ 1 \end{bmatrix}.$$

Choosing $x_3 = -1$ yields $X = [\,3 \quad -1 \quad -1\,]^T$ and $\{1 - 2x - x^2\}$ as a basis of $\mathbf{V}_{-1}$.
These results are summarized as follows:

Eigenvalue	Multiplicity		Eigenspace
	Geometric	Algebraic	
-2	2	2	$\mathbf{V}_{-2} = \langle 1 - 2x, 1 - x - x^2 \rangle$
-1	1	1	$\mathbf{V}_{-1} = \langle 1 - 2x - x^2 \rangle$

7.2 Exercises

In Problems 1–6, let T be the linear operator on $\mathbf{R}^n$ that is represented by the given matrix A relative to the standard basis of $\mathbf{R}^n$. Find the algebraic multiplicity and the geometric multiplicity of each eigenvalue of T. The matrices here are taken from Problems 1–12 in Exercises 7.1.

1. $A = \begin{bmatrix} -1 & 2 \\ -2 & 3 \end{bmatrix}$

2. $A = \begin{bmatrix} 1 & 2 \\ 2 & -2 \end{bmatrix}$

3. $A = \begin{bmatrix} 1 & -1 & -1 \\ 1 & 3 & 2 \\ -1 & -1 & 0 \end{bmatrix}$ **4.** $A = \begin{bmatrix} 3 & 1 & -1 \\ -1 & 1 & 1 \\ 1 & 1 & 1 \end{bmatrix}$

5. $A = \begin{bmatrix} 3 & 0 & 4 & 4 \\ 0 & -1 & 0 & 0 \\ 0 & -4 & -1 & -4 \\ 0 & 4 & 0 & 3 \end{bmatrix}$ **6.** $A = \begin{bmatrix} 2 & 0 & 0 & 0 \\ 0 & 1 & -1 & 1 \\ 1 & 0 & 1 & 0 \\ 1 & 0 & -1 & 2 \end{bmatrix}$

7. Let T be the linear operator on $\mathbf{R}^2$ that has the matrix

$$A = \begin{bmatrix} 1 & 2 \\ 2 & -2 \end{bmatrix}$$

relative to the standard basis of $\mathbf{R}^2$.

(a) Find eigenvectors $\mathbf{v}_1$, $\mathbf{v}_2$ of T such that $\{\mathbf{v}_1, \mathbf{v}_2\}$ is a basis of $\mathbf{R}^2$.

(b) Find the matrix of T relative to this basis.

8. Let T be the linear operator on $\mathbf{R}^3$ defined by

$$T(x_1, x_2, x_3) = (x_1 + x_2 + x_3, x_1 + x_2, x_1 - x_2).$$

(a) Find eigenvectors $\mathbf{v}_1$, $\mathbf{v}_2$, $\mathbf{v}_3$ of T such that $\{\mathbf{v}_1, \mathbf{v}_2, \mathbf{v}_3\}$ is a basis of $\mathbf{R}^3$.

(b) Find the matrix of T relative to this basis.

9. Suppose that the basis $\mathcal{A} = \{\mathbf{v}_1, \mathbf{v}_2, \ldots, \mathbf{v}_n\}$ of $\mathbf{V}$ consists entirely of eigenvectors of T. Determine the matrix of T relative to $\mathcal{A}$.

10. The linear operator T on $\mathbf{R}^2$ has the matrix

$$\begin{bmatrix} 5 & -2 \\ -2 & 2 \end{bmatrix}$$

relative to the basis $\mathcal{A} = \{(3, 3), (1, -1)\}$. Find the eigenvalues of T and obtain an eigenvector of T corresponding to each eigenvalue.

11. For each matrix A below, let T be the linear operator on $\mathbf{R}^3$ that has matrix A relative to the basis $\mathcal{A} = \{(1, 0, 0), (1, 1, 0), (1, 1, 1)\}$. Find the algebraic and geometric multiplicities of each eigenvalue, and a basis for each eigenspace.

(a) $A = \begin{bmatrix} 8 & 5 & -5 \\ 5 & 8 & -5 \\ 15 & 15 & -12 \end{bmatrix}$ **(b)** $A = \begin{bmatrix} -4 & -3 & -1 \\ -4 & 0 & -4 \\ 8 & 4 & 5 \end{bmatrix}$

(c) $A = \begin{bmatrix} 3 & 2 & 2 \\ 1 & 4 & 1 \\ -2 & -4 & -1 \end{bmatrix}$ **(d)** $A = \begin{bmatrix} 8 & 5 & 6 \\ 0 & -2 & 0 \\ -10 & -5 & -8 \end{bmatrix}$

12. Let T be a linear operator on $\mathcal{P}_2$ that has matrix

$$\begin{bmatrix} -2 & 0 & 1 \\ 0 & -3 & 0 \\ -1 & 0 & -4 \end{bmatrix}$$

relative to the basis $\mathcal{A} = \{3 + x, 1 + x^2, x - x^2\}$.

(a) Find the algebraic and geometric multiplicities of each eigenvalue.

(b) Find a basis of each eigenspace.

13. Let T be a linear operator on $\mathbf{R}_{2\times 2}$ that has matrix

$$\begin{bmatrix} -1 & 1 & -4 & -5 \\ -3 & 3 & -4 & -5 \\ 0 & 0 & 2 & 0 \\ 3 & -3 & 0 & 5 \end{bmatrix}$$

relative to the basis

$$\mathcal{A} = \left\{ \begin{bmatrix} 1 & 0 \\ 0 & 0 \end{bmatrix}, \begin{bmatrix} 1 & 0 \\ 1 & 0 \end{bmatrix}, \begin{bmatrix} 1 & 1 \\ 1 & 0 \end{bmatrix}, \begin{bmatrix} 1 & 1 \\ 1 & 1 \end{bmatrix} \right\}.$$

(a) Find the algebraic and geometric multiplicities of each eigenvalue.

(b) Find a basis of each eigenspace.

14. Let $\mathcal{C}$ denote the field of complex numbers, and let $\mathcal{C}^n$ be as defined in Example 1 of Section 4.1. Let $\mathcal{E}_n$ be the basis $\mathcal{E}_n = \{\mathbf{e}_1, \mathbf{e}_2, \ldots, \mathbf{e}_n\}$ of $\mathcal{C}^n$, where $\mathbf{e}_1 = (1, 0, 0, \ldots, 0)$, $\mathbf{e}_2 = (0, 1, 0, \ldots, 0)$, etc. If T is the linear operator that has the given matrix A relative to $\mathcal{E}_n$, find the algebraic and geometric multiplicities of each eigenvalue, and a basis for each eigenspace.

(a) $A = \begin{bmatrix} 2 & 1-i \\ 1+i & 3 \end{bmatrix}$ **(b)** $A = \begin{bmatrix} 5 & i \\ -i & 2 \end{bmatrix}$

(c) $A = \begin{bmatrix} 3 & 4 & 2 \\ 1 & 3 & 1 \\ 1 & 2 & 2 \end{bmatrix}$ **(d)** $A = \begin{bmatrix} i & 0 & 0 \\ -2i & i & -2+i \\ 0 & 0 & -2 \end{bmatrix}$

(e) $A = \begin{bmatrix} 1+i & 0 & 0 \\ -2i & 1+i & 2i \\ i & 0 & 1 \end{bmatrix}$ **(f)** $A = \begin{bmatrix} -2+2i & 0 & -2+i \\ 0 & -i & 0 \\ 4-2i & 0 & 4-i \end{bmatrix}$

*In Problems 15–18, find the eigenvalues of the given linear operator T on $\mathcal{P}_2$ over $\mathbf{R}$. For each eigenvalue, (**a**) state the algebraic multiplicity, (**b**) state the geometric multiplicity, and (**c**) find a basis for each eigenspace.*

15. $T(a_0 + a_1 x + a_2 x^2) = (3a_0 + a_1 + a_2) + 2a_1 x + 2a_2 x^2$

16. $T(a_0 + a_1 x + a_2 x^2) = 2a_0 + (3a_0 + a_1 + 2a_2)x + (3a_0 - a_1 + 4a_2)x^2$

17. $T(a_0 + a_1 x + a_2 x^2) = (a_0 + 3a_1 + 2a_2) + 2a_1 x + (a_1 + 2a_2)x^2$

18. $T(a_0 + a_1 x + a_2 x^2) = (2a_0 + a_1) + 2a_1 x + (2a_0 + 3a_1 + a_2)x^2$

19. Prove that the relation of similarity over $\mathcal{F}$ is an equivalence relation on the set of all $n \times n$ matrices over $\mathcal{F}$.

20. Which $n \times n$ matrices over $\mathcal{F}$ are similar to the given matrix?

(a) I_n **(b)** cI_n, where $c \in \mathcal{F}$ **(c)** $\mathbf{0}_n$

21. If A and B are invertible matrices and B is similar to A, prove that B^{-1} is similar to A^{-1}.

22. Prove that if B is similar to A, then B^T is similar to A^T.

23. Prove or disprove: A and A^T have the same eigenspaces.

24. Prove that if A and B are $n \times n$ matrices over $\mathcal{F}$ with A invertible, then BA is similar over $\mathcal{F}$ to AB.

25. Prove that if B is similar to A over $\mathcal{F}$, then $p(B) = \sum_{i=0}^{k} a_i B^i$ is similar to $p(A) = \sum_{i=0}^{k} a_i A^i$ for any $a_0, a_1, \ldots, a_k \in \mathcal{F}$.

26. Prove that similar matrices have the same rank.

27. Prove that if B is similar to A, then $\det(B) = \det(A)$.

28. Let $B = P^{-1}AP$ and suppose that X is an eigenvector of A corresponding to the eigenvalue λ. Show that λ is an eigenvalue of B, and find a corresponding eigenvector.

29. For any square matrix $A = [a_{ij}]_n$, the **trace** of A, $t(A)$, is defined by $t(A) = \sum_{i=1}^{n} a_{ii}$. That is, $t(A)$ is the sum of the diagonal elements of A. Prove that if B is similar to A, then $t(B) = t(A)$.

7.3 Representation by a Diagonal Matrix

The simplest form that the matrix of a linear operator can have is that of a scalar matrix. For if cI_n is the matrix of T relative to the basis $\mathcal{A} = \{\mathbf{v}_1, \mathbf{v}_2, \ldots, \mathbf{v}_n\}$ of $\mathbf{V}$, then $T(\mathbf{v}_i) = c\mathbf{v}_i$ for each i, and this implies that for any $\mathbf{v} = \sum_{i=1}^{n} a_i \mathbf{v}_i$ in $\mathbf{V}$,

$$T(\mathbf{v}) = \sum_{i=1}^{n} a_i T(\mathbf{v}_i) = \sum_{i=1}^{n} a_i c \mathbf{v}_i = c \sum_{i=1}^{n} a_i \mathbf{v}_i = c\mathbf{v}.$$

That is, c is an eigenvalue of T and every nonzero vector in $\mathbf{V}$ is an eigenvector corresponding to c. If $c > 0$, T can be described geometrically as an expansion about the origin if $c \geq 1$ and as a contraction about the origin if $c \leq 1$. If $c < 0$, T can be described as a reflection through the origin followed by an expansion or a contraction.

The next simplest form for a matrix is a diagonal matrix (of which the scalar matrix is a special case). If a linear operator has a diagonal matrix relative to a certain basis, this diagonal matrix displays at a glance the essential features of the transformation. With $\mathbf{v}$ confined to an eigenspace $\mathbf{V}_\lambda$, $T(\mathbf{v}) = \lambda \mathbf{v}$ so that T maps $\mathbf{V}_\lambda$ in the same fashion as a scalar linear operator maps the entire space $\mathbf{V}$.

Although the class of linear operators that can be represented by a diagonal matrix is quite large, not all linear operators can be represented in this way. Our primary objective in this section is to characterize those linear operators that can be represented by a diagonal matrix. This characterization is given in the next theorem.

7.16
THEOREM

The linear operator T on $\mathbf{V}$ can be represented by a diagonal matrix if and only if there is a basis of $\mathbf{V}$ that consists entirely of eigenvectors of T.

Proof Suppose that $\mathcal{A} = \{\mathbf{v}_1, \mathbf{v}_2, \ldots, \mathbf{v}_n\}$ is a basis of $\mathbf{V}$ such that each $\mathbf{v}_i$ is an eigenvector of T with λ_i as the corresponding eigenvalue. Then $T(\mathbf{v}_i) = \lambda_i \mathbf{v}_i$ and the matrix of T relative to $\mathcal{A}$ is

$$[T]_{\mathcal{A}} = \begin{bmatrix} \lambda_1 & 0 & \cdots & 0 \\ 0 & \lambda_2 & \cdots & 0 \\ \vdots & \vdots & \ddots & \vdots \\ 0 & 0 & \cdots & \lambda_n \end{bmatrix}.$$

Thus T is represented by a diagonal matrix relative to $\mathcal{A}$.

On the other hand, if T has a diagonal matrix

$$D = \begin{bmatrix} d_1 & 0 & \cdots & 0 \\ 0 & d_2 & \cdots & 0 \\ \vdots & \vdots & \ddots & \vdots \\ 0 & 0 & \cdots & d_n \end{bmatrix},$$

relative to the basis $\{\mathbf{v}_1, \mathbf{v}_2, \ldots, \mathbf{v}_n\}$ of $\mathbf{V}$, then $T(\mathbf{v}_i) = d_i \mathbf{v}_i$ so that each d_i is an eigenvalue of T with $\mathbf{v}_i$ as an associated eigenvector. ∎

There are several corollaries that are worthy of mention.

7.17
COROLLARY

If T is represented by a diagonal matrix, the elements on the diagonal are the eigenvalues of T.

Proof This follows at once from the last part of the proof of the theorem. ∎

7.18
COROLLARY

If T has n distinct eigenvalues in $\mathcal{F}$, then T can be represented by a diagonal matrix.

Proof Suppose that T has n distinct eigenvalues $\lambda_1, \lambda_2, \ldots, \lambda_n$ in $\mathcal{F}$. Consider a set $\mathcal{A} = \{\mathbf{v}_1, \mathbf{v}_2, \ldots, \mathbf{v}_n\}$ of n vectors in $\mathbf{V}$ that contains exactly one eigenvector corresponding to each λ_i. The set $\mathcal{A}$ is linearly independent by Theorem 7.8 and therefore forms a basis of the n-dimensional vector space $\mathbf{V}$. ∎

7.19
COROLLARY

If the $n \times n$ matrix A over $\mathcal{F}$ has n distinct eigenvalues in $\mathcal{F}$, then A is similar over $\mathcal{F}$ to a diagonal matrix.

Proof If A is an $n \times n$ matrix over $\mathcal{F}$ with n distinct eigenvalues in $\mathcal{F}$, then any linear operator that A represents has n distinct eigenvalues, by Theorem 7.4. But such a linear operator can be represented by a diagonal matrix, and this diagonal matrix is similar to A (by Definition 7.11 and Theorem 5.15). ■ ■ ■

7.20
THEOREM

Suppose that all eigenvalues of T are in $\mathcal{F}$. Then T can be represented by a diagonal matrix if and only if the geometric multiplicity of each eigenvalue of T is equal to the algebraic multiplicity.

Proof Let T be a linear operator on the n-dimensional vector space $\mathbf{V}$. In view of Theorem 7.16, it is sufficient to show that there exists a basis of eigenvectors of T if and only if the geometric and algebraic multiplicities of each eigenvalue are equal. As stated in the theorem, all eigenvalues of T are assumed to be in $\mathcal{F}$.

Let $\lambda_1, \lambda_2, \ldots, \lambda_r$ be the distinct eigenvalues of T, let $n_i = \dim(\mathbf{V}_{\lambda_i})$ be the geometric multiplicity of λ_i, and let m_i be the algebraic multiplicity of λ_i. Since m_i is the multiplicity of λ_i as a zero of the nth-degree polynomial $\det(A - xI)$, $\sum_{i=1}^{r} m_i = n$. According to Theorem 7.15, $0 < n_i \leq m_i$ for each i. Hence $\sum_{i=1}^{r} n_i = n$ if and only if $n_i = m_i$ for each i.

Now let $\mathcal{B}_i = \{\mathbf{u}_{i1}, \mathbf{u}_{i2}, \ldots, \mathbf{u}_{in_i}\}$ be a basis of $\mathbf{V}_{\lambda_i}$ for each i, and put

$$\mathcal{B} = \{\mathbf{u}_{11}, \ldots, \mathbf{u}_{1n_1}, \mathbf{u}_{21}, \ldots, \mathbf{u}_{2n_2}, \ldots, \mathbf{u}_{r1}, \ldots, \mathbf{u}_{rn_r}\}.$$

The set $\mathcal{B}$ contains $\sum_{i=1}^{r} n_i$ vectors and clearly spans

$$\mathbf{V}_{\lambda_1} + \mathbf{V}_{\lambda_2} + \cdots + \mathbf{V}_{\lambda_r}.$$

The sum $\sum_{i=1}^{r} \mathbf{V}_{\lambda_i}$ is direct by Corollary 7.10, and therefore

$$\dim \left(\sum_{i=1}^{r} \mathbf{V}_{\lambda_i} \right) = \sum_{i=1}^{r} \dim \left(\mathbf{V}_{\lambda_i} \right) = \sum_{i=1}^{r} n_i.$$

Hence $\mathcal{B}$ is a basis of $\sum_{i=1}^{r} \mathbf{V}_{\lambda_i}$.

Assume that there exists a basis $\{\mathbf{v}_1, \mathbf{v}_2, \ldots, \mathbf{v}_n\}$ of eigenvectors of T. Each $\mathbf{v}_j$ is in some $\mathbf{V}_{\lambda_j}$ and therefore dependent on $\mathcal{B}$. This means that $\mathcal{B}$ spans $\mathbf{V}$ and consequently has n elements since it is linearly independent. Thus $\sum_{i=1}^{r} n_i = n$ and $n_i = m_i$ for $i = 1, 2, \ldots, r$.

Assume now that $n_i = m_i$ for $i = 1, 2, \ldots, r$. Then $\sum_{i=1}^{r} n_i = n$ so that $\mathcal{B}$ has n vectors. Since $\mathcal{B}$ is linearly independent, $\mathcal{B}$ must be a basis of $\mathbf{V}$. And since $\mathcal{B}$ is composed of eigenvectors of T, the proof is complete. ■ ■ ■

In the remainder of this chapter, the frequent references to diagonal matrices make it desirable to have a more compact notation for this type of matrix. This notational convenience is provided in the next definition.

7.21
DEFINITION

The diagonal matrix $D = [d_{ij}]_n$ with $d_{ij} = 0$ for $i \neq j$ and $d_{ii} = \lambda_i$ will be denoted by $D = \mathrm{diag}\{\lambda_1, \lambda_2, \ldots, \lambda_n\}$.

We have seen that the problem of finding a diagonal matrix and a basis such that a given linear operator is represented by the diagonal matrix is one type of eigenvalue problem. Since we have a systematic method for finding the eigenvalues and eigenvectors of a linear operator, we are already equipped to solve this type of problem. We also have available from Chapter 5 a method for finding an invertible matrix P such that $P^{-1}AP$ is diagonal. For, with any convenient choice of basis $\mathcal{A}$, P is the matrix of transition from $\mathcal{A}$ to a basis $\mathcal{A}'$ of eigenvectors of T, so we can write

$$D = [T]_{\mathcal{A}'} = P^{-1}AP.$$

In most eigenvalue problems, the linear operator T is not given explicitly. Instead, one encounters the matrix A and is confronted with the problem of finding an invertible P such that $P^{-1}AP$ is diagonal. In such a situation, the formulation of the problem in terms of linear operators, vectors, and bases is only an encumbrance. It is more efficient to proceed directly to the problem of finding the columns of P. For this procedure, it is desirable to formulate the problem $P^{-1}AP = D = \mathrm{diag}\{\lambda_1, \lambda_2, \ldots, \lambda_n\}$ in the form $AP = PD$. With $A = [a_{ij}]_n$ and $P = [p_{ij}]_n$, the element in the ith row and jth column of AP is $\sum_{k=1}^{n} a_{ik} p_{kj}$, whereas the corresponding element in PD is $p_{ij}\lambda_j$. With j fixed, we have

$$\begin{bmatrix} \sum_{k=1}^{n} a_{1k} p_{kj} \\ \sum_{k=1}^{n} a_{2k} p_{kj} \\ \vdots \\ \sum_{k=1}^{n} a_{nk} p_{kj} \end{bmatrix} = \begin{bmatrix} p_{1j}\lambda_j \\ p_{2j}\lambda_j \\ \vdots \\ p_{nj}\lambda_j \end{bmatrix} = \lambda_j \begin{bmatrix} p_{1j} \\ p_{2j} \\ \vdots \\ p_{nj} \end{bmatrix}.$$

But the left-hand member of this equation is the same as the product

$$\begin{bmatrix} a_{11} & a_{12} & \cdots & a_{1n} \\ a_{21} & a_{22} & \cdots & a_{2n} \\ \vdots & \vdots & \ddots & \vdots \\ a_{n1} & a_{n2} & \cdots & a_{nn} \end{bmatrix} \begin{bmatrix} p_{1j} \\ p_{2j} \\ \vdots \\ p_{nj} \end{bmatrix} = AP_j,$$

where P_j is the jth column of P. Thus, we find that the equation

$$P^{-1}AP = \text{diag}\{\lambda_1, \lambda_2, \ldots, \lambda_n\}$$

is equivalent to the system of equations

$$AP_j = \lambda_j P_j, \quad j = 1, 2, \ldots, n,$$

where P_j is the jth column of P. But this system says precisely that the jth column of P is an eigenvector of A corresponding to λ_j. The requirement that P be invertible is equivalent to the requirement that these eigenvectors of A be linearly independent. Eigenvectors that are associated with distinct eigenvalues automatically form a linearly independent set (as in Theorem 7.8), but care must be taken to ensure independence whenever the geometric multiplicity of an eigenvalue exceeds 1. The procedure is illustrated in our next example.

EXAMPLE 1 Consider the problem of finding a real invertible matrix P such that $P^{-1}AP$ is diagonal, where

$$A = \begin{bmatrix} 7 & 3 & 3 & 2 \\ 0 & 1 & 2 & -4 \\ -8 & -4 & -5 & 0 \\ 2 & 1 & 2 & 3 \end{bmatrix}.$$

As the initial step, we find the characteristic equation of A, given by

$$(x-3)^2(x-1)(x+1) = 0.$$

We consider first the repeated eigenvalue $\lambda_1 = \lambda_2 = 3$, for if the geometric multiplicity of this eigenvalue is less than 2, then no matrix P of the required type exists (Theorem 7.20). The equation $(A - 3I)X = \mathbf{0}$ is given by

$$\begin{bmatrix} 4 & 3 & 3 & 2 \\ 0 & -2 & 2 & -4 \\ -8 & -4 & -8 & 0 \\ 2 & 1 & 2 & 0 \end{bmatrix} \begin{bmatrix} x_1 \\ x_2 \\ x_3 \\ x_4 \end{bmatrix} = \begin{bmatrix} 0 \\ 0 \\ 0 \\ 0 \end{bmatrix}.$$

Straightforward computations show that the solutions here are given by

$$\begin{bmatrix} x_1 \\ x_2 \\ x_3 \\ x_4 \end{bmatrix} = x_3 \begin{bmatrix} -\frac{3}{2} \\ 1 \\ 1 \\ 0 \end{bmatrix} + x_4 \begin{bmatrix} 1 \\ -2 \\ 0 \\ 1 \end{bmatrix}.$$

Hence the eigenvalue 3 has geometric multiplicity 2, and

$$P_1 = \begin{bmatrix} 3 \\ -2 \\ -2 \\ 0 \end{bmatrix} \quad \text{and} \quad P_2 = \begin{bmatrix} 1 \\ -2 \\ 0 \\ 1 \end{bmatrix}$$

provide two linearly independent columns of P. Repetition of the same procedure yields the solutions

$$P_3 = \begin{bmatrix} 1 \\ -2 \\ 0 \\ 0 \end{bmatrix} \quad \text{for } \lambda_3 = 1 \quad \text{and} \quad P_4 = \begin{bmatrix} 1 \\ -6 \\ 4 \\ -1 \end{bmatrix} \quad \text{for } \lambda_4 = -1.$$

Thus the matrix

$$P = [\,P_1 \quad P_2 \quad P_3 \quad P_4\,] = \begin{bmatrix} 3 & 1 & 1 & 1 \\ -2 & -2 & -2 & -6 \\ -2 & 0 & 0 & 4 \\ 0 & 1 & 0 & -1 \end{bmatrix}$$

is an invertible matrix such that $P^{-1}AP = \text{diag}\{3, 3, 1, -1\}$.

As a check on this solution, the student may verify that $AP = P \cdot \text{diag}\{3, 3, 1, -1\}$ or that

$$P^{-1} = \begin{bmatrix} 1 & \frac{1}{2} & \frac{1}{2} & 0 \\ \frac{1}{2} & \frac{1}{4} & \frac{1}{2} & 1 \\ -3 & -2 & -\frac{5}{2} & -1 \\ \frac{1}{2} & \frac{1}{4} & \frac{1}{2} & 0 \end{bmatrix}$$

and $P^{-1}AP = \text{diag}\{3, 3, 1, -1\}$. ∎

7.3 Exercises

*In Problems 1–6, (a) determine whether the given matrix A is similar over **R** to a diagonal matrix, and (b) whenever possible, find an invertible matrix P over **R** such that $P^{-1}AP$ is a diagonal matrix. The matrices here are the same as in Problems 1–6 of Exercises 7.2.*

1. $A = \begin{bmatrix} -1 & 2 \\ -2 & 3 \end{bmatrix}$

2. $A = \begin{bmatrix} 1 & 2 \\ 2 & -2 \end{bmatrix}$

3. $A = \begin{bmatrix} 1 & -1 & -1 \\ 1 & 3 & 2 \\ -1 & -1 & 0 \end{bmatrix}$

4. $A = \begin{bmatrix} 3 & 1 & -1 \\ -1 & 1 & 1 \\ 1 & 1 & 1 \end{bmatrix}$

5. $A = \begin{bmatrix} 3 & 0 & 4 & 4 \\ 0 & -1 & 0 & 0 \\ 0 & -4 & -1 & -4 \\ 0 & 4 & 0 & 3 \end{bmatrix}$

6. $A = \begin{bmatrix} 2 & 0 & 0 & 0 \\ 0 & 1 & -1 & 1 \\ 1 & 0 & 1 & 0 \\ 1 & 0 & -1 & 2 \end{bmatrix}$

In Problems 7–10, (a) determine whether the given linear operator T can be represented by a diagonal matrix, and (b) whenever possible, find a diagonal matrix and a basis such that T is represented by the diagonal matrix relative to the basis. The linear operators are the same as those in Problems 13–16 in Exercises 7.1.

7. $T(x_1, x_2) = (x_1 + 2x_2, 2x_1 - 2x_2)$ on $\mathbf{R}^2$

8. $T(x_1, x_2) = (2x_1 + x_2, x_1 - x_2)$ on $\mathbf{R}^2$

9. $T(x_1, x_2, x_3) = (x_1 + x_2 - x_3, -x_1 + 3x_2 - x_3, -x_1 + 2x_2)$ on $\mathbf{R}^3$

10. $T(x_1, x_2, x_3, x_4) = (x_1+x_3+x_4, 2x_1+x_2+3x_3+x_4, -x_1-x_3-x_4, 3x_1+2x_2+5x_3+x_4)$ on $\mathbf{R}^4$

In Problems 11–14, let T be the linear operator on $\mathbf{R}^3$ that has the given matrix A relative to the basis $\mathcal{A} = \{(1, 0, 0), (1, 1, 0), (1, 1, 1)\}$. (a) Determine whether T can be represented by a diagonal matrix, and (b) whenever possible, find a diagonal matrix and a basis of $\mathbf{R}^3$ such that T is represented by the diagonal matrix relative to the basis. These linear operators are the same as those in Problem 11 of Exercises 7.2.

11. $A = \begin{bmatrix} 8 & 5 & -5 \\ 5 & 8 & -5 \\ 15 & 15 & -12 \end{bmatrix}$

12. $A = \begin{bmatrix} -4 & -3 & -1 \\ -4 & 0 & -4 \\ 8 & 4 & 5 \end{bmatrix}$

13. $A = \begin{bmatrix} 3 & 2 & 2 \\ 1 & 4 & 1 \\ -2 & -4 & -1 \end{bmatrix}$

14. $A = \begin{bmatrix} 8 & 5 & 6 \\ 0 & -2 & 0 \\ -10 & -5 & -8 \end{bmatrix}$

In Problems 15–20, (a) determine whether the given matrix A is similar over C to a diagonal matrix, and (b) whenever possible, find an invertible matrix P over C such that $P^{-1}AP$ is a diagonal matrix. The matrices here are the same as in Problem 14 of Exercises 7.2.

15. $A = \begin{bmatrix} 2 & 1-i \\ 1+i & 3 \end{bmatrix}$

16. $A = \begin{bmatrix} 5 & i \\ -i & 2 \end{bmatrix}$

17. $A = \begin{bmatrix} 3 & 4 & 2 \\ 1 & 3 & 1 \\ 1 & 2 & 2 \end{bmatrix}$

18. $A = \begin{bmatrix} i & 0 & 0 \\ -2i & i & -2+i \\ 0 & 0 & -2 \end{bmatrix}$

19. $A = \begin{bmatrix} 1+i & 0 & 0 \\ -2i & 1+i & 2i \\ i & 0 & 1 \end{bmatrix}$

20. $A = \begin{bmatrix} -2+2i & 0 & -2+i \\ 0 & -i & 0 \\ 4-2i & 0 & 4-i \end{bmatrix}$

21. Whenever possible, perform a check on the work in the indicated problem by verifying that $AP = PD$, where D is the diagonal matrix that is similar to A.

(a) Problem 1   (b) Problem 2   (c) Problem 3   (d) Problem 4
(e) Problem 5   (f) Problem 6   (g) Problem 15   (h) Problem 16
(i) Problem 17   (j) Problem 18   (k) Problem 19   (l) Problem 20

22. Whenever possible, perform a check on the work in the indicated problem by computing P^{-1} and verifying that $P^{-1}AP$ is indeed a diagonal matrix.

(a) Problem 1   (b) Problem 2   (c) Problem 3   (d) Problem 4

(e) Problem 5   (f) Problem 6   (g) Problem 15   (h) Problem 16

(i) Problem 17   (j) Problem 18   (k) Problem 19   (l) Problem 20

23. Give an example of a 2×2 matrix over $\mathbf{R}$ that is not similar over $\mathbf{R}$ to a diagonal matrix.

24. Give an example of two 2×2 matrices that have the same characteristic equation but are not similar.

25. Show that the characteristic polynomial of the matrix

$$C = \begin{bmatrix} 0 & 1 & 0 & \cdots & 0 \\ 0 & 0 & 1 & \cdots & 0 \\ \vdots & \vdots & \vdots & \ddots & \vdots \\ 0 & 0 & 0 & \cdots & 1 \\ -c_0 & -c_1 & -c_2 & \cdots & -c_{n-1} \end{bmatrix}$$

is $p(x) = (-1)^n(x^n + c_{n-1}x^{n-1} + \cdots + c_1 x + c_0)$. The matrix C is called the **companion matrix** of the polynomial $p(x)$. (*Hint:* Expand $\det(C - xI)$ about the last row.)

26. Use the result of Problem 25 to write down a matrix with the given polynomial $p(x)$ as its characteristic polynomial.

(a) $p(x) = -x^3 + 5x^2 - 2$ (b) $p(x) = x^2 - 3x + 2$

(c) $p(x) = x^4 + 5x^2 + 4$ (d) $p(x) = -x^5 + 1$

27. If A is similar to a diagonal matrix D, then show that there exists an invertible matrix P such that $A^k = PD^kP^{-1}$ and use this result to evaluate each of the following.

(a) A^5 where $A = \begin{bmatrix} 1 & 0 \\ 4 & -1 \end{bmatrix}$ (b) A^3 where $A = \begin{bmatrix} 8 & 1 \\ 2 & 7 \end{bmatrix}$

(c) A^{10} where $A = \begin{bmatrix} 1 & 0 & 3 \\ 2 & -1 & 3 \\ 0 & 0 & -2 \end{bmatrix}$ (d) A^6 where $A = \begin{bmatrix} -7 & -2 & -8 \\ 5 & 3 & 5 \\ 5 & 2 & 6 \end{bmatrix}$

28. If an invertible matrix A is similar to a diagonal matrix D, then show that there exists an invertible matrix P such that $A^{-1} = PD^{-1}P^{-1}$ and use this result to evaluate A^{-1} for each of the following matrices.

(a) $\begin{bmatrix} 3 & -2 & 0 \\ -1 & 2 & 0 \\ -1 & 0 & 2 \end{bmatrix}$ (b) $\begin{bmatrix} -2 & -3 & 0 & 3 \\ 3 & 4 & 0 & -3 \\ 0 & 0 & 2 & 0 \\ 3 & 3 & 0 & -2 \end{bmatrix}$

29. Prove or disprove each of the following statements.

(a) Every matrix that is similar to a diagonal matrix is invertible.

(b) Every invertible matrix is similar to a diagonal matrix.

(c) Every triangular matrix is similar to a diagonal matrix.

30. If A and B are each similar to a diagonal matrix, and have the same eigenvectors (not necessarily the same eigenvalues), then prove that A and B commute.

31. Suppose the only eigenvalues of A are ± 1 and A is similar to a diagonal matrix. Prove that $A^{-1} = A$.

32. Suppose that $\lambda_1, \ldots, \lambda_r$ are the distinct eigenvalues of T, and that each λ_i is in $\mathcal{F}$. Prove that T can be represented by a diagonal matrix if and only if

$$\mathbf{V} = V_{\lambda_1} \oplus \cdots \oplus \mathbf{V}_{\lambda_r}.$$

8 Functions of Vectors

There are several standard types of functions defined on a vector space that have found widespread application. The linear transformation, which we have already studied, is probably the most important of these, but there are others that are of great value. The linear functional is central to the study of linear programming. The quadratic form is frequently useful in statistics, engineering, and physics. We shall encounter each of these types of functions in this chapter.

8.1 Linear Functionals

We recall from Chapter 5 that whenever $\mathbf{U}$ and $\mathbf{V}$ are vector spaces over the same field $\mathcal{F}$, the set of all linear transformations of $\mathbf{U}$ into $\mathbf{V}$ is a vector space over $\mathcal{F}$ (Theorem 5.3). Also, we have seen in Example 1 of Section 4.1 that a field $\mathcal{F}$ may be regarded as a vector space over itself. Thus, for any vector space $\mathbf{V}$ over $\mathcal{F}$, the set of all linear transformations of $\mathbf{V}$ into $\mathcal{F}$ is a vector space over $\mathcal{F}$. This type of linear transformation is of such importance that it has a special name.

8.1 DEFINITION	Let $\mathbf{V}$ be a vector space over the field $\mathcal{F}$. A linear transformation of $\mathbf{V}$ into $\mathcal{F}$ is called a **linear functional** on $\mathbf{V}$. The set of all linear functionals on $\mathbf{V}$ is denoted by $\mathbf{V}^*$.

Thus, a linear functional on $\mathbf{V}$ is a scalar-valued function f defined on $\mathbf{V}$ that has the property

$$f(a\mathbf{u} + b\mathbf{v}) = af(\mathbf{u}) + bf(\mathbf{v})$$

for all $\mathbf{u}, \mathbf{v} \in \mathbf{V}$. As mentioned above, $\mathbf{V}^*$ is a vector space over $\mathcal{F}$. There are several interesting relations between $\mathbf{V}$ and $\mathbf{V}^*$ that we will investigate later, but first let us consider some examples illustrating Definition 8.1.

EXAMPLE 1 Let $\mathbf{V} = \mathbf{R}^n$, and let $\mathbf{c} = (c_1, c_2, \ldots, c_n)$ be a fixed vector in $\mathbf{V}$. For each

$$\mathbf{v} = (x_1, x_2, \ldots, x_n) \in \mathbf{V},$$

define $f(\mathbf{v})$ to be

$$f(\mathbf{v}) = \mathbf{c} \cdot \mathbf{v} = c_1 x_1 + c_2 x_2 + \cdots + c_n x_n.$$

The function f so defined is scalar-valued, and

$$f(a\mathbf{u} + b\mathbf{v}) = \mathbf{c} \cdot (a\mathbf{u} + b\mathbf{v}) = a(\mathbf{c} \cdot \mathbf{u}) + b(\mathbf{c} \cdot \mathbf{v}) = af(\mathbf{u}) + bf(\mathbf{v}).$$

Thus f is a linear functional on $\mathbf{R}^n$.   ■

EXAMPLE 2   Let $\mathbf{V} = \mathbf{R}^3$, and define $f(\mathbf{v}) = \|\mathbf{v}\|$ for each $\mathbf{v} \in \mathbf{V}$. Then f is scalar-valued, but f is not a linear transformation. For example,

$$f(\mathbf{e}_1) + f(\mathbf{e}_2) = f(1, 0, 0) + f(0, 1, 0) = 2$$

and

$$f(\mathbf{e}_1 + \mathbf{e}_2) = f(1, 1, 0) = \sqrt{2}.$$

Thus f is *not* a linear functional on $\mathbf{R}^3$.   ■

EXAMPLE 3   For a fixed positive integer n, let $\mathbf{V}$ be the vector space $\mathcal{P}_n$ consisting of all polynomials in x with coefficients in the field $\mathcal{F}$ and degree $\leq n$. For each polynomial $p(x) = \sum_{i=0}^{n} a_i x^i$, define $f(p(x)) = p(0)$. The mapping f is clearly scalar-valued. For any $p(x) = \sum_{i=0}^{n} a_i x^i$ and $q(x) = \sum_{i=0}^{n} b_i x^i$ in $\mathcal{P}_n$ and any $a, b \in \mathcal{F}$,

$$f(ap(x) + bq(x)) = f\left(\sum_{i=0}^{n}(aa_i + bb_i)x^i\right)$$

$$= aa_0 + bb_0$$

$$= af(p(x)) + bf(q(x)).$$

This shows that f is a linear functional on $\mathcal{P}_n$.   ■

EXAMPLE 4   Let $\mathbf{V}$ be the vector space $\mathbf{R}_{n \times n}$ of all $n \times n$ matrices over $\mathbf{R}$ as defined in Chapter 4. For each $A = [a_{ij}]_n \in \mathbf{V}$, the trace of A, denoted by $t(A)$, is given by $t(A) = \sum_{i=1}^{n} a_{ii}$. It is left as an exercise (Problem 10) to verify that t is a linear functional on $\mathbf{R}_{n \times n}$.   ■

EXAMPLE 5   The set of all convergent sequences of real numbers is a subspace $\mathbf{W}$ of the vector space in Example 3 in Section 4.1. The function f defined by $f(\{a_n\}) = \lim_{n \to \infty} a_n$ is a linear functional on $\mathbf{W}$.   ■

As mentioned earlier, it is already known that $\mathbf{V}^*$ is a vector space over $\mathcal{F}$, just as $\mathbf{V}$ is. The following theorem shows that $\mathbf{V}^*$ has the same dimension as $\mathbf{V}$ whenever $\mathbf{V}$ is of finite dimension.

8.2 THEOREM

If $\mathbf{V}$ is a vector space of finite dimension n over $\mathcal{F}$, then $\mathbf{V}^*$ is also of dimension n over $\mathcal{F}$.

Proof Suppose that $\mathcal{A} = \{\mathbf{u}_1, \mathbf{u}_2, \ldots, \mathbf{u}_n\}$ is a basis of **V**. For each j, $(j = 1, 2, \ldots, n)$, let $\mathbf{p}_j$ be defined at $\mathbf{u} = \sum_{i=1}^{n} x_i \mathbf{u}_i$ in **V** by

$$\mathbf{p}_j(\mathbf{u}) = \mathbf{p}_j\left(\sum_{i=1}^{n} x_i \mathbf{u}_i\right) = x_j.$$

Since each $\mathbf{u}$ can be written uniquely as $\mathbf{u} = \sum_{i=1}^{n} x_i \mathbf{u}_i$, the value $\mathbf{p}_j(\mathbf{u})$ is well-defined, and $\mathbf{p}_j$ is a mapping of **V** into $\mathcal{F}$. For any a, b in $\mathcal{F}$ and $\mathbf{u} = \sum_{i=1}^{n} x_i \mathbf{u}_i$, $\mathbf{v} = \sum_{i=1}^{n} y_i \mathbf{u}_i$ in **V**,

$$\mathbf{p}_j(a\mathbf{u} + b\mathbf{v}) = \mathbf{p}_j\left(\sum_{i=1}^{n} (ax_i + by_i)\mathbf{u}_i\right)$$

$$= ax_j + by_j$$

$$= a\mathbf{p}_j(\mathbf{u}) + b\mathbf{p}_j(\mathbf{v}).$$

Thus each $\mathbf{p}_j$ is contained in $\mathbf{V}^*$.

The contention is that the set $\mathcal{A}^* = \{\mathbf{p}_1, \mathbf{p}_2, \ldots, \mathbf{p}_n\}$ is a basis of $\mathbf{V}^*$. To show that $\mathcal{A}^*$ spans $\mathbf{V}^*$, let f be any linear functional on **V**. If we put $a_i = f(\mathbf{u}_i)$, then for any $\mathbf{u} = \sum_{i=1}^{n} x_i \mathbf{u}_i$ in **V**,

$$f(\mathbf{u}) = \sum_{i=1}^{n} x_i f(\mathbf{u}_i)$$

$$= \sum_{i=1}^{n} x_i a_i$$

$$= \sum_{i=1}^{n} a_i \mathbf{p}_i(\mathbf{u})$$

$$= \left(\sum_{i=1}^{n} a_i \mathbf{p}_i\right)(\mathbf{u}).$$

Hence $f = \sum_{i=1}^{n} a_i \mathbf{p}_i$, and $\mathcal{A}^*$ spans $\mathbf{V}^*$. To see that $\mathcal{A}^*$ is linearly independent, suppose that $c_1, c_2, \ldots, c_n$ is a set of scalars such that

$$c_1\mathbf{p}_1 + c_2\mathbf{p}_2 + \cdots + c_n\mathbf{p}_n = Z,$$

where Z is the zero linear functional. Then $\left(\sum_{i=1}^{n} c_i \mathbf{p}_i\right)(\mathbf{u}) = 0$ for each $\mathbf{u} \in \mathbf{V}$. In particular, for each $\mathbf{u}_j$,

$$0 = \left(\sum_{i=1}^{n} c_i \mathbf{p}_i\right)(\mathbf{u}_j) = \sum_{i=1}^{n} c_i \mathbf{p}_i(\mathbf{u}_j) = \sum_{i=1}^{n} c_i \delta_{ij} = c_j.$$

That is, each $c_j = 0$, and this completes the proof. ∎ ■ ■ ■

The basis $\mathcal{A}^*$ in the proof of Theorem 8.2 is called the **dual basis** of $\mathcal{A}$. The linear functionals $\mathbf{p}_j$ in $\mathcal{A}^*$ are called the **coordinate projections** relative to $\mathcal{A}$.

EXAMPLE 6   The set $\mathcal{A} = \{\mathbf{u}_1 = 1, \mathbf{u}_2 = 1 + x, \mathbf{u}_3 = 1 + x + x^2\}$ is a basis of $\mathcal{P}_2$ over $\mathbf{R}$. To find the dual basis $\mathcal{A}^*$, of $\mathcal{P}_2^*$, we follow the first part of the proof of Theorem 8.2 and write an arbitrary $\mathbf{u} = a_0 + a_1 x + a_2 x^2$ in $\mathcal{P}_2$ as a linear combination of the base vectors.

$$a_0 + a_1 x + a_2 x^2 = c_1(1) + c_2(1 + x) + c_3(1 + x + x^2)$$

Solving

$$
\begin{aligned}
a_0 &= c_1 + c_2 + c_3 \\
a_1 &= \phantom{c_1 + {}} c_2 + c_3 \\
a_2 &= \phantom{c_1 + c_2 + {}} c_3
\end{aligned}
$$

for c_1, c_2 and c_3 yields

$$c_1 = a_0 - a_1$$

$$c_2 = a_1 - a_2$$

$$c_3 = a_2.$$

Since the elements of the dual basis are given by

$$\mathbf{p}_j(\mathbf{u}) = \mathbf{p}_j \left(\sum_{i=1}^{3} c_i \mathbf{u}_i \right) = c_j$$

then

$$\mathbf{p}_1(a_0 + a_1 x + a_2 x^2) = a_0 - a_1$$

$$\mathbf{p}_2(a_0 + a_1 x + a_2 x^2) = a_1 - a_2$$

$$\mathbf{p}_3(a_0 + a_1 x + a_2 x^2) = a_2$$

and the dual basis $\mathcal{A}^*$ of $\mathcal{P}_2^*$ is $\mathcal{A}^* = \{\mathbf{p}_1, \mathbf{p}_2, \mathbf{p}_3\}$.   ∎

For later use we note that the defining property of the coordinate projections $\mathbf{p}_j$ is that $\mathbf{p}_j(\mathbf{u}_i) = \delta_{ij}$, for each base vector $\mathbf{u}_i$.

Whenever $\mathbf{V}$ is finite-dimensional, $\mathbf{V}^*$ is called the **dual space** of $\mathbf{V}$. If $\mathbf{V}$ is of infinite dimension, then $\mathbf{V}^*$ is not necessarily isomorphic to $\mathbf{V}$, and the term "dual space" is not ordinarily used.

For the remainder of this section, $\mathbf{V}$ will denote an n-dimensional vector space over a field $\mathcal{F}$. As a linear transformation of $\mathbf{V}$ into $\mathcal{F}$, each linear functional has a $1 \times n$ matrix relative to each basis of $\mathbf{V}$. According to Definition 5.10, the matrix of f relative to the basis $\mathcal{A} = \{\mathbf{u}_1, \mathbf{u}_2, \ldots, \mathbf{u}_n\}$ of $\mathbf{V}$ is $A = [\, a_1 \quad a_2 \quad \cdots \quad a_n \,]$ where $a_j = f(\mathbf{u}_j)$. (We are using the basis $\mathcal{E}_1 = \{1\}$ of $\mathcal{F}$ here, and will adhere to this choice consistently.) The matrix of f provides a convenient method of computing the values $f(\mathbf{u})$. For if $\mathbf{u}$ has coordinate matrix $X = [\, x_1 \quad x_2 \quad \cdots \quad x_n \,]^T$, then

$$f(\mathbf{u}) = \sum_{i=1}^{n} x_i f(\mathbf{u}_i)$$

$$= \sum_{i=1}^{n} x_i a_i$$

$$= \begin{bmatrix} a_1 & a_2 & \cdots & a_n \end{bmatrix} \begin{bmatrix} x_1 \\ x_2 \\ \vdots \\ x_n \end{bmatrix}$$

$$= AX,$$

where we identify the 1×1 matrix $[\sum_{i=1}^{n} x_i a_i]$ with the element $\sum_{i=1}^{n} x_i a_i$ in $\mathcal{F}$. Actually, this result is nothing new, but merely a special case of Theorem 5.12, and we can write

$$f(\mathbf{u}) = [f(\mathbf{u})]_{\mathcal{E}_1} = [f]_{\mathcal{E}_1, \mathcal{A}}[\mathbf{u}]_{\mathcal{A}} = AX.$$

The set $\{1\}$ is clearly the simplest choice of basis for $\mathcal{F}$, so we do not propose to make any changes here. But there is no reason to restrict ourselves in the choice of basis in $\mathbf{V}$, and Theorem 5.15 describes completely the results of such a change. (The space $\mathbf{V}$ here is playing the role of $\mathbf{U}$ in Theorem 5.15.) If f has matrix A relative to the basis $\mathcal{A}$ of $\mathbf{V}$, and if Q is the matrix of transition from $\mathcal{A}$ to the basis $\mathcal{B}$, then f has matrix $B = AQ$ relative to $\mathcal{B}$, or

$$B = [f]_{\mathcal{E}_1, \mathcal{B}} = [f]_{\mathcal{E}_1, \mathcal{A}} Q_{\mathcal{A} \to \mathcal{B}} = AQ.$$

Any basis of $\mathbf{V}$ has a dual basis in $\mathbf{V}^*$, so a change of basis from $\mathcal{A}$ to $\mathcal{B}$ in $\mathbf{V}$ induces a corresponding change of basis from $\mathcal{A}^*$ to $\mathcal{B}^*$ in $\mathbf{V}^*$. Our next theorem describes the relation between these changes of bases.

8.4
THEOREM

If Q is the transition matrix from the basis $\mathcal{A} = \{\mathbf{u}_1, \mathbf{u}_2, \ldots, \mathbf{u}_n\}$ to the basis $\mathcal{B} = \{\mathbf{v}_1, \mathbf{v}_2, \ldots, \mathbf{v}_n\}$ of $\mathbf{V}$, then $(Q^T)^{-1}$ is the matrix of transition from $\mathcal{A}^*$ to $\mathcal{B}^*$ in $\mathbf{V}^*$.

Proof Rather than prove the statement in the conclusion, we shall prove the equivalent assertion that Q^T is the transition matrix from $\mathcal{B}^*$ to $\mathcal{A}^*$.

Let $\mathcal{A}^* = \{\mathbf{p}_1, \mathbf{p}_2, \ldots, \mathbf{p}_n\}$, and let $\mathcal{B}^* = \{\mathbf{g}_1, \mathbf{g}_2, \ldots, \mathbf{g}_n\}$. Now $\mathbf{p}_j = \sum_{k=1}^{n} c_{kj} \mathbf{g}_k$, where $C = [c_{ij}]_n$ is the matrix of transition from $\mathcal{B}^*$ to $\mathcal{A}^*$. Then

$$\mathbf{p}_j(\mathbf{v}_i) = \sum_{k=1}^{n} c_{kj} \mathbf{g}_k(\mathbf{v}_i) = \sum_{k=1}^{n} c_{kj} \delta_{ki} = c_{ij}.$$

But $\mathbf{v}_i = \sum_{k=1}^{n} q_{ki}\mathbf{u}_k$ since $Q = [q_{ij}]_n$ is the transition matrix from $\mathcal{A}$ to $\mathcal{B}$. Hence

$$\mathbf{p}_j(\mathbf{v}_i) = \sum_{k=1}^{n} q_{ki}\mathbf{p}_j(\mathbf{u}_k) = \sum_{k=1}^{n} q_{ki}\delta_{jk} = q_{ji},$$

so we have $c_{ij} = \mathbf{p}_j(\mathbf{v}_i) = q_{ji}$, and $C = Q^T$. ■■■

The ideas developed in this section are illustrated in the following example.

EXAMPLE 7 Let $\mathbf{V}$ be the vector space $\mathcal{C}^3$ of all ordered triples of complex numbers.[1] The mapping f given by $f(c_1, c_2, c_3) = ic_1 - ic_2 + c_3$ is a linear functional on $\mathcal{C}^3$. Relative to the basis $\mathcal{A} = \{\mathbf{u}_1 = (1, 0, 0), \mathbf{u}_2 = (0, i, 0), \mathbf{u}_3 = (0, 1, i)\}$, f has matrix $A = [\,i \quad 1 \quad 0\,]$. The coordinates x_i of $\mathbf{u} = (c_1, c_2, c_3)$ relative to $\mathcal{A}$ are given by $x_1 = c_1$, $x_2 = -ic_2 + c_3$, $x_3 = -ic_3$, and

$$AX = [f]_{\mathcal{E}_1, \mathcal{A}}[\mathbf{u}]_{\mathcal{A}} = \begin{bmatrix} i & 1 & 0 \end{bmatrix} \begin{bmatrix} c_1 \\ -ic_2 + c_3 \\ -ic_3 \end{bmatrix} = ic_1 - ic_2 + c_3 = f(\mathbf{u}).$$

The elements of the dual basis $\mathcal{A}^* = \{\mathbf{p}_1, \mathbf{p}_2, \mathbf{p}_3\}$ are given by

$$\mathbf{p}_1(c_1, c_2, c_3) = c_1,$$
$$\mathbf{p}_2(c_1, c_2, c_3) = -ic_2 + c_3,$$
$$\mathbf{p}_3(c_1, c_2, c_3) = -ic_3.$$

The matrix

$$Q = \begin{bmatrix} 0 & -i & 0 \\ 1 & 0 & 1 \\ 0 & 0 & -i \end{bmatrix}$$

is the matrix of transition from $\mathcal{A}$ to $\mathcal{B} = \{\mathbf{v}_1, \mathbf{v}_2, \mathbf{v}_3\}$, where $\mathbf{v}_1 = (0, i, 0)$, $\mathbf{v}_2 = (-i, 0, 0)$, $\mathbf{v}_3 = (0, 0, 1)$. The matrix of f relative to $\mathcal{B}$ is

$$B = [f]_{\mathcal{E}_1, \mathcal{B}} = [f]_{\mathcal{E}_1, \mathcal{A}} Q_{\mathcal{A} \to \mathcal{B}} = AQ = \begin{bmatrix} 1 & 1 & 1 \end{bmatrix}.$$

The coordinates y_i of $\mathbf{u} = (c_1, c_2, c_3)$ relative to $\mathcal{B}$ are $y_1 = -ic_2$, $y_2 = ic_1$, $y_3 = c_3$, and

$$BY = [f]_{\mathcal{E}_1, \mathcal{B}}[\mathbf{u}]_{\mathcal{B}} = \begin{bmatrix} 1 & 1 & 1 \end{bmatrix} \begin{bmatrix} -ic_2 \\ ic_1 \\ c_3 \end{bmatrix} = -ic_2 + ic_1 + c_3 = f(\mathbf{u}).$$

The dual basis is $\mathcal{B}^* = \{\mathbf{g}_1, \mathbf{g}_2, \mathbf{g}_3\}$, where

$$\mathbf{g}_1(c_1, c_2, c_3) = -ic_2,$$
$$\mathbf{g}_2(c_1, c_2, c_3) = ic_1,$$
$$\mathbf{g}_3(c_1, c_2, c_3) = c_3.$$

---

[1] The symbol i here denotes the square root of -1.

The matrix $(Q^T)^{-1}$ of transition from $\mathcal{A}^*$ to $\mathcal{B}^*$ is given by

$$(Q^T)^{-1} = \begin{bmatrix} 0 & i & 0 \\ 1 & 0 & 0 \\ -i & 0 & i \end{bmatrix},$$

so that $\mathbf{g}_1 = \mathbf{p}_2 - i\mathbf{p}_3, \mathbf{g}_2 = i\mathbf{p}_1$, and $\mathbf{g}_3 = i\mathbf{p}_3$. The reader may verify that these last equalities are correct by evaluating both members for an arbitrary $\mathbf{u} \in \mathcal{C}^3$. ∎

8.1 Exercises

In Problems 1–8, determine whether or not the given function f is a linear functional on the given vector space. If f is not a linear functional, give a justification.

1. Define f on $\mathcal{P}_n$ over $\mathbf{R}$ by $f(p(x)) = p(c)$ for a constant c.

2. Define f on $\mathbf{R}^2$ by $f(x_1, x_2) = |x_1 + x_2|$.

3. Define f on $\mathbf{R}_{n\times n}$ by $f(A) = \det(A)$.

4. Define f on $\mathbf{R}_{n\times n}$ by $f(A) = \prod_{i=1}^n a_{ii}$, where $A = [a_{ij}]$.

5. Define f on $\mathbf{R}_{n\times n}$ by $f(A) = n\sum_{i=1}^n a_{ii}$, where $A = [a_{ij}]$.

6. Define f on $\mathbf{R}_{n\times 1}$ by

$$f\left(\begin{bmatrix} a_1 \\ a_2 \\ \vdots \\ a_n \end{bmatrix}\right) = \sum_{i=1}^n a_i.$$

7. Define f on $\mathbf{R}^3$ by $f(x_1, x_2, x_3) = x_2$.

8. Define f on $\mathbf{R}^3$ by $f(x_1, x_2, x_3) = 1 + x_2$.

In Problems 9–16, determine whether or not the given function f is a linear functional on the given vector space and justify your answer.

9. Define f on $\mathbf{R}^2$ by $f(x_1, x_2) = x_1 - x_2$.

10. Define f on $\mathbf{R}_{n\times n}$ by $f(A) = t(A)$, where $t(A)$ is the trace of A.

11. Define f on $\mathcal{P}_2$ over $\mathbf{R}$ by $f(a_0 + a_1 x + a_2 x^2) = a_0 + a_1 + a_2$.

12. Define f on the vector space $\mathcal{P}_n$ over $\mathbf{R}$ by $f(a_0 + a_1 x + \cdots + a_n x^n) = n$.

13. Define f on the vector space $\mathbf{V}$ of all real-valued continuous functions on the closed interval $[0, 1]$ by $f(g) = \int_0^1 g(t)dt$.

14. Define f on the vector space $\mathbf{V}$ of all real-valued continuous functions on the closed interval $[0, 1]$ by $f(g) = \int_0^1 h(t)g(t)dt$, for fixed $h(t)$ in $\mathbf{V}$.

15. Define f on the vector space **V** of all real-valued differentiable functions g on **R** by $f(g) = g'(2)$.

16. Define f on the vector space **V** of all real-valued differentiable functions g on **R** by $f(g(t)) = 2g'(0)$.

17. If $f(x_1, x_2, x_3) = 2x_1 - x_2 + 4x_3$, find a vector $\mathbf{c} \in \mathbf{R}^3$ such that $f(\mathbf{u}) = \mathbf{c} \cdot \mathbf{u}$ for all $\mathbf{u} \in \mathbf{R}^3$.

18. Find the matrix of the given linear functional relative to the given basis of $\mathbf{R}^n$.
    (a) $f(x_1, x_2, x_3) = 3x_1 - 2x_2 + 7x_3$, $\mathcal{A} = \{(1, 0, 1), (1, 1, 0), (0, 1, 1)\}$
    (b) $f(x_1, x_2, x_3) = 6x_1 + 5x_2 - 8x_3$, $\mathcal{A} = \{(4, -3, 1), (5, -3, 1), (3, -2, 1)\}$
    (c) $f(x_1, x_2, x_3, x_4) = 2x_1 + 4x_3 + 12x_4$,
        $\mathcal{A} = \{(1, 0, 0, 0), (1, 1, 0, 0), (1, 1, 1, 0), (1, 1, 1, 1)\}$
    (d) $f(x_1, x_2, x_3, x_4) = 9x_1 - 6x_2 + 3x_4$,
        $\mathcal{A} = \{(1, 0, 1, 0), (1, 0, 0, 1), (0, 0, 1, 1), (0, 1, 1, 0)\}$

19. Use the matrices found in Problem 18 to compute $f(\mathbf{e}_i)$ for each $\mathbf{e}_i$ in the standard basis $\mathcal{E}_n$.

20. Find the matrix A of the linear functional f relative to the basis $\mathcal{A}$ of $\mathcal{P}_2$ over **R** and use A to find $f(p(x))$ for each of the following.
    (a) $f(a_0 + a_1 x + a_2 x^2) = a_0 + a_1 + a_2$, $\mathcal{A} = \{x, 1 - x, 1 - x^2\}$, $p(x) = 2 - x^2$
    (b) $f(a_0 + a_1 x + a_2 x^2) = 3a_0 - 2a_1 + a_2$, $\mathcal{A} = \{1 - x, 1 - x - x^2, 1 - x^2\}$,
        $p(x) = 2 - x + 2x^2$

21. Let $\mathcal{A} = \{\mathbf{u}_1, \mathbf{u}_2, \mathbf{u}_3\}$ where $\mathbf{u}_1 = (1, 2, 0)$, $\mathbf{u}_2 = (1, 0, 2)$, $\mathbf{u}_3 = (0, 1, 2)$, and let $\mathcal{A}^* = \{\mathbf{p}_1, \mathbf{p}_2, \mathbf{p}_3\}$. Find the three expressions that give the values of $\mathbf{p}_i(x_1, x_2, x_3)$ for $i = 1, 2, 3$.

22. For the given basis $\mathcal{A}$ of $\mathbf{R}^3$, find the dual basis $\mathcal{A}^*$.
    (a) $\mathcal{A} = \{(0, 0, 1), (0, 1, 1), (1, 1, 1)\}$ (b) $\mathcal{A} = \{(0, 1, 1), (1, 1, 0), (1, 0, 1)\}$
    (c) $\mathcal{A} = \{(2, 0, 2), (0, 1, 1), (2, -3, 1)\}$ (d) $\mathcal{A} = \{(1, -1, 0), (2, -1, 1), (1, 1, 3)\}$

23. Let $\mathcal{E}_3^* = \{\mathbf{g}_1, \mathbf{g}_2, \mathbf{g}_3\}$ denote the dual basis of $\mathcal{E}_3 = \{\mathbf{e}_1, \mathbf{e}_2, \mathbf{e}_3\}$. In each part of Problem 22, find the coordinates of the elements $\mathbf{p}_i$ of $\mathcal{A}^*$ relative to $\mathcal{E}_3^*$.

24. For the given basis $\mathcal{A}$ of $\mathcal{P}_n$ over **R**, find the dual basis of $\mathcal{A}$.
    (a) $\mathcal{A} = \{1, 1 - x, 1 + x - x^2\}$,
    (b) $\mathcal{A} = \{2 + x, x + x^2, 1 + x + x^2\}$
    (c) $\mathcal{A} = \{x, x + x^2, 1 - x + x^3, x^3\}$
    (d) $\mathcal{A} = \{1 + x^2, 1 + x^3, 1 + x, 1 + x^2 - x^3\}$

25. Suppose that the linear functional f has matrix $A = [\, a_1 \quad a_2 \quad \cdots \quad a_n \,]$ relative to the basis $\mathcal{A}$ of **V**. Prove that the coordinate matrix of f relative to $\mathcal{A}^*$ is A^T.

26. Let $\mathcal{A} = \{(1, 1, 0), (1, 0, 0), (1, 1, 1)\}$ and $\mathcal{A}^* = \{\mathbf{p}_1, \mathbf{p}_2, \mathbf{p}_3\}$, and let f be the linear functional that has coordinates $[\, 1 \quad 2 \quad 3 \,]^T$ relative to $\mathcal{A}^*$. Find the values of $f(5, 4, 3)$ and $f(x_1, x_2, x_3)$.

27. Let $\mathcal{A} = \{x, 1 - x, 1 - x^2\}$ and $\mathcal{A}^* = \{\mathbf{p}_1, \mathbf{p}_2, \mathbf{p}_3\}$, and let f be the linear functional that has coordinates $[\,2 \quad -1 \quad 1\,]^T$ relative to $\mathcal{A}^*$. Find the values of $f(1 + x - x^2)$ and $f(a_0 + a_1 x + a_2 x^2)$.

28. Let $\mathbf{u}$ be a fixed vector in the n-dimensional vector space $\mathbf{V}$.

    (a) Prove that there is a nonzero $f \in \mathbf{V}^*$ such that $f(\mathbf{u}) = 0$.

    (b) Let $\mathbf{u} = (1, 2, 3) \in \mathbf{R}^3$. Find a linear functional $f \neq Z$ such that $f(\mathbf{u}) = 0$.

29. Use the matrix of transition from $\mathcal{E}_n^*$ to $\mathcal{A}^*$ to find $\mathcal{A}^* = \{\mathbf{p}_1, \mathbf{p}_2, \ldots, \mathbf{p}_n\}$ for the given basis $\mathcal{A}$ of $\mathbf{R}^n$. Write each $\mathbf{p}_i$ as a linear combination of the elements $\mathbf{g}_i$ of $\mathcal{E}_n^*$.

    (a) $\mathcal{A} = \{(1, -3, 2), (0, 1, 1), (1, 7, 13)\}$

    (b) $\mathcal{A} = \{(1, 1, 3), (2, -1, 1), (1, -1, 0)\}$

    (c) $\mathcal{A} = \{(0, 1, 0, 0), (1, -2, 0, -1), (0, 1, 0, 1), (2, -7, 1, -2)\}$

    (d) $\mathcal{A} = \{(3, 6, 0, 2), (2, 4, 1, 1), (2, 5, -1, 1), (1, 3, 0, 1)\}$

30. Suppose that the linear functional f on $\mathbf{V}$ has matrix $A = [\,a_1 \quad a_2 \quad \cdots \quad a_n\,]$ relative to the basis $\mathcal{A}$ of $\mathbf{V}$, and that Q is the matrix of transition from $\mathcal{A}$ to $\mathcal{B}$. Without using Theorem 5.15, prove that the matrix of f relative to $\mathcal{B}$ is AQ.

31. Let $\mathbf{V}$ be a finite-dimensional vector space. For any nonempty subset $\mathcal{M}$ of $\mathbf{V}$, the **annihilator** of $\mathcal{M}$ is the set $\mathcal{M}^0$ of linear functionals given by

$$\mathcal{M}^0 = \{f \in \mathbf{V}^* \mid f(\mathbf{u}) = 0 \text{ for all } \mathbf{u} \in \mathcal{M}\}.$$

    (a) Prove that $\mathcal{M}^0$ is a subspace of $\mathbf{V}^*$.

    (b) Prove that if $\mathbf{W}$ is a subspace of $\mathbf{V}$, then $\dim(\mathbf{W}) + \dim(\mathbf{W}^0) = \dim(\mathbf{V})$.

32. Find a basis for $\mathcal{M}^0$.

    (a) $\mathcal{M} = \{(1, 2, -2, 4), (1, 1, 1, 6), (2, 3, -1, 10)\}$

    (b) $\mathcal{M} = \{(2, 0, -3, 6), (2, 1, 0, 4), (0, 1, 3, -2)\}$

    (c) $\mathcal{M} = \{(2, 2, 1, 0), (0, 4, 1, 0), (4, 8, 3, 0)\}$

    (d) $\mathcal{M} = \{(1, 0, 1, 0), (2, -3, -4, 1), (1, 1, -3, 0)\}$

33. For any nonempty subset $\mathcal{T}$ of $\mathbf{V}^*$,

$$\mathcal{T}^0 = \{\mathbf{u} \in \mathbf{V} \mid f(\mathbf{u}) = 0 \text{ for all } f \in \mathcal{T}\}.$$

    (a) Prove that $\mathcal{T}^0$ is a subspace of $\mathbf{V}$.

    (b) Prove that if $\mathcal{T}$ is a subspace of $\mathbf{V}^*$, that $\dim(\mathcal{T}) + \dim(\mathcal{T}^0) = \dim(\mathbf{V})$.

34. Prove that $(\mathbf{W}^0)^0 = \mathbf{W}$ for any subspace $\mathbf{W}$ of $\mathbf{V}$, and $(\mathcal{T}^0)^0 = \mathcal{T}$ for any subspace $\mathcal{T}$ of $\mathbf{V}^*$.

35. Prove that if $\mathcal{M}_1 \subseteq \mathcal{M}_2$, then $\mathcal{M}_1^0 \supseteq \mathcal{M}_2^0$.

36. Let $\mathbf{W}_1$ and $\mathbf{W}_2$ be subspaces of $\mathbf{V}$.

    (a) Prove that $(\mathbf{W}_1 + \mathbf{W}_2)^0 = \mathbf{W}_1^0 \cap \mathbf{W}_2^0$.

    (b) Prove that $(\mathbf{W}_1 \cap \mathbf{W}_2)^0 = \mathbf{W}_1^0 + \mathbf{W}_2^0$.

37. Let $\mathbf{V}$ be of dimension n over $\mathcal{F}$. It follows from Theorem 8.2 that $(\mathbf{V}^*)^* = \mathbf{V}^{**}$ is an n-dimensional vector space over $\mathcal{F}$.

(a) For each $\mathbf{u} \in \mathbf{V}$, define the function $h_\mathbf{u}$ on $\mathbf{V}^*$ by $h_\mathbf{u}(f) = f(\mathbf{u})$ for all f in $\mathbf{V}^*$. Prove that $h_\mathbf{u} \in \mathbf{V}^{**}$.

(b) Prove that the mapping from $\mathbf{V}$ to $\mathbf{V}^{**}$ defined by $\phi(\mathbf{u}) = h_\mathbf{u}$ is an isomorphism from $\mathbf{V}$ to $\mathbf{V}^{**}$.

(*Comment*: Since $\mathbf{V}$ and $\mathbf{V}^{**}$ are n-dimensional spaces over $\mathcal{F}$, each is isomorphic to $\mathcal{F}^n$. The isomorphism ϕ is such a natural one, however, that it is ordinarily used to identify $\mathbf{V}$ and $\mathbf{V}^{**}$ as being the same space. That is, $\mathbf{u}$ and $h_\mathbf{u}$ are regarded as the same entity. This point of view is advantageous in certain instances in linear programming.)

8.2 Real Quadratic Forms

In this section we turn our attention to a second type of vector function, the *quadratic form*. For the time being, we are concerned only with *real quadratic forms*. We consider quadratic forms in a more general setting in Section 8.8.

**8.5
DEFINITION**

Let q be a mapping of $\mathbf{R}^n$ into $\mathbf{R}$. Then q is a **real quadratic form** if there exist constants c_{ij} in $\mathbf{R}$ such that

$$q(\mathbf{v}) = \sum_{i=1}^{n}\sum_{j=1}^{n} c_{ij}x_ix_j$$

for each $\mathbf{v} = (x_1, x_2, \ldots, x_n)$ in $\mathbf{R}^n$.

For convenience, we shall refer to a real quadratic form in this section simply as a "quadratic form." And since $q(\mathbf{v})$ is a polynomial in $x_1, x_2, \ldots, x_n$, we refer to q as a quadratic form in the variables $x_1, x_2, \ldots, x_n$. The use of parentheses in both $q(\mathbf{v})$ and $\mathbf{v} = (x_1, x_2, \ldots, x_n)$ leads to the clumsy expression $q(\mathbf{v}) = q((x_1, x_2, \ldots, x_n))$, so we shall drop one set of parentheses from this notation. Thus a quadratic form in x_1, x_2 is given by

$$q(x_1, x_2) = ax_1^2 + bx_2^2 + cx_1x_2.$$

The student has no doubt encountered such expressions as

$$q(x, y) = ax^2 + by^2 + cxy$$

in analytic geometry or the calculus. For example, the left-hand member of

$$9x^2 + 16y^2 = 144$$

defines a quadratic form in x and y, as does the left-hand member of

$$xy = 1.$$

Typically, the set of all points (x, y) in $\mathbf{R}^2$ for which a given quadratic form has a constant value is a conic section with center at the origin.

The value of a real quadratic form $q(x, y, z)$ in x, y, z is a polynomial

$$q(x, y, z) = ax^2 + by^2 + cz^2 + dxy + exz + fyz,$$

where the coefficients are real numbers. Typically, the set of all points (x, y, z) in $\mathbf{R}^3$ for which a given quadratic form has a constant value is a quadric surface.

EXAMPLE 1   Consider the real quadratic form q defined on $\mathbf{R}^3$ by

$$q(x_1, x_2, x_3) = x_1^2 - 2x_2^2 + 4x_3^2 + 4x_1x_2 - 3x_2x_3.$$

If we let $X = [\,x_1 \quad x_2 \quad x_3\,]^T$ and

$$A = \begin{bmatrix} 1 & 4 & 0 \\ 0 & -2 & -3 \\ 0 & 0 & 4 \end{bmatrix},$$

a simple computation shows that $X^T A X$ is a matrix with a single element $q(x_1, x_2, x_3)$:

$$X^T A X = [x_1^2 - 2x_2^2 + 4x_3^2 + 4x_1x_2 - 3x_2x_3].$$

When $q(x_1, x_2, x_3)$ and this matrix of order 1 are identified as being the same, we have

$$q(x_1, x_2, x_3) = X^T A X.$$

In choosing this matrix A, we simply entered the coefficient a_{ij} of $x_i x_j$ in the ith row and jth column of A. Since the cross-product terms can be written in many ways, the matrix A used in $q(x_1, x_2, x_3) = X^T A X$ is not unique. For instance, the term $4x_1x_2$ can be written as $4x_2x_1$ or as $3x_1x_2 + x_2x_1$. These variations would lead one to use

$$A = \begin{bmatrix} 1 & 0 & 0 \\ 4 & -2 & -3 \\ 0 & 0 & 4 \end{bmatrix} \quad \text{or} \quad A = \begin{bmatrix} 1 & 3 & 0 \\ 1 & -2 & -3 \\ 0 & 0 & 4 \end{bmatrix}.$$

With the cross-product terms split into equal parts, the symmetric matrix

$$A = \begin{bmatrix} 1 & 2 & 0 \\ 2 & -2 & -\frac{3}{2} \\ 0 & -\frac{3}{2} & 4 \end{bmatrix}$$

would be used.   ∎

For any $q(x_1, x_2, \ldots, x_n) = \sum_{i=1}^n \sum_{j=1}^n c_{ij} x_i x_j$, it is always possible to write

$$q(x_1, x_2, \ldots, x_n) = X^T A X$$

with $X = [\, x_1 \quad x_2 \quad \cdots \quad x_n \,]^T$ and with A a symmetric $n \times n$ matrix as in the foregoing example. For any $r \neq s$, the sum of the two terms $c_{rs}x_r x_s + c_{sr}x_s x_r$ can be split into two parts $a_{rs}x_r x_s$ and $a_{sr}x_s x_r$ with

$$a_{rs} = a_{sr} = \frac{c_{rs} + c_{sr}}{2}.$$

Together with $a_{rr} = c_{rr}$, this yields

$$q(x_1, x_2, \ldots, x_n) = \sum_{i=1}^{n} \sum_{j=1}^{n} a_{ij} x_i x_j = X^T A X,$$

where $A = [a_{ij}]_n$ is a symmetric matrix.

8.6
DEFINITION

Whenever $q(x_1, x_2, \ldots, x_n)$ is written as

$$q(x_1, x_2, \ldots, x_n) = X^T A X$$

with

$$X = \begin{bmatrix} x_1 \\ x_2 \\ \vdots \\ x_n \end{bmatrix}$$

and with $A = [a_{ij}]_n$ a symmetric matrix, we say that A **represents** the quadratic form q, or that A is the **matrix** of q relative to $x_1, x_2, \ldots, x_n$.

According to this definition, a matrix must be symmetric in order to be called *the matrix* of a quadratic form, or to say that the matrix *represents* the quadratic form.

Unless stated otherwise, it will be assumed from now on that each $X^T A X$ is written with A symmetric. Under this restriction, the matrix that represents a certain quadratic form is unique in a given set of variables $x_1, x_2, \ldots, x_n$. As a first step in establishing this result, we prove the following lemma.

8.7
LEMMA

If A is a real symmetric matrix of order n and $X^T A X = 0$ for all choices of $X = [\, x_1 \quad x_2 \quad \cdots \quad x_n \,]^T$, then A is a zero matrix.

Proof Suppose that A is a matrix that satisfies the given conditions. Let $x_r = 1$ and all other $x_i = 0$. Then $0 = X^T A X = a_{rr}$. Now let $r \neq s$, and put $x_r = 1$ and $x_s = 1$, with all other $x_i = 0$. Then

$$0 = X^T A X$$

$$= \sum_{i=1}^{n} \sum_{j=1}^{n} a_{ij} x_i x_j$$

$$= a_{rs} x_r x_s + a_{sr} x_s x_r$$

$$= a_{rs} + a_{sr}.$$

But $a_{rs} + a_{sr} = 2a_{rs}$ since A is symmetric, so we have $a_{rs} = 0$. Since r and s were arbitrary, A is a zero matrix. ∎ ∎ ∎

8.8
THEOREM

> The matrix of the real quadratic form q relative to $x_1, x_2, \ldots, x_n$ is unique.

Proof Suppose that A and B both represent q relative to $x_1, x_2, \ldots, x_n$. Then

$$X^T A X = q(x_1, x_2, \ldots, x_n)$$
$$= X^T B X,$$

and this implies that

$$X^T (A - B) X = 0$$

for all choices of $X = [\, x_1 \ \ x_2 \ \ \cdots \ \ x_n \,]^T$. Now $A - B$ is symmetric since A and B are symmetric (see Problem 6), and therefore $A = B$ by the lemma. ∎ ∎ ∎

Let us now reexamine the definition of a real quadratic form. As stated in Definition 8.5, $q(\mathbf{v})$ is a scalar that is uniquely determined by the vector $\mathbf{v}$ in $\mathbf{R}^n$. Although the discussion up to this point has been primarily in terms of the components x_i of $\mathbf{v}$, this should not obscure the fact that q is a function of the vector variable $\mathbf{v}$. Now the vector $\mathbf{v}$ is uniquely determined by its coordinates relative to any given basis of $\mathbf{R}^n$, and so the value $q(\mathbf{v})$ should also be uniquely determined by these coordinates. Our next theorem gives $q(\mathbf{v})$ explicitly in terms of these coordinates. Theorem 5.14 provides the key to the proof.

8.9
THEOREM

> Suppose that, for each $\mathbf{v} = (x_1, x_2, \ldots, x_n)$ in $\mathbf{R}^n$,
>
> $$q(\mathbf{v}) = X^T A X,$$
>
> where A is the matrix of q relative to $x_1, x_2, \ldots, x_n$. Let P be the matrix of transition from $\mathcal{E}_n$ to the basis $\mathcal{A}$ and let $Y = [\, y_1 \ \ y_2 \ \ \cdots \ \ y_n \,]^T$ be the coordinate matrix of

v relative to $\mathcal{A}$. Then

$$q(\mathbf{v}) = Y^T BY,$$

where $B = P^T A P$.

Proof Since $X = [\,x_1 \quad x_2 \quad \cdots \quad x_n\,]^T$ is the coordinate matrix of $\mathbf{v} = (x_1, x_2, \ldots, x_n)$ relative to $\mathcal{E}_n$, $X = PY$ by Theorem 5.14. Hence

$$q(\mathbf{v}) = X^T A X = (PY)^T A(PY) = Y^T (P^T A P)Y. \qquad \blacksquare\blacksquare\blacksquare$$

EXAMPLE 2 Let q be the quadratic form given by

$$q(x_1, x_2, x_3) = 13x_1^2 + 5x_2^2 + 26x_3^2 - 16x_1x_2 - 40x_1x_3 + 24x_2x_3,$$

and let $Y = [\,y_1 \quad y_2 \quad y_3\,]^T$ be the coordinate matrix of $\mathbf{v} = (x_1, x_2, x_3)$ relative to the basis $\mathcal{A} = \{(1, 2, 0), (0, -2, 1), (2, 5, -1)\}$. We shall find the expression for $q(\mathbf{v})$ in terms of y_1, y_2, y_3.

Written in matrix form, $q(\mathbf{v})$ appears as

$$q(x_1, x_2, x_3) = \begin{bmatrix} x_1 & x_2 & x_3 \end{bmatrix} \begin{bmatrix} 13 & -8 & -20 \\ -8 & 5 & 12 \\ -20 & 12 & 26 \end{bmatrix} \begin{bmatrix} x_1 \\ x_2 \\ x_3 \end{bmatrix}.$$

The matrix of transition from $\mathcal{E}_3$ to $\mathcal{A}$ is

$$P = \begin{bmatrix} 1 & 0 & 2 \\ 2 & -2 & 5 \\ 0 & 1 & -1 \end{bmatrix},$$

so Theorem 8.9 assures us that $q(\mathbf{v}) = Y^T BY$, where

$$B = P^T A P$$

$$= \begin{bmatrix} 1 & 2 & 0 \\ 0 & -2 & 1 \\ 2 & 5 & -1 \end{bmatrix} \begin{bmatrix} 13 & -8 & -20 \\ -8 & 5 & 12 \\ -20 & 12 & 26 \end{bmatrix} \begin{bmatrix} 1 & 0 & 2 \\ 2 & -2 & 5 \\ 0 & 1 & -1 \end{bmatrix}$$

$$= \begin{bmatrix} 1 & 0 & 0 \\ 0 & -2 & 0 \\ 0 & 0 & 3 \end{bmatrix}.$$

Thus the expression for $q(\mathbf{v})$ in terms of y_1, y_2, y_3 is given by the simple form

$$q(\mathbf{v}) = y_1^2 - 2y_2^2 + 3y_3^2. \qquad \blacksquare$$

Theorem 8.9 makes available many different expressions for the value $q(\mathbf{v})$ of a quadratic form. From one point of view, it describes the effect of a change of variables from the set $x_1, x_2, \ldots, x_n$ to the set $y_1, y_2, \ldots, y_n$ according to the rule that $X = PY$. Such a change of variables is called a **linear** change of variables. The terminology of Definition 8.6 can be applied to the variables y_i as well as the x_i.

8.10
COROLLARY

> Suppose that $\mathbf{v} = (x_1, x_2, \ldots, x_n)$ and A represents the quadratic form q relative to $x_1, x_2, \ldots, x_n$. If the variables $y_1, y_2, \ldots, y_n$ are related to the x_i by $X = PY$ with P invertible, then the matrix $B = P^T A P$ represents q relative to $y_1, y_2, \ldots, y_n$.

Proof With $B = P^T A P$, we have $q(\mathbf{v}) = Y^T B Y$ from the theorem. This expression is valid for any $\mathbf{v} = (x_1, x_2, \ldots, x_n)$ since $Y = P^{-1} X$ can take on any prescribed value. Now B is symmetric, since

$$B^T = (P^T A P)^T = P^T A^T (P^T)^T = P^T A P = B.$$

Hence B represents q relative to $y_1, y_2, \ldots, y_n$. ■ ■ ■

The results of the preceding theorem and corollary motivate the introduction of the relation of congruence on matrices.

8.11
DEFINITION

> Let A and B be matrices over the field $\mathcal{F}$. Then B is **congruent** to A over $\mathcal{F}$ if there is an invertible matrix P over $\mathcal{F}$ such that $B = P^T A P$.

It is left as an exercise (Problem 8) to prove that congruence over $\mathcal{F}$ is an equivalence relation on the set of all square matrices over $\mathcal{F}$.

The principal objective of this and the next two sections is to show that the polynomial expression for the value of a quadratic form q takes on the particularly simple form of a "sum of squares"

$$q(\mathbf{v}) = \lambda_1 y_1^2 + \lambda_2 y_2^2 + \cdots + \lambda_n y_n^2$$

whenever the basis $\mathcal{A}$ in Theorem 8.9 is chosen appropriately. Example 2 illustrates this simple form, but we have no method for finding an appropriate basis $\mathcal{A}$ at this time.

8.2 Exercises

1. Write $q(x_1, x_2, \ldots, x_n)$ as $X^T A X$ with two different matrices A, one of which is symmetric.

(a) $q(x_1, x_2) = 4x_1^2 + 6x_1x_2 - 2x_2^2$

(b) $q(x_1, x_2) = 3x_1^2 + 5x_2^2$

(c) $q(x_1, x_2, x_3) = -2x_1^2 - 3x_1x_3 + 8x_2x_3 - 10x_2x_1 - 7x_3^2$

(d) $q(x_1, x_2, x_3) = 12x_1^2 - 8x_2x_1 + 4x_1x_3 - 5x_2^2$

(e) $q(x_1, x_2, x_3) = 6x_2^2 - 9x_1x_2 - 7x_1^2$

(f) $q(x_1, x_2, x_3, x_4) = -3x_2^2 + 4x_1^2 - 11x_1x_4 + 5x_2x_4 + 18x_1x_2 + 16x_4^2$

(g) $q(x_1, x_2, x_3) = \sum_{i=1}^{3} \sum_{j=1}^{3} a_{ij}x_i x_j$ with $a_{ij} = i + j$

(h) $q(x_1, x_2, x_3) = \sum_{i=1}^{3} \sum_{j=1}^{3} a_{ij}x_i x_j$ with $a_{ij} = i - j$

2. Suppose that for each $\mathbf{v} = (x_1, x_2, \ldots, x_n)$ in $\mathbf{R}^n$, $q(\mathbf{v}) = X^T A X$ for the given matrix A. For the given basis $\mathcal{A}$ of $\mathbf{R}^n$, find the expression for $q(\mathbf{v})$ in terms of the coordinates y_i of $\mathbf{v}$ relative to $\mathcal{A}$.

(a) $A = \begin{bmatrix} \frac{3}{2} & \sqrt{2} & -\frac{1}{2} \\ \sqrt{2} & 1 & -\sqrt{2} \\ -\frac{1}{2} & -\sqrt{2} & -\frac{5}{2} \end{bmatrix}$, $\mathcal{A} = \{(1, 0, 1), (3, \sqrt{2}, 1), (3\sqrt{2}, -4, \sqrt{2})\}$

(b) $A = \begin{bmatrix} 1 & -2 & 0 \\ -2 & 2 & -2 \\ 0 & -2 & 3 \end{bmatrix}$, $\mathcal{A} = \{(1, 0, 1), (0, 1, 1), (1, 1, 0)\}$

(c) $A = \begin{bmatrix} 1 & -1 & -1 \\ -1 & 1 & -1 \\ -1 & -1 & 1 \end{bmatrix}$, $\mathcal{A} = \{(1, 0, 0), (1, 1, 0), (1, 1, 1)\}$

(d) $A = \begin{bmatrix} 1 & 1 & 2 \\ 1 & 3 & \frac{1}{2} \\ 2 & \frac{1}{2} & 7 \end{bmatrix}$, $\mathcal{A} = \{(1, 0, 0), (1, -1, 0), (-11, 3, 4)\}$

3. Find the matrix that represents the given quadratic form relative to the variables y_i.

(a) $q(x_1, x_2, x_3) = 4x_1x_2 - 2x_1x_3 + x_2^2 + 2x_2x_3 - 2x_3^2$,
$x_1 = y_1$
$x_2 = -y_1 + y_2 + 2y_3$
$x_3 = -y_1 + 2y_2 + 2y_3$

(b) $q(x_1, x_2, x_3) = x_1^2 + 4x_3^2 - 2x_1x_2 + 4x_1x_3 - 6x_2x_3$,
$x_1 = -4y_1 - 5y_2 + 3y_3$
$x_2 = 3y_1 + 3y_2 - 2y_3$
$x_3 = -y_1 - y_2 + y_3$

(c) $q(x_1, x_2, x_3) = 4x_1x_2 + 4x_1x_3 + 4x_2x_3$,
$x_1 = 2y_1 + y_2 - 2y_3$
$x_2 = 3y_1 + y_3$
$x_3 = 6y_1 - 3y_2 + 2y_3$

(d) $q(x_1, x_2, x_3) = -x_1^2 + 2x_2^2 - 6x_3^2 + 4x_1x_2 - x_1x_3 - 4x_2x_3$,
$x_1 = y_1 + 2y_2 + y_3$
$x_2 = -y_1 - y_2 + y_3$
$x_3 = y_2 + 3y_3$

4. Verify that if $a_{rs} = a_{sr} = \frac{1}{2}(c_{rs}+c_{sr})$ for $r, s = 1, 2, \ldots, n$, then $\sum_{i=1}^{n} \sum_{j=1}^{n} c_{ij}x_ix_j = \sum_{i=1}^{n} \sum_{j=1}^{n} a_{ij}x_ix_j$.

5. Recall that a matrix A is *skew-symmetric* if and only if $A^T = -A$. Prove that if A is skew-symmetric, then $X^T A X = 0$ for all $X = [\, x_1 \quad x_2 \quad \cdots \quad x_n \,]^T$.

6. Prove that the sum and difference of two symmetric matrices of the same order are symmetric.

7. Prove that any square matrix can be written uniquely as the sum of a symmetric and a skew-symmetric matrix. (*Hint:* $A + A^T$ is symmetric.)

8. Prove that the relation of congruence over $\mathcal{F}$ is an equivalence relation on the set of all square matrices over $\mathcal{F}$.

9. Prove that two symmetric matrices A and B over $\mathbf{R}$ are congruent if and only if they represent the same real quadratic form relative to two sets of n variables that are related by an invertible linear change of variables.

10. (a) Show that the set of symmetric matrices of order n over $\mathbf{R}$ is not closed under multiplication.

   (b) Under what condition will the product of two symmetric matrices be symmetric?

11. Show that AA^T and $A^T A$ are symmetric for any matrix A.

12. Show that if A is an $n \times n$ matrix over $\mathbf{R}$ such that $X^T A X = 0$ for all $X = [\, x_1 \quad x_2 \quad \cdots \quad x_n \,]^T$, then A is skew-symmetric.

13. Let q be a quadratic form represented by the symmetric matrix A relative to $x_1, x_2, \ldots, x_n$. Show that q can be expressed as $q(x_1, x_2, \ldots, x_n) = (AX)^T X$.

14. Let q be a real quadratic form defined on $\mathbf{R}^n$.

   (a) Express $q(k\mathbf{v})$ in terms of $q(\mathbf{v})$, for $k \in \mathbf{R}$ and $\mathbf{v} \in \mathbf{R}^n$.

   (b) Prove or disprove: $q(\mathbf{v} + \mathbf{w}) = q(\mathbf{v}) + q(\mathbf{w})$ for $\mathbf{v}, \mathbf{w} \in \mathbf{R}^n$.

8.3 Orthogonal Matrices

Corollary 8.10 shows that the problem of reducing the expression for $q(\mathbf{v})$ by a linear change of variable to the form

$$q(\mathbf{v}) = \lambda_1 y_1^2 + \lambda_2 y_2^2 + \cdots + \lambda_n y_n^2$$

is equivalent to that of finding an invertible matrix P such that

$$P^T A P = \text{diag}\{\lambda_1, \lambda_2, \ldots, \lambda_n\}.$$

This problem has something of the same flavor as the diagonalization problem studied in Section 7.3. The difference is that $P^T A P$ has taken the place of $P^{-1} A P$. It is not unnatural, then, to consider the possibility of finding a matrix P over $\mathbf{R}$ such that $P^T A P$ is diagonal and $P^T = P^{-1}$. These are clearly strong restrictions to be placed on P, but oddly enough it is true that such a P *always* exists (whenever A is real and symmetric). Those matrices P that have the property that $P^T = P^{-1}$ are quite useful in many instances, and have a special name.

8.12 DEFINITION

A matrix P over $\mathbf{R}$ is **orthogonal** if $P^T = P^{-1}$.

It would seem natural to expect a connection between the use of the word orthogonal to describe a matrix and the use of the same word to describe a set of vectors in Chapter 1. We recall that an orthogonal set of vectors is a set $\{\mathbf{u}_\lambda \mid \lambda \in \mathcal{L}\}$ such that $\mathbf{u}_{\lambda_1} \cdot \mathbf{u}_{\lambda_2} = 0$ whenever $\lambda_1 \neq \lambda_2$. An **orthogonal basis** of a subspace $\mathbf{W}$ of $\mathbf{R}^n$, then, is a basis of $\mathbf{W}$ that is an orthogonal set of vectors. The two uses of the word orthogonal are related, but not exactly in the way that one might expect. Definition 8.13 and Theorem 8.14 explain the relation completely.

8.13 DEFINITION

A set of vectors $\{\mathbf{u}_\lambda \mid \lambda \in \mathcal{L}\}$ is **orthonormal** if the set is orthogonal and if each $\mathbf{u}_\lambda$ has length 1.

The word "orthonormal" is a fusion of the words "orthogonal" and "normal." A **normalized** set of vectors is a set in which each vector has unit length.

8.14 THEOREM

Let P be an $r \times r$ matrix over $\mathbf{R}$. Then P is orthogonal if and only if P is the matrix of transition from one orthonormal set of r vectors in $\mathbf{R}^n$ to another orthonormal set of r vectors in $\mathbf{R}^n$.

Proof Let $P = [p_{ij}]_{r \times r}$ over $\mathbf{R}$, and let $\mathcal{A} = \{\mathbf{u}_1, \mathbf{u}_2, \ldots, \mathbf{u}_r\}$ be an orthonormal set in $\mathbf{R}^n$. Then P is the matrix of transition from $\mathcal{A}$ to a set $\mathcal{B} = \{\mathbf{v}_1, \mathbf{v}_2, \ldots, \mathbf{v}_r\}$ of vectors in $\mathbf{R}^n$ such that $\mathbf{v}_j = \sum_{k=1}^r p_{kj}\mathbf{u}_k$. Now $\mathcal{B}$ is orthonormal if and only if $\mathbf{v}_i \cdot \mathbf{v}_j = \delta_{ij}$ for all pairs i, j. We have

$$\mathbf{v}_i \cdot \mathbf{v}_j = \left(\sum_{k=1}^r p_{ki}\mathbf{u}_k\right)\left(\sum_{m=1}^r p_{mj}\mathbf{u}_m\right)$$

$$= \sum_{k=1}^r \sum_{m=1}^r p_{ki} p_{mj}\mathbf{u}_k \cdot \mathbf{u}_m$$

$$= \sum_{k=1}^r \sum_{m=1}^r p_{ki} p_{mj}\delta_{km} = \sum_{k=1}^r p_{ki} p_{kj},$$

where $\mathbf{u}_k \cdot \mathbf{u}_m = \delta_{km}$ since $\mathcal{A}$ is orthonormal. But this last sum is precisely the element in the ith row and jth column of $P^T P$. Thus $\mathbf{v}_i \cdot \mathbf{v}_j = \delta_{ij}$ if and only if $P^T P = I_r$, and the proof is complete.   ■ ■ ■

Thus a matrix P is orthogonal if and only if it preserves the property of being orthonormal from one basis to another. Whether the bases are orthonormal or not, we say that there is an **orthogonal change of basis** whenever the matrix of transition from one basis to the other is orthogonal. The associated linear change of variable $X = PY$ is called an **orthogonal change of variable**.

There is a question lurking in the background here that should be brought out into the open and answered. The question is this: Does every nonzero subspace of $\mathbf{R}^n$ have an orthonormal basis? Theorem 8.15 shows that the answer is affirmative, and the proof describes a procedure for obtaining an orthonormal basis from any given basis. This procedure is known as the **Gram-Schmidt Orthogonalization Process**.

8.15 THEOREM

Let $\mathcal{A} = \{\mathbf{u}_1, \mathbf{u}_2, \ldots, \mathbf{u}_r\}$ be a basis of the subspace $\mathbf{W}$ of $\mathbf{R}^n$. There exists an orthonormal basis $\mathcal{N} = \{\mathbf{v}_1, \mathbf{v}_2, \ldots, \mathbf{v}_r\}$ of $\mathbf{W}$ such that each $\mathbf{v}_i$ is a linear combination of $\mathbf{u}_1, \mathbf{u}_2, \ldots, \mathbf{u}_i$.

Proof   The proof is by induction on the dimension r of $\mathbf{W}$. In order to describe the procedure clearly, the routine is presented in full for $r = 1, 2, 3$.

Since $\mathcal{A}$ is linearly independent, each $\mathbf{u}_i \neq \mathbf{0}$ and $\|\mathbf{u}_i\| > 0$. Let $\mathbf{v}_1 = (\mathbf{u}_1/\|\mathbf{u}_1\|)$. Then $\mathbf{v}_1$ is a unit vector, and the proof is complete if $r = 1$.

For $r > 1$, let

$$\mathbf{w}_2 = \mathbf{u}_2 - (\mathbf{v}_1 \cdot \mathbf{u}_2)\mathbf{v}_1.$$

Then $\mathbf{w}_2 \neq \mathbf{0}$ since $\{\mathbf{v}_1, \mathbf{u}_2\}$ is linearly independent. We have

$$\mathbf{v}_1 \cdot \mathbf{w}_2 = \mathbf{v}_1 \cdot \mathbf{u}_2 - (\mathbf{v}_1 \cdot \mathbf{u}_2)\mathbf{v}_1 \cdot \mathbf{v}_1$$
$$= \mathbf{v}_1 \cdot \mathbf{u}_2 - \mathbf{v}_1 \cdot \mathbf{u}_2$$
$$= 0,$$

so $\mathbf{w}_2$ is orthogonal to $\mathbf{v}_1$. Put $\mathbf{v}_2 = (\mathbf{w}_2/\|\mathbf{w}_2\|)$. Then $\{\mathbf{v}_1, \mathbf{v}_2\}$ is an orthonormal set and $\mathbf{v}_i$ is a linear combination of $\mathbf{u}_1, \ldots, \mathbf{u}_i$ for $i = 1, 2$.

If $r > 2$, let

$$\mathbf{w}_3 = \mathbf{u}_3 - (\mathbf{v}_1 \cdot \mathbf{u}_3)\mathbf{v}_1 - (\mathbf{v}_2 \cdot \mathbf{u}_3)\mathbf{v}_2.$$

Then $\mathbf{w}_3 \neq \mathbf{0}$, since $\mathbf{w}_3 = \mathbf{0}$ would require that $\mathbf{u}_3$ be dependent on $\{\mathbf{v}_1, \mathbf{v}_2\}$ and hence on $\{\mathbf{u}_1, \mathbf{u}_2\}$. Since

$$\mathbf{v}_1 \cdot \mathbf{w}_3 = \mathbf{v}_1 \cdot \mathbf{u}_3 - (\mathbf{v}_1 \cdot \mathbf{u}_3)\mathbf{v}_1 \cdot \mathbf{v}_1 - (\mathbf{v}_2 \cdot \mathbf{u}_3)\mathbf{v}_1 \cdot \mathbf{v}_2$$

$$= \mathbf{v}_1 \cdot \mathbf{u}_3 - \mathbf{v}_1 \cdot \mathbf{u}_3$$

$$= 0,$$

$\mathbf{w}_3$ is orthogonal to $\mathbf{v}_1$. Similarly, $\mathbf{w}_3$ is orthogonal to $\mathbf{v}_2$. The vector $\mathbf{v}_3 = (\mathbf{w}_3/\|\mathbf{w}_3\|)$ is a unit vector and $\{\mathbf{v}_1, \mathbf{v}_2, \mathbf{v}_3\}$ is an orthonormal set such that $\mathbf{v}_i$ is a linear combination of $\mathbf{u}_1, \ldots, \mathbf{u}_i$ for $i = 1, 2, 3$.

Assume the theorem is true for all subspaces with $r = k$. Let $\mathcal{A} = \{\mathbf{u}_1, \ldots, \mathbf{u}_k, \mathbf{u}_{k+1}\}$ be a basis of the $(k + 1)$-dimensional subspace $\mathbf{W}$. By the induction hypothesis, the subspace $\langle \mathbf{u}_1, \mathbf{u}_2, \ldots, \mathbf{u}_k \rangle$ has an orthonormal basis $\{\mathbf{v}_1, \mathbf{v}_2, \ldots, \mathbf{v}_k\}$ such that $\mathbf{v}_i$ is a linear combination of $\mathbf{u}_1, \ldots, \mathbf{u}_i$ for each i. Let

$$\mathbf{w}_{k+1} = \mathbf{u}_{k+1} - (\mathbf{v}_1 \cdot \mathbf{u}_{k+1})\mathbf{v}_1 - (\mathbf{v}_2 \cdot \mathbf{u}_{k+1})\mathbf{v}_2 - \cdots - (\mathbf{v}_k \cdot \mathbf{u}_{k+1})\mathbf{v}_k$$

$$= \mathbf{u}_{k+1} - \sum_{j=1}^{k} (\mathbf{v}_j \cdot \mathbf{u}_{k+1})\mathbf{v}_j.$$

Then $\mathbf{w}_{k+1} \neq \mathbf{0}$, since otherwise $\mathbf{u}_{k+1}$ would be dependent on $\{\mathbf{v}_1, \mathbf{v}_2, \ldots, \mathbf{v}_k\}$ and hence on $\{\mathbf{u}_1, \mathbf{u}_2, \ldots, \mathbf{u}_k\}$. Since

$$\mathbf{v}_i \cdot \mathbf{w}_{k+1} = \mathbf{v}_i \cdot \mathbf{u}_{k+1} - \sum_{j=1}^{k} (\mathbf{v}_j \cdot \mathbf{u}_{k+1})\mathbf{v}_i \cdot \mathbf{v}_j$$

$$= \mathbf{v}_i \cdot \mathbf{u}_{k+1} - \mathbf{v}_i \cdot \mathbf{u}_{k+1}$$

$$= 0,$$

$\mathbf{w}_{k+1}$ is orthogonal to $\mathbf{v}_i$ for $i = 1, 2, \ldots, k$. Thus

$$\mathbf{v}_{k+1} = \frac{\mathbf{w}_{k+1}}{\|\mathbf{w}_{k+1}\|}$$

is a unit vector and $\{\mathbf{v}_1, \ldots, \mathbf{v}_k, \mathbf{v}_{k+1}\}$ is an orthonormal set such that $\mathbf{v}_i$ is a linear combination of $\{\mathbf{u}_1, \ldots, \mathbf{u}_i\}$ for $i = 1, 2, \ldots, k + 1$. Since $\{\mathbf{v}_1, \mathbf{v}_2, \ldots, \mathbf{v}_{k+1}\}$ is a linearly independent set of $k + 1$ vectors in $\mathbf{W}$, it is a basis of $\mathbf{W}$ by Theorem 1.34, and this completes the proof. ■■■

EXAMPLE 1 We shall use the Gram-Schmidt Orthogonalization Process to obtain an orthonormal basis of $\langle \mathcal{A} \rangle$, where $\mathcal{A}$ is the linearly independent set given by

$$\mathcal{A} = \{\mathbf{u}_1 = (2, 0, 2, 1), \mathbf{u}_2 = (0, 0, 4, 1), \mathbf{u}_3 = (8, 0, 3, 5)\}.$$

Using the same notation as in the proof of Theorem 8.15, we begin by writing

$$\mathbf{v}_1 = \frac{\mathbf{u}_1}{\|\mathbf{u}_1\|} = \tfrac{1}{3}(2, 0, 2, 1) = \left(\tfrac{2}{3}, 0, \tfrac{2}{3}, \tfrac{1}{3}\right).$$

Next we let

$$\begin{aligned}
\mathbf{w}_2 &= \mathbf{u}_2 - (\mathbf{v}_1 \cdot \mathbf{u}_2)\mathbf{v}_1 \\
&= (0, 0, 4, 1) - 3\left(\tfrac{2}{3}, 0, \tfrac{2}{3}, \tfrac{1}{3}\right) \\
&= (-2, 0, 2, 0)
\end{aligned}$$

and

$$\mathbf{v}_2 = \frac{\mathbf{w}_2}{\|\mathbf{w}_2\|} = \tfrac{\sqrt{2}}{4}(-2, 0, 2, 0) = \left(-\tfrac{\sqrt{2}}{2}, 0, \tfrac{\sqrt{2}}{2}, 0\right).$$

To obtain the third vector in our orthonormal basis of $\langle \mathcal{A} \rangle$, we write

$$\begin{aligned}
\mathbf{w}_3 &= \mathbf{u}_3 - (\mathbf{v}_1 \cdot \mathbf{u}_3)\mathbf{v}_1 - (\mathbf{v}_2 \cdot \mathbf{u}_3)\mathbf{v}_2 \\
&= (8, 0, 3, 5) - 9\left(\tfrac{2}{3}, 0, \tfrac{2}{3}, \tfrac{1}{3}\right) - \left(-\tfrac{5\sqrt{2}}{2}\right)\left(-\tfrac{\sqrt{2}}{2}, 0, \tfrac{\sqrt{2}}{2}, 0\right) \\
&= (8, 0, 3, 5) - (6, 0, 6, 3) + \left(-\tfrac{5}{2}, 0, \tfrac{5}{2}, 0\right) \\
&= \left(-\tfrac{1}{2}, 0, -\tfrac{1}{2}, 2\right).
\end{aligned}$$

Finally, we let

$$\mathbf{v}_3 = \frac{\mathbf{w}_3}{\|\mathbf{w}_3\|} = \tfrac{\sqrt{2}}{6}(-1, 0, -1, 4).$$

Thus the set $\{\mathbf{v}_1, \mathbf{v}_2, \mathbf{v}_3\}$ is an orthonormal basis of $\langle \mathcal{A} \rangle$, where

$$\mathbf{v}_1 = \tfrac{1}{3}(2, 0, 2, 1), \quad \mathbf{v}_2 = \tfrac{\sqrt{2}}{2}(-1, 0, 1, 0), \quad \mathbf{v}_3 = \tfrac{\sqrt{2}}{6}(-1, 0, -1, 4). \qquad\blacksquare$$

In the next example, we use the Gram-Schmidt Orthogonalization Process to extend an orthonormal set of 2 vectors to an orthonormal basis of $\mathbf{R}^3$.

EXAMPLE 2 The vectors $\mathbf{v}_1 = \tfrac{1}{\sqrt{3}}(1, 1, 1)$ and $\mathbf{v}_2 = \tfrac{1}{\sqrt{2}}(1, -1, 0)$ are orthonormal and the set $\mathcal{A} = \{\mathbf{v}_1, \mathbf{v}_2\}$ can be extended to an orthonormal basis of $\mathbf{R}^3$. Since any set of orthonormal vectors is linearly independent, we first extend $\mathcal{A}$ to a linearly independent set of three vectors. The vector $\mathbf{u}_3 = (0, 0, 1)$ does not depend on $\mathcal{A}$. Applying the Gram-Schmidt Orthogonalization Process (the last step only) to the set $\{\mathbf{v}_1, \mathbf{v}_2, \mathbf{u}_3\}$ we find:

$$\begin{aligned}
\mathbf{w}_3 &= \mathbf{u}_3 - (\mathbf{v}_1 \cdot \mathbf{u}_3)\mathbf{v}_1 - (\mathbf{v}_2 \cdot \mathbf{u}_3)\mathbf{v}_2 \\
&= (0, 0, 1) - \left(\tfrac{1}{\sqrt{3}}(1, 1, 1) \cdot (0, 0, 1)\right)\tfrac{1}{\sqrt{3}}(1, 1, 1) \\
&\quad - \left(\tfrac{1}{\sqrt{2}}(1, -1, 0) \cdot (0, 0, 1)\right)\tfrac{1}{\sqrt{2}}(1, -1, 0) \\
&= (0, 0, 1) - \tfrac{1}{\sqrt{3}} \cdot \tfrac{1}{\sqrt{3}}(1, 1, 1) - 0 \cdot \tfrac{1}{\sqrt{2}}(1, -1, 0) \\
&= (0, 0, 1) - \tfrac{1}{3}(1, 1, 1) \\
&= \tfrac{1}{3}(-1, -1, 2).
\end{aligned}$$

Finally we normalize $\mathbf{w}_3$,

$$\mathbf{v}_3 = \frac{\mathbf{w}_3}{\|\mathbf{w}_3\|} = \tfrac{1}{\sqrt{6}}(-1, -1, 2)$$

and $\left\{ \frac{1}{\sqrt{3}}(1, 1, 1), \frac{1}{\sqrt{2}}(1, -1, 0), \frac{1}{\sqrt{6}}(-1, -1, 2) \right\}$ is an orthonormal basis of $\mathbf{R}^3$.   ■

8.3   Exercises

1. Let

$$A = \left\{ \left(\tfrac{2}{3}, \tfrac{2}{3}, \tfrac{1}{3} \right), \left(-\tfrac{2}{3}, \tfrac{1}{3}, \tfrac{2}{3} \right) \right\} \quad \text{and} \quad P = \begin{bmatrix} \tfrac{1}{2} & -\tfrac{\sqrt{3}}{2} \\ \tfrac{\sqrt{3}}{2} & \tfrac{1}{2} \end{bmatrix}.$$

(a) Verify that A is an orthonormal set.

(b) Verify that P is an orthogonal matrix.

(c) Suppose that P is the transition matrix from A to $\{\mathbf{v}_1, \mathbf{v}_2\}$. Find $\mathbf{v}_1$ and $\mathbf{v}_2$, and verify that $\{\mathbf{v}_1, \mathbf{v}_2\}$ is orthonormal.

(d) Extend A to an orthonormal basis of $\mathbf{R}^3$.

2. Given that $\{\mathbf{u}_1 = (1, -1, 0, 1), \mathbf{u}_2 = (4, 1, 3, 0), \mathbf{u}_3 = (2, -3, 9, 4)\}$ is linearly independent, find an orthonormal set $\{\mathbf{v}_1, \mathbf{v}_2, \mathbf{v}_3\}$ such that each $\mathbf{v}_i$ is a linear combination of $\mathbf{u}_1, \ldots, \mathbf{u}_i$.

3. Write $\mathbf{v}_3$ in Problem 2 as a linear combination of $\mathbf{u}_1, \mathbf{u}_2, \mathbf{u}_3$.

4. Given that the set A is linearly independent, use the Gram-Schmidt Orthogonalization Process to obtain an orthonormal basis of $\langle A \rangle$.

(a) $A = \{(1, 2, -2, 4), (1, 1, 1, 6), (5, 2, 2, 5)\}$

(b) $A = \{(2, 0, -3, 6), (2, 1, 0, 4), (1, 7, -1, 3)\}$

(c) $A = \{(2, 2, 1), (0, 4, 1), (8, 3, 5)\}$

(d) $A = \{(1, 0, 1, 0), (1, 1, -3, 0), (2, -3, -4, 1), (2, 3, -2, -3)\}$

5. Extend each of the following orthonormal sets to an orthonormal basis of $\mathbf{R}^n$.

(a) $\left\{ \tfrac{1}{2}\left(1, \sqrt{3}, 0\right), \tfrac{1}{2}\left(-\sqrt{3}, 1, 0\right) \right\}$ (b) $\left\{ \tfrac{1}{\sqrt{3}}(1, 1, 1), \tfrac{1}{\sqrt{2}}(0, 1, -1) \right\}$

(c) $\left\{ (1, 0, 0, 0), \tfrac{1}{\sqrt{2}}(0, 0, 1, -1), \tfrac{1}{\sqrt{3}}(0, 1, 1, 1) \right\}$

(d) $\left\{ \tfrac{1}{\sqrt{3}}(1, 0, 1, 1), \tfrac{1}{\sqrt{15}}(1, 3, -2, 1), \tfrac{1}{\sqrt{35}}(1, 3, 3, -4) \right\}$

6. For each of the following orthogonal sets A, find an orthonormal basis of $\langle A \rangle$ and extend that basis to an orthonormal basis of $\mathbf{R}^4$.

(a) $A = \{(-1, 1, 1, 0), (0, -1, 1, 1)\}$ (b) $A = \{(1, 1, 1, 1), (0, -1, 1, 0)\}$

(c) $A = \{(1, 2, 0, -1), (1, -1, 0, -1)\}$ (d) $A = \{(-1, -1, 1, 3), (0, 3, 0, 1)\}$

7. Prove that the set of all orthogonal matrices of order n is closed under multiplication.

8. Prove that P^{-1} is orthogonal whenever P is orthogonal.

9. Prove that if P is orthogonal, then $\det(P) = \pm 1$.

10. Prove that, given any unit vector $\mathbf{v}_1$ in $\mathbf{R}^n$, there is an orthonormal basis $\mathbf{R}^n$ that has $\mathbf{v}_1$ for its first element.

11. A linear operator T on $\mathbf{R}^n$ is called **orthogonal** if and only if $\|T(\mathbf{u})\| = \|\mathbf{u}\|$ for every $\mathbf{u} \in \mathbf{R}^n$. Prove that T is orthogonal if and only if $T(\mathbf{u}) \cdot T(\mathbf{v}) = \mathbf{u} \cdot \mathbf{v}$ for all $\mathbf{u}, \mathbf{v} \in \mathbf{R}^n$. (*Hint:* $\|\mathbf{u} + \mathbf{v}\|^2 - \|\mathbf{u}\|^2 - \|\mathbf{v}\|^2 = 2\mathbf{u} \cdot \mathbf{v}$.)

12. (See Problem 11.) Prove that T is orthogonal if and only if T maps an orthonormal basis $\{\mathbf{v}_1, \mathbf{v}_2, \ldots, \mathbf{v}_n\}$ of $\mathbf{R}^n$ onto an orthonormal basis $\{T(\mathbf{v}_1), T(\mathbf{v}_2), \ldots, T(\mathbf{v}_n)\}$ of $\mathbf{R}^n$.

13. (See Problem 12.) Let $\mathcal{N} = \{\mathbf{v}_1, \mathbf{v}_2, \ldots, \mathbf{v}_n\}$ be an arbitrary orthonormal basis of $\mathbf{R}^n$. Prove that T is orthogonal if and only if the matrix of T relative to $\mathcal{N}$ is an orthogonal matrix.

14. Determine whether each of the following linear operators is orthogonal.
    (a) $T : \mathbf{R}^3 \to \mathbf{R}^3$ defined by $T(x_1, x_2, x_3) = (x_3, x_2, x_1)$
    (b) $T : \mathbf{R}^3 \to \mathbf{R}^3$ defined by $T(x_1, x_2, x_3) = (-x_1, -x_2, -x_3)$
    (c) $T : \mathbf{R}^4 \to \mathbf{R}^4$ defined by $T(x_1, x_2, x_3, x_4) = (x_1, 0, 0, x_4)$
    (d) $T : \mathbf{R}^4 \to \mathbf{R}^4$ defined by $T(x_1, x_2, x_3, x_4) = (x_1, x_2 + x_3, 0, x_4)$
    (e) $T : \mathbf{R}^4 \to \mathbf{R}^4$ defined by $T(x_1, x_2, x_3, x_4) = \left(2x_1, \frac{1}{2}x_2, x_3, x_4\right)$
    (f) $T : \mathbf{R}^4 \to \mathbf{R}^4$ defined by $T(x_1, x_2, x_3, x_4) = (x_1, x_1, x_1, x_1)$

15. (See Problem 13.) Show that the product of two orthogonal linear operators on $\mathbf{R}^n$ is an orthogonal linear operator.

16. It follows from Problem 13 that any orthogonal linear operator is invertible. Prove that if T is orthogonal, then T^{-1} is orthogonal.

17. For each of the linear operators T in Problem 14 that is orthogonal, find T^{-1}.

8.4 Reduction of Real Quadratic Forms

We return now to the problem posed at the beginning of Section 8.3, that of reducing a real quadratic form $q(\mathbf{v}) = \sum_{i=1}^{n} \sum_{j=1}^{n} a_{ij} x_i x_j$ to the form

$$q(\mathbf{v}) = \lambda_1 y_1^2 + \lambda_2 y_2^2 + \cdots + \lambda_n y_n^2$$

by an *orthogonal* linear change of variables $X = PY$. This is equivalent to finding an orthonormal basis $\mathcal{A}$ of $\mathbf{R}^n$ such that the linear operator T with matrix $A = [a_{ij}]_n$ relative to $\mathcal{E}_n$ is represented by $P^{-1}AP = \mathrm{diag}\{\lambda_1, \lambda_2, \ldots, \lambda_n\}$ relative to $\mathcal{A}$. This is the same situation as that considered in Section 7.3, except that now P must be orthogonal. If the eigenvalues λ_j are real, then the solutions P_j to the equation

$$(A - \lambda_j I)P_j = 0$$

can be taken to be real since A is real. We show in our next theorem that the λ_j's are indeed real because A is symmetric. Before proceeding to this theorem, we introduce some needed terminology and notation.

8.16 DEFINITION

Let $C = [c_{rs}]_{m \times n}$ be a matrix over the field $\mathcal{C}$ of complex numbers. The **conjugate** of C is the matrix $\overline{C} = [\overline{c}_{rs}]_{m \times n}$, where $\overline{c}_{rs} = a_{rs} - ib_{rs}$ is the conjugate[2] of the complex number $c_{rs} = a_{rs} + ib_{rs}$.

If a and b are real numbers, the notation $\overline{z}$ for the conjugate $a - bi$ of the complex number $z = a + bi$ is a standard one, and we have merely extended this notation to matrices. The basic properties $\overline{z_1 z_2} = \overline{z}_1 \overline{z}_2$ and $\overline{z_1 + z_2} = \overline{z}_1 + \overline{z}_2$ are valid for matrices: $\overline{AB} = \overline{A}\,\overline{B}$ and $\overline{A + B} = \overline{A} + \overline{B}$. (See Problem 5 of the Exercises.)

8.17 DEFINITION

If $C = [c_{rs}]_{m \times n}$ is a matrix over $\mathcal{C}$, then the **conjugate transpose** of C is the matrix $C^* = (\overline{C})^T$.

The proofs of the equalities $C^* = \overline{(C^T)}$, $(A_1 + A_2 + \cdots + A_n)^* = A_1^* + A_2^* + \cdots + A_n^*$, and $(A_1 A_2 \cdots A_n)^* = A_n^* \cdots A_2^* A_1^*$ are left as exercises.

8.18 THEOREM

The spectrum of a real symmetric matrix is a set of real numbers. That is, all of the eigenvalues of a real symmetric matrix are real.

Proof Let A be a real symmetric matrix, let λ be any eigenvalue[3] of A in the field $\mathcal{C}$, and let X be an eigenvector over $\mathcal{C}$ associated with λ. Then $AX = \lambda X$, and this implies that

$$X^* A X = \lambda X^* X.$$

We regard this equality of 1 by 1 matrices as an equality of numbers. Now

$$(X^* A X)^* = X^* A^* (X^*)^* = X^* A X$$

since $\overline{A} = A$ and $A^T = A$ imply that $A^* = A$. Hence $X^* A X$, is a real number. Also,

$$X^* X = \sum_{k=1}^{n} \overline{x}_k x_k = \sum_{k=1}^{n} |x_k|^2,$$

---

[2] The symbol i here denotes the square root of -1; a_{rs} and b_{rs} are real numbers.
[3] We do not prove it here, but every polynomial of degree n with coefficients in $\mathcal{C}$ has exactly n zeros, all of which are in $\mathcal{C}$. In particular, the polynomial $\det(A - \lambda I)$ has n zeros in $\mathcal{C}$ if A has order n.

which is a positive real number since $X \neq \mathbf{0}$. This means that

$$\lambda = \frac{X^* A X}{X^* X},$$

a quotient of real numbers.   ∎

We proceed now to the main result of this section.

8.19 THEOREM

If A is a real symmetric matrix of order n, there exists an orthogonal matrix P over $\mathbf{R}$ such that $P^{-1} A P$ is a diagonal matrix.

Proof The proof is by induction on the order n of A. The theorem is trivially true for $n = 1$.

Assume that the theorem is true for all $k \times k$ matrices, and let A be a real symmetric matrix of order $k + 1$. Let T be the linear operator on $\mathbf{R}^{k+1}$ that has matrix A relative to $\mathcal{E}_{k+1}$. Let λ_1 be any eigenvalue of T (and A). Then λ_1 is real and there is a corresponding eigenvector $\mathbf{v}_1$ in $\mathbf{R}^{k+1}$. Since any multiple of $\mathbf{v}_1$ is in the eigenspace $\mathbf{V}_{\lambda_1}$, we may assume that $\mathbf{v}_1$ is a unit vector. It follows from Theorem 8.15 that $\{\mathbf{v}_1\}$ can be extended to an orthonormal basis

$$\mathcal{N} = \{\mathbf{v}_1, \mathbf{v}_2, \ldots, \mathbf{v}_{k+1}\}$$

of $\mathbf{R}^{k+1}$ (See Problem 10 of Exercises 8.3). The matrix of transition P_1 from $\mathcal{E}_{k+1}$ to $\mathcal{N}$ is orthogonal (Theorem 8.14), and T is represented by

$$A_1 = P_1^{-1} A P_1 = \left[\begin{array}{c|c} \lambda_1 & \mathbf{0} \\ \hline \mathbf{0} & A_2 \end{array} \right]$$

relative to $\mathcal{N}$. (The elements to the right of λ_1 in the first row must be zero because of symmetry.) The $k \times k$ matrix A_2 is real and symmetric, so it follows from our induction hypotheses that there is an orthogonal $k \times k$ matrix Q such that $Q^{-1} A_2 Q = \text{diag}\{\lambda_2, \ldots, \lambda_{k+1}\}$. It is readily verified that

$$P_2 = \left[\begin{array}{c|c} 1 & \mathbf{0} \\ \hline \mathbf{0} & Q \end{array} \right]$$

is orthogonal and that

$$P_2^{-1} A_1 P_2 = P_2^{-1} P_1^{-1} A P_1 P_2 = \text{diag}\{\lambda_1, \lambda_2, \ldots, \lambda_{k+1}\}.$$

Now $P_1 P_2$ is orthogonal since

$$(P_1 P_2)^T = P_2^T P_1^T = P_2^{-1} P_1^{-1} = (P_1 P_2)^{-1},$$

and thus $P = P_1 P_2$ is the desired matrix.   ∎

It is important to note that the diagonal elements of the matrix

$$P^{-1}AP = \operatorname{diag}\{\lambda_1, \lambda_2, \ldots, \lambda_n\}$$

are the eigenvalues of A.

8.20
COROLLARY

Any real quadratic form q with $q(\mathbf{v}) = X^T A X$ can be reduced to a diagonalized representation

$$q(\mathbf{v}) = Y^T (P^T A P) Y = \lambda_1 y_1^2 + \lambda_2 y_2^2 + \cdots + \lambda_n y_n^2$$

by an orthogonal change of variables $X = PY$. For each $\mathbf{v} = (x_1, x_2, \ldots, x_n) \in \mathbf{R}^n$, $Y = [\, y_1 \quad y_2 \quad \cdots \quad y_n \,]^T$ is the coordinate matrix of $\mathbf{v}$ relative to $\mathcal{A}$, where P is the matrix of transition from $\mathcal{E}_n$ to $\mathcal{A}$.

Proof   The proof is left as an exercise in Problem 17.   ■ ■ ■

The proof of Theorem 8.19 is not exactly constructive, although it does suggest an iterative procedure to obtain P by beginning with the selection of λ_1, $\mathbf{v}_1$, and P_1. It is not advantageous for us to pursue this lead, as the next theorem leads to a much more efficient procedure.

We observe that if $\mathbf{u}$ and $\mathbf{v}$ are vectors in $\mathbf{R}^n$ with coordinate matrices

$$X = [\, x_1 \quad x_2 \quad \cdots \quad x_n \,]^T \quad \text{and} \quad Y = [\, y_1 \quad y_2 \quad \cdots \quad y_n \,]^T$$

relative to $\mathcal{E}_n$, then $\mathbf{u} \cdot \mathbf{v} = \sum_{k=1}^n x_k y_k = X^T Y$. In particular, $\mathbf{u}$ and $\mathbf{v}$ are orthogonal if and only if $X^T Y = 0$.

8.21
THEOREM

Let A be a real symmetric matrix. If λ_r and λ_s are distinct eigenvalues of A with associated eigenvectors P_r and P_s, respectively, then $P_r^T P_s = 0$.

Proof   Suppose that λ_r and λ_s are distinct eigenvalues of A with P_r and P_s as corresponding eigenvectors. Then $A P_r = \lambda_r P_r$ and $A P_s = \lambda_s P_s$. Now

$$P_r^T A P_s = P_r^T (\lambda_s P_s) = \lambda_s P_r^T P_s$$

and

$$(P_r^T A P_s)^T = P_s^T A P_r = P_s^T (\lambda_r P_r) = \lambda_r P_s^T P_r = \lambda_r (P_r^T P_s)^T.$$

But $(P_r^T A P_s)^T = P_r^T A P_s$ and $(P_r^T P_s)^T = P_r^T P_s$ since $P_r^T A P_s$ and $P_r^T P_s$ are matrices of order 1. Hence

$$\lambda_r P_r^T P_s = \lambda_s P_r^T P_s$$

and

$$(\lambda_r - \lambda_s)P_r^T P_s = 0.$$

Since $\lambda_r - \lambda_s \neq 0$, it must be that $P_r^T P_s = 0$. ∎ ∎ ∎

In Section 7.3, we found that $P = [\, P_1 \quad P_2 \quad \cdots \quad P_n \,]$ was an invertible matrix such that $P^{-1}AP$ was diagonal if and only if the columns P_j of P were the coordinate matrices relative to $\mathcal{E}_n$ of a basis of eigenvectors $\mathbf{v}_j$ of the associated linear transformation T. Since $P_r^T P_s$ is the element in row r and column s of $P^T P$, the requirement that $P^T = P^{-1}$ is satisfied if and only if $P_r^T P_s = \delta_{rs}$. Since $P_r^T P_s = \mathbf{v}_r \cdot \mathbf{v}_s$, this is equivalent to requiring that the basis of eigenvectors be orthonormal. Theorem 8.21 assures us that eigenvectors from distinct eigenspaces are automatically orthogonal. Thus, the only modification of the procedure in Section 7.3 that is necessary to make P orthogonal is to choose orthonormal bases of the eigenspaces $\mathbf{V}_{\lambda_j}$. This is illustrated in the next example.

EXAMPLE 1 Consider the problem of finding an orthogonal matrix P such that $P^T A P$ is diagonal, where

$$A = \begin{bmatrix} 1 & -1 & -1 \\ -1 & 1 & -1 \\ -1 & -1 & 1 \end{bmatrix}.$$

As explained above, the basic problem is to find an orthonormal basis of eigenvectors of the linear transformation T that has matrix A relative to $\mathcal{E}_3$. The characteristic equation of A is $-(x+1)(x-2)^2 = 0$. By solving the systems

$$(A - (-1)I)X = \mathbf{0} \quad \text{and} \quad (A - 2I)X = \mathbf{0},$$

we find that $\{(1, 1, 1)\}$ is a basis of the eigenspace $\mathbf{V}_{-1}$ and $\{(1, -1, 0), (1, 0, -1)\}$ is a basis of $\mathbf{V}_2$. Applying the Gram-Schmidt process, we obtain

$$\left\{ \tfrac{1}{\sqrt{3}}(1, 1, 1) \right\}$$

and

$$\left\{ \tfrac{1}{\sqrt{2}}(1, -1, 0),\ \tfrac{1}{\sqrt{6}}(1, 1, -2) \right\}$$

as orthonormal bases of $\mathbf{V}_{-1}$ and $\mathbf{V}_2$, respectively. Hence

$$P = \begin{bmatrix} \frac{1}{\sqrt{3}} & \frac{1}{\sqrt{2}} & \frac{1}{\sqrt{6}} \\ \frac{1}{\sqrt{3}} & -\frac{1}{\sqrt{2}} & \frac{1}{\sqrt{6}} \\ \frac{1}{\sqrt{3}} & 0 & -\frac{2}{\sqrt{6}} \end{bmatrix} = \frac{1}{\sqrt{6}} \begin{bmatrix} \sqrt{2} & \sqrt{3} & 1 \\ \sqrt{2} & -\sqrt{3} & 1 \\ \sqrt{2} & 0 & -2 \end{bmatrix}$$

is an orthogonal matrix such that $P^T A P = \mathrm{diag}\{-1, 2, 2\}$. ∎

Since the introduction of quadratic forms was partially motivated by references to the conic sections and quadric surfaces, it is of interest to relate our results here to these geometric quantities.

A conic section always has an equation of the form

$$ax^2 + bxy + cy^2 + dx + ey + f = 0$$

in rectangular coordinates x, y. According to Corollary 8.20, the quadratic form q with $q(x, y) = ax^2 + bxy + cy^2$ can be reduced by an orthogonal change of variables

$$x = p_{11}x' + p_{12}y'$$

$$y = p_{21}x' + p_{22}y'$$

to a diagonalized form

$$\lambda_1(x')^2 + \lambda_2(y')^2.$$

It is shown in analytic geometry that such a change of variables corresponds to a rotation of the coordinate axes about the origin. The different possibilities for the signs of the eigenvalues correspond to different types of conic sections, with degenerate cases possible in each instance. If λ_1 and λ_2 are nonzero and of the same sign, the conic section is a circle or an ellipse. If λ_1 and λ_2 are nonzero and of opposite sign, the conic section is a hyperbola. If exactly one of λ_1, λ_2 is zero, the conic section is a parabola. If both λ_1 and λ_2 are zero, the graph is a straight line.

EXAMPLE 2   Consider the conic section

$$3x^2 + 4xy + 3y^2 = 1$$

determined by the quadratic form $q(x, y) = 3x^2 + 4xy + 3y^2$. According to Corollary 8.20, this quadratic form can be reduced to the diagonalized form $q(x, y) = \lambda_1(x')^2 + \lambda_2(y')^2$ by an orthogonal change of variables. The symmetric matrix of q relative to $\mathcal{E}_2$ is given by

$$A = \begin{bmatrix} 3 & 2 \\ 2 & 3 \end{bmatrix}$$

whose characteristic polynomial is $|A - xI| = (3 - x)^2 - 4$. Hence the eigenvalues and corresponding eigenspaces of A are

Eigenvalue	Eigenspace
$\lambda_1 = 1$	$\mathbf{V}_1 = \langle (-1, 1) \rangle$
$\lambda_2 = 5$	$\mathbf{V}_5 = \langle (1, 1) \rangle.$

Normalizing each of the eigenvectors gives eigenspaces

$$\mathbf{V}_1 = \left\langle \tfrac{1}{\sqrt{2}}(-1, 1) \right\rangle \quad \text{and} \quad \mathbf{V}_5 = \left\langle \tfrac{1}{\sqrt{2}}(1, 1) \right\rangle.$$

Then the orthogonal matrix

$$P = \tfrac{1}{\sqrt{2}} \begin{bmatrix} -1 & 1 \\ 1 & 1 \end{bmatrix}$$

is the matrix of transition from $\mathcal{E}_2$ to the orthonormal basis $\left\{ \tfrac{1}{\sqrt{2}}(-1, 1), \tfrac{1}{\sqrt{2}}(1, 1) \right\}$ of eigenvectors and

$$P^T A P = \tfrac{1}{\sqrt{2}} \begin{bmatrix} -1 & 1 \\ 1 & 1 \end{bmatrix} \begin{bmatrix} 3 & 2 \\ 2 & 3 \end{bmatrix} \tfrac{1}{\sqrt{2}} \begin{bmatrix} -1 & 1 \\ 1 & 1 \end{bmatrix} = \begin{bmatrix} 1 & 0 \\ 0 & 5 \end{bmatrix}.$$

The diagonalized form of q is

$$q(x', y') = [x' \quad y'] \begin{bmatrix} 1 & 0 \\ 0 & 5 \end{bmatrix} \begin{bmatrix} x' \\ y' \end{bmatrix} = (x')^2 + 5(y')^2$$

where

$$\begin{bmatrix} x \\ y \end{bmatrix} = \tfrac{1}{\sqrt{2}} \begin{bmatrix} -1 & 1 \\ 1 & 1 \end{bmatrix} \begin{bmatrix} x' \\ y' \end{bmatrix}$$

gives the orthogonal change of variables

$$x = -\tfrac{1}{2}\sqrt{2}x' + \tfrac{1}{2}\sqrt{2}y'$$
$$y = \tfrac{1}{2}\sqrt{2}x' + \tfrac{1}{2}\sqrt{2}y'.$$

Therefore the conic section

$$3x^2 + 4xy + 3y^2 = 1$$

has been rotated by the orthogonal change of variables and is now recognized as the ellipse

$$(x')^2 + 5(y')^2 = 1. \qquad\blacksquare$$

As mentioned in Section 8.2, the quadric surfaces can be related to quadratic forms in three variables. This relation can be analyzed in a manner analogous to that for the conic sections. However, this analysis becomes somewhat involved, and it is omitted here for this reason.

8.4 Exercises

1. Find an orthogonal matrix P such that $P^T A P$ is diagonal.

(a) $A = \begin{bmatrix} 0 & 2 & 2 \\ 2 & 0 & 2 \\ 2 & 2 & 0 \end{bmatrix}$

(b) $A = \begin{bmatrix} 1 & 2 & -4 \\ 2 & -2 & -2 \\ -4 & -2 & 1 \end{bmatrix}$

(c) $A = \begin{bmatrix} 17 & 2 & -2 \\ 2 & 14 & 4 \\ -2 & 4 & 14 \end{bmatrix}$
$\qquad$
(d) $A = \begin{bmatrix} 4 & -1 & 0 & 1 \\ -1 & 5 & -1 & 0 \\ 0 & -1 & 4 & -1 \\ 1 & 0 & -1 & 5 \end{bmatrix}$

2. Identify each quadratic form as one of the conic sections using a suitable orthogonal change of variables. In each case, give the orthogonal matrix P and the equation of the conic after the change of variables.

   (a) $2x^2 + 18xy + 2y^2 = 8$
   $\qquad\qquad$
   (b) $5x^2 + 6xy - 3y^2 = 24$

   (c) $5x^2 + 4xy + 2y^2 = 6$
   $\qquad\qquad$
   (d) $3x^2 - 4xy + 3y^2 = 25$

3. Diagonalize each of the following quadratic forms using an orthogonal change of variables. In each case, give the diagonalized form and the orthogonal matrix P.

   (a) $q(x_1, x_2, x_3) = 2x_1^2 + 2x_1 x_2 + 2x_2^2 + 3x_3^2$

   (b) $q(x_1, x_2, x_3) = 19x_1^2 + 20x_1 x_2 - 10x_1 x_3 + 4x_2^2 + 20x_2 x_3 + 19x_3^2$

   (c) $q(x_1, x_2, x_3) = \frac{17}{2}x_1^2 + \frac{17}{2}x_2^2 + x_3^2 - x_1 x_2 + 4x_1 x_3 + 4x_2 x_3$

   (d) $q(x_1, x_2, x_3, x_4) = -x_1^2 + 4x_1 x_3 - x_2^2 + 4x_1 x_4 + x_3^2 + x_4^2$

   (e) $q(x_1, x_2, x_3, x_4) = -x_1^2 - 2x_2^2 - x_3^2 + 4x_1 x_2 + 2x_1 x_3 + 4x_1 x_4 + 4x_2 x_3 - 4x_3 x_4$

   (f) $q(x_1, x_2, x_3, x_4) = 21x_1^2 + 13x_2^2 + 37x_3^2 + 13x_4^2 + 6x_1 x_2 - 30x_1 x_3 + 6x_1 x_4$
   $$- 10x_2 x_3 + 2x_2 x_4 - 10x_3 x_4$$

   (*Hint:* The eigenvalues of A are 12 and 48.)

4. Let $z_1, z_2, \ldots, z_n$ and $a_1, a_2, \ldots, a_n$ be complex numbers.

   (a) Verify that $\overline{z_1 z_2} = \bar{z}_1 \bar{z}_2$ and $\overline{z_1 + z_2} = \bar{z}_1 + \bar{z}_2$.

   (b) By induction, extend the results of part (a) to $\overline{z_1 z_2 \cdots z_n} = \bar{z}_1 \bar{z}_2 \cdots \bar{z}_n$ and $\overline{z_1 + z_2 + \cdots + z_n} = \bar{z}_1 + \bar{z}_2 + \cdots + \bar{z}_n$.

   (c) Prove that $\overline{\sum_{k=1}^{n} a_k z_k} = \sum_{k=1}^{n} \bar{a}_k \bar{z}_k$.

5. Let $A = [a_{rs}]_{m \times n}$ and $B = [b_{rs}]_{n \times p}$ over $\mathcal{C}$.

   (a) Prove that $\overline{AB} = \bar{A}\,\bar{B}$.

   (b) Prove that $\overline{A + B} = \bar{A} + \bar{B}$ whenever $A + B$ is defined.

6. Prove that $C^* = \overline{(C^T)}$.

7. Prove that $(A^*)^* = A$.

8. Prove that $(cA)^* = \bar{c}\,A^*$ for any $c \in \mathcal{C}$.

9. Prove that $(A_1 + A_2 + \cdots + A_n)^* = A_1^* + A_2^* + \cdots + A_n^*$.

10. Prove that $(A_1 A_2 \cdots A_n)^* = A_n^* \cdots A_2^* A_1^*$.

11. Prove that $(A^*)^{-1} = (A^{-1})^*$, if A is invertible.

12. In the proof of Theorem 8.19, verify that the matrix P_2 is orthogonal and that $P_2^{-1} A_1 P_2 = \text{diag}\{\lambda_1, \lambda_2, \ldots, \lambda_{k+1}\}$.

13. Let $\mathbf{u}$ and $\mathbf{v}$ be vectors in $\mathbf{R}^n$ with coordinate matrices $X = [\,x_1 \quad x_2 \quad \cdots \quad x_n\,]^T$ and $Y = [\,y_1 \quad y_2 \quad \cdots \quad y_n\,]^T$, respectively, relative to an orthonormal basis $\{\mathbf{v}_1, \mathbf{v}_2, \dots, \mathbf{v}_n\}$ of $\mathbf{R}^n$. Prove that $\mathbf{u} \cdot \mathbf{v} = X^T Y$.

14. Let $\mathcal{N} = \{\mathbf{v}_1, \mathbf{v}_2, \dots, \mathbf{v}_n\}$ be an orthonormal basis of $\mathbf{R}^n$, and $X = [\,x_1 \quad x_2 \quad \cdots \quad x_n\,]^T$ the coordinate matrix of $\mathbf{u}$ relative to $\mathcal{N}$. Prove that $x_k = \mathbf{u} \cdot \mathbf{v}_k$ for $k = 1, 2, \dots, n$.

15. Prove that if $A = B^T B$ for some matrix B over $\mathbf{R}$, then $X^T A X \geq 0$ for all $X = [\,x_1 \quad x_2 \quad \cdots \quad x_n\,]^T$ over $\mathbf{R}$.

16. Assume that the linear operator T of $\mathbf{R}^n$ has a symmetric matrix relative to an orthonormal basis of $\mathbf{R}^n$. Prove that the eigenvectors of T which correspond to distinct eigenvalues are orthogonal.

17. Prove Corollary 8.20.

8.5 Classification of Real Quadratic Forms

We have seen in Section 8.4 that an arbitrary real quadratic form q with $q(\mathbf{v}) = X^T A X$ can be reduced by an invertible (even orthogonal) linear change of variables $X = PY$ to a diagonalized representation

$$q(\mathbf{v}) = \lambda_1 y_1^2 + \lambda_2 y_2^2 + \cdots + \lambda_n y_n^2,$$

where

$$P^T A P = D = \text{diag}\{\lambda_1, \lambda_2, \dots, \lambda_n\}.$$

The matrix $Y = [\,y_1 \quad y_2 \quad \cdots \quad y_n\,]^T$ is the coordinate matrix of $\mathbf{v} = (x_1, x_2, \dots, x_n)$ relative to the basis $\mathcal{A} = \{\mathbf{u}_1, \mathbf{u}_2, \dots, \mathbf{u}_n\}$, where P is the transition matrix from $\mathcal{E}_n$ to $\mathcal{A}$. An interchange of two vectors $\mathbf{u}_i$ and $\mathbf{u}_j$ in $\mathcal{A}$ amounts to an interchange of the variables y_i and y_j, and such an interchange is reflected in the matrix D by an interchange of λ_i and λ_j. The diagonalized representation can thus be written as

$$q(\mathbf{v}) = \lambda_1 y_1^2 + \lambda_2 y_2^2 + \cdots + \lambda_r y_r^2,$$

where each of $\lambda_1, \lambda_2, \dots, \lambda_r$ is nonzero. The matrices A and D have the same rank since they are equivalent. Consequently, the number of nonzero λ_i's is always the same as the rank r of A.

> **8.22** **DEFINITION**
>
> The **rank** of the real quadratic form q with $q(\mathbf{v}) = X^T A X$ is the rank of the matrix A.

The discussion above shows that the rank of q is the same as the number of variables having nonzero coefficients in a diagonalized representation of q. We shall examine these nonzero terms in more detail.

In any two diagonalized representations of a real quadratic form q, the number of variables with positive coefficients is the same and the number of variables with negative coefficients is the same.

Proof   Suppose that q has rank r, and let

$$q(\mathbf{v}) = d_1 y_1^2 + d_2 y_2^2 + \cdots + d_r y_r^2$$

and

$$q(\mathbf{v}) = d_1' z_1^2 + d_2' z_2^2 + \cdots + d_r' z_r^2$$

be any two diagonalized representations of q.[4] We may assume without loss of generality that the diagonal elements in both $D = \text{diag}\{d_1, d_2, \ldots, d_r, 0, \ldots, 0\}$ and $D' = \text{diag}\{d_1', d_2', \ldots, d_r', 0, \ldots, 0\}$ are arranged so that the positive elements come first, followed by the negative elements. That is,

$$D = \text{diag}\{d_1, \ldots, d_p, d_{p+1}, \ldots, d_r, 0, \ldots, 0\}$$

with the first p elements d_i positive, and

$$D' = \text{diag}\{d_1', \ldots, d_k', d_{k+1}', \ldots, d_r', 0, \ldots, 0\}$$

with the first k elements d_i' positive.

Let $\mathcal{A} = \{\mathbf{u}_1, \mathbf{u}_2, \ldots, \mathbf{u}_n\}$ and $\mathcal{B} = \{\mathbf{v}_1, \mathbf{v}_2, \ldots, \mathbf{v}_n\}$ be bases of $\mathbf{R}^n$ chosen so that $[y_1 \ y_2 \ \cdots \ y_n]^T$ and $[z_1 \ z_2 \ \cdots \ z_n]^T$ are the coordinate matrices of $\mathbf{v} = (x_1, x_2, \ldots, x_n)$ relative to $\mathcal{A}$ and $\mathcal{B}$, respectively (see Corollary 8.20). Now $q(\mathbf{u}_i) = d_i > 0$ for $i = 1, 2, \ldots, p$. Therefore, for any $\mathbf{w} = \sum_{i=1}^{p} y_i \mathbf{u}_i \in \mathbf{W}_1 = \langle \mathbf{u}_1, \mathbf{u}_2, \ldots, \mathbf{u}_p \rangle$,

$$q(\mathbf{w}) = \sum_{i=1}^{p} d_i y_i^2 \geq 0.$$

Also, $q(\mathbf{v}_i) = d_i' < 0$ for $i = k+1, k+2, \ldots, n$, so that for each $\mathbf{w} = \sum_{i=k+1}^{n} z_i \mathbf{v}_i \in \mathbf{W}_2 = \langle \mathbf{v}_{k+1}, \mathbf{v}_{k+2}, \ldots, \mathbf{v}_n \rangle$,

$$q(\mathbf{w}) = \sum_{i=k+1}^{n} d_i' z_i^2 \leq 0.$$

Thus $\mathbf{W}_1 \cap \mathbf{W}_2 = \{\mathbf{0}\}$ and

$$\dim(\mathbf{W}_1 + \mathbf{W}_2) = \dim(\mathbf{W}_1) + \dim(\mathbf{W}_2).$$

---

[4] These two diagonalized representations correspond to two diagonal matrices D and D' that are congruent to the original matrix A of q : $D = P_1^T A P_1$ and $D' = P_2^T A P_2$ for invertible P_1 and P_2. That is, we are dealing here with congruence of matrices, not with orthogonal similarity.

But $\dim(\mathbf{W}_1 + \mathbf{W}_2) \le n$ and

$$\dim(\mathbf{W}_1) + \dim(\mathbf{W}_2) = p + n - k.$$

Therefore, $p + n - k \le n$ and $p \le k$.

From the symmetry of the conditions on p and k, it follows that $k \le p$ and $k = p$. The second part of the conclusion follows immediately from $r - k = r - p$. ∎

8.24 COROLLARY

Let A be a real symmetric matrix. Any two diagonal matrices that are congruent to A over $\mathbf{R}$ have the same number of positive elements and the same number of negative elements on the diagonal.

Proof See Problem 6. ∎

8.25 DEFINITION

The **index** of the quadratic form q is the number p of positive coefficients appearing in a diagonalized representation of q. The difference s between p and the number of negative coefficients in a diagonalized representation of q is the **signature** of q. That is, the signature of q is the number $s = p - (r - p) = 2p - r$, where r is the rank of q.

Theorem 8.23 shows that the index and signature of q are well-defined terms. The signature of q is a measure of the "positiveness" or "negativeness" of q.

8.26 THEOREM

A quadratic form q on $\mathbf{R}^n$ with rank r and index p can be represented as

$$q(\mathbf{v}) = z_1^2 + \cdots + z_p^2 - z_{p+1}^2 - \cdots - z_r^2$$

by a suitable invertible linear change of variables.

Proof Let

$$q(\mathbf{v}) = d_1 y_1^2 + \cdots + d_p y_p^2 + d_{p+1} y_{p+1}^2 + \cdots + d_r y_r^2$$

be a diagonalized representation of q, with $d_1, \ldots, d_p$ positive and $d_{p+1}, \ldots, d_r$ negative. For $i = 1, 2, \ldots, p$, d_i has a positive real square root $\sqrt{d_i}$, and we put $z_i = \sqrt{d_i}\, y_i$. For $i = p+1, p+2, \ldots, r$, $-d_i$ has a positive real square root $\sqrt{|d_i|}$, and we put $z_i = \sqrt{|d_i|}\, y_i$. For $i = r+1, \ldots, n$, we put $z_i = y_i$. The linear change of variables

$$
\begin{bmatrix} z_1 \\ z_2 \\ \vdots \\ z_n \end{bmatrix} = \mathrm{diag}\left\{ \sqrt{d_1}, \ldots, \sqrt{d_p}, \sqrt{|d_{p+1}|}, \ldots, \sqrt{|d_r|}, 1, \ldots, 1 \right\} \begin{bmatrix} y_1 \\ y_2 \\ \vdots \\ y_n \end{bmatrix}
$$

is clearly invertible, and

$$
\begin{aligned}
q(\mathbf{v}) &= d_1 y_1^2 + \cdots + d_p y_p^2 - (-d_{p+1} y_{p+1}^2) - \cdots - (-d_r y_r^2) \\
&= \left(\sqrt{d_1} y_1 \right)^2 + \cdots + \left(\sqrt{d_p} y_p \right)^2 - \left(\sqrt{|d_{p+1}|} y_{p+1} \right)^2 - \cdots - \left(\sqrt{|d_r|} y_r \right)^2 \\
&= z_1^2 + \cdots + z_p^2 - z_{p+1}^2 - \cdots - z_r^2.
\end{aligned}
$$

■ ■ ■

The form $q(\mathbf{v}) = z_1^2 + \cdots + z_p^2 - z_{p+1}^2 - \cdots - z_r^2$ in Theorem 8.26 is called the **canonical form** for q. There are two corollaries to the theorem concerning symmetric matrices. Proofs are requested in the Exercises.

8.27 COROLLARY

Any real symmetric matrix A is congruent over $\mathbf{R}$ to a unique matrix of the form

$$
C = \left[\begin{array}{c|c|c}
I_p & 0 & 0 \\
\hline
0 & -I_{r-p} & 0 \\
\hline
0 & 0 & 0
\end{array} \right],
$$

where r is the rank of A.

The number p of positive 1's in C is called the **index** of A. Since the linear change of variable in the proof of Theorem 8.26 is clearly not necessarily orthogonal, the matrix P used to obtain $P^T A P = C$ in Corollary 8.27 may not be orthogonal.

8.28 COROLLARY

Two symmetric $n \times n$ matrices over $\mathbf{R}$ are congruent over $\mathbf{R}$ if and only if they have the same rank and the same index.

EXAMPLE 1 In Example 1 of Section 8.4, we obtained the orthogonal matrix

$$
P_1 = \frac{1}{\sqrt{6}} \begin{bmatrix} \sqrt{2} & \sqrt{3} & 1 \\ \sqrt{2} & -\sqrt{3} & 1 \\ \sqrt{2} & 0 & -2 \end{bmatrix}
$$

such that $P_1^T A P_1 = \text{diag}\{-1, 2, 2\}$, where

$$A = \begin{bmatrix} 1 & -1 & -1 \\ -1 & 1 & -1 \\ -1 & -1 & 1 \end{bmatrix}.$$

Thus the change of variables

$$x_1 = \tfrac{1}{\sqrt{6}}(\sqrt{2}y_1 + \sqrt{3}y_2 + y_3)$$

$$x_2 = \tfrac{1}{\sqrt{6}}(\sqrt{2}y_1 - \sqrt{3}y_2 + y_3)$$

$$x_3 = \tfrac{1}{\sqrt{6}}(\sqrt{2}y_1 - 2y_3)$$

reduces the quadratic form

$$q(\mathbf{v}) = q(x_1, x_2, x_3) = x_1^2 + x_2^2 + x_3^2 - 2x_1x_2 - 2x_1x_3 - 2x_2x_3$$

to the form

$$q(\mathbf{v}) = -y_1^2 + 2y_2^2 + 2y_3^2.$$

In the terminology of the first paragraph of this section, an interchange of the variables y_1 and y_3 corresponds to an interchange of $\mathbf{u}_1$ and $\mathbf{u}_3$, or an interchange of columns one and three in P_1. Thus

$$P_2 = \tfrac{1}{\sqrt{6}} \begin{bmatrix} 1 & \sqrt{3} & \sqrt{2} \\ 1 & -\sqrt{3} & \sqrt{2} \\ -2 & 0 & \sqrt{2} \end{bmatrix}$$

is an orthogonal matrix such that $P_2^T A P_2 = \text{diag}\{2, 2, -1\}$. This means that the orthogonal change of variable $[\, x_1 \;\; x_2 \;\; x_3 \,]^T = P_2[\, y_1' \;\; y_2' \;\; y_3' \,]^T$ reduces $q(\mathbf{v})$ to the form

$$q(\mathbf{v}) = 2(y_1')^2 + 2(y_2')^2 - (y_3')^2.$$

Following the method of the proof of Theorem 8.26, we write $y_1' = \tfrac{1}{\sqrt{2}}z_1$, $y_2' = \tfrac{1}{\sqrt{2}}z_2$, and $y_3' = z_3$. This reduces q to the canonical form

$$q(\mathbf{v}) = z_1^2 + z_2^2 - z_3^2.$$

The matrix of transition P_3 corresponding to this last change of variables $[\, y_1' \;\; y_2' \;\; y_3' \,]^T = P_3[\, z_1 \;\; z_2 \;\; z_3 \,]^T$ is

$$P_3 = \begin{bmatrix} \tfrac{1}{\sqrt{2}} & 0 & 0 \\ 0 & \tfrac{1}{\sqrt{2}} & 0 \\ 0 & 0 & 1 \end{bmatrix}.$$

Combining, we have the change of variable $[\, x_1 \;\; x_2 \;\; x_3 \,]^T = P_2 P_3[\, z_1 \;\; z_2 \;\; z_3 \,]^T$, and

$$P = P_2 P_3 = \frac{1}{2\sqrt{3}} \begin{bmatrix} 1 & \sqrt{3} & 2 \\ 1 & -\sqrt{3} & 2 \\ -2 & 0 & 2 \end{bmatrix}$$

is a matrix such that $P^T A P = \text{diag}\{1, 1, -1\}$. We note that P is not orthogonal. It is clear that q has rank 3, index 2, and signature 1. ∎

8.29
DEFINITION

Let q be a real quadratic form on $\mathbf{R}^n$ with rank r and index p.

1. If $p = r = n$, q is called **positive definite**.
2. If $p = r$, q is called **positive semidefinite**.
3. If $p = 0$ and $r = n$, q is called **negative definite**.
4. If $p = 0$, q is called **negative semidefinite**.

If none of the descriptions in Definition 8.29 apply to a quadratic form q, then q is said to be **indefinite**. The quadratic form q in Example 1 is an example of an indefinite form since the rank r of q is 3 and the index p is 2.

Each of the conditions in Definition 8.29 can be formulated in terms of the range of values of q. These formulations are given in the next theorem, with proofs requested in Problem 9.

8.30
THEOREM

Let q be a real quadratic form on $\mathbf{R}^n$.

1. q is positive definite if and only if $q(\mathbf{v}) > 0$ for all $\mathbf{v} \neq \mathbf{0}$ in $\mathbf{R}^n$.
2. q is positive semidefinite if and only if $q(\mathbf{v}) \geq 0$ for all $\mathbf{v} \in \mathbf{R}^n$.
3. q is negative definite if and only if $q(\mathbf{v}) < 0$ for all $\mathbf{v} \neq \mathbf{0}$ in $\mathbf{R}^n$.
4. q is negative semidefinite if and only if $q(\mathbf{v}) \leq 0$ for all $\mathbf{v} \in \mathbf{R}^n$.

EXAMPLE 2 The quadratic form q defined by

$$q(x_1, x_2, x_3, x_4) = x_1^2 + x_2^2 + x_3^2 + x_4^2 - 2x_1 x_3 - 2x_2 x_4$$

can be described by $q(\mathbf{v}) = X^T A X$ where $X = [x_1 \quad x_2 \quad x_3 \quad x_4]^T$ is the coordinate matrix of $\mathbf{v}$ relative to $\mathcal{E}_4$ and

$$A = \begin{bmatrix} 1 & 0 & -1 & 0 \\ 0 & 1 & 0 & -1 \\ -1 & 0 & 1 & 0 \\ 0 & -1 & 0 & 1 \end{bmatrix}.$$

The characteristic polynomial of A is

$$|A - xI| = x^4 - 4x^3 + 4x^2$$
$$= x^2(x-2)^2.$$

Since the eigenvalues of A are $\lambda = 2$ and $\lambda = 0$, the diagonalized form for q is

$$q(y_1, y_2, y_3, y_4) = 2y_1^2 + 2y_2^2.$$

Thus the rank r of q is 2, the index p is 2, the signature is $s = 2p - r = 2$, and q is a positive semidefinite real quadratic form. ∎

8.5 Exercises

1. For each $\mathbf{v} = (x_1, x_2, \ldots, x_n)$ in $\mathbf{R}^n$, a quadratic form q is defined by $q(\mathbf{v}) = X^T A X$ for the given A. Find the rank, index, and signature of q.

   (a) $A = \begin{bmatrix} \frac{3}{2} & \sqrt{2} & -\frac{1}{2} \\ \sqrt{2} & 1 & -\sqrt{2} \\ -\frac{1}{2} & -\sqrt{2} & -\frac{5}{2} \end{bmatrix}$ **(b)** $A = \begin{bmatrix} 1 & -2 & 0 \\ -2 & 2 & -2 \\ 0 & -2 & 3 \end{bmatrix}$

   (c) $A = \begin{bmatrix} 1 & 1 & 1 \\ 1 & 1 & 1 \\ 1 & 1 & 1 \end{bmatrix}$ **(d)** $A = \begin{bmatrix} -2 & 2 & 2 \\ 2 & 1 & 4 \\ 2 & 4 & 1 \end{bmatrix}$

2. Find the canonical form for each of the quadratic forms referred to in Problem 1.

3. For each of the following matrices A, find an invertible real matrix P such that $P^T A P$ is of the form C given in Corollary 8.27.

   (a) $A = \begin{bmatrix} 0 & 2 & 2 \\ 2 & 0 & 2 \\ 2 & 2 & 0 \end{bmatrix}$ **(b)** $A = \begin{bmatrix} 1 & 2 & -4 \\ 2 & -2 & -2 \\ -4 & -2 & 1 \end{bmatrix}$

   (c) $A = \begin{bmatrix} 17 & 2 & -2 \\ 2 & 14 & 4 \\ -2 & 4 & 14 \end{bmatrix}$ **(d)** $A = \begin{bmatrix} 4 & -1 & 0 & 1 \\ -1 & 5 & -1 & 0 \\ 0 & -1 & 4 & -1 \\ 1 & 0 & -1 & 5 \end{bmatrix}$

4. For each of the matrices A in Problem 3, let q be the quadratic form on $\mathbf{R}^n$ with $q(\mathbf{v}) = X^T A X$. Find a basis $\mathcal{B}$ of $\mathbf{R}^n$ such that $q(\mathbf{v})$ has the form

$$q(\mathbf{v}) = z_1^2 + \cdots + z_p^2 - z_{p+1}^2 - \cdots - z_r^2$$

   with $[\, z_1 \quad z_2 \quad \cdots \quad z_n \,]^T$ as the coordinate matrix of $\mathbf{v}$ relative to $\mathcal{B}$.

5. Identify each of the following real quadratic forms as positive definite, positive semidefinite, negative definite, negative semidefinite or indefinite. For each, state the rank, index, and signature.

(a) $q(x_1, x_2, x_3) = -4x_1^2 - x_2^2 - 4x_3^2 - 4x_1x_2 + 8x_1x_3 + 4x_2x_3$

(b) $q(x_1, x_2, x_3) = 2x_1^2 + 3x_2^2 + 2x_3^2 + 2x_1x_3$

(c) $q(x_1, x_2, x_3) = -3x_1^2 - 4x_2^2 - 3x_3^2 - 2x_1x_3$

(d) $q(x_1, x_2, x_3) = x_1^2 + x_2^2 + x_3^2 + 6x_1x_2 + 6x_1x_3 + 6x_2x_3$

(e) $q(x_1, x_2, x_3) = 2x_1x_2 + 2x_1x_3 + 2x_2x_3$

(f) $q(x_1, x_2, x_3) = 2x_1x_2 - 2x_1x_3 + 2x_2x_3$

(g) $q(x_1, x_2, x_3, x_4) = 2x_1^2 + 2x_2^2 + 2x_3^2 + 2x_4^2 + 4x_1x_2 + 4x_1x_3 + 4x_1x_4 + 4x_2x_3$
$+ 4x_2x_4 + 4x_3x_4$

(h) $q(x_1, x_2, x_3, x_4) = 2x_1^2 + 2x_2^2 + 2x_3^2 + 2x_4^2 + 2x_1x_2 + 2x_1x_3 + 2x_1x_4 + 2x_2x_3$
$+ 2x_2x_4 + 2x_3x_4$

(i) $q(x_1, x_2, x_3, x_4) = -9x_1^2 - 25x_2^2 - x_3^2 - x_4^2 + 30x_1x_2 - 6x_1x_3 - 6x_1x_4$
$+ 10x_2x_3 + 10x_2x_4 - 2x_3x_4$

(j) $q(x_1, x_2, x_3, x_4) = 2x_1x_2 + 2x_1x_3 - 2x_1x_4 + 2x_2x_3 - 2x_2x_4 - 2x_3x_4$

6. Prove Corollary 8.24.

7. Prove Corollary 8.27.

8. Prove Corollary 8.28.

9. Prove Theorem 8.30.

10. A real symmetric matrix A is defined to be **positive definite** if the quadratic form $q(x_1, x_2, \ldots, x_n) = X^T A X$ is positive definite. The terms **positive semidefinite,** etc., are defined similarly.

(a) Prove that A is positive definite if and only if $A = B^T B$ for some real invertible matrix B.

(b) Prove that A is positive semidefinite if and only if there exists a (possibly singular) real matrix Q such that $A = Q^T Q$.

(c) Prove that A is positive definite if and only if all of the eigenvalues of A are positive.

(d) Prove that A is negative definite if and only if all of the eigenvalues of A are negative.

(e) Prove that if A is positive definite, then A is invertible and A^{-1} is positive definite.

(f) Prove that if A and B are positive definite, then $A + B$ is positive definite.

11. Prove that if A is positive definite, then there exists a positive definite matrix S such that $S^2 = A$. (The matrix S is called the **square root** of A.)

12. Find the square root of the each of the following positive definite matrices.

(a) $A = \begin{bmatrix} 2 & 1 & 1 \\ 1 & 2 & 1 \\ 1 & 1 & 2 \end{bmatrix}$
(b) $A = \begin{bmatrix} 2 & 0 & 1 \\ 0 & 3 & 0 \\ 1 & 0 & 2 \end{bmatrix}$

8.6 Bilinear Forms

The *Cartesian product* $\mathcal{S} \times \mathcal{T}$ of two sets $\mathcal{S}, \mathcal{T}$ is the set of all ordered pairs of elements from $\mathcal{S}$ and $\mathcal{T}$:

$$\mathcal{S} \times \mathcal{T} = \{(s, t) \mid s \in \mathcal{S} \text{ and } t \in \mathcal{T}\}.$$

Few concepts of such simplicity have been put to such a wide range of uses in mathematics. For this is the basic concept behind the coordinate systems, the definitions of binary relation and function, and the constructions of several number systems. Although not stated explicitly in our development, the idea is implicit in the familiar vector spaces $\mathcal{F}^n$. For the next three sections, we shall be concerned with Cartesian products of sets of vectors, or rather, with functions defined on such sets of vectors.

8.31 DEFINITION
Let $\mathbf{U}$ and $\mathbf{V}$ be vector spaces over the same field $\mathcal{F}$. A **bilinear form** on $\mathbf{U}$ and $\mathbf{V}$ is a mapping f of pairs of vectors $(\mathbf{u}, \mathbf{v})$ in $\mathbf{U} \times \mathbf{V}$ onto scalars $f(\mathbf{u}, \mathbf{v})$ in $\mathcal{F}$ that has the properties

$$f(a_1\mathbf{u}_1 + a_2\mathbf{u}_2, \mathbf{v}) = a_1 f(\mathbf{u}_1, \mathbf{v}) + a_2 f(\mathbf{u}_2, \mathbf{v}) \tag{i}$$

and

$$f(\mathbf{u}, b_1\mathbf{v}_1 + b_2\mathbf{v}_2) = b_1 f(\mathbf{u}, \mathbf{v}_1) + b_2 f(\mathbf{u}, \mathbf{v}_2). \tag{ii}$$

The conditions (i) and (ii) are described by saying that f is a linear function of each of the vector variables. These conditions (i) and (ii) are together equivalent to the single requirement that

$$f(a_1\mathbf{u}_1 + a_2\mathbf{u}_2, b_1\mathbf{v}_1 + b_2\mathbf{v}_2) = a_1 b_1 f(\mathbf{u}_1, \mathbf{v}_1) + a_1 b_2 f(\mathbf{u}_1, \mathbf{v}_2)$$
$$+ a_2 b_1 f(\mathbf{u}_2, \mathbf{v}_1) + a_2 b_2 f(\mathbf{u}_2, \mathbf{v}_2). \tag{iii}$$

These linearity conditions extend readily to

$$f\left(\sum_{i=1}^{m} a_i\mathbf{u}_i, \sum_{j=1}^{n} b_j\mathbf{v}_j\right) = \sum_{i=1}^{m} \sum_{j=1}^{n} a_i b_j f(\mathbf{u}_i, \mathbf{v}_j) \tag{iv}$$

for all positive integers m and n. Thus, a function f from $\mathbf{U} \times \mathbf{V}$ to $\mathcal{F}$ is a bilinear form on $\mathbf{U}$ and $\mathbf{V}$ if and only if (iv) is satisfied as an identity.

We are primarily interested in the case where the vector spaces $\mathbf{U}$ and $\mathbf{V}$ in Definition 8.31 are finite-dimensional. For the remainder of this section, $\mathbf{U}$ and $\mathbf{V}$ will denote vector spaces over $\mathcal{F}$ with dimensions m and n, respectively.

8.32
DEFINITION

Let $\mathcal{A} = \{\mathbf{u}_1, \mathbf{u}_2, \ldots, \mathbf{u}_m\}$ and $\mathcal{B} = \{\mathbf{v}_1, \mathbf{v}_2, \ldots, \mathbf{v}_n\}$ be bases of **U** and **V**, respectively, and let f be a bilinear form on **U** and **V**. The **matrix** of f relative to $\mathcal{A}$ and $\mathcal{B}$ is the matrix $A = [a_{ij}]_{m \times n}$, where

$$a_{ij} = f(\mathbf{u}_i, \mathbf{v}_j)$$

for $i = 1, 2, \ldots, m; \; j = 1, 2, \ldots, n.$

We say that the matrix A **represents** the bilinear form f.

EXAMPLE 1   The mapping $f : \mathbf{R}^4 \times \mathbf{R}^2 \to \mathbf{R}$ defined by

$$f((x_1, x_2, x_3, x_4), (y_1, y_2)) = 3x_1y_1 + x_1y_2 - x_2y_1 + 2x_2y_2 - x_3y_2 + x_4y_1$$

is a bilinear form. The matrix of f relative to the standard bases $\mathcal{E}_4$ and $\mathcal{E}_2$ of $\mathbf{R}^4$ and $\mathbf{R}^2$, respectively, is determined by $a_{ij} = f(\mathbf{u}_i, \mathbf{v}_j)$ for each $\mathbf{u}_i \in \mathcal{E}_3$ and $\mathbf{v}_j \in \mathcal{E}_2$.

$$a_{11} = f((1, 0, 0, 0), (1, 0)) = 3 \qquad a_{12} = f((1, 0, 0, 0), (0, 1)) = 1$$
$$a_{21} = f((0, 1, 0, 0), (1, 0)) = -1 \qquad a_{22} = f((0, 1, 0, 0), (0, 1)) = 2$$
$$a_{31} = f((0, 0, 1, 0), (1, 0)) = 0 \qquad a_{32} = f((0, 0, 1, 0), (0, 1)) = -1$$
$$a_{41} = f((0, 0, 0, 1), (1, 0)) = 1 \qquad a_{42} = f((0, 0, 0, 1), (0, 1)) = 0$$

Thus the matrix representing f relative to $\mathcal{E}_4$ and $\mathcal{E}_2$ is given by

$$A = \begin{bmatrix} 3 & 1 \\ -1 & 2 \\ 0 & -1 \\ 1 & 0 \end{bmatrix}.$$
∎

As with linear functionals, the function values $f(\mathbf{u}, \mathbf{v})$ can be expressed compactly by use of the matrix of f.

8.33
THEOREM

Let $X = [\, x_1 \quad x_2 \quad \cdots \quad x_m \,]^T$ denote the coordinate matrix of a vector $\mathbf{u} \in \mathbf{U}$ relative to the basis $\mathcal{A}$ of **U**, and let $Y = [\, y_1 \quad y_2 \quad \cdots \quad y_n \,]^T$ be the coordinate matrix of a vector $\mathbf{v} \in \mathbf{V}$ relative to the basis $\mathcal{B}$ of **V**. If $A = [a_{ij}]_{m \times n}$, then A is the matrix of the bilinear form f relative to $\mathcal{A}$ and $\mathcal{B}$ if and only if the equation

$$f(\mathbf{u}, \mathbf{v}) = X^T A Y$$

is satisfied for all choices of $\mathbf{u} \in \mathbf{U}, \mathbf{v} \in \mathbf{V}.$

Proof Assume first that $f(\mathbf{u}, \mathbf{v}) = X^T A Y$ is satisfied for all $\mathbf{u} \in \mathbf{U}, \mathbf{v} \in \mathbf{V}$. For the particular choices $\mathbf{u} = \mathbf{u}_i$ and $\mathbf{v} = \mathbf{v}_j$, we have $X = [\delta_{1i} \ \delta_{2i} \ \cdots \ \delta_{mi}]^T$ and $Y = [\delta_{1j} \ \delta_{2j} \ \cdots \ \delta_{nj}]^T$. Hence

$$f(\mathbf{u}_i, \mathbf{v}_j) = X^T A Y$$

$$= \begin{bmatrix} \delta_{1i} & \delta_{2i} & \cdots & \delta_{mi} \end{bmatrix} \begin{bmatrix} \sum_{k=1}^{n} a_{1k}\delta_{kj} \\ \sum_{k=1}^{n} a_{2k}\delta_{kj} \\ \vdots \\ \sum_{k=1}^{n} a_{mk}\delta_{kj} \end{bmatrix}$$

$$= \begin{bmatrix} \delta_{1i} & \delta_{2i} & \cdots & \delta_{mi} \end{bmatrix} \begin{bmatrix} a_{1j} \\ a_{2j} \\ \vdots \\ a_{mj} \end{bmatrix}$$

$$= \sum_{k=1}^{m} \delta_{ki} a_{kj}$$

$$= a_{ij},$$

and $A = [a_{ij}]_{m \times n}$ is the matrix of f relative to $\mathcal{A}$ and $\mathcal{B}$.

Assume, on the other hand, that $A = [a_{ij}]_{m \times n}$ is the matrix of f relative to $\mathcal{A}$ and $\mathcal{B}$. This means that $a_{ij} = f(\mathbf{u}_i, \mathbf{v}_j)$ for $i = 1, 2, \ldots, m; j = 1, 2, \ldots, n$. For any $\mathbf{u} = \sum_{i=1}^{m} x_i \mathbf{u}_i$, and $\mathbf{v} = \sum_{j=1}^{n} y_j \mathbf{v}_j$, the condition (iv) stated previously implies that

$$f(\mathbf{u}, \mathbf{v}) = \sum_{i=1}^{m} \sum_{j=1}^{n} x_i y_j f(\mathbf{u}_i, \mathbf{v}_j)$$

$$= \sum_{i=1}^{m} \sum_{j=1}^{n} x_i a_{ij} y_j$$

$$= \sum_{i=1}^{m} x_i \left(\sum_{j=1}^{n} a_{ij} y_j \right).$$

Now $\sum_{j=1}^{n} a_{ij} y_j$ is the element in the ith row of the $m \times 1$ matrix AY, and thus

$$f(\mathbf{u}, \mathbf{v}) = \sum_{i=1}^{m} x_i \left(\sum_{j=1}^{n} a_{ij} y_j \right) = X^T A Y. \quad\blacksquare\blacksquare\blacksquare$$

EXAMPLE 2 We can use Theorem 8.33 to compute $f((2, 0, 7, -1), (3, -1))$ for the bilinear form in Example 1. Since the elements of the coordinate matrices for $(2, 0, 7, -1)$ and $(3, -1)$

relative to the standard bases are the same as their components, we have

$$f((2, 0, 7, -1), (3, -1)) = \begin{bmatrix} 2 & 0 & 7 & -1 \end{bmatrix} \begin{bmatrix} 3 & 1 \\ -1 & 2 \\ 0 & -1 \\ 1 & 0 \end{bmatrix} \begin{bmatrix} 3 \\ -1 \end{bmatrix} = 20.$$ ∎

It follows from Definition 8.32 and Theorem 8.33 that the two bilinear forms f and g on **U** and **V** are equal if and only if they have the same matrix relative to fixed bases $\mathcal{A}$ of **U** and $\mathcal{B}$ of **V**.

As we have in similar situations previously, we ask about the effects of changes of bases in **U** and **V**. The answer is obtained quite easily.

8.34
THEOREM

Let A be the matrix of the bilinear form f relative to the bases $\mathcal{A}$ of **U** and $\mathcal{B}$ of **V**. If Q is the matrix of transition from $\mathcal{A}$ to the basis $\mathcal{A}'$ of **U** and P is the matrix of transition from $\mathcal{B}$ to the basis $\mathcal{B}'$ of **V**, then $Q^T A P$ is the matrix of f relative to $\mathcal{A}'$ and $\mathcal{B}'$.

Proof Suppose that $\mathbf{u} \in \mathbf{U}$ has coordinate matrix X relative to $\mathcal{A}$ and X' relative to $\mathcal{A}'$. Let $\mathbf{v} \in \mathbf{V}$ have coordinate matrix Y relative to $\mathcal{B}$ and Y' relative to $\mathcal{B}'$. By Theorem 5.14, $X = QX'$ and $Y = PY'$. Combined with the "only if" part of Theorem 8.33, this yields

$$f(\mathbf{u}, \mathbf{v}) = X^T A Y$$
$$= (QX')^T A (PY')$$
$$= (X')^T (Q^T A P) Y'.$$

But, by the "if" part of Theorem 8.33, this means that $Q^T A P$ is the matrix of f relative to $\mathcal{A}'$ and $\mathcal{B}'$. ∎∎∎

8.35
COROLLARY

Two $m \times n$ matrices A and B over $\mathcal{F}$ represent the same bilinear form on **U** and **V** relative to certain (not necessarily different) choices of bases of **U** and **V** if and only if A and B are equivalent over $\mathcal{F}$.

Proof See Problem 5. ∎∎∎

The **rank** of a bilinear form f is defined to be the rank r of any matrix that represents f.

8.36	Let f be a bilinear form on **U** and **V**. With suitable choices of bases in **U** and **V**, f
COROLLARY	can be represented by a matrix D_r that has the first r diagonal elements equal to 1 and all other elements zero.

Proof See Problem 6.   ■ ■ ■

The next example illustrates how bases are chosen in order for the matrix representing a bilinear form to be of the form described in Corollary 8.36.

EXAMPLE 3 Let f be the bilinear form on $\mathbf{U} = \mathbf{R}^3$ and $\mathbf{V} = \mathbf{R}^2$ that is defined by

$$f((x_1, x_2, x_3), (y_1, y_2)) = -7x_1y_1 - 10x_1y_2 - 2x_2y_1 - 3x_2y_2 + 12x_3y_1 + 17x_3y_2.$$

We consider the following problems.

1. Write the matrix A of f relative to $\mathcal{A} = \{(1, 0, 0), (1, 1, 0), (1, 1, 1)\}$ and $\mathcal{B} = \{(1, -1), (2, -1)\}$.
2. Use the matrix A to compute the value of $f((2, 3, 1), (0, -1))$.
3. Determine bases $\mathcal{A}'$ of **U** and $\mathcal{B}'$ of **V** such that the matrix of f relative to $\mathcal{A}'$ and $\mathcal{B}'$ is of the form D_r described in Corollary 8.36.

We obtain the matrix A by computing the values $a_{ij} = f(\mathbf{u}_i, \mathbf{v}_j)$. For example,

$$a_{11} = f((1, 0, 0), (1, -1))$$
$$= -7(1)(1) - 10(1)(-1) - 2(0)(1) - 3(0)(-1) + 12(0)(1) + 17(0)(-1)$$
$$= 3.$$

Similarly,

$$a_{21} = f((1, 1, 0), (1, -1)) = 4,$$
$$a_{31} = f((1, 1, 1), (1, -1)) = -1,$$

and so on. Thus

$$A = \begin{bmatrix} 3 & -4 \\ 4 & -5 \\ -1 & 2 \end{bmatrix}.$$

In order to use A to compute the value of $f((2, 3, 1), (0, -1))$, we first write

$$(2, 3, 1) = (-1)(1, 0, 0) + (2)(1, 1, 0) + (1)(1, 1, 1)$$

and

$$(0, -1) = 2(1, -1) + (-1)(2, -1).$$

Then the equation $f(\mathbf{u}, \mathbf{v}) = X^T A Y$ yields

$$f((2, 3, 1), (0, -1)) = \begin{bmatrix} -1 & 2 & 1 \end{bmatrix} \begin{bmatrix} 3 & -4 \\ 4 & -5 \\ -1 & 2 \end{bmatrix} \begin{bmatrix} 2 \\ -1 \end{bmatrix} = 12.$$

By the method of Section 3.7, we find that

$$Q^T = \begin{bmatrix} 1 & 0 & 0 \\ 0 & 1 & 0 \\ 3 & -2 & 1 \end{bmatrix} \quad \text{and} \quad P = \begin{bmatrix} -5 & 4 \\ -4 & 3 \end{bmatrix}$$

are invertible matrices such that

$$Q^T A P = \begin{bmatrix} 1 & 0 \\ 0 & 1 \\ 0 & 0 \end{bmatrix}.$$

Using Q as the matrix of transition from A to A' and P as the matrix of transition from B to B', we obtain

$$A' = \{(1, 0, 0), (1, 1, 0), (2, -1, 1)\} \quad \text{and} \quad B' = \{(-13, 9), (10, -7)\}. \qquad \blacksquare$$

8.6 Exercises

1. Find the matrix of the given bilinear form f on $\mathbf{U}$ and $\mathbf{V}$ with respect to the given bases A and B.

   (a) $\mathbf{U} = \mathbf{R}^2, \mathbf{V} = \mathbf{R}^3, A = \mathcal{E}_2, B = \mathcal{E}_3,$
   $f((x_1, x_2), (y_1, y_2, y_3)) = 2x_1 y_2 - x_1 y_3 + 2x_2 y_1 + x_2 y_2 + x_2 y_3$

   (b) $\mathbf{U} = \mathbf{V} = \mathbf{R}^3, A = B = \mathcal{E}_3,$
   $f((x_1, x_2, x_3), (y_1, y_2, y_3)) = x_1 y_1 - x_1 y_2 + 2x_2 y_1 + x_2 y_3 + 2x_3 y_2 + x_3 y_3$

   (c) $\mathbf{U} = \mathbf{V} = \mathbf{R}^3, A = B = \{(1, 1, 1), (1, 1, 0), (1, 0, 0)\},$
   $f((x_1, x_2, x_3), (y_1, y_2, y_3)) = 2x_1 y_2 - x_1 y_3 + 2x_2 y_1 + x_2 y_2 + x_2 y_3 - x_3 y_1$
   $\qquad\qquad + x_3 y_2 - 2x_3 y_3$

   (d) $\mathbf{U} = \mathbf{R}^3, \mathbf{V} = \mathbf{R}^2, A = \{(1, 1, 0), (1, -1, 1), (0, 1, 0)\}, B = \{(1, 2), (2, 1)\},$
   $f((x_1, x_2, x_3), (y_1, y_2)) = 2x_1 y_1 + 4x_1 y_2 - 6x_2 y_1 + 3x_2 y_2 + x_3 y_1$

   (e) $\mathbf{U} = \mathcal{C}^3, \mathbf{V} = \mathcal{C}^2, A = \{(i, 0, 0), (1, i, 0), (0, 0, 2i)\}, B = \{(1 - i, i), (i, -i)\},$
   $f((x_1, x_2, x_3), (y_1, y_2)) = 5x_1 y_1 + ix_1 y_2 - ix_2 y_1 + 2x_2 y_2 + 2x_3 y_1 - x_3 y_2$

   (f) $\mathbf{U} = \mathcal{C}^3, \mathbf{V} = \mathcal{C}^2, A = \{(1, 0, 0), (1, 1, 0), (1, 1, 1)\}, B = \{(1, -1), (2, -1)\},$
   $f((x_1, x_2, x_3), (y_1, y_2)) = x_1 y_2 + x_2 y_1 + 2x_2 y_2 - 2x_3 y_1 + 2x_3 y_2$

2. Use the matrix obtained in the corresponding part of Problem 1 to compute $f(\mathbf{u}, \mathbf{v})$ for the given vectors.

(a) $\mathbf{u} = (3, -1)$, $\mathbf{v} = (0, 4, -1)$ **(b)** $\mathbf{u} = (5, -6, 3)$, $\mathbf{v} = (2, -1, 0)$

(c) $\mathbf{u} = (1, -2, 2)$, $\mathbf{v} = (0, 0, 1)$ **(d)** $\mathbf{u} = (0, 3, -1)$, $\mathbf{v} = (8, 7)$

(e) $\mathbf{u} = (i, 0, i)$, $\mathbf{v} = (2, 0)$ **(f)** $\mathbf{u} = (1, 0, -1)$, $\mathbf{v} = (1, 0)$

3. Each of parts (a)–(f) below relate to the corresponding part of Problem 1 above. In each case, a new pair of bases $\mathcal{A}'$, $\mathcal{B}'$ is given for the vector spaces $\mathbf{U}$, $\mathbf{V}$. Find the matrix of f relative to $\mathcal{A}'$ and $\mathcal{B}'$ by use of Theorem 8.34.

(a) $\mathcal{A}' = \{(4, -1), (-1, 5)\}$, $\mathcal{B}' = \{(1, 2, -4), (1, -1, -1), (-4, -2, 1)\}$

(b) $\mathcal{A}' = \mathcal{B}' = \{(17, 2, -2), (1, 7, 2), (-1, 2, 7)\}$

(c) $\mathcal{A}' = \mathcal{B}' = \{(5, 4, 2), (1, -1, -2), (1, -1, 1)\}$

(d) $\mathcal{A}' = \{(3, 1, 2), (3, -1, 1), (0, 5, -2)\}$, $\mathcal{B}' = \{(1, -1), (0, 3)\}$

(e) $\mathcal{A}' = \{(-1, 0, 0), (-1, -i, 0), (0, 0, 1)\}$, $\mathcal{B}' = \{(2, 1 + i), (1, -1)\}$

(f) $\mathcal{A}' = \{(1, 0, 0), (0, 1, 0), (-6, 2, 1)\}$, $\mathcal{B}' = \{(-2, 1), (1, 0)\}$

4. Let f be the bilinear form on $\mathbf{R}^4$ and $\mathbf{R}^3$ that has the given matrix A relative to $\mathcal{E}_4$ and $\mathcal{E}_3$. Find bases $\mathcal{A}'$ of $\mathbf{R}^4$ and $\mathcal{B}'$ of $\mathbf{R}^3$ such that, relative to $\mathcal{A}'$ and $\mathcal{B}'$, f has a matrix of the form D_r described in Corollary 8.36.

(a) $A = \begin{bmatrix} 1 & 0 & 2 \\ 4 & 1 & 3 \\ 3 & 1 & 1 \\ 2 & 1 & -1 \end{bmatrix}$ **(b)** $A = \begin{bmatrix} 1 & 2 & 3 \\ -2 & -4 & -6 \\ 1 & 0 & -1 \\ -1 & 0 & 1 \end{bmatrix}$

(c) $A = \begin{bmatrix} 0 & 0 & 0 \\ 2 & 1 & 3 \\ 4 & 2 & 6 \\ 1 & 0 & 1 \end{bmatrix}$ **(d)** $A = \begin{bmatrix} 0 & 1 & 2 \\ 3 & -2 & -1 \end{bmatrix}$

5. Prove Corollary 8.35.

6. Prove Corollary 8.36.

7. Prove that the rank of a bilinear form as defined in this section is unique.

8. If f and g are two bilinear forms on $\mathbf{U}$ and $\mathbf{V}$, then $f + g$ is defined by

$$(f + g)(\mathbf{u}, \mathbf{v}) = f(\mathbf{u}, \mathbf{v}) + g(\mathbf{u}, \mathbf{v})$$

for all $(\mathbf{u}, \mathbf{v}) \in \mathbf{U} \times \mathbf{V}$. Also, for any $a \in \mathcal{F}$, the product af is defined by

$$(af)(\mathbf{u}, \mathbf{v}) = a \cdot f(\mathbf{u}, \mathbf{v}).$$

Prove that the set of all bilinear forms on $\mathbf{U}$ and $\mathbf{V}$ is a vector space with respect to these operations of addition and scalar multiplication.

9. (See Problem 8.) Let $\mathbf{W}$ be the vector space of all bilinear forms on $\mathbf{R}^m$ and $\mathbf{R}^n$. Prove that the mapping $m(f) = A$ that sends a bilinear form onto its matrix A relative to $\mathcal{E}_m$ and $\mathcal{E}_n$ is an isomorphism of the vector space $\mathbf{W}$ onto the vector space $\mathbf{R}_{m \times n}$ as defined in Section 4.1.

8.7 Symmetric Bilinear Forms

As in the case of linear transformations, there is a great deal of interest in those bilinear forms for the special case where $\mathbf{V} = \mathbf{U}$. In this case, we shorten the phrase "bilinear form on $\mathbf{V}$ and $\mathbf{V}$" to "bilinear form on $\mathbf{V}$." We also agree that, without a statement to the contrary, the bases $\mathcal{A}$ and $\mathcal{B}$ are the same whenever $\mathbf{V} = \mathbf{U}$. This means that $a_{ij} = f(\mathbf{u}_i, \mathbf{u}_j)$ in the matrix of f relative to $\mathcal{A}$. Throughout this section, we shall have $\mathbf{V} = \mathbf{U}$ and $\mathbf{V}$ an n-dimensional vector space over $\mathcal{F}$.

The imposition of the conditions that $\mathbf{U} = \mathbf{V}$ and $\mathcal{A} = \mathcal{B}$ means that our bilinear forms are now represented by square matrices only. As a more interesting consequence of these conditions, we have the next theorem. This result should be compared with Corollary 8.35 in the preceding section.

8.37 THEOREM

Let A and B be square matrices of order n over $\mathcal{F}$. Then B is congruent to A over $\mathcal{F}$ if and only if B and A represent the same bilinear form on $\mathbf{V}$.

Proof Let A and B be matrices of order n over $\mathcal{F}$, and suppose that A represents the bilinear form f on $\mathbf{V}$ relative to the basis $\mathcal{A}$.

Assume first that B is congruent to A over $\mathcal{F}$. Then $B = P^T A P$ for some invertible P over $\mathcal{F}$. Since P is invertible with elements in $\mathcal{F}$, P is the transition matrix from $\mathcal{A}$ to a basis $\mathcal{A}'$ of $\mathbf{V}$. With $\mathbf{U} = \mathbf{V}$, $\mathcal{A} = \mathcal{B}$, $Q = P$, and $\mathcal{A}' = \mathcal{B}'$ in Theorem 8.34, we have $B = P^T A P$ is the matrix of f relative to $\mathcal{A}'$.

On the other hand, suppose that B represents f relative to some basis $\mathcal{A}'$ of $\mathbf{V}$. If P is the matrix of transition from $\mathcal{A}$ to $\mathcal{A}'$, then Theorem 8.34 asserts that $P^T A P$ is the matrix of f relative to $\mathcal{A}'$. But this matrix is unique, so it must be that $B = P^T A P$. Now P is invertible with elements in $\mathcal{F}$ since it is the transition matrix from one basis of $\mathbf{V}$ to another. Hence B is congruent to A over $\mathcal{F}$. ∎ ∎ ∎

The connection established between bilinear forms and matrices in Theorem 8.33 suggests the possibility of describing properties of bilinear forms in terms of their matrices. Several interesting results along these lines can be obtained whenever the matrices are square.

8.38 DEFINITION

A bilinear form f on $\mathbf{V}$ is **symmetric** if $f(\mathbf{u}, \mathbf{v}) = f(\mathbf{v}, \mathbf{u})$ for all $\mathbf{u}, \mathbf{v} \in \mathbf{V}$.

8.39 THEOREM

A bilinear form f on $\mathbf{V}$ is symmetric if and only if every matrix that represents f is symmetric.

Proof Suppose that f is symmetric, and let $A = [a_{ij}]_n$ be a matrix that represents f. Then $a_{ij} = f(\mathbf{u}_i, \mathbf{u}_j) = f(\mathbf{u}_j, \mathbf{u}_i) = a_{ji}$, and A is symmetric.

Suppose now that f is represented by a symmetric matrix $A = [a_{ij}]_n$ relative to the basis $\mathcal{A}$ of $\mathbf{V}$. Let $\mathbf{u}$ and $\mathbf{v}$ be arbitrary vectors with coordinate matrices X and Y, respectively, relative to $\mathcal{A}$. Then $f(\mathbf{u}, \mathbf{v}) = X^T A Y$ and $f(\mathbf{v}, \mathbf{u}) = Y^T A X$. Since $Y^T A X$ is a 1 by 1 matrix, $Y^T A X = (Y^T A X)^T$. Thus

$$f(\mathbf{v}, \mathbf{u}) = (Y^T A X)^T = X^T A^T (Y^T)^T = X^T A Y = f(\mathbf{u}, \mathbf{v}),$$

where $A^T = A$, since A is symmetric.   ■ ■ ■

Our next theorem has an immediate corollary concerning symmetric bilinear forms.

8.40 THEOREM

If $1 + 1 \neq 0$ in $\mathcal{F}$, every symmetric matrix A of order n over $\mathcal{F}$ is congruent over $\mathcal{F}$ to a diagonal matrix.

Proof The proof is by induction on the order n of A. The theorem is trivially true for $n = 1$.

Assume that the theorem is true for all symmetric matrices of order k over $\mathcal{F}$, and let A be a symmetric matrix of order $k + 1$ over $\mathcal{F}$. Let $\mathcal{A} = \{\mathbf{u}_1, \mathbf{u}_2, \ldots, \mathbf{u}_{k+1}\}$ be a basis of $\mathcal{F}^{k+1}$, and let f be the symmetric bilinear form on $\mathcal{F}^{k+1}$ that has matrix A relative to $\mathcal{A}$. If A is the zero matrix, there is nothing to prove. Assume, then, that $a_{rs} = f(\mathbf{u}_r, \mathbf{u}_s) \neq 0$ for the pair r, s. If $f(\mathbf{u}_r, \mathbf{u}_r) = 0$ and $f(\mathbf{u}_s, \mathbf{u}_s) = 0$, then

$$f(\mathbf{u}_r + \mathbf{u}_s, \mathbf{u}_r + \mathbf{u}_s) = 2f(\mathbf{u}_r, \mathbf{u}_s) \neq 0.$$

Thus there is a vector $\mathbf{v}_1 \in \mathcal{F}^{k+1}$ such that $d_1 = f(\mathbf{v}_1, \mathbf{v}_1) \neq 0$. The set $\{\mathbf{v}_1\}$ can be extended to a basis $\mathcal{B} = \{\mathbf{v}_1, \mathbf{v}_2, \ldots, \mathbf{v}_{k+1}\}$ of $\mathcal{F}^{k+1}$. The set $\mathcal{A}' = \{\mathbf{u}_1', \mathbf{u}_2', \ldots, \mathbf{u}_{k+1}'\}$ is obtained from the basis $\mathcal{B}$ as follows:

$$\mathbf{u}_1' = \mathbf{v}_1 \quad \text{and} \quad \mathbf{u}_j' = \mathbf{v}_j - \frac{f(\mathbf{u}_1', \mathbf{v}_j)}{d_1} \mathbf{u}_1'$$

for $j = 2, \ldots, k + 1$. That is,

$$\mathbf{u}_1' = \mathbf{v}_1$$

$$\mathbf{u}_2' = \mathbf{v}_2 - \frac{f(\mathbf{u}_1', \mathbf{v}_2)}{d_1} \mathbf{u}_1'$$

$$\vdots$$

$$\mathbf{u}_{k+1}' = \mathbf{v}_{k+1} - \frac{f(\mathbf{u}_1', \mathbf{v}_{k+1})}{d_1} \mathbf{u}_1'.$$

It is clear, then, that $\mathcal{A}'$ can be obtained from $\mathcal{B}$ by a sequence of elementary operations. Therefore, $\mathcal{A}'$ is a basis of $\mathcal{F}^{k+1}$. Let A_1 denote the matrix of f relative to $\mathcal{A}'$. The matrix

A_1 is symmetric since f is symmetric. Now $f(\mathbf{u}_1', \mathbf{u}_1') = f(\mathbf{v}_1, \mathbf{v}_1) = d_1$, and if $j > 1$,

$$f(\mathbf{u}_1', \mathbf{u}_j') = f\left(\mathbf{u}_1', \mathbf{v}_j - \frac{f(\mathbf{u}_1', \mathbf{v}_j)}{d_1}\mathbf{u}_1'\right)$$

$$= f(\mathbf{u}_1', \mathbf{v}_j) - \frac{f(\mathbf{u}_1', \mathbf{v}_j)}{d_1} f(\mathbf{u}_1', \mathbf{u}_1')$$

$$= 0.$$

Thus A_1 is of the form

$$A_1 = \left[\begin{array}{c|c} d_1 & \mathbf{0} \\ \hline \mathbf{0} & A_2 \end{array}\right],$$

where A_2 is a symmetric $k \times k$ matrix. If Q_1 denotes the matrix of transition from $\mathcal{A}$ to $\mathcal{B}$ and Q_2 denotes the matrix of transition from $\mathcal{B}$ to $\mathcal{A}'$, then $P_1 = Q_1 Q_2$ denotes the invertible matrix of transition from $\mathcal{A}$ to $\mathcal{A}'$, and $A_1 = P_1^T A P_1$ by Theorem 8.34.

By the induction hypothesis, there is an invertible matrix Q of order k such that $Q^T A_2 Q = \text{diag}\{d_2, \ldots, d_{k+1}\}$. Hence

$$P_2 = \left[\begin{array}{c|c} 1 & \mathbf{0} \\ \hline \mathbf{0} & Q \end{array}\right]$$

is an invertible matrix, and

$$(P_1 P_2)^T A(P_1 P_2) = P_2^T P_1^T A P_1 P_2$$

$$= P_2^T A_1 P_2$$

$$= \left[\begin{array}{c|c} 1 & \mathbf{0} \\ \hline \mathbf{0} & Q^T \end{array}\right]\left[\begin{array}{c|c} d_1 & \mathbf{0} \\ \hline \mathbf{0} & A_2 \end{array}\right]\left[\begin{array}{c|c} 1 & \mathbf{0} \\ \hline \mathbf{0} & Q \end{array}\right]$$

$$= \left[\begin{array}{c|c} d_1 & \mathbf{0} \\ \hline \mathbf{0} & Q^T A_2 Q \end{array}\right]$$

$$= \text{diag}\{d_1, d_2, \ldots, d_{k+1}\}.$$

Thus $P = P_1 P_2$ is an invertible matrix such that $P^T A P$ is diagonal. The theorem follows by induction. ∎

8.41 COROLLARY If $1 + 1 \neq 0$ in $\mathcal{F}$, then every symmetric bilinear form on $\mathbf{V}$ can be represented by a diagonal matrix.

Proof The proof is left as an exercise (Problem 3).   ■ ■ ■

> **8.42**
> **COROLLARY**
>
> Every symmetric matrix over $\mathcal{C}$ is congruent over $\mathcal{C}$ to a diagonal matrix D_r with the first r diagonal elements equal to 1 and all other elements zero.

Proof See Problem 4.   ■ ■ ■

Much the same as with Theorem 8.19, the proof of Theorem 8.40 provides a basis for an approach to the problem of finding an invertible matrix P over $\mathcal{F}$ such that $P^T A P$ is diagonal, but this approach is not very efficient. We can proceed more directly in the following manner to obtain such a P. The first column of P is simply the coordinate matrix

$$P_1 = \begin{bmatrix} p_{11} \\ p_{21} \\ \vdots \\ p_{n1} \end{bmatrix}$$

relative to $\mathcal{A}$ of a vector $\mathbf{u}'_1$ chosen such that

$$d_1 = f(\mathbf{u}'_1, \mathbf{u}'_1) \neq 0.$$

The vector $\mathbf{u}'_2$ must be chosen so that $\{\mathbf{u}'_1, \mathbf{u}'_2\}$ is linearly independent and $f(\mathbf{u}'_1, \mathbf{u}'_2) = 0$. This latter condition means that the coordinate matrix

$$P_2 = \begin{bmatrix} p_{12} \\ p_{22} \\ \vdots \\ p_{n2} \end{bmatrix}$$

of $\mathbf{u}'_2$ relative to $\mathcal{A}$ must satisfy the equation

$$P_1^T A P_2 = 0.$$

That is, $x_1 = p_{12}, x_2 = p_{22}, \ldots, x_n = p_{n2}$ must be a solution to the system $(P_1^T A) X = \mathbf{0}$. Similarly, the requirements that $f(\mathbf{u}'_1, \mathbf{u}'_3) = 0$ and $f(\mathbf{u}'_2, \mathbf{u}'_3) = 0$ are reflected by the conditions $(P_1^T A) P_3 = 0$ and $(P_2^T A) P_3 = 0$ on the coordinate matrix P_3 of $\mathbf{u}'_3$ relative to $\mathcal{A}$, and so on. The method is illustrated in the following example.

EXAMPLE 1 The problem is to determine a real invertible matrix P such that $P^T A P$ is diagonal, where

$$A = \begin{bmatrix} 0 & 1 & 2 \\ 1 & 0 & 0 \\ 2 & 0 & 0 \end{bmatrix}.$$

Let f be the symmetric bilinear form on $\mathbf{R}^3$ that has matrix A relative to $\mathcal{E}_3$. Since $f(\mathbf{e}_i, \mathbf{e}_i) = 0$ for $i = 1, 2, 3$ and $f(\mathbf{e}_1, \mathbf{e}_2) \neq 0$, we choose $\mathbf{u}'_1 = \mathbf{e}_1 + \mathbf{e}_2$ such that $f(\mathbf{u}'_1, \mathbf{u}'_1) \neq 0$. Then $P_1 = [\,1 \quad 1 \quad 0\,]^T$ and

$$P_1^T A = [\,1 \quad 1 \quad 2\,].$$

The choice $P_2 = [\,1 \quad 1 \quad -1\,]^T$ satisfies $(P_1^T A)P_2 = 0$ and makes $\{\mathbf{u}'_1, \mathbf{u}'_2\}$ linearly independent. The matrix $P_2^T A$ is given by

$$P_2^T A = [\,-1 \quad 1 \quad 2\,].$$

The choice $P_3 = [\,0 \quad -2 \quad 1\,]^T$ satisfies $(P_1^T A)P_3 = 0$ and $(P_2^T A)P_3 = 0$, and makes $\{\mathbf{u}'_1, \mathbf{u}'_2, \mathbf{u}'_3\}$ linearly independent. The matrix $P = [\,P_1 \quad P_2 \quad P_3\,]$ is given by

$$P = \begin{bmatrix} 1 & 1 & 0 \\ 1 & 1 & -2 \\ 0 & -1 & 1 \end{bmatrix},$$

and $P^T A P = \mathrm{diag}\{2, -2, 0\}$.   ∎

The following definition and theorem apply to infinite-dimensional vector spaces as well as those of finite dimension, although our main interest is in the latter case.

8.43 DEFINITION   Let $\mathbf{V}$ be a vector space over $\mathcal{F}$. A mapping q of $\mathbf{V}$ into $\mathcal{F}$ is a **quadratic form** on $\mathbf{V}$ if and only if there is a symmetric bilinear form f on $\mathbf{V}$ such that $q(\mathbf{u}) = f(\mathbf{u}, \mathbf{u})$ for all $\mathbf{u} \in \mathbf{V}$.

It should be observed that the definition of a real quadratic form that was given earlier is entirely consistent with this definition.

It is clear that each symmetric bilinear form f determines a unique associated quadratic form q by the rule that $q(\mathbf{u}) = f(\mathbf{u}, \mathbf{u})$. The following theorem shows that this correspondence between symmetric bilinear forms and quadratic forms is one-to-one if $1 + 1 \neq 0$ in $\mathcal{F}$.

8.44 THEOREM   Let $\mathbf{V}$ be a vector space over the field $\mathcal{F}$ in which $1 + 1 \neq 0$. If the quadratic form q on $\mathbf{V}$ is determined by the symmetric bilinear form f on $\mathbf{V}$, then

$$f(\mathbf{u}, \mathbf{v}) = \tfrac{1}{2}[q(\mathbf{u} + \mathbf{v}) - q(\mathbf{u}) - q(\mathbf{v})]$$

for all $\mathbf{u}, \mathbf{v}$ in $\mathbf{V}$.

Proof The theorem follows from the simplification

$$
\begin{aligned}
\tfrac{1}{2}[q(\mathbf{u}+\mathbf{v}) - q(\mathbf{u}) - q(\mathbf{v})] &= \tfrac{1}{2}[f(\mathbf{u}+\mathbf{v}, \mathbf{u}+\mathbf{v}) - f(\mathbf{u}, \mathbf{u}) - f(\mathbf{v}, \mathbf{v})] \\
&= \tfrac{1}{2}[f(\mathbf{u}, \mathbf{u}) + f(\mathbf{u}, \mathbf{v}) + f(\mathbf{v}, \mathbf{u}) + f(\mathbf{v}, \mathbf{v}) \\
&\quad - f(\mathbf{u}, \mathbf{u}) - f(\mathbf{v}, \mathbf{v})] \\
&= \tfrac{1}{2}[2f(\mathbf{u}, \mathbf{v})] \\
&= f(\mathbf{u}, \mathbf{v}).
\end{aligned}
$$

■ ■ ■

8.45
DEFINITION

Let $\mathbf{V}$ be a finite-dimensional vector space over the field $\mathcal{F}$ in which $1 + 1 \neq 0$. The **matrix** of the quadratic form q relative to the basis $\mathcal{A}$ of $\mathbf{V}$ is the same as the matrix of the symmetric bilinear form f that determines q.

We note that the matrix of a quadratic form is required to be symmetric.

For the remainder of this chapter, we restrict our attention to those vector spaces that have a field of scalars $\mathcal{F}$ in which $1 + 1 \neq 0$.

The intimate connection between quadratic forms and symmetric bilinear forms means that many of the results obtained for symmetric bilinear forms translate immediately into statements about quadratic forms. The most important of these are listed next.

1. If q has matrix A relative to the basis $\mathcal{A}$ of $\mathbf{V}$ and $\mathbf{u}$ has coordinate matrix X relative to $\mathcal{A}$, then $q(\mathbf{u}) = X^T A X$ (Theorem 8.33).

2. If q has matrix A relative to the basis $\mathcal{A}$ of $\mathbf{V}$ and P is the matrix of transition from $\mathcal{A}$ to $\mathcal{A}'$, then $P^T A P$ is the matrix of q relative to $\mathcal{A}'$ (Theorem 8.34).

3. Every quadratic form on $\mathbf{V}$ can be represented by a diagonal matrix (Corollary 8.41).

4. Every quadratic form on a vector space over $\mathcal{C}$ can be represented by a matrix D_r with the first r diagonal elements equal to 1 and all other elements zero (Problem 6).

8.7 Exercises

1. For each matrix A, find an invertible matrix P such that $P^T A P$ is diagonal.

(a) $A = \begin{bmatrix} 0 & 2 & -1 \\ 2 & 1 & 1 \\ -1 & 1 & -2 \end{bmatrix}$
(b) $A = \begin{bmatrix} 2 & -1 & 1 \\ -1 & 3 & 0 \\ 1 & 0 & 0 \end{bmatrix}$

(c) $A = \begin{bmatrix} 0 & 2 & 3 \\ 2 & 0 & -2 \\ 3 & -2 & 0 \end{bmatrix}$
(d) $A = \begin{bmatrix} 1 & 2 & 0 \\ 2 & -2 & -3 \\ 0 & -3 & 4 \end{bmatrix}$

$$\textbf{(e)} \ A = \begin{bmatrix} 0 & -1 & 2 \\ -1 & 0 & -3 \\ 2 & -3 & 0 \end{bmatrix} \qquad\qquad \textbf{(f)} \ A = \begin{bmatrix} 0 & 2 & 3 \\ 2 & 1 & -1 \\ 3 & -1 & 1 \end{bmatrix}$$

2. For each matrix A in Problem 1, let f be the bilinear form on $\mathbf{R}^3$ that has matrix A relative to the basis $\mathcal{A} = \{(1, 1, 1), (1, 0, 1), (0, 1, -1)\}$. Use the matrix P from Problem 1 to find a basis of $\mathbf{R}^3$ relative to which f is represented by a diagonal matrix. Write out the corresponding diagonalized expression for f.

3. Prove Corollary 8.41.

4. Prove Corollary 8.42.

5. Translate Corollary 8.42 into a statement concerning symmetric bilinear forms on vector spaces over $\mathcal{C}$.

6. Use Corollary 8.42 to prove (4) above in the list of properties of quadratic forms over $\mathcal{C}$.

7. Let q represent an arbitrary quadratic form.

   (a) Prove or disprove: $q(a\mathbf{u} + b\mathbf{v}) = aq(\mathbf{u}) + bq(\mathbf{v})$.

   (b) Prove that $q(\mathbf{u} + \mathbf{v}) + q(\mathbf{u} - \mathbf{v}) = 2[q(\mathbf{u}) + q(\mathbf{v})]$.

8. Let q be a quadratic form on $\mathbf{R}^n$ with symmetric matrix A relative to the standard basis $\mathcal{E}_n : q(x_1, x_2, \ldots, x_n) = X^T A X$. If $\lambda_1 \geq \lambda_2 \geq \cdots \geq \lambda_n$ are the (necessarily real) eigenvalues of A, prove that $\lambda_1 \geq q(\mathbf{v}) \geq \lambda_n$ for all $\mathbf{v} = (x_1, x_2, \ldots, x_n)$ with length 1.

9. A bilinear form f on $\mathbf{V}$ is by definition **skew-symmetric** if and only if

$$f(\mathbf{u}, \mathbf{v}) = -f(\mathbf{v}, \mathbf{u})$$

   for all $\mathbf{u}, \mathbf{v} \in \mathbf{V}$. Prove that a bilinear form on $\mathbf{V}$ is skew-symmetric if and only if any matrix that represents f is skew-symmetric.

10. Let $\mathbf{V}$ be a vector space over a field $\mathcal{F}$ in which $1 + 1 \neq 0$, and let f be a bilinear form on $\mathbf{V}$. Prove that f is skew-symmetric if and only if $f(\mathbf{v}, \mathbf{v}) = 0$ for all $\mathbf{v} \in \mathbf{V}$.

11. Prove that if $1 + 1 \neq 0$ in $\mathcal{F}$, every bilinear form f on $\mathbf{V}$ can be written uniquely as the sum of a symmetric and a skew-symmetric bilinear form. (See Problem 8 of Exercises 8.6.)

12. Express each of the following bilinear forms f as the sum of a symmetric and a skew-symmetric bilinear form.

   (a) $f((x_1, x_2), (y_1, y_2)) = 2x_1y_1 + x_1y_2 - 4x_2y_2$

   (b) $f((x_1, x_2, x_2), (y_1, y_2, y_3)) = x_1y_1 + x_1y_2 + 2x_1y_3 - 4x_2y_2 + 6x_2y_3 - x_3y_1 + 2x_3y_3$

13. Assume that $1 + 1 \neq 0$ in $\mathcal{F}$. The definition of a quadratic form on $\mathbf{V}$ can be "extended" so as to include those mappings q of $\mathbf{V}$ into $\mathcal{F}$ such that $q(\mathbf{v}) = f(\mathbf{v}, \mathbf{v})$ for some (not necessarily symmetric) bilinear form on $\mathbf{V}$.

   (a) Prove that f is skew-symmetric if and only if f determines the zero quadratic form on $\mathbf{V}$.

(b) Prove that the two bilinear forms f_1 and f_2 on **V** define the same quadratic form on **V** if and only if $f_1 - f_2$ is skew-symmetric.

(c) Does this "extension" actually enlarge the set of quadratic forms on **V**? Why or why not?

8.8 Hermitian Forms

We restrict our attention now to those vector spaces with a scalar field $\mathcal{F}$ that is either **R** or the field $\mathcal{C}$ of complex numbers. Throughout this section, each statement involving $\mathcal{F}$ may be read in only two ways: with $\mathcal{F} = \mathbf{R}$, and with $\mathcal{F} = \mathcal{C}$.

The development of Chapter 9 is based on the interesting and important concept of an inner product on a vector space. We shall see there that, so long as our work is with real vector spaces, symmetric bilinear forms are sufficient for our needs. But with complex vector spaces, we shall have need for another type of function, the *Hermitian form*. A Hermitian form has somewhat the same relation to a complex bilinear form as a symmetric bilinear form has to a bilinear form.

8.46 DEFINITION

Let **U** and **V** be vector spaces over the field $\mathcal{F}$. A **complex bilinear form** on **U** and **V** is a mapping f of pairs of vectors $(\mathbf{u}, \mathbf{v}) \in \mathbf{U} \times \mathbf{V}$ onto scalars $f(\mathbf{u}, \mathbf{v}) \in \mathcal{F}$ that has the properties

$$f(a_1\mathbf{u}_1 + a_2\mathbf{u}_2, \mathbf{v}) = \bar{a}_1 f(\mathbf{u}_1, \mathbf{v}) + \bar{a}_2 f(\mathbf{u}_2, \mathbf{v}) \tag{i}$$

and

$$f(\mathbf{u}, b_1\mathbf{v}_1 + b_2\mathbf{v}_2) = b_1 f(\mathbf{u}, \mathbf{v}_1) + b_2 f(\mathbf{u}, \mathbf{v}_2), \tag{ii}$$

where $\bar{a}_i$ denotes the complex conjugate of a_i.

We describe the conditions (i) and (ii) by saying that f is complex linear in the first variable and linear in the second variable. It is readily seen that a function f from $\mathbf{U} \times \mathbf{V}$ into $\mathcal{F}$ is a complex bilinear form if and only if

$$f\left(\sum_{i=1}^{r} a_i\mathbf{u}_i, \sum_{j=1}^{s} b_j\mathbf{v}_j\right) = \sum_{i=1}^{r}\sum_{j=1}^{s} \bar{a}_i b_j f(\mathbf{u}_i, \mathbf{v}_j)$$

for all positive integers r, s.

Whenever the field of scalars is real, a complex bilinear form reduces to a bilinear form. In this sense, a complex bilinear form is a generalization of a bilinear form. We must keep in mind, however, that the two concepts are quite distinct whenever $\mathcal{F}$ is the field $\mathcal{C}$.

Throughout the remainder of this section, **U** and **V** will denote vector spaces over $\mathcal{F}$ with dimensions m and n, respectively, and f will denote a complex bilinear form on **U** and **V**.

8.47
DEFINITION

Let $\mathcal{A} = \{\mathbf{u}_1, \mathbf{u}_2, \ldots, \mathbf{u}_m\}$ and $\mathcal{B} = \{\mathbf{v}_1, \mathbf{v}_2, \ldots, \mathbf{v}_n\}$ be bases of $\mathbf{U}$ and $\mathbf{V}$, respectively. The **matrix of a complex bilinear form** f relative to $\mathcal{A}$ and $\mathcal{B}$ is the matrix $A = [a_{ij}]_{m \times n}$, where $a_{ij} = f(\mathbf{u}_i, \mathbf{v}_j)$ for $i = 1, 2, \ldots, m$; $j = 1, 2, \ldots, n$.

At this point, a parallel can be perceived between the properties of bilinear forms and those of complex bilinear forms. Actually, the parallel is so strong that it is quite repetitious to develop the properties of complex bilinear forms in as much detail as was done with bilinear forms. At the same time, the adjustments that are necessitated by complex linearity in the first variable are not altogether obvious. Consequently, we shall more or less outline the development here with statements of the major results as theorems, and leave the proofs of most of these theorems as exercises. In all cases, a proof can be obtained by a suitable modification of the proof of the corresponding result for bilinear forms.

We recall that the conjugate transpose $(\overline{A})^T$ of a matrix A is denoted by A^*.

8.48
THEOREM

Let $X = [\,x_1 \;\; x_2 \;\; \cdots \;\; x_m\,]^T$ denote the coordinate matrix of $\mathbf{u} \in \mathbf{U}$ relative to the basis $\mathcal{A}$ of $\mathbf{U}$, and let $Y = [\,y_1 \;\; y_2 \;\; \cdots \;\; y_n\,]^T$ be the coordinate matrix of $\mathbf{v} \in \mathbf{V}$ relative to the basis $\mathcal{B}$ of $\mathbf{V}$. If $A = [a_{ij}]_{m \times n}$, then A is the matrix of the complex bilinear form f on $\mathbf{U}$ and $\mathbf{V}$ if and only if the equation

$$f(\mathbf{u}, \mathbf{v}) = X^* A Y$$

is satisfied for all choices of $\mathbf{u} \in \mathbf{U}$, $\mathbf{v} \in \mathbf{V}$.

Proof See Problem 8. ∎

EXAMPLE 1 Consider the mapping f defined from $\mathcal{C}^2 \times \mathcal{C}^3$ into $\mathcal{C}$ by

$$f(\mathbf{u}, \mathbf{v}) = 3i\overline{u}_1 v_2 + (-1 - 3i)\overline{u}_1 v_3 + i\overline{u}_2 v_1 + (4 - i)\overline{u}_2 v_2 + (-4 + 3i)\overline{u}_2 v_3,$$

where $\mathbf{u} = (u_1, u_2)$ and $\mathbf{v} = (v_1, v_2, v_3)$. Guided by the equation $a_{ij} = f(\mathbf{e}_i, \mathbf{e}_j)$ in Definition 8.47, we see that the value of $f(\mathbf{u}, \mathbf{v})$ can be written in the form

$$f(\mathbf{u}, \mathbf{v}) = f((u_1, u_2), (v_1, v_2, v_3))$$

$$= \begin{bmatrix} \overline{u}_1 & \overline{u}_2 \end{bmatrix} \begin{bmatrix} 0 & 3i & -1 - 3i \\ i & 4 - i & -4 + 3i \end{bmatrix} \begin{bmatrix} v_1 \\ v_2 \\ v_3 \end{bmatrix}.$$

Hence f is the complex bilinear form on $\mathcal{C}^2$ and $\mathcal{C}^3$ that has the matrix

$$A = \begin{bmatrix} 0 & 3i & -1 - 3i \\ i & 4 - i & -4 + 3i \end{bmatrix}$$

relative to the standard bases $\mathcal{E}_2$ of $\mathcal{C}^2$ and $\mathcal{E}_3$ of $\mathcal{C}^3$. ∎

It follows from Definition 8.47 and Theorem 8.48 that two complex bilinear forms f and g on **U** and **V** are equal if and only if they have the same matrix relative to bases $\mathcal{A}$ of **U** and $\mathcal{B}$ of **V**.

8.49
THEOREM

Let A be the matrix of f relative to the bases $\mathcal{A}$ of **U** and $\mathcal{B}$ of **V**. If Q is the matrix of transition from $\mathcal{A}$ to the basis $\mathcal{A}'$ of **U** and P is the matrix of transition from $\mathcal{B}$ to the basis $\mathcal{B}'$ of **V**, then Q^*AP is the matrix of f relative to $\mathcal{A}'$ and $\mathcal{B}'$.

Proof See Problem 9.
■ ■ ■

EXAMPLE 2 Let f be the same complex bilinear form as in Example 1, and consider the problem of finding the matrix of f relative to the bases $\mathcal{A}' = \{(1, i), (i, 0)\}$ of $\mathcal{C}^2$ and $\mathcal{B}' = \{(1, 1, 1), (1, 1, 0), (1, 0, 0)\}$ of $\mathcal{C}^3$.

The matrix of transition from $\mathcal{A}$ to $\mathcal{A}'$ is

$$Q = \begin{bmatrix} 1 & i \\ i & 0 \end{bmatrix},$$

and the matrix of transition from $\mathcal{B}$ to $\mathcal{B}'$ is

$$P = \begin{bmatrix} 1 & 1 & 1 \\ 1 & 1 & 0 \\ 1 & 0 & 0 \end{bmatrix}.$$

By Theorem 8.49, the matrix of f relative to $\mathcal{A}'$ and $\mathcal{B}'$ is

$$Q^*AP = \begin{bmatrix} 1 & -i \\ -i & 0 \end{bmatrix} \begin{bmatrix} 0 & 3i & -1-3i \\ i & 4-i & -4+3i \end{bmatrix} \begin{bmatrix} 1 & 1 & 1 \\ 1 & 1 & 0 \\ 1 & 0 & 0 \end{bmatrix}$$

$$= \begin{bmatrix} 2 & -i & 1 \\ i & 3 & 0 \end{bmatrix}.$$

Thus, relative to the bases $\mathcal{A}'$ and $\mathcal{B}'$ we have

$$f(\mathbf{u}, \mathbf{v}) = 2\bar{u}'_1 v'_1 - i\,\bar{u}'_1 v'_2 + \bar{u}'_1 v'_3 + i\,\bar{u}'_2 v'_1 + 3\bar{u}'_2 v'_2.$$

■

Our main interest in complex bilinear forms is with the case where $\mathbf{V} = \mathbf{U}$. As with bilinear forms and linear transformations, we assume that the bases $\mathcal{A}$ and $\mathcal{B}$ of $\mathbf{V} = \mathbf{U}$ are the same unless it is stated otherwise. For the remainder of the section, we shall have $\mathbf{V} = \mathbf{U}$ and $\mathbf{V}$ an n-dimensional vector space over $\mathcal{F}$.

The equivalence relation that we need to replace congruence of matrices over $\mathcal{F}$ is given in our next definition.

| **8.50**
DEFINITION | Let A and B be matrices over the field $\mathcal{F}$. Then B is **conjunctive** (or **Hermitian congruent**) to A over $\mathcal{F}$ if there is an invertible matrix P over $\mathcal{F}$ such that $B = P^*AP$. |

Theorem 8.49 and Definition 8.50 lead to the following result.

| **8.51**
THEOREM | Let A and B be matrices of order n over $\mathcal{F}$. Then B is conjunctive to A over $\mathcal{F}$ if and only if B and A represent the same complex bilinear form on **V**. |

The term "Hermitian" applies to complex bilinear forms in about the same way as the term "symmetric" applies to bilinear forms. As a matter of fact, when $\mathcal{F}$ is the field of real numbers, the two terms are coincident, just as "bilinear" and "complex bilinear" are. In this sense, a Hermitian complex bilinear form is a generalization of a symmetric bilinear form.

| **8.52**
DEFINITION | A complex bilinear form f on **V** is **Hermitian** if and only if

$$f(\mathbf{u}, \mathbf{v}) = \overline{f(\mathbf{v}, \mathbf{u})}$$

for all $\mathbf{u}, \mathbf{v} \in \mathbf{V}$. |

| **8.53**
DEFINITION | A matrix H over the field $\mathcal{F}$ is **Hermitian** if and only if $H^* = H$. |

Thus a real matrix A is Hermitian if and only if it is symmetric. The relation between Hermitian complex bilinear forms and Hermitian matrices is exactly what one would expect.

| **8.54**
THEOREM | A complex bilinear form f on **V** is Hermitian if and only if every matrix that represents f is Hermitian. |

Proof See Problem 11.   ■ ■ ■

Our next theorem corresponds to the result in Theorem 8.40 for symmetric matrices. We include a proof for this theorem, since it is quite important and its proof is a bit more difficult than the others in this section.

8.55 **THEOREM**	Every Hermitian matrix H of order n over $\mathcal{C}$ is conjunctive over $\mathcal{C}$ to a diagonal matrix.

Proof The proof is by induction on the order n of H. The theorem is trivially valid for $n = 1$.

Assume that the theorem is true for all Hermitian matrices of order k over $\mathcal{C}$, and let H be a Hermitian matrix of order $k + 1$ over $\mathcal{C}$. Let $\mathcal{A} = \{\mathbf{u}_1, \mathbf{u}_2, \ldots, \mathbf{u}_{k+1}\}$ be a basis of $\mathcal{C}^{k+1}$, and let h be the Hermitian complex bilinear form on $\mathcal{C}^{k+1}$ that has matrix H relative to $\mathcal{A}$. If $H = \mathbf{0}$, H is a diagonal matrix already. Assume now that $H \neq \mathbf{0}$ and $h(\mathbf{u}_r, \mathbf{u}_s) = a + bi \neq 0$ for the pair r, s. If $h(\mathbf{u}_r, \mathbf{u}_r) = 0$ and $h(\mathbf{u}_s, \mathbf{u}_s) = 0$, then

$$h(\mathbf{u}_r + i\mathbf{u}_s, \mathbf{u}_r + i\mathbf{u}_s) = h(\mathbf{u}_r, \mathbf{u}_r) + ih(\mathbf{u}_r, \mathbf{u}_s) - ih(\mathbf{u}_s, \mathbf{u}_r) + h(\mathbf{u}_s, \mathbf{u}_s)$$

$$= i[h(\mathbf{u}_r, \mathbf{u}_s) - h(\mathbf{u}_s, \mathbf{u}_r)]$$

$$= i[h(\mathbf{u}_r, \mathbf{u}_s) - \overline{h(\mathbf{u}_r, \mathbf{u}_s)}]$$

$$= -2b.$$

A similar computation shows that

$$h(\mathbf{u}_r + \mathbf{u}_s, \mathbf{u}_r + \mathbf{u}_s) = h(\mathbf{u}_r, \mathbf{u}_s) + \overline{h(\mathbf{u}_r, \mathbf{u}_s)}$$

$$= 2a.$$

Thus, there is a vector $\mathbf{v}_1 \in \mathcal{C}^{k+1}$ such that $d_1 = h(\mathbf{v}_1, \mathbf{v}_1) \neq 0$. The set $\{\mathbf{v}_1\}$ can be extended to a basis $\mathcal{B} = \{\mathbf{v}_1, \mathbf{v}_2, \ldots, \mathbf{v}_{k+1}\}$ of $\mathcal{C}^{k+1}$. The set $\mathcal{A}' = \{\mathbf{u}_1', \mathbf{u}_2', \ldots, \mathbf{u}_{k+1}'\}$ is obtained from $\mathcal{B}$ as follows:

$$\mathbf{u}_1' = \mathbf{v}_1$$

and

$$\mathbf{u}_j' = \mathbf{v}_j - \frac{h(\mathbf{u}_1', \mathbf{v}_j)}{d_1}\mathbf{u}_1',$$

for $j = 2, \ldots, k + 1$. Since $\mathcal{A}'$ is obtained from $\mathcal{B}$ by elementary operations, $\mathcal{A}'$ is a basis of $\mathcal{C}^{k+1}$. The matrix H_1 of h relative to $\mathcal{A}'$ is Hermitian, since h is Hermitian. Now $h(\mathbf{u}_1', \mathbf{u}_1') = d_1$, and a simple computation shows that $h(\mathbf{u}_1', \mathbf{u}_j') = 0$ for $j = 2, \ldots, k + 1$. Hence H_1 is of the form

$$H_1 = \left[\begin{array}{c|c} d_1 & \mathbf{0} \\ \hline \mathbf{0} & H_2 \end{array}\right],$$

where H_2 is a $k \times k$ Hermitian matrix. The matrix of transition P_1 from $\mathcal{A}$ to $\mathcal{A}'$ is invertible over $\mathcal{C}$, and $H_1 = P_1^* H P_1$ by Theorem 8.49.

By the induction hypothesis, there is an invertible matrix Q of order k over $\mathcal{C}$ such that $Q^* H_2 Q = \text{diag}\{d_2, \ldots, d_{k+1}\}$. Now

$$P_2 = \left[\begin{array}{c|c} 1 & 0 \\ \hline 0 & Q \end{array} \right]$$

is an invertible matrix over $\mathcal{C}$, and

$$(P_1 P_2)^* H (P_1 P_2) = P_2^* P_1^* H P_1 P_2$$
$$= P_2^* H_1 P_2$$
$$= \left[\begin{array}{c|c} d_1 & 0 \\ \hline 0 & Q^* H_2 Q \end{array} \right]$$
$$= \text{diag}\{d_1, d_2, \ldots, d_{k+1}\}.$$

Thus, $P = P_1 P_2$ is an invertible matrix over $\mathcal{C}$ such that $P^* H P$ is diagonal. The theorem follows from the principle of mathematical induction. ∎ ∎ ∎

**8.56
DEFINITION**

A Hermitian complex bilinear form on **V** is called a **Hermitian form** on **V**.

It follows from Theorems 8.55 and 8.51 that any Hermitian form h on **V** can be represented by a diagonal matrix. Now the condition $H^* = H$ requires that the diagonal elements of a Hermitian matrix must be real. Hence the diagonal elements in a diagonalized representation of h are always real.

The proof of Theorem 8.23 can be modified so as to obtain a proof of the following theorem. With $q(\mathbf{v})$ replaced by $h(\mathbf{v}, \mathbf{v})$ and $\mathbf{R}^n$ replaced by $\mathcal{C}^n$, the only other changes necessary are in the expressions for $h(\mathbf{v}, \mathbf{v})$. For example, the two diagonalized representations would appear as $h(\mathbf{v}, \mathbf{v}) = \sum_{k=1}^{r} d_k |y_k|^2$ and $h(\mathbf{v}, \mathbf{v}) = \sum_{k=1}^{r} c_k |z_k|^2$.

**8.57
THEOREM**

In any two diagonalized representations of a Hermitian form h, the number of positive terms is the same and the number of negative terms is the same.

The definitions of *index*, *signature*, *positive definite*, etc., apply to Hermitian forms as they are given in Definitions 8.25 and 8.29. The positive definite Hermitian forms are those that are fundamental to Chapter 9. The index of a Hermitian matrix is by definition the same as that of a Hermitian form that it represents.

8.58 COROLLARY

Any Hermitian matrix H is conjunctive over $\mathcal{C}$ to a unique matrix of the form

$$C = \left[\begin{array}{c|c|c} I_p & 0 & 0 \\ \hline 0 & -I_{r-p} & 0 \\ \hline 0 & 0 & 0 \end{array}\right],$$

where r is the rank of H. Two $n \times n$ Hermitian matrices are conjunctive over $\mathcal{C}$ if and only if they have the same rank and index.

Proof See Problem 13.　■ ■ ■

8.59 COROLLARY

With a suitable choice of basis in $\mathbf{V}$, any Hermitian form on $\mathbf{V}$ can be represented by a matrix of the form C in Corollary 8.58.

Proof See Problem 14.　■ ■ ■

8.60 COROLLARY

A Hermitian form h is positive definite if and only if

$$h(\mathbf{v}, \mathbf{v}) > 0$$

for all $\mathbf{v} \neq \mathbf{0}$.

Proof See Problem 15.　■ ■ ■

　　Just after Corollary 8.42 in Section 8.7, we described a method for obtaining an invertible matrix P that would reduce a given symmetric matrix A to a diagonal matrix $P^T A P$, and we illustrated the method in Example 1 of that section. In order to obtain a method for finding an invertible matrix P that will reduce a Hermitian matrix H to a diagonal matrix $P^* H P$, the only changes that are necessary in that description are that each $P_i^T A$ be replaced by $P_i^* H$. The method for obtaining P such that $P^* H P$ is of the form C in Corollary 8.58 is entirely analogous to the techniques of Section 8.7. Our next example gives a demonstration of this method.

EXAMPLE 3 For the Hermitian matrix

$$H = \left[\begin{array}{rrr} 5 & 4i & -4 \\ -4i & 3 & 5i \\ -4 & -5i & 3 \end{array}\right],$$

we shall find an invertible matrix P such that

$$P^*HP = \left[\begin{array}{c|c|c} I_p & 0 & 0 \\ \hline 0 & -I_{r-p} & 0 \\ \hline 0 & 0 & 0 \end{array}\right].$$

We first find an invertible matrix P_1 such that $P_1^*HP_1$ is diagonal. Since $h_{11} = 5$ is not zero in $H = [h_{ij}]$, we can use

$$C_1 = \begin{bmatrix} 1 \\ 0 \\ 0 \end{bmatrix}$$

as the first column of P_1. The second column C_2 of P_1 needs to make $\{C_1, C_2\}$ linearly independent and satisfy $(C_1^*H)C_2 = 0$, which appears as

$$\begin{bmatrix} 5 & 4i & -4 \end{bmatrix} C_2 = 0.$$

The choice

$$C_2 = \begin{bmatrix} 4 \\ 0 \\ 5 \end{bmatrix}$$

satisfies both conditions. The third column C_3 of P_1 needs to make $\{C_1, C_2, C_3\}$ linearly independent and satisfy $(C_1^*H)C_3 = 0$ and $(C_2^*H)C_3 = 0$. That is, we need a column C_3 that is not a linear combination of C_1 and C_2 and that satisfies both equations

$$\begin{bmatrix} 5 & 4i & -4 \end{bmatrix} C_3 = 0,$$
$$\begin{bmatrix} 0 & -9i & -1 \end{bmatrix} C_3 = 0.$$

The choice

$$C_3 = \begin{bmatrix} -8i \\ 1 \\ -9i \end{bmatrix}$$

satisfies all the conditions. Thus

$$P_1 = \begin{bmatrix} 1 & 4 & -8i \\ 0 & 0 & 1 \\ 0 & 5 & -9i \end{bmatrix}$$

is a matrix such that $P_1^*HP_1$ is diagonal. Performing the multiplication, we find that

$$P_1^*HP_1 = \begin{bmatrix} 5 & 0 & 0 \\ 0 & -5 & 0 \\ 0 & 0 & 16 \end{bmatrix}.$$

Interchanging the second and third columns in P_1 yields

$$P_2 = \begin{bmatrix} 1 & -8i & 4 \\ 0 & 1 & 0 \\ 0 & -9i & 5 \end{bmatrix}$$

such that $P_2^* H P_2 = \text{diag}\{5, 16, -5\}$. If we now put

$$P_3 = \begin{bmatrix} \frac{1}{\sqrt{5}} & 0 & 0 \\ 0 & \frac{1}{4} & 0 \\ 0 & 0 & \frac{1}{\sqrt{5}} \end{bmatrix}$$

and $P = P_2 P_3$, we get

$$P^* H P = (P_2 P_3)^* H (P_2 P_3)$$

$$= P_3^* (P_2^* H P_2) P_3$$

$$= \begin{bmatrix} \frac{1}{\sqrt{5}} & 0 & 0 \\ 0 & \frac{1}{4} & 0 \\ 0 & 0 & \frac{1}{\sqrt{5}} \end{bmatrix} \begin{bmatrix} 5 & 0 & 0 \\ 0 & 16 & 0 \\ 0 & 0 & -5 \end{bmatrix} \begin{bmatrix} \frac{1}{\sqrt{5}} & 0 & 0 \\ 0 & \frac{1}{4} & 0 \\ 0 & 0 & \frac{1}{\sqrt{5}} \end{bmatrix}$$

$$= \begin{bmatrix} 1 & 0 & 0 \\ 0 & 1 & 0 \\ 0 & 0 & -1 \end{bmatrix},$$

and this matrix has the form described in Corollary 8.58. ■

8.8 Exercises

1. Which of the following matrices are Hermitian?

(a) $\begin{bmatrix} 5 & i \\ -i & 2 \end{bmatrix}$ **(b)** $\begin{bmatrix} 2 & 1-i \\ 1+i & 3 \end{bmatrix}$ **(c)** $\begin{bmatrix} 1 & 1 \\ -1 & 1 \end{bmatrix}$

(d) $\begin{bmatrix} 0 & 1 & 1 \\ 1 & 0 & 1 \\ 1 & 1 & 0 \end{bmatrix}$ **(e)** $\begin{bmatrix} 5 & i & 2 \\ -i & 2 & -1 \\ 2 & -1 & 2 \end{bmatrix}$ **(f)** $\begin{bmatrix} 0 & 2 & 1 \\ -2 & 1 & -1 \\ -1 & 1 & -2 \end{bmatrix}$

(g) $\begin{bmatrix} \frac{11}{9} & -\frac{2}{3}i & -\frac{1}{9} \\ \frac{2}{3}i & 2 & -\frac{1}{3}i \\ -\frac{1}{9} & \frac{1}{3}i & \frac{5}{9} \end{bmatrix}$ **(h)** $\begin{bmatrix} 3i & 0 & 4i \\ 0 & 2i & -2 \\ -4i & 2 & i \end{bmatrix}$

2. For each Hermitian matrix H in Problem 1, answer the following questions.

(a) Find an invertible matrix P such that $P^* H P$ is diagonal.

(b) State the rank and index of H.

(c) Which of these matrices are conjuctive over $\mathcal{C}$?

3. For each Hermitian matrix H in Problem 1, find an invertible matrix P such that P^*HP is of the form C given in Corollary 8.58.

4. In each of parts (a)–(h) of Problem 1, let f be the complex bilinear form on C^n that has the given matrix relative to $\mathcal{E}_n$.

   (a) Which of these functions f are Hermitian forms?
   (b) For each f that is a Hermitian form, find a basis $\mathcal{A}'$ of C^n with respect to which f is represented by a diagonal matrix.

5. Prove that the relation "conjunctive" is an equivalence relation on the set of all square matrices of order n over the field $\mathcal{F}$.

6. Prove the following corollary to Theorem 8.49. Two $m \times n$ matrices over $\mathcal{F}$ represent the same complex bilinear form on $\mathbf{U}$ and $\mathbf{V}$ if and only if A and B are equivalent over $\mathcal{F}$.

7. Let f be a complex bilinear form on $\mathbf{U}$ and $\mathbf{V}$ over C. Prove that, with suitable choices of bases in $\mathbf{U}$ and $\mathbf{V}$, f can be represented by a matrix D_r that has the first r diagonal elements equal to 1 and all other elements zero.

8. Prove Theorem 8.48.

9. Prove Theorem 8.49.

10. Prove Theorem 8.51.

11. Prove Theorem 8.54.

12. Write out the details of the proof of Theorem 8.57.

13. Prove Corollary 8.58.

14. Prove Corollary 8.59.

15. Prove Corollary 8.60.

16. **(a)** Prove that the eigenvalues of a Hermitian matrix H are all real.
    (b) Prove that eigenvectors associated with distinct eigenvalues are orthogonal; that is, if X_1 is an eigenvector of H associated with λ_1 and X_2 is an eigenvector of H associated with λ_2 ($\lambda_1 \neq \lambda_2$), then $X_1^*X_2 = 0$.

17. Prove that a Hermitian matrix H is positive definite if and only if there exists an invertible matrix P such that $H = P^*P$.

18. Given that the matrix H in Problem 1(e) is a positive definite Hermitian matrix, find an invertible matrix P such that $P^*P = H$.

19. Let A and B be Hermitian matrices of the same dimensions.

    (a) Prove that $A + B$ and $A - B$ are Hermitian.
    (b) Prove that if $AB = BA$, then AB is Hermitian.
    (c) Prove that A^T is Hermitian.
    (d) Prove that if A is invertible, then A^{-1} is Hermitian.

(e) Prove that A^k is Hermitian, for k any integer.

(f) Prove that $\det(A)$ is real.

20. A matrix A is called **skew-Hermitian** if $A^* = -A$. Prove that any square matrix A can be written uniquely as $A = B + C$ with B Hermitian and C skew-Hermitian.

21. Prove that if A is skew-Hermitian, then iA is Hermitian.

22. Prove that any square matrix A that is written in the form $A = B + iC$, where B and C are real, is skew-Hermitian if and only if B is skew-symmetric and C is symmetric.

23. Characterize the diagonal elements of a skew-Hermitian matrix.

24. Prove that if A is skew-Hermitian then any eigenvalue of A is of the form $\lambda = ib$, where b is real.

9 Inner Product Spaces

The vector spaces $\mathbf{R}^n$ have been an intuitive guide in our development thus far, and we have extended most of the concepts introduced there to more general settings. The outstanding exceptions are the basic concepts of length and inner product in $\mathbf{R}^n$. In this chapter, we generalize these concepts and study those properties of vector spaces that are based on an inner product. It is possible to define an inner product for vector spaces over fields other than the field $\mathbf{R}$ of real numbers or the field $\mathcal{C}$ of complex numbers, but our interest is restricted to these cases. Accordingly, throughout this chapter, we shall always have either $\mathcal{F} = \mathbf{R}$ or $\mathcal{F} = \mathcal{C}$, and $\mathbf{V}$ shall denote a finite-dimensional vector space over $\mathcal{F}$.

9.1 Inner Products

We begin with the definition of an inner product. As mentioned above, $\mathcal{F}$ denotes a field that is either $\mathbf{R}$ or $\mathcal{C}$, and $\mathbf{V}$ is a finite-dimensional vector space over $\mathcal{F}$. The conjugate of the complex number z is denoted by $\bar{z}$.

9.1 DEFINITION

An **inner product** on $\mathbf{V}$ is a mapping f of pairs of vectors $(\mathbf{u}, \mathbf{v})$ in $\mathbf{V} \times \mathbf{V}$ onto scalars $f(\mathbf{u}, \mathbf{v}) \in \mathcal{F}$ that has the following properties:

(i) $f(a_1\mathbf{u}_1 + a_2\mathbf{u}_2, \mathbf{v}) = \bar{a}_1 f(\mathbf{u}_1, \mathbf{v}) + \bar{a}_2 f(\mathbf{u}_2, \mathbf{v})$
(ii) $f(\mathbf{u}, \mathbf{v}) = \overline{f(\mathbf{v}, \mathbf{u})}$
(iii) $f(\mathbf{v}, \mathbf{v}) > 0$ if $\mathbf{v} \neq \mathbf{0}$ and $f(\mathbf{0}, \mathbf{0}) = 0$.

We note that property (ii) requires that $f(\mathbf{v}, \mathbf{v})$ be real and property (iii) requires that this real number be positive except when $\mathbf{v} = \mathbf{0}$.

If $\mathcal{F} = \mathbf{R}$ in Definition 9.1, $\mathbf{V}$ is called a **real inner product space**, or a **Euclidean space**. If $\mathcal{F} = \mathcal{C}$, $\mathbf{V}$ is called a **complex inner product space**, or a **unitary space**. The term **inner product space** is used to refer collectively to Euclidean spaces and unitary spaces.

The properties listed in Definition 9.1 invite a comparison between inner products and Hermitian forms. This comparison yields the following theorem, which is important even though the proof is trivial.

9.2 THEOREM

A mapping f of $\mathbf{V} \times \mathbf{V}$ into $\mathcal{F}$ is an inner product on $\mathbf{V}$ if and only if f is a positive definite Hermitian form.

Proof Suppose first that f is a positive definite Hermitian form. Then f is a complex bilinear form, so it follows from Definition 8.46 that f has property (i) of Definition 9.1. By Definition 8.52, the mapping f has property (ii) of Definition 9.1. Finally, f has property (iii) by Corollary 8.60. Hence f is an inner product on **V**.

Assume, on the other hand, that f is an inner product on **V**. The references cited in the preceding paragraph show that f satisfies all of the requirements of a positive definite Hermitian form except possibly the condition that

$$f(\mathbf{u}, b_1\mathbf{v}_1 + b_2\mathbf{v}_2) = b_1 f(\mathbf{u}, \mathbf{v}_1) + b_2 f(\mathbf{u}, \mathbf{v}_2)$$

in Definition 8.46. But

$$
\begin{aligned}
f(\mathbf{u}, b_1\mathbf{v}_1 + b_2\mathbf{v}_2) &= \overline{f(b_1\mathbf{v}_1 + b_2\mathbf{v}_2, \mathbf{u})} \\
&= \overline{\bar{b}_1 f(\mathbf{v}_1, \mathbf{u}) + \bar{b}_2 f(\mathbf{v}_2, \mathbf{u})} \\
&= b_1 \overline{f(\mathbf{v}_1, \mathbf{u})} + b_2 \overline{f(\mathbf{v}_2, \mathbf{u})} \\
&= b_1 f(\mathbf{u}, \mathbf{v}_1) + b_2 f(\mathbf{u}, \mathbf{v}_2),
\end{aligned}
$$

so that f is indeed a positive definite Hermitian form. ■■■

The next example presents the most commonly used inner products in $\mathbf{R}^n$ and $\mathcal{C}^n$.

EXAMPLE 1 For any two vectors $\mathbf{u} = (u_1, u_2, \ldots, u_n)$ and $\mathbf{v} = (v_1, v_2, \ldots, v_n)$ in $\mathbf{R}^n$, let the value $f(\mathbf{u}, \mathbf{v})$ be given by

$$f(\mathbf{u}, \mathbf{v}) = u_1 v_1 + u_2 v_2 + \cdots + u_n v_n = \sum_{k=1}^{n} u_k v_k.$$

Then $f(\mathbf{u}, \mathbf{v})$ is the familiar scalar product $\mathbf{u} \cdot \mathbf{v}$ of Chapter 1. The properties required in Definition 9.1 follow easily from Theorem 1.20.

The corresponding inner product on $\mathcal{C}^n$ is given by

$$g(\mathbf{u}, \mathbf{v}) = \bar{u}_1 v_1 + \bar{u}_2 v_2 + \cdots + \bar{u}_n v_n = \sum_{k=1}^{n} \bar{u}_k v_k,$$

where $\mathbf{u} = (u_1, u_2, \ldots, u_n)$ and $\mathbf{v} = (v_1, v_2, \ldots, v_n)$ in $\mathcal{C}^n$. The verification that this mapping is an inner product is left as an exercise (Problem 3). ■

The inner products in Example 1 will be referred to hereafter as the **standard inner products** on $\mathbf{R}^n$ and $\mathcal{C}^n$, respectively. Whenever an inner product on $\mathbf{R}^n$ or $\mathcal{C}^n$ is not specified, it is understood to be the standard inner product.

EXAMPLE 2 Let $\mathbf{u} = (u_1, u_2)$ and $\mathbf{v} = (v_1, v_2)$ in $\mathbf{R}^2$. We shall determine if the following rule determines an inner product on $\mathbf{R}^2$:

$$f(\mathbf{u}, \mathbf{v}) = 2u_1 v_1 + u_1 v_2 + u_2 v_1 + u_2 v_2.$$

We see that the value of f can be written as

$$f((u_1, u_2), (v_1, v_2)) = \begin{bmatrix} u_1 & u_2 \end{bmatrix} \begin{bmatrix} 2 & 1 \\ 1 & 1 \end{bmatrix} \begin{bmatrix} v_1 \\ v_2 \end{bmatrix},$$

and since we are dealing with a real vector space, f is a Hermitian form. Thus f is an inner product if and only if f is positive definite; that is, if and only if the third property in Definition 9.1 is satisfied. Since

$$f(\mathbf{v}, \mathbf{v}) = 2v_1^2 + 2v_1 v_2 + v_2^2$$
$$= v_1^2 + (v_1 + v_2)^2,$$

we see that $f(\mathbf{v}, \mathbf{v}) \geq 0$ and $f(\mathbf{v}, \mathbf{v}) = 0$ if and only if $v_1 = 0$ and $v_1 + v_2 = 0$, that is, if and only if $\mathbf{v} = \mathbf{0}$. Thus f is an inner product on $\mathbf{R}^2$, one that looks quite different from the standard inner product. ∎

The connection established in Theorem 9.2 makes available the results of Section 8.8 for use with inner products. In particular, an inner product f on $\mathbf{V}$ has a unique matrix $A = [a_{ij}]_{n \times n}$ relative to each basis $\mathcal{A} = \{\mathbf{v}_1, \mathbf{v}_2, \ldots, \mathbf{v}_n\}$ of $\mathbf{V}$. This matrix A is determined by the conditions $a_{ij} = f(\mathbf{v}_i, \mathbf{v}_j)$, and $f(\mathbf{u}, \mathbf{v}) = X^* A Y$ where $\mathbf{u}$ and $\mathbf{v}$ have coordinate matrices X and Y, respectively, relative to $\mathcal{A}$. If P is the matrix of transition from $\mathcal{A}$ to $\mathcal{A}'$, then f has matrix $P^* A P$ relative to $\mathcal{A}'$, by Theorem 8.49. Since f is positive definite, Corollary 8.59 implies that there is a basis $\mathcal{A}'$ of $\mathbf{V}$ such that the inner product f has matrix I_n relative to $\mathcal{A}'$. Relative to this basis $\mathcal{A}'$, $f(\mathbf{u}, \mathbf{v})$ is given by $f(\mathbf{u}, \mathbf{v}) = X^* Y = \sum_{k=1}^{n} \bar{x}_k y_k$. Thus the standard inner products in Example 1 furnish typical examples provided the choice of basis is appropriate.

The results obtained in this chapter are valid for all finite-dimensional inner-product spaces, in that they are not dependent on a particular choice of f. For this reason, it is customary to replace the notation $f(\mathbf{u}, \mathbf{v})$ by a more convenient one. We choose to drop the f from the notation, and simply write $(\mathbf{u}, \mathbf{v})$ instead of $f(\mathbf{u}, \mathbf{v})$. This change of notation has an additional advantage in that it reminds us that we are dealing with an inner product, and not just a complex bilinear form.

According to the result in Problem 17 of Exercises 8.8, a Hermitian matrix H is positive definite if and only if $H = P^* P$ for some invertible matrix P. This result provides an easy way to construct inner products on $\mathcal{C}^n$. For $\mathbf{u} = (u_1, u_2, \ldots, u_n)$ and $\mathbf{v} = (v_1, v_2, \ldots, v_n)$ in $\mathcal{C}^n$, let

$$U = \begin{bmatrix} u_1 \\ u_2 \\ \vdots \\ u_n \end{bmatrix}, \quad V = \begin{bmatrix} v_1 \\ v_2 \\ \vdots \\ v_n \end{bmatrix}.$$

We can choose an invertible P of order n, put $H = P^* P$, and then the rule

$$(\mathbf{u}, \mathbf{v}) = U^* H V$$

defines an inner product on $\mathcal{C}^n$. We say that the invertible matrix P **generates** this inner product.

EXAMPLE 3 As a specific example of the preceding discussion, the matrix

$$P = \begin{bmatrix} 1 & 0 & 0 \\ -i & 1 & i \\ 0 & i & 1 \end{bmatrix}$$

generates the inner product on $\mathcal{C}^3$ defined as follows. For $\mathbf{u} = (u_1, u_2, u_3)$ and $\mathbf{v} = (v_1, v_2, v_3)$ in $\mathcal{C}^3$,

$$(\mathbf{u}, \mathbf{v}) = \begin{bmatrix} \bar{u}_1 & \bar{u}_2 & \bar{u}_3 \end{bmatrix} P^* P \begin{bmatrix} v_1 \\ v_2 \\ v_3 \end{bmatrix}$$

$$= \begin{bmatrix} \bar{u}_1 & \bar{u}_2 & \bar{u}_3 \end{bmatrix} \begin{bmatrix} 2 & i & -1 \\ -i & 2 & 0 \\ -1 & 0 & 2 \end{bmatrix} \begin{bmatrix} v_1 \\ v_2 \\ v_3 \end{bmatrix}$$

$$= 2\bar{u}_1 v_1 + i\bar{u}_1 v_2 - \bar{u}_1 v_3 - i\bar{u}_2 v_1 + 2\bar{u}_2 v_2 - \bar{u}_3 v_1 + 2\bar{u}_3 v_3. \quad \blacksquare$$

9.1 Exercises

1. Using the standard inner product in $\mathcal{C}^3$, compute $(\mathbf{u}, \mathbf{v})$ and $(\mathbf{v}, \mathbf{u})$ for the given vectors.

   (a) $\mathbf{u} = (2 - i, 1 - 5i, 1 + 2i)$, $\mathbf{v} = (2 + i, -3, -5 + i)$
   (b) $\mathbf{u} = (11, -6i, -1)$, $\mathbf{v} = (2i, 6, -i)$

2. Repeat Problem 1 using the inner product in Example 3 of this section.

3. Let $f : \mathbf{R}^n \times \mathbf{R}^n \to \mathbf{R}$ and $g : \mathcal{C}^n \times \mathcal{C}^n \to \mathcal{C}$ be the standard inner products as defined in Example 1.

   (a) Prove that f is an inner product on $\mathbf{R}^n$.
   (b) Prove that g is an inner product on $\mathcal{C}^n$.

4. Let $\mathbf{u} = (u_1, u_2, \ldots, u_n)$ and $\mathbf{v} = (v_1, v_2, \ldots, v_n)$ in $\mathbf{R}^n$. Prove or disprove that the given rule defines an inner product on $\mathbf{R}^n$.

   (a) $(\mathbf{u}, \mathbf{v}) = 2u_1 v_1 + u_1 v_2 + u_2 v_1 + 2u_2 v_2$ on $\mathbf{R}^2$
   (b) $(\mathbf{u}, \mathbf{v}) = u_1 v_2 + u_2 v_1$ on $\mathbf{R}^2$
   (c) $(\mathbf{u}, \mathbf{v}) = u_1 v_1 + u_1 v_2 + u_2 v_1 + u_2 v_2$ on $\mathbf{R}^2$
   (d) $(\mathbf{u}, \mathbf{v}) = 2u_1 v_1 + u_2 v_2$ on $\mathbf{R}^2$
   (e) $(\mathbf{u}, \mathbf{v}) = u_1 v_1 + u_3 v_3$ on $\mathbf{R}^3$
   (f) $(\mathbf{u}, \mathbf{v}) = u_1^2 v_1^2 + u_2^2 v_2^2 + u_3^2 v_3^2$ on $\mathbf{R}^3$

5. In each of parts (a)–(h) of Problem 1 of Exercises 8.8, let f be the complex bilinear form on $\mathcal{C}^n$ that has the given matrix relative to $\mathcal{E}_n$. Which of these functions are inner products on $\mathcal{C}^n$?

6. Assume that the function f defined on $\mathcal{C}^2 \times \mathcal{C}^2$ by

$$f((a_1, a_2), (b_1, b_2)) = 5\bar{a}_1 b_1 + i\bar{a}_1 b_2 - i\bar{a}_2 b_1 + 2\bar{a}_2 b_2$$

   is an inner product on $\mathcal{C}^2$.
   (a) Find the matrix of f relative to the basis $\mathcal{E}_2 = \{(1, 0), (0, 1)\}$.
   (b) Find a basis $\mathcal{A} = \{\mathbf{v}_1, \mathbf{v}_2\}$ of $\mathcal{C}^2$ such that $(\mathbf{v}_i, \mathbf{v}_j) = \delta_{ij}$ for all i, j.

7. Assume that the matrix

$$H = \begin{bmatrix} 5 & i & 2 \\ -i & 2 & -1 \\ 2 & -1 & 2 \end{bmatrix}$$

   is a positive definite Hermitian matrix, and let $(\mathbf{u}, \mathbf{v})$ be the inner product on $\mathcal{C}^3$ that has matrix H relative to the basis

$$\mathcal{A} = \{(1, 0, 0), (1, 1, 0), (1, 1, 1)\}.$$

   Write out the value of $(\mathbf{u}, \mathbf{v})$ for arbitrary vectors $\mathbf{u} = (u_1, u_2, u_3)$ and $\mathbf{v} = (v_1, v_2, v_3)$ in $\mathcal{C}^3$.

8. Find a basis $\mathcal{A}'$ of $\mathcal{C}^3$ for which the inner product in Problem 7 has the form $(\mathbf{u}, \mathbf{v}) = X^* Y$, where $\mathbf{u}$ and $\mathbf{v}$ have coordinate matrices X and Y, respectively, relative to $\mathcal{A}'$.

9. Recall that the *trace* of a matrix $A = [a_{ij}]_{n \times n}$ is the number $t(A) = \sum_{i=1}^{n} a_{ii}$. Prove that $(A, B) = t(A^T B)$ is an inner product on the vector space $\mathbf{R}_{n \times n}$ of all $n \times n$ matrices over $\mathbf{R}$.

10. Let $A = [a_{ij}]_{n \times n}$, $B = [b_{ij}]_{n \times n}$ and prove that $(A, B) = \sum_{i=1}^{n} \sum_{j=1}^{n} a_{ij} b_{ij}$ defines an inner product on the vector space $\mathbf{R}_{n \times n}$ of all square matrices of order n over $\mathbf{R}$. (This inner product is called the **Frobenius** inner product.)

11. Prove or disprove: $(a + bi, c + di) = ac + bd$, for a, b, c, d real numbers, defines an inner product on $\mathcal{C}^1$.

12. **(a)** Let $p(x) = a_0 + a_1 x + \cdots + a_n x^n$ and $q(x) = b_0 + b_1 x + \cdots + b_n x^n$ be polynomials in $\mathcal{P}_n$ over $\mathbf{R}$. Show that $(p, q) = \sum_{i=1}^{n+1} p(a_i)q(a_i)$ for $n+1$ distinct real numbers a_i, defines an inner product on $\mathcal{P}_n$ over $\mathbf{R}$.
    (b) Show that $(p, q) = \sum_{i=1}^{n} p(a_i)q(a_i)$ does not define an inner product on $\mathcal{P}_n$.

13. Let $p(x) = a_0 + a_1 x + \cdots + a_n x^n$ and $q(x) = b_0 + b_1 x + \cdots + b_n x^n$ be polynomials in $\mathcal{P}_n$ over $\mathcal{C}$. Prove that $(p, q) = \bar{a}_0 b_0 + \bar{a}_1 b_1 + \cdots + \bar{a}_n b_n$ defines an inner product on the vector space $\mathcal{P}_n$ over $\mathcal{C}$.

14. The definition of an inner product in Definition 9.1 applies equally well to infinite-dimensional vector spaces $\mathbf{V}$. Assuming the usual properties of the definite integral, prove that $(f, g) = \int_0^1 f(x)g(x)dx$ defines an inner product on the vector space of all continuous real-valued functions on $[0, 1]$.

15. Show that $(\mathbf{v}, \mathbf{0}) = 0$ for all $\mathbf{v} \in \mathbf{V}$.

16. Prove that if $(\mathbf{u}, \mathbf{v}) = 0$ for all $\mathbf{v} \in \mathbf{V}$, then $\mathbf{u} = \mathbf{0}$.

17. Prove that if $\mathbf{u}$ and $\mathbf{v}$ are two vectors in $\mathbf{V}$ such that $(\mathbf{u}, \mathbf{w}) = (\mathbf{v}, \mathbf{w})$ for all $\mathbf{w}$ in $\mathbf{V}$, then $\mathbf{u} = \mathbf{v}$.

18. Let $\mathcal{A} = \{\mathbf{v}_1, \mathbf{v}_2, \ldots, \mathbf{v}_n\}$ be a basis of the inner product space $\mathbf{V}$, and let $\mathbf{u}$ and $\mathbf{v}$ be arbitrary vectors with coordinate matrices X and Y, respectively, relative to $\mathcal{A}$. Prove that $(\mathbf{u}, \mathbf{v}) = X^*Y$ for all $\mathbf{u}, \mathbf{v} \in \mathbf{V}$ if and only if $(\mathbf{v}_i, \mathbf{v}_j) = \delta_{ij}$ for all pairs i, j.

9.2 Norms and Distances

The concept of an inner product leads to the notion of the *length* or *norm* of a vector.

**9.3
DEFINITION**

For a given inner product $(\mathbf{u}, \mathbf{v})$ on $\mathbf{V}$, the **length** or **norm** of a vector $\mathbf{v}$ in $\mathbf{V}$ is the real number $\|\mathbf{v}\|$ given by $\|\mathbf{v}\| = \sqrt{(\mathbf{v}, \mathbf{v})}$.

EXAMPLE 1   We shall compute $\|\mathbf{v}\|$ for $\mathbf{v} = (2i, 6, -i)$ in $\mathcal{C}^3$ by using the standard inner product and also by using the inner product in Example 3 of Section 9.1.

Using the standard inner product, we have

$$(\mathbf{v}, \mathbf{v}) = (-2i)(2i) + (6)(6) + (i)(-i) = 41$$

and

$$\|\mathbf{v}\| = \sqrt{41}.$$

Using the inner product from Example 3 of Section 9.1, we have

$$(\mathbf{v}, \mathbf{v}) = \begin{bmatrix} -2i & 6 & i \end{bmatrix} \begin{bmatrix} 2 & i & -1 \\ -i & 2 & 0 \\ -1 & 0 & 2 \end{bmatrix} \begin{bmatrix} 2i \\ 6 \\ -i \end{bmatrix} = 110$$

and

$$\|\mathbf{v}\| = \sqrt{110}. \qquad \blacksquare$$

The norm of a vector as given in Definition 9.3 is readily seen to be a generalization of Definition 1.19. Our last example illustrates that the number obtained as the length of a vector depends on the particular inner product that is being used. However, we shall see in Theorem 9.5 that this length has the properties that one would naturally expect. In the derivation of these properties, the famous Cauchy–Schwarz inequality is of invaluable aid.

9.4 THEOREM

The Cauchy–Schwarz Inequality   For any vectors **u**, **v** in **V**,

$$|(\mathbf{u}, \mathbf{v})| \leq \|\mathbf{u}\| \cdot \|\mathbf{v}\|.$$

Proof If $\mathbf{v} = \mathbf{0}$, the equality holds with both members zero. Consider the case where $\mathbf{v} \neq \mathbf{0}$. Since the inner product is a positive definite Hermitian form, we have

$$0 \leq (\mathbf{u} - z\mathbf{v}, \mathbf{u} - z\mathbf{v})$$

$$= (\mathbf{u}, \mathbf{u}) - z(\mathbf{u}, \mathbf{v}) - \overline{z}(\mathbf{v}, \mathbf{u}) + z\overline{z}(\mathbf{v}, \mathbf{v})$$

$$= (\mathbf{u}, \mathbf{u}) - z(\mathbf{u}, \mathbf{v}) - \overline{z}\,\overline{(\mathbf{u}, \mathbf{v})} + z\overline{z}(\mathbf{v}, \mathbf{v})$$

for any complex number z. Since $\mathbf{v} \neq \mathbf{0}$, we may let

$$z = \frac{\overline{(\mathbf{u}, \mathbf{v})}}{(\mathbf{v}, \mathbf{v})}.$$

This yields

$$0 \leq (\mathbf{u}, \mathbf{u}) - \frac{\overline{(\mathbf{u}, \mathbf{v})}(\mathbf{u}, \mathbf{v})}{(\mathbf{v}, \mathbf{v})} - \frac{(\mathbf{u}, \mathbf{v})\overline{(\mathbf{u}, \mathbf{v})}}{(\mathbf{v}, \mathbf{v})} + \frac{(\mathbf{u}, \mathbf{v})\overline{(\mathbf{u}, \mathbf{v})}}{(\mathbf{v}, \mathbf{v})^2}(\mathbf{v}, \mathbf{v})$$

$$= (\mathbf{u}, \mathbf{u}) - \frac{(\mathbf{u}, \mathbf{v})\overline{(\mathbf{u}, \mathbf{v})}}{(\mathbf{v}, \mathbf{v})}.$$

Multiplication of both members by $(\mathbf{v}, \mathbf{v})$ yields to

$$(\mathbf{u}, \mathbf{v})\overline{(\mathbf{u}, \mathbf{v})} \leq (\mathbf{u}, \mathbf{u})(\mathbf{v}, \mathbf{v})$$

or

$$|(\mathbf{u}, \mathbf{v})|^2 \leq \|\mathbf{u}\|^2\|\mathbf{v}\|^2.$$

The statement of the theorem follows from taking the positive square root of both members of the last inequality. ∎

    It is left as an exercise to prove that the equality holds in the Cauchy–Schwarz inequality if and only if $\{\mathbf{u}, \mathbf{v}\}$ is linearly dependent (Problem 9).

EXAMPLE 2 We shall verify the Cauchy–Schwarz inequality using the inner product from Example 3 of the last section with

$$\mathbf{u} = (1 + i, -3i, -2), \quad \mathbf{v} = (2i, 6, -i).$$

We have $\|\mathbf{v}\| = \sqrt{110}$ from our last example. Performing the other required computations, we get

$$(\mathbf{u}, \mathbf{v}) = \begin{bmatrix} 1-i & 3i & -2 \end{bmatrix} \begin{bmatrix} 2 & i & -1 \\ -i & 2 & 0 \\ -1 & 0 & 2 \end{bmatrix} \begin{bmatrix} 2i \\ 6 \\ -i \end{bmatrix} = 11 + 61i,$$

$$|(\mathbf{u}, \mathbf{v})| = \sqrt{121 + 3721} = \sqrt{3842},$$

$$(\mathbf{u}, \mathbf{u}) = \begin{bmatrix} 1-i & 3i & -2 \end{bmatrix} \begin{bmatrix} 2 & i & -1 \\ -i & 2 & 0 \\ -1 & 0 & 2 \end{bmatrix} \begin{bmatrix} 1+i \\ -3i \\ -2 \end{bmatrix} = 40,$$

$$\|\mathbf{u}\| = \sqrt{40}.$$

The inequality

$$\sqrt{3842} \le \sqrt{40}\sqrt{110} = \sqrt{4400}$$

is a valid one, verifying the Cauchy–Schwarz inequality in this case. ■

We now derive the fundamental properties of the norm.

9.5 THEOREM

The norm of a vector has the following properties:

(i) $\|\mathbf{v}\| > 0$ if $\mathbf{v} \ne \mathbf{0}$, and $\|\mathbf{0}\| = 0$;

(ii) $\|a\mathbf{v}\| = |a| \cdot \|\mathbf{v}\|$;

(iii) $\|\mathbf{u} + \mathbf{v}\| \le \|\mathbf{u}\| + \|\mathbf{v}\|$. (The **Triangle Inequality**)

Proof The first of these follows immediately from $\|\mathbf{v}\| = \sqrt{(\mathbf{v}, \mathbf{v})}$, since the inner product $(\mathbf{v}, \mathbf{v})$ is a positive definite Hermitian form.

For any $a \in \mathcal{F}$ and any $\mathbf{v} \in \mathbf{V}$,

$$\|a\mathbf{v}\| = \sqrt{(a\mathbf{v}, a\mathbf{v})} = \sqrt{\bar{a}a(\mathbf{v}, \mathbf{v})} = |a|\sqrt{(\mathbf{v}, \mathbf{v})}.$$

This proves (ii).

For any complex number $z = a + bi$ with $a, b \in \mathbf{R}$, $z + \bar{z} = 2a \le 2|z|$. Hence

$$\|\mathbf{u} + \mathbf{v}\|^2 = (\mathbf{u} + \mathbf{v}, \mathbf{u} + \mathbf{v})$$
$$= (\mathbf{u}, \mathbf{u}) + (\mathbf{u}, \mathbf{v}) + \overline{(\mathbf{u}, \mathbf{v})} + (\mathbf{v}, \mathbf{v})$$
$$\le \|\mathbf{u}\|^2 + 2|(\mathbf{u}, \mathbf{v})| + \|\mathbf{v}\|^2.$$

But $|(\mathbf{u}, \mathbf{v})| \le \|\mathbf{u}\| \cdot \|\mathbf{v}\|$ by the Cauchy–Schwarz inequality, so

$$\|\mathbf{u} + \mathbf{v}\|^2 \le \|\mathbf{u}\|^2 + 2\|\mathbf{u}\| \cdot \|\mathbf{v}\| + \|\mathbf{v}\|^2$$
$$= (\|\mathbf{u}\| + \|\mathbf{v}\|)^2.$$

This result is equivalent to property (iii), and the proof is complete. ■■■

There is one more basic concept to be introduced in an inner product space, that of a distance function.

**9.6
DEFINITION**

For a given inner product $(\mathbf{u}, \mathbf{v})$ on $\mathbf{V}$, the **distance** $d(\mathbf{u}, \mathbf{v})$ between vectors $\mathbf{u}$ and $\mathbf{v}$ in $\mathbf{V}$ is defined by $d(\mathbf{u}, \mathbf{v}) = \|\mathbf{u} - \mathbf{v}\|$.

The distance thus defined has the properties listed in the next theorem. The proof is left as an exercise.

**9.7
THEOREM**

The distance function d has the following properties:

(i) $d(\mathbf{u}, \mathbf{v}) > 0$ if $\mathbf{u} \neq \mathbf{v}$, and $d(\mathbf{v}, \mathbf{v}) = 0$;

(ii) $d(\mathbf{u}, \mathbf{v}) = d(\mathbf{v}, \mathbf{u})$;

(iii) $d(\mathbf{u}, \mathbf{v}) + d(\mathbf{v}, \mathbf{w}) \geq \mathbf{d(u, w)}$.

Proof See Problem 10. ∎ ∎ ∎

A set in which there is defined a distance function with the properties (i), (ii), (iii) of Theorem 9.7 is called a **metric space**, and the distance function is called a **metric** or **norm** for the space. The only norms that we are interested in are those connected with an inner product, but there are more general norms.

9.2 Exercises

1. Compute the norm of the given $\mathbf{v} \in \mathcal{C}^3$, using the standard inner product.

(a) $\mathbf{v} = (3, -i, 1)$ **(b)** $\mathbf{v} = (0, 2i, 1)$

(c) $\mathbf{v} = (2i, i, 0)$ **(d)** $\mathbf{v} = (5 + i, 5, 0)$

2. Using the inner product from Problem 7 of Exercises 9.1, compute the norm of each $\mathbf{v}$ in Problem 1 above.

3. Let $\mathbf{u} = (u_1, u_2)$ and $\mathbf{v} = (v_1, v_2)$. The rule

$$(\mathbf{u}, \mathbf{v}) = 2u_1v_1 + u_1v_2 + u_2v_1 + 2u_2v_2$$

defines an inner product on $\mathbf{R}^2$.

(a) Use this inner product and verify the Cauchy–Schwarz inequality for $\mathbf{u} = (1, 2)$ and $\mathbf{v} = (3, -5)$.

(b) Use this inner product and compute $\|\mathbf{w}\|$ for $\mathbf{w} = (-1, 3)$.

(c) Using this inner product and the vectors from part (a), compute $d(\mathbf{u}, \mathbf{v})$.

4. Use the inner product on $\mathbf{R}^2$ generated by

$$\begin{bmatrix} 1 & 1 \\ -1 & 2 \end{bmatrix}$$

and verify the Cauchy–Schwarz inequality for $\mathbf{u} = (3, -1)$ and $\mathbf{v} = (2, 4)$.

5. Assume that the trace function $(U, V) = t(U^T V)$ is an inner product on $\mathbf{R}_{2\times 2}$.

(a) Find the norm of

$$\begin{bmatrix} 1 & 2 \\ 0 & 2 \end{bmatrix}.$$

(b) Find the distance between

$$\begin{bmatrix} 1 & 2 \\ 0 & 2 \end{bmatrix} \quad \text{and} \quad \begin{bmatrix} 1 & 0 \\ -2 & -2 \end{bmatrix}.$$

6. Assume that the rule $(f, g) = \int_0^1 f(x)g(x)\,dx$ defines an inner product on the vector space of all continuous real-valued functions on $[0, 1]$.

(a) Find the norm of $f(x) = x^3$.

(b) Find the distance between $f(x) = x$ and $g(x) = x^3$.

7. Using the standard inner product on $\mathcal{C}^2$, write out the value of $d(\mathbf{u}, \mathbf{v})$ for arbitrary vectors $\mathbf{u}, \mathbf{v} \in \mathcal{C}^2$.

8. Find the value $d(\mathbf{u}, \mathbf{v})$ for the distance function determined by the inner product in Problem 7 of Exercises 9.1.

9. Prove that the equality sign holds in Theorem 9.4 if and only if $\{\mathbf{u}, \mathbf{v}\}$ is linearly dependent. (*Hint:* Consider $((\mathbf{u}, \mathbf{u})\mathbf{v} - (\mathbf{u}, \mathbf{v})\mathbf{u}, (\mathbf{u}, \mathbf{u})\mathbf{v} - (\mathbf{u}, \mathbf{v})\mathbf{u})$ in case $\mathbf{u} \neq \mathbf{0}$.)

10. Prove Theorem 9.7.

11. Use the Cauchy–Schwarz inequality to prove the following statements.

(a) For any real numbers $a_1, a_2, \ldots, a_n$ and $b_1, b_2, \ldots, b_n$,

$$\left(\sum_{k=1}^{n} a_k b_k \right)^2 \leq \left(\sum_{k=1}^{n} a_k^2 \right) \left(\sum_{k=1}^{n} b_k^2 \right)$$

and, in particular, $(a_1 + a_2 + \cdots + a_n)^2 \leq n(a_1^2 + a_2^2 + \cdots + a_n^2)$. When does equality hold in this last statement?

(b) For any complex numbers $a_1, a_2, \ldots, a_n$ and $b_1, b_2, \ldots, b_n$,

$$\left| \sum_{k=1}^{n} \bar{a}_k b_k \right|^2 \leq \left(\sum_{k=2}^{n} |a_k|^2 \right) \left(\sum_{k=1}^{n} |b_k|^2 \right).$$

12. Let $\mathbf{V}$ be a Euclidean space.

(a) Show that the Cauchy–Schwarz inequality implies that

$$-1 \leq \frac{(\mathbf{u}, \mathbf{v})}{\|\mathbf{u}\| \cdot \|\mathbf{v}\|} \leq 1$$

for any nonzero $\mathbf{u}$, $\mathbf{v}$ in $\mathbf{V}$.

(b) We define the angle θ between two nonzero vectors $\mathbf{u}$ and $\mathbf{v}$ in $\mathbf{V}$ by

$$\cos \theta = \frac{(\mathbf{u}, \mathbf{v})}{\|\mathbf{u}\| \cdot \|\mathbf{v}\|},$$

with $0 \leq \theta \leq \pi$. Prove the *law of cosines* in $\mathbf{V}$:

$$\|\mathbf{u} - \mathbf{v}\|^2 = \|\mathbf{u}\|^2 + \|\mathbf{v}\|^2 - 2\|\mathbf{u}\| \cdot \|\mathbf{v}\| \cos \theta.$$

13. For $\mathbf{u} = (1, -4)$ and $\mathbf{v} = (2, -3)$, use the inner product from Problem 3 to compute $\cos \theta$, where θ is the angle between $\mathbf{u}$ and $\mathbf{v}$.

14. Let

$$U = \begin{bmatrix} u_1 & u_2 \\ u_3 & u_4 \end{bmatrix} \quad \text{and} \quad V = \begin{bmatrix} v_1 & v_2 \\ v_3 & v_4 \end{bmatrix}.$$

Given that the rule

$$(U, V) = u_1 v_1 + u_2 v_2 + u_3 v_3 + u_4 v_4$$

defines an inner product on $\mathbf{R}_{2 \times 2}$, use this inner product to compute $\cos \theta$, where θ is the angle between

$$A = \begin{bmatrix} 1 & 2 \\ -2 & 0 \end{bmatrix} \quad \text{and} \quad B = \begin{bmatrix} 0 & 3 \\ 6 & 2 \end{bmatrix}.$$

15. Prove that for any $\mathbf{u}$, $\mathbf{v}$ in a Euclidean space $\mathbf{V}$,

$$\|\mathbf{u} + \mathbf{v}\|^2 + \|\mathbf{u} - \mathbf{v}\|^2 = 2(\|\mathbf{u}\|^2 + \|\mathbf{v}\|^2).$$

16. Prove that for any $\mathbf{u}$, $\mathbf{v}$ in a Euclidean space $\mathbf{V}$,

$$(\mathbf{u}, \mathbf{v}) = \tfrac{1}{4}\|\mathbf{u} + \mathbf{v}\|^2 - \tfrac{1}{4}\|\mathbf{u} - \mathbf{v}\|^2.$$

17. Prove that for any $\mathbf{u}$, $\mathbf{v}$ in a Euclidean space $\mathbf{V}$,

$$\|\mathbf{u} - \mathbf{v}\| \geq |\|\mathbf{u}\| - \|\mathbf{v}\||.$$

18. Prove that for any $\mathbf{u}$, $\mathbf{v}$ in a unitary space $\mathbf{V}$,

$$(\mathbf{u}, \mathbf{v}) = \tfrac{1}{4}\|\mathbf{u} + \mathbf{v}\|^2 - \tfrac{1}{4}\|\mathbf{u} - \mathbf{v}\|^2 + \tfrac{i}{4}\|\mathbf{u} + i\mathbf{v}\|^2 - \tfrac{i}{4}\|\mathbf{u} - i\mathbf{v}\|^2.$$

9.3 Orthonormal Bases

With the generalizations of the concepts of inner product and length of a vector, there are corresponding generalizations of many of the results in Chapter 8.

9.8 **DEFINITION**	A set $\{\mathbf{v}_\lambda \mid \lambda \in \mathcal{L}\}$ of vectors in $\mathbf{V}$ is **orthogonal** if $(\mathbf{v}_{\lambda_1}, \mathbf{v}_{\lambda_2}) = 0$ whenever $\lambda_1 \neq \lambda_2$. The set $\{\mathbf{v}_\lambda \mid \lambda \in \mathcal{L}\}$ is **orthonormal** if it is orthogonal and if each $\mathbf{v}_\lambda$ has norm 1.

It is easily proved that an orthogonal set of nonzero vectors in $\mathbf{V}$ is necessarily linearly independent (see Problem 9).

The definition of an orthogonal matrix over $\mathbf{R}$ remains unchanged from Definition 8.12: P is orthogonal if and only if $P^T = P^{-1}$.

The proof of Theorem 8.14 modifies easily to yield the following result and Theorem 9.11 as well.

9.9 **THEOREM**	Let P be an $r \times r$ matrix over $\mathbf{R}$, and let $\mathbf{V}$ be a real inner product space. Then P is orthogonal if and only if P is the transition matrix from one orthonormal set of r vectors in $\mathbf{V}$ to another orthonormal set of r vectors in $\mathbf{V}$.

Proof See Problem 19.   ■ ■ ■

9.10 **DEFINITION**	A matrix U over $\mathcal{C}$ is **unitary** if and only if $U^* = U^{-1}$.

As transition matrices, unitary matrices play the same role for complex inner product spaces as orthogonal matrices do for Euclidean spaces.

9.11 **THEOREM**	Let U be an $r \times r$, matrix over $\mathcal{C}$, and let $\mathbf{V}$ be a complex inner product space. Then U is unitary if and only if U is the transition matrix from one orthonormal set of r vectors in $\mathbf{V}$ to another orthonormal set of r vectors in $\mathbf{V}$.

Proof See Problem 21.   ■ ■ ■

If $\mathbf{v}_j \cdot \mathbf{u}_{k+1}$ is replaced by $(\mathbf{v}_j, \mathbf{u}_{k+1})$ in the Gram–Schmidt orthogonalization process (Theorem 8.15), there results a proof of the following theorem. The verification of this is requested in Problem 22.

9.12 THEOREM	Let **V** be an inner product space, and let $\mathcal{A} = \{\mathbf{u}_1, \mathbf{u}_2, \ldots, \mathbf{u}_r\}$ be a basis of the subspace **W** of **V**. There exists an orthonormal basis $\mathcal{N} = \{\mathbf{v}_1, \mathbf{v}_2, \ldots, \mathbf{v}_r\}$ of **W** such that each $\mathbf{v}_i$ is a linear combination of $\mathbf{u}_1, \mathbf{u}_2, \ldots, \mathbf{u}_i$.

As an immediate corollary, every finite-dimensional inner product space has an orthonormal basis.

9.13 COROLLARY	Any orthonormal set $\{\mathbf{v}_1, \mathbf{v}_2, \ldots, \mathbf{v}_r\}$ of vectors in **V** can be extended to an orthonormal basis $\{\mathbf{v}_1, \mathbf{v}_2, \ldots, \mathbf{v}_r, \mathbf{v}_{r+1}, \ldots, \mathbf{v}_n\}$ of **V**.

Proof See Problem 23. ∎

With a slight adjustment of Problem 14 of Exercises 9.1, it can be shown that the rule $(p, q) = \int_{-1}^{1} p(x)q(x)\, dx$ defines an inner product on $\mathcal{P}_n$ over **R**, the vector space of all polynomials of degree n or less over **R**. Beginning with the standard basis $\mathcal{A} = \{1, x, \ldots, x^n\}$ of $\mathcal{P}_n$, the Gram–Schmidt orthogonalization process can be used with this inner product to produce an orthonormal basis of $\mathcal{P}_n$. However, if an orthogonal basis is all that is required, the Gram–Schmidt process can be adapted in the following way.

Gram–Schmidt Process Let $\mathcal{A} = \{\mathbf{u}_1, \mathbf{u}_2, \ldots, \mathbf{u}_r\}$ be a basis of the subspace **W** of the inner product space **V**. There exists an orthogonal basis $\mathcal{N} = \{\mathbf{v}_1, \mathbf{v}_2, \ldots, \mathbf{v}_r\}$ of **W** such that

$$\mathbf{v}_i = \mathbf{u}_i - \frac{(\mathbf{v}_1, \mathbf{u}_i)}{(\mathbf{v}_1, \mathbf{v}_1)}\mathbf{v}_1 - \frac{(\mathbf{v}_2, \mathbf{u}_i)}{(\mathbf{v}_2, \mathbf{v}_2)}\mathbf{v}_2 - \cdots - \frac{(\mathbf{v}_{r-1}, \mathbf{u}_i)}{(\mathbf{v}_{r-1}, \mathbf{v}_{r-1})}\mathbf{v}_{r-1}, \quad \text{for } i = 2, 3, \ldots, r.$$

If an orthonormal set is required then each $\mathbf{v}_i$ can then be *normalized*, yielding an orthonormal basis

$$\mathcal{N}' = \left\{ \frac{\mathbf{v}_1}{\|\mathbf{v}_1\|}, \frac{\mathbf{v}_2}{\|\mathbf{v}_2\|}, \ldots, \frac{\mathbf{v}_r}{\|\mathbf{v}_r\|} \right\}.$$

EXAMPLE 1 The set $\{1, x, x^2\}$ is a basis of $\mathcal{P}_2$ over **R**, however, not an orthogonal basis since

$$(1, x^2) = \int_{-1}^{1} 1 \cdot x^2\, dx = \left(\frac{x^3}{3} \right)_{-1}^{1} = \frac{2}{3} \neq 0.$$

The vectors 1 and x are orthogonal since

$$(1, x) = \int_{-1}^{1} 1 \cdot x\, dx = \left(\frac{x^2}{2} \right)_{-1}^{1} = 0$$

and we need only to extend $\{1, x\}$ to an orthogonal basis of $\mathcal{P}_2$. Utilizing the Gram–Schmidt procedure, we set $\mathbf{v}_1 = 1$, $\mathbf{v}_2 = x$, $\mathbf{u}_3 = x^2$ and

$$\mathbf{v}_3 = \mathbf{u}_3 - \frac{(\mathbf{v}_1, \mathbf{u}_3)}{(\mathbf{v}_1, \mathbf{v}_1)} \mathbf{v}_1 - \frac{(\mathbf{v}_2, \mathbf{u}_3)}{(\mathbf{v}_2, \mathbf{v}_2)} \mathbf{v}_2$$

$$= x^2 - \frac{\int_{-1}^{1} 1 \cdot x^2 \, dx}{\int_{-1}^{1} 1 \cdot 1 \, dx} 1 - \frac{\int_{-1}^{1} x \cdot x^2 \, dx}{\int_{-1}^{1} x \cdot x \, dx} x$$

$$= x^2 - \left(\frac{2/3}{2}\right) \cdot 1 - \frac{0}{2/3} \cdot x$$

$$= x^2 - \frac{1}{3}.$$

Thus the set $\mathcal{N} = \{1, x, x^2 - \frac{1}{3}\}$ is an orthogonal basis of $\mathcal{P}_2$. The vectors in $\mathcal{N}$ are called **Legendre polynomials**. ∎

A basis $\mathcal{A} = \{\mathbf{v}_1, \mathbf{v}_2, \ldots, \mathbf{v}_n\}$ of $\mathbf{V}$ is orthonormal if and only if $(\mathbf{v}_i, \mathbf{v}_j) = \delta_{ij}$. This condition is equivalent to the requirement that $(\mathbf{u}, \mathbf{v}) = X^*Y$, where $\mathbf{u}$ and $\mathbf{v}$ have coordinate matrices X and Y, respectively, relative to $\mathcal{A}$ (see Problem 18 of Exercises 9.1). Thus, the orthonormal bases of $\mathbf{V}$ relative to a certain inner product are precisely those bases for which the inner product has I_n as its matrix (see Section 9.1).

EXAMPLE 2 With the inner product on $\mathcal{C}^3$ that has the matrix

$$H = \begin{bmatrix} \frac{11}{9} & -\frac{2i}{3} & -\frac{1}{9} \\ \frac{2i}{3} & 2 & -\frac{i}{3} \\ -\frac{1}{9} & \frac{i}{3} & \frac{5}{9} \end{bmatrix}$$

relative to the standard basis $\mathcal{E}_3$ of $\mathcal{C}^3$, the inner product of $\mathbf{u} = (u_1, u_2, u_3)$ and $\mathbf{v} = (v_1, v_2, v_3)$ in $\mathcal{C}^3$ is given by

$$(\mathbf{u}, \mathbf{v}) = \begin{bmatrix} \bar{u}_1 & \bar{u}_2 & \bar{u}_3 \end{bmatrix} H \begin{bmatrix} v_1 \\ v_2 \\ v_3 \end{bmatrix}.$$

If we let $\mathbf{v}_1 = \frac{1}{\sqrt{3}}(1, 0, 2)$, $\mathbf{v}_2 = \frac{1}{\sqrt{2}}(-1, i, 1)$, $\mathbf{v}_3 = \frac{1}{\sqrt{6}}(1, i, -1)$, it is readily verified that $\mathcal{A} = \{\mathbf{v}_1, \mathbf{v}_2, \mathbf{v}_3\}$ is an orthonormal basis of $\mathcal{C}^3$ relative to this inner product. The matrix of transition from $\mathcal{E}_3$ to $\mathcal{A}$ is given by

$$P = \begin{bmatrix} \frac{1}{\sqrt{3}} & -\frac{1}{\sqrt{2}} & \frac{1}{\sqrt{6}} \\ 0 & \frac{i}{\sqrt{2}} & \frac{i}{\sqrt{6}} \\ \frac{2}{\sqrt{3}} & \frac{1}{\sqrt{2}} & -\frac{1}{\sqrt{6}} \end{bmatrix}$$

and $P^*HP = I_3$. We note that the matrix P is not unitary, but it should not be since $\mathcal{E}_3$ is not an orthonormal basis relative to this inner product. ∎

9.3 Exercises

1. Show that, with the standard inner product in $\mathcal{C}^3$, each of the following sets is orthogonal.

   (a) $\left\{(\frac{3}{2}, \sqrt{2}, -\frac{1}{2}), (\sqrt{2}, -2, -\sqrt{2}), (-3, \sqrt{2}, -5)\right\}$

   (b) $\left\{(3i, 2i\sqrt{2}, -i), (-i, 2i\sqrt{2}, 5i), (6i\sqrt{2}, -7i, 4i\sqrt{2})\right\}$

2. Normalize the sets in Problem 1.

3. Show that the set $\{(1, 0, 0), (1 + 2i, 1 + 2i, 1), (3 + 4i, 3 + 6i, 1 + 4i)\}$ is orthogonal with respect to the inner product of Problem 7 in Exercises 9.1.

4. Using the inner product as in Problem 3, find another orthogonal set of three nonzero vectors in $\mathcal{C}^3$.

5. With the standard inner product in $\mathcal{C}^3$, use the Gram–Schmidt process to find an orthonormal basis of $\langle \mathcal{A} \rangle$.

   (a) $\mathcal{A} = \{(1, -i, 1), (2, 0, 1 - i)\}$ **(b)** $\mathcal{A} = \{(0, -i, 1), (1 + i, 2, 1)\}$

6. Let the inner product (A, B) be defined on $\mathbf{R}_{2\times 2}$ as in Problem 9 of Exercises 9.1: $(A, B) = t(A^T B)$. Obtain an orthonormal basis of $\mathbf{R}_{2\times 2}$ by applying the Gram–Schmidt process to the basis

$$\left\{ \begin{bmatrix} 1 & 0 \\ 0 & 0 \end{bmatrix}, \begin{bmatrix} 1 & 1 \\ 0 & 0 \end{bmatrix}, \begin{bmatrix} 1 & 1 \\ 1 & 0 \end{bmatrix}, \begin{bmatrix} 1 & 1 \\ 1 & 1 \end{bmatrix} \right\}.$$

7. Find an orthonormal basis of $\mathbf{R}^3$ that has $(\frac{1}{3}, -\frac{2}{3}, \frac{2}{3})$ as its first element.

8. Given that

$$\mathcal{A} = \left\{ \left(\frac{1}{2}, -\frac{i}{\sqrt{2}}, \frac{1}{2}\right), \left(-\frac{1}{2}i, \frac{1}{\sqrt{2}}, -\frac{i}{2}\right), \left(\frac{1+i}{2}, 0, \frac{-1-i}{2}\right) \right\}$$

   is an orthonormal basis of $\mathcal{C}^3$ relative to the standard inner product, extend each of the orthonormal sets $\mathcal{B}$ below to an orthonormal basis of $\mathcal{C}^3$.

   (a) $\mathcal{B} = \left\{ \left(\frac{i}{2}, \frac{1}{\sqrt{2}}, \frac{i}{2}\right) \right\}$

   (b) $\mathcal{B} = \left\{ \left(\frac{1}{2}, \frac{i}{\sqrt{2}}, \frac{1}{2}\right), \left(\frac{2+i}{2\sqrt{2}}, -\frac{i}{2}, -\frac{i}{2\sqrt{2}}\right) \right\}$

9. Prove that every orthogonal set $\{\mathbf{v}_\lambda \mid \lambda \in \mathcal{L}\}$ of nonzero vectors in $\mathbf{V}$ is linearly independent.

10. Let $\mathbf{V}$ be a Euclidean space, and let $\mathbf{u}, \mathbf{v} \in \mathbf{V}$.

   (a) Prove the Pythagorean theorem in $\mathbf{V}$: $\|\mathbf{u}\|^2 + \|\mathbf{v}\|^2 = \|\mathbf{u} + \mathbf{v}\|^2$ if and only if $\mathbf{u}$ is orthogonal to $\mathbf{v}$.

   (b) Prove that $\mathbf{u} + \mathbf{v}$ and $\mathbf{u} - \mathbf{v}$ are orthogonal if and only if $\|\mathbf{u}\| = \|\mathbf{v}\|$.

11. Let $\{\mathbf{u}, \mathbf{v}, \mathbf{w}\}$ be an orthonormal set of vectors in a Euclidean space. Find the value of $\|\mathbf{u} + \mathbf{v} + \mathbf{w}\|$.

12. Let $\{\mathbf{u}, \mathbf{v}\}$ be an orthonormal set of vectors in an inner product space. Prove that $\|a\mathbf{u} + b\mathbf{v}\|^2 = |a|^2 + |b|^2$ for any a, b in $\mathcal{C}$.

13. Prove or disprove: The relation defined by "orthogonal" is an equivalence relation on the set of all vectors in an inner product space.

14. Suppose $\mathbf{u}$, $\mathbf{v}$, and $\mathbf{w}$ are vectors in an inner product space such that $\mathbf{u}$ is orthogonal to both $\mathbf{v}$ and $\mathbf{w}$. Prove that $\mathbf{u}$ is orthogonal to $a\mathbf{v} + b\mathbf{w}$ for all scalars a and b.

15. Let $\mathcal{A} = \{\mathbf{u}_1, \mathbf{u}_2, \ldots, \mathbf{u}_n\}$ be an orthonormal basis for an inner product space $\mathbf{V}$. If $\mathbf{v}$ and $\mathbf{w}$ are in $\mathbf{V}$, then show that

$$(\mathbf{v}, \mathbf{w}) = \overline{(\mathbf{u}_1, \mathbf{v})}(\mathbf{u}_1, \mathbf{w}) + \overline{(\mathbf{u}_2, \mathbf{v})}(\mathbf{u}_2, \mathbf{w}) + \cdots + \overline{(\mathbf{u}_n, \mathbf{v})}(\mathbf{u}_n, \mathbf{w}).$$

16. Let $\mathcal{A} = \{\mathbf{u}_1, \mathbf{u}_2, \ldots, \mathbf{u}_r\}$ is a basis of the subspace $\mathbf{W}$ of an inner product space $\mathbf{V}$. Show that $\mathcal{N} = \{\mathbf{v}_1, \mathbf{v}_2, \ldots, \mathbf{v}_r\}$ where

$$\mathbf{v}_i = \mathbf{u}_i - \frac{(\mathbf{v}_1, \mathbf{u}_i)}{(\mathbf{v}_1, \mathbf{v}_1)}\mathbf{v}_1 - \frac{(\mathbf{v}_2, \mathbf{u}_i)}{(\mathbf{v}_2, \mathbf{v}_2)}\mathbf{v}_2 - \cdots - \frac{(\mathbf{v}_{r-1}, \mathbf{u}_i)}{(\mathbf{v}_{r-1}, \mathbf{v}_{r-1})}\mathbf{v}_{r-1},$$

for $i = 2, 3, \ldots, r$, is an orthogonal basis of $\mathbf{W}$.

17. **(a)** Extend the orthogonal basis $\mathcal{N}$ of $\mathcal{P}_2$ found in Example 1 to an orthogonal basis of $\mathcal{P}_3$ over $\mathbf{R}$.

    (b) Normalize the vectors in the orthogonal basis of $\mathcal{P}_3$. The resulting vectors in the orthonormal basis are called the **normalized Legendre polynomials**.

18. **(a)** Use the inner product defined by $(p, q) = \int_0^1 p(x)q(x)\,dx$ on $\mathcal{P}_n$ to find an orthogonal basis $\mathcal{N}$ of $\mathcal{P}_2$ over $\mathbf{R}$.

    (b) Normalize the base vectors found in **(a)**.

    (c) Use $\mathcal{N}$ to find an orthogonal basis $\{p_1, p_2, p_3\}$ of $\mathcal{P}_2$ such that $p_i(1) = 1$ for $i = 1, 2, 3$.

19. Prove Theorem 9.9.

20. Prove that the set of all unitary matrices of order n is closed under multiplication.

21. Prove Theorem 9.11.

22. Write out a detailed proof of Theorem 9.12.

23. Prove Corollary 9.13.

9.4 Orthogonal Complements

The inner product in $\mathbf{V}$ forms the basis for a special type of decomposition of $\mathbf{V}$ as a direct sum with a given subspace as a summand. This type of decomposition involves the *orthogonal complement* of a subspace, a term that we shall define shortly.

9.14 DEFINITION

Let $\mathcal{A}$ be a nonempty subset of **V**. The set $\mathcal{A}^\perp$ (read $\mathcal{A}$ perp) is defined by

$$\mathcal{A}^\perp = \{\mathbf{v} \in \mathbf{V} \mid (\mathbf{u}, \mathbf{v}) = 0 \quad \text{for all } \mathbf{u} \in \mathcal{A}\}.$$

That is, $\mathcal{A}^\perp$ consists precisely of all those vectors in **V** that are orthogonal to every vector in $\mathcal{A}$.

EXAMPLE 1 Let $\mathbf{V} = \mathbf{R}^3$ with the standard inner product, and let us consider several possibilities for $\mathcal{A}$.

If $\mathcal{A} = \{(0, 0, 0)\}$, then $\mathcal{A}^\perp = \mathbf{R}^3$.

If $\mathcal{A}$ consists of the single nonzero vector (a_1, a_2, a_3), then $\mathcal{A}^\perp$ is the set of all vectors (x_1, x_2, x_3) in the plane $a_1x_1 + a_2x_2 + a_3x_3 = 0$.

If $\mathcal{A}$ consists of two linearly independent vectors, $\mathcal{A} = \{(a_1, a_2, a_3), (b_1, b_2, b_3)\}$, then $\mathcal{A}^\perp$ is the line of intersection of the two planes $a_1x_1 + a_2x_2 + a_3x_3 = 0$ and $b_1x_1 + b_2x_2 + b_3x_3 = 0$.

If $\mathcal{A}$ is a basis of $\mathbf{R}^3$, then $\mathcal{A}^\perp$ is the zero subspace. ∎

In each of the cases considered in the preceding example, $\mathcal{A}^\perp$ was a subspace of $\mathbf{R}^3$. It is not difficult to show that this is always the case.

9.15 THEOREM

For any nonempty subset $\mathcal{A}$ of **V**, $\mathcal{A}^\perp$ is a subspace of **V**.

Proof See Problem 4. ∎∎∎

9.16 THEOREM

For any nonempty subset $\mathcal{A}$ of **V**, $\mathcal{A}^\perp = \langle \mathcal{A} \rangle^\perp$.

Proof Let $\mathbf{v} \in \langle \mathcal{A} \rangle^\perp$. Then $(\mathbf{u}, \mathbf{v}) = 0$ for all $\mathbf{u} \in \langle \mathcal{A} \rangle$, and surely $(\mathbf{u}, \mathbf{v}) = 0$ for all $\mathbf{u} \in \mathcal{A}$ since $\mathcal{A} \subseteq \langle \mathcal{A} \rangle$, Thus $\langle \mathcal{A} \rangle^\perp \subseteq \mathcal{A}^\perp$.

Now let $\mathbf{v} \in \mathcal{A}^\perp$. Any vector $\mathbf{u} \in \langle \mathcal{A} \rangle$ can be written as $\mathbf{u} = \sum_{k=1}^r a_k \mathbf{u}_k$ with $\mathbf{u}_k \in \mathcal{A}$, and

$$(\mathbf{u}, \mathbf{v}) = \left(\sum_{k=1}^r a_k \mathbf{u}_k, \mathbf{v} \right) = \sum_{k=1}^r \overline{a}_k (\mathbf{u}_k, \mathbf{v}).$$

But $(\mathbf{u}_k, \mathbf{v}) = 0$ for each k since $\mathbf{v} \in \mathcal{A}^\perp$. Hence $(\mathbf{u}, \mathbf{v}) = 0$ and $\mathbf{v} \in \langle \mathcal{A} \rangle^\perp$ since $\mathbf{u}$ was arbitrary in $\langle \mathcal{A} \rangle$. This gives $\mathcal{A}^\perp \subseteq \langle \mathcal{A} \rangle^\perp$, and the proof is complete. ∎∎∎

Theorem 9.16 shows that there is no loss of generality if we restrict our attention to those $\mathbf{W}^\perp$ where $\mathbf{W}$ is a subspace of $\mathbf{V}$. In this case, $\mathbf{W}^\perp$ is called the **orthogonal complement** of $\mathbf{W}$. The justification for this terminology is contained in the next theorem.

9.17 THEOREM

If $\mathbf{W}$ is any subspace of $\mathbf{V}$, then the sum $\mathbf{W} + \mathbf{W}^\perp$ is direct and $\mathbf{W} \oplus \mathbf{W}^\perp = \mathbf{V}$.

Proof Now $\mathbf{v} \in \mathbf{W} \cap \mathbf{W}^\perp$ must satisfy $(\mathbf{v}, \mathbf{v}) = 0$, so $\mathbf{W} \cap \mathbf{W}^\perp = \{\mathbf{0}\}$, and the sum $\mathbf{W} + \mathbf{W}^\perp$ is direct.

If $\mathbf{W} = \{\mathbf{0}\}$ or $\mathbf{W} = \mathbf{V}$, the theorem is trivial. Suppose then that $\{\mathbf{v}_1, \mathbf{v}_2, \ldots, \mathbf{v}_r\}$ is an orthonormal basis of $\mathbf{W}$ and $r < n$. By Corollary 9.13, the set $\{\mathbf{v}_1, \mathbf{v}_2, \ldots, \mathbf{v}_r\}$ can be extended to an orthonormal basis

$$\mathcal{A} = \{\mathbf{v}_1, \mathbf{v}_2, \ldots, \mathbf{v}_r, \mathbf{v}_{r+1}, \ldots, \mathbf{v}_n\}$$

of $\mathbf{V}$. We shall show that $\mathcal{B} = \{\mathbf{v}_{r+1}, \ldots, \mathbf{v}_n\}$ is a basis of $\mathbf{W}^\perp$. As part of a basis, $\mathcal{B}$ is linearly independent, so it is necessary only to show that $\mathcal{B}$ spans $\mathbf{W}^\perp$. To do this, let $\mathbf{v} = \sum_{k=1}^{n} a_k \mathbf{v}_k$ be any vector in $\mathbf{W}^\perp$. Since $\mathcal{A}$ is orthonormal, $(\mathbf{v}_i, \mathbf{v}_k) = \delta_{ik}$. Thus, for $i = 1, 2, \ldots, r$, we have

$$0 = (\mathbf{v}_i, \mathbf{v})$$

$$= \left(\mathbf{v}_i, \sum_{k=1}^{n} a_k \mathbf{v}_k \right)$$

$$= \sum_{k=1}^{n} a_k \delta_{ik}$$

$$= a_i.$$

This means that $\mathbf{v} = \sum_{k=r+1}^{n} a_k \mathbf{v}_k$, and therefore $\mathcal{B}$ spans $\mathbf{W}^\perp$. ∎

9.18 COROLLARY

If $\mathbf{W}$ has dimension r, then $\mathbf{W}^\perp$ has dimension $n - r$.[1]

Proof In the proof of the theorem, $\{\mathbf{v}_1, \mathbf{v}_2, \ldots, \mathbf{v}_r\}$ is a basis of $\mathbf{W}$ and $\{\mathbf{v}_{r+1}, \ldots, \mathbf{v}_n\}$ is a basis of $\mathbf{W}^\perp$. ∎

Whenever $\mathbf{v} \in \mathbf{V}$ is written as $\mathbf{v} = \mathbf{v}_1 + \mathbf{v}_2$ with $\mathbf{v}_1 \in \mathbf{W}$ and $\mathbf{v}_2 \in \mathbf{W}^\perp$, $\mathbf{v}_1$ is called the **orthogonal projection** of $\mathbf{v}$ onto $\mathbf{W}$.

Since the orthogonal complement $\mathbf{W}^\perp$ of a subspace $\mathbf{W}$ is again a subspace, it in turn has an orthogonal complement $(\mathbf{W}^\perp)^\perp = \mathbf{W}^{\perp\perp}$. This procedure of forming orthogonal

---

[1] Compare this result with Problem 30 of Section 1.6.

complements can be repeated endlessly, but the next theorem shows that such repetition leads only back and forth between the same two subspaces.

9.19 THEOREM

For any subspace $\mathbf{W}$ of $\mathbf{V}$, $\mathbf{W}^{\perp\perp} = \mathbf{W}$.

Proof Since $\mathbf{W} \oplus \mathbf{W}^{\perp} = \mathbf{V}$, any $\mathbf{v} \in \mathbf{V}$ can be written uniquely as $\mathbf{v} = \mathbf{v}_1 + \mathbf{v}_2$ with $\mathbf{v}_1 \in \mathbf{W}$ and $\mathbf{v}_2 \in \mathbf{W}^{\perp}$. To prove that $\mathbf{W}^{\perp\perp} = \mathbf{W}$, it is sufficient to show that $\mathbf{v} \in \mathbf{W}^{\perp\perp}$ if and only if $\mathbf{v}_2 = \mathbf{0}$.

Now $\mathbf{v} \in \mathbf{W}^{\perp\perp}$ if and only if

$$(\mathbf{v}_1, \mathbf{u}) + (\mathbf{v}_2, \mathbf{u}) = (\mathbf{v}_1 + \mathbf{v}_2, \mathbf{u}) = (\mathbf{v}, \mathbf{u}) = 0$$

for all $\mathbf{u} \in \mathbf{W}^{\perp}$. Since $\mathbf{v}_1 \in \mathbf{W}$, $(\mathbf{v}_1, \mathbf{u}) = 0$ for all $\mathbf{u} \in \mathbf{W}^{\perp}$. This means that $\mathbf{v} \in \mathbf{W}^{\perp\perp}$ if and only if $(\mathbf{v}_2, \mathbf{u}) = 0$ for all $\mathbf{u} \in \mathbf{W}^{\perp}$. But $\mathbf{v}_2 \in \mathbf{W}^{\perp}$, so the last condition holds if and only if $(\mathbf{v}_2, \mathbf{v}_2) = 0$. Hence $\mathbf{v} \in \mathbf{W}^{\perp\perp}$ if and only if $\mathbf{v}_2 = \mathbf{0}$. ∎

9.4 Exercises

1. Let $\mathbf{V} = \mathbf{R}^4$, and let $\mathbf{W} = \langle (1, -2, 2, -7), (1, -2, 3, -9) \rangle$.
   (a) Find an orthonormal basis of $\mathbf{W}^{\perp}$.
   (b) Show how to write an arbitrary vector $\mathbf{v} \in \mathbf{V}$ as $\mathbf{v} = \mathbf{v}_1 + \mathbf{v}_2$, with $\mathbf{v}_1 \in \mathbf{W}$ and $\mathbf{v}_2 \in \mathbf{W}^{\perp}$.

2. Find an orthonormal basis of $\mathbf{W}^{\perp}$ in $\mathcal{C}^3$.
   (a) $\mathbf{W} = \langle (1, i, 1 - i), (i, -1, 0) \rangle$ (b) $\mathbf{W} = \langle (1, i, i - 1), (1, 1 + i, 2i - 1) \rangle$

3. Use the inner product defined by $(p, q) = \int_0^1 p(x)q(x)\,dx$ on $\mathcal{P}_n$ over $\mathbf{R}$ and let $\mathbf{W} = \langle 1, x \rangle$ in $\mathcal{P}_3$.
   (a) Find $\mathbf{W}^{\perp}$.
   (b) Write an arbitrary polynomial $a_0 + a_1 x + a_2 x^2 + a_3 x^3$ as $p(x) + q(x)$ with $p(x)$ in $\mathbf{W}$ and $q(x)$ in $\mathbf{W}^{\perp}$.

4. Prove Theorem 9.15.

5. Prove that $(\mathbf{W}_1 + \mathbf{W}_2)^{\perp} = \mathbf{W}_1^{\perp} \cap \mathbf{W}_2^{\perp}$, for subspaces $\mathbf{W}_1$ and $\mathbf{W}_2$ of $\mathbf{V}$.

6. Prove that $(\mathbf{W}_1 \cap \mathbf{W}_2)^{\perp} = \mathbf{W}_1^{\perp} + \mathbf{W}_2^{\perp}$, for subspaces $\mathbf{W}_1$ and $\mathbf{W}_2$ of $\mathbf{V}$.

7. Let $\mathbf{V}$ be Euclidean, and let $\mathbf{W}$ be a subspace of $\mathbf{V}$. Prove that whenever $\mathbf{v} \in \mathbf{V}$ is written as $\mathbf{v}_1 + \mathbf{v}_2$ with $\mathbf{v}_1 \in \mathbf{W}$ and $\mathbf{v}_2 \in \mathbf{W}^{\perp}$, then $\|\mathbf{v}\|^2 = \|\mathbf{v}_1\|^2 + \|\mathbf{v}_2\|^2$.

8. The row space of an $n \times n$ matrix A is defined to be the column space of A^T. Prove that the orthogonal complement of the row space of A is the nullspace of A; that is, if $\mathbf{S}$ is the row space of A and $\mathbf{W}$ is the nullspace of A, then $\mathbf{S}^{\perp} = \mathbf{W}$.

9. Let f be a fixed linear functional on $\mathbf{V}$ with matrix $A = [\, a_1 \;\; a_2 \;\; \cdots \;\; a_n \,]$ relative to the orthonormal basis $\mathcal{A}$ of $\mathbf{V}$, and let $\phi(f)$ be the vector in $\mathbf{V}$ that has coordinate matrix A^* relative to $\mathcal{A}$. Prove that $f(\mathbf{v}) = (\phi(f), \mathbf{v})$ for all $\mathbf{v} \in \mathbf{V}$.

10. Prove that the mapping ϕ that maps each $f \in \mathbf{V}^*$ onto $\phi(f)$ in $\mathbf{V}$ is a bijective mapping from $\mathbf{V}^*$ to $\mathbf{V}$, but that ϕ is not an isomorphism except when $\mathbf{V}$ is Euclidean.

11. Prove that, for any subspace $\mathbf{W}$ of $\mathbf{V}$, f is in the annihilator $\mathbf{W}^0$ if and only if $\phi(f) \in \mathbf{W}^\perp$.

12. With ϕ as in Problem 10, define (f, g) on $\mathbf{V}^*$ by $(f, g) = (\phi(f), \phi(g))$. Prove that (f, g) defines an inner product on $\mathbf{V}^*$.

9.5 Isometries

For the remainder of the chapter, $(\mathbf{u}, \mathbf{v})$ will denote an arbitrarily chosen but fixed inner product on $\mathbf{V}$. With this fixed inner product, we direct our attention to those linear operators on $\mathbf{V}$ that preserve distances, or leave the lengths of vectors unchanged. These linear operators are called *isometries*. More precisely, we have the following definition.

9.20 DEFINITION

A linear operator T on $\mathbf{V}$ is an **isometry** if $\|T(\mathbf{v})\| = \|\mathbf{v}\|$ for all $\mathbf{v} \in \mathbf{V}$. An isometry on a Euclidean space is called an **orthogonal operator**, and an isometry on a unitary space is called a **unitary operator**.

Since $\|\mathbf{v}\| = \sqrt{(\mathbf{v}, \mathbf{v})}$, one would expect that a linear operator would preserve the norm if and only if it preserved the inner product. Our next theorem shows that this is actually the case.

9.21 THEOREM

A linear operator T on $\mathbf{V}$ is an isometry if and only if

$$(\mathbf{u}, \mathbf{v}) = (T(\mathbf{u}), T(\mathbf{v}))$$

for all $\mathbf{u}, \mathbf{v}$ in $\mathbf{V}$.

Proof If $(\mathbf{u}, \mathbf{v}) = (T(\mathbf{u}), T(\mathbf{v}))$ for all $\mathbf{u}, \mathbf{v}$ in $\mathbf{V}$, then

$$\|T(\mathbf{v})\| = \sqrt{(T(\mathbf{v}), T(\mathbf{v}))} = \sqrt{(\mathbf{v}, \mathbf{v})} = \|\mathbf{v}\|,$$

and T is an isometry.

Suppose conversely that T is an isometry. We first observe that

$$\|\mathbf{u} + \mathbf{v}\|^2 = \|\mathbf{u}\|^2 + (\mathbf{u}, \mathbf{v}) + (\mathbf{v}, \mathbf{u}) + \|\mathbf{v}\|^2. \tag{i}$$

If $\mathbf{V}$ is a Euclidean space, then $(\mathbf{v}, \mathbf{u}) = (\mathbf{u}, \mathbf{v})$, and

$$(\mathbf{u}, \mathbf{v}) = \tfrac{1}{2}\{\|\mathbf{u} + \mathbf{v}\|^2 - \|\mathbf{u}\|^2 - \|\mathbf{v}\|^2\}.$$

Hence

$$(T(\mathbf{u}), T(\mathbf{v})) = \tfrac{1}{2}\{\|T(\mathbf{u}) + T(\mathbf{v})\|^2 - \|T(\mathbf{u})\|^2 - \|T(\mathbf{v})\|^2\}$$

$$= \tfrac{1}{2}\{\|\mathbf{u} + \mathbf{v}\|^2 - \|\mathbf{u}\|^2 - \|\mathbf{v}\|^2\}$$

$$= (\mathbf{u}, \mathbf{v}),$$

and the proof is complete for this case.

If $\mathbf{V}$ is a unitary space, we have

$$(\mathbf{u}, \mathbf{v}) + \overline{(\mathbf{u}, \mathbf{v})} = \|\mathbf{u} + \mathbf{v}\|^2 - \|\mathbf{u}\|^2 - \|\mathbf{v}\|^2 \tag{ii}$$

from equation (i), and another relation is required. We find that

$$\|\mathbf{u} + i\mathbf{v}\|^2 = \|\mathbf{u}\|^2 + i(\mathbf{u}, \mathbf{v}) - i\overline{(\mathbf{u}, \mathbf{v})} + \|\mathbf{v}\|^2,$$

and therefore

$$(\mathbf{u}, \mathbf{v}) - \overline{(\mathbf{u}, \mathbf{v})} = -i\{\|\mathbf{u} + i\mathbf{v}\|^2 - \|\mathbf{u}\|^2 - \|\mathbf{v}\|^2\}. \tag{iii}$$

Equations (ii) and (iii) combine to yield

$$(\mathbf{u}, \mathbf{v}) = \tfrac{1}{2}\{\|\mathbf{u} + \mathbf{v}\|^2 - i\|\mathbf{u} + i\mathbf{v}\|^2 + (i - 1)(\|\mathbf{u}\|^2 + \|\mathbf{v}\|^2)\}. \tag{iv}$$

It is left as an exercise (Problem 5) to show that this equation implies $(T(\mathbf{u}), T(\mathbf{v})) = (\mathbf{u}, \mathbf{v})$. ■ ■ ■

Another characterization of an isometry can be given in terms of the images of orthonormal bases.

9.22 THEOREM

A linear operator T on $\mathbf{V}$ is an isometry if and only if T maps an orthonormal basis of $\mathbf{V}$ onto an orthonormal basis of $\mathbf{V}$.

Proof Let $\mathcal{A} = \{\mathbf{v}_1, \mathbf{v}_2, \ldots, \mathbf{v}_n\}$ be an orthonormal basis of $\mathbf{V}$, and consider the set $T(\mathcal{A}) = \{T(\mathbf{v}_1), T(\mathbf{v}_2), \ldots, T(\mathbf{v}_n)\}$.

If T is an isometry, then by Theorem 9.21 we have $(T(\mathbf{v}_i), T(\mathbf{v}_j)) = (\mathbf{v}_i, \mathbf{v}_j) = \delta_{ij}$, and therefore $T(\mathcal{A})$ is an orthonormal basis of $\mathbf{V}$.

Assume now that $T(\mathcal{A})$ is an orthonormal basis of $\mathbf{V}$. For any $\mathbf{u} = \sum_{i=1}^{n} a_i \mathbf{v}_i$, and $\mathbf{v} = \sum_{j=1}^{n} b_j \mathbf{v}_j$ in $\mathbf{V}$,

$$(T(\mathbf{u}), T(\mathbf{v})) = \left(T\left(\sum_{i=1}^{n} a_i \mathbf{v}_i\right), T\left(\sum_{j=1}^{n} b_j \mathbf{v}_j\right) \right)$$

$$= \sum_{i=1}^{n}\sum_{j=1}^{n} \overline{a}_i b_j (T(\mathbf{v}_i), T(\mathbf{v}_j))$$

$$= \sum_{i=1}^{n}\sum_{j=1}^{n} \overline{a}_i b_j \delta_{ij}$$

$$= \sum_{i=1}^{n}\sum_{j=1}^{n} \overline{a}_i b_j (\mathbf{v}_i, \mathbf{v}_j)$$

$$= \left(\sum_{i=1}^{n} a_i \mathbf{v}_i, \sum_{j=1}^{n} b_j \mathbf{v}_j \right)$$

$$= (\mathbf{u}, \mathbf{v}).$$

Therefore T is an isometry by Theorem 9.21.   ■ ■ ■

Yet another characterization of an isometry can be made in terms of its matrix relative to an orthonormal basis. This time it is necessary to separate the real and complex cases.

9.23 THEOREM   Let T be a linear operator on the unitary space $\mathbf{V}$. Then T is unitary if and only if every matrix that represents T relative to an orthonormal basis of $\mathbf{V}$ is a unitary matrix.

Proof Let $A = [a_{ij}]_{n\times n}$ be the matrix of T relative to an orthonormal basis $\mathcal{A} = \{\mathbf{v}_1, \mathbf{v}_2, \ldots, \mathbf{v}_n\}$ of $\mathbf{V}$. Then $T(\mathbf{v}_j) = \sum_{i=1}^{n} a_{ij}\mathbf{v}_i$, so that A is the matrix of transition from $\mathcal{A}$ to $T(\mathcal{A})$. By Theorem 9.22, T is unitary if and only if $T(\mathcal{A})$ is orthonormal. But $T(\mathcal{A})$ is orthonormal if and only if A is unitary by Theorem 9.11.   ■ ■ ■

A similar application of Theorems 9.22 and 9.9 yields the corresponding result for orthogonal operators.

9.24 THEOREM   Let T be a linear operator on the Euclidean space $\mathbf{V}$. Then T is orthogonal if and only if every matrix that represents T relative to an orthonormal basis of $\mathbf{V}$ is an orthogonal matrix.

EXAMPLE 1 Let T be the linear operator on $\mathcal{C}^3$ given by

$$T(\mathbf{v}_1, \mathbf{v}_2, \mathbf{v}_3) = \left(\tfrac{1}{2}\mathbf{v}_1 - \tfrac{i}{2}\mathbf{v}_2 + \tfrac{1+i}{2}\mathbf{v}_3, -\tfrac{i}{\sqrt{2}}\mathbf{v}_1 + \tfrac{1}{\sqrt{2}}\mathbf{v}_2, \tfrac{1}{2}\mathbf{v}_1 - \tfrac{i}{2}\mathbf{v}_2 - \tfrac{1+i}{2}\mathbf{v}_3\right).$$

The standard basis $\mathcal{E}_3$ is an orthonormal basis relative to the standard inner product on $\mathcal{C}^3$. The matrix of T with respect to $\mathcal{E}_3$ is

$$A = \begin{bmatrix} \dfrac{1}{2} & -\dfrac{i}{2} & \dfrac{1+i}{2} \\ -\dfrac{i}{\sqrt{2}} & \dfrac{1}{\sqrt{2}} & 0 \\ \dfrac{1}{2} & -\dfrac{i}{2} & \dfrac{-1-i}{2} \end{bmatrix}.$$

According to Theorem 9.23, T is a unitary operator if and only if A is a unitary matrix. Checking to see $A^* = A^{-1}$, we find that

$$A^*A = \begin{bmatrix} \dfrac{1}{2} & \dfrac{i}{\sqrt{2}} & \dfrac{1}{2} \\ \dfrac{i}{2} & \dfrac{1}{\sqrt{2}} & \dfrac{i}{2} \\ \dfrac{1-i}{2} & 0 & \dfrac{-1+i}{2} \end{bmatrix} \begin{bmatrix} \dfrac{1}{2} & -\dfrac{i}{2} & \dfrac{1+i}{2} \\ -\dfrac{i}{\sqrt{2}} & \dfrac{1}{\sqrt{2}} & 0 \\ \dfrac{1}{2} & -\dfrac{i}{2} & \dfrac{-1-i}{2} \end{bmatrix} = \begin{bmatrix} 1 & 0 & 0 \\ 0 & 1 & 0 \\ 0 & 0 & 1 \end{bmatrix}.$$

Therefore A is a unitary matrix, and T is a unitary operator. Illustrating Theorem 9.22, we see that T maps the orthonormal basis $\mathcal{E}_3$ onto the orthonormal basis

$$T(\mathcal{E}_3) = \left\{ \left(\tfrac{1}{2}, -\tfrac{i}{\sqrt{2}}, \tfrac{1}{2}\right), \left(-\tfrac{i}{2}, \tfrac{1}{\sqrt{2}}, -\tfrac{i}{2}\right), \left(\tfrac{1+i}{2}, 0, \tfrac{-1-i}{2}\right) \right\}. \qquad \blacksquare$$

9.5 Exercises

1. Determine whether or not each of the matrices below represents an orthogonal operator relative to an orthonormal basis of $\mathbf{R}^3$.

(a) $\begin{bmatrix} 1 & 1 & 0 \\ 1 & -1 & 0 \\ 0 & 0 & 1 \end{bmatrix}$ (b) $\begin{bmatrix} \frac{1}{2} & -\frac{\sqrt{3}}{2} & 0 \\ 0 & 0 & 1 \\ \frac{\sqrt{3}}{2} & \frac{1}{2} & 0 \end{bmatrix}$

(c) $\frac{1}{3}\begin{bmatrix} 2 & -2 & 1 \\ 2 & 1 & -2 \\ 1 & 2 & 2 \end{bmatrix}$ (d) $\begin{bmatrix} 1 & 1 & 1 \\ -1 & 1 & 1 \\ 0 & -1 & 2 \end{bmatrix}$

2. Determine whether or not each of the given matrices represents a unitary operator relative to an orthonormal basis of $\mathcal{C}^3$.

(a) $\frac{1}{6}\begin{bmatrix} 3-4i & 3+i & i \\ (i-2)\sqrt{2} & 2(i+1)\sqrt{2} & (2i-1)\sqrt{2} \\ -1 & 1+3i & 4-3i \end{bmatrix}$

(b) $\begin{bmatrix} 2 & 1-i & 0 \\ 1+i & 3 & 0 \\ 0 & 0 & 1 \end{bmatrix}$ **(c)** $\begin{bmatrix} 1 & 2 & 0 \\ i & -i & i \\ -1 & 1 & 1 \end{bmatrix}$

(d) $\frac{1}{\sqrt{6}}\begin{bmatrix} \sqrt{2} & \sqrt{3} & 1 \\ \sqrt{2} & -\sqrt{3} & 1 \\ \sqrt{2} & 0 & -2 \end{bmatrix}$

3. (a) Given that $\left\{\left(\frac{1}{\sqrt{2}}, \frac{1}{\sqrt{2}}\right), \left(\frac{1}{\sqrt{2}}, -\frac{1}{\sqrt{2}}\right)\right\}$ is an orthonormal basis of $\mathbf{R}^2$, find an isometry that maps $\left(\frac{1}{\sqrt{2}}, \frac{1}{\sqrt{2}}\right)$ onto $(0, -1)$.

(b) Verify: $\|T(x, y)\| = \|(x, y)\|$ for all (x, y) in $\mathbf{R}^2$.

(c) Verify: $(T(x, y), T(w, z)) = ((x, y), (w, z))$ for all $(x, y), (w, z)$ in $\mathbf{R}^2$.

4. Given that $\left\{\left(\frac{1}{3}, -\frac{2}{3}, \frac{2}{3}\right), \left(-\frac{2}{3}, -\frac{2}{3}, -\frac{1}{3}\right), \left(\frac{2}{3}, -\frac{1}{3}, -\frac{2}{3}\right)\right\}$ is an orthonormal basis of $\mathbf{R}^3$, find an isometry T that maps $\left(\frac{1}{3}, -\frac{2}{3}, \frac{2}{3}\right)$ onto $(1, 0, 0)$ and $\left(-\frac{2}{3}, -\frac{2}{3}, -\frac{1}{3}\right)$ onto $\left(0, \frac{1}{\sqrt{2}}, \frac{1}{\sqrt{2}}\right)$.

5. Prove that equation (iv) in the proof of Theorem 9.21 implies that $(T(\mathbf{u}), T(\mathbf{v})) = (\mathbf{u}, \mathbf{v})$.

6. Prove Theorem 9.24.

7. Prove that if A is either orthogonal or unitary, then $|\det(A)| = 1$.

8. Prove that the product of two isometries is an isometry.

9. Prove that T is an isometry if and only if the image of a unit vector under T is always a unit vector.

10. Prove or disprove, where U and V are unitary matrices.

(a) $U + V$ is unitary.   **(b)** $\overline{U}$ is unitary.   **(c)** U^T is unitary.

(d) U^* is unitary.   **(e)** U^{-1} is unitary.

11. Let $\mathbf{U}$ and $\mathbf{V}$ be inner product spaces with the inner product and norm in $\mathbf{U}$ designated by $(\mathbf{u}_1, \mathbf{u}_2)_\mathbf{U}$ and $\|\mathbf{u}\|_\mathbf{U}$, and that in $\mathbf{V}$ by $(\mathbf{v}_1, \mathbf{v}_2)_\mathbf{V}$ and $\|\mathbf{v}\|_\mathbf{V}$, respectively. A linear transformation $T : \mathbf{U} \to \mathbf{V}$ is defined to be an **isometry** if $\|T(\mathbf{u})\|_\mathbf{V} = \|\mathbf{u}\|_\mathbf{U}$ for all $\mathbf{u}$ in $\mathbf{U}$. Prove that if T is an isometry then $(T(\mathbf{u}_1, \mathbf{u}_2))_\mathbf{V} = (\mathbf{u}_1, \mathbf{u}_2)_\mathbf{U}$, for all $\mathbf{u}_1, \mathbf{u}_2$ in $\mathbf{U}$.

12. Let $\mathbf{U}$ and $\mathbf{V}$ be inner product spaces as described in Problem 11. The spaces $\mathbf{U}$ and $\mathbf{V}$ are defined to be **isometric** if there exists a linear transformation $T : \mathbf{U} \to \mathbf{V}$ such that T is an isometry and an isomorphism. Prove that the relation "isometric" is an equivalence relation on the set of all inner product spaces.

13. Prove that a linear transformation $T : \mathbf{U} \to \mathbf{V}$ is an isometry if and only if T maps an orthonormal basis of $\mathbf{U}$ onto an orthonormal basis of $\mathbf{V}$.

14. (a) Find an isometry T from $\mathbf{R}^2$ to $\mathcal{P}_1$ over $\mathbf{R}$ using the standard inner product in $\mathbf{R}^2$ and the inner product defined by $(p, q) = \int_0^1 p(x)q(x)\,dx$ in $\mathcal{P}_1$. (*Hint:* An orthonormal basis for $\mathcal{P}_1$ can be found in Problem 18 of Exercises 9.3.)

   (b) Verify: $\|T(a, b)\|_{\mathcal{P}_1} = \|(a, b)\|_{\mathbf{R}^2}$ for all (a, b) in $\mathbf{R}^2$.

   (c) Verify: $(T(a, b), T(c, d))_{\mathcal{P}_1} = ((a, b), (c, d))_{\mathbf{R}^2}$ for all (a, b), (c, d) in $\mathbf{R}^2$.

15. Find an isometry T from $\mathbf{R}^3$ to $\mathcal{P}_2$ over $\mathbf{R}$, using the standard inner product in $\mathbf{R}^3$ and the inner product defined by $\int_{-1}^1 p(x)q(x)\,dx$ in $\mathcal{P}_2$. (*Hint:* An orthonormal basis for $\mathcal{P}_2$ can be found in Problem 17 of Exercises 9.3.)

9.6 Normal Matrices

In Chapter 7, we formulated necessary and sufficient conditions in order for a linear operator on $\mathbf{V}$ to be represented by a diagonal matrix. At that time, our investigation was concerned with vector spaces in general, whereas now we are concerned only with inner product spaces. This prompts us to pose a restricted form of the original question by asking which linear operators can be represented by a diagonal matrix relative to an *orthonormal* basis of $\mathbf{V}$.

A given linear operator T on $\mathbf{V}$ will have a unique matrix A relative to a particular orthonormal basis of $\mathbf{V}$. If P is the transition matrix from this basis to a new basis of $\mathbf{V}$, the new matrix of T is $P^{-1}AP$. When $\mathbf{V}$ is unitary, the new basis is orthonormal if and only if P is a unitary matrix. Thus, for a unitary space $\mathbf{V}$, T can be represented by a diagonal matrix relative to an orthonormal basis if and only if there exists a unitary matrix P such that $P^{-1}AP$ is diagonal. Similarly, for a Euclidean space $\mathbf{V}$, T can be represented by a diagonal matrix relative to an orthonormal basis if and only if there exists an orthogonal matrix P such that $P^{-1}AP$ is diagonal.

9.25 DEFINITION   If A and B are square matrices over $\mathcal{C}$, then B is **unitarily similar** to A if and only if there exists a unitary matrix U such that $B = U^{-1}AU$. If B and A are square matrices over $\mathbf{R}$, then B is **orthogonally similar** to A if and only if $B = P^{-1}AP$ for some orthogonal matrix P.

Unitary similarity is an equivalence relation on the square matrices over $\mathcal{C}$, and orthogonal similarity is an equivalence relation on the square matrices over $\mathbf{R}$ (Problems 4 and 5).

9.26 DEFINITION   A matrix $B = [b_{ij}]_{m \times n}$ is **upper triangular** if $b_{ij} = 0$ for all $i > j$.

That is, an upper triangular matrix is one that has only zero elements below the main diagonal. Similarly, a **lower triangular** matrix is one that has only zero elements above the main diagonal.

The next proof should be compared with that of Theorem 8.19, for the technique of the proof is much the same.

9.27 THEOREM

Every $n \times n$ matrix A over $\mathcal{C}$ is unitarily similar to an upper triangular matrix B, and the diagonal elements of B are the eigenvalues of A.

Proof   The theorem is trivial for $n = 1$. We proceed by induction on n.

Assume that the theorem is true for all $k \times k$ matrices over $\mathcal{C}$, and let A be a matrix of order $k + 1$ over $\mathcal{C}$. Let $\mathbf{V}$ be a unitary space of dimension $k + 1$, and let $\mathcal{A}$ be an orthonormal basis of $\mathbf{V}$. The matrix A represents a unique linear operator T on $\mathbf{V}$ relative to the basis $\mathcal{A}$. Let λ be an eigenvalue of T with $\mathbf{v}$ a corresponding eigenvector of norm 1. By Corollary 9.13, there is an orthonormal basis

$$\mathcal{B} = \{\mathbf{v}_1, \mathbf{v}_2, \ldots, \mathbf{v}_n\}$$

of $\mathbf{V}$ with $\mathbf{v}_1 = \mathbf{v}$. The matrix of transition U_1 from $\mathcal{A}$ to $\mathcal{B}$ is unitary, and the matrix of T relative to $\mathcal{B}$ is of the form

$$A_1 = U_1^{-1} A U_1 = \begin{bmatrix} \lambda & R_1 \\ 0 & A_2 \end{bmatrix},$$

where R_1 is 1 by k and A_2 is of order k. By the induction hypothesis, there is a unitary matrix Q such that $Q^{-1} A_2 Q$ is upper triangular. The matrix

$$U_2 = \begin{bmatrix} 1 & 0 \\ 0 & Q \end{bmatrix}$$

is unitary, and

$$U_2^{-1} A_1 U_2 = \begin{bmatrix} 1 & 0 \\ 0 & Q^{-1} \end{bmatrix} \begin{bmatrix} \lambda & R_1 \\ 0 & A_2 \end{bmatrix} \begin{bmatrix} 1 & 0 \\ 0 & Q \end{bmatrix}$$

$$= \begin{bmatrix} \lambda & R_1 Q \\ 0 & Q^{-1} A_2 Q \end{bmatrix}.$$

Thus, $U_2^{-1} A_1 U_2$ is upper triangular since $Q^{-1} A_2 Q$ is upper triangular. The matrix $U = U_1 U_2$ is unitary since U_1 and U_2 are unitary, and

$$B = U^{-1} A U$$

is upper triangular since

$$B = U_2^{-1} U_1^{-1} A U_1 U_2 = U_2^{-1} A_1 U_2.$$

It is clear that the diagonal elements of B are the eigenvalues of $B = [b_{ij}]$ since

$$\det(B - xI) = (b_{11} - x)(b_{22} - x) \cdots (b_{nn} - x).$$

But B and A have the same eigenvalues since they are similar.   ■ ■ ■

9.28
COROLLARY

Every linear operator on a unitary space **V** can be represented by an upper triangular matrix relative to an orthonormal basis of **V**.

On the surface, it would appear that there should be a result for real matrices that corresponds to Theorem 9.27. That is, it would seem likely that every $n \times n$ real matrix would be orthogonally similar to an upper triangular matrix over **R**. But a closer examination shows that this is not the case at all. For the diagonal elements of a triangular matrix are the eigenvalues of that matrix, and the eigenvalues of a real matrix are not necessarily real (see Problem 1 for an example). However, for real matrices that have only real eigenvalues, the proof of Theorem 9.27 can be modified so as to prove the following theorem.

9.29
THEOREM

Let A be a real matrix of order n. Then A is orthogonally similar to an upper triangular matrix if and only if all the eigenvalues of A are real.

Proof See Problem 8.   ■ ■ ■

9.30
COROLLARY

A linear operator T on a Euclidean space **V** can be represented by an upper triangular matrix relative to an orthonormal basis of **V** if and only if all the eigenvalues of T are real.

Now we are ready to establish the criterion for a matrix to be unitarily similar to a diagonal matrix.

9.31
THEOREM

A square matrix A over $\mathcal{C}$ is unitarily similar to a diagonal matrix if and only if $AA^* = A^*A$.

Proof Assume that there exists a unitary matrix U such that

$$U^{-1}AU = D = \text{diag}\{d_1, d_2, \ldots, d_n\}.$$

We first note that

$$DD^* = D^*D = \operatorname{diag}\{|d_1|^2, |d_2|^2, \ldots, |d_n|^2\}.$$

Since $A = UDU^{-1}$ and $U^{-1} = U^*$, we have

$$
\begin{aligned}
AA^* &= (UDU^*)(UD^*U^*) \\
&= UDD^*U^* \\
&= UD^*DU^* \\
&= UD^*U^*UDU^* \\
&= A^*A.
\end{aligned}
$$

Assume now that $AA^* = A^*A$. By Theorem 9.27, there is a unitary matrix U such that $B = U^{-1}AU = U^*AU$ is upper triangular. Now

$$BB^* = U^*AUU^*A^*U = U^*AA^*U,$$

and a similar calculation shows that

$$B^*B = U^*A^*AU.$$

Therefore, $BB^* = B^*B$, since $AA^* = A^*A$. Since $b_{rs} = 0$ in $B = [b_{ij}]$ whenever $r > s$, the element in the ith row and jth column of BB^* is $\sum_{k=1}^{n} b_{ik}\bar{b}_{jk} = \sum_{k=i}^{n} b_{ik}\bar{b}_{jk}$. Similarly, the element in the ith row and jth column of B^*B is $\sum_{k=1}^{n} \bar{b}_{ki}b_{kj} = \sum_{k=1}^{i} \bar{b}_{ki}b_{kj}$. Equating the diagonal elements of BB^* and B^*B, we have

$$\sum_{k=1}^{i} |b_{ki}|^2 = \sum_{k=i}^{n} |b_{ik}|^2. \tag{v}$$

For $i = 1$, this yields

$$|b_{11}|^2 = |b_{11}|^2 + |b_{12}|^2 + \cdots + |b_{1n}|^2.$$

Therefore $b_{1j} = 0$ for all $j > 1$, and all elements in the first row of B except b_{11} are zero. In particular, $b_{12} = 0$. With $i = 2$ in equation (v), we now have

$$|b_{22}|^2 = |b_{12}|^2 + |b_{22}|^2 = |b_{22}|^2 + |b_{23}|^2 + \cdots + |b_{2n}|^2.$$

Therefore $b_{2j} = 0$ for all $j > 2$, and all elements except b_{22} in the second row of B are zero. This procedure can be repeated with equation (v) to obtain

$$|b_{ii}|^2 = |b_{ii}|^2 + |b_{i,i+1}|^2 + \cdots + |b_{in}|^2$$

for $i = 1, 2, \ldots, n$. Therefore, $b_{ij} = 0$ for all $j > i$, and B is a diagonal matrix. This completes the proof. ∎

9.32
DEFINITION

A square matrix A over a field $\mathcal{F} \subseteq \mathcal{C}$ is called **normal** if and only if $AA^* = A^*A$.

9.33
COROLLARY

A linear operator T on a unitary space $\mathbf{V}$ can be represented by a diagonal matrix relative to an orthonormal basis of $\mathbf{V}$ if and only if any matrix representing T relative to an orthonormal basis of $\mathbf{V}$ is normal.

The last theorem says that those matrices over $\mathcal{C}$ that are unitarily similar to a diagonal matrix are precisely the normal matrices. These include the symmetric real matrices, the Hermitian matrices, the orthogonal matrices, and the unitary matrices. However, there are normal matrices that do not fall into any of these categories. Such an example is found in Problem 2.

Our next theorem gives a simple characterization of those real matrices of order n that are orthogonally similar to a diagonal matrix.

9.34
THEOREM

A real matrix of order n is orthogonally similar to a diagonal matrix if and only if it is symmetric.

Proof   If A is real and symmetric, then A is orthogonally similar to a diagonal matrix by Theorem 8.19.

Assume that the real matrix A is orthogonally similar to a diagonal matrix D, and let P be an orthogonal matrix such that $P^{-1}AP = P^TAP = D$. Then $A = PDP^T$, so that

$$A^T = PD^TP^T = PDP^T = A,$$

and A is symmetric.   ■ ■ ■

9.35
COROLLARY

A linear operator T on a Euclidean space $\mathbf{V}$ can be represented by a diagonal matrix relative to an orthonormal basis of $\mathbf{V}$ if and only if every matrix that represents T relative to an orthonormal basis of $\mathbf{V}$ is symmetric.

The results established in this section give a practical way to determine whether or not a matrix is unitarily similar or orthogonally similar to a diagonal matrix, but a systematic procedure for finding a matrix that will accomplish the diagonalization is yet lacking. Such a procedure will be developed at the end of the next section.

9.6 Exercises

1. Determine which of the following real matrices are orthogonally similar to an upper triangular matrix.

   (a) $\begin{bmatrix} 1 & 2 \\ 3 & 4 \end{bmatrix}$ (b) $\begin{bmatrix} 1 & 1 \\ -1 & 1 \end{bmatrix}$ (c) $\begin{bmatrix} 1 & 1 \\ 1 & -1 \end{bmatrix}$ (d) $\begin{bmatrix} 7 & -6 \\ -6 & -2 \end{bmatrix}$

2. Determine which of the following matrices are orthogonal, which are unitary, and which are normal.

   (a) $\begin{bmatrix} -1 & -\frac{3}{\sqrt{2}} & \frac{3}{\sqrt{2}} \\ -\frac{3}{\sqrt{2}} & \frac{1}{2} & \frac{3}{2} \\ \frac{3}{\sqrt{2}} & \frac{3}{2} & \frac{1}{2} \end{bmatrix}$ (b) $\begin{bmatrix} \frac{1}{2} & -\frac{\sqrt{3}}{2} & 0 \\ \frac{\sqrt{3}}{2} & \frac{1}{2} & 0 \\ 0 & 0 & 1 \end{bmatrix}$

   (c) $\begin{bmatrix} i & i-1 \\ i+1 & 0 \end{bmatrix}$ (d) $\begin{bmatrix} 2 & i \\ i & 2 \end{bmatrix}$

   (e) $\frac{1}{2}\begin{bmatrix} 1 & i & 1+i \\ -i & 1 & -1+i \\ -1+i & 1+i & 0 \end{bmatrix}$ (f) $\begin{bmatrix} 5 & i & 2 \\ -i & 2 & -1 \\ 2 & -1 & 2 \end{bmatrix}$

3. For what values of x, y, z is the matrix

$$A = \begin{bmatrix} a & x & y \\ 0 & b & z \\ 0 & 0 & c \end{bmatrix}$$

   a normal matrix?

4. Prove that unitary similarity is an equivalence relation on the square matrices over $\mathcal{C}$.

5. Prove that orthogonal similarity is an equivalence relation on the square matrices over **R**.

6. Prove that every Hermitian matrix is normal.

7. Prove that every unitary matrix is normal.

8. Prove Theorem 9.29.

9. Prove that if a matrix U is unitary, then all eigenvalues of U have absolute value 1.

10. Prove that a square matrix A over $\mathcal{C}$ is normal if and only if every matrix that is unitarily similar to A is normal.

11. Let $B = [b_{ij}]$ be an upper triangular matrix that is square. Prove that B^2 is an upper triangular matrix.

12. Let C be any square matrix over $\mathcal{C}$.

    (a) Prove that C can be written as $C = A + Bi$ with A and B Hermitian.

    (b) Prove that C is normal if and only if $AB = BA$.

9.7 Normal Linear Operators

In the preceding section, we found that a linear operator T on $\mathbf{V}$ can be represented by a diagonal matrix relative to an orthonormal basis if and only if every matrix that represents T relative to an orthonormal basis is normal. Our main purpose in this section is to interpret this requirement as a condition on the linear operator T. This turns naturally to an investigation of the linear operator that has matrix A^* whenever T has matrix A relative to an orthonormal basis $\mathcal{A}$.

Let $\mathcal{A} = \{\mathbf{v}_1, \mathbf{v}_2, \ldots, \mathbf{v}_n\}$ be an orthonormal basis of $\mathbf{V}$. For an arbitrary vector $\mathbf{v} = \sum_{i=1}^{n} x_i \mathbf{v}_i$ in $\mathbf{V}$, we have $(\mathbf{v}_j, \mathbf{v}) = \sum_{i=1}^{n} x_i (\mathbf{v}_j, \mathbf{v}_i) = x_j$. That is, $\mathbf{v} = \sum_{i=1}^{n} (\mathbf{v}_i, \mathbf{v}) \mathbf{v}_i$. Thus a vector $\mathbf{v}$ is determined uniquely whenever the inner product of $\mathbf{v}$ with each base vector $\mathbf{v}_j$ is specified.

Suppose now that T is a certain fixed linear operator on $\mathbf{V}$. For a fixed $\mathbf{v}$ in $\mathbf{V}$, the set of equations

$$(\mathbf{v}_j, \mathbf{u}) = (T(\mathbf{v}_j), \mathbf{v}), \quad j = 1, 2, \ldots, n, \tag{vi}$$

determines a unique vector $\mathbf{u}$ in $\mathbf{V}$ since it specifies the inner product of $\mathbf{u}$ with each $\mathbf{v}_j$ in $\mathcal{A}$. Thus the rule $T^*(\mathbf{v}) = \mathbf{u}$ defines a mapping T^* of $\mathbf{V}$ into $\mathbf{V}$. For an arbitrary $\mathbf{w} = \sum_{j=1}^{n} b_j \mathbf{v}_j$, in $\mathbf{V}$,

$$(\mathbf{u}, \mathbf{w}) = \sum_{j=1}^{n} b_j (\mathbf{u}, \mathbf{v}_j)$$

and

$$(\mathbf{v}, T(\mathbf{w})) = \sum_{j=1}^{n} b_j (\mathbf{v}, T(\mathbf{v}_j)).$$

Thus the set of equations (vi) is equivalent to the requirement that $(\mathbf{u}, \mathbf{w}) = (\mathbf{v}, T(\mathbf{w}))$ for all $\mathbf{w} \in \mathbf{V}$. In other words, for each $\mathbf{v} \in \mathbf{V}$, the value $T^*(\mathbf{v})$ is determined by the equation

$$(T^*(\mathbf{v}), \mathbf{w}) = (\mathbf{v}, T(\mathbf{w})) \quad \text{for all } \mathbf{w} \in \mathbf{V}. \tag{vii}$$

This leads to the following definition.

**9.36
DEFINITION**

For each linear operator T on $\mathbf{V}$, the **adjoint** of T is the mapping T^* of $\mathbf{V}$ into $\mathbf{V}$ that is defined by the equation

$$(T^*(\mathbf{v}), \mathbf{w}) = (\mathbf{v}, T(\mathbf{w}))$$

for all $\mathbf{v}, \mathbf{w} \in \mathbf{V}$.

> **9.37**
> **THEOREM**
>
> For any linear operator T on $\mathbf{V}$, the adjoint of T is a linear operator on $\mathbf{V}$.

Proof See Problem 7. ■■■

Our next theorem provides a basis for the desired interpretation of the results in Section 9.6.

> **9.38**
> **THEOREM**
>
> If the linear operator T on $\mathbf{V}$ has matrix $A = [a_{ij}]$ relative to the orthonormal basis $\mathcal{A}$ of $\mathbf{V}$, then T^* has matrix A^* relative to $\mathcal{A}$.

Proof Let $\mathcal{A} = \{\mathbf{v}_1, \mathbf{v}_2, \ldots, \mathbf{v}_n\}$. Then $T(\mathbf{v}_i) = \sum_{k=1}^{n} a_{ki}\mathbf{v}_k$ since A is the matrix of T relative to $\mathcal{A}$. Suppose $B = [b_{ij}]_n$ is the matrix of T^* relative to $\mathcal{A}$, so that $T^*(\mathbf{v}_j) = \sum_{k=1}^{n} b_{kj}\mathbf{v}_k$. From the definition of T^*, we have $(T(\mathbf{v}_i), \mathbf{v}_j) = (\mathbf{v}_i, T^*(\mathbf{v}_j))$. But

$$(T(\mathbf{v}_i), \mathbf{v}_j) = \left(\sum_{k=1}^{n} a_{ki}\mathbf{v}_k, \mathbf{v}_j \right)$$

$$= \sum_{k=1}^{n} \overline{a}_{ki}(\mathbf{v}_k, \mathbf{v}_j)$$

$$= \sum_{k=1}^{n} \overline{a}_{ki}\delta_{kj}$$

$$= \overline{a}_{ji}$$

and

$$(\mathbf{v}_i, T^*(\mathbf{v}_j)) = \left(\mathbf{v}_i, \sum_{k=1}^{n} b_{kj}\mathbf{v}_k \right)$$

$$= \sum_{k=1}^{n} b_{kj}(\mathbf{v}_i, \mathbf{v}_k)$$

$$= \sum_{k=1}^{n} b_{kj}\delta_{ik}$$

$$= b_{ij}.$$

Thus $b_{ij} = \overline{a}_{ji}$ and $B = A^*$. ■■■

It is left as an exercise to prove that $(T^*)^* = T$ for any linear operator T on $\mathbf{V}$. A linear operator T is called **self-adjoint** if $T^* = T$. It follows from the preceding theorem that if $\mathbf{V}$

is unitary, then T is self-adjoint if and only if every matrix that represents T relative to an orthonormal basis of V is Hermitian. If V is Euclidean, then T is self-adjoint if and only if every matrix that represents T relative to an orthonormal basis is symmetric. A self-adjoint linear operator of a unitary space is called a **Hermitian** operator, and a self-adjoint linear operator on a Euclidean space is a **symmetric** operator.

9.39 DEFINITION

A linear operator T on V is **normal** if and only if $TT^* = T^*T$.

If T has matrix A relative to an orthonormal basis $\mathcal{A}$ of V, then TT^* has matrix AA^* and T^*T has matrix A^*A relative to $\mathcal{A}$. Thus T is a normal linear transformation of V if and only if every matrix of T relative to an orthonormal basis is a normal matrix.

9.40 THEOREM

A linear operator T on the unitary space V can be represented by a diagonal matrix relative to an orthonormal basis of V if and only if T is normal.

Proof This is a restatement of Corollary 9.33. ∎ ∎ ∎

9.41 THEOREM

A linear operator T on the unitary space V is normal if and only if there exists an orthonormal basis of V that consists entirely of eigenvectors of T.

Proof By the preceding theorem, T is normal if and only if T can be represented by a diagonal matrix relative to an orthonormal basis. But, according to the proof of Theorem 7.16, $\mathcal{A} = \{\mathbf{v}_1, \mathbf{v}_2, \ldots, \mathbf{v}_n\}$ is a basis of eigenvectors of T if and only if the matrix of T relative to $\mathcal{A}$ is diagonal. ∎ ∎ ∎

As was promised in the last section, we proceed now to develop a systematic method for finding a unitary matrix that will accomplish a desired diagonalization.

Many of the results in Section 7.3 are helpful here, even though we are presently restricted to orthonormal bases. If T is represented by a diagonal matrix, the elements on the diagonal are the eigenvalues of T (Corollary 7.17). If T is normal, the geometric multiplicity of each eigenvalue is equal to the algebraic multiplicity (Theorem 7.20).

For orthogonal similarity with a real matrix A, Theorem 9.34 shows that the treatment in Section 8.4 is complete, and the methods developed there apply unchanged. For unitary similarity, some changes are necessary. We must first obtain the result for normal matrices that corresponds to Theorem 8.21 for real symmetric matrices. The first step in this direction is to relate the eigenvalues of T to those of T^*.

9.42
THEOREM

A number λ is an eigenvalue of the normal linear operator T with $\mathbf{v}$ as an associated eigenvector if and only if $\bar{\lambda}$ is an eigenvalue of T^* with the same $\mathbf{v}$ as an associated eigenvector.

Proof See Problem 12.

9.43
THEOREM

Let A be a normal matrix of order n. If λ_r and λ_s are distinct eigenvalues of A with associated eigenvectors U_r and U_s, then $U_r^* U_s = 0$.

Proof Let $\mathbf{V}$ be an n-dimensional inner product space, and let $\mathcal{A}$ be an orthonormal basis of $\mathbf{V}$. Let T be the linear operator on $\mathbf{V}$ that has matrix A relative to $\mathcal{A}$. An eigenvector of A corresponding to a certain eigenvalue is the coordinate matrix of an eigenvector of T corresponding to the same eigenvalue. Hence U_r and U_s are the coordinate matrices of eigenvectors $\mathbf{v}_r$ and $\mathbf{v}_s$ that correspond to the eigenvalues λ_r and λ_s, respectively. By Problem 18 of Exercises 9.1, $U_r^* U_s = 0$ if and only if $(\mathbf{v}_r, \mathbf{v}_s) = 0$.

According to Theorem 9.42, $\bar{\lambda}_r$ is an eigenvalue of T^* and $T^*(\mathbf{v}_r) = \bar{\lambda}_r \mathbf{v}_r$. Hence

$$(T^*(\mathbf{v}_r), \mathbf{v}_s) = (\bar{\lambda}_r \mathbf{v}_r, \mathbf{v}_s) = \lambda_r (\mathbf{v}_r, \mathbf{v}_s).$$

We also have

$$(\mathbf{v}_r, T(\mathbf{v}_s)) = (\mathbf{v}_r, \lambda_s \mathbf{v}_s) = \lambda_s (\mathbf{v}_r, \mathbf{v}_s).$$

But $(T^*(\mathbf{v}_r), \mathbf{v}_s) = (\mathbf{v}_r, T(\mathbf{v}_s))$, so this means that $\lambda_r (\mathbf{v}_r, \mathbf{v}_s) = \lambda_s (\mathbf{v}_r, \mathbf{v}_s)$ and

$$(\lambda_r - \lambda_s)(\mathbf{v}_r, \mathbf{v}_s) = 0.$$

Since $\lambda_r - \lambda_s \neq 0$, it must be that $(\mathbf{v}_r, \mathbf{v}_s) = 0$, and the proof is complete.

We can now formulate our method for finding a unitary matrix U that will diagonalize a given normal matrix $A = [a_{ij}]_n$. Let U_j denote the jth column of $U = [u_{ij}]_n$, so that $U = [\, U_1 \quad U_2 \quad \cdots \quad U_n \,]$. The requirement that $U^{-1} A U = \text{diag}\{\lambda_1, \lambda_2, \ldots, \lambda_n\}$ is equivalent to the system of equations

$$AU_j = \lambda_j U_j, \quad j = 1, 2, \ldots, n.$$

That is, each U_j must be an eigenvector of A corresponding to λ_j. Since $U_r^* U_s$ is the element in row r and column s of $U^* U$, the requirement that $U^* = U^{-1}$ is satisfied if and only if $U_r^* U_s = \delta_{rs}$. With the same notation as in the proof of Theorem 9.43, $U_r^* U_s = (\mathbf{v}_r, \mathbf{v}_s)$. Thus, the columns U_r of U must be the coordinates of an orthonormal basis of eigenvectors of T. Theorem 9.43 assures us that eigenvectors from different eigenspaces are automatically orthogonal. Thus the only modification of the procedure in Section 7.3 that is necessary to make U unitary is to choose orthonormal bases of the eigenspaces $\mathbf{V}_{\lambda_j}$.

The Gram–Schmidt process can be used to obtain orthonormal bases of those $\mathbf{V}_{\lambda_j}$ that have dimension greater than 1.

EXAMPLE 1 Consider the problem of finding a unitary matrix U such that $U^{-1}AU$ is diagonal, where A is the normal matrix

$$A = \begin{bmatrix} \dfrac{2+i}{2} & 0 & \dfrac{-2+i}{2} \\ 0 & i & 0 \\ \dfrac{-2+i}{2} & 0 & \dfrac{2+i}{2} \end{bmatrix}.$$

The characteristic polynomial is found to be

$$\det(A - xI) = -(x - i)^2(x - 2).$$

The augmented matrix $[A - 2I \mid \mathbf{0}]$ reduces by row operations to

$$\begin{bmatrix} 1 & 0 & 1 & 0 \\ 0 & 1 & 0 & 0 \\ 0 & 0 & 0 & 0 \end{bmatrix},$$

so the corresponding eigenvectors in $\mathbf{V}_2 \subseteq \mathcal{C}^3$ are of the form $x_3(-1, 0, 1)$. This means that $\mathbf{V}_2$ has an orthonormal basis consisting of a real unit vector. But in order to exhibit a solution with a unitary U that is not orthogonal, we choose $x_3 = -\sqrt{3} + i$ and obtain the orthonormal basis

$$\left\{ \left(\frac{\sqrt{3} - i}{2\sqrt{2}}, 0, \frac{-\sqrt{3} + i}{2\sqrt{2}} \right) \right\}$$

for $\mathbf{V}_2$. The matrix $A - iI$ leads to solutions of the form $x_2(0, 1, 0) + x_3(1, 0, 1)$. With $x_2 = 1, x_3 = i$, we obtain $(i, 1, i)$, and with $x_2 = i, x_3 = 1$, we obtain $(1, i, 1)$. Application of the Gram–Schmidt process to the basis $\{(i, 1, i), (1, i, 1)\}$ leads to the orthonormal basis

$$\left\{ \left(\frac{i}{\sqrt{3}}, \frac{1}{\sqrt{3}}, \frac{i}{\sqrt{3}} \right), \left(\frac{1}{\sqrt{6}}, \frac{2i}{\sqrt{6}}, \frac{1}{\sqrt{6}} \right) \right\}$$

of $\mathbf{V}_i$. Thus

$$U = \begin{bmatrix} \dfrac{i}{\sqrt{3}} & \dfrac{1}{\sqrt{6}} & \dfrac{\sqrt{3} - i}{2\sqrt{2}} \\ \dfrac{1}{\sqrt{3}} & \dfrac{2i}{\sqrt{6}} & 0 \\ \dfrac{i}{\sqrt{3}} & \dfrac{1}{\sqrt{6}} & \dfrac{-\sqrt{3} + i}{2\sqrt{2}} \end{bmatrix}$$

is a unitary matrix such that $U^{-1}AU = \text{diag}\{i, i, 2\}$. ∎

9.7 Exercises

1. For each of the following linear operators on $\mathcal{C}^2$, write out the value of $T^*(a_1, a_2)$.

   (a) $T(a_1, a_2) = (a_1 + (1 - i)a_2, (1 + i)a_1 + 2a_2)$

   (b) $T(a_1, a_2) = (a_1 + ia_2, a_1 - a_2)$

   (c) $T(a_1, a_2) = (ia_1 - ia_2, a_1)$

   (d) $T(a_1, a_2) = (ia_1 + (i - 1)a_2, (1 + i)a_1)$

2. For each linear operator T in Problem 1, find an orthonormal basis of eigenvectors whenever such a basis exists.

3. Whenever possible, find a unitary matrix U such that $U^{-1}AU$ is diagonal. The matrices in parts (a)–(e) are from Problem 2 in Exercises 9.6.

   (a) $A = \begin{bmatrix} -1 & -\frac{3}{\sqrt{2}} & \frac{3}{\sqrt{2}} \\ -\frac{3}{\sqrt{2}} & \frac{1}{2} & \frac{3}{2} \\ \frac{3}{\sqrt{2}} & \frac{3}{2} & \frac{1}{2} \end{bmatrix}$ (b) $A = \begin{bmatrix} \frac{1}{2} & -\frac{\sqrt{3}}{2} & 0 \\ \frac{\sqrt{3}}{2} & \frac{1}{2} & 0 \\ 0 & 0 & 1 \end{bmatrix}$

   (c) $A = \begin{bmatrix} i & i - 1 \\ i + 1 & 0 \end{bmatrix}$ (d) $A = \begin{bmatrix} 2 & i \\ i & 2 \end{bmatrix}$

   (e) $A = \frac{1}{2}\begin{bmatrix} 1 & i & 1 + i \\ -i & 1 & -1 + i \\ -1 + i & 1 + i & 0 \end{bmatrix}$ (f) $A = \begin{bmatrix} 3 & 1 & 0 \\ 1 & \frac{5}{2} & 1 - i \\ 0 & 1 + i & 3 \end{bmatrix}$

4. Consider the following matrices.

   $$A = \begin{bmatrix} \frac{1}{2} & \frac{\sqrt{3}}{2} \\ \frac{\sqrt{3}}{2} & -\frac{1}{2} \end{bmatrix}, \qquad B = \begin{bmatrix} \frac{1}{\sqrt{2}} & \frac{1}{\sqrt{2}} \\ \frac{i}{\sqrt{2}} & -\frac{i}{\sqrt{2}} \end{bmatrix}, \qquad C = \begin{bmatrix} 1 & 2 \\ 2 & -2 \end{bmatrix},$$

   $$D = \begin{bmatrix} 2 + i & -2 + i \\ -2 + i & 2 + i \end{bmatrix}, \qquad E = \begin{bmatrix} 1 & 2 \\ 0 & 3 \end{bmatrix}$$

   (a) Determine which of these matrices are orthogonal, which are unitary, and which are normal.

   (b) Which of these matrices are unitarily similar to an upper triangular matrix?

   (c) Which of these matrices are unitarily similar to a diagonal matrix?

   (d) Which of these matrices are similar over $\mathcal{C}$ to a diagonal matrix?

5. Let T be a linear operator on $\mathbf{V}$ over $\mathcal{F}$.

   (a) Prove that $(T^*)^* = T$.

   (b) Prove that $(aT)^* = \bar{a}T^*$, for all $a \in \mathcal{F}$.

6. Let S and T be two linear operators on $\mathbf{V}$.

   (a) Prove that $(S + T)^* = S^* + T^*$.   (b) Prove that $(ST)^* = T^*S^*$.

7. Prove Theorem 9.37.

8. Show that $T + T^*$ is self-adjoint.

9. Let T be a linear operator on the unitary space $\mathbf{V}$. Prove that T is Hermitian if and only if $(T(\mathbf{v}), \mathbf{v})$ is real for all $\mathbf{v} \in \mathbf{V}$.

10. Show that T is an isometry if and only if $T^* = T^{-1}$.

11. Prove that a linear operator T on the unitary space $\mathbf{V}$ is normal if and only if $\|T(\mathbf{v})\| = \|T^*(\mathbf{v})\|$ for all $\mathbf{v} \in \mathbf{V}$.

12. Prove Theorem 9.42.

13. Prove that a normal linear operator T is self-adjoint if and only if all the eigenvalues of T are real.

14. Prove that if T is normal, then T and T^* have the same kernel.

15. Prove that if T is an invertible linear operator on $\mathbf{V}$, then T^* is invertible and $(T^*)^{-1} = (T^{-1})^*$.

16. A linear operator T on an inner product space $\mathbf{V}$ is said to be **skew-adjoint** if $T^* = -T$. (For $\mathbf{V}$ unitary, T is called **skew-Hermitian**, and for $\mathbf{V}$ Euclidean, T is called **skew-symmetric**.)

(a) Prove that, for any linear operator T on $\mathbf{V}$, $T - T^*$ is skew-adjoint.

(b) Show that any linear operator T on $\mathbf{V}$ can be written uniquely as the sum of a self-adjoint linear operator and a skew-adjoint linear operator.

10 Spectral Decompositions

In this chapter we consider once again the question of diagonalization of a linear operator on a finite-dimensional vector space $\mathbf{V}$. We have seen in Chapter 7 that a linear operator T on $\mathbf{V}$ is diagonalizable (i.e., can be represented by a diagonal matrix) if and only if there exists a basis of $\mathbf{V}$ that consists of eigenvectors of T. For a unitary vector space $\mathbf{V}$, those linear operators that can be represented by a diagonal matrix relative to an orthonormal basis of $\mathbf{V}$ are the same as the normal linear operators on $\mathbf{V}$. One of our main objectives now is to describe the diagonalizable linear operators in terms that are free of any reference to an inner product. The characterization that we obtain is in terms of a spectral decomposition. A certain acquaintance with projections is essential to a formulation of the concept of a spectral decomposition.

For the entire chapter, $\mathbf{V}$ shall denote an n-dimensional vector space over a field $\mathcal{F}$, and T shall denote a linear operator on $\mathbf{V}$.

10.1 Projections and Direct Sums

In Section 9.4, we have seen that a finite-dimensional inner product space $\mathbf{V}$ can be decomposed as a direct sum $\mathbf{V} = \mathbf{W} \oplus \mathbf{W}^\perp$, where $\mathbf{W}$ is an arbitrary subspace of $\mathbf{V}$ and $\mathbf{W}^\perp$ is the orthogonal complement of $\mathbf{W}$. Each vector $\mathbf{v} \in \mathbf{V}$ can be written uniquely as $\mathbf{v} = \mathbf{v}_1 + \mathbf{v}_2$ with $\mathbf{v}_1 \in \mathbf{W}$ and $\mathbf{v}_2 \in \mathbf{W}^\perp$, and the linear operator P defined by $P(\mathbf{v}) = \mathbf{v}_1$ is called the **projection of $\mathbf{V}$ onto $\mathbf{W}$ along $\mathbf{W}^\perp$** (see Problem 3).

The situation in the preceding paragraph generalizes to any direct sum decomposition $\mathbf{V} = \mathbf{W}_1 \oplus \mathbf{W}_2$ of an arbitrary (not necessarily an inner product space) $\mathbf{V}$. For each $\mathbf{v}$ in $\mathbf{V}$ can be written uniquely as $\mathbf{v} = \mathbf{v}_1 + \mathbf{v}_2$ with $\mathbf{v}_i \in \mathbf{W}_i$, and the mapping $P(\mathbf{v}) = \mathbf{v}_1$ is again a linear operator on $\mathbf{V}$. Corresponding to the situation above, P is called the **projection of $\mathbf{V}$ onto $\mathbf{W}_1$ along $\mathbf{W}_2$**. (In order to make a distinction for the case where $\mathbf{W}_2 = \mathbf{W}_1^\perp$, P is frequently called the **orthogonal projection of $\mathbf{V}$ onto $\mathbf{W}_1$** when $\mathbf{W}_2 = \mathbf{W}_1^\perp$.) For any $\mathbf{v} = \mathbf{v}_1 + \mathbf{v}_2$ in $\mathbf{V}$,

$$P^2(\mathbf{v}_1 + \mathbf{v}_2) = P(\mathbf{v}_1) = \mathbf{v}_1 = P(\mathbf{v}_1 + \mathbf{v}_2),$$

so P has the property that $P^2 = P$.

10.1 DEFINITION A linear operator T on $\mathbf{V}$ is called **idempotent** if $T^2 = T$.

Our discussion above shows that the projection P of $\mathbf{V}$ onto $\mathbf{W}_1$ along $\mathbf{W}_2$ is idempotent, and that $\mathbf{W}_1 = P(\mathbf{V})$, $\mathbf{W}_2 = P^{-1}(\mathbf{0})$. The converse is also true: An idempotent linear operator T is always a projection of $\mathbf{V}$ onto $T(\mathbf{V})$ along $T^{-1}(\mathbf{0})$.

If T is an idempotent linear operator on $\mathbf{V}$, then T is the projection of $\mathbf{V}$ onto $T(\mathbf{V})$ along $T^{-1}(\mathbf{0})$.

Proof Assume that $T^2 = T$. For any $\mathbf{u} \in T(\mathbf{V})$, $\mathbf{u} = T(\mathbf{v})$ for some $\mathbf{v} \in \mathbf{V}$. Hence

$$T(\mathbf{u}) = T^2(\mathbf{v}) = T(\mathbf{v}) = \mathbf{u},$$

and T acts as the identity transformation on $T(\mathbf{V})$. Thus, for any $\mathbf{v}$ in $T(\mathbf{V}) \cap T^{-1}(\mathbf{0})$, we have $\mathbf{v} = T(\mathbf{v}) = \mathbf{0}$, and the sum $T(\mathbf{V}) + T^{-1}(\mathbf{0})$ is direct. Let $\mathbf{v}$ be an arbitrary vector, and let $\mathbf{v}_1 = T(\mathbf{v})$, $\mathbf{v}_2 = (1 - T)(\mathbf{v})$, where 1 denotes the identity transformation. Now $\mathbf{v}_1$ is clearly in $T(\mathbf{V})$ and $\mathbf{v}_2$ is in $T^{-1}(\mathbf{0})$ since[1] $T(\mathbf{v}_2) = (T - T^2)(\mathbf{v}) = Z(\mathbf{v}) = \mathbf{0}$. Since

$$\mathbf{v} = T(\mathbf{v}) + (1 - T)(\mathbf{v}) = \mathbf{v}_1 + \mathbf{v}_2,$$

we have $\mathbf{V} = T(\mathbf{V}) \oplus T^{-1}(\mathbf{0})$, and T is the projection of $\mathbf{V}$ onto $T(\mathbf{V})$ along $T^{-1}(\mathbf{0})$. ∎

In view of Theorem 10.2, an idempotent linear operator is referred to simply as a projection: T is a **projection** if and only if $T^2 = T$.

The relation between projections and direct sums that was described in the first two paragraphs of this section can be extended to direct sums that involve more than two terms.

Let $\mathbf{V} = \mathbf{W}_1 \oplus \mathbf{W}_2 \oplus \cdots \oplus \mathbf{W}_r$, and let $\mathbf{v} = \mathbf{v}_1 + \mathbf{v}_2 + \cdots + \mathbf{v}_r$ denote the decomposition of each $\mathbf{v} \in \mathbf{V}$ with $\mathbf{v}_i \in \mathbf{W}_i$. For $i = 1, 2, \ldots, r$, let P_i be defined by

$$P_i(\mathbf{v}_1 + \mathbf{v}_2 + \cdots + \mathbf{v}_r) = \mathbf{v}_i.$$

Then the set $\{P_1, P_2, \ldots, P_r\}$ has the following properties:

(a) each P_i is a projection;
(b) $P_i P_j = Z$ whenever $i \neq j$;
(c) $P_1 + P_2 + \cdots + P_r = 1$.

Each P_i is a projection since $P_i^2(\mathbf{v}) = P_i(\mathbf{v}_i) = \mathbf{v}_i = P_i(\mathbf{v})$. If $i \neq j$, $P_i P_j(\mathbf{v}) = P_i(\mathbf{v}_j) = \mathbf{0} = Z(\mathbf{v})$ for all $\mathbf{v} \in \mathbf{V}$, so (b) holds. For any $\mathbf{v} = \mathbf{v}_1 + \mathbf{v}_2 + \cdots + \mathbf{v}_r$ in $\mathbf{V}$, we have

$$\mathbf{v} = P_1(\mathbf{v}) + P_2(\mathbf{v}) + \cdots + P_r(\mathbf{v})$$

$$= (P_1 + P_2 + \cdots + P_r)(\mathbf{v}),$$

and hence $P_1 + P_2 + \cdots + P_r = 1$.

---

[1] The symbol Z denotes the zero linear transformation.

The three properties of the set $\{P_1, P_2, \ldots, P_r\}$ in the preceding paragraph motivate the following definition.

10.3 DEFINITION

A set of projections $\{P_1, P_2, \ldots, P_r\}$ is an **orthogonal** set if $P_i P_j = Z$ whenever $i \neq j$. A **complete** set of projections for $\mathbf{V}$ is an orthogonal set $\{P_1, P_2, \ldots, P_r\}$ of nonzero projections with the property that $P_1 + P_2 + \cdots + P_r = 1$.

Two projections P_1, P_2 are called **orthogonal** if $\{P_1, P_2\}$ forms an orthogonal set, that is, if $P_1 P_2 = Z = P_2 P_1$.

We have seen above that every direct sum decomposition $\mathbf{V} = \mathbf{W}_1 \oplus \cdots \oplus \mathbf{W}_r$ determines a complete set of projections $\{P_1, \ldots, P_r\}$ for $\mathbf{V}$ such that $P_j(\mathbf{v}) = \mathbf{v}_j$ whenever $\mathbf{v} = \mathbf{v}_1 + \cdots + \mathbf{v}_r$ with $\mathbf{v}_i \in \mathbf{W}_i$. The converse is also true, in that if $\{P_1, \ldots, P_r\}$ is a complete set of projections for $\mathbf{V}$, then $\mathbf{V} = P_1(\mathbf{V}) \oplus \cdots \oplus P_r(\mathbf{V})$ and $P_j(\mathbf{v}) = \mathbf{v}_j$ whenever $\mathbf{v} = \mathbf{v}_1 + \cdots + \mathbf{v}_r$ with $\mathbf{v}_i \in P_i(\mathbf{V})$. Now $\mathbf{v} = P_1(\mathbf{v}) + \cdots + P_r(\mathbf{v})$ for each $\mathbf{v} \in \mathbf{V}$ since $P_1 + \cdots + P_r = 1$, and therefore $\mathbf{V} = P_1(\mathbf{V}) + \cdots + P_r(\mathbf{V})$. To see that the sum is direct, let

$$\mathbf{w} = P_i(\mathbf{u}_1) = \sum_{\substack{j=1 \\ j \neq i}}^{r} P_j(\mathbf{u}_2)$$

be in

$$P_i(\mathbf{V}) \cap \sum_{\substack{j=1 \\ j \neq i}}^{r} P_j(\mathbf{V}).$$

Then

$$\mathbf{w} = P_i(\mathbf{u}_1) = P_i^2(\mathbf{u}_1) = P_i\left(\sum_{\substack{j=1 \\ j \neq i}}^{r} P_j(\mathbf{u}_2)\right)$$

$$= \sum_{\substack{j=1 \\ j \neq i}}^{r} P_i P_j(\mathbf{u}_2) = \sum_{\substack{j=1 \\ j \neq i}}^{r} Z(\mathbf{u}_2) = \mathbf{0}.$$

It is clear that $P_j(\mathbf{v}) = \mathbf{v}_j$ whenever $\mathbf{v} = \mathbf{v}_1 + \cdots + \mathbf{v}_r$ with $\mathbf{v}_i \in P_i(\mathbf{V})$.

10.4 DEFINITION

A complete set of projections $\{P_1, P_2, \ldots, P_r\}$ for $\mathbf{V}$ and a direct sum decomposition $\mathbf{V} = \mathbf{W}_1 \oplus \mathbf{W}_2 \oplus \cdots \oplus \mathbf{W}_r$ are said to **correspond** to each other, or to be **associated** with each other if $P_j(\mathbf{v}) = \mathbf{v}_j$ whenever $\mathbf{v} \in \mathbf{V}$ is written in the unique form $\mathbf{v} = \mathbf{v}_1 + \mathbf{v}_2 + \cdots + \mathbf{v}_r$ with $\mathbf{v}_i \in \mathbf{W}_i$.

Thus, $\mathbf{V} = \mathbf{W}_1 \oplus \mathbf{W}_2 \oplus \cdots \oplus \mathbf{W}_r$ and $\{P_1, P_2, \ldots, P_r\}$ correspond if and only if $P_j(\mathbf{V}) = \mathbf{W}_j$ and P_j acts as the identity on $\mathbf{W}_j$.

Let us consider now some simple examples of complete sets of projections. The reader should note in each case the corresponding direct sum decomposition.

EXAMPLE 1   Let $\mathbf{V} = \mathbf{R}^n$, and define P_i by

$$P_i(a_1, a_2, \ldots, a_n) = a_i \mathbf{e}_i = (0, \ldots, 0, a_i, 0, \ldots, 0)$$

for $i = 1, 2, \ldots, n$. Then $\{P_1, P_2, \ldots, P_n\}$ is a complete set of projections for $\mathbf{R}^n$. This situation generalizes readily to an arbitrary n-dimensional vector space $\mathbf{V}$. For any given basis $\mathcal{A} = \{\mathbf{v}_1, \mathbf{v}_2, \ldots, \mathbf{v}_n\}$, let $P_i(\sum_{j=1}^n a_j \mathbf{v}_j) = a_i \mathbf{v}_i$. Then $\{P_1, P_2, \ldots, P_n\}$ is a complete set of projections for $\mathbf{V}$. ∎

Variations in the number of projections in a complete set can easily be made. For example, for each $\mathbf{v} = (a_1, a_2, a_3)$ in $\mathbf{R}^3$, let the mappings T_1 and T_2 be defined by $T_1(\mathbf{v}) = (a_1, 0, 0)$ and $T_2(\mathbf{v}) = (0, a_2, a_3)$. Then $\{T_1, T_2\}$, is a complete set of projections for $\mathbf{R}^3$.

10.5
THEOREM

> Each eigenvalue of a projection is either 0 or 1.

Proof   Suppose that λ is an eigenvalue of the projection P, and let $\mathbf{v}$ be an associated eigenvector. Then $(P - \lambda)(\mathbf{v}) = \mathbf{0}$, and

$$\begin{aligned}
(1 - \lambda)\lambda \mathbf{v} &= (1 - \lambda)P(\mathbf{v}) \\
&= (P - \lambda P)(\mathbf{v}) \\
&= (P^2 - \lambda P)(\mathbf{v}) \\
&= P((P - \lambda)(\mathbf{v})) \\
&= P(\mathbf{0}) \\
&= \mathbf{0}.
\end{aligned}$$

Since $\mathbf{v} \neq \mathbf{0}$, it must be that $(1 - \lambda)\lambda = 0$, and λ is either 0 or 1. ■ ■ ■

The fact that a projection P has the property $P^2 = P$ and also acts as the identity transformation on its range might lead one to expect the matrix of a projection to look somehow like the identity matrix. This is not necessarily the case, however. For example, the matrix

$$A = \begin{bmatrix} 3 & -3 & -2 & 3 \\ -4 & 6 & 4 & -5 \\ 3 & -3 & -2 & 3 \\ -4 & 6 & 4 & -5 \end{bmatrix}$$

is such that $A^2 = A$, and hence represents a projection. But with an appropriate choice of basis, the matrix of a projection P can be made to take on a form very much like I_n. Now $P(\mathbf{V})$ is the same as the eigenspace of P corresponding to the eigenvalue 1, and $\mathbf{V} = P(\mathbf{V}) \oplus P^{-1}(\mathbf{0})$. Thus, if a basis $\{\mathbf{v}_1, \ldots, \mathbf{v}_r\}$ of $P(\mathbf{V})$ is extended to a basis $\mathcal{A} = \{\mathbf{v}_1, \ldots, \mathbf{v}_r, \mathbf{v}_{r+1}, \ldots, \mathbf{v}_n\}$ of $\mathbf{V}$ with $\{\mathbf{v}_{r+1}, \ldots, \mathbf{v}_n\}$ a basis of $P^{-1}(\mathbf{0})$, then the matrix of P relative to $\mathcal{A}$ is

$$D_r = \begin{bmatrix} I_r & \mathbf{0} \\ \mathbf{0} & \mathbf{0} \end{bmatrix}.$$

An interchange of the $\mathbf{v}_i$ in $\mathcal{A}$ produces an interchange of the elements on the diagonal of D_r, so the 1's on the diagonal can be placed in any desired diagonal positions.

We illustrate these ideas with the matrix

$$A = \begin{bmatrix} 3 & -3 & -2 & 3 \\ -4 & 6 & 4 & -5 \\ 3 & -3 & -2 & 3 \\ -4 & 6 & 4 & -5 \end{bmatrix}$$

that represents a projection P.

EXAMPLE 2 Since A represents a projection P, its eigenvalues are 0 and 1. The eigenspaces are given by

$$\mathbf{V}_1 = \langle (1, 0, 1, 0), (0, 1, 0, 1) \rangle \quad \text{and} \quad \mathbf{V}_0 = \langle (-1, 1, 0, 2), (-\tfrac{2}{3}, 0, 1, \tfrac{4}{3}) \rangle.$$

So $\mathbf{R}^4$ can be decomposed into the direct sum $\mathbf{R}^4 = \mathbf{V}_1 \oplus \mathbf{V}_0$, where $\mathbf{V}_1 = P(\mathbf{V})$ and $\mathbf{V}_0 = P^{-1}(\mathbf{0})$. Hence the matrix of P relative to the basis

$$\{(1, 0, 1, 0), (0, 1, 0, 1), (-1, 1, 0, 2), (-\tfrac{2}{3}, 0, 1, \tfrac{4}{3})\}$$

has the diagonal form $D = \text{diag}\{1, 1, 0, 0\}$. ■

10.1 Exercises

1. Show that $\{P_1, P_2\}$ is a complete set of projections for $\mathbf{R}^3$ where

$$P_1(x_1, x_2, x_3) = (\tfrac{1}{2}(x_1 + x_3), x_2, \tfrac{1}{2}(x_1 + x_3))$$

and

$$P_2(x_1, x_2, x_3) = (\tfrac{1}{2}(x_1 - x_3), 0, \tfrac{1}{2}(-x_1 + x_3)).$$

2. Show that each of the following matrices represents a projection P. Find $P(\mathbf{V})$ and $P^{-1}(\mathbf{0})$ such that $\mathbf{V} = P(\mathbf{V}) \oplus P^{-1}(\mathbf{0})$. Also give the diagonalized form for P.

(a) $\frac{1}{3} \begin{bmatrix} 2 & 1 & -1 \\ 1 & 2 & 1 \\ -1 & 1 & 2 \end{bmatrix}$ **(b)** $\frac{1}{2} \begin{bmatrix} 1 & 0 & 1 \\ 0 & 2 & 0 \\ 1 & 0 & 1 \end{bmatrix}$

(c) $\frac{1}{2} \begin{bmatrix} 1 & 0 & -1 & 0 \\ 0 & 1 & 0 & -1 \\ -1 & 0 & 1 & 0 \\ 0 & -1 & 0 & 1 \end{bmatrix}$ **(d)** $\frac{1}{5} \begin{bmatrix} 2 & -2 & -1 & -1 \\ -2 & 2 & 1 & 1 \\ -1 & 1 & 3 & -2 \\ -1 & 1 & -2 & 3 \end{bmatrix}$

3. If $V = W_1 \oplus W_2$, prove that the projection of V onto W_1 along W_2 is a linear operator on V.

4. Let T be a linear operator on V. Prove that T is a projection if and only if $1 - T$ is a projection.

5. Prove that if P is the projection of V onto W_1 along W_2, then $1 - P$ is the projection of V onto W_2 along W_1.

6. Let P_1, P_2, P_3 be nonzero projections of V such that $P_1 + P_2 + P_3 = 1$, and assume that $1 + 1 \neq 0$ in $\mathcal{F}$. Prove that $\{P_1, P_2, P_3\}$ is a complete set of projections for V.

7. Give an example of two projections P_1, P_2 such that $P_1 P_2 = Z$ but $P_2 P_1 \neq Z$.

8. Prove that if P is a projection on the inner product space V, then P^* is a projection on V.

9. Prove that any complete set of projections for V is linearly independent.

10. Let $V = W_1 + W_2 + \cdots + W_r$, where each W_i is a subspace of V. Prove that $V = W_1 \oplus W_2 \oplus \cdots \oplus W_r$ if and only if $v_1 + v_2 + \cdots + v_r = 0$ and $v_i \in W_i$ imply that each $v_j = 0$.

11. Let P_1 be the projection of V onto S_1 along S_2, and let P_2 be the projection of V onto W_1 along W_2, and assume that $1 + 1 \neq 0$ in $\mathcal{F}$.

    (a) Prove that $P_1 + P_2$ is a projection if and only if $\{P_1, P_2\}$ is orthogonal.

    (b) Prove that if P_1 and P_2 are orthogonal projections, then $P_1 + P_2$ is the projection of V onto $S_1 \oplus W_1$ along $S_2 \cap W_2$.

12. Let V be an inner product space. Prove that if P is the projection of V onto W_1 along W_2, then P^* is the projection of V onto $W_2^\perp$ along $W_1^\perp$.

13. According to the definition in the second paragraph of this section, a projection P on an inner product space V is called an orthogonal projection if and only if $P(V)$ and $P^{-1}(0)$ are orthogonal subspaces. Prove that a projection P is an orthogonal projection if and only if P is self-adjoint.

14. Let P be a projection on the inner product space V. Prove that if $\|P(v)\| \leq \|v\|$ for all $v \in V$, then P is an orthogonal projection.

15. Prove that a projection on an inner product space V is self-adjoint if and only if it is normal.

10.2 Spectral Decompositions

With the contents of the preceding section at hand, we are prepared to formulate a new characterization of the diagonalizable linear operators on $\mathbf{V}$. It is clear that in order for T to be diagonalizable, all of its eigenvalues must be in $\mathcal{F}$. That is, the characteristic polynomial of T must factor into linear factors over $\mathcal{F}$. The problem, then, is to determine what additional conditions are necessary in order that T be diagonalizable.

The characterization of diagonalizable operators that we shall obtain is phrased in terms of projections. The projections are a very special type of operator, so it is natural to ask how the set of projections fits into the vector space that consists of all linear operators on $\mathbf{V}$. The sum of two projections is usually *not* a projection (see Problem 1), so the projections do not form a subspace. However, our main interest is not with the set of all projections, but rather with those subspaces that are spanned by a complete set of projections. That is, we are interested primarily in those linear operators that are linear combinations of a complete set of projections. Our first result concerns the eigenvectors of this type of operator.

10.6 THEOREM

Let $\{P_1, P_2, \ldots, P_r\}$ be a complete set of projections for $\mathbf{V}$, and suppose that $T = c_1 P_1 + c_2 P_2 + \cdots + c_r P_r$ for some scalars c_i. Then each c_i is an eigenvalue of T, and each eigenvector $\mathbf{v}_i$ associated with the eigenvalue 1 of P_i is an eigenvector of T associated with c_i.

Proof Let $\mathbf{v}_i$ be an eigenvector associated with the eigenvalue 1 of P_i, so that $P_i(\mathbf{v}_i) = \mathbf{v}_i$. For any $j \neq i$, we have $P_j(\mathbf{v}_i) = P_j(P_i(\mathbf{v}_i)) = \mathbf{0}$. Therefore

$$T(\mathbf{v}_i) = c_1 P_1(\mathbf{v}_i) + c_2 P_2(\mathbf{v}_i) + \cdots + c_r P_r(\mathbf{v}_i)$$

$$= c_i P_i(\mathbf{v}_i)$$

$$= c_i \mathbf{v}_i,$$

and c_i is an eigenvalue of T with $\mathbf{v}_i$ as an associated eigenvector. ∎

10.7 THEOREM

The eigenvectors of a complete set of projections span $\mathbf{V}$.

Proof Let $\{P_1, P_2, \ldots, P_r\}$ be a complete set of projections for $\mathbf{V}$. For an arbitrary $\mathbf{v} \in \mathbf{V}$, consider the vector $P_i(\mathbf{v})$. Since

$$P_i(P_i(\mathbf{v})) = P_i^2(\mathbf{v}) = P_i(\mathbf{v}),$$

$P_i(\mathbf{v})$ is an eigenvector of P_i unless $P_i(\mathbf{v}) = \mathbf{0}$. Since $1 = P_1 + P_2 + \cdots + P_r$,

$$\mathbf{v} = P_1(\mathbf{v}) + P_1(\mathbf{v}) + \cdots + P_r(\mathbf{v}).$$

But $P_1(\mathbf{v}) + P_2(\mathbf{v}) + \cdots + P_r(\mathbf{v})$ is a linear combination of eigenvectors of $P_1, P_2, \ldots, P_r$, and the proof is complete.   ■ ■ ■

We are now in a position to prove the main theorem of this section.

10.8 THEOREM

Let T be a linear operator on $\mathbf{V}$ with distinct eigenvalues $\lambda_1, \lambda_2, \ldots, \lambda_r$. Then T is diagonalizable if and only if

$$T = \lambda_1 P_1 + \lambda_2 P_2 + \cdots + \lambda_r P_r,$$

where $\{P_1, P_2, \ldots, P_r\}$ is a complete set of projections for $\mathbf{V}$.

Proof Suppose first that $T = \lambda_1 P_1 + \lambda_2 P_2 + \cdots + \lambda_r P_r$, where $\{P_1, P_2, \ldots, P_r\}$ is a complete set of projections for $\mathbf{V}$. Let $\mathcal{A} = \{\mathbf{v}_1, \mathbf{v}_2, \ldots, \mathbf{v}_n\}$ be a basis of $\mathbf{V}$, and consider the set of nr vectors

$$\mathcal{B} = \{P_1(\mathbf{v}_1), \ldots, P_1(\mathbf{v}_n), P_2(\mathbf{v}_1), \ldots, P_2(\mathbf{v}_n), \ldots, P_r(\mathbf{v}_1), \ldots, P_r(\mathbf{v}_n)\}.$$

For any $\mathbf{v} = \sum_{i=1}^{n} a_i \mathbf{v}_i$, we have

$$\mathbf{v} = (P_1 + P_2 + \cdots + P_r)(\mathbf{v})$$

$$= (P_1 + P_2 + \cdots + P_r)\left(\sum_{i=1}^{n} a_i \mathbf{v}_i\right)$$

$$= \sum_{i=1}^{n} \sum_{j=1}^{r} a_i P_j(\mathbf{v}_i).$$

Thus, $\mathcal{B}$ spans $\mathbf{V}$, and therefore contains a basis $\mathcal{C} = \{\mathbf{u}_1, \mathbf{u}_2, \ldots, \mathbf{u}_n\}$ of $\mathbf{V}$. Each $\mathbf{u}_k$ is a nonzero vector of the form $\mathbf{u}_k = P_j(\mathbf{v}_i)$, and hence is an eigenvector of P_j associated with the eigenvalue 1. By Theorem 10.6, $\mathbf{u}_k$ is an eigenvector of T associated with the eigenvalue λ_j. Thus $\mathcal{C}$ is a basis of eigenvectors of T, and T is diagonalizable by Theorem 7.16.

Suppose now that T is diagonalizable, and let $\mathcal{B} = \{\mathbf{w}_1, \mathbf{w}_2, \ldots, \mathbf{w}_n\}$ be a basis of $\mathbf{V}$ such that T is represented by a diagonal matrix D relative to $\mathcal{B}$. As in the proof of Theorem 7.16, the $\mathbf{w}_i$ are eigenvectors of T, and the diagonal elements of D are the eigenvalues λ_i of T. By Theorem 7.20, the geometric multiplicity of λ_j is the same as the algebraic multiplicity m_j. That is, $\mathbf{V}_{\lambda_j}$ is of dimension m_j. By Corollary 7.10, the sum $\sum_{j=1}^{r} \mathbf{V}_{\lambda_j}$ is direct. Since $\sum_{j=1}^{r} m_j = n$, it must be that $\mathbf{V} = \mathbf{V}_{\lambda_1} \oplus \mathbf{V}_{\lambda_2} \oplus \cdots \oplus \mathbf{V}_{\lambda_r}$. With $\mathbf{W}_j = \mathbf{V}_{\lambda_j}$ in Definition 10.4, the set of projections P_j defined there is a complete set of projections for $\mathbf{V}$. For each $\mathbf{v} \in \mathbf{V}$, we have

$$\mathbf{v} = \mathbf{v}_1 + \mathbf{v}_2 + \cdots + \mathbf{v}_r$$

with $\mathbf{v}_j = P_j(\mathbf{v})$. And since $\mathbf{v}_j \in \mathbf{V}_{\lambda_j}$, we have $T(\mathbf{v}_j) = \lambda_j \mathbf{v}_j$. Thus

$$\begin{aligned}
T(\mathbf{v}) &= T(\mathbf{v}_1) + T(\mathbf{v}_2) + \cdots + T(\mathbf{v}_r) \\
&= \lambda_1 \mathbf{v}_1 + \lambda_2 \mathbf{v}_2 + \cdots + \lambda_r \mathbf{v}_r \\
&= \lambda_1 P_1(\mathbf{v}) + \lambda_2 P_2(\mathbf{v}) + \cdots + \lambda_r P_r(\mathbf{v}) \\
&= (\lambda_1 P_1 + \lambda_2 P_2 + \cdots + \lambda_r P_r)(\mathbf{v}),
\end{aligned}$$

and $T = \lambda_1 P_1 + \lambda_2 P_2 + \cdots + \lambda_r P_r$. ∎

10.9 DEFINITION

If the linear operator T on $\mathbf{V}$ can be written in the form

$$T = \lambda_1 P_1 + \lambda_2 P_2 + \cdots + \lambda_r P_r,$$

where $\{P_1, P_2, \ldots, P_r\}$ is a complete set of projections for $\mathbf{V}$ and $\lambda_1, \lambda_2, \ldots, \lambda_r$ are the distinct eigenvalues of T, then the expression $\lambda_1 P_1 + \lambda_2 P_2 + \cdots + \lambda_r P_r$ is called a **spectral decomposition** of T.

Theorem 10.8 asserts that T is diagonalizable if and only if T has a spectral decomposition.

10.10 THEOREM

If T has a spectral decomposition $T = \lambda_1 P_1 + \lambda_2 P_2 + \cdots + \lambda_r P_r$, then $f(T) = f(\lambda_1) P_1 + f(\lambda_2) P_2 + \cdots + f(\lambda_r) P_r$ for every polynomial $f(x)$.

Proof We first show that $T^i = \sum_{j=1}^{r} \lambda_j^i P_j$. This is trivial for $i = 0$ and $i = 1$. Assuming that $T^k = \sum_{j=1}^{r} \lambda_j^k P_j$, we have

$$\begin{aligned}
T^{k+1} &= \left(\sum_{j=1}^{r} \lambda_j^k P_j \right) \left(\sum_{m=1}^{r} \lambda_m P_m \right) \\
&= \sum_{j=1}^{r} \sum_{m=1}^{r} \lambda_j^k \lambda_m P_j P_m \\
&= \sum_{j=1}^{r} \lambda_j^{k+1} P_j.
\end{aligned}$$

By induction, $T^i = \sum_{j=1}^{r} \lambda_j^i P_j$ for each nonnegative integer i. For any polynomial $f(x) = \sum_{i=0}^{s} c_i x^i$, we thus have

$$f(T) = \sum_{i=0}^{s} c_i \left(\sum_{j=1}^{r} \lambda_j^i P_j \right)$$

$$= \sum_{j=1}^{r} \left(\sum_{i=0}^{s} c_i \lambda_j^i \right) P_j$$

$$= \sum_{j=1}^{r} f(\lambda_j) P_j.$$ ∎∎∎

Theorem 10.10 will prove to be useful in the next section.

10.11 THEOREM

If $T = \lambda_1 P_1 + \lambda_2 P_2 + \cdots + \lambda_r P_r$ is a spectral decomposition of T, then $\mathbf{V} = \mathbf{V}_{\lambda_1} \oplus \mathbf{V}_{\lambda_2} \oplus \cdots \oplus \mathbf{V}_{\lambda_r}$, and P_j is the projection of $\mathbf{V}$ onto $\mathbf{V}_{\lambda_j}$ along $\sum_{\substack{j=1 \\ i \neq j}}^{r} \mathbf{V}_{\lambda_i}$.

Proof Let $T = \lambda_1 P_1 + \lambda_2 P_2 + \cdots + \lambda_r P_r$ be a spectral decomposition of T. Then T is diagonalizable by Theorem 10.8, and $\mathbf{V} = \mathbf{V}_{\lambda_1} \oplus \mathbf{V}_{\lambda_2} \oplus \cdots \oplus \mathbf{V}_{\lambda_r}$ as in the proof of that theorem. Since $1 = P_1 + P_2 + \cdots + P_r$, we have $\mathbf{v} = P_1(\mathbf{v}) + P_2(\mathbf{v}) + \cdots + P_r(\mathbf{v})$ for any $\mathbf{v}$ in $\mathbf{V}$. But $P_j(\mathbf{v})$ is in $\mathbf{V}_{\lambda_j}$ since

$$T(P_j(\mathbf{v})) = \left(\sum_{i=1}^{r} \lambda_i P_i \right) (P_j(\mathbf{v})) = \lambda_j (P_j(\mathbf{v})).$$

Thus we have $P_j(\mathbf{V}) \subseteq \mathbf{V}_{\lambda_j}$ and hence

$$\dim(P_j(\mathbf{V})) \leq \dim(\mathbf{V}_{\lambda_j}).$$

This implies the inequalities

$$n = \sum_{j=1}^{r} \dim(P_j(\mathbf{V})) \leq \sum_{j=1}^{r} \dim(\mathbf{V}_{\lambda_j}) = n$$

and therefore

$$\dim(P_j(\mathbf{V})) = \dim(\mathbf{V}_{\lambda_j}),$$

for $j = 1, 2, \ldots, r$. It follows that $P_j(\mathbf{V}) = \mathbf{V}_{\lambda_j}$ for $j = 1, 2, \ldots, r$. Hence P_j is the projection of $\mathbf{V}$ onto $\mathbf{V}_{\lambda_j}$ along the sum of the remaining eigenspaces.   ∎∎∎

As might be expected, a restriction to orthonormal bases of an inner product space $\mathbf{V}$ corresponds to a restriction on the projections in the spectral decomposition. The exact description is as follows.

10.12 THEOREM	A linear operator T on an n-dimensional inner product space $\mathbf{V}$ can be represented by a diagonal matrix relative to an orthonormal basis of $\mathbf{V}$ if and only if T has a spectral decomposition in which the projections are self-adjoint.

Proof With the proofs of Theorems 7.16 and 10.8 in mind, it is sufficient to prove that an orthonormal basis of eigenvectors of T exists if and only if T has a spectral decomposition in which each projection is self-adjoint.

Suppose first that $T = \lambda_1 P_1 + \lambda_2 P_2 + \cdots + \lambda_r P_r$, with the P_i self-adjoint. Let $\mathcal{B}_j = \{\mathbf{u}_{j1}, \mathbf{u}_{j2}, \ldots, \mathbf{u}_{jn_j}\}$ be an orthonormal basis of the eigenspace $\mathbf{V}_{\lambda_j}$ for $j = 1, 2, \ldots, r$. As in the proof of Theorem 7.20, the set

$$\mathcal{B} = \{\mathbf{u}_{11}, \ldots, \mathbf{u}_{1n_1}, \mathbf{u}_{21}, \ldots, \mathbf{u}_{2n_2}, \ldots, \mathbf{u}_{r1}, \ldots, \mathbf{u}_{rn_r}\}$$

is a basis of $\mathbf{V}$. This basis is orthonormal if and only if $(\mathbf{u}_{it}, \mathbf{u}_{js}) = 0$ whenever $i \neq j$. According to Theorem 10.11, P_j is the projection of $\mathbf{V}$ onto $\mathbf{V}_{\lambda_j}$ along $\sum_{i \neq j} \mathbf{V}_{\lambda_j}$. Hence $P_j(\mathbf{v}) = \mathbf{v}$ for each $\mathbf{v}$ in $\mathbf{V}_{\lambda_j}$. In other words, each nonzero vector of $\mathbf{V}_{\lambda_j}$ is an eigenvector of P_j associated with 1. Consequently, we have

$$(\mathbf{u}_{it}, \mathbf{u}_{js}) = (P_i(\mathbf{u}_{it}), P_j(\mathbf{u}_{js}))$$
$$= (\mathbf{u}_{it}, P_i^* P_j(\mathbf{u}_{js}))$$
$$= (\mathbf{u}_{it}, P_i P_j(\mathbf{u}_{js}))$$
$$= (\mathbf{u}_{it}, \mathbf{0})$$
$$= 0$$

whenever $i \neq j$.

Conversely, suppose that there exists an orthonormal basis of eigenvectors of T. With the same notation as used in the proof of Theorem 10.8, $\mathbf{V} = \mathbf{V}_{\lambda_1} \oplus \mathbf{V}_{\lambda_2} \oplus \cdots \oplus \mathbf{V}_{\lambda_r}$. The set of projections $\{P_1, P_2, \ldots, P_r\}$ is a complete set for $\mathbf{V}$, and $T = \lambda_1 P_1 + \lambda_2 P_2 + \cdots + \lambda_r P_r$. The proof will be complete if we show that each P_i is self-adjoint. Let $\mathbf{u} = P_1(\mathbf{u}) + P_2(\mathbf{u}) + \cdots + P_r(\mathbf{u})$ and $\mathbf{v} = P_1(\mathbf{v}) + P_2(\mathbf{v}) + \cdots + P_r(\mathbf{v})$ be any two vectors in $\mathbf{V}$. For $i \neq j$, $P_i(\mathbf{u})$ and $P_j(\mathbf{v})$ are in distinct eigenspaces $\mathbf{V}_{\lambda_i}$ and $\mathbf{V}_{\lambda_j}$, and so are orthogonal. This implies that

$$(\mathbf{u}, P_j(\mathbf{v})) = \left(\sum_{i=1}^{r} P_i(\mathbf{u}), P_j(\mathbf{v}) \right)$$
$$= (P_j(\mathbf{u}), P_j(\mathbf{v}))$$
$$= \left(P_j(\mathbf{u}), \sum_{i=1}^{r} P_i(\mathbf{v}) \right)$$
$$= (P_j(\mathbf{u}), \mathbf{v}),$$

and therefore $P_j^* = P_j$. ∎

We shall devise a method for obtaining the projections involved in a spectral decomposition near the end of the next section.

10.2 Exercises

1. Verify that the mappings of $\mathbf{R}^2$ into $\mathbf{R}^2$ defined by $P_1(x_1, x_2) = (x_1 + x_2, 0)$ and $P_2(x_1, x_2) = (x_2, x_2)$ are projections, and show that $P_1 + P_2$ is not a projection.

2. Let P_1 and P_2 be the projections defined on $\mathcal{C}^2$ by $P_1(x_1, x_2) = (x_1, x_1)$ and $P_2(x_1, x_2) = (x_2, x_2)$.

   (a) Let $T = 3P_1 + 4P_2$ and determine if T is diagonalizable.
   (b) Find the eigenvalues and associated eigenvectors of T.
   (c) Determine whether or not $P_1 + P_2$ is a projection.

3. Let P_1 and P_2 be the projections defined on $\mathbf{R}^3$ by

$$P_1(x_1, x_2, x_3) = \left(\tfrac{1}{2}(x_1 + x_3), x_2, \tfrac{1}{2}(x_1 + x_3) \right)$$

and

$$P_2(x_1, x_2, x_3) = \left(\tfrac{1}{2}(x_1 - x_3), 0, \tfrac{1}{2}(-x_1 + x_3) \right).$$

   (a) Let $T = 5P_1 - 2P_2$ and determine if T is diagonalizable.
   (b) State the eigenvalues and associated eigenvectors of T.

4. Let P_1 and P_2 be the projections defined on $\mathbf{R}^4$ by

$$P_1(x_1, x_2, x_3, x_4) = \left(\tfrac{1}{2}(x_1 + x_3), \tfrac{1}{2}(x_2 + x_4), \tfrac{1}{2}(x_1 + x_3), \tfrac{1}{2}(x_2 + x_4) \right)$$

and

$$P_2(x_1, x_2, x_3, x_4) = \left(\tfrac{1}{2}(x_1 - x_3), \tfrac{1}{2}(x_2 - x_4), \tfrac{1}{2}(-x_1 + x_3), \tfrac{1}{2}(-x_2 + x_4) \right).$$

   (a) Let $T = 4P_1 + 5P_2$ and determine if T is diagonalizable.
   (b) State the eigenvalues and associated eigenvectors of T.

5. Prove that if T is invertible and has a spectral decomposition, then T^{-1} has a spectral decomposition with the same complete set of projections.

6. Let $\{P_1, P_2, \ldots, P_r\}$ be a complete set of projections for $\mathbf{V}$, and suppose that $T = c_1 P_1 + c_2 P_2 + \cdots + c_r P_r$.

   (a) Prove that each eigenvalue of T is equal to at least one of the scalars c_j.
   (b) Give an example which shows that there may be a vector $\mathbf{v}$ such that $\mathbf{v}$ is an eigenvector of T associated with c_j, but $\mathbf{v}$ is not an eigenvector of any P_i.

7. Prove that each projection P_i in a spectral decomposition

$$T = \lambda_1 P_1 + \lambda_2 P_2 + \cdots + \lambda_r P_r$$

of T has rank equal to the geometric multiplicity of the eigenvalue λ_i of T.

8. Prove that the spectral decomposition of T is unique if it exists.

9. Let T be a normal linear transformation of the inner product space $\mathbf{V}$. Prove that T^* has a spectral decomposition.

10. Let $T = \sum_{i=1}^{r} \lambda_i P_i$ be a spectral decomposition for the linear operator T on the inner product space $\mathbf{V}$. Suppose that $\{f_1(x), f_2(x), \ldots, f_r(x)\}$ is a set of polynomials with real coefficients such that $f_i(\lambda_j) = \delta_{ij}$. Prove that $f_i(T) = P_i$ for $i = 1, 2, \ldots, r$.

11. A linear operator T on an inner product space $\mathbf{V}$ is called **nonnegative** if T is self-adjoint and $(T(\mathbf{v}), \mathbf{v}) \geq 0$ for every $\mathbf{v} \in \mathbf{V}$.

    (a) Prove that any nonnegative linear operator T on $\mathbf{V}$ has a spectral decomposition $T = \sum_{i=1}^{r} \lambda_i P_i$ with each $\lambda_i \geq 0$.

    (b) Show that $S = \sqrt{\lambda_1} P_1 + \sqrt{\lambda_2} P_2 + \cdots + \sqrt{\lambda_r} P_r$ is a nonnegative linear operator such that $S^2 = T$ (i.e., S is a nonnegative square root of T).

    (c) Show that the nonnegative square root of S in (b) is unique.

10.3 Minimal Polynomials and Spectral Decompositions

In this section, we shall need to use certain properties of addition and multiplication of polynomials in x over a field $\mathcal{F}$ (i.e., polynomials in x with coefficients in $\mathcal{F}$). The set of all polynomials in x over $\mathcal{F}$ is denoted by $\mathcal{F}[x]$. A derivation of the information that we need about $\mathcal{F}[x]$ would constitute a major digression at this point. For this reason, we shall simply state without proofs the results that are needed.

If $p(x)$ and $m(x)$ are in $\mathcal{F}[x]$ with $p(x)$ nonzero, then there exist polynomials $q(x), r(x)$ in $\mathcal{F}[x]$ such that

$$m(x) = p(x)q(x) + r(x)$$

with $r(x)$ either the zero polynomial or a polynomial of degree less than that of $p(x)$. This statement is known as the **division algorithm** for elements of $\mathcal{F}[x]$. If $r(x)$ is the zero polynomial, we say that $p(x)$ **divides** $m(x)$, and that $p(x)$ is a **divisor** of $m(x)$.

A nonzero polynomial $p(x)$ in $\mathcal{F}[x]$ is called **monic** if the coefficient of the highest degree term in $p(x)$ is 1. That is, the highest power of x that appears in $p(x)$ has 1 as its coefficient.

A monic polynomial $d(x)$ in $\mathcal{F}[x]$ is called the **greatest common divisor** of a set of nonzero polynomials $q_1(x), q_2(x), \ldots, q_r(x)$ in $\mathcal{F}[x]$ if

1. $d(x)$ is a divisor of each of the polynomials $q_i(x)$, and
2. every polynomial $p(x)$ that divides each $q_i(x)$ is also a divisor of $d(x)$.

Every set of nonzero polynomials $q_1(x), q_2(x), \ldots, q_r(x)$ in $\mathcal{F}[x]$ has a unique monic greatest common divisor $d(x)$ in $\mathcal{F}[x]$. Moreover, there exists a set of polynomials $g_1(x), g_2(x), \ldots, g_r(x)$ in $\mathcal{F}[x]$ such that

$$d(x) = g_1(x)q_1(x) + g_2(x)q_2(x) + \cdots + g_r(x)q_r(x).$$

In some of the examples and exercises, we assume that the student is familiar with the partial fraction decomposition of a quotient of two nonzero polynomials.

We have already encountered polynomials $p(T) = \sum_{i=0}^{s} c_i T^i$ in a linear operator T on $\mathbf{V}$ and the corresponding polynomials $p(A) = \sum_{i=0}^{s} c_i A^i$ in a matrix A that represents T. It has been noted that T and A satisfy the same polynomial equations.

From one point of view, a polynomial $p(A) = \sum_{i=0}^{s} c_i A^i$ in the square matrix A can be thought of as being obtained from a polynomial $p(x) = \sum_{i=0}^{s} c_i x^i$ by replacing the powers x^i of the indeterminate x by the corresponding powers A^i of the matrix A. This suggests the construction of other types of polynomials involving matrices by making other replacements in $p(x)$. One might treat x as an indeterminate scalar and replace the scalar coefficients c_i by matrix coefficients C_i, or one might replace both the c_i and the x^i by matrix quantities. The algebra connected with this last type of polynomial is quite involved, and we shall not be concerned with them here. Our interest is confined to only two types of polynomials:

1. Those obtained by leaving the scalar coefficients c_i unchanged and replacing x by a square matrix A.
2. Those obtained by treating x as an indeterminate scalar and replacing the scalar coefficients c_i by matrix coefficients C_i.

There is another point of view from which the polynomials of the second type may be regarded. Any polynomial $\sum_{i=0}^{s} C_i x^i$ of this type can be considered to be a matrix with elements in $\mathcal{F}[x]$. For example,

$$\begin{bmatrix} 2 & -1 \\ 0 & 4 \end{bmatrix} x^3 + \begin{bmatrix} -1 & 0 \\ 8 & 3 \end{bmatrix} x + \begin{bmatrix} 4 & 5 \\ -7 & 0 \end{bmatrix} = \begin{bmatrix} 2x^3 - x + 4 & -x^3 + 5 \\ 8x - 7 & 4x^3 + 3x \end{bmatrix}.$$

In either form, this type of matrix is called a **matrix polynomial**. In our development, we have considered only matrices with elements in a field, and $\mathcal{F}[x]$ is not a field. But $\mathcal{F}[x]$ is contained in the field $\mathcal{F}(x)$ of all rational functions in x over $\mathcal{F}$, and we can consider matrix polynomials in this context. For future use, we note that two matrix polynomials are equal if and only if they have equal coefficients of each power of x.

Since powers of A commute with scalars and with each other, multiplication is commutative for polynomials of the first type: $p(A)q(A) = q(A)p(A)$ for any $p(x), q(x)$ in $\mathcal{F}[x]$. In this case, factorizations in $\mathcal{F}[x]$ remain valid: If $p(x) = g(x)h(x)$ in $\mathcal{F}[x]$, then $p(A) = g(A)h(A)$. A certain amount of care must be exercised when working with polynomials of the second type, since multiplication is not commutative then. Factorizations in $\mathcal{F}[x]$ no longer remain valid here. For example, if $AB \neq BA$, then

$$(Ax - B)(Ax + B) = A^2 x^2 + (AB - BA)x - B^2$$
$$\neq A^2 x^2 - B^2.$$

We recall from Chapter 4 that the set $\mathcal{F}_{n\times n}$ of all $n \times n$ matrices over $\mathcal{F}$ is a vector space over $\mathcal{F}$. If E_{ij} denotes the $n \times n$ matrix that has the element in row i, column j equal to 1 and all other elements 0, then an arbitrary $A = [a_{ij}]$ in $\mathcal{F}_{n\times n}$ can be written uniquely as

$$A = \sum_{i=1}^{n} \sum_{j=1}^{n} a_{ij} E_{ij},$$

and the set $\{E_{ij}\}$ is a basis of $\mathcal{F}_{n\times n}$. Hence $\mathcal{F}_{n\times n}$ has dimension n^2, and the set

$$I = A^0, A, A^2, \ldots, A^{n^2}$$

is linearly dependent for any $n \times n$ matrix A over $\mathcal{F}$. That is, there is a polynomial $p(x) = \sum_{i=0}^{n^2} c_i x^i$ such that $p(A) = \mathbf{0}$. This means that there exists a monic polynomial $m(x)$ of smallest degree such that $m(A) = \mathbf{0}$.

| 10.13 THEOREM | Let A be an $n \times n$ matrix, and let $m(x)$ be a monic polynomial over $\mathcal{F}$ of smallest degree such that $m(A) = \mathbf{0}$. For any $p(x)$ in $\mathcal{F}[x]$, $p(A) = \mathbf{0}$ if and only if $m(x)$ is a factor of $p(x)$. |

Proof If $p(x) = m(x)h(x)$, then $p(A) = m(A)h(A) = \mathbf{0}$. Suppose conversely that $p(A) = \mathbf{0}$. Let $q(x)$ and $r(x)$ be the quotient and remainder upon division of $p(x)$ by $m(x)$:

$$p(x) = m(x)q(x) + r(x),$$

where either $r(x)$ is zero, or the degree of $r(x)$ is less than that of $m(x)$. Then since $m(A) = \mathbf{0}$,

$$r(A) = m(A)q(A) + r(A) = p(A) = \mathbf{0}.$$

This means that $r(x)$ must be the zero polynomial, since otherwise $r(A) = \mathbf{0}$ would contradict the choice of $m(x)$ as having the smallest possible degree such that $m(A) = \mathbf{0}$. Hence $m(x)$ is a factor of $p(x)$. ■ ■ ■

This theorem makes it easy to prove that the polynomial $m(x)$ in the hypothesis is unique (see Problem 6).

| 10.14 DEFINITION | Let A be an $n \times n$ matrix over $\mathcal{F}$. The unique monic polynomial $m(x)$ of smallest degree such that $m(A) = \mathbf{0}$ is called the **minimal polynomial** of A. The **minimal polynomial** of a linear operator T on $\mathbf{V}$ is by definition the same as the minimal polynomial of any matrix that represents T. |

Since a linear operator T and any matrix that represents it satisfy the same polynomial equations, the minimal polynomial of T is well-defined.

The discussion preceding Theorem 10.13 is an efficient argument for the existence of the minimal polynomial $m(x)$ of A, but it suggests no practical method for finding $m(x)$. The next theorem is a great help in this direction. It is known as the **Hamilton–Cayley theorem**.

10.15
THEOREM

Let $A = [a_{ij}]$ be an arbitrary $n \times n$ matrix over $\mathcal{F}$, and let $f(x)$ be the characteristic polynomial of A. Then $f(A) = \mathbf{0}$.

Proof Since the characteristic matrix

$$A - xI = [a_{ij} - x\delta_{ij}]$$

is a matrix with polynomials as elements, the minor of each $a_{ij} - x\delta_{ij}$ is the determinant of an $(n-1) \times (n-1)$ matrix with polynomial elements. Hence each element in $B = \mathrm{adj}(A - xI)$ is a polynomial in x. Moreover,

$$
\begin{aligned}
B(A - xI) &= \mathrm{adj}(A - xI) \cdot (A - xI) \\
&= \det(A - xI)I \\
&= f(x)I.
\end{aligned}
$$

If $f(x) = c_n x^n + \cdots + c_1 x + c_0$, then

$$
\begin{aligned}
f(A) - f(x)I &= (c_n A^n + \cdots + c_1 A + c_0 I) - (c_n I x^n + \cdots + c_1 I x + c_0 I) \\
&= c_n (A^n - I x^n) + \cdots + c_2 (A^2 - I x^2) + c_1 (A - I x).
\end{aligned}
$$

The usual properties of matrix multiplication and matrix addition yield the factorization $A^k - I x^k = P_k (A - Ix)$, where

$$P_k = I x^{k-1} + A x^{k-2} + \cdots + A^{k-2} x + A^{k-1}$$

is a matrix polynomial of degree $k - 1$. Hence

$$f(A) - f(x)I = \sum_{k=1}^{n} c_k P_k (A - Ix),$$

and

$$f(A) = f(x)I + \sum_{k=1}^{n} c_k P_k (A - Ix) = \mathrm{adj}(A - Ix) \cdot (A - Ix) + \sum_{k=1}^{n} c_k P_k (A - Ix)$$

$$= \left\{ \mathrm{adj}(A - Ix) + \sum_{k=1}^{n} c_k P_k \right\} (A - Ix).$$

The matrix inside the braces is a matrix polynomial, say $B_t x^t + \cdots + B_1 x + B_0$. This gives

$$f(A) = (B_t x^t + \cdots + B_1 x + B_0)(A - Ix)$$
$$= -B_t x^{t+1} + (B_t A - B_{t-1}) x^t + \cdots + (B_1 A - B_0) x + B_0 A.$$

Since two matrix polynomials are equal if and only if they have corresponding coefficients that are equal, this requires that $f(A) = B_0 A$ and the coefficients of each positive power of x on the right side be zero:

$$B_t = 0$$
$$B_t A - B_{t-1} = 0$$
$$\vdots$$
$$B_2 A - B_1 = 0$$
$$B_1 A - B_0 = 0.$$

Solving for $B_t, B_{t-1}, \ldots, B_1, B_0$ in order we find

$$0 = B_t = B_{t-1} = \cdots = B_1 = B_0.$$

But this gives $f(A) = B_0 A = 0 \cdot A = 0$, and the theorem is proved. ∎

The minimal polynomial of a diagonalizable operator T can be used to obtain the projections for a spectral decomposition. Since the existence of a spectral decomposition requires that the eigenvalues of T be in $\mathcal{F}$, we assume for the remainder of this chapter that the characteristic polynomial $f(x)$ of T factors into linear factors over $\mathcal{F}$:

$$f(x) = (-1)^n (x - \lambda_1)^{m_1} (x - \lambda_2)^{m_2} \cdots (x - \lambda_r)^{m_r}, \tag{i}$$

where the m_j are the algebraic multiplicities of the distinct eigenvalues λ_j. Since the minimal polynomial $m(x)$ divides $f(x)$, this means that

$$m(x) = (x - \lambda_1)^{t_1} (x - \lambda_2)^{t_2} \cdots (x - \lambda_r)^{t_r}. \tag{ii}$$

We shall adopt the notation of this paragraph for the characteristic polynomial $f(x)$ and the minimal polynomial $m(x)$ of T throughout the remainder of the chapter.

There are some further notational conventions that are essential. For each factor $(x - \lambda_j)^{t_j}$ of $m(x)$, let $q_j(x)$ be the polynomial

$$q_j(x) = \frac{m(x)}{(x - \lambda_j)^{t_j}}$$
$$= (x - \lambda_1)^{t_1} \cdots (x - \lambda_{j-1})^{t_{j-1}} (x - \lambda_{j+1})^{t_{j+1}} \cdots (x - \lambda_r)^{t_r}.$$

Any nonconstant common divisor of $q_1(x), q_2(x), \ldots, q_r(x)$ would necessarily have a linear factor of the form $x - \lambda_i$ since the only linear factors of each $q_j(x)$ are of this type. But $x - \lambda_i$ is not a factor of $q_i(x)$, so there are no nonconstant common divisors of

$q_1(x), q_2(x), \ldots, q_r(x)$. That is, the greatest common divisor of $q_1(x), q_2(x), \ldots, q_r(x)$ is 1. Hence there are polynomials $g_1(x), g_2(x), \ldots, g_r(x)$ such that

$$1 = g_1(x)q_1(x) + g_2(x)q_2(x) + \cdots + g_r(x)q_r(x).$$

Since

$$q_j(x) = \frac{m(x)}{(x - \lambda_j)^{t_j}},$$

this yields

$$\frac{1}{m(x)} = \frac{g_1(x)}{(x - \lambda_1)^{t_1}} + \frac{g_2(x)}{(x - \lambda_2)^{t_2}} + \cdots + \frac{g_r(x)}{(x - \lambda_r)^{t_r}}.$$

Let $p_j(x) = g_j(x)q_j(x)$ for $j = 1, 2, \ldots, r$. The set of polynomials $p_1(x), p_2(x), \ldots, p_r(x)$ has the following properties:

(i) $p_i(x)p_j(x)$ has $m(x)$ as a factor if $i \neq j$, say $p_i(x)p_j(x) = h_{ij}(x)m(x)$
(ii) $1 = p_1(x) + p_2(x) + \cdots + p_r(x)$.

> **10.16**
> **THEOREM**
>
> Let $F_j = p_j(T)$, and let $\mathcal{K}_j$ denote the kernel of $(T - \lambda_j)^{t_j}$ for $j = 1, 2, \ldots, r$. Then $\{F_1, F_2, \ldots, F_r\}$ is a complete set of projections for **V**, and F_j is the projection of **V** onto $\mathcal{K}_j$ along $\sum_{i \neq j} \mathcal{K}_i$.

Proof Since $p_i(x)p_j(x) = h_{ij}(x)m(x)$ for $i \neq j$ and $m(T) = Z$, we have

$$F_i F_j = p_i(T)p_j(T) = h_{ij}(T)m(T) = Z$$

whenever $i \neq j$. It follows from (ii) above that $1 = F_1 + F_2 + \cdots + F_r$. Thus

$$F_j = F_j \left(\sum_{i=1}^{r} F_i \right) = \sum_{i=1}^{r} F_j F_i = F_j^2,$$

and $\{F_1, F_2, \ldots, F_r\}$ is an orthogonal set of projections such that $1 = F_1 + F_2 + \cdots + F_r$. We shall show that F_j is the projection of **V** onto $\mathcal{K}_j$ along $\sum_{i \neq j} \mathcal{K}_i$, and it will follow that $F_j \neq Z$ since $\mathcal{K}_j$ clearly contains the eigenspace $\mathbf{V}_{\lambda_j}$ of T.

Now

$$(x - \lambda_j)^{t_j} p_j(x) = (x - \lambda_j)^{t_j} g_j(x)q_j(x)$$

$$= (x - \lambda_j)^{t_j} g_j(x) \frac{m(x)}{(x - \lambda_j)^{t_j}}$$

$$= g_j(x)m(x),$$

so $(T - \lambda_j)^{t_j} F_j = (T - \lambda_j)^{t_j} p_j(T) = g_j(T)m(T) = Z$. Hence $(T - \lambda_j)^{t_j} F_j(\mathbf{v}) = \mathbf{0}$ for all $\mathbf{v}$, and $F_j(\mathbf{V}) \subseteq \mathcal{K}_j$. Now let $\mathbf{v} \in \mathcal{K}_j$, so that $(T - \lambda_j)^{t_j}(\mathbf{v}) = \mathbf{0}$. For $i \neq j$, $(T - \lambda_j)^{t_j}$ is a factor of $F_i = p_i(T)$ since $(x - \lambda_j)^{t_j}$ is a factor of $p_i(x)$. Hence $F_i(\mathbf{v}) = \mathbf{0}$ for all $i \neq j$, and

$$\mathbf{v} = (F_1 + F_2 + \cdots + F_r)(\mathbf{v}) = F_j(\mathbf{v}).$$

Thus, $\mathbf{v} \in F_j(\mathbf{V})$, and $\mathcal{K}_j = F_j(\mathbf{V})$. This completes the proof of the theorem.   ■ ■ ■

10.17
DEFINITION

> A subspace $\mathbf{W}$ of $\mathbf{V}$ is **invariant** under T, or T-**invariant**, if
>
> $$T(\mathbf{W}) \subseteq \mathbf{W}.$$

That is, a subspace $\mathbf{W}$ of $\mathbf{V}$ is T-invariant if and only if $T(\mathbf{v}) \in \mathbf{W}$ for all $\mathbf{v} \in \mathbf{W}$. Every linear operator T on $\mathbf{V}$ has invariant subspaces. The eigenspaces $\mathbf{V}_\lambda$ of T are invariant under T, as are the zero subspace and $\mathbf{V}$.

If $\mathbf{W}$ is a subspace of $\mathbf{V}$ that is invariant under T, then T induces a linear transformation $T_\mathbf{W}$ of $\mathbf{W}$ into $\mathbf{W}$ defined by $T_\mathbf{W}(\mathbf{v}) = T(\mathbf{v})$ for all $\mathbf{v} \in \mathbf{W}$. That is, as long as $\mathbf{v} \in \mathbf{W}$, $T_\mathbf{W}$ and T map $\mathbf{v}$ onto the same vector. The distinction between $T_\mathbf{W}$ and T is that $T_\mathbf{W}$ is defined only on $\mathbf{W}$. The transformation $T_\mathbf{W}$ is called the **restriction of T to $\mathbf{W}$**.

10.18
THEOREM

> Let $\mathcal{K}_j$ be the kernel of $(T - \lambda_j)^{t_j}$ for $j = 1, 2, \ldots, r$. Then
>
> **(a)** $\mathbf{V} = \mathcal{K}_1 \oplus \mathcal{K}_2 \oplus \cdots \oplus \mathcal{K}_r$;
> **(b)** each $\mathcal{K}_j$ is invariant under T;
> **(c)** if T_j denotes the restriction of T to $\mathcal{K}_j$, then the minimal polynomial of T_j is $(x - \lambda_j)^{t_j}$.

Proof Consider the complete set of projections $\{F_1, F_2, , \ldots, F_r\}$ where $F_j = p_j(T)$. By Theorem 10.16, F_j is the projection of $\mathbf{V}$ onto $\mathcal{K}_j = F_j(\mathbf{V})$ along $\sum_{i \neq j} \mathcal{K}_i$. Thus, $\mathbf{V} = \mathcal{K}_1 \oplus \mathcal{K}_2 \oplus \cdots \oplus \mathcal{K}_r$ is the direct sum decomposition corresponding to this complete set of projections.

To establish (b), let $\mathbf{v} \in \mathcal{K}_j$ and consider $T(\mathbf{v})$. Since $(T - \lambda_j)^{t_j}$ is a polynomial in T, it commutes with T to yield

$$(T - \lambda_j)^{t_j}(T(\mathbf{v})) = T((T - \lambda_j)^{t_j}(\mathbf{v})) = T(\mathbf{0}) = \mathbf{0}.$$

Hence $T(\mathbf{v})$ is in $\mathcal{K}_j$, and $\mathcal{K}_j$ is invariant under T.

Now $(T_j - \lambda_j)^{t_j}$ is the zero transformation of $\mathcal{K}_j$ since $(T - \lambda_j)^{t_j}(\mathbf{v}) = \mathbf{0}$ for all $\mathbf{v} \in \mathcal{K}_j$. Therefore, the minimal polynomial of T_j divides $(x - \lambda_j)^{t_j}$. That is, the minimal polynomial of T_j is $(x - \lambda_j)^s$ for some $s \leq t_j$. Suppose that $s < t_j$. Now each $\mathbf{v} \in \mathbf{V}$ can be written uniquely as

$$\mathbf{v} = \mathbf{v}_1 + \mathbf{v}_2 + \cdots + \mathbf{v}_r$$

with $\mathbf{v}_i \in \mathcal{K}_i$. Since

$$(T - \lambda_i)^{t_i}(\mathbf{v}_i) = \mathbf{0}$$

for each i, then

$$q_j(T)(\mathbf{v}) = q_j(T)(\mathbf{v}_j).$$

And since

$$(T - \lambda_j)^s(\mathbf{v}_j) = (T_j - \lambda_j)^s(\mathbf{v}_j) = \mathbf{0}$$

for all $\mathbf{v}_j \in \mathcal{K}_j$, this means that $p(T) = (T - \lambda_j)^s q_j(T)$ is the zero transformation on $\mathbf{V}$. But this is a contradiction, since $p(x) = (x - \lambda_j)^s q_j(x)$ has degree less than that of $m(x) = (x - \lambda_j)^{t_j} q_j(x)$. Therefore, $s = t_j$. ■■■

10.19 THEOREM

The linear operator T on $\mathbf{V}$ has a spectral decomposition if and only if the minimal polynomial $m(x)$ of T has the form

$$m(x) = (x - \lambda_1)(x - \lambda_2) \cdots (x - \lambda_r),$$

where $\lambda_1, \lambda_2, \ldots, \lambda_r$ are the distinct eigenvalues of T.

Proof If the minimal polynomial $m(x)$ has the given form, then each $t_i = 1$ in equation (ii). Hence $\mathcal{K}_i = \mathbf{V}_{\lambda_i}$ for $i = 1, 2, \ldots, r$, and

$$\mathbf{V} = \mathbf{V}_{\lambda_1} \oplus \mathbf{V}_{\lambda_2} \oplus \cdots \oplus \mathbf{V}_{\lambda_r}.$$

It follows from the proof of Theorem 7.20 that T is diagonalizable, and therefore T has a spectral decomposition by Theorem 10.8.

Conversely, suppose that T has a spectral decomposition

$$T = \lambda_1 P_1 + \lambda_2 P_2 + \cdots + \lambda_r P_r.$$

For any polynomial $p(x)$,

$$p(T) = p(\lambda_1) P_1 + p(\lambda_2) P_2 + \cdots + p(\lambda_r) P_r$$

by Theorem 10.10. If $p(\lambda_i) \neq 0$ for some i, $p(T)$ has $p(\lambda_i)$ as a nonzero eigenvalue by Theorem 10.6. Hence $p(T) = Z$ if and only if $p(\lambda_i) = 0$ for $i = 1, 2, \ldots, r$. This implies at once that

$$m(x) = (x - \lambda_1)(x - \lambda_2) \cdots (x - \lambda_r).$$ ■■■

The result that we have been working toward is now at hand.

10.20 THEOREM

If $T = \lambda_1 P_1 + \lambda_2 P_2 + \cdots + \lambda_r P_r$ is a spectral decomposition, then $P_j = p_j(T)$.

Proof Let $T = \lambda_1 P_1 + \lambda_2 P_2 + \cdots + \lambda_r P_r$ be a spectral decomposition of T. By Theorem 10.19,

$$m(x) = (x - \lambda_1)(x - \lambda_2) \cdots (x - \lambda_r).$$

That is, each t_j in equation (ii) is 1, and $\mathcal{K}_j = \mathbf{V}_{\lambda_j}$ for each j. With $F_j = p_j(T)$, then $\{F_1, F_2, \ldots, F_r\}$ is a complete set of projections for $\mathbf{V}$, and F_j is the projection of $\mathbf{V}$ onto $\mathcal{K}_j$ along $\sum_{i \neq j} \mathcal{K}_i$ by Theorem 10.16. By Theorem 10.11, P_j is the projection of $\mathbf{V}$ onto $\mathbf{V}_{\lambda_j}$ along $\sum_{i \neq j} \mathbf{V}_{\lambda_i}$. Since $\mathcal{K}_j = \mathbf{V}_{\lambda_j}$ for each j, $P_j = F_j = p_j(T)$ for each j, and the proof is complete. ∎ ∎ ∎

The next corollary follows readily from the fact that each projection in a spectral decomposition of T is a polynomial in T.

10.21 COROLLARY

If $T = \lambda_1 P_1 + \lambda_2 P_2 + \cdots + \lambda_r P_r$ is a spectral decomposition of T, then a linear operator S on $\mathbf{V}$ commutes with T if and only if S commutes with each P_j.

Proof See Problem 10. ∎ ∎ ∎

One immediate and desirable consequence of Theorem 10.20 is that we now have a systematic procedure available for finding the projections in a spectral decomposition. This procedure, of course, is to determine the P_j by use of $P_j = p_j(T)$.

EXAMPLE 1 Let T be the linear transformation of $\mathbf{R}^3$ that has the matrix A relative to the standard basis where

$$A = \begin{bmatrix} 1 & -1 & -1 \\ -1 & 1 & -1 \\ -1 & -1 & 1 \end{bmatrix}.$$

We shall determine whether or not T has a spectral decomposition, and shall obtain such a decomposition if there is one. The characteristic polynomial $f(x)$ is given by

$$f(x) = -(x + 1)(x - 2)^2.$$

It follows from Theorem 10.19 that T has a spectral decomposition if and only if

$$m(x) = (x + 1)(x - 2).$$

A simple calculation shows that

$$m(A) = A^2 - A - 2I = \mathbf{0},$$

so T does indeed have a spectral decomposition. With $\lambda_1 = -1$, $\lambda_2 = 2$, the polynomials $q_j(x)$ are given by $q_1(x) = x - 2$, $q_2(x) = x + 1$. The partial fraction decomposition

$$\frac{1}{m(x)} = \frac{g_1(x)}{x + 1} + \frac{g_2(x)}{x - 2}$$

leads to $g_1(x) = -\frac{1}{3}$, $g_2(x) = \frac{1}{3}$. Hence $p_1(x) = -\frac{1}{3}(x - 2)$ and $p_2(x) = \frac{1}{3}(x + 1)$. The projections P_1, P_2 thus have respective matrices E_1, E_2 relative to $\mathcal{E}_3$ given by

$$E_1 = -\tfrac{1}{3}(A - 2I) = \tfrac{1}{3}\begin{bmatrix} 1 & 1 & 1 \\ 1 & 1 & 1 \\ 1 & 1 & 1 \end{bmatrix},$$

$$E_2 = \tfrac{1}{3}(A + I) = \tfrac{1}{3}\begin{bmatrix} 2 & -1 & -1 \\ -1 & 2 & -1 \\ -1 & -1 & 2 \end{bmatrix}.$$

It is easily checked that $E_1^2 = E_1$, $E_2^2 = E_2$, and $A = \lambda_1 E_1 + \lambda_2 E_2$. ■

In case T does not have a spectral decomposition, a somewhat similar decomposition can be obtained that in many cases is quite useful. We have seen that the mappings $P_j = F_j = p_j(T)$ form a complete set of projections $\{P_1, P_2, \ldots, P_r\}$ for $\mathbf{V}$. Consider the linear operator

$$D = \lambda_1 P_1 + \lambda_2 P_2 + \cdots + \lambda_r P_r,$$

where $\lambda_1, \lambda_2, \ldots, \lambda_r$ are the distinct eigenvalues of T. Since D has a spectral decomposition, it is diagonalizable. Let $N = T - D$. Using the representation

$$T = T(P_1 + P_2 + \cdots + P_r) = TP_1 + TP_2 + \cdots + TP_r$$

for T, N is given by

$$N = (T - \lambda_1)P_1 + (T - \lambda_2)P_2 + \cdots + (T - \lambda_r)P_r.$$

It is left as an exercise (Problem 11) to prove that

$$N^k = (T - \lambda_1)^k P_1 + (T - \lambda_2)^k P_2 + \cdots + (T - \lambda_r)^k P_r$$

for all positive integers k. Recall from the proof of Theorem 10.16 that $(T - \lambda_j)^{t_j} P_j = Z$, and thus $N^k = Z$ if $k \geq t_j$ for all j.

10.22
DEFINITION

A linear operator T is called **nilpotent** if $T^k = Z$ for some positive integer k. The smallest positive integer t such that $T^t = Z$ is the **index** of nilpotency of T.

10.23
THEOREM

If the minimal polynomial of T factors into a product of linear factors (not necessarily distinct) over $\mathcal{F}$, then T can be written as the sum $T = D + N$ of a diagonalizable transformation D and a nilpotent transformation N, where D and N are polynomials in T, and consequently commute.

Proof Let D and N be as given in the paragraph preceding Definition 10.22. From the expressions there for D and N and the fact that $P_j = p_j(T)$, it is reasonably clear that D and N are polynomials in T and hence commute. Nevertheless, we furnish some additional details. The projections P_j commute with T since $P_j = p_j(T)$. Now

$$ND = \left(\sum_{i=1}^{r} (T - \lambda_i) P_i \right) \left(\sum_{j=1}^{r} \lambda_j P_j \right)$$

$$= \sum_{i=1}^{r} \sum_{j=1}^{r} (T - \lambda_i) \lambda_j P_i P_j,$$

and

$$DN = \left(\sum_{j=1}^{r} \lambda_j P_j \right) \left(\sum_{i=1}^{r} (T - \lambda_i) P_i \right)$$

$$= \sum_{i=1}^{r} \sum_{j=1}^{r} \lambda_j P_j (T - \lambda_i) P_i.$$

But

$$\lambda_j P_j (T - \lambda_i) P_i = (T - \lambda_i) \lambda_j P_i P_j,$$

so $DN = ND$. ∎

In case $\mathbf{V}$ is a unitary space, the hypothesis of Theorem 10.23 is automatically satisfied, and every linear operator on $\mathbf{V}$ is the sum of a diagonalizable and a nilpotent transformation.

Spectral decompositions of linear operators have the usual parallel statements for matrices. If T has matrix A relative to a certain basis and the projections P_i in $T = \lambda_1 P_1 + \lambda_2 P_2 + \cdots + \lambda_r P_r$ have matrices E_i for $i = 1, 2, \ldots, r$, then

$$A = \lambda_1 E_1 + \lambda_2 E_2 + \cdots + \lambda_r E_r$$

is called a **spectral decomposition** for A. The E_i are called **projection matrices** or **principal idempotents** for A. It is left as an exercise to prove that $A E_i = \lambda_i E_i$.

If T, D, and N in $T = D + N$ have matrices A, B, and C, respectively, then $A = B + C$. The matrix B is diagonalizable, and C is a nilpotent matrix.

In the next example, we illustrate the situation when the minimal polynomial of T does not factor into distinct linear factors.

EXAMPLE 2 Suppose T is the linear transformation of $\mathbf{R}^3$ with matrix relative to the standard basis given by

$$A = \begin{bmatrix} 0 & -3 & 4 \\ 1 & 2 & -2 \\ 2 & 3 & -2 \end{bmatrix}.$$

The characteristic polynomial is $f(x) = (-1)^3(x - 2)(x + 1)^2$ and since the minimal polynomial is $m(x) = (x - 2)(x + 1)^2$, then T does not have a spectral decomposition. However, A can be expressed as the sum of a diagonalizable matrix and a nilpotent matrix. Define

$$q_1(x) = \frac{m(x)}{x - 2} = (x + 1)^2 \quad \text{and} \quad q_2(x) = \frac{m(x)}{(x + 1)^2} = x - 2.$$

Then we must determine $g_1(x)$ and $g_2(x)$ such that

$$\frac{1}{m(x)} = \frac{g_1(x)}{x - 2} + \frac{g_2(x)}{(x + 1)^2}.$$

The partial fraction decomposition of

$$\frac{1}{(x - 2)(x + 1)^2} = \frac{a}{x - 2} + \frac{b}{x + 1} + \frac{c}{(x + 1)^2}$$

yields $a = \frac{1}{9}$, $b = -\frac{1}{9}$, and $c = -\frac{1}{3}$. Combining the last two terms in the partial fraction decomposition gives

$$\frac{1}{(x - 2)(x + 1)^2} = \frac{\frac{1}{9}}{x - 2} + \frac{-\frac{1}{9}(x + 4)}{(x + 1)^2}$$

so that

$$g_1(x) = \frac{1}{9} \quad \text{and} \quad g_2(x) = -\frac{1}{9}(x + 4).$$

Setting $p_1(x) = g_1(x)q_1(x) = \frac{1}{9}(x + 1)^2$ and $p_2(x) = g_2(x)q_2(x) = -\frac{1}{9}(x + 4)(x - 2)$ leads to the complete set of projections with matrices

$$p_1(A) = \tfrac{1}{9}(A + I)^2 = \tfrac{1}{9}\begin{bmatrix} 1 & -3 & 4 \\ 1 & 3 & -2 \\ 2 & 3 & -1 \end{bmatrix}^2 = \tfrac{1}{3}\begin{bmatrix} 2 & 0 & 2 \\ 0 & 0 & 0 \\ 1 & 0 & 1 \end{bmatrix}$$

and

$$p_2(A) = -\tfrac{1}{9}(A + 4I)(A - 2I)$$

$$= -\tfrac{1}{9}\begin{bmatrix} 4 & -3 & 4 \\ 1 & 6 & -2 \\ 2 & 3 & 2 \end{bmatrix}\begin{bmatrix} -2 & -3 & 4 \\ 1 & 0 & -2 \\ 2 & 3 & -4 \end{bmatrix} = \tfrac{1}{3}\begin{bmatrix} 1 & 0 & -2 \\ 0 & 3 & 0 \\ -1 & 0 & 2 \end{bmatrix}.$$

Then the diagonalizable matrix B is given by

$$B = 2p_1(A) + (-1)p_2(A)$$

$$= 2 \left(\frac{1}{3} \begin{bmatrix} 2 & 0 & 2 \\ 0 & 0 & 0 \\ 1 & 0 & 1 \end{bmatrix} \right) + (-1) \left(\frac{1}{3} \begin{bmatrix} 1 & 0 & -2 \\ 0 & 3 & 0 \\ -1 & 0 & 2 \end{bmatrix} \right) = \begin{bmatrix} 1 & 0 & 2 \\ 0 & -1 & 0 \\ 1 & 0 & 0 \end{bmatrix}$$

and the nilpotent matrix is

$$C = A - B = \begin{bmatrix} 0 & -3 & 4 \\ 1 & 2 & -2 \\ 2 & 3 & -2 \end{bmatrix} - \begin{bmatrix} 1 & 0 & 2 \\ 0 & -1 & 0 \\ 1 & 0 & 0 \end{bmatrix} = \begin{bmatrix} -1 & -3 & 2 \\ 1 & 3 & -2 \\ 1 & 3 & -2 \end{bmatrix}. \quad\blacksquare$$

10.3 Exercises

1. Verify the Hamilton–Cayley theorem for each matrix A.

(a) $A = \begin{bmatrix} -4 & -3 & -1 \\ -4 & 0 & -4 \\ 8 & 4 & 5 \end{bmatrix}$

(b) $A = \begin{bmatrix} 8 & 5 & -5 \\ 5 & 8 & -5 \\ 15 & 15 & -12 \end{bmatrix}$

(c) $A = \begin{bmatrix} 8 & 5 & 6 & 0 \\ 0 & -2 & 0 & 0 \\ -10 & -5 & -8 & 0 \\ 2 & 1 & 1 & 2 \end{bmatrix}$

(d) $A = \begin{bmatrix} 7 & 3 & 3 & 2 \\ 0 & 1 & 2 & -4 \\ -8 & -4 & -5 & 0 \\ 2 & 1 & 1 & 3 \end{bmatrix}$

2. Find the minimal polynomial of each matrix A in Problem 1.

3. Determine all real 2×2 matrices A such that $A^2 = -I$.

4. Let T be the linear operator on $\mathbf{R}^n$ that has the given matrix A relative to the standard basis $\mathcal{E}_n$. Find the spectral decomposition of T.

(a) $A = \begin{bmatrix} 8 & 5 & -5 \\ 5 & 8 & -5 \\ 15 & 15 & -12 \end{bmatrix}$

(b) $A = \begin{bmatrix} 3 & 2 & 2 \\ 1 & 4 & 1 \\ -2 & -4 & -1 \end{bmatrix}$

(c) $A = \begin{bmatrix} 7 & 3 & 3 & 2 \\ 0 & 1 & 2 & -4 \\ -8 & -4 & -5 & 0 \\ 2 & 1 & 2 & 3 \end{bmatrix}$

(d) $A = \begin{bmatrix} 4 & -1 & 0 & 1 \\ -1 & 5 & -1 & 0 \\ 0 & -1 & 4 & -1 \\ 1 & 0 & -1 & 5 \end{bmatrix}$

5. Write the given matrix A as the sum of a diagonalizable matrix and a nilpotent matrix.

(a) $A = \begin{bmatrix} 7 & 3 & 3 & 2 \\ 0 & 1 & 2 & -4 \\ -8 & -4 & -5 & 0 \\ 2 & 1 & 1 & 3 \end{bmatrix}$

(b) $A = \begin{bmatrix} 8 & 5 & 6 & 0 \\ 0 & -2 & 0 & 0 \\ -10 & -5 & -8 & 0 \\ 2 & 1 & 1 & 2 \end{bmatrix}$

(c) $A = \begin{bmatrix} 1 & 1 & 0 & 0 \\ 0 & 1 & 1 & 0 \\ 0 & 0 & 1 & 1 \\ -1 & 0 & 2 & 1 \end{bmatrix}$
   (d) $A = \begin{bmatrix} 0 & 3 & 2 & -2 \\ 0 & 3 & 2 & -2 \\ -8 & 6 & -2 & 3 \\ -7 & 7 & -1 & 2 \end{bmatrix}$

6. Prove that the polynomial $m(x)$ in the hypothesis of Theorem 10.13 is unique.

7. Prove that, for an arbitrary square matrix A, A and A^T satisfy the same polynomial equations with scalar coefficients.

8. Let $f(x) = c_n x^n + c_{n-1} x^{n-1} + \cdots + c_1 x + c_0$ be the characteristic polynomial of A. If $\det(A) \neq 0$, prove that $A^{-1} = -(1/c_0)(c_n A^{n-1} + c_{n-1} A^{n-2} + \cdots + c_1 I)$.

9. Prove that similar matrices have the same minimal polynomial.

10. Prove Corollary 10.21.

11. Let $N = (T - \lambda_1)P_1 + (T - \lambda_2)P_2 + \cdots + (T - \lambda_r)P_r$, where $\{P_1, P_2, \ldots, P_r\}$ is a complete set of projections. Prove that

$$N^k = (T - \lambda_1)^k P_1 + (T - \lambda_2)^k P_2 + \cdots + (T - \lambda_r)^k P_r$$

for all positive integers k.

12. Let $\{P_1, P_2, \ldots, P_r\}$ be the complete set of projections determined by $P_j = p_j(T)$. With $T = TP_1 + TP_2 + \cdots + TP_r$, show that $(TP_i) \cdot (TP_j) = Z$ if $i \neq j$, and that $p(T) = p(TP_1) + p(TP_2) + \cdots + p(TP_r)$ for any polynomial $p(x)$ with zero constant term.

13. Let $T = \lambda_1 P_1 + \lambda_2 P_2 + \cdots + \lambda_r P_r$ be a spectral decomposition for the linear operator T on $\mathbf{V}$. Prove that a linear operator S on $\mathbf{V}$ commutes with T if and only if every $\mathcal{K}_j$ is invariant under S.

14. Prove that if $A = \lambda_1 E_1 + \lambda_2 E_2 + \cdots + \lambda_r E_r$ is a spectral decomposition for A, then $A E_i = \lambda_i E_i$ for each i.

10.4 Nilpotent Transformations

We have seen in the preceding section that if the minimum polynomial of T factors into linear factors over $\mathcal{F}$, then T can be written as $T = D + N$, where D is diagonalizable and N is nilpotent. The degree of simplicity that can be obtained in a matrix representation of T thus depends on the type of representation that is possible for N. The representation of T that is accepted as being simplest, or nearest to a diagonal form, is the Jordan canonical form. The derivation of the Jordan canonical form is the principal objective of the remainder of this chapter. This derivation hinges on an investigation of the invariant subspaces of an operator.

Suppose that $\mathbf{W}$ is an invariant subspace of $\mathbf{V}$ under T. Let $\{\mathbf{v}_1, \ldots, \mathbf{v}_k\}$ be a basis of $\mathbf{W}$, and extend this set to a basis $\mathcal{A} = \{\mathbf{v}_1, \ldots, \mathbf{v}_k, \mathbf{v}_{k+1}, \ldots, \mathbf{v}_n\}$ of $\mathbf{V}$. Since $\mathbf{W}$ is invariant under T, $T(\mathbf{v}_j) = \sum_{i=1}^{k} a_{ij}\mathbf{v}_i$ for $j = 1, 2, \ldots, k$. Hence the matrix of T relative to $\mathcal{A}$ is of the form

$$A = \begin{bmatrix} A_1 & A_3 \\ 0 & A_2 \end{bmatrix},$$

where A_1 is the matrix of $T_{\mathbf{W}}$ relative to $\{\mathbf{v}_1, \ldots, \mathbf{v}_k\}$.

If $\mathbf{V} = \mathbf{W}_1 \oplus \mathbf{W}_2$, where each of $\mathbf{W}_1$ and $\mathbf{W}_2$ is invariant under T, and if the basis $\mathcal{A} = \{\mathbf{v}_1, \ldots, \mathbf{v}_k, \mathbf{v}_{k+1}, \ldots, \mathbf{v}_n\}$ of $\mathbf{V}$ is chosen so that $\{\mathbf{v}_1, \ldots, \mathbf{v}_k\}$ is a basis of $\mathbf{W}_1$ and $\{\mathbf{v}_{k+1}, \ldots, \mathbf{v}_n\}$ is a basis of $\mathbf{W}_2$, then we also have $T(\mathbf{v}_j) = \sum_{i=k+1}^{n} a_{ij}\mathbf{v}_i$ for $j = k + 1,$ $\ldots, n$. Hence $A_3 = \mathbf{0}$, and the matrix of T relative to $\mathcal{A}$ is of the form

$$A = \begin{bmatrix} A_1 & 0 \\ 0 & A_2 \end{bmatrix},$$

where A_i is the matrix of the restriction of T to $\mathbf{W}_i$.

The results of the preceding paragraph generalize readily to direct sums with more than two terms. If $\mathbf{W}_1, \mathbf{W}_2, \ldots, \mathbf{W}_r$ are T-invariant subspaces of $\mathbf{V}$ such that $\mathbf{V}$ is the direct sum $\mathbf{V} = \mathbf{W}_1 \oplus \mathbf{W}_2 \oplus \cdots \oplus \mathbf{W}_r$, and if the basis

$$\mathcal{B} = \{\mathbf{u}_{11}, \ldots, \mathbf{u}_{1n_1}, \mathbf{u}_{21}, \ldots, \mathbf{u}_{2n_2}, \ldots, \mathbf{u}_{r1}, \ldots, \mathbf{u}_{rn_r}\}$$

of $\mathbf{V}$ is such that $\{\mathbf{u}_{i1}, \ldots, \mathbf{u}_{in_i}\}$ is a basis of $\mathbf{W}_i$, then the matrix of T relative to $\mathcal{B}$ is of the form

$$A = \begin{bmatrix} A_1 & 0 & \cdots & 0 \\ 0 & A_2 & \cdots & 0 \\ \vdots & \vdots & \ddots & \vdots \\ 0 & 0 & \cdots & A_r \end{bmatrix},$$

where A_i is the matrix of the restriction of T to $\mathbf{W}_i$. A matrix such as this A is called a **diagonal block matrix**.

The type of invariant subspace that we shall be mainly concerned with is the cyclic subspace, to be defined shortly.

Let $\mathbf{v}$ be any nonzero vector in $\mathbf{V}$. Since a linearly independent subset of $\mathbf{V}$ can have at most n elements, there is a unique positive integer k such that $\{\mathbf{v}, T(\mathbf{v}), \ldots, T^{k-1}(\mathbf{v})\}$ is linearly independent and $T^k(\mathbf{v})$ is dependent on $\{\mathbf{v}, T(\mathbf{v}), \ldots, T^{k-1}(\mathbf{v})\}$. Then there are scalars $a_0, a_1, \ldots, a_{k-1}$ such that $T^k(\mathbf{v}) = \sum_{i=0}^{k-1} a_i T^i(\mathbf{v})$. It follows from this that the subspace

$$\langle \mathbf{v}, T(\mathbf{v}), \ldots, T^{k-1}(\mathbf{v}) \rangle$$

is T-invariant (Problem 3).

10.24 DEFINITION Let $\mathbf{v}$ be a nonzero vector in $\mathbf{V}$, and let k be the unique positive integer such that $\{\mathbf{v}, T(\mathbf{v}), \ldots, T^{k-1}(\mathbf{v})\}$ is linearly independent and $T^k(\mathbf{v})$ is dependent on $\{\mathbf{v}, T(\mathbf{v}), \ldots, T^{k-1}(\mathbf{v})\}$. The subspace $\langle \mathbf{v}, T(\mathbf{v}), \ldots, T^{k-1}(\mathbf{v}) \rangle$ is denoted by $\mathbf{C}(\mathbf{v}, T)$

and is called the **cyclic subspace of v relative to** T, or the **cyclic subspace generated by v under** T. The particular basis $\{T^{k-1}(\mathbf{v}), \ldots, T(\mathbf{v}), \mathbf{v}\}$ of $\mathbf{C}(\mathbf{v}, T)$ is called the **cyclic basis** generated by $\mathbf{v}$ under T and is denoted by $\mathcal{B}(\mathbf{v}, T)$.

The arrangement of the elements of $\mathcal{B}(\mathbf{v}, T)$ according to descending rather than ascending powers of T seems a bit unnatural, but this is done so that the restriction of T to $\mathbf{C}(\mathbf{v}, T)$ will have a special form of matrix relative to $\mathcal{B}(\mathbf{v}, T)$. Since

$$T(T^{k-1}(\mathbf{v})) = T^k(\mathbf{v}) = a_{k-1}T^{k-1}(\mathbf{v}) + \cdots + a_1 T(\mathbf{v}) + a_0 \mathbf{v}$$

and $T(T^j(\mathbf{v})) = T^{j+1}(\mathbf{v})$ for $j = 0, 1, \ldots, k-2$, the restriction of T to $\mathbf{C}(\mathbf{v}, T)$ has the matrix

$$A = \begin{bmatrix} a_{k-1} & 1 & 0 & \cdots & 0 \\ a_{k-2} & 0 & 1 & \cdots & 0 \\ \vdots & \vdots & \vdots & \ddots & \vdots \\ a_1 & 0 & 0 & \cdots & 1 \\ a_0 & 0 & 0 & \cdots & 0 \end{bmatrix}$$

relative to $\mathcal{B}(\mathbf{v}, T)$. The special form that this matrix A takes on for a nilpotent transformation T is especially useful.

10.25 THEOREM   If T is a nilpotent transformation of $\mathbf{V}$, then the restriction of T to $\mathbf{C}(\mathbf{v}, T)$ has the matrix

$$\begin{bmatrix} 0 & 1 & 0 & \cdots & 0 \\ 0 & 0 & 1 & \cdots & 0 \\ \vdots & \vdots & \vdots & \ddots & \vdots \\ 0 & 0 & 0 & \cdots & 1 \\ 0 & 0 & 0 & \cdots & 0 \end{bmatrix}$$

relative to $\mathcal{B}(\mathbf{v}, T)$.

Proof With the notation of the paragraph just before the theorem, we shall show that $T^k(\mathbf{v}) = \mathbf{0}$. Our proof of this fact is an "overkill": We show that the restriction $T_{\mathbf{c}}$ of T to $\mathbf{C}(\mathbf{v}, T)$ has minimal polynomial x^k.

Since T is nilpotent on $\mathbf{V}$, $T_{\mathbf{c}}$ is nilpotent, say with index s. The minimal polynomial of $T_{\mathbf{c}}$ must therefore divide x^s. Let x^r denote this minimal polynomial. Since $T^{k-1}(\mathbf{v})$ is in the basis $\mathcal{B}(\mathbf{v}, T)$, then $r \geq k$.

Consider the polynomial $p(x) = a_0 + a_1 x + \cdots + a_{k-1}x^{k-1} - x^k$. We shall show that

$$S = p(T_{\mathbf{c}}) = a_0 + a_1 T_{\mathbf{c}} + \cdots + a_{k-1}T_{\mathbf{c}}^{k-1} - T_{\mathbf{c}}^k$$

is the zero transformation of $\mathbf{C}(\mathbf{v}, T)$. To do this, it is sufficient to show that S takes on the value $\mathbf{0}$ at each vector in $\mathcal{B}(\mathbf{v}, T)$. Since $T_{\mathbf{c}}(\mathbf{v}) = T(\mathbf{v})$,

$$S(\mathbf{v}) = a_0\mathbf{v} + a_1 T(\mathbf{v}) + \cdots + a_{k-1}T^{k-1}(\mathbf{v}) - T^k(\mathbf{v}) = \mathbf{0}$$

by the choice of the scalars a_i. For $j = 1, \ldots, k - 1$, we have

$$\begin{aligned} S(T^j(\mathbf{v})) &= a_0 T^j(\mathbf{v}) + \cdots + a_{k-1}T^{k-1}(T^j(\mathbf{v})) - T^k(T^j(\mathbf{v})) \\ &= T^j(a_0\mathbf{v} + \cdots + a_{k-1}T^{k-1}(\mathbf{v}) - T^k(\mathbf{v})) \\ &= T^j(\mathbf{0}) \\ &= \mathbf{0}. \end{aligned}$$

Hence $S = p(T_{\mathbf{c}})$ is the zero transformation, and x^r divides $p(x)$. This requires that $r \leq k$. Therefore $r = k$ and $p(x) = x^r = x^k$. ∎

Our principal concern is with the connection between cyclic subspaces and the kernels of the powers of a nilpotent transformation. Before proceeding to our main result in this direction, some preliminary lemmas are in order. These lemmas by themselves are of little consequence, but they are essential to the proof of Theorem 10.28.

10.26 LEMMA

Suppose that T is a nilpotent transformation of index t on $\mathbf{V}$, and let $\mathbf{W}_j$ be the kernel of T^j. Then $\{\mathbf{0}\} = \mathbf{W}_0 \subseteq \mathbf{W}_1 \subseteq \cdots \subseteq \mathbf{W}_t = \mathbf{V}$, and $\mathbf{W}_{j-1} \neq \mathbf{W}_j$ for $j = 1, 2, \ldots, t$.

Proof The equality $\{\mathbf{0}\} = \mathbf{W}_0$ follows from the fact that T^0 is the identity transformation, and $\mathbf{W}_t = \mathbf{V}$ since T has index t on $\mathbf{V}$.

Since $T^{j-1}(\mathbf{v}) = \mathbf{0}$ implies $T^j(\mathbf{v}) = T(\mathbf{0}) = \mathbf{0}$, $\mathbf{W}_{j-1} \subseteq \mathbf{W}_j$ for $j = 1, 2, \ldots, t$. Let $\mathbf{u}$ be a vector in $\mathbf{V}$ such that $T^{t-1}(\mathbf{u}) \neq \mathbf{0}$, and let $1 \leq j \leq t$. Then $T^{t-j}(\mathbf{u})$ is in $\mathbf{W}_j$ since $T^j(T^{t-j}(\mathbf{u})) = \mathbf{0}$, and $T^{t-j}(\mathbf{u})$ is not in $\mathbf{W}_{j-1}$ since $T^{j-1}(T^{t-j}(\mathbf{u})) = T^{t-1}(\mathbf{u}) \neq \mathbf{0}$. Thus $\mathbf{W}_{j-1} \neq \mathbf{W}_j$. ∎

10.27 LEMMA

Let the subspaces $\mathbf{W}_i$ be as in Lemma 10.26, with $j \geq 2$. Suppose that $\{\mathbf{u}_1, \ldots, \mathbf{u}_k\}$ is a basis of $\mathbf{W}_{j-2}$, and that $\{\mathbf{w}_1, \ldots, \mathbf{w}_r\}$ is a linearly independent set of vectors in $\mathbf{W}_j$ such that

$$\langle \mathbf{w}_1, \ldots, \mathbf{w}_r \rangle \cap \mathbf{W}_{j-1} = \{\mathbf{0}\}.$$

Then $\{\mathbf{u}_1, \ldots, \mathbf{u}_k, T(\mathbf{w}_1), \ldots, T(\mathbf{w}_r)\}$ is a linearly independent subset of $\mathbf{W}_{j-1}$.

Proof Since $T^{j-1}(T(\mathbf{w}_i)) = T^j(\mathbf{w}_i) = \mathbf{0}$ for each i, the set

$$\{\mathbf{u}_1, \ldots, \mathbf{u}_k, T(\mathbf{w}_1), \ldots, T(\mathbf{w}_r)\}$$

is contained in $\mathbf{W}_{j-1}$. Assume that

$$\sum_{i=1}^{k} b_i \mathbf{u}_i + \sum_{m=1}^{r} c_m T(\mathbf{w}_m) = \mathbf{0}.$$

Then

$$\sum_{m=1}^{r} c_m T(\mathbf{w}_m) = -\sum_{i=1}^{k} b_i \mathbf{u}_i \quad \text{is in } \mathbf{W}_{j-2},$$

so that

$$T^{j-2}\left(\sum_{m=1}^{r} c_m T(\mathbf{w}_m)\right) = \mathbf{0}.$$

This implies that $T^{j-1}(\sum_{m=1}^{r} c_m \mathbf{w}_m) = \mathbf{0}$, and therefore $\sum_{m=1}^{r} c_m \mathbf{w}_m$ is in $\mathbf{W}_{j-1}$. Since

$$\langle \mathbf{w}_1, \ldots, \mathbf{w}_r \rangle \cap \mathbf{W}_{j-1} = \{\mathbf{0}\},$$

it must be that $\sum_{m=1}^{r} c_m \mathbf{w}_m = \mathbf{0}$. Hence each $c_m = 0$, and this implies $\sum_{i=1}^{k} b_i \mathbf{u}_i = \mathbf{0}$. This requires in turn that all $b_i = 0$ since $\{\mathbf{u}_1, \ldots, \mathbf{u}_k\}$ is linearly independent. Therefore, $\{\mathbf{u}_1, \ldots, \mathbf{u}_k, T(\mathbf{w}_1), \ldots, T(\mathbf{w}_r)\}$ is linearly independent.   ■ ■ ■

The **superdiagonal elements** in a matrix $A = [a_{ij}]_{m \times n}$ are those elements $a_{i,i+1}$ for $i = 1, 2, \ldots, m - 1$. That is, the superdiagonal elements are those immediately above the diagonal elements.

10.28 THEOREM

Let T be a nilpotent operator of index t on $\mathbf{V}$, let $\mathbf{W}_j$ be the kernel of T^j for $j = 1, 2, \ldots, t$, and let $s = \text{nullity}(T)$. There exists a basis $\mathcal{B}$ of $\mathbf{V}$ such that the matrix of T relative to $\mathcal{B}$ is given by

$$A = \begin{bmatrix} A_1 & \mathbf{0} & \cdots & \mathbf{0} \\ \mathbf{0} & A_2 & \cdots & \mathbf{0} \\ \vdots & \vdots & \ddots & \vdots \\ \mathbf{0} & \mathbf{0} & \cdots & A_s \end{bmatrix},$$

where each A_i is a square matrix that has all superdiagonal elements 1 and all other elements 0, A_1 is of order t, and $\text{order}(A_i) \geq \text{order}(A_{i+1})$ for each i. The matrix A is uniquely determined by T.

Proof By Lemma 10.27, there is a basis

$$\mathcal{A} = \{\mathbf{v}_{11}, \ldots, \mathbf{v}_{1s_1}, \mathbf{v}_{21}, \ldots, \mathbf{v}_{2s_2}, \ldots, \mathbf{v}_{t1}, \ldots, \mathbf{v}_{ts_t}\}$$

of $\mathbf{V}$ such that $\{\mathbf{v}_{11}, \ldots, \mathbf{v}_{js_j}\}$ is a basis of $\mathbf{W}_j$ for $j = 1, \ldots, t$. Such a basis can be obtained by extending a basis $\{\mathbf{v}_{11}, \ldots, \mathbf{v}_{1s_1}\}$ of $\mathbf{W}_1$ to a basis

$$\{\mathbf{v}_{11}, \ldots, \mathbf{v}_{1s_1}, \mathbf{v}_{21}, \ldots, \mathbf{v}_{2s_2}\}$$

of $\mathbf{W}_2$, then extending this basis to a basis of $\mathbf{W}_3$, and so on. For each j, the elements of $\mathcal{A}$ that are in $\mathbf{W}_j$ but not in $\mathbf{W}_{j-1}$ are precisely those in the segment $\mathbf{v}_{j1}, \ldots, \mathbf{v}_{js_j}$. We shall replace each of these segments by a new set of vectors in order to obtain a basis $\mathcal{A}'$ with certain properties. Roughly, the idea is this. The vectors in segment $j - 1$ will be replaced by the images of the vectors in the next segment to the right, and then this set will be extended to a basis of $\mathbf{W}_{j-1}$.

No change is required in the last s_t vectors of $\mathcal{A}$. That is, we put

$$\mathbf{v}'_{t1} = \mathbf{v}_{t1}, \quad \mathbf{v}'_{t2} = \mathbf{v}_{t2}, \quad \ldots, \quad \mathbf{v}'_{ts_t} = \mathbf{v}_{ts_t}.$$

If $\sum_{m=1}^{s_t} c_m \mathbf{v}'_{tm}$ is in $\mathbf{W}_{t-1}$, then

$$\sum_{m=1}^{s_t} c_m \mathbf{v}'_{tm} = \sum_{i=1}^{t-1} \sum_{j=1}^{s_i} a_{ij} \mathbf{v}_{ij}.$$

The linear independence of $\mathcal{A}$ requires that each $a_{ij} = 0$ and each $c_m = 0$. Hence $\langle \mathbf{v}'_{t1}, \ldots, \mathbf{v}'_{ts_t} \rangle \cap \mathbf{W}_{t-1} = \{\mathbf{0}\}$. Let

$$\mathbf{v}'_{t-1,1} = T(\mathbf{v}_{t1}), \quad \mathbf{v}'_{t-1,2} = T(\mathbf{v}_{t2}), \quad \ldots, \quad \mathbf{v}'_{t-1,s_t} = T(\mathbf{v}_{ts_t}).$$

By Lemma 10.27, the set

$$\{\mathbf{v}_{11}, \ldots, \mathbf{v}_{1s_1}, \ldots, \mathbf{v}_{t-2,1}, \ldots, \mathbf{v}_{t-2,s_{t-2}}, \mathbf{v}'_{t-1,1}, \ldots, \mathbf{v}'_{t-1,s_t}\}$$

is linearly independent, and so can be extended to a basis

$$\{\mathbf{v}_{11}, \ldots, \mathbf{v}_{1s_1}, \ldots, \mathbf{v}_{t-2,1}, \ldots, \mathbf{v}'_{t-1,1}, \ldots, \mathbf{v}'_{t-1,s_t}, \ldots, \mathbf{v}'_{t-1,s_{t-1}}\}$$

of $\mathbf{W}_{t-1}$. At this point, the last two segments in the basis

$$\{\mathbf{v}_{11}, \ldots, \mathbf{v}_{1s_1}, \ldots, \mathbf{v}_{t-2,s_{t-2}}, \mathbf{v}'_{t-1,1}, \ldots, \mathbf{v}'_{t-1,s_{t-1}}, \mathbf{v}'_{t1}, \ldots, \mathbf{v}'_{ts_t}\}$$

of $\mathbf{V}$ have the desired form.

The procedure is repeated, replacing the segment $\mathbf{v}_{j-1,1}, \ldots, \mathbf{v}_{j-1,s_{j-1}}$ by the images

$$\mathbf{v}'_{j-1,1} = T(\mathbf{v}'_{j1}), \quad \mathbf{v}'_{j-1,2} = T(\mathbf{v}'_{j2}), \quad \ldots, \quad \mathbf{v}'_{j-1,s_j} = T(\mathbf{v}'_{js_j}),$$

and then adjoining elements $\mathbf{v}'_{j-1,s_j+1}, \ldots, \mathbf{v}'_{j-1,s_{j-1}}$ so as to form a basis of $\mathbf{W}_{j-1}$. This repetition leads finally to a basis

$$\mathcal{A}' = \{\mathbf{v}'_{11}, \ldots, \mathbf{v}'_{1s_1}, \mathbf{v}'_{21}, \ldots, \mathbf{v}'_{2s_2}, \ldots, \mathbf{v}'_{t1}, \ldots, \mathbf{v}'_{ts_t}\}$$

such that

$$\mathbf{v}'_{j-1,1} = T(\mathbf{v}'_{j1}), \ldots, \mathbf{v}'_{j-1,s_j} = T(\mathbf{v}'_{js_j})$$

for each j.

The vectors in $\mathcal{A}'$ are now rearranged so that the first vectors from each segment are written first, followed in order by the second vectors from each segment, and so on until the vectors in the last segment are exhausted. Whenever a segment of vectors is exhausted, we continue the same procedure with the remaining segments, if there are any. This process leads to the basis

$$\mathcal{B} = \{\mathbf{v}'_{11}, \mathbf{v}'_{21}, \ldots, \mathbf{v}'_{t1}, \ldots, \mathbf{v}'_{1s_t}, \ldots, \mathbf{v}'_{ts_t}, \mathbf{v}'_{1,s_t+1}, \ldots, \mathbf{v}'_{t-1,s_t+1}, \ldots, \mathbf{v}'_{1s_1}, \ldots, \mathbf{v}'_{ks_1}\}.$$

Since

$$\mathbf{v}'_{11} = T(\mathbf{v}'_{21}), \quad \mathbf{v}'_{21} = T(\mathbf{v}'_{31}), \quad \ldots, \quad \mathbf{v}'_{t-1,1} = T(\mathbf{v}'_{t1}),$$

we have

$$\mathbf{v}'_{11} = T^{t-1}(\mathbf{v}'_{t1}), \quad \mathbf{v}'_{21} = T^{t-2}(\mathbf{v}'_{t1}), \quad \ldots, \quad \mathbf{v}'_{t-1,1} = T(\mathbf{v}'_{t1}),$$

and $\{\mathbf{v}'_{11}, \mathbf{v}'_{21}, \ldots, \mathbf{v}'_{t1}\}$ is the same as the cyclic basis $\mathcal{B}(\mathbf{v}'_{t1}, T)$ of $\mathbf{C}(\mathbf{v}'_{t1}, T)$. Similarly, those $\mathbf{v}'_{ij}$ with the same second subscript (i.e., with the same position in the segments of $\mathcal{A}'$) are the same as the vectors in the cyclic basis $\mathcal{B}(\mathbf{v}'_{mj}, T)$ of $\mathbf{C}(\mathbf{v}'_{mj}, T)$:

$$\mathcal{B}(\mathbf{v}'_{mj}, T) = \{\mathbf{v}'_{1j}, \mathbf{v}'_{2j}, \ldots, \mathbf{v}'_{mj}\},$$

where m is the number of the last segment in $\mathcal{A}'$ that has at least j elements. Thus

$$\mathcal{B} = \{\mathcal{B}(\mathbf{v}'_{t1}, T), \ldots, \mathcal{B}(\mathbf{v}'_{ts_t}, T), \mathcal{B}(\mathbf{v}'_{t-1,s_t+1}, T), \ldots, \mathcal{B}(\mathbf{v}'_{1,s_2+1}, T), \ldots, \mathcal{B}(\mathbf{v}'_{1s_1}, T)\},$$

where $\mathcal{B}(\mathbf{v}'_{ij}, T)$ is the cyclic basis of $\mathbf{C}(\mathbf{v}'_{ij}, T)$. By Theorem 10.25, the matrix of the restriction of T to $\mathbf{C}(\mathbf{v}'_{ij}, T)$ is of the form

$$A_j = \begin{bmatrix} 0 & 1 & 0 & \cdots & 0 \\ 0 & 0 & 1 & \cdots & 0 \\ \vdots & \vdots & \vdots & \ddots & \vdots \\ 0 & 0 & 0 & \cdots & 1 \\ 0 & 0 & 0 & \cdots & 0 \end{bmatrix}.$$

Since each $\mathbf{C}(\mathbf{v}'_{ij}, T)$ is T-invariant and $\mathbf{V}$ is the direct sum of the $\mathbf{C}(\mathbf{v}'_{ij}, T)$, the matrix of T with respect to $\mathcal{B}$ is of the form A given in the statement of the theorem. There are s_1 submatrices A_j in A, one for each $\mathbf{C}(\mathbf{v}'_{ij}, T)$, and $s_1 = \dim(\mathbf{W}_1) = \text{nullity}(T)$. The matrix A_1 is of order t, since $\mathcal{B}(\mathbf{v}'_{t1}, T)$ has t elements. The inequality

$$\text{order}(A_i) \geq \text{order}(A_{i+1})$$

is clear from the construction of the basis $\mathcal{B}$.

Now T determines the subspaces $\mathbf{W}_j$, and the dimensions of the $\mathbf{W}_j$ determine the matrices A_j. Hence A is uniquely determined by T. ■ ■ ■

EXAMPLE 1 Let T be the linear transformation of $\mathbf{R}^5$ that has matrix

$$M = \begin{bmatrix} 3 & 0 & -2 & -3 & 1 \\ 4 & -1 & -1 & -4 & 0 \\ 4 & -1 & -1 & -4 & 0 \\ 1 & 0 & -1 & -1 & 1 \\ 2 & -1 & 0 & -2 & 0 \end{bmatrix}$$

relative to the standard basis. It is easily verified that T is nilpotent of index 3. We follow the proof of Theorem 10.28 to obtain a basis $\mathcal{B}$ of $\mathbf{R}^5$ such that T has a matrix A of the form described in the theorem.

The row-echelon form of M leads to the basis $\{(1, 0, 0, 1, 0), (1, 2, 2, 0, 1)\}$ of the kernel $\mathbf{W}_1$ of T. The row-echelon form for M^2 shows that the kernel $\mathbf{W}_2$ of T^2 is the set of all $(x_1, x_2, x_3, x_4, x_5)$ in $\mathbf{R}^5$ such that $x_2 = x_3$. The basis of $\mathbf{W}_1$ can then be extended to the basis

$$\{(1, 0, 0, 1, 0), (1, 2, 2, 0, 1); (0, 1, 1, 0, 0), (0, 1, 1, 0, 1)\}$$

of $\mathbf{W}_2$. (The semicolon separates the segments in the basis.) This basis can in turn be extended to the basis

$$\mathcal{A} = \{(1, 0, 0, 1, 0), (1, 2, 2, 0, 1); (0, 1, 1, 0, 0), (0, 1, 1, 0, 1); (0, 0, 1, 0, 0)\}$$

of the desired type.

As the first step in modifying the basis, we find $T(0, 0, 1, 0, 0) = (-2, -1, -1, -1, 0)$ and replace the segment $(0, 1, 1, 0, 0), (0, 1, 1, 0, 1)$ in the basis of $\mathbf{W}_2$ by the vector $(-2, -1, -1, -1, 0)$. The vector $(0, 1, 1, 0, 0)$ can be used to extend this set to the basis

$$\{(1, 0, 0, 1, 0), (1, 2, 2, 0, 1); (-2, -1, -1, -1, 0), (0, 1, 1, 0, 0)\}$$

of $\mathbf{W}_2$. [Incidentally, the vector $(0, 1, 1, 0, 1)$ cannot be used.] Next we replace the segment $(1, 0, 0, 1, 0), (1, 2, 2, 0, 1)$ by $T(-2, -1, -1, -1, 0), T(0, 1, 1, 0, 0)$ to obtain the basis

$$\mathcal{A}' = \{(-1, -2, -2, 0, -1), (-2, -2, -2, -1, -1);$$
$$(-2, -1, -1, -1, 0), (0, 1, 1, 0, 0); (0, 0, 1, 0, 0)\}$$

as described in the proof of the theorem. By rearrangement of the vectors according to their positions in the segments, we obtain the basis

$$\mathcal{B} = \{(-1, -2, -2, 0, -1), (-2, -1, -1, -1, 0), (0, 0, 1, 0, 0);$$
$$(-2, -2, -2, -1, -1), (0, 1, 1, 0, 0)\}.$$

It is easily verified that the matrix of T relative to $\mathcal{B}$ is

$$A = \begin{bmatrix} A_1 & \mathbf{0} \\ \mathbf{0} & A_2 \end{bmatrix}, \quad \text{where} \quad A_1 = \begin{bmatrix} 0 & 1 & 0 \\ 0 & 0 & 1 \\ 0 & 0 & 0 \end{bmatrix}, \quad A_2 = \begin{bmatrix} 0 & 1 \\ 0 & 0 \end{bmatrix}. \quad \blacksquare$$

10.4 Exercises

1. Let T be the nilpotent transformation of $\mathbf{R}^n$ that has the given matrix relative to the standard basis. Find a basis $\mathcal{B}$ of $\mathbf{R}^n$ that satisfies the conditions of Theorem 10.28, and exhibit the matrix A described in that theorem.

(a) $\begin{bmatrix} 4 & -1 & -19 & 3 \\ -3 & 0 & 12 & -2 \\ 1 & 0 & -4 & 1 \\ 0 & 0 & 0 & 0 \end{bmatrix}$ (b) $\begin{bmatrix} 2 & 0 & -1 & -2 \\ 4 & -1 & -1 & -4 \\ 4 & -1 & -1 & -4 \\ 0 & 0 & 0 & 0 \end{bmatrix}$

(c) $\begin{bmatrix} -1 & -1 & -2 & 0 & 1 \\ 0 & 1 & 3 & 0 & -2 \\ 2 & 1 & 1 & 0 & 0 \\ 1 & 0 & 0 & 0 & 1 \\ 1 & 1 & 2 & 0 & -1 \end{bmatrix}$ (d) $\begin{bmatrix} 3 & 0 & -2 & -2 & 1 \\ 4 & -1 & -1 & -2 & 0 \\ 4 & -1 & -1 & -3 & 0 \\ 1 & 0 & -1 & -1 & 1 \\ 2 & -1 & 0 & -2 & 0 \end{bmatrix}$

2. In each part below, let T be the nilpotent transformation in the corresponding part of Problem 1. Find $\mathcal{B}(\mathbf{v}, T)$ for the given $\mathbf{v}$, and find the matrix relative to $\mathcal{B}(\mathbf{v}, T)$ of the restriction of T to $\mathbf{C}(\mathbf{v}, T)$.

(a) $\mathbf{v} = (1, 1, 0, -1)$ (b) $\mathbf{v} = (1, 1, 1, 1)$

(c) $\mathbf{v} = (-1, -1, 1, 0, 1)$ (d) $\mathbf{v} = (1, 2, -1, 1, 2)$

3. Prove that the subspace $\mathbf{C}(\mathbf{v}, T)$ in Definition 10.24 is T-invariant.

4. Prove that T is nilpotent of index t if and only if the minimum polynomial of T is x^t.

5. Prove that T is nilpotent if and only if the characteristic polynomial of T is x^n.

6. The **index** of nilpotency of a nilpotent matrix A is the least positive integer t such that $A^t = \mathbf{0}$. Prove that if A is nilpotent of index t, then any matrix that is similar to A is also nilpotent of index t.

7. Show that the zero operator Z is the only normal linear operator on a unitary space $\mathbf{V}$ that is nilpotent.

8. Prove that the cyclic subspace of $\mathbf{v}$ relative to T is the set of all vectors of the form $g(T)(\mathbf{v})$, where $g(T)$ is a polynomial in T.

9. Prove that $\mathbf{C}(\mathbf{v}, T)$ is the intersection of all T-invariant subspaces that contain $\mathbf{v}$.

10.5 The Jordan Canonical Form

Throughout the development, a good portion of our effort has been connected in one way or another with representations of linear operators by diagonal matrices, and this is in keeping with the importance of the topic. A person's knowledge of this area might be regarded as a good rough measure of his knowledge of the fundamentals of linear algebra.

Not all linear operators can be represented by a diagonal matrix, but a standard form can be obtained that comes very close to being a diagonal matrix. This standard form is known as the *Jordan canonical form.* In a sense, it is the ultimate result concerning diagonalization, and this is the main reason for its inclusion here. There are situations in which it is of great value.

We begin with the definition of a Jordan matrix.

10.29 DEFINITION

The **Jordan matrix** of order k with eigenvalue λ is the $k \times k$ matrix

$$
\begin{bmatrix}
\lambda & 1 & 0 & \cdots & 0 & 0 \\
0 & \lambda & 1 & \cdots & 0 & 0 \\
0 & 0 & \lambda & \cdots & 0 & 0 \\
\vdots & \vdots & \vdots & \ddots & \vdots & \vdots \\
0 & 0 & 0 & \cdots & \lambda & 1 \\
0 & 0 & 0 & \cdots & 0 & \lambda
\end{bmatrix}.
$$

10.30 THEOREM

Let T have characteristic polynomial

$$
f(x) = (-1)^n (x - \lambda_1)^{m_1} \cdots (x - \lambda_r)^{m_r}
$$

and minimal polynomial

$$
m(x) = (x - \lambda_1)^{t_1} \cdots (x - \lambda_r)^{t_r}
$$

over $\mathcal{F}$, and let n_i be the geometric multiplicity of λ_i. Then there exists a basis of **V** such that the matrix of T relative to this basis has the following form:

$$
J =
\begin{bmatrix}
J_1 & \mathbf{0} & \cdots & \mathbf{0} \\
\mathbf{0} & J_2 & \cdots & \mathbf{0} \\
\vdots & \vdots & \ddots & \vdots \\
\mathbf{0} & \mathbf{0} & \cdots & J_r
\end{bmatrix},
$$

with J_i a square matrix of order m_i given by

$$
J_i =
\begin{bmatrix}
J_{i1} & \mathbf{0} & \cdots & \mathbf{0} \\
\mathbf{0} & J_{i2} & \cdots & \mathbf{0} \\
\vdots & \vdots & \ddots & \vdots \\
\mathbf{0} & \mathbf{0} & \cdots & J_{in_i}
\end{bmatrix},
$$

where each J_{ik} is a Jordan matrix with eigenvalue λ_i such that J_{i1} has order t_i and order $(J_{ik}) \geq$ order $(J_{i,k+1})$ for all k. For a prescribed ordering of the eigenvalues, the matrix J is uniquely determined by T and is called the **Jordan canonical matrix** for T.

Proof Let $T = D + N$ be the expression of T as the sum of a diagonalizable transformation D and a nilpotent transformation N as in the proof of Theorem 10.23. With the same notation as used in Section 10.3, let $\mathcal{K}_i$ denote the kernel of $(T - \lambda_i)^{t_i}$, and let $F_i = p_i(T)$. Then $\mathbf{V} = \mathcal{K}_1 \oplus \cdots \oplus \mathcal{K}_r$, and F_i is the projection of $\mathbf{V}$ onto $\mathcal{K}_i$ along $\sum_{j \neq i} \mathcal{K}_j$, by Theorem 10.16. Hence any nonzero vector in $\mathcal{K}_i$ is an eigenvector of F_i corresponding to the eigenvalue 1, and the restriction of F_i to $\mathcal{K}_j$ is the zero transformation if $i \neq j$. For the time being, let d_i denote the dimension of $\mathcal{K}_i$. With every choice of bases $\mathcal{B}_i = \{\mathbf{u}_{i1}, \dots, \mathbf{u}_{id_i}\}$ for the subspaces $\mathcal{K}_i$, the set

$$\mathcal{B} = \{\mathbf{u}_{11}, \dots, \mathbf{u}_{1d_1}, \mathbf{u}_{21}, \dots, \mathbf{u}_{2d_2}, \dots, \mathbf{u}_{r1}, \dots, \mathbf{u}_{rd_r}\}$$

is a basis of $\mathbf{V}$. Since $F_i(\mathcal{K}_i) = \mathcal{K}_i$ and $F_i(\mathcal{K}_j) = \{\mathbf{0}\}$ if $i \neq j$, each $\mathcal{K}_i$ is invariant under $D = \lambda_1 F_1 + \cdots + \lambda_r F_r$, and the restriction of D to $\mathcal{K}_i$ has matrix $\lambda_i I_{d_i}$ relative to $\mathcal{B}_i$. It follows from the discussion at the beginning of Section 10.4 that D has the matrix

$$K = \begin{bmatrix} \lambda_1 I_{d_1} & \mathbf{0} & \cdots & \mathbf{0} \\ \mathbf{0} & \lambda_2 I_{d_2} & \cdots & \mathbf{0} \\ \vdots & \vdots & \ddots & \vdots \\ \mathbf{0} & \mathbf{0} & \cdots & \lambda_r I_{d_r} \end{bmatrix}$$

relative to $\mathcal{B}$.

Consider now the matrix of N relative to a basis of this type. Since

$$(T - \lambda_i) F_i(\mathcal{K}_i) \subseteq (T - \lambda_i)(\mathcal{K}_i) \subseteq \mathcal{K}_i$$

and $(T - \lambda_i) F(\mathcal{K}_j) = \{\mathbf{0}\}$ if $i \neq j$, each $\mathcal{K}_i$ is invariant under

$$N = (T - \lambda_1) F_1 + \cdots + (T - \lambda_r) F_r.$$

Now the restriction N_i of N to $\mathcal{K}_i$ is the same as the restriction of $T - \lambda_i$ to $\mathcal{K}_i$. By part (c) of Theorem 10.18, the minimal polynomial of the restriction of T to $\mathcal{K}_i$ is $(x - \lambda_i)^{t_i}$. Therefore, the restriction of $T - \lambda_i$ to $\mathcal{K}_i$ has minimal polynomial x^{t_i}. This means that the minimal polynomial of N_i is x^{t_i}, and that N_i is nilpotent of index t_i. According to Theorem 10.28, there exists a basis $\mathcal{B}_i' = \{\mathbf{u}_{i1}', \dots, \mathbf{u}_{id_i}'\}$ of $\mathcal{K}_i$ such that N_i has a $d_i \times d_i$ matrix of the form

$$L_i = \begin{bmatrix} L_{i1} & \mathbf{0} & \cdots & \mathbf{0} \\ \mathbf{0} & L_{i2} & \cdots & \mathbf{0} \\ \vdots & \vdots & \ddots & \vdots \\ \mathbf{0} & \mathbf{0} & \cdots & L_{in_i} \end{bmatrix},$$

where each L_{ij} has the form

$$L_{ij} = \begin{bmatrix} 0 & 1 & 0 & \cdots & 0 \\ 0 & 0 & 1 & \cdots & 0 \\ \vdots & \vdots & \vdots & \ddots & \vdots \\ 0 & 0 & 0 & \cdots & 1 \\ 0 & 0 & 0 & \cdots & 0 \end{bmatrix},$$

L_{i1} has order t_i and order $(L_{ij}) \geq$ order $(L_{i,j+1})$ for all j. The number of diagonal blocks L_{ij} in L_i is the nullity of N_i. Since N_i is the same as the restriction of $T - \lambda_i$ to $\mathcal{K}_i$, and since the kernel of $T - \lambda_i$ is contained in $\mathcal{K}_i$, we have

$$\text{nullity } (N_i) = \text{nullity } (T - \lambda_i)$$
$$= \dim(T - \lambda_i)^{-1}(\mathbf{0})$$
$$= n_i.$$

Relative to the basis

$$\mathcal{B}' = \{\mathbf{u}'_{11}, \ldots, \mathbf{u}'_{1d_1}, \mathbf{u}'_{21}, \ldots, \mathbf{u}'_{2d_2}, \ldots, \mathbf{u}'_{r1}, \ldots, \mathbf{u}'_{rd_r}\},$$

N has the matrix

$$L = \begin{bmatrix} L_1 & \mathbf{0} & \cdots & \mathbf{0} \\ \mathbf{0} & L_2 & \cdots & \mathbf{0} \\ \vdots & \vdots & \ddots & \vdots \\ \mathbf{0} & \mathbf{0} & \cdots & L_r \end{bmatrix}$$

since each $\mathcal{K}_i$ is invariant under N. Thus T has the matrix

$$J = K + L = \begin{bmatrix} \lambda_1 I_{d_1} + L_1 & \mathbf{0} & \cdots & \mathbf{0} \\ \mathbf{0} & \lambda_2 I_{d_2} + L_2 & \cdots & \mathbf{0} \\ \vdots & \vdots & \ddots & \vdots \\ \mathbf{0} & \mathbf{0} & \cdots & \lambda_r I_{d_r} + L_r \end{bmatrix}$$

relative to $\mathcal{B}'$. Now J is an upper triangular matrix, so

$$\det(J - xI) = (-1)^n (x - \lambda_1)^{d_1} \cdots (x - \lambda_r)^{d_r}.$$

But $\det(J - xI) = f(x)$ since J represents T. Therefore $d_i = m_i$ for each i. Letting $J_i = \lambda_i I_{m_i} + L_i$, the matrix J is in the required form.

Now K is uniquely determined by $f(x)$ whenever the ordering of the eigenvalues is designated. By Theorem 10.28, N_i determines the matrix L_i, and a prescribed ordering of the eigenvalues then determines L. Hence $J = K + L$ is unique if the order of the eigenvalues is specified. ■ ■ ■

10.31
THEOREM

If an $n \times n$ matrix A over $\mathcal{F}$ has characteristic polynomial $f(x)$ and minimal polynomial $m(x)$ that factor over $\mathcal{F}$ as given in Theorem 10.30, then A is similar over $\mathcal{F}$ to a unique matrix J as described in that theorem. The matrix J is called the **Jordan canonical form** for A.

Proof Let A be a matrix that satisfies the given conditions. With any given choices of an n-dimensional vector space $\mathbf{V}$ over $\mathcal{F}$ and a basis of $\mathbf{V}$, A determines a unique linear operator T on $\mathbf{V}$. Any such linear operator has $f(x)$ as its characteristic polynomial and $m(x)$

as its minimal polynomial. Since any n-dimensional vector space over $\mathcal{F}$ is isomorphic to $\mathcal{F}^n$, we may assume without loss of generality that T is a linear operator on $\mathcal{F}^n$. The projections F_i, the subspaces $\mathcal{K}_i$, and the operators D and N are determined independently of the choice of basis in $\mathcal{F}^n$. Thus the matrix J is uniquely determined by A. ■ ■ ■

The proof of Theorem 10.30 furnishes a method for obtaining the basis and the matrix J described in that theorem, but several short-cuts can be made in the procedure. This is illustrated in the following example.

EXAMPLE 1 Consider the linear operator T on $\mathbf{R}^6$ that has the matrix

$$A = \begin{bmatrix} -1 & 1 & 0 & 0 & 0 & 0 \\ -1 & -1 & 1 & 0 & 0 & 0 \\ 0 & 1 & -1 & 0 & 0 & 0 \\ -2 & 1 & 1 & -1 & 1 & 0 \\ 3 & 0 & -1 & -1 & -3 & 0 \\ -1 & 1 & 1 & 0 & 0 & -2 \end{bmatrix}$$

relative to $\mathcal{E}_6$. We shall (a) find a basis $\mathcal{B}'$ of $\mathbf{R}^6$ such that T has the Jordan canonical matrix J relative to $\mathcal{B}'$, (b) determine the matrix J, and (c) find a matrix P such that $P^{-1}AP = J$.

The characteristic polynomial of A is $f(x) = (x + 1)^3(x + 2)^3$, so $\lambda_1 = -1$ and $\lambda_2 = -2$ are the distinct eigenvalues of T. It is actually not necessary to find the minimal polynomial.

Our first step is to find a basis $\mathcal{B}'_1$ for the kernel $\mathcal{K}_1$ of $(T + 1)^{t_1}$ that is of the type in the proof of Theorem 10.30. We have seen that the restriction N_1 of N to $\mathcal{K}_1$ is the same as the restriction of $T - \lambda_1$ to $\mathcal{K}_1$. It follows that $(N_1)^j$ is the restriction of $(T - \lambda_1)^j$ to $\mathcal{K}_1$ for each positive integer j. Since the kernel $\mathbf{W}_j$ of $(T - \lambda_1)^j$ is contained in the kernel $\mathbf{W}_{j+1}$ of $(T - \lambda_1)^{j+1}$, we begin by finding a basis of the kernel $\mathbf{W}_1$ of $T - \lambda_1$, extending to a basis of $\mathbf{W}_2$, and so on. Since the kernel of $T - \lambda_1$ is contained in $\mathcal{K}_1$, the kernel of the restriction of $T - \lambda_1$ to $\mathcal{K}_1$ is the same as the kernel of $T - \lambda_1$.

The reduction of $A - \lambda_1 I = A + I$ to row-echelon form yields $\{(1, 0, 1, 0, 1, 0)\}$ the basis of the kernel $\mathbf{W}_1$. By use of the row-echelon form of $(A + I)^2$, this extends to the basis $\{(1, 0, 1, 0, 1, 0); (1, 1, 1, 1, 0, 1)\}$ of $\mathbf{W}_2$. Repetition of this procedure with $(A + I)^3$ produces the basis

$$\{(1, 0, 1, 0, 1, 0); (1, 1, 1, 1, 0, 1); (0, 0, 1, 0, 0, 0)\}$$

of $\mathbf{W}_3$. Upon finding the dimension of $\mathbf{W}_4$, we discover that $\mathbf{W}_3 = \mathbf{W}_4$. By Lemma 10.26, this indicates that the index of N_1 is $t_1 = 3$. Following the procedure of Theorem 10.28, we replace $(1, 1, 1, 1, 0, 1)$ by

$$N_1(0, 0, 1, 0, 0, 0) = (T - \lambda_1)(0, 0, 1, 0, 0, 0)$$

and

$$(1, 0, 1, 0, 1, 0) \quad \text{by} \quad N_1^2(0, 0, 1, 0, 0, 0)$$

to obtain

$$\mathcal{B}_1' = \{(1,0,1,0,1,0), (0,1,0,1,-1,1), (0,0,1,0,0,0)\}.$$

The matrix of N_1 relative to this basis is

$$L_1 = \begin{bmatrix} 0 & 1 & 0 \\ 0 & 0 & 1 \\ 0 & 0 & 0 \end{bmatrix}.$$

Following the same procedure with the eigenvalue $\lambda_2 = -2$, and letting $\mathbf{W}_j = \ker(N_2^j) = \ker(T+2)^j$, we find

$$\mathbf{W}_1 = \langle (0,0,0,1,1,0), (0,0,0,0,0,1) \rangle,$$

$$\mathbf{W}_2 = \langle (0,0,0,1,1,0), (0,0,0,0,0,1); (0,0,0,0,1,0) \rangle.$$

We find that $\mathbf{W}_2 = \mathbf{W}_3$, and this indicates that $t_2 = 2$. We then replace the first vector in the basis of $\mathbf{W}_2$ by

$$N_2(0,0,0,0,1,0) = (T+2)(0,0,0,0,1,0) = (0,0,0,1,-1,0)$$

and extend this to a basis of $\mathbf{W}_1$ to obtain the basis

$$\{(0,0,0,1,-1,0), (0,0,0,0,0,1); (0,0,0,0,1,0)\}$$

of $\mathbf{W}_2$. Rearrangement of the vectors in this basis yields

$$\mathcal{B}_2' = \{(0,0,0,1,-1,0), (0,0,0,0,1,0), (0,0,0,0,0,1)\}.$$

The matrix of N_2 relative to $\mathcal{B}_2'$ is

$$L_2 = \begin{bmatrix} 0 & 1 & 0 \\ 0 & 0 & 0 \\ 0 & 0 & 0 \end{bmatrix}.$$

The desired basis $\mathcal{B}'$ is thus given by

$$\mathcal{B}' = \{(1,0,1,0,1,0), (0,1,0,1,-1,1), (0,0,1,0,0,0),$$
$$(0,0,0,1,-1,0), (0,0,0,0,1,0), (0,0,0,0,0,1)\},$$

and the matrix of T relative to this basis is

$$J = \begin{bmatrix} \lambda_1 I_{m_1} + L_1 & \mathbf{0} \\ \mathbf{0} & \lambda_2 I_{m_2} + L_2 \end{bmatrix}$$

$$= \begin{bmatrix} -1 & 1 & 0 & 0 & 0 & 0 \\ 0 & -1 & 1 & 0 & 0 & 0 \\ 0 & 0 & -1 & 0 & 0 & 0 \\ 0 & 0 & 0 & -2 & 1 & 0 \\ 0 & 0 & 0 & 0 & -2 & 0 \\ 0 & 0 & 0 & 0 & 0 & -2 \end{bmatrix}.$$

The matrix of transition from $\mathcal{E}_6$ to $\mathcal{B}'$ is

$$P = \begin{bmatrix} 1 & 0 & 0 & 0 & 0 & 0 \\ 0 & 1 & 0 & 0 & 0 & 0 \\ 1 & 0 & 1 & 0 & 0 & 0 \\ 0 & 1 & 0 & 1 & 0 & 0 \\ 1 & -1 & 0 & -1 & 1 & 0 \\ 0 & 1 & 0 & 0 & 0 & 1 \end{bmatrix}.$$

It is easily verified that $P^{-1}AP = J$. ∎

10.5 Exercises

1. Write the transformation T in Example 1 of this section as $T = D + N$, where D is diagonalizable and N is nilpotent.

2. Suppose a matrix A has characteristic polynomial $f(x)$ and minimal polynomial $m(x)$. For each of the following, list all possible Jordan canonical forms for A.
   (a) $f(x) = (-1)^7(x - 2)^4(x - 7)^3$, $m(x) = (x - 2)^2(x - 7)^2$, $\lambda_1 = 2, \lambda_2 = 7$
   (b) $f(x) = (-1)^6(x + 1)^3(x - 1)^2(x + 6)$, $m(x) = (x + 1)(x - 1)^2(x + 6)$, $\lambda_1 = 1$, $\lambda_2 = -1, \lambda_3 = -6$
   (c) $f(x) = (-1)^6(x + 3)^5(x - 4)$, $m(x) = (x + 3)^3(x - 4)$, $\lambda_1 = -3, \lambda_2 = 4$
   (d) $f(x) = (-1)^7(x + 5)^5(x + 4)^2$, $m(x) = (x + 5)^4(x + 4)^2$, $\lambda_1 = -5, \lambda_2 = -4$
   (e) $f(x) = (-1)^6(x - 8)^6$, $m(x) = (x - 8)^3$, $\lambda_1 = 8$
   (f) $f(x) = (-1)^7(x - 2)^4(x + 2)^3$, $m(x) = (x - 2)^3(x + 2)^2$, $\lambda_1 = 2, \lambda_2 = -2$

3. For each of the following matrices A, its minimum polynomial $m(x)$ is given. Find an invertible matrix P such that $P^{-1}AP$ is in Jordan canonical form.
   (a) $A = \begin{bmatrix} 2 & 1 & 0 \\ 0 & 2 & 0 \\ 2 & 3 & 1 \end{bmatrix}$, $m(x) = (x - 2)^2(x - 1)$
   (b) $A = \begin{bmatrix} 1 & -4 & 0 & 0 \\ 1 & -3 & 0 & 0 \\ -1 & 2 & -1 & 0 \\ -3 & 6 & 0 & 2 \end{bmatrix}$, $m(x) = (x + 1)^2(x - 2)$

4. Find the Jordan canonical form for each matrix A.
   (a) $A = \begin{bmatrix} 7 & 3 & 3 & 2 \\ 0 & 1 & 2 & -4 \\ -8 & -4 & -5 & 0 \\ 2 & 1 & 1 & 3 \end{bmatrix}$
   (b) $A = \begin{bmatrix} 8 & 5 & 6 & 0 \\ 0 & -2 & 0 & 0 \\ -10 & -5 & -8 & 0 \\ 2 & 1 & 1 & 2 \end{bmatrix}$
   (c) $A = \begin{bmatrix} 1 & 1 & 0 & 0 \\ 0 & 1 & 1 & 0 \\ 0 & 0 & 1 & 1 \\ -1 & 0 & 2 & 1 \end{bmatrix}$
   (d) $A = \begin{bmatrix} 3 & 1 & 0 & 0 \\ 0 & 0 & 1 & 0 \\ -1 & -3 & 3 & 0 \\ 1 & 3 & -1 & 2 \end{bmatrix}$

5. For each part of Problem 4, let T be the linear transformation of $\mathbf{R}^4$ that has the matrix A relative to $\mathcal{E}_4$. Find a basis of $\mathbf{R}^4$ such that the matrix of T relative to this basis is the Jordan canonical matrix J for T, and write down a matrix P such that $P^{-1}AP = J$.

6. Use the results of Problems 4 and 5 to write each matrix A in Problem 4 as the sum of a diagonalizable matrix and a nilpotent matrix.

11 Numerical Methods

Most of the applications of linear algebra involve either the solution of systems of linear equations or the determination of the eigenvalues of a matrix. In many of these situations, it is not practical to attempt an exact solution to the problem, but rather to obtain an approximate solution within certain bounds of accuracy. Accordingly, numerical methods have been devised to obtain these approximate solutions.

A great variety of numerical methods has been developed. Those presented in this chapter are in many cases not the most practical methods available. They are chosen because they depend on the elementary theory and illustrate the fundamental techniques. Procedures which are quite sophisticated have been developed in connection with the use of computers, but no attempt is made here to reach such an advanced level. For example, a great deal of work has been done on error analysis and estimates of rates of convergence, but these topics are not touched on here.

In this chapter we consider only vectors in $\mathbf{R}^n$. A more general setting for the work is possible in most instances, but there is no advantage for us in that setting. In some of our work, there is a geometric rationale behind a method that is of much intuitive value. In order to strengthen this geometric intuition, we shall use the words *point* and *vector* interchangeably for elements in $\mathbf{R}^n$.

It is desirable to use the term *norm* here in a more general sense than was done in Chapter 9. The more general meaning of the term includes functions that are not derived from an inner product.

11.1 Sequences and Series of Vectors

We begin with the generalization of the term *norm*.

11.1 DEFINITION

A **norm** for $\mathbf{R}^n$ is a mapping of vectors $\mathbf{v}$ in $\mathbf{R}^n$ onto scalars $\|\mathbf{v}\|$ in $\mathbf{R}$ that has the following properties:

(i) $\|\mathbf{v}\| > 0$ if $\mathbf{v} \neq \mathbf{0}$, and $\|\mathbf{0}\| = 0$;
(ii) $\|a\mathbf{v}\| = |a| \cdot \|\mathbf{v}\|$ for any $a \in \mathbf{R}$;
(iii) $\|\mathbf{u} + \mathbf{v}\| \leq \|\mathbf{u}\| + \|\mathbf{v}\|$ (**The Triangle Inequality**).

These three properties are the same as those given in Theorem 9.5. Thus the use of the term *norm* for the function $\|\mathbf{v}\| = \sqrt{(\mathbf{v}, \mathbf{v})}$ on an inner product space is consistent with this definition. Moreover, the examples and exercises in Section 9.2 furnish familiar examples of norms on $\mathbf{R}^n$. It should be kept in mind, however, that there is not necessarily any connection between a certain norm on $\mathbf{R}^n$ and the inner products on $\mathbf{R}^n$ (see Problem 16). We shall use $d(\mathbf{v})$ to denote the standard norm on $\mathbf{R}^n$: If $\mathbf{v} = (x_1, x_2, \ldots, x_n)$, then $d(\mathbf{v}) = (x_1^2 + x_2^2 + \cdots + x_n^2)^{1/2}$.

Two important consequences of the properties in Definition 11.1 are that

$$\|\mathbf{u} - \mathbf{v}\| \geq \|\mathbf{u}\| - \|\mathbf{v}\|$$

and

$$\left\| \sum_{k=1}^{r} \mathbf{v}_k \right\| \leq \sum_{k=1}^{r} \|\mathbf{v}_k\|.$$

The proofs are requested in Problems 4 and 5.

The next theorem furnishes two important examples of norms on $\mathbf{R}^n$.

11.2 THEOREM

Each of the functions defined below is a norm for $\mathbf{R}^n$:

1. $\|(x_1, x_2, \ldots, x_n)\| = \max_j\{|x_j|\}$;
2. $\|(x_1, x_2, \ldots, x_n)\| = \sum_{j=1}^{n} |x_j|$.

In (1), $\max_j\{|x_j|\}$ denotes the largest number in the set $\{|x_1|, |x_2|, \ldots, |x_n|\}$.

Proof See Problem 6. ∎

EXAMPLE 1 Consider the vector $\mathbf{v} = (1, 4, -1)$ in $\mathbf{R}^3$. With the standard norm on $\mathbf{R}^3$, we have

$$\|\mathbf{v}\| = \sqrt{x_1^2 + x_2^2 + x_3^2} = \sqrt{18} = 3\sqrt{2}.$$

Using the norm given by (1) in Theorem 11.2, we get

$$\|\mathbf{v}\| = \max\{|x_1|, |x_2|, |x_3|\} = 4,$$

while the norm given by (2) in Theorem 11.2 yields

$$\|\mathbf{v}\| = |x_1| + |x_2| + |x_3| = 6.$$ ∎

11.3 DEFINITION

A **sequence** of vectors in $\mathbf{R}^n$ is a set $\{\mathbf{v}_k\}$ of vectors in $\mathbf{R}^n$ that is indexed by the set of all positive integers k.

That is, a sequence of vectors in $\mathbf{R}^n$ is a function that has values in $\mathbf{R}^n$ and has the set of all positive integers as its domain. As with sequences of numbers, we shall also consider sets of vectors $\mathbf{v}_0, \mathbf{v}_1, \mathbf{v}_2, \ldots$ that are indexed by the set of all nonnegative integers as sequences.

**11.4
DEFINITION**

Let $\mathbf{v}_k = (x_{1k}, x_{2k}, \ldots, x_{nk})$, $k = 1, 2, \ldots$ be a sequence of vectors in $\mathbf{R}^n$. The sequence $\{\mathbf{v}_k\}$ **converges** to $\mathbf{b} = (b_1, b_2, \ldots, b_n)$ in $\mathbf{R}^n$ if and only if

$$\lim_{k \to \infty} x_{ik} = b_i \quad \text{for } i = 1, 2, \ldots, n.$$

When $\{\mathbf{v}_k\}$ converges to $\mathbf{b}$, the vector $\mathbf{b}$ is called the **limit** of the sequence $\{\mathbf{v}_k\}$, and we write $\lim_{k \to \infty} \mathbf{v}_k = \mathbf{b}$ or $\mathbf{v}_k \to \mathbf{b}$.

A sequence of vectors that has a limit is called a **convergent** sequence. It is left as an exercise to prove that the limit of a convergent sequence of vectors in $\mathbf{R}^n$ is unique (Problem 7). A sequence that does not converge is said to **diverge**, or to be **divergent**.

EXAMPLE 2 For the sequence $\{\mathbf{v}_k\}$ in $\mathbf{R}^3$ with

$$\mathbf{v}_k = \left(e^{-k}, \frac{3k^2 + 2k}{k^2 - k + 1}, \cos(k^{-1}) \right),$$

we have

$$\lim_{k \to \infty} \mathbf{v}_k = (0, 3, 1)$$

since

$$\lim_{k \to \infty} e^{-k} = \lim_{k \to \infty} \frac{1}{e^k} = 0,$$

$$\lim_{k \to \infty} \frac{3k^2 + 2k}{k^2 - k + 1} = \lim_{k \to \infty} \frac{6k + 2}{2k - 1} = 3$$

and

$$\lim_{k \to \infty} \cos(k^{-1}) = \lim_{k \to \infty} \cos\left(\frac{1}{k}\right) = \cos 0 = 1. \qquad \blacksquare$$

The proof of the next theorem is left as an exercise in Problem 8.

**11.5
THEOREM**

Let $\mathcal{A} = \{\mathbf{u}_1, \mathbf{u}_2, \ldots, \mathbf{u}_n\}$ be a basis of $\mathbf{R}^n$. A sequence $\{\mathbf{v}_k\}$ of vectors $\mathbf{v}_k = \sum_{i=1}^{n} y_{ik}\mathbf{u}_i$ converges to the vector $\mathbf{a} = \sum_{i=1}^{n} a_i\mathbf{u}_i$ if and only if $\lim_{k \to \infty} y_{ik} = a_i$ for $i = 1, 2, \ldots, n$.

Our next result establishes part of an important connection between norms and convergence of sequences in $\mathbf{R}^n$. The notation $\|\mathbf{v}\|$ will indicate an arbitrary norm on $\mathbf{R}^n$ for the remainder of this chapter.

**11.6
THEOREM**

If the sequence $\{\mathbf{v}_k\}$ converges to the zero vector in $\mathbf{R}^n$, then

$$\lim_{k \to \infty} \|\mathbf{v}_k\| = 0.$$

Proof Assume that $\{\mathbf{v}_k\}$ converges to $\mathbf{0}$ in $\mathbf{R}^n$. Let $\mathbf{v}_k = (x_{1k}, x_{2k}, \dots, x_{nk}) = \sum_{i=1}^{n} x_{ik}\mathbf{e}_i$, where $\mathcal{E}_n = \{\mathbf{e}_1, \mathbf{e}_2, \dots, \mathbf{e}_n\}$ denotes the standard basis of $\mathbf{R}^n$. Then we have

$$\|\mathbf{v}_k\| \leq \sum_{i=1}^{n} \|x_{ik}\mathbf{e}_i\| \qquad \text{by Problem 5}$$

$$= \sum_{i=1}^{n} |x_{ik}| \cdot \|\mathbf{e}_i\| \qquad \text{by Definition 11.1 (ii)}$$

$$\leq b \sum_{i=1}^{n} |x_{ik}|$$

where $b = \max_i\{\|\mathbf{e}_i\|\}$. From Theorem 11.5, we know that $\lim_{k \to \infty} x_{ik} = 0$ for $i = 1, 2, \dots, n$. Thus

$$\lim_{k \to \infty} b \sum_{i=1}^{n} |x_{ik}| = b \sum_{i=1}^{n} \lim_{k \to \infty} |x_{ik}| = 0,$$

and this implies that $\lim_{k \to \infty} \|\mathbf{v}_k\| = 0$. ■■■

The next objective in this section is to prove the converse of Theorem 11.6. Our proof makes use of the concept of continuity of a mapping from $\mathbf{R}^n$ into $\mathbf{R}$: A mapping f of $\mathbf{R}^n$ into $\mathbf{R}$ is **continuous** at $(a_1, a_2, \dots, a_n)$ in $\mathbf{R}^n$ if, for every given $\epsilon > 0$, there is a corresponding $\delta > 0$ such that

$$|f(x_1, x_2, \dots, x_n) - f(a_1, a_2, \dots, a_n)| < \epsilon$$

whenever $\sqrt{(x_1 - a_1)^2 + (x_2 - a_2)^2 + \cdots + (x_n - a_n)^2} < \delta$.

We shall need an important result concerning the extreme values (maxima and minima) of a function that is continuous on a closed and bounded subset of $\mathbf{R}^n$. By a **closed** subset of $\mathbf{R}^n$, we mean a subset $\mathcal{S}$ of $\mathbf{R}^n$ that contains all of its limit points. That is, if there is a sequence of elements $\mathbf{v}_k$ of $\mathcal{S}$ such that $\mathbf{v}_k \to \mathbf{v}$, then $\mathbf{v}$ is also in $\mathcal{S}$. A subset $\mathcal{S}$ of $\mathbf{R}^n$ is **bounded** if there exists a fixed number r such that $d(\mathbf{v}) \leq r$ for every $\mathbf{v}$ in $\mathcal{S}$. The result referred to before states that a function which is continuous at each point of a closed and bounded subset $\mathcal{S}$ of $\mathbf{R}^n$ takes on a minimum value and a maximum value on $\mathcal{S}$. In other words, there are points $\mathbf{u}_1$ and $\mathbf{u}_2$ in $\mathcal{S}$ such that $f(\mathbf{u}_1) \leq f(\mathbf{v}) \leq f(\mathbf{u}_2)$ for all $\mathbf{v}$ in $\mathcal{S}$.

| 11.7 LEMMA | For any fixed norm $\|\mathbf{v}\|$ on $\mathbf{R}^n$, the function f defined on $\mathbf{R}^n$ by $$f(x_1, x_2, \ldots, x_n) = \|(x_1, x_2, \ldots, x_n)\|$$ is continuous at each point of $\mathbf{R}^n$. |

Proof Let $(a_1, a_2, \ldots, a_n)$ be in $\mathbf{R}^n$, and let $\epsilon > 0$ be given. Let $b = \max_i\{\|\mathbf{e}_i\|\}$, and choose δ such that $\delta < \epsilon/(bn)$. Whenever

$$\sqrt{(x_1 - a_1)^2 + (x_2 - a_2)^2 + \cdots + (x_n - a_n)^2} < \delta,$$

we have

$$|f(x_1, x_2, \ldots, x_n) - f(a_1, a_2, \ldots, a_n)|$$

$$= \left\| \left\| \sum_{i=1}^{n} x_i \mathbf{e}_i \right\| - \left\| \sum_{i=1}^{n} a_i \mathbf{e}_i \right\| \right\|$$

$$\leq \left\| \sum_{i=1}^{n} (x_i - a_i) \mathbf{e}_i \right\| \qquad \text{(Prob. 4)}$$

$$\leq |x_1 - a_1| \|\mathbf{e}_1\| + \cdots + |x_n - a_n| \|\mathbf{e}_n\| \qquad \text{(Prob. 5)}$$

$$\leq b(|x_1 - a_1| + |x_2 - a_2| + \cdots + |x_n - a_n|)$$

$$\leq bn \cdot \max_i\{|x_i - a_i|\}$$

$$\leq bn\sqrt{(x_1 - a_1)^2 + (x_2 - a_2)^2 + \cdots + (x_n - a_n)^2}$$

$$< bn\left(\frac{\epsilon}{bn}\right)$$

$$= \epsilon$$

Thus, f is continuous at $(a_1, a_2, \ldots, a_n)$. ∎

| 11.8 LEMMA | For any fixed norm $\|\mathbf{v}\|$ on $\mathbf{R}^n$, there exists a number $b > 0$ such that $$b \cdot d(\mathbf{v}) \leq \|\mathbf{v}\|$$ for all $\mathbf{v}$ in $\mathbf{R}^n$. |

Proof If $\mathbf{v} = \mathbf{0}$, the inequality is true for every choice of b.

Consider next the set S which consists of all those vectors $\mathbf{v}$ in $\mathbf{R}^n$ such that $d(\mathbf{v}) = 1$. The proofs that S is closed and bounded are left as exercises (Problems 10 and 11). The function f defined by $f(\mathbf{v}) = \|\mathbf{v}\|$ is continuous on S by Lemma 11.7, and therefore has a minimum value $b \geq 0$ on the set $\mathcal{S}$. Moreover, $b > 0$ since $b = 0$ would imply $\|\mathbf{v}\| = 0$ for some $\mathbf{v} \neq \mathbf{0}$, contradicting (i) in Definition 11.1. Thus we have a positive b such that

$$b \cdot d(\mathbf{v}) = b \leq \|\mathbf{v}\|$$

for all $\mathbf{v}$ in $\mathcal{S}$.

Now let $\mathbf{v}$ be an arbitrary nonzero vector in $\mathbf{R}^n$. Since the vector $\mathbf{v}/d(\mathbf{v})$ is in $\mathcal{S}$,

$$b = b \cdot d\left(\frac{\mathbf{v}}{d(\mathbf{v})}\right) \leq \left\|\frac{\mathbf{v}}{d(\mathbf{v})}\right\| = \frac{1}{d(\mathbf{v})}\|\mathbf{v}\|.$$

Multiplication of both sides of this inequality by $d(\mathbf{v})$ yields $b \cdot d(\mathbf{v}) \leq \|\mathbf{v}\|$, and the proof is complete. ∎ ∎ ∎

11.9 THEOREM

Let $\|\mathbf{v}\|$ denote an arbitrary norm on $\mathbf{R}^n$. If $\lim_{k \to \infty} \|\mathbf{v}_k\| = 0$, then $\{\mathbf{v}_k\}$ converges to the zero vector in $\mathbf{R}^n$.

Proof Let $\mathbf{v}_k = (x_{1k}, x_{2k}, \ldots, x_{nk})$, and assume that $\|\mathbf{v}_k\| \to 0$. Let $\epsilon > 0$ be given. By Lemma 11.8, there is a positive number b such that $b \cdot d(\mathbf{v}) \leq \|\mathbf{v}\|$ for each $\mathbf{v} = (x_1, x_2, \ldots, x_n)$ in $\mathbf{R}^n$. Since $\|\mathbf{v}_k\| \to 0$, there is a positive number m such that $k > m$ implies $\|\mathbf{v}_k\| < b\epsilon$. For $k > m$ and $i = 1, 2, \ldots, n$, we have

$$b|x_{ik}| \leq b\sqrt{x_{1k}^2 + x_{2k}^2 + \cdots + x_{nk}^2} = b \cdot d(\mathbf{v}_k) \leq \|\mathbf{v}_k\| < b\epsilon.$$

Thus $|x_{ik}| < \epsilon$ for $k > m$, and it follows that $\lim_{k \to \infty} x_{ik} = 0$ for each i. ∎ ∎ ∎

Theorems 11.6 and 11.9 together yield the following relation between norms and the convergence of sequences. This is the major result of this section.

11.10 THEOREM

If $\{\mathbf{v}_k\}$ is a sequence of vectors in $\mathbf{R}^n$, then $\lim_{k \to \infty} \mathbf{v}_k = \mathbf{v}$ if and only if

$$\lim_{k \to \infty} \|\mathbf{v}_k - \mathbf{v}\| = 0.$$

Proof The statement $\lim_{k \to \infty} \mathbf{v}_k = \mathbf{v}$ is equivalent to $\lim_{k \to \infty}(\mathbf{v}_k - \mathbf{v}) = 0$, and this in turn is equivalent to $\lim_{k \to \infty} \|\mathbf{v}_k - \mathbf{v}\| = 0$, by Theorems 11.6 and 11.9. ∎ ∎ ∎

The striking thing about Theorem 11.10 is that it shows that $\|\mathbf{v}_k - \mathbf{v}\| \to 0$ for an *arbitrary* norm on $\mathbf{R}^n$ if and only if the same condition holds with the *standard* norm on $\mathbf{R}^n$. In other words, the test for convergence $\|\mathbf{v}_k - \mathbf{v}\| \to 0$ is independent of the choice of the norm.

11.11
DEFINITION

Given a sequence $\{\mathbf{u}_k\}$ of vectors in $\mathbf{R}^n$, the sequence $\{\mathbf{v}_k\}$ with $\mathbf{v}_r = \mathbf{u}_1 + \mathbf{u}_2 + \cdots + \mathbf{u}_r$ is called a **series**, and is denoted by $\sum_{k=1}^{\infty} \mathbf{u}_k$ or $\sum_k \mathbf{u}_k$. The terms $\mathbf{v}_r = \sum_{k=1}^{r} \mathbf{u}_k$ are called the **partial sums** of the series $\sum_{k=1}^{\infty} \mathbf{u}_k$. If $\mathbf{v}_k \to \mathbf{v}$, then $\mathbf{v}$ is called the **sum** of the series. We say that $\sum_{k=1}^{\infty} \mathbf{u}_k$ converges to $\mathbf{v}$, and write $\sum_{k=1}^{\infty} \mathbf{u}_k = \mathbf{v}$.

11.1 Exercises

1. Let $\mathbf{v} = (-2, 1, 4)$ in $\mathbf{R}^3$.
   (a) Compute the standard norm $d(\mathbf{v})$ for this $\mathbf{v}$.
   (b) Compute $\|\mathbf{v}\|$ using the norm in part 1 of Theorem 11.2.
   (c) Compute $\|\mathbf{v}\|$ using the norm in part 2 of Theorem 11.2.

2. Work Problem 1 using the vector $\mathbf{v} = (2, 4, -2, -5)$ in $\mathbf{R}^4$.

3. Find $\lim_{k \to \infty} \mathbf{v}_k$ for each of the following $\{\mathbf{v}_k\}$ that converge.

   (a) $\mathbf{v}_k = \left(\dfrac{\ln k}{k^2 - 1}, \dfrac{\cos k}{e^k} \right)$
           (b) $\mathbf{v}_k = \left(k \sin \dfrac{1}{k}, e^{-1/k^2} \right)$

   (c) $\mathbf{v}_k = \left(ke^{-k}, \dfrac{k^2 - 1}{4k^2 - k}, k \tan \dfrac{1}{k} \right)$
       (d) $\mathbf{v}_k = \left(\left(\dfrac{2}{3} \right)^k, \dfrac{k^2}{3k^2 + 2}, \dfrac{\sin k}{k}, k^{-2} \right)$

   (e) $\mathbf{v}_k = \left(k^{-1/2}, \sin k, \dfrac{1}{k-2} \right)$
          (f) $\mathbf{v}_k = \left(e^{-k}, \dfrac{k+1}{k-1}, k \ln k \right)$

4. Prove that $\|\mathbf{u} - \mathbf{v}\| \geq |\|\mathbf{u}\| - \|\mathbf{v}\||$ for an arbitrary norm on $\mathbf{R}^n$.

5. Prove that $\| \sum_{k=1}^{r} \mathbf{v}_k \| \leq \sum_{k=1}^{r} \|\mathbf{v}_k\|$ for any positive integer r.

6. Prove Theorem 11.2.

7. Prove that the limit of a convergent sequence of vectors in $\mathbf{R}^n$ is unique.

8. Prove Theorem 11.5.

9. Let $\{\mathbf{u}_k\}$ and $\{\mathbf{v}_k\}$ be convergent sequences of vectors in $\mathbf{R}^n$, and let $a \in \mathbf{R}$.
   (a) Prove that $\lim_{k \to \infty} a\mathbf{v}_k = a \lim_{k \to \infty} \mathbf{v}_k$.
   (b) Prove that $\lim_{k \to \infty} (\mathbf{u}_k + \mathbf{v}_k) = \lim_{k \to \infty} \mathbf{u}_k + \lim_{k \to \infty} \mathbf{v}_k$.

10. Prove that the set S in the proof of Lemma 11.8 is closed.

11. Prove that the set S in the proof of Lemma 11.8 is bounded.

12. Let $a \in \mathbf{R}$, and let $\sum_{k=1}^{\infty} \mathbf{u}_k$ and $\sum_{k=1}^{\infty} \mathbf{v}_k$ be convergent series with sums $\mathbf{u}$ and $\mathbf{v}$, respectively.
    (a) Prove that $\sum_{k=1}^{\infty} a\mathbf{v}_k$ converges to $a\mathbf{v}$.
    (b) Prove that $\sum_{k=1}^{\infty} (\mathbf{u}_k + \mathbf{v}_k)$ converges to $\mathbf{u} + \mathbf{v}$.

13. The term *norm* as given in Definition 11.1 can be extended to an arbitrary n-dimensional vector space $\mathbf{V}$ over $\mathbf{R}$ by simply replacing $\mathbf{R}^n$ by $\mathbf{V}$ in that definition. Let $\mathbf{V}$ be a two-dimensional vector space over $\mathbf{R}$, and let $\{\mathbf{u}_1, \mathbf{u}_2\}$ be a basis of $\mathbf{V}$. Let $\{\mathbf{v}_k\}$ be a sequence of vectors in $\mathbf{V}$, with $\mathbf{v}_k = x_{1k}\mathbf{u}_1 + x_{2k}\mathbf{u}_2$. Prove that $\|\mathbf{v}_k\| \to 0$ if and only if $x_{ik} \to 0$ for $i = 1, 2$.

14. Let P be an invertible $n \times n$ matrix over $\mathbf{R}$. Suppose that for $k = 1, 2, \ldots, n$, the real variables $y_{1k}, y_{2k}, \ldots, y_{nk}$ and $x_{1k}, x_{2k}, \ldots, x_{nk}$ are related by

$$[\, y_{1k} \quad y_{2k} \quad \cdots \quad y_{nk} \,]^T = P[\, x_{1k} \quad x_{2k} \quad \cdots \quad x_{nk} \,]^T.$$

Prove that $\lim_{k \to \infty} y_{ik} = 0$ for $i = 1, 2, \ldots, n$ if and only if $\lim_{k \to \infty} x_{ik} = 0$ for $i = 1, 2, \ldots, n$.

15. Explain the connection between Problem 13 and Theorem 11.5.

16. For any norm $\|\mathbf{v}\|$ on $\mathbf{R}^n$ derived from an inner product $\|\mathbf{v}\| = \sqrt{(\mathbf{v}, \mathbf{v})}$, Problem 15 of Exercises 9.2 asserts that

$$\|\mathbf{u} + \mathbf{v}\|^2 + \|\mathbf{u} - \mathbf{v}\|^2 = 2(\|\mathbf{u}\|^2 + \|\mathbf{v}\|^2).$$

This property is known as the **parallelogram rule**. Show that the parallelogram rule does not hold on $\mathbf{R}^2$ for the norm in part 2 of Theorem 11.2. Hence conclude that this norm is not derived from any inner product on $\mathbf{R}^2$.

11.2 Sequences and Series of Matrices

Many of the iterative methods for the solution of systems of equations are based on the convergence of a sequence of matrices. A treatment of convergence of sequences of matrices depends of the properties of norms of matrices.

> **11.12 DEFINITION**
>
> For any given norm $\|\mathbf{v}\|$ on $\mathbf{R}^n$, the **associated** or **corresponding norm** $\|X\|$ is defined on $\mathbf{R}_{n \times 1}$ as follows: if $X = [\, x_1 \quad x_2 \quad \cdots \quad x_n \,]^T$, then $\|X\| = \|(x_1, x_2, \ldots, x_n)\|$.

The properties listed in Definition 11.1 translate directly into the following statements concerning the norm $\|X\|$ on $\mathbf{R}_{n \times 1}$:

(i) $\|X\| > 0$ if $X \neq \mathbf{0}$, and $\|\mathbf{0}\| = 0$;

(ii) $\|aX\| = |a| \cdot \|X\|$ for any $a \in \mathbf{R}$;

(iii) $\|X + Y\| \leq \|X\| + \|Y\|$.

Thus the norm $\|X\|$, considered as a mapping of the vector space $\mathbf{R}_{n \times 1}$ into $\mathbf{R}$, has the same properties as a norm on $\mathbf{R}^n$. It is possible to extend the definition of a norm to arbitrary vector spaces $\mathbf{V}$ over $\mathbf{R}$ by requiring a mapping of $\mathbf{V}$ into $\mathbf{R}$ with the three properties in Definition 11.1. We shall *not* move in this direction, however. To the contrary, we shall confine our attention exclusively to those norms of the type in Definition 11.12.

Our main objective here is to obtain a norm $\|A\|$ on the set of $n \times n$ matrices A over $\mathbf{R}$, and to do this in such a way that $\|A\|$ is related in a certain way to the norm in Definition 11.12. This relation is stated explicitly in part 4 of Theorem 11.15.

If $Y = [\, y_1 \quad y_2 \quad \cdots \quad y_n \,]^T$, we have seen that $\|Y\| = \|(y_1, y_2, \ldots, y_n)\|$ is a continuous function of $y_1, y_2, \ldots, y_n$. If $Y = AX$ for $X = [\, x_1 \quad x_2 \quad \cdots \quad x_n \,]^T$ and a fixed matrix $A = [a_{ij}]_n$, then each $y_i = \sum_{j=1}^{n} a_{ij} x_j$ is clearly a continuous function of $x_1, x_2, \ldots, x_n$. Hence the composite function $g(x_1, x_2, \ldots, x_n) = \|AX\|$ is a continuous function of $x_1, x_2, \ldots, x_n$. It follows from Lemmas 11.7 and 11.8 that the set S of all $\mathbf{v} = (x_1, x_2, \ldots, x_n)$ in $\mathbf{R}^n$ such that $\|\mathbf{v}\| = 1$ is closed and bounded, and therefore the function g attains a maximum value on S. That is, there is a matrix $X_0 = [\, x_1^{(0)} \quad x_2^{(0)} \quad \cdots \quad x_n^{(0)} \,]^T$ such that $\|X_0\| = 1$ and

$$g\left(x_1^{(0)}, x_2^{(0)}, \ldots, x_n^{(0)}\right) = \|AX_0\| = \sup_{\|X\|=1} \{\|AX\|\},$$

where $\sup_{\|X\|=1}\{\|AX\|\}$ denotes the least upper bound of the set of function values $\|AX\|$ obtained with $\|X\| = 1$. This result enables us to define the desired norm on $\mathbf{R}_{n \times n}$.

11.13
DEFINITION

With $\|X\|$ as given in Definition 11.12, the **associated** or **corresponding norm** of each $n \times n$ matrix A over $\mathbf{R}$ is defined by

$$\|A\| = \sup_{\|X\|=1} \{\|AX\|\}.$$

11.14
THEOREM

Each of the following functions is the associated norm on $\mathbf{R}_{n \times n}$ for the norm on $\mathbf{R}^n$ that is given in the corresponding part of Theorem 11.2.

1. $\|A\|_r = \max_i \left\{ \sum_{j=1}^{n} |a_{ij}| \right\}$

2. $\|A\|_c = \max_j \left\{ \sum_{i=1}^{n} |a_{ij}| \right\}$

Proof We shall prove the statement for (1) and leave the proof of the statement (2) as an exercise in Problem 6.

Let $\|A\|_r$ denote the norm on $\mathbf{R}_{n \times n}$ corresponding to $\|X\| = \max_i\{|x_i|\}$. For any X such that $\|X\| = 1$, we have

$$\|A\|_r = \sup_{\|X\|=1} \{\|AX\|\}$$

$$= \sup_{\substack{i \\ \|X\|=1}} \left\{ \left| \sum_{j=1}^{n} a_{ij} x_j \right| \right\}$$

$$\leq \sup_{\substack{i \\ \|X\|=1}} \left\{ \sum_{j=1}^{n} |a_{ij}| \cdot |x_j| \right\}$$

$$\leq \max_{i} \left\{ \sum_{j=1}^{n} |a_{ij}| \right\}.$$

Suppose the maximum value of $\sum_{j=1}^{n} |a_{ij}|$ occurs in row k. Let $X = [\,x_1 \quad x_2 \quad \cdots \quad x_n\,]^T$, where $x_j = 1$ if $a_{kj} \geq 0$ and $x_j = -1$ if $a_{kj} < 0$. Then $\|X\| = 1$, and $a_{kj} x_j = |a_{kj}|$ for each j. Thus we have

$$\max_{i} \left\{ \sum_{j=1}^{n} |a_{ij}| \right\} = \sum_{j=1}^{n} |a_{kj}|$$

$$= \sum_{j=1}^{n} a_{kj} x_j$$

$$= \left| \sum_{j=1}^{n} a_{kj} x_j \right|$$

$$\leq \sup_{\substack{i \\ \|X\|=1}} \left\{ \left| \sum_{j=1}^{n} a_{ij} x_j \right| \right\}$$

$$= \|A\|_r.$$

This means that $\|A\|_r = \max_i \left\{ \sum_{j=1}^{n} |a_{ij}| \right\}$.   ■■■

The norms $\|A\|_r$ and $\|A\|_c$ are known respectively as the **row norm** and **column norm** for n by n matrices over $\mathbf{R}$.

EXAMPLE 1   We have three specific examples of norms that we can exhibit for square matrices: the row norm, the column norm, and the norm that corresponds to the standard norm

$$d(\mathbf{v}) = d(x_1, x_2, \ldots, x_n) = \sqrt{x_1^2 + x_2^2 + \cdots + x_n^2}$$

on $\mathbf{R}^n$. We will illustrate each of these using a simple example in $\mathbf{R}_{2\times 2}$ with

$$A = \begin{bmatrix} 4 & -3 \\ -4 & 3 \end{bmatrix}.$$

For the row norm $\|A\|_r$, we have

$$\|A\|_r = \max_i \left\{ \sum_{j=1}^{2} |a_{ij}| \right\} = 7,$$

and the column norm is given by

$$\|A\|_c = \max_j \left\{ \sum_{i=1}^{2} |a_{ij}| \right\} = 8.$$

The computation of $\|A\|$ that corresponds to the standard norm on $\mathbf{R}^2$ is more involved. For an arbitrary $X = \begin{bmatrix} x & y \end{bmatrix}^T$ in $\mathbf{R}^2$,

$$AX = \begin{bmatrix} 4x - 3y \\ -4x + 3y \end{bmatrix}$$

and

$$\|A\| = \sup_{\|X\|=1} \{ \|AX\| \}$$

$$= \sup_{\|X\|=1} \left\{ \sqrt{(4x - 3y)^2 + (-4x + 3y)^2} \right\}$$

$$= \sup_{\|X\|=1} \left\{ \sqrt{2(4x - 3y)^2} \right\}.$$

Now $\sqrt{2(4x - 3y)^2}$ has a maximum value at the same point where $h = (4x - 3y)^2$ has a maximum value. Thus we need to solve the problem of maximizing $h = (4x - 3y)^2$ subject to the constraint $x^2 + y^2 = 1$. The method of Lagrange multipliers can be used to advantage on this problem, but the methods of an introductory calculus course are adequate. Using implicit differentiation with the equations $h = (4x - 3y)^2$ and $x^2 + y^2 = 1$, we get

$$\frac{dh}{dx} = 2(4x - 3y)(4 - 3y') \quad \text{and} \quad 2x + 2yy' = 0.$$

Solving for y', we obtain $y' = -x/y$ and

$$\frac{dh}{dx} = 2(4x - 3y)\left(4 + \frac{3x}{y}\right).$$

Thus h has critical points where $x^2 + y^2 = 1$ and one of the equations $4x - 3y = 0$, $3x + 4y = 0$, $y = 0$ hold. We find that

the solutions to the system $x^2 + y^2 = 1$, $4x - 3y = 0$ are $\left(\frac{3}{5}, \frac{4}{5}\right)$ and $\left(-\frac{3}{5}, -\frac{4}{5}\right)$;

the solutions to the system $x^2 + y^2 = 1$, $3x + 4y = 0$ are $\left(\frac{4}{5}, -\frac{3}{5}\right)$ and $\left(-\frac{4}{5}, \frac{3}{5}\right)$;

the solutions to the system $x^2 + y^2 = 1$, $y = 0$ are $(1, 0)$ and $(-1, 0)$.

The values of h at the critical points are as follows:

$$\text{at } \left(\tfrac{3}{5}, \tfrac{4}{5}\right) \quad \text{and} \quad \left(-\tfrac{3}{5}, -\tfrac{4}{5}\right), \quad h = 0;$$

$$\text{at } \left(\tfrac{4}{5}, -\tfrac{3}{5}\right) \quad \text{and} \quad \left(-\tfrac{4}{5}, \tfrac{3}{5}\right), \quad h = 25;$$

$$\text{at } (1, 0) \quad \text{and} \quad (-1, 0), \quad h = 16.$$

Thus

$$\|A\| = \sup_{\|X\|=1} \left\{ \sqrt{2(4x - 3y)^2} \right\}$$

$$= \sup_{\|X\|=1} \left\{ \sqrt{2h} \right\}$$

$$= \sup \left\{ 0, 5\sqrt{2}, 4\sqrt{2} \right\}$$

$$= 5\sqrt{2} \qquad \blacksquare$$

It was shown just before Definition 11.13 that there always exists an $n \times 1$ matrix X_0 with $\|X_0\| = 1$ such that $\|A\| = \|AX_0\|$. The other basic properties of $\|A\|$ are contained in our next theorem.

11.15 THEOREM

Each norm $\|A\|$ on $\mathbf{R}_{n \times n}$ has the following properties:

1. $\|A\| > 0$ if $A \neq \mathbf{0}$, and $\|\mathbf{0}\| = 0$;
2. $\|aA\| = |a| \cdot \|A\|$ for any a in $\mathbf{R}$;
3. $\|A + B\| \leq \|A\| + \|B\|$ for any A, B in $\mathbf{R}_{n \times n}$;
4. $\|AY\| \leq \|A\| \cdot \|Y\|$ for any A in $\mathbf{R}_{n \times n}$ and any Y in $\mathbf{R}_{n \times 1}$;
5. $\|AB\| \leq \|A\| \cdot \|B\|$ for any A, B in $\mathbf{R}_{n \times n}$.

Proof The statement (1) is obvious. For any a in $\mathbf{R}$ and any A in $\mathbf{R}_{n \times n}$, we have

$$\|aA\| = \sup_{\|X\|=1} \{\|aAX\|\}$$

$$= \sup_{\|X\|=1} \{|a| \cdot \|AX\|\}$$

$$= |a| \cdot \|A\|.$$

Hence property (2) is true.

Since $\|AX + BX\| \le \|AX\| + \|BX\|$,

$$\|A + B\| = \sup_{\|X\|=1} \{\|AX + BX\|\}$$

$$\le \sup_{\|X\|=1} \{\|AX\|\} + \sup_{\|X\|=1} \{\|BX\|\}$$

$$= \|A\| + \|B\|.$$

The statement (4) is obvious for $Y = \mathbf{0}$. For any nonzero n by 1 matrix Y over $\mathbf{R}$, $X = (1/\|Y\|)Y$ has norm 1, and

$$\|A\| = \sup_{\|X\|=1} \{\|AX\|\}$$

$$\ge \left\| A\left(\frac{1}{\|Y\|}Y\right) \right\|$$

$$= \frac{1}{\|Y\|} \cdot \|AY\|.$$

Thus $\|A\| \cdot \|Y\| \ge \|AY\|$.

Let X_0 be an n by 1 matrix such that $\|X_0\| = 1$ and $\|AB\| = \|ABX_0\|$. By part (4), we have

$$\|AB\| = \|ABX_0\| \le \|A\| \cdot \|BX_0\| \le \|A\| \cdot \|B\| \cdot \|X_0\| = \|A\| \cdot \|B\|.$$

This completes the proof. ■ ■ ■

11.16
DEFINITION

A **sequence** of r by s matrices over $\mathbf{R}$ is a set $\{A_k\}$ of r by s matrices $A_k = [a_{ij}^{(k)}]_{r \times s}$ over $\mathbf{R}$ that is indexed by the set of all positive integers k. The sequence $\{A_k\}$ converges to the limit $A = [a_{ij}]_{r \times s}$ if $\lim_{k\to\infty} a_{ij}^{(k)} = a_{ij}$ for all pairs i, j.

As with vectors, we write $\lim_{k\to\infty} A_k = A$ or $A_k \to A$ to indicate that $\{A_k\}$ converges to A.

EXAMPLE 2 Consider the problem of evaluating $\lim_{k\to\infty} A_k$ where

$$A_k = \begin{bmatrix} 2^{-k} & \dfrac{5}{k} & \dfrac{\ln k}{k} \\[2ex] \dfrac{3k^2}{k^2 + 3} & k \sin \dfrac{1}{k} & \dfrac{\cos k}{k} \end{bmatrix}.$$

Using L'Hospital's rule when needed, we find that

$$\lim_{k\to\infty} 2^{-k} = \lim_{k\to\infty} \frac{1}{2^k} = 0, \quad \lim_{k\to\infty} \frac{5}{k} = 0, \quad \lim_{k\to\infty} \frac{\ln k}{k} = \lim_{k\to\infty} \frac{1/k}{1} = 0,$$

$$\lim_{k\to\infty} \frac{3k^2}{k^2+3} = 3, \quad \lim_{k\to\infty} k \sin\frac{1}{k} = \lim_{k\to\infty} \frac{\sin(1/k)}{(1/k)} = 1, \quad \lim_{k\to\infty} \frac{\cos k}{k} = 0.$$

Thus

$$\lim_{k\to\infty} A_k = \begin{bmatrix} 0 & 0 & 0 \\ 3 & 1 & 0 \end{bmatrix}. \qquad\blacksquare$$

Let $\{\mathbf{v}_k\}$ be a sequence of vectors in $\mathbf{R}^n$ and let $X_k = [\, x_{1k} \quad x_{2k} \quad \cdots \quad x_{nk}\,]^T$ denote the coordinate matrix of $\mathbf{v}_k$ relative to the basis $\mathcal{A}$ of $\mathbf{R}^n$. Theorem 11.5 shows that $\{\mathbf{v}_k\}$ converges to the vector $\mathbf{b}$ with coordinates $B = [\, b_1 \quad b_2 \quad \cdots \quad b_n\,]^T$ relative to $\mathcal{A}$ if and only if the sequence $\{X_k\}$ converges to B.

We shall be interested mainly in the convergence of sequences of $n \times n$ matrices. Convergence of this type of sequences is related to norms in much the same way as in the case of vectors.

11.17 **THEOREM**	Let $\{A_k\}$ be a sequence of $n \times n$ matrices over $\mathbf{R}$. Then $A_k \to \mathbf{0}$ if and only if $\|A_k\| \to 0$.

Proof Suppose first that $\{A_k\}$ is a sequence of $n \times n$ matrices $A_k = [a_{ij}^{(k)}]$ such that $\lim_{k\to\infty} a_{ij}^{(k)} = 0$ for all pairs i, j. For each A_k let $X_k = [\, x_{1k} \quad x_{2k} \quad \cdots \quad x_{nk}\,]^T$ be a matrix such that $\|X_k\| = 1$ and $\|A_k\| = \|A_k X_k\|$. Let

$$Y_k = [\, y_{1k} \quad y_{2k} \quad \cdots \quad y_{nk}\,]^T = A_k X_k.$$

Then $y_{ik} = \sum_{j=1}^{n} a_{ij}^{(k)} x_{jk}$, and

$$\lim_{k\to\infty} y_{ik} = \sum_{j=1}^{n} \lim_{k\to\infty} a_{ij}^{(k)} x_{jk} = 0$$

since $a_{ij}^{(k)} \to 0$ and $\{x_{jk}\}$ is bounded. Thus, $Y_k \to \mathbf{0}$, and $\|A_k\| = \|Y_k\| \to 0$ by Theorem 11.6.

Assume now that

$$\|A_k\| = \sup_{\|X\|=1}\{\|A_k X\|\} \to 0.$$

Then $\|A_k X\| \to 0$ for each fixed X such that $\|X\| = 1$. For $j = 1, 2, \ldots, n$, let $E_j = [\, \delta_{1j} \quad \delta_{2j} \quad \cdots \quad \delta_{nj}\,]^T$ and

$$X_j = \frac{1}{\|E_j\|} E_j.$$

Each matrix X_j has norm 1, and

$$A_k X_j = \frac{1}{\|E_j\|} A_k E_j = \frac{1}{\|E_j\|} \left[\, a_{1j}^{(k)} \quad a_{2j}^{(k)} \quad \cdots \quad a_{nj}^{(k)}\,\right]^T$$

is a sequence such that $\lim_{k \to \infty} \|A_k X_j\| = 0$ for each j. Since

$$\|A_k X_j\| = \frac{1}{\|E_j\|} \left\| \left(a_{1j}^{(k)}, a_{2j}^{(k)}, \ldots, a_{nj}^{(k)} \right) \right\|,$$

this implies that $\lim_{k \to \infty} a_{ij}^{(k)} = 0$ for each i, by Theorem 11.9. Therefore, $A_k \to \mathbf{0}$, and the proof is complete. ■ ■ ■

**11.18
COROLLARY**

If $\{A_k\}$ is a sequence of $n \times n$ matrices over $\mathbf{R}$, then $A_k \to A$ if and only if

$$\|A_k - A\| \to 0.$$

Proof See Problem 8. ■ ■ ■

**11.19
COROLLARY**

If A is an $n \times n$ matrix over $\mathbf{R}$ such that $\|A\| < 1$, then $\lim_{k \to \infty} A^k = \mathbf{0}$.

Proof See Problem 9. ■ ■ ■

**11.20
THEOREM**

Let A be a square matrix over $\mathbf{R}$. The absolute value of any eigenvalue of A does not exceed the value of any norm of A.

Proof Let λ be an eigenvalue of A, and let Y be a corresponding eigenvector with $\|Y\| = 1$. We have

$$\|A\| = \sup_{\|X\|=1} \{\|AX\|\}$$

$$\geq \|AY\|$$

$$= \|\lambda Y\|$$

$$= |\lambda| \cdot \|Y\|$$

$$= |\lambda|.$$ ■ ■ ■

In combination with Theorem 11.23, the next theorem is of key importance in Section 11.3. In the proof of this theorem, we make use of the Jordan canonical form. The derivation of the Jordan canonical form for a matrix A proceeded under the assumption that the characteristic polynomial of A factored completely over the field $\mathcal{F}$ (see Theorem 10.31). We are presently considering only matrices over $\mathbf{R}$, and such a factorization is not assured over $\mathbf{R}$. However, a complete factorization over the field $\mathcal{C}$ is available, and it establishes the existence of the Jordan canonical form. In the proof of Theorem 11.21, we consider the real matrix A as a matrix over $\mathcal{C}$. This puts the proof in its proper context.

11.21
THEOREM

Let A be a square matrix over **R**. Then $\lim_{k \to \infty} A^k = \mathbf{0}$ if and only if $|\lambda| < 1$ for every eigenvalue λ of A.

Proof We shall show first that if A has an eigenvalue λ with $|\lambda| \geq 1$, then $\{A^k\}$ does not converge to $\mathbf{0}$. Suppose that λ is an eigenvalue of A such that $|\lambda| \geq 1$. For each positive integer k, λ^k is an eigenvalue of A^k, and $|\lambda^k| \geq 1$. Hence $\|A^k\| \geq |\lambda^k| \geq 1$ by Theorem 11.20. This means that $\{\|A^k\|\}$ does not converge to 0, and therefore $\{A^k\}$ does not converge to $\mathbf{0}$.

Assume now that $|\lambda| < 1$ for every eigenvalue λ of A. Let the Jordan canonical form J for A be given by

$$
J = \begin{bmatrix}
\lambda_1 I_{m_1} + N_1 & 0 & \cdots & 0 \\
0 & \lambda_2 I_{m_2} + N_2 & \cdots & 0 \\
\vdots & \vdots & \ddots & \vdots \\
0 & 0 & \cdots & \lambda_r I_{m_r} + N_r
\end{bmatrix},
$$

where N_i in the diagonal block $J_i = \lambda_i I_{m_i} + N_i$ is nilpotent of index t_i. There exists an invertible matrix P such that $A = P^{-1}JP$. Since $A^k = P^{-1}J^kP$ for each positive integer k, it is sufficient to show that $\lim_{k \to \infty} J^k = \mathbf{0}$. Now

$$
J^k = \begin{bmatrix}
J_1^k & 0 & \cdots & 0 \\
0 & J_2^k & \cdots & 0 \\
\vdots & \vdots & \ddots & \vdots \\
0 & 0 & \cdots & J_r^k
\end{bmatrix}
$$

for each k, so we need only show that $J_i^k \to \mathbf{0}$ for each i. Since $N_i^j = \mathbf{0}$ for $j \geq t_i$, we have

$$
J_i^k = \sum_{j=0}^{t_i - 1} \binom{k}{j} \lambda_i^{k-j} I_{m_i} N_i^j,
$$

where

$$
\binom{k}{j} = \frac{k!}{j!(k-j)!}.
$$

For a fixed j, let $a_k = \binom{k}{j} \lambda_i^{k-j}$. Now

$$
\left| \frac{a_{k+1}}{a_k} \right| = \frac{k+1}{k+1-j} |\lambda_i| \to |\lambda_i|
$$

as $k \to \infty$, and $|\lambda_i| < 1$. Hence $\lim_{k \to \infty} a_k = 0$, and therefore

$$\lim_{k \to \infty} \binom{k}{j} \lambda_i^{k-j} I_{m_i} N_i^j = \mathbf{0}$$

for $j = 0, 1, \ldots, t_i - 1$. It follows that $J_i^k \to \mathbf{0}$ for each i, and the proof is complete. ∎

**11.22
DEFINITION**

Given a sequence $\{A_k\}$ of r by s matrices over $\mathbf{R}$, the sequence $\{S_k\}$ with $S_k = A_1 + A_2 + \cdots + A_k$ is called a **series**, and is denoted by $\sum_{k=1}^{\infty} A_k$. The terms S_k are called the **partial sums** of the series $\sum_{k=1}^{\infty} A_k$. If $S_k \to S$, then S is called the **sum** of the series. We say that $\sum_{k=1}^{\infty} A_k$ **converges** to S, and write $\sum_{k=1}^{\infty} A_k = S$.

Our interest in series is confined to power series. The series $\sum_{k=0}^{\infty} A^k$ is of particular importance.

**11.23
THEOREM**

The series $\sum_{k=0}^{\infty} A^k$ converges if and only if $A^k \to \mathbf{0}$ as $k \to \infty$. In the case of convergence, the sum is $(I - A)^{-1}$.

Proof Let $S_r = \sum_{k=0}^{r} A^k$ denote the partial sums of the series $\sum_{k=0}^{\infty} A^k$. Suppose first that the series $\sum_{k=0}^{\infty} A^k$ converges to S. Then

$$\lim_{k \to \infty} A^k = \lim_{k \to \infty} (S_k - S_{k-1})$$
$$= \lim_{k \to \infty} S_k - \lim_{k \to \infty} S_{k-1}$$
$$= S - S$$
$$= \mathbf{0}.$$

Suppose, conversely, that $A^k \to \mathbf{0}$ as $k \to \infty$. This implies that all eigenvalues of A are less than 1 in absolute value, and consequently $\det(A - I) \neq 0$. Hence $I - A$ is invertible. Since

$$(I + A + \cdots + A^k)(I - A) = I - A^{k+1},$$

we have

$$S_k = I + A + \cdots + A^k$$
$$= (I - A^{k+1})(I - A)^{-1}.$$

Therefore $S_k \to (I - A)^{-1}$ as $k \to \infty$. ∎

| 11.24 | The series $\sum_{k=0}^{\infty} A^k$ converges to $(I - A)^{-1}$ if and only if $|\lambda| < 1$ for every eigenvalue |
| COROLLARY | λ of A. |

Proof This follows immediately from Theorems 11.21 and 11.23.   ■ ■ ■

11.2 Exercises

1. Let

$$A = \begin{bmatrix} 4 & -4 \\ 3 & -3 \end{bmatrix}.$$

(a) Find the value of the row norm $\|A\|_r$.

(b) Find the value of the column norm $\|A\|_c$.

(c) Find the value of $\|A\|$ that corresponds to the standard norm $d(x, y)$ on $\mathbf{R}^2$.

2. Work Problem 1 using

$$A = \begin{bmatrix} 3 & -4 \\ -3 & 4 \end{bmatrix}.$$

3. Work Problem 1 using

$$A = \begin{bmatrix} 4 & 3 \\ 3 & 4 \end{bmatrix}.$$

4. Find $\lim_{k \to \infty} A_k$ for each of the following $\{A_k\}$ that converge.

(a) $A_k = \begin{bmatrix} \dfrac{1}{4^k} & \dfrac{k^2 - 1}{k^2 + 1} \\ \tan^{-1} k & \tan \dfrac{\pi}{2k} \end{bmatrix}$

(b) $A_k = \begin{bmatrix} \tan^{-1}\left(\dfrac{k}{k+1}\right) & \dfrac{\cos k}{\sqrt{k}} \\ \dfrac{\ln k}{k} & \dfrac{2^k}{e^k} \end{bmatrix}$

(c) $A_k = \begin{bmatrix} \dfrac{\ln(1/k)}{k} & \sqrt{k} \\ \dfrac{\cos k}{k+1} & \dfrac{(-1)^k}{k} \end{bmatrix}$

(d) $A_k = \begin{bmatrix} \dfrac{k}{e^k} & \sqrt{k+1} - \sqrt{k} \\ \dfrac{\ln(e^k + 1)}{5k} & k \ln k \end{bmatrix}$

(e) $A_k = \begin{bmatrix} \dfrac{(-1)^k}{k^2} & (-1)^k \sin \dfrac{1}{k} \\ \ln \dfrac{k}{k+1} & \cos \dfrac{2\pi}{k} \end{bmatrix}$

(f) $A_k = \begin{bmatrix} (-1)^k \tan \dfrac{1}{k} & \tan \dfrac{2}{k} \\ \cos^{-1} \left(\dfrac{k^2}{2k^2 + 1} \right) & \dfrac{k^2}{e^k} \end{bmatrix}$

5. Use Lemmas 11.7 and 11.8 to prove that the set $\mathcal{S}$ of all $\mathbf{v}$ in $\mathbf{R}^n$ such that $\|\mathbf{v}\| = 1$ is closed and bounded.

6. Prove that the function given in part (2) of Theorem 11.14 is the associated norm on $\mathbf{R}_{n \times n}$ for the norm on $\mathbf{R}^n$ in part (2) of Theorem 11.2.

7. Verify directly that $\|A\|_r$ and $\|A\|_c$ in Theorem 11.14 have the properties listed in Theorem 11.15.

8. Prove Corollary 11.18.

9. Prove Corollary 11.19.

11.3 The Standard Method of Iteration

Consider a system of equations $CX = B$, where $C = [c_{ij}]_{n \times n}$, $X = [\, x_1 \quad x_2 \quad \cdots \quad x_n \,]^T$, and $B = [\, b_1 \quad b_2 \quad \cdots \quad b_n \,]^T$ are matrices over $\mathbf{R}$. The matrix C is assumed to be invertible.

The system $CX = B$ may be rewritten as $X = AX + B$, where $A = I - C$. Let X_0 be an arbitrarily chosen $n \times 1$ matrix over $\mathbf{R}$, and consider the sequence $\{X_k\}$ given by

$$X_1 = AX_0 + B, \, X_2 = AX_1 + B, \ldots, X_k = AX_{k-1} + B, \ldots.$$

That is,

$$X_k = A^k X_0 + (I + A + \cdots + A^{k-1})B.$$

In the standard method of iteration, the matrices X_k are taken to be the successive approximations to the solution of the system.

11.25 THEOREM

The standard method of iteration converges to the solution of the system if $|\lambda| < 1$ for all eigenvalues λ of A.

Proof If all the eigenvalues of A have absolute value less than 1, then $A^k \to \mathbf{0}$ by Theorem 11.21, and $\{I + A + \cdots + A^k\}$ converges to $(I - A)^{-1}$ by Theorem 11.23. Hence $\{X_k\}$ converges to $(I - A)^{-1}B$. But $(I - A)^{-1}B$ is the solution of the system, since

$$A(I - A)^{-1}B + B = (A(I - A)^{-1} + I)B$$
$$= (A(I - A)^{-1} + (I - A)(I - A)^{-1})B$$
$$= (A + (I - A))(I - A)^{-1}B = (I - A)^{-1}B. \quad \blacksquare\blacksquare\blacksquare$$

11.26
COROLLARY

If some norm of A is less than 1, the standard method of iteration converges to the solution of the system.

Proof If some norm of A is less than 1, then all eigenvalues of A are less than 1, by Theorem 11.20. According to Theorem 11.25, this implies that the sequence $\{X_k\}$ converges to the solution. ■ ■ ■

EXAMPLE 1 Consider the following system of equations.

$$0.69x_1 - 0.17x_2 - 0.01x_3 = 0$$

$$-0.23x_1 + 0.94x_2 - 0.06x_3 = 1$$

$$-0.25x_1 + 0.40x_2 - 0.75x_3 = 1$$

We shall solve this system by using the standard method of iteration, obtaining our results to two decimal places. To write the system in the form $X = AX + B$, we put

$$A = I - C = \begin{bmatrix} 1 & 0 & 0 \\ 0 & 1 & 0 \\ 0 & 0 & 1 \end{bmatrix} - \begin{bmatrix} 0.69 & -0.17 & -0.01 \\ -0.23 & 0.94 & -0.06 \\ -0.25 & 0.40 & 0.75 \end{bmatrix} = \begin{bmatrix} 0.31 & 0.17 & 0.01 \\ 0.23 & 0.06 & 0.06 \\ 0.25 & -0.40 & 0.25 \end{bmatrix}$$

Using the row norm $\|A\|_r$, we find that $\|A\|_r = \max\{0.49, 0.35, 0.90\} = 0.90$. Since $\|A\|_r < 1$, we are assured that the standard method of iteration will converge to the solution of the system.

As an arbitrary choice of the initial matrix, we take $X_0 = [\,1\quad 0\quad 0\,]^T$. We then compute the matrices $X_k = AX_{k-1} + B$ in the sequence $\{X_k\}$, rounding to two decimal places.[1] This yields the following results.

$$X_0 = \begin{bmatrix} 1 \\ 0 \\ 0 \end{bmatrix}, \quad X_1 = AX_0 + B = \begin{bmatrix} 0.31 \\ 1.23 \\ 1.25 \end{bmatrix}, \quad X_2 = AX_1 + B = \begin{bmatrix} 0.32 \\ 1.22 \\ 0.90 \end{bmatrix},$$

$$X_3 = AX_2 + B = \begin{bmatrix} 0.32 \\ 1.20 \\ 0.82 \end{bmatrix}, \quad X_4 = AX_3 + B = \begin{bmatrix} 0.31 \\ 1.19 \\ 0.81 \end{bmatrix}, \quad X_5 = AX_4 + B = \begin{bmatrix} 0.31 \\ 1.19 \\ 0.80 \end{bmatrix}.$$

In the computation of $X_6 = AX_5 + B$, we get $X_6 = X_5$ when rounded to two decimal places. Thus an approximate solution to the system is given by $x_1 = 0.31$, $x_2 = 1.19$, $x_3 = 0.80$. ■

The following example illustrates that the standard method of iteration may converge even though $\|A\| > 1$ for some $\|A\|$.

---

[1] This rounding does not guarantee that the second decimal place is correct. To be certain that the second decimal place is correct, we would need to carry at least three decimal places in our calculations.

EXAMPLE 2 Consider the system

$$0.42x_1 + 0.21x_2 + 0.93x_3 = 0.33$$
$$-0.44x_1 + 1.18x_2 + 0.94x_3 = 0.03$$
$$-0.20x_1 + 0.10x_2 + 1.30x_3 = 0.41$$

that has

$$A = I - C = \begin{bmatrix} 0.58 & -0.21 & -0.93 \\ 0.44 & -0.18 & -0.94 \\ 0.20 & -0.10 & -0.30 \end{bmatrix}.$$

Computing the row and column norms of A, we obtain

$$\|A\|_r = \max\{1.72, 1.56, 0.60\} = 1.72$$

and

$$\|A\|_c = \max\{1.22, 0.49, 2.17\} = 2.17.$$

Thus both $\|A\|_r$ and $\|A\|_c$ are greater than 1.

However, the computer software *Scientific WorkPlace* gives the eigenvalues of A as $-0.2, 0.2$, and 0.1. This means, of course, that the standard method of iteration will converge in this case. Beginning with

$$X_0 = \begin{bmatrix} 1 \\ 0 \\ 0 \end{bmatrix},$$

we obtain

$$X_1 = \begin{bmatrix} 0.91 \\ 0.47 \\ 0.61 \end{bmatrix}, \quad X_2 = \begin{bmatrix} 0.19 \\ -0.23 \\ 0.36 \end{bmatrix}, \quad X_3 = \begin{bmatrix} 0.15 \\ -0.18 \\ 0.36 \end{bmatrix}, \quad X_4 = \begin{bmatrix} 0.12 \\ -0.21 \\ 0.35 \end{bmatrix} = X_5.$$

Thus an approximate solution to the system is $x_1 = 0.12$, $x_2 = -0.21$, $x_3 = 0.35$. ■

11.3 Exercises

1. Prove by induction that $X_k = A^k X_0 + (I + A + \cdots + A^{k-1})B$ in the standard method of iteration.

2. Prove that if the standard method of iteration converges to the solution of $X = AX + B$ for all choices of X_0 and B, then all eigenvalues of A have absolute value less than 1.

3. Write the given system in the form $X = AX + B$, and show that the standard method of iteration will converge for the system.

(a) $1.01x_1 - 0.51x_2 + 0.35x_3 = \quad 0.11$
$\quad\quad 0.21x_1 + 1.19x_2 - 0.13x_3 = -0.02$
$\quad -0.33x_1 - 0.33x_2 + 1.25x_3 = \quad 0.10$

(b) $1.27x_1 + 0.37x_2 - 0.21x_3 = \quad 0.10$
$\quad\quad 0.45x_1 + 0.55x_2 + 0.27x_3 = -0.10$
$\quad -0.09x_1 - 0.09x_2 + 1.07x_3 = \quad\quad 0$

(c) $1.10x_1 - 0.36x_2 + 0.06x_3 = \quad 0.17$
$\quad -0.32x_1 + 1.25x_2 + 0.21x_3 = \quad 0.25$
$\quad\quad 0.12x_1 - 0.13x_2 + 0.95x_3 = -0.24$

(d) $0.68x_1 + 0.21x_2 + 0.31x_3 = \quad 0.15$
$\quad -0.14x_1 + 1.18x_2 + 0.31x_3 = -0.16$
$\quad -0.20x_1 + 0.10x_2 + 0.53x_3 = \quad 0.15$

4. Use the standard method of iteration to obtain approximate solutions to the systems in Problem 3. Round off the successive approximations to two decimal places and continue until two successive approximations agree.

5. If $\|A\| < 1$ and S denotes the solution to $X = AX + B$, show that

$$\|S - X_k\| \leq \|X_0\| \cdot \|A\|^k + \frac{\|B\| \cdot \|A\|^k}{1 - \|A\|}.$$

6. Use Problem 5 with the norms in part 1 of Theorems 11.2 and 11.14 to estimate $\|S - X_4\|$ in Problem 3 with $X_0 = \mathbf{0}$.

11.4 Cimmino's Method

Cimmino's method first appeared in an article by Gianfranco Cimmino in 1938 entitled *Calcolo approssimato per le soluzioni dei sistemi di equazioni lineari*. It is an iterative method for solving certain types of systems of linear equations. Let $AX = B$ be a system of linear equations with $A = [a_{ij}]_{n \times n}$, $X = [x_1 \ x_2 \ \cdots \ x_n]^T$, and $B = [b_1 \ b_2 \ \cdots \ b_n]^T$ over $\mathbf{R}$. For $i = 1, 2, \ldots, n$, let $\mathbf{a}_i = (a_{i1}, a_{i2}, \ldots, a_{in})$. The ith equation in the system can be written as $\mathbf{a}_i \cdot \mathbf{v} = b_i$, where $\mathbf{v} = (x_1, x_2, \ldots, x_n)$. We assume that A has rank n, and that the system is normalized so that each $\mathbf{a}_i$ has norm 1.

The set of all points $(x_1, x_2, \ldots, x_n)$ in $\mathbf{R}^n$ that satisfy a fixed linear equation $a_1x_1 + a_2x_2 + \cdots + a_nx_n = b$ is called a **hyperplane**. If $\mathbf{a} = (a_1, a_2, \ldots, a_n)$ and $\mathbf{v} = (x_1, x_2, \ldots, x_n)$, then the point $(x_1, x_2, \ldots, x_n)$ is in the hyperplane if and only if $\mathbf{a} \cdot \mathbf{v} = b$. This is, of course, in analogy with the plane in $\mathbf{R}^3$. Two points $\mathbf{b}, \mathbf{c}$ in $\mathbf{R}^n$ are **symmetric** with respect to the hyperplane if (i) $\mathbf{c} - \mathbf{b}$ is a multiple of $\mathbf{a}$ and (ii) $\frac{1}{2}(\mathbf{b} + \mathbf{c})$ is in the hyperplane.

In Cimmino's method, a sequence $\{\mathbf{b}_k\}$ that converges to the solution $\mathbf{v}$ of the system is generated from an arbitrarily chosen point $\mathbf{b}_0$. As the initial step, a set of positive masses $m_1, m_2, \ldots, m_n$ is selected. The point $\mathbf{b}_{j+1}$ is obtained from $\mathbf{b}_j$ as follows: For $i = 1, 2, \ldots, n$, the point $\mathbf{c}_i$ which is symmetric to $\mathbf{b}_j$ with respect to the ith hyperplane $a_{i1}x_1 + a_{i2}x_2 + \cdots + a_{in}x_n = b_i$ is determined. The point $\mathbf{c}_i$ is assigned mass m_i, and the center of mass for this system is determined. This center of mass is the next approximation $\mathbf{b}_{j+1}$. (See Figure 11.1.)

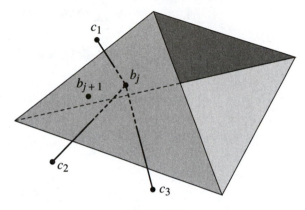

Figure 11.1

Using the fact that $\mathbf{a}_i \cdot \mathbf{a}_i = 1$, it is easily verified that $\mathbf{c}_i$ is given by the equation $\mathbf{c}_i = \mathbf{b}_j - 2(\mathbf{a}_i \cdot \mathbf{b}_j - b_i)\mathbf{a}_i$. If we let m be the sum of the masses, then $\mathbf{b}_{j+1}$ is given by this expression:

$$\mathbf{b}_{j+1} = \mathbf{b}_j - \frac{2}{m} \sum_{i=1}^{n} m_i (\mathbf{a}_i \cdot \mathbf{b}_j - b_i)\mathbf{a}_i.$$

Since the rank of A is n, the set of vectors $\mathcal{A} = \{\mathbf{a}_1, \mathbf{a}_2, \ldots, \mathbf{a}_n\}$ forms a basis of $\mathbf{R}^n$. With this particular choice of basis, we are able to obtain a recursion relation that leads to an explicit expression for the coordinates of the jth approximation. These coordinates can then be used to prove that Cimmino's method always converges. The recursion relation is given in the next lemma.

11.27 LEMMA

Suppose that $\mathbf{b}_j$ and $\mathbf{v}$ have coordinates relative to $\mathcal{A}$ given respectively by the matrices $Y_j = [\, y_{1j} \quad y_{2j} \quad \cdots \quad y_{nj}\,]^T$ and $S = [\, s_1 \quad s_2 \quad \cdots \quad s_n\,]^T$. Then

$$Y_{j+1} = \left(I - \frac{2}{m}R\right) Y_j + \frac{2}{m} RS,$$

where $R = [r_{ij}]_{n \times n} = [m_i \mathbf{a}_i \cdot \mathbf{a}_j]_{n \times n}$.

Proof With j fixed, let

$$d_i = m_i \mathbf{a}_i \cdot (\mathbf{b}_j - \mathbf{v})$$

$$= \sum_{k=1}^{n} m_i (y_{kj} - s_k) \mathbf{a}_i \cdot \mathbf{a}_k$$

$$= \sum_{k=1}^{n} r_{ik} (y_{kj} - s_k).$$

Then $D = [\, d_1 \quad d_2 \quad \cdots \quad d_n \,]^T = R(Y_j - S)$. Now

$$\mathbf{b}_{j+1} = \mathbf{b}_j - \frac{2}{m} \sum_{i=1}^{n} m_i (\mathbf{a}_i \cdot \mathbf{b}_j - b_i) \mathbf{a}_i$$

$$= \mathbf{b}_j - \frac{2}{m} \sum_{i=1}^{n} m_i (\mathbf{a}_i \cdot \mathbf{b}_j - \mathbf{a}_i \cdot \mathbf{v}) \mathbf{a}_i$$

$$= \sum_{i=1}^{n} y_{ij} \mathbf{a}_i - \frac{2}{m} \sum_{i=1}^{n} (m_i \mathbf{a}_i \cdot (\mathbf{b}_j - \mathbf{v})) \mathbf{a}_i$$

Equating coefficients, we have

$$y_{i,j+1} = y_{ij} - \frac{2}{m} \sum_{i=1}^{n} m_i \mathbf{a}_i \cdot (\mathbf{b}_j - \mathbf{v}) = y_{ij} - \frac{2}{m} d_i,$$

and therefore

$$Y_{j+1} = Y_j - \frac{2}{m} D$$

$$= Y_j - \frac{2}{m} R(Y_j - S)$$

$$= \left(I - \frac{2}{m} R \right) Y_j + \frac{2}{m} RS. \qquad \blacksquare\blacksquare\blacksquare$$

The explicit expression for Y_j in terms of Y_0 is given in the following lemma. A proof is requested in Problem 2.

11.28 LEMMA

The coordinates Y_j are given by

$$Y_j = \left(I - \frac{2}{m} R \right)^j Y_0 + \frac{2}{m} \left(\sum_{t=0}^{j-1} \left(I - \frac{2}{m} R \right)^t \right) RS.$$

As one might expect, our proof of convergence is based on an examination of the eigenvalues of $I - (2/m)R$. The first step in this direction is in the next lemma.

11.29 LEMMA

The eigenvalues λ of $(1/m)R$ satisfy $0 < \lambda < 1$.

Proof Using the norm $\|A\|_c$ as given in Theorem 11.14 together with the fact that $|\mathbf{a}_i \cdot \mathbf{a}_j| \leq 1$, it is easily shown that $\|(1/m)R\|_c < 1$. Therefore the eigenvalues of $(1/m)R$ are less than 1 in absolute value, by Theorem 11.20.

To show that the eigenvalues are positive, we let $M = \text{diag}\{\sqrt{m_1}, \sqrt{m_2}, \ldots, \sqrt{m_n}\}$, $p_{ij} = \mathbf{a}_i \cdot \mathbf{a}_j$, and $P = [p_{ij}] = AA^T$, where $A = [a_{ij}]_{n \times n}$. With this notation,

$$|R - \lambda I| = \begin{vmatrix} m_1 p_{11} - \lambda & m_1 p_{12} & \cdots & m_1 p_{1n} \\ m_2 p_{21} & m_2 p_{22} - \lambda & \cdots & m_2 p_{2n} \\ \vdots & \vdots & \ddots & \vdots \\ m_n p_{n1} & m_n p_{n2} & \cdots & m_n p_{nn} - \lambda \end{vmatrix}$$

$$= m_1 m_2 \cdots m_n \left| P - \text{diag}\left\{ \frac{\lambda}{m_1}, \frac{\lambda}{m_2}, \ldots, \frac{\lambda}{m_n} \right\} \right|.$$

Some straightforward calculations show that

$$|M| \cdot |R - \lambda I| \cdot |M| = m_1 m_2 \cdots m_n |MPM - \lambda I|.$$

It follows that $|R - \lambda I| = 0$ if and only if $|MPM - \lambda I| = 0$. That is, R and MPM have the same eigenvalues. But $MPM = MAA^T M^T = (MA)(MA)^T$, where MA is invertible. Consequently, MPM is positive definite. Therefore all eigenvalues of R and $(1/m)R$ are positive. ■■■

The proof of the following lemma is requested in Problem 4.

11.30 LEMMA

The eigenvalues of $I - (2/m)R$ have absolute value less than 1.

We are now able to show that $\{\mathbf{b}_j\}$ converges to the solution $\mathbf{v}$ of the system.

11.31 THEOREM

The sequence $\{\mathbf{b}_j\}$ converges to $\mathbf{v}$.

Proof From Lemma 11.28, we have the equation

$$Y_j = \left(I - \frac{2}{m}R\right)^j Y_0 + \frac{2}{m}\left(\sum_{t=0}^{j-1}\left(I - \frac{2}{m}R\right)^t\right)RS.$$

Since the eigenvalues of $I - (2/m)R$ are less than 1 in absolute value,

$$\lim_{j\to\infty}\left(I - \frac{2}{m}R\right)^j = 0$$

and

$$\sum_{t=0}^{\infty}\left(I - \frac{2}{m}R\right)^t = \left(I - \left(I - \frac{2}{m}R\right)\right)^{-1} = \left(\frac{2}{m}R\right)^{-1}.$$

Thus $\{Y_j\}$ converges to

$$0\cdot Y_0 + \frac{2}{m}\left(\frac{2}{m}R\right)^{-1}RS = S. \qquad \blacksquare\blacksquare\blacksquare$$

It is interesting to note that each expression in the lemmas that involves Y_j also involves S. As a consequence, none of these can be used in order to obtain the successive approximations. However, it is easy to obtain such an expression from

$$\mathbf{b}_{j+1} = \mathbf{b}_j - \frac{2}{m}\sum_{i=1}^{n}m_i(\mathbf{a}_i\cdot\mathbf{b}_j - b_i)\mathbf{a}_i.$$

If we let A_i and X_j denote the coordinate matrices relative to $\mathcal{E}_n$ of $\mathbf{a}_i$ and $\mathbf{b}_j$, respectively, then

$$X_{j+1} = X_j - \frac{2}{m}\sum_{i=1}^{n}m_i(A_i^T X_j - b_i)A_i.$$

EXAMPLE 1 We shall illustrate Cimmino's method with the following system, rounding to two decimal places.

$$\begin{aligned}
5x_1 + 6x_2 + 3x_3 &= 6 \\
-6x_1 + 3x_2 + 5x_3 &= 7 \\
3x_1 - 5x_2 + 6x_3 &= 39
\end{aligned}$$

As an initial step, we normalize the system so that each $\mathbf{a}_i$ has norm 1. This gives the equivalent system

$$\begin{aligned}
0.60x_1 + 0.72x_2 + 0.36x_3 &= 0.72 \\
-0.72x_1 + 0.36x_2 + 0.60x_3 &= 0.84 \\
0.36x_1 - 0.60x_2 + 0.72x_3 &= 4.66
\end{aligned}$$

that has

$$\mathbf{a}_1 = (0.60, 0.72, 0.36),$$

$$\mathbf{a}_2 = (-0.72, 0.36, 0.60),$$

$$\mathbf{a}_3 = (0.36, -0.60, 0.72),$$

$$b_1 = 0.72, \quad b_2 = 0.84, \quad b_3 = 4.66.$$

We choose $\mathbf{b}_0 = (1, 0, 0)$ and $m_i = \frac{1}{3}$ for $i = 1, 2, 3$. The coordinate matrices A_i and X_0 relative to $\mathcal{E}_n$ are given by

$$A_1 = \begin{bmatrix} 0.60 \\ 0.72 \\ 0.36 \end{bmatrix}, \quad A_2 = \begin{bmatrix} -0.72 \\ 0.36 \\ 0.60 \end{bmatrix}, \quad A_3 = \begin{bmatrix} 0.36 \\ -0.60 \\ 0.72 \end{bmatrix}, \quad X_0 = \begin{bmatrix} 1 \\ 0 \\ 0 \end{bmatrix}.$$

The coordinate matrices X_j of the successive approximations relative to $\mathcal{E}_n$ are obtained from the equation

$$X_{j+1} = X_j - \frac{2}{m} \sum_{i=1}^{n} m_i (A_i^T X_j - b_i) A_i$$

$$= X_j - \frac{2}{3} \sum_{i=1}^{3} (A_i^T X_j - b_i) A_i.$$

With the result of each iteration rounded to two decimal places, the successive approximations are given by

$$X_1 = \begin{bmatrix} 1.33 \\ -1.29 \\ 2.72 \end{bmatrix}, \quad X_2 = \begin{bmatrix} 1.32 \\ -1.78 \\ 3.64 \end{bmatrix}, \quad X_3 = \begin{bmatrix} 1.32 \\ -1.93 \\ 3.95 \end{bmatrix}, \quad X_4 = \begin{bmatrix} 1.27 \\ -2.03 \\ 4.06 \end{bmatrix},$$

$$X_5 = \begin{bmatrix} 1.24 \\ -2.07 \\ 4.10 \end{bmatrix}, \quad X_6 = \begin{bmatrix} 1.23 \\ -2.08 \\ 4.11 \end{bmatrix}, \quad X_7 = \begin{bmatrix} 1.23 \\ -2.08 \\ 4.12 \end{bmatrix}, \quad X_8 = \begin{bmatrix} 1.23 \\ -2.08 \\ 4.12 \end{bmatrix}.$$

Since we have $X_7 = X_8$, the iteration terminates and the approximate solution that we have obtained by Cimmino's method is

$$x_1 = 1.23, x_2 = -2.08, x_3 = 4.12.$$

In a footnote to Example 1 in Section 11.3, it was stated that the rounding of successive approximations to two decimal places does not ensure that the second decimal place is correct. This example provides an illustration of that fact. The solution provided by *Scientific WorkPlace* software is

$$x_1 = 1.2329, x_2 = -2.0959, x_3 = 4.1370,$$

correct to four decimal places. Thus both values $x_2 = -2.09$ and $x_3 = 4.13$ are off by 1 in the second decimal place. ∎

11.4 Exercises

1. Show that the expression for c_i in the fourth paragraph of Section 11.4 is correct.

2. Prove Lemma 11.28.

3. Prove that

$$\left\| \frac{1}{m} R \right\|_c < 1$$

   in the proof of Lemma 11.29. (*Hint:* Use the Schwarz inequality on r_{ij}.)

4. Prove Lemma 11.30. (*Hint:* Show that if λ is an eigenvalue of $I - (2/m)R$, then $\frac{1}{2}(1-\lambda)$ is an eigenvalue of $(1/m)R$.)

5. Use Cimmino's method to obtain approximate solutions to the following systems. Round off the successive approximations to two decimal places and continue until two successive approximations agree. Take each $m_i = 1/n$.

   (a) $7x_1 + 7x_2 = 20$
       $7x_1 - 7x_2 = 10$

   (b) $5x_1 - 6x_2 = 6$
       $6x_1 + 7x_2 = 21$

   (c) $\quad 7x_1 + 7x_2 + 3x_3 = 10$
       $-7x_1 + 3x_2 + 7x_3 = 0$
       $\quad 3x_1 - 7x_2 + 7x_3 = 20$

   (d) $6x_1 + 7x_2 + 4x_3 = -10$
       $6x_1 - 7x_2 + 4x_3 = 10$
       $6x_1 - 8x_3 = 0$

11.5 An Iterative Method for Determining Eigenvalues

We shall consider now a standard method of iteration for obtaining the eigenvalues of a matrix that resembles the iterative method for solving systems of equations that was presented in Section 11.3. This method is applicable only to $n \times n$ matrices over **R** that have n distinct real eigenvalues, and we proceed here under the assumption that this is the case. This assumption imposes a severe theoretical limitation on the method, but this restriction is met in a great many physical problems.

Let A be an $n \times n$ real matrix with n distinct real eigenvalues $\lambda_1, \lambda_2, \ldots, \lambda_n$. We assume that the eigenvalues are indexed so that $|\lambda_1| \geq |\lambda_2| \geq \cdots \geq |\lambda_n|$. In many problems it is only necessary to find the eigenvalue with largest absolute value, and the standard method of iteration is especially appropriate then. However, the method can be modified so as to obtain the entire set of eigenvalues if that is desired.

We proceed now on the assumption that $|\lambda_1| > |\lambda_2|$. A treatment of the case where $|\lambda_1| = |\lambda_2|$ is given later in this section. We first develop a procedure for obtaining a single eigenvalue of A, and then present a continuation that will obtain the remaining eigenvalues if they are desired.

The standard method of iteration is based on a sequence $\{V_k\}$ of $n \times 1$ matrices over **R** generated by the rule

$$V_1 = AV_0, \quad V_2 = AV_1, \quad \ldots, \quad V_{k+1} = AV_k, \quad \ldots$$

where V_0 is an arbitrarily chosen nonzero vector in $\mathbf{R}_{n \times 1}$. It is clear that V_k is given by $V_k = A^k V_0$.

In many instances where iteration might be used to obtain an eigenvalue of A, it would be known that A is invertible. That is, it would be known that 0 is not an eigenvalue of A. We wish to admit the possibility that 0 be an eigenvalue, and this means that some V_k may be **0**, even though $V_0 \neq \mathbf{0}$. If this possibility occurs, it is clear that an eigenvalue and a corresponding eigenvector have been found. Even though this is obvious, it is important enough to be stated as a theorem.

11.32
THEOREM

If some $V_k = \mathbf{0}$ in $\{V_k\}$, let r be the smallest integer such that $V_r \neq \mathbf{0}$ and $A V_r = \mathbf{0}$. Then 0 is an eigenvalue of A with V_r as a corresponding eigenvector.

Until stated differently, the conditions that $V_k \neq \mathbf{0}$ and $|\lambda_1| > |\lambda_2|$ will be assumed to be satisfied. For $i = 1, 2, \ldots, n$, X_i will denote an eigenvector of A associated with λ_i. Since the eigenvalues are distinct, the set $\{X_1, X_2, \ldots, X_n\}$ is linearly independent, and therefore there are scalars a_i such that $V_0 = \sum_{i=1}^{n} a_i X_i$.

11.33
THEOREM

Suppose that $a_1 \neq 0$ in $V_0 = \sum_{i=1}^{n} a_i X_i$. With i fixed and $v_{i,k+1}/v_{ik}$ restricted to nonzero v_{ik} in $V_k = (v_{1k}, v_{2k}, \ldots, v_{nk})$,

$$\lim_{k \to \infty} \frac{v_{i,k+1}}{v_{ik}} = \lambda_1.$$

Proof   Since

$$V_k = A^k V_0 = \sum_{i=1}^{n} a_i A^k X_i = \sum_{i=1}^{n} a_i \lambda_i^k X_i$$

for each positive integer k, we have

$$V_k = a_1 \lambda_1^k \left(X_1 + \sum_{i=2}^{n} \frac{a_i}{a_1} \left(\frac{\lambda_i}{\lambda_1} \right)^k X_i \right).$$

Since $|\lambda_i/\lambda_1| < 1$ for each $i > 1$,

$$\lim_{k \to \infty} \left(\frac{\lambda_i}{\lambda_1} \right)^k = 0.$$

Hence $\|V_k - a_1 \lambda_1^k X_1\| \to 0$ as $k \to \infty$. But

$$\|V_{k+1} - \lambda_1 V_k\| = \|V_{k+1} - a_1 \lambda_1^{k+1} X_1 + a_1 \lambda_1^{k+1} X_1 - \lambda_1 V_k\|$$

$$\leq \|V_{k+1} - a_1 \lambda_1^{k+1} X_1\| + |\lambda_1| \cdot \|V_k - a_1 \lambda_1^k X_1\|,$$

so this implies that $\|V_{k+1} - \lambda_1 V_k\| \to 0$ as $k \to \infty$. That is, $\lim(v_{i,k+1} - \lambda_1 v_{ik}) = 0$ for each i. With v_{ik} restricted to nonzero values, we thus have

$$\lim_{k \to \infty} \frac{v_{i,k+1}}{v_{ik}} = \lambda_1.$$ ∎

One variation of the iteration process consists of a modification of the matrices in the sequence $\{V_k\}$. With $\|V_k\| = \max_i \{|v_{ik}|\}$, the successive approximations Y_k are obtained according to

$$Y_0 = \frac{1}{\|V_0\|} V_0, \quad Y_1 = \frac{1}{\|AY_0\|} AY_0, \quad \ldots, \quad Y_{k+1} = \frac{1}{\|AY_k\|} AY_k, \ldots.$$

Each Y_k is a scalar multiple of V_k for a positive scalar, and $\|Y_k\| = 1$ for every k. Hence

$$Y_k = \frac{1}{\|V_k\|} V_k.$$

There are two techniques for obtaining λ_1 from the sequences $\{V_k\}$ and $\{Y_k\}$. In one method, the ratios $v_{i,k+1}/v_{ik}$ are computed and compared for the nonzero v_{ik} in V_k. Whenever these agree to a certain degree of accuracy, this indicates that their common value gives the eigenvalue λ_1 to that degree of accuracy. An associated eigenvector can then be obtained from $Y_k = V_k/\|V_k\|$ if it is needed. With the other method, the sequence $\{Y_k\}$ is computed until repetition is obtained to the desired accuracy. The eigenvalue λ_1 is then found by computing ratios of the corresponding elements of AY_k and Y_k.

The method using the ratios $v_{i,k+1}/v_{ik}$ is illustrated in the following example, and the other method is used in Example 2 near the end of this section.

EXAMPLE 1 Let

$$A = \begin{bmatrix} 2 & -2 & 3 \\ 1 & 1 & 1 \\ 1 & 3 & -1 \end{bmatrix}.$$

We shall use the standard method of iteration to find the eigenvalue λ_1 of A that has the greatest absolute value. Our technique will be to compute the ratios $v_{i,k+1}/v_{ik}$ for nonzero v_{ik} until two consecutive values agree to two decimal places and then take this common value as our approximation to λ_1.

With $V_0 = [\, 1 \quad 0 \quad 0 \,]^T$, we obtain the following elements of $\{V_k\}$.

$$V_0 = \begin{bmatrix} 1 \\ 0 \\ 0 \end{bmatrix}, \qquad V_1 = AV_0 = \begin{bmatrix} 2 \\ 1 \\ 1 \end{bmatrix}, \qquad V_2 = AV_1 = \begin{bmatrix} 5 \\ 4 \\ 4 \end{bmatrix},$$

$$V_3 = AV_2 = \begin{bmatrix} 14 \\ 13 \\ 13 \end{bmatrix}, \qquad V_4 = AV_3 = \begin{bmatrix} 41 \\ 40 \\ 40 \end{bmatrix}, \qquad V_5 = AV_4 = \begin{bmatrix} 122 \\ 121 \\ 121 \end{bmatrix},$$

$$V_6 = AV_5 = \begin{bmatrix} 365 \\ 364 \\ 364 \end{bmatrix}, \qquad V_7 = AV_6 = \begin{bmatrix} 1094 \\ 1093 \\ 1093 \end{bmatrix}, \qquad V_8 = AV_7 = \begin{bmatrix} 3281 \\ 3280 \\ 3280 \end{bmatrix}.$$

Computing the ratios $v_{1,k+1}/v_{1,k}$, we get

$$\frac{v_{1,1}}{v_{1,0}} = \frac{2}{1} = 2, \qquad \frac{v_{1,2}}{v_{1,1}} = \frac{5}{2} = 2.5, \qquad \frac{v_{1,3}}{v_{1,2}} = \frac{14}{5} = 2.8,$$

$$\frac{v_{1,4}}{v_{1,3}} = \frac{41}{14} = 2.9286, \qquad \frac{v_{1,5}}{v_{1,4}} = \frac{122}{41} = 2.9756, \qquad \frac{v_{1,6}}{v_{1,5}} = \frac{365}{122} = 2.9918,$$

$$\frac{v_{1,7}}{v_{1,6}} = \frac{1094}{365} = 2.9973, \qquad \frac{v_{1,8}}{v_{1,7}} = \frac{3281}{1094} = 2.9991.$$

Since each of the last two calculations round to 3.00, we take $\lambda_1 = 3.00$ as our approximation. By inspection, we can see that the choices $i = 2$ and $i = 3$ in this example would not change our estimate of

$$\lim_{k \to \infty} \frac{v_{i,k+1}}{v_{ik}}.$$

Scientific WorkPlace gives the eigenvalues of A as 1, 3, and -2. ■

11.34 THEOREM

Suppose that $a_1 \neq 0$ in $V_0 = \sum_{i=1}^{n} a_i X_i$. If $\lambda_1 > 0$, $\{Y_k\}$ converges to an eigenvector of A corresponding to λ_1. If $\lambda_1 < 0$, $\{Y_{2k}\}$ converges to an eigenvector of A corresponding to λ_1, and $\{Y_{2k+1}\}$ converges to the negative of this eigenvector.

Proof With the same notation as in the proof of Theorem 11.33, we have

$$V_k = a_1 \lambda_1^k X_1 + a_2 \lambda_2^k X_2 + \cdots + a_n \lambda_n^k X_n,$$

and hence

$$\frac{V_k}{\lambda_1^k} = a_1 X_1 + a_2 \left(\frac{\lambda_2}{\lambda_1} \right)^k X_2 + \cdots + a_n \left(\frac{\lambda_n}{\lambda_1} \right)^k X_n.$$

Since $|\lambda_i| < |\lambda_1|$ for all $i \neq 1$,

$$\lim_{k \to \infty} \frac{V_k}{\lambda_1^k} = a_1 X_1.$$

This implies that

$$\lim_{k \to \infty} \frac{\|V_k\|}{|\lambda_1|^k} = \|a_1 X_1\|, \quad \text{and} \quad \lim_{k \to \infty} \frac{|\lambda_1|^k}{\|V_k\|} = \frac{1}{\|a_1 X_1\|}.$$

Now Y_k can be written as

$$Y_k = \frac{\lambda_1^k}{\|V_k\|} \cdot \frac{V_k}{\lambda_1^k}.$$

If $\lambda_1 > 0$, then

$$\lim_{k \to \infty} \frac{\lambda_1^k}{\|V_k\|} = \frac{1}{\|a_1 X_1\|}, \quad \text{and} \quad \lim_{k \to \infty} Y_k = \frac{a_1 X_1}{\|a_1 X_1\|}.$$

If $\lambda_1 < 0$, then

$$\lim_{k \to \infty} \frac{\lambda_1^{2k}}{\|V_{2k}\|} = \frac{1}{\|a_1 X_1\|}, \quad \text{and} \quad \lim_{k \to \infty} Y_{2k} = \frac{a_1 X_1}{\|a_1 X_1\|}.$$

For the sequence of odd-numbered terms, we have

$$\lim_{k \to \infty} \frac{\lambda_1^{2k+1}}{\|V_{2k+1}\|} = -\frac{1}{\|a_1 X_1\|}, \quad \text{and} \quad \lim_{k \to \infty} Y_{2k+1} = -\frac{a_1 X_1}{\|a_1 X_1\|}. \qquad \blacksquare\blacksquare\blacksquare$$

Consider now the case where $|\lambda_1| = |\lambda_2|$. Since all eigenvalues of A are distinct and real, this means that $\lambda_2 = -\lambda_1$. We may assume that $\lambda_1 > 0$. In this case,

$$\frac{V_k}{\lambda_1^k} = a_1 X_1 + (-1)^k a_2 X_2 + \left(\frac{\lambda_3}{\lambda_1}\right)^k a_3 X_3 + \cdots + \left(\frac{\lambda_n}{\lambda_1}\right)^k a_n X_n.$$

Thus we have

$$\lim_{k \to \infty} \frac{V_{2k}}{\lambda_1^{2k}} = a_1 X_1 + a_2 X_2 \quad \text{and} \quad \lim_{k \to \infty} \frac{V_{2k+1}}{\lambda_1^{2k+1}} = a_1 X_1 - a_2 X_2.$$

This implies that

$$\lim_{k \to \infty} Y_{2k} = \frac{a_1 X_1 + a_2 X_2}{\|a_1 X_1 + a_2 X_2\|} \quad \text{and} \quad \lim_{k \to \infty} Y_{2k+1} = \frac{a_1 X_1 - a_2 X_2}{\|a_1 X_1 - a_2 X_2\|}.$$

Since $\{Y_{2k}\}$ and $\{Y_{2k+1}\}$ converge to two different limits that are not negatives of each other, the case where $\lambda_2 = -\lambda_1$ is easily recognized. The next theorem gives a method for obtaining λ_1 and λ_2 in this case. The proof is similar to that of Theorem 11.33.

**11.35
THEOREM**

Consider the case where $\lambda_2 = -\lambda_1$ and $\lambda_1 > 0$. With i fixed and $v_{i,2k+2}/v_{i,2k}$ restricted to nonzero $v_{i,2k}$,

$$\lim_{k \to \infty} \frac{v_{i,2k+2}}{v_{i,2k}} = \lambda_1^2.$$

Proof See Problem 5. $\blacksquare\blacksquare\blacksquare$

The same type of statement as that in Theorem 11.35 can be made for the quotients $v_{i,2k+1}/v_{i,2k-1}$.

From the expression

$$V_{k+1} + \lambda_1 V_k = 2a_1 \lambda_1^{k+1} X_1 + (\lambda_1 + \lambda_3)\lambda_3^k a_3 X_3 + \cdots + (\lambda_1 + \lambda_n)\lambda_n^k a_n X_n,$$

it is clear that

$$\lim_{k \to \infty} \frac{V_{k+1} + \lambda_1 V_k}{\lambda_1^{k+1}} = 2a_1 X_1.$$

This yields an eigenvector corresponding to λ_1.

A similar analysis shows that $(V_{k+1} - \lambda_1 V_k)/(-\lambda_1)^{k+1}$ converges to $2a_2 X_2$.

We turn our attention now to the problem of determining the remaining eigenvalues of A after one of them is known.

We have seen in Chapter 7 that if λ is an eigenvalue of A with X as an associated eigenvector, then $p(A)X = p(\lambda)X$ for any polynomial $p(x)$. That is, $p(\lambda)$ is an eigenvalue of $p(A)$ with X as an associated eigenvector. With $V_0 = \sum_{i=1}^{n} a_i X_i$ as before, the sequence $\{V_k\}$ for the matrix $p(A)$ is given by

$$V_k = (p(\lambda_1))^k a_1 X_1 + (p(\lambda_2))^k a_2 X_2 + \cdots + (p(\lambda_n))^k a_n X_n.$$

Let λ_j be an eigenvalue of A which has an intermediate absolute value. If $|p(\lambda_j)|$ is greatest for the eigenvalues of $p(A)$, then the ratios $v_{i,k+1}/v_{ik}$ will converge to $p(\lambda_j)$, and $\{Y_k\}$ will converge to a multiple of X_j.

Suppose now that X_1 and λ_1 have been obtained. If $p(x) = x - \lambda_1$, the term $(p(\lambda_1))^k a_1 X_1$ disappears from V_k. That is, we have

$$V_k = (\lambda_2 - \lambda_1)^k a_2 X_2 + (\lambda_3 - \lambda_1)^k a_3 X_3 + \cdots + (\lambda_n - \lambda_1)^k a_n X_n.$$

In most instances, the term $a_1(p(\lambda_1))^k X_1$ will not actually disappear since the value λ_1' obtained for λ_1 is not entirely accurate. But the term $(\lambda_1 - \lambda_1')^k a_1 X_1$ tends to 0 as k increases. Thus, iteration with the matrix $A - \lambda_1 I$ will converge to a multiple of the eigenvector X_j that has the eigenvalue $\lambda_j - \lambda_1$ of greatest absolute value. Once $\lambda_j - \lambda_1$ is known, λ_j can be found by adding λ_1. Whenever λ_j has been obtained, the iteration procedure may be repeated with the matrix $(A - \lambda_1 I)(A - \lambda_j I)$, and the sequence of vectors Y_k will converge to another eigenvector of the original matrix A. This procedure can be repeated until each eigenvalue of A has been obtained.

The discussion in the preceding paragraph applies when λ_1 has been obtained. There is, of course, the possibility that $\lambda_n = 0$ is the first eigenvalue obtained. If λ_1 is replaced by λ_n in the discussion, the resulting argument shows that the method of iteration then yields the eigenvalue λ_1.

EXAMPLE 2   To illustrate the foregoing procedure, we will find the eigenvalue λ_2 for the matrix

$$A = \begin{bmatrix} 2 & -2 & 3 \\ 1 & 1 & 1 \\ 1 & 3 & -1 \end{bmatrix}$$

in Example 1 of this section. We will be iterating with the matrix

$$A - \lambda_1 I = A - 3I = \begin{bmatrix} -1 & -2 & 3 \\ 1 & -2 & 1 \\ 1 & 3 & -4 \end{bmatrix},$$

and the iteration will converge to the eigenvalue $\lambda_j - 3$ having greatest absolute value. For variety, this time we will use the sequence $\{Y_k\}$ where $B = A - \lambda_1 I = A - 3I$ and

$$Y_k = \frac{1}{\|BY_{k-1}\|} BY_{k-1}$$

so that $\|Y_k\| = 1$ for all k. With $Y_k = [\, y_{1k} \quad y_{2k} \quad y_{3k} \,]^T$, we will use the norm $\|Y_k\| = \max_i \{|y_{ik}|\}$. Beginning with the arbitrary choice $Y_0 = [\,1 \quad 1 \quad 0\,]^T$, we obtain the following matrices in $\{Y_k\}$.

$$Y_1 = \begin{bmatrix} -0.75 \\ -0.25 \\ 1.00 \end{bmatrix}, \qquad Y_2 = \begin{bmatrix} 0.77273 \\ 0.13636 \\ -1.00000 \end{bmatrix}, \quad Y_3 = \begin{bmatrix} -0.78071 \\ -0.09649 \\ 1.00000 \end{bmatrix},$$

$$Y_4 = \begin{bmatrix} 0.78374 \\ 0.08131 \\ -1.00000 \end{bmatrix}, \quad Y_5 = \begin{bmatrix} -0.78493 \\ -0.07536 \\ 1.00000 \end{bmatrix}, \quad Y_6 = \begin{bmatrix} 0.78541 \\ 0.07300 \\ -1.00000 \end{bmatrix},$$

$$Y_7 = \begin{bmatrix} -0.78559 \\ -0.07206 \\ 1.00000 \end{bmatrix}, \quad Y_8 = \begin{bmatrix} 0.78566 \\ 0.07168 \\ -1.00000 \end{bmatrix}, \quad Y_9 = \begin{bmatrix} -0.78569 \\ -0.07153 \\ 1.00000 \end{bmatrix}.$$

Rounding our results to two decimal places, it seems clear that the eigenvalue $\lambda_2 - 3$ of B is negative, the sequence $\{Y_{2k}\}$ is converging to the eigenvector of B given by

$$X = \begin{bmatrix} 0.79 \\ 0.07 \\ -1.00 \end{bmatrix},$$

and the sequence $\{Y_{2k+1}\}$ is converging to $-X$. To find the value of $\lambda_2 - 3$, we compute

$$BY_9 = \begin{bmatrix} 3.9288 \\ 0.3574 \\ -5.0003 \end{bmatrix}$$

and the ratios of corresponding elements of BY_9 and Y_9. The ratios are

$$\frac{3.9288}{-0.7857} = -5.0004, \qquad \frac{0.3574}{-0.0715} = -4.9986 \quad \text{and} \quad \frac{-5.0003}{1.0000} = -5.0003.$$

Thus we conclude that -5 is the eigenvalue of B with largest absolute value, $\lambda_2 - 3 = -5$, and $\lambda_2 = -2$ is the eigenvalue of A with second largest absolute value.

As an alternative to the use of BY_9 and Y_9, we could have used

$$AY_9 = \begin{bmatrix} 2 & -2 & 3 \\ 1 & 1 & 1 \\ 1 & 3 & -1 \end{bmatrix} \begin{bmatrix} -0.78569 \\ -0.07153 \\ 1.00000 \end{bmatrix} = \begin{bmatrix} 1.5717 \\ 0.1428 \\ -2.0003 \end{bmatrix}$$

and Y_9 with the corresponding ratios

$$\frac{1.5717}{-0.7857} = -2.0004, \quad \frac{0.1428}{-0.0715} = -1.9972, \quad \text{and} \quad \frac{-2.0003}{1.0000} = -2.0003.$$

These computations lead directly to the eigenvalue $\lambda_2 = -2$ of A without finding the eigenvalue of B.

To find the remaining eigenvalue of A, we iterate with the matrix

$$\begin{aligned} C &= (A - \lambda_1 I)(A - \lambda_2 I) \\ &= (A - 3I)(A + 2I) \\ &= \begin{bmatrix} -3 & 5 & -2 \\ 3 & -5 & 2 \\ 3 & -5 & 2 \end{bmatrix}, \end{aligned}$$

using

$$Y_k = \frac{1}{\|CY_{k-1}\|} CY_{k-1}.$$

Starting with $Y_0 = [\,1 \quad 0 \quad 0\,]^T$, we obtain

$$Y_0 = \begin{bmatrix} 1 \\ 0 \\ 0 \end{bmatrix}, \quad Y_1 = \begin{bmatrix} -1 \\ 1 \\ 1 \end{bmatrix}, \quad Y_2 = \begin{bmatrix} 1 \\ -1 \\ -1 \end{bmatrix}.$$

Since $Y_2 = -Y_1$, we see that the sequence will continue to alternate between Y_1 and $-Y_1$. Thus Y_1 is an eigenvector of C corresponding to the eigenvalue -1 of C, and therefore Y_1 is an eigenvector of A. To find the associated eigenvalue of A, we compute

$$AY_1 = \begin{bmatrix} 2 & -2 & 3 \\ 1 & 1 & 1 \\ 1 & 3 & -1 \end{bmatrix} \begin{bmatrix} -1 \\ 1 \\ 1 \end{bmatrix} = \begin{bmatrix} -1 \\ 1 \\ 1 \end{bmatrix} = Y_1.$$

Hence the eigenvalue of A corresponding to Y_1 is 1. We have found the eigenvalues of A to be $\lambda_1 = 3$, $\lambda_2 = -2$, and $\lambda_3 = 1$.   ∎

In practice, the remaining eigenvalues are not usually found by the procedure used in Example 2. Our example is intended to illustrate the theory, and refined techniques that give better accuracy are normally used. These refinements are too complicated for presentation here.

11.5 Exercises

1. Use the sequence $\{V_k\}$ and Theorem 11.33 to find λ_1 for the given matrix A. Obtain an answer correct to two digits.

   (a) $A = \begin{bmatrix} 28 & -27 \\ 18 & -17 \end{bmatrix}$,

   (b) $A = \begin{bmatrix} -36 & 57 \\ -38 & 59 \end{bmatrix}$

   (c) $A = \begin{bmatrix} 23 & 22 & 25 \\ 19 & 20 & 19 \\ -22 & -22 & -24 \end{bmatrix}$

   (d) $A = \begin{bmatrix} 5 & 56 & 31 \\ 3 & 58 & 31 \\ -6 & -56 & -32 \end{bmatrix}$

2. Use the method described in this section with the sequence $\{Y_k\}$ to obtain the value of λ_2, correct to two digits, for each matrix A in Problem 1.

3. For each A in Problem 1, use the sequence $\{Y_k\}$ and Theorem 11.34 to find an eigenvector corresponding to λ_1. Obtain an answer correct to two digits in each element.

4. Use the eigenvector found in Problem 3 to obtain λ_1 to two digits.

5. Prove Theorem 11.35.

6. State and prove a result for the quotients $v_{i,2k+1}/v_{i,2k-1}$ that corresponds to the statement in Theorem 11.35.

7. For the case where $\lambda_2 = -\lambda_1$, prove that $(V_{k+1} - \lambda_1 V_k)/(-\lambda_1)^{k+1}$ converges to $2a_2 X_2$.

Answers to Selected Exercises

Exercises 1.1, page 7

3. (a) Yes, $\mathbf{v} = -2(1, 1, 1) + 3(0, 1, 1) + 3(0, 0, 1)$ **(c)** No

(e) Yes, $\mathbf{v} = -(1, 0, -1, 1) + 0(-2, 0, 2, -2) + 1(1, 1, 1, 1) + 0(2, 1, 0, 2)$

5. $(a_1, a_2, a_3) = (a_1 - a_2)\mathbf{u}_1 + (a_2 - a_3)\mathbf{u}_2 + a_3\mathbf{u}_3$

7. $(a_1, a_2, a_3) = a_1\mathbf{u}_1 + (a_2 - a_1)\mathbf{u}_2 + (a_3 - a_2 + a_1)\mathbf{u}_3 + (a_4 - a_3 + a_2 - a_1)\mathbf{u}_4$

8. (a) Yes **(c)** No **9.** Yes **14. (a)** 12 **(c)** Minimum of m and n

Exercises 1.2, page 11

1. (a) Yes **(c)** No

2. (a) $(1, 1, 0) + (0, 1, 1) - (-1, 1, 1) - (2, 1, 0) = (0, 0, 0)$
$(2, 1, 0) = (1, 1, 0) + (0, 1, 1) - (-1, 1, 1)$

3. $(1, 0, 1)$ is such a vector **11.** No

Exercises 1.3, page 16

1. No, property (4) of Definition 1.9 fails.

3. No, property (1) of Definition 1.9 fails.

5. Yes **7.** Yes **9.** No, property (4) of Definition 1.9 fails.

11. Yes **13.** No, property (1) of Definition 1.9 fails.

17. Definition 1.4 Let $\mathcal{A} = \{\mathbf{u}_\lambda \mid \lambda \in \mathcal{L}\}$ be a nonempty set of vectors in $\mathbf{R}^n$. A vector $\mathbf{v}$ is *linearly dependent* on $\mathcal{A}$ if there exist vectors $\mathbf{u}_{\lambda_1}, \mathbf{u}_{\lambda_2}, \ldots, \mathbf{u}_{\lambda_k}$ in $\mathcal{A}$ and $a_1, a_2, \ldots, a_k$ in $\mathbf{R}$ such that $\mathbf{v} = a_1\mathbf{u}_{\lambda_1} + a_2\mathbf{u}_{\lambda_2} + \cdots + a_k\mathbf{u}_{\lambda_k}$. A vector of the form $\sum_{i=1}^{k} a_i\mathbf{u}_{\lambda_i}$ is called a *linear combination* of the vectors in $\mathcal{A}$.

Definition 1.7 Let $\mathcal{A} = \{\mathbf{u}_\lambda \mid \lambda \in \mathcal{L}\}$ be a nonempty set of vectors in $\mathbf{R}^n$. Then $\mathcal{A}$ is *linearly dependent* if there exist vectors $\mathbf{u}_{\lambda_1}, \mathbf{u}_{\lambda_2}, \ldots, \mathbf{u}_{\lambda_k}$ and scalars $a_1, a_2, \ldots, a_k$, not all zero, such that $a_1\mathbf{u}_{\lambda_1} + a_2\mathbf{u}_{\lambda_2} + \cdots + a_k\mathbf{u}_{\lambda_k} = 0$. If $\mathcal{A}$ is not linearly dependent, it is *linearly independent*.

18. (a) Yes **(c)** No **19. (a)** $\{(0, 0)\}$

20. (a) Set of all vectors of the form $(2a, a, -2a)$ where $a \in \mathbf{R}$.

Exercises 1.4, page 21

1. (a) $\{(1, 0, 0), (0, 1, 0), (0, 0, 1)\}$

2. (a) Yes, $(2, 3) = 2(1, 1) + (0, 1)$ **(c)** Yes, $(-4, 4) = -7(0, -1) - 1(4, 3)$

(e) Yes, $(-1, 2, 5) = 2(0, 1, 3) - (1, 0, 1)$ **(g)** No

3. (a) $(1, 0, 0)$ **(c)** $(0, 1, 0)$ **4. (a)** Yes   **(c)** No

5. (a) Yes   **(c)** Yes   **(e)** Yes   **6. (a)** Yes   **7. (a)** No   **(c)** Yes   **(e)** No

8. (a) $\{(1, 0), (0, 0)\}$ **(c)** $\{(1, -1, 0), (1, -1, 1)\}$
   (e) $\{(1, -1, 0), (1, -1, 1), (0, 0, 0), (-1, 1, 0)\}$

9. $\{(1, -3)\}$

Exercises 1.5, page 30

1.

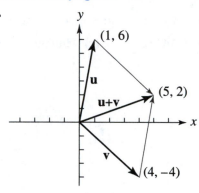

3. $\bigcap_{\lambda \in \mathcal{L}} \mathcal{M}_\lambda$ is the origin.

$\bigcup_{\lambda \in \mathcal{L}} \mathcal{M}_\lambda$ is the set of all points except those with coordinates $(0, y)$, where $y \neq 0$.

4. (a) $3x + 4y = 0$ **5. (a)** $2x + 3y - z = 0$ **(c)** $x - 2y + z = 0$

6. (a) 13 **(c)** 5 **(e)** $\sqrt{30}$ **(g)** $9\sqrt{2}$ **(i)** $\sqrt{145}$

7. (a) $x = 6$ **(c)** $x = \frac{1}{2}$ **8. (a)** $\frac{1}{13}(3, -4, 12)$ **9. (a)** $-4\sqrt{21}/21$

10. (a) $(3, 6)$ **(c)** $\frac{3}{11}(-2, 5, 2)$ **11. (a)** $2\sqrt{5}$

13. $(-26, 39) = 2(5, 12) - 3(12, -5)$

15. (a) $\{(1, 2, 1)\}$ **(b)** $\{(3, 1, -2), (-2, 1, 3), (0, 1, 0)\}$ **23. (a)** 5 **(c)** $\sqrt{19}$

Exercises 1.6, page 42

1. (a) Linearly independent   **(c)** Linearly dependent   **(e)** Linearly independent

2. (a) Does not span $\mathbf{R}^3$ **(c)** Does not span $\mathbf{R}^3$ **(e)** Spans $\mathbf{R}^3$

3. (a) Not a basis for $\mathbf{R}^3$ **(c)** Basis for $\mathbf{R}^3$ **(e)** Not a basis for $\mathbf{R}^3$

4. (a) Not a basis of $\mathbf{R}^4$ **(c)** Basis of $\mathbf{R}^4$ **8. (a)** $\frac{3}{5}, \frac{1}{5}$ **(c)** $4, 2$

9. (a) $1, -1, 0$ **(c)** $1, 1, 1$ **10. (a)** $\frac{5}{7}, \frac{10}{7}$ **11. (a)** $1, -1, 1$

12. (a) $-2x + y, x$ **13. (a)** $\frac{1}{2}(x + y - z), \frac{1}{2}(x - y + z), -\frac{1}{2}(x - y - z)$

14. (a) $\{(1, 2)\}$ **(c)** $\{(1, 0, 3)\}$ **(e)** $\{(3, 1, 0), (0, -1, 2)\}$
    (g) $\{(1, 0, 0, 0), (0, 1, 0, 0), (-1, 1, 1, -1)\}$

15. (a) $\{(3, -4)\}$

16. (a) $\{(-2, 3, 0), (1, 0, 3)\}$ **(c)** $\{(-1, 0, 7), (0, 1, 0)\}$

17. (a) $\{(1, 2, 0), (1, 0, 0), (0, 0, 1)\}$

18. (a) $\{(0, 0, 4, -5), (1, 0, 0, 0), (0, 1, 0, 0), (0, 0, 1, 0)\}$
    (c) $\{(-2, 1, 0, 0), (0, 1, 1, 1), (1, 0, 0, 0), (0, 0, 1, 0)\}$

19. (a) $\{(2, 6, -3), (5, 15, -8), (5, 3, -2)\}$
    (c) $\{(1, 1, 0), (2, 4, 1), (1, 2, 1)\}$

20. (a) $\{(1, 0, 1, -1), (3, -2, 3, 5)\}$
    (c) $\{(2, -1, 0, 1), (1, 2, 1, 0), (5, 3, 2, 1)\}$

21. (a) 2   **(c)** 2

22. (a) $\langle (1, 0, 0, 0), (0, 1, 0, 0) \rangle$
    (c) $\langle (-2, 2, -6, 4), (2, 1, 1, -1) \rangle$

23. (a) $\langle (1, 0, 0, 0), (0, 1, 0, 0), (0, 0, 1, 0) \rangle$
    (c) $\langle (0, 1, 4, 0, 0), (0, 0, 1, 3, 0), (0, 2, 7, -6, 0) \rangle$

25. (a) $\{(1, 0, 2, 3), (0, 1, -2, -3), (0, 1, 1, 0), (0, 0, 0, 1)\}$
    (c) $\{(1, 1, 0, 0), (0, 0, 1, 1), (1, 0, 1, 0), (0, 1, 0, -1)\}$

Exercises 2.1, page 52

1. $\{(1, 0, 0, 0, 0), (0, 1, 0, 0, 0), (0, 0, 1, 0, 0), (0, 0, 0, 1, 0), (0, 0, 0, 0, 1)\}$

3. Replace the second vector by the sum of the second vector and (-2) times the first vector.

5. Apply the sequence E_1, E_2, E_3, E_4 where

E_1: Replace the second vector by the sum of the second and (-1) times the first,

E_2: Interchange the two vectors,

E_3: Multiply the second vector by 3,

E_4: Replace the second vector by the sum of the second and the first.

6. (a) Apply the sequence E_1, E_2, E_3, E_4, E_5, E_6, E_7, E_8, E_9 where
    E_1: Replace the first vector by the sum of the first and 3 times the second,
    E_2: Replace the third vector by the sum of the third and the second,
    E_3: Multiply the first vector by -1,
    E_4: Replace the second vector by the sum of the second and 2 times the first,
    E_5: Replace the third vector by the sum of the third and 4 times the first,
    E_6: Multiply the third vector by $-\frac{1}{11}$,

E_7: Replace the first vector by the sum of the first and 3 times the third,

E_8: Replace the second vector by the sum of the second and 6 times the third,

E_9: Interchange the first and second vectors.

7. (a) Apply the sequence E_1, E_2, E_3, E_4, E_5, E_6, E_7, E_8, E_9, E_{10}, E_{11}, E_{12}, E_{13} where

E_1: Replace the second vector by the sum of the second and -1 times the third,

E_2: Replace the third vector by the sum of the third and -3 times the second,

E_3: Replace the first vector by the sum of the first and -2 times the fourth,

E_4: Replace the second vector by the sum of the second and -1 times the fourth,

E_5: Replace the third vector by the sum of the third and 3 times the fourth,

E_6: Multiply the third vector by $\frac{1}{19}$,

E_7: Replace the first vector by the sum of the first and 9 times the third,

E_8: Replace the second vector by the sum of the second and 6 times the third,

E_9: Replace the fourth vector by the sum of the fourth and -5 times the third,

E_{10}: Multiply the first vector by $\frac{1}{2}$,

E_{11}: Replace the fourth vector by the sum of the fourth and -1 times the first,

E_{12}: Interchange the first and second vectors,

E_{13}: Interchange the second and fourth vectors.

8. (a) No   **9. (a)** No

11. E_1: Replace the fourth vector by the sum of the fourth and the third vector.

E_2: Replace the third vector by the sum of the third and the second vector.

E_3: Replace the second vector by the sum of the second and the first vector.

13. Apply the sequence F_1, F_2, F_3 of elementary operations where

$F_1 = E_3^{-1}$: replace the third vector by the sum of the third vector and 2 times the second vector,

$F_2 = E_2^{-1}$: replace the second vector by the sum of the second vector and -2 times the first vector,

$F_3 = E_1^{-1}$: multiply the first vector by $-\frac{1}{3}$.

15. Apply the sequence F_1, F_2, F_3 of elementary operations where

$F_1 = E_3^{-1}$: replace the second vector by the sum of the second and -1 times the first vector,

$F_2 = E_2^{-1}$: replace the third vector by the sum of the third and -1 times the second vector,

$F_3 = E_1^{-1}$: replace the fourth vector by the sum of the fourth and -1 times the third vector.

17. Apply the sequence F_1, F_2, F_3, F_4 to $\mathcal{A}$ with the F_i as follows.

F_1: Interchange the second and third vectors to obtain from $\mathcal{A}$ the set $\mathcal{A}_1' = \{(1, 0, 2), (0, 1, 5), (2, 3, 19)\}$.

F_2: Replace the third vector by the sum of the third vector and -2 times the first vector to obtain the set $\mathcal{A}_2' = \{(1, 0, 2), (0, 1, 5), (0, 3, 15)\}$ from $\mathcal{A}_1'$.

F_3: Replace the third vector by the sum of the third vector and -3 times the second vector to obtain $\mathcal{A}_3' = \{(1, 0, 2), (0, 1, 5), (0, 0, 0)\}$ from $\mathcal{A}_2'$.

F_4: Multiply the first vector by 2 to obtain the set $\mathcal{A}_4 = \{(2, 0, 4), (0, 1, 5), (0, 0, 0)\}$.

Exercises 2.2, page 56

6. (a) Linearly independent   **(c)** Linearly dependent   **(e)** Linearly dependent

Exercises 2.3, page 65

1. E_1: Interchange the first and third vectors.

E_2: Multiply the first vector by $\frac{1}{2}$.

E_3: Replace the second vector by the sum of the second and (-3) times the first.

E_4: Replace the fourth vector by the sum of the fourth and the first.

E_5: Multiply the second vector by $(-\frac{1}{2})$.

E_6: Replace the first vector by the sum of the first and (-1) times the second.

E_7: Replace the third vector by the sum of the third and the second.

E_8: Replace the fourth vector by the sum of the fourth and (-3) times the second.

2. (a) $\mathcal{A}' = \{(1, 0, 5, 0, 2), (0, 1, 2, 0, 1), (0, 0, 0, 1, 1), (0, 0, 0, 0, 0)\}$

3. (a) 3, so the set is linearly dependent.   **4. (a)** 2   **(c)** 3

5. (a) $\{(1, 0, 0, 1), (0, 1, 0, -1), (0, 0, 1, 1)\}$ **(c)** $\{(1, 0, 1, 1), (0, 1, 1, -1)\}$

6. (a) $\{(1, -\frac{1}{2}, 0)\}$ **(c)** $\{(1, 0, -\frac{1}{3}), (0, 1, 1)\}$ **(e)** $\{(0, 0, 1, 0), (0, 0, 0, 1)\}$
   (g) $\{(1, 0, 0, 0, \frac{1}{4}), (0, 1, 0, 0, -1), (0, 0, 0, 1, 1)\}$

7. (a) $\{(1, -\frac{2}{3})\}$ **8. (a)** $\{(1, 0, 1), (0, 1, -1)\}$ **(c)** $\{(1, 0, 0), (0, 1, 2)\}$

9. (a) $\langle \mathcal{A} \rangle = \langle \mathcal{B} \rangle$ **(c)** $\langle \mathcal{A} \rangle \neq \langle \mathcal{B} \rangle$ **(e)** $\langle \mathcal{A} \rangle = \langle \mathcal{B} \rangle$ **10. (a)** 3

11. (a) E_1: Replace the second vector by the sum of the second vector and (-1) times the first.

E_2: Replace the first vector by the sum of the first and second.

E_3: Multiply the second vector by (-1).

E_4: Replace the first vector by the sum of the first and $(-\frac{1}{2})$ times the second.

E_5: Multiply the first vector by 2.

(c) E_1: Interchange the first and second vectors.

E_2: Replace the second vector by the sum of the second vector and (-4) times the first vector.

E_3: Replace the first vector by the sum of the first vector and (-1) times the second vector.

E_4: Replace the first vector by the sum of the first vector and (-1) times the second vector.

E_5: Multiply the second vector by 2.

Exercises 3.1, page 72

1. $x = 2y$ **2. (a)** $(7, 14)$ **(c)** $(2, 4, -14, 1)$ **3. (a)** $\begin{bmatrix} 6 \\ 2 \end{bmatrix}$ **(c)** $\begin{bmatrix} 2 \\ -2 \\ -1 \end{bmatrix}$

5. $d_1 = x_3, d_2 = x_2 - x_3, d_3 = x_1 - x_2$ **7.** $\begin{bmatrix} 3 \\ -1 \end{bmatrix}$ **9.** $\begin{bmatrix} 1 \\ 2 \\ -1 \end{bmatrix}$

11. $\begin{bmatrix} a \\ b - a \\ c - b \end{bmatrix}$ **15. (a)** $\begin{bmatrix} 2 & 1 \\ 3 & 2 \end{bmatrix}$ **(c)** $\begin{bmatrix} 1 & -1 \\ 2 & 6 \end{bmatrix}$ **(e)** $\begin{bmatrix} 1 & 1 \\ 1 & 0 \\ 1 & -1 \\ 1 & 1 \end{bmatrix}$

16. (a) $\begin{bmatrix} 2 & -1 \\ -3 & 2 \end{bmatrix}$ **(c)** $\begin{bmatrix} \frac{3}{4} & \frac{1}{8} \\ -\frac{1}{4} & \frac{1}{8} \end{bmatrix}$ **(e)** Does not exist.   **17.** p

18. (a) $\{(4, 4), (-1, 1)\}$
   (c) $\{(0, 0, 0, 0), (0, 1, 4, 4), (8, -2, 8, -16), (5, 3, 10, -4), (5, 2, 6, -8)\}$
   (e) $\{(0, 1, -4), (-1, 2, 4), (1, -2, 0)\}$

19. (a) $\begin{bmatrix} 3 & -8 & -5 \\ 0 & 4 & 4 \end{bmatrix}$ is a matrix of transition from the first set to the second set.
   (c) No matrix of transition exists.

21. $\{(1, 0, -1, 2), (0, 1, 2, -1)\}$

Exercises 3.2, page 80

1. (a) $\begin{bmatrix} 29 & -5 \\ 6 & -6 \\ 32 & -5 \end{bmatrix}$ **(c)** $\begin{bmatrix} 9 & -18 \\ 29 & -2 \end{bmatrix}$ **(e)** $\begin{bmatrix} 11 & 12 \\ -75 & -7 \end{bmatrix}$

2. (a) BA does not exist.   **(c)** $\begin{bmatrix} 1 & -6 & -5 \\ 12 & -2 & -11 \\ 18 & 32 & 8 \end{bmatrix}$ **(e)** $\begin{bmatrix} 4 & -3 & -4 & 2 \\ -10 & 32 & -25 & 2 \\ -36 & 69 & -24 & -6 \\ -16 & 12 & 16 & -8 \end{bmatrix}$

3. (a) $\begin{bmatrix} 1 & -1 \\ -1 & 1 \\ 1 & -1 \end{bmatrix}$ **(c)** $\begin{bmatrix} 0 & 0 & 0 \\ 1 & 0 & -1 \\ 4 & 2 & 0 \\ 9 & 6 & 3 \end{bmatrix}$

4. (a) $n = r$ **(c)** $n = r$ and $m = t$ **5.** $r = s$

7. $A = \begin{bmatrix} 1 & 2 \\ 2 & 1 \end{bmatrix}$, $B = \begin{bmatrix} 2 & 1 \\ 1 & 2 \end{bmatrix}$

9. $A = \begin{bmatrix} 1 & 2 \\ 2 & 4 \end{bmatrix}, B = \begin{bmatrix} 3 & 2 \\ 2 & 4 \end{bmatrix}, C = \begin{bmatrix} 1 & 4 \\ 3 & 3 \end{bmatrix}$ **11. (a)** $\begin{bmatrix} \pm\sqrt{x} & 0 \\ 0 & \pm\sqrt{y} \end{bmatrix}$

13. (a) $c_{ij} = 28 + 12i - 6j - 3ij$ **(c)** $c_{ij} = 6i - 3j + 2ij - 10$

15. $\begin{bmatrix} 1 & 4 & 3 & 2 \\ 0 & 1 & 1 & 7 \\ 2 & -3 & 1 & -1 \end{bmatrix} \begin{bmatrix} x_1 \\ x_2 \\ x_3 \\ x_4 \end{bmatrix} = \begin{bmatrix} 3 \\ -5 \\ 0 \end{bmatrix}$ **17. (a)** $\begin{bmatrix} 2 & 1 \\ -1 & -2 \end{bmatrix}$ **(b)** $\begin{bmatrix} 5 & 1 \\ 1 & 2 \end{bmatrix}$

Exercises 3.3, page 91

1. (a) $\mathcal{A} = \{(2, -5, -2), (-6, 13, 4), (6, 1, 10)\}$ Singular

(c) $\mathcal{A} = \{(5, 2, 2), (2, 1, 9), (7, 0, 3)\}$ Nonsingular

2. (a) Elementary **(c)** Not elementary **(e)** Elementary **(g)** Elementary

3. (a) $\begin{bmatrix} 1 & 0 & 0 \\ 0 & 1 & 0 \\ 0 & 0 & -3 \end{bmatrix}, \begin{bmatrix} 1 & 0 & 0 \\ 0 & 1 & 0 \\ 0 & 0 & -\frac{1}{3} \end{bmatrix}$ **(c)** $\begin{bmatrix} 1 & 0 & 0 \\ 0 & 0 & 1 \\ 0 & 1 & 0 \end{bmatrix}, \begin{bmatrix} 1 & 0 & 0 \\ 0 & 0 & 1 \\ 0 & 1 & 0 \end{bmatrix}$

(e) $\begin{bmatrix} 0 & 0 & -2 \\ 0 & 1 & 0 \\ 1 & 0 & 0 \end{bmatrix}, \begin{bmatrix} 0 & 0 & 1 \\ 0 & 1 & 0 \\ -\frac{1}{2} & 0 & 0 \end{bmatrix}$

4. (a) $\begin{bmatrix} 1 & -4 & 0 \\ 0 & 1 & 0 \\ 0 & 0 & 1 \end{bmatrix}$ **(c)** Not elementary **(e)** $\begin{bmatrix} 0 & 0 & 1 \\ 0 & 1 & 0 \\ 1 & 0 & 0 \end{bmatrix}$ **(g)** $\begin{bmatrix} 1 & 0 & 0 \\ 0 & 1 & 0 \\ 0 & 0 & \frac{1}{2} \end{bmatrix}$

5. (a) $A^{-1} = \begin{bmatrix} 0 & 1 \\ 1 & 0 \end{bmatrix}\begin{bmatrix} 1 & 0 \\ 0 & \frac{1}{3} \end{bmatrix}\begin{bmatrix} 1 & 0 \\ -2 & 1 \end{bmatrix}$

(c) $A^{-1} = \begin{bmatrix} 1 & 0 & 0 \\ 0 & 1 & -3 \\ 0 & 0 & 1 \end{bmatrix}\begin{bmatrix} 1 & 0 & 2 \\ 0 & 1 & 0 \\ 0 & 0 & 1 \end{bmatrix}\begin{bmatrix} \frac{1}{4} & 0 & 0 \\ 0 & 1 & 0 \\ 0 & 0 & 1 \end{bmatrix}$

6. (a) $\begin{bmatrix} -2 & -3 \\ 1 & 1 \end{bmatrix}$ **(c)** $\begin{bmatrix} \frac{2}{5} & -\frac{1}{5} \\ \frac{1}{5} & \frac{2}{5} \end{bmatrix}$

7. (a) $\begin{bmatrix} -4 & 5 \\ -3 & 4 \end{bmatrix}$ **(b)** $\mathcal{C} = \{(-7, -4), (9, 5)\}$

(c) $(AB)^{-1}$ exists because AB is the matrix of transition from the basis $\mathcal{A}$ to the basis $\mathcal{C}$ in $\mathbf{R}^2$.

8. (a) $\begin{bmatrix} 1 & 0 \\ 0 & 4 \end{bmatrix}\begin{bmatrix} 1 & 4 \\ 0 & 1 \end{bmatrix}\begin{bmatrix} -1 & 0 \\ 0 & 1 \end{bmatrix}\begin{bmatrix} 1 & 0 \\ \frac{3}{4} & 1 \end{bmatrix}$ **(c)** $\begin{bmatrix} -5 & 0 \\ 0 & 1 \end{bmatrix}\begin{bmatrix} 1 & -\frac{4}{5} \\ 0 & 1 \end{bmatrix}\begin{bmatrix} 1 & 0 \\ 2 & 1 \end{bmatrix}$

(e) $\begin{bmatrix} 1 & 0 & 0 \\ 0 & 1 & 0 \\ 0 & 0 & 3 \end{bmatrix}\begin{bmatrix} 1 & 0 & 0 \\ 0 & -1 & 0 \\ 0 & 0 & 1 \end{bmatrix}\begin{bmatrix} 2 & 0 & 0 \\ 0 & 1 & 0 \\ 0 & 0 & 1 \end{bmatrix}$

9. (a) $\begin{bmatrix} -1 & 1 \\ \frac{3}{4} & -\frac{1}{2} \end{bmatrix}$ **(c)** $\begin{bmatrix} -\frac{1}{5} & \frac{4}{5} \\ \frac{2}{5} & -\frac{3}{5} \end{bmatrix}$ **(e)** $\begin{bmatrix} \frac{1}{2} & 0 & 0 \\ 0 & -1 & 0 \\ 0 & 0 & \frac{1}{3} \end{bmatrix}$

11. $A^{-1} = \begin{bmatrix} 0 & -2 \\ 6 & 5 \end{bmatrix}$ **13.** $X = A^{-1}B$ **14. (a)** $X = \begin{bmatrix} -1 \\ -1 \end{bmatrix}$ **(c)** $X = \begin{bmatrix} 2 & -4 \\ 1 & -2 \end{bmatrix}$

16. (c) $B = A^{-1}CA$ **(e)** $A^{-1} = B(AB)^{-1}$ **(g)** $C^{-1} = BA$

Exercises 3.4, page 102

1. (a) Interchange the first and second columns.

(c) Replace the second column by the sum of the second column and 2 times the third column.

(e) Multiply the third column by -3.

2. (a) In reduced column-echelon form.

(c) Not in reduced column-echelon form.

(e) Not in reduced column-echelon form.

(g) In reduced column-echelon form.

3. (a) $\begin{bmatrix} 1 & -2 \\ 0 & 1 \end{bmatrix}$ **(c)** $\begin{bmatrix} 1 & 0 \\ 1 & 1 \end{bmatrix}$ **4. (a)** $\begin{bmatrix} -3 & 4 & 2 \\ 0 & 0 & 1 \\ -8 & 1 & 3 \end{bmatrix}$ **(c)** $\begin{bmatrix} 1 & 2 & 0 \\ 2 & 7 & -4 \end{bmatrix}$

5. Not always. Let

$$A = \begin{bmatrix} 1 & 0 \\ 0 & 0 \\ 1 & 1 \end{bmatrix} \quad \text{and} \quad M = \begin{bmatrix} 1 & 0 & 0 \\ -2 & 1 & 0 \\ 0 & 0 & 1 \end{bmatrix}.$$

Then A is the matrix of transition from $\mathcal{E}_3$ to $\mathcal{A} = \{(1, 0, 1), (0, 0, 1)\}$ and

$$MA = \begin{bmatrix} 1 & 0 \\ -2 & 0 \\ 1 & 1 \end{bmatrix}$$

is the matrix of transition from $\mathcal{E}_3$ to $\mathcal{B} = \{(1, -2, 1), (0, 0, 1)\}$. Now $(1, -2, 1)$ is in $\mathcal{B}$ but not in $\langle \mathcal{A} \rangle$ since it has a nonzero second component. Thus $\langle \mathcal{B} \rangle \neq \langle \mathcal{A} \rangle$ in this case.

6. (a) $\begin{bmatrix} 0 & 0 & 0 \\ 1 & 0 & 0 \\ 2 & 0 & 0 \\ 0 & 1 & 0 \end{bmatrix}$ **(c)** $\begin{bmatrix} 1 & 0 & 0 & 0 \\ 0 & 1 & 0 & 0 \\ -1 & 0 & 0 & 0 \\ -1 & 2 & 0 & 0 \end{bmatrix}$ **(e)** $\begin{bmatrix} 1 & 0 & 0 & 0 \\ 2 & 0 & 0 & 0 \\ 0 & 1 & 0 & 0 \\ 1 & 2 & 0 & 0 \\ 0 & 0 & 1 & 0 \\ 1 & 1 & 1 & 0 \end{bmatrix}$

(g) $\begin{bmatrix} 1 & 0 & 0 & 0 & 0 \\ 0 & 1 & 0 & 0 & 0 \\ 2 & 0 & 0 & 0 & 0 \\ 0 & 0 & 1 & 0 & 0 \end{bmatrix}$

7. (a) $\begin{bmatrix} -1 & 2 & -1 \\ 1 & -1 & 0 \\ 0 & 0 & 1 \end{bmatrix}$ **(c)** $\begin{bmatrix} 1 & 0 & -1 & -1 \\ -2 & 1 & -1 & 1 \\ 0 & 0 & 1 & 0 \\ 0 & 0 & 0 & 1 \end{bmatrix}$ **(e)** $\begin{bmatrix} 1 & -\frac{7}{8} & -\frac{3}{8} & \frac{41}{8} \\ 0 & \frac{1}{4} & \frac{1}{4} & -\frac{7}{4} \\ 0 & \frac{3}{8} & -\frac{1}{8} & -\frac{5}{8} \\ 0 & 0 & 0 & 1 \end{bmatrix}$

(g) $\begin{bmatrix} \frac{1}{2} & -\frac{1}{2} & -2 & -1 & 0 \\ \frac{1}{2} & \frac{1}{2} & -1 & -1 & 1 \\ 0 & 0 & 1 & 0 & 1 \\ 0 & 0 & 0 & 1 & -2 \\ \frac{3}{2} & -\frac{3}{2} & 1 & 0 & 0 \end{bmatrix}$

8. (a) $\{(0, 1, 2, 0), (0, 0, 0, 1)\}$ **(c)** $\{(1, 0, -1, -1), (0, 1, 0, 2)\}$

(e) $\{(1, 2, 0, 1, 0, 1), (0, 0, 1, 2, 0, 1), (0, 0, 0, 0, 1, 1)\}$

(g) $\{(1, 0, 2, 0), (0, 1, 0, 0), (0, 0, 0, 1)\}$

9. Let M_i, $i = 1, 2, \ldots, n$, be an elementary matrix of type I formed by multiplying the ith column of I_n by a. Then $aI_n = M_1 M_2 \cdots M_n$.

Exercises 3.5, page 112

1. (a) Multiply the first row by 4.   **(c)** Interchange the first and third rows.

(e) Multiply the second row by -3.

2. (a) $\begin{bmatrix} 5 & -6 & 1 \\ -6 & 9 & 0 \\ 1 & 0 & 1 \end{bmatrix}$ **(c)** $\begin{bmatrix} -9 & 12 & -1 \\ -4 & 6 & 0 \end{bmatrix}$ **(e)** $\begin{bmatrix} -9 & 12 & -1 \\ -4 & 6 & 0 \end{bmatrix}$

(g) $\begin{bmatrix} 0 & -3 \\ 2 & 8 \\ 8 & -10 \end{bmatrix}$

3. $A^T = \begin{bmatrix} 1 & 0 \\ 0 & 3 \end{bmatrix} \begin{bmatrix} 0 & 1 \\ 1 & 0 \end{bmatrix} \begin{bmatrix} 1 & 0 \\ 2 & 1 \end{bmatrix}$

4. (a) Not in reduced row-echelon form.   **(c)** Not in reduced row-echelon form.

(e) Not in reduced row-echelon form.

5. (a) $\begin{bmatrix} 1 & 0 & 1 \\ 0 & 1 & 0 \\ 0 & 0 & 0 \\ 0 & 0 & 0 \end{bmatrix}$ **(c)** $\begin{bmatrix} 1 & 0 & 1 & 1 \\ 0 & 1 & 1 & -1 \\ 0 & 0 & 0 & 0 \\ 0 & 0 & 0 & 0 \end{bmatrix}$ **(e)** $\begin{bmatrix} 1 & 0 & 0 & -\frac{41}{8} \\ 0 & 1 & 0 & \frac{7}{4} \\ 0 & 0 & 1 & \frac{5}{8} \\ 0 & 0 & 0 & 0 \\ 0 & 0 & 0 & 0 \\ 0 & 0 & 0 & 0 \end{bmatrix}$

(g) $\begin{bmatrix} 1 & 0 & 2 & 1 & 0 \\ 0 & 1 & 1 & 1 & 0 \\ 0 & 0 & 0 & 0 & 1 \\ 0 & 0 & 0 & 0 & 0 \end{bmatrix}$

6. (a) $\begin{bmatrix} 0 & -1 & 0 & 2 \\ 0 & 1 & 0 & -1 \\ 0 & -2 & 1 & 0 \\ 1 & 0 & 0 & 0 \end{bmatrix}$ **(c)** $\begin{bmatrix} 1 & 0 & 0 & 0 \\ -2 & 1 & 0 & 0 \\ 1 & 0 & 1 & 0 \\ 1 & -2 & 0 & 1 \end{bmatrix}$

(e) $\begin{bmatrix} 1 & 0 & -\frac{7}{8} & 0 & -\frac{3}{8} & 0 \\ 0 & 0 & \frac{1}{4} & 0 & \frac{1}{4} & 0 \\ 0 & 0 & \frac{3}{8} & 0 & -\frac{1}{8} & 0 \\ -1 & 0 & -1 & 0 & -1 & 1 \\ -1 & 0 & -2 & 1 & 0 & 0 \\ -2 & 1 & 0 & 0 & 0 & 0 \end{bmatrix}$ **(g)** $\begin{bmatrix} \frac{1}{2} & -\frac{1}{2} & 0 & 0 \\ \frac{1}{2} & \frac{1}{2} & 0 & 0 \\ \frac{3}{2} & -\frac{3}{2} & 0 & 1 \\ -2 & 0 & 1 & 0 \end{bmatrix}$

7. (a) $\begin{bmatrix} -\frac{1}{2} & \frac{3}{2} & 1 \\ \frac{1}{2} & -\frac{1}{2} & 0 \\ \frac{1}{2} & -\frac{1}{2} & -1 \end{bmatrix}$ **(c)** $\begin{bmatrix} \frac{10}{9} & -\frac{4}{9} & \frac{1}{9} \\ -\frac{8}{9} & \frac{5}{9} & \frac{1}{9} \\ -\frac{1}{9} & \frac{4}{9} & -\frac{1}{9} \end{bmatrix}$

8. (a) $\begin{bmatrix} -1 & -2 \\ \frac{1}{2} & \frac{3}{2} \end{bmatrix}$ **(c)** $\begin{bmatrix} -2 & -1 & 3 \\ -1 & 0 & 1 \\ -2 & -1 & 2 \end{bmatrix}$ **(e)** Not invertible **(g)** Not invertible

9. (a) $\mathcal{A} = \{(1, -1, 0), (0, 1, 1), (2, 0, -1)\}, A^{-1} = \begin{bmatrix} \frac{1}{3} & -\frac{2}{3} & \frac{2}{3} \\ \frac{1}{3} & \frac{1}{3} & \frac{2}{3} \\ \frac{1}{3} & \frac{1}{3} & -\frac{1}{3} \end{bmatrix}$

(c) $\mathcal{A} = \{(0, -1, 0, -1), (1, 2, 1, 0), (1, 0, 1, 1), (0, 0, 1, 1)\}$,

$A^{-1} = \begin{bmatrix} 0 & -\frac{1}{3} & \frac{2}{3} & -\frac{2}{3} \\ 0 & \frac{1}{3} & \frac{1}{3} & -\frac{1}{3} \\ 1 & -\frac{1}{3} & -\frac{1}{3} & \frac{1}{3} \\ -1 & 0 & 1 & 0 \end{bmatrix}$

13. Those of type I or type III

Exercises 3.6, page 122

1. (a) A and B are column-equivalent. **(c)** A and B are not column-equivalent.

2. (a) A and B are not row-equivalent. **(c)** A and B are row-equivalent.

3. $\begin{bmatrix} 0 & -2 & 3 \\ 1 & 3 & -2 \\ 0 & -1 & 1 \end{bmatrix}$

5. (a) $Q = \begin{bmatrix} 1 & 0 & \frac{9}{4} \\ 0 & -\frac{2}{3} & 0 \\ 1 & -2 & \frac{19}{8} \end{bmatrix}$ **(c)** A and B are not column-equivalent.

6. (a) A solution is given by

$$B = A \begin{bmatrix} 1 & 0 & 0 \\ 0 & 1 & 0 \\ 1 & 0 & 1 \end{bmatrix} \begin{bmatrix} 1 & 0 & 0 \\ 0 & -\frac{2}{3} & 0 \\ 0 & 0 & 1 \end{bmatrix} \begin{bmatrix} 1 & 0 & 0 \\ 0 & 1 & 0 \\ 0 & -2 & 1 \end{bmatrix} \begin{bmatrix} 1 & 0 & 0 \\ 0 & 1 & 0 \\ 0 & 0 & \frac{1}{8} \end{bmatrix} \begin{bmatrix} 1 & 0 & \frac{9}{4} \\ 0 & 1 & 0 \\ 0 & 0 & 1 \end{bmatrix}.$$

(c) A and B are not column-equivalent.

7. $\begin{bmatrix} 4 & 14 & -7 \\ 6 & 29 & -14 \\ 5 & 25 & -12 \end{bmatrix}$

9. $PA = \begin{bmatrix} 1 & 2 & 0 \\ 0 & 0 & 1 \\ 0 & 0 & 0 \\ 0 & 0 & 0 \end{bmatrix}$, where

$$P = \begin{bmatrix} 1 & 0 & 0 & 0 \\ 0 & 1 & 0 & 0 \\ 0 & 0 & 1 & 0 \\ 0 & -9 & 0 & 1 \end{bmatrix} \begin{bmatrix} 1 & 0 & 0 & 0 \\ 0 & 1 & 0 & 0 \\ 0 & 4 & 1 & 0 \\ 0 & 0 & 0 & 1 \end{bmatrix} \begin{bmatrix} 1 & 3 & 0 & 0 \\ 0 & 1 & 0 & 0 \\ 0 & 0 & 1 & 0 \\ 0 & 0 & 0 & 1 \end{bmatrix} \begin{bmatrix} 1 & 0 & 0 & 0 \\ 0 & 1 & 0 & 0 \\ 0 & 0 & 1 & 0 \\ -3 & 0 & 0 & 1 \end{bmatrix} \begin{bmatrix} 1 & 0 & 0 & 0 \\ 0 & 1 & 0 & 0 \\ 2 & 0 & 1 & 0 \\ 0 & 0 & 0 & 1 \end{bmatrix}$$

or

$$P = \begin{bmatrix} 1 & 3 & 0 & 0 \\ 0 & 1 & 0 & 0 \\ 0 & 0 & 1 & 0 \\ 0 & 0 & 0 & 1 \end{bmatrix} \begin{bmatrix} 1 & 0 & 0 & 0 \\ 0 & 1 & 0 & 0 \\ 0 & 4 & 1 & 0 \\ 0 & 0 & 0 & 1 \end{bmatrix} \begin{bmatrix} 1 & 0 & 0 & 0 \\ 0 & 1 & 0 & 0 \\ 0 & 0 & 1 & 0 \\ 0 & -9 & 0 & 1 \end{bmatrix} \begin{bmatrix} 1 & 0 & 0 & 0 \\ 0 & 1 & 0 & 0 \\ 0 & 0 & 1 & 0 \\ -3 & 0 & 0 & 1 \end{bmatrix} \begin{bmatrix} 1 & 0 & 0 & 0 \\ 0 & 1 & 0 & 0 \\ 2 & 0 & 1 & 0 \\ 0 & 0 & 0 & 1 \end{bmatrix}$$

11. (a) A and B are not row-equivalent.   **(c)** $P = \begin{bmatrix} 0 & 0 & 0 & -1 \\ \frac{1}{2} & 0 & 0 & -\frac{7}{2} \\ \frac{1}{2} & 0 & 1 & -\frac{3}{2} \\ \frac{5}{2} & 1 & 0 & -\frac{3}{2} \end{bmatrix}$

12. (a) $\langle A \rangle = \langle B \rangle = \langle (1, 0), (0, 1) \rangle$
   (c) $\langle A \rangle = \langle (1, -2, 0, 0), (0, 0, 1, -1) \rangle$, $\langle B \rangle = \langle (1, 0, -1, -2), (0, 1, 1, 1) \rangle$, $\langle A \rangle \neq \langle B \rangle$

13. (a) All nonsingular $n \times n$ matrices.   **(b)** All nonsingular $n \times n$ matrices.

Exercises 3.7, page 130

1. (a) 2   **(c)** 3   **(e)** 2   **2. (a)** 2   **(c)** 2

3. (a) $\{(1, 0, -1, 2), (0, 1, 2, -1)\}$ **(c)** $\{(1, 0, -1, -2), (0, 1, 1, 1)\}$

4. (a) $\{(2, 1, 0, 3), (1, 2, 3, 0)\}$ **(c)** $\{(1, 4, 3, 2), (0, 1, 1, 1)\}$

5. (a) $\begin{bmatrix} 1 & 2 & 4 & 1 \\ 2 & 0 & 1 & 1 \\ 0 & 1 & 0 & 1 \\ 3 & 4 & 0 & 1 \end{bmatrix}$, $\mathcal{E}_4, 4$ **(c)** $\begin{bmatrix} 1 & 0 \\ 0 & 0 \\ 0 & 1 \\ 0 & 0 \end{bmatrix}$, $\{(1, 0, 0, 0), (0, 0, 1, 0)\}$, 2

(e) $\begin{bmatrix} 1 \\ -1 \\ -1 \end{bmatrix}, \{(1, -1, -1)\}, 1$ **(g)** $\begin{bmatrix} 1 \\ 2 \end{bmatrix}, \{(1, 2)\}, 1$

(i) $\begin{bmatrix} -1 & 1 \\ 1 & 0 \\ 0 & 1 \end{bmatrix}, \{(1, 0, 1), (0, 1, 1)\}, 2$

6. (a) 3 **(c)** 3 **(e)** 5

11. Those in part (a) and part (b) are equivalent, and those in part (c) and part (d) are equivalent.

13. (a) Not column-equivalent, row-equivalent, equivalent

(c) Not column-equivalent, row-equivalent, equivalent

15. (a) $\mathcal{A} = \{(1, 1, 0, 0), (0, 1, 1, 0), (0, 0, 1, 1), (1, 2, 2, 1)\}$

$\mathcal{A}' = \{(1, 0, 0, 1), (0, 1, 0, -1), (0, 0, 1, 1), (0, 0, 0, 0)\}$

$\mathcal{B} = \{(1, 0, 0, 1), (0, 1, 0, -1), (0, 0, 1, 1), (0, 0, 0, 1)\}$

$$P = \begin{bmatrix} 1 & 0 & 0 & 0 \\ 0 & 1 & 0 & 0 \\ 0 & 0 & 1 & 0 \\ -1 & 1 & -1 & 1 \end{bmatrix}, Q = \begin{bmatrix} 1 & 0 & 0 & -1 \\ -1 & 1 & 0 & -1 \\ 1 & -1 & 1 & -1 \\ 0 & 0 & 0 & 1 \end{bmatrix}$$

16. (a) $P = \begin{bmatrix} 1 & 0 & 0 \\ 0 & 1 & 0 \\ -1 & -1 & 1 \end{bmatrix}, Q = \begin{bmatrix} 0 & 0 & 0 & 1 \\ 0 & 1 & -2 & 0 \\ 0 & 0 & 1 & 0 \\ 1 & -2 & 0 & 0 \end{bmatrix}$

17. $P = \begin{bmatrix} 1 & 0 & 0 & 0 \\ -1 & -1 & 1 & 0 \\ 0 & 1 & 0 & 0 \\ 2 & -1 & 0 & 1 \end{bmatrix}, Q = \begin{bmatrix} 1 & 2 & 1 \\ -4 & -6 & -3 \\ 0 & 0 & 1 \end{bmatrix}$

Exercises 3.8, page 137

1. $L = \begin{bmatrix} 1 & 0 \\ 4 & 1 \end{bmatrix}, U \begin{bmatrix} 1 & 3 \\ 0 & -14 \end{bmatrix}$ **3.** $L = \begin{bmatrix} 1 & 0 & 0 \\ 3 & 1 & 0 \\ -1 & \frac{1}{11} & 1 \end{bmatrix}, U = \begin{bmatrix} 1 & -3 & -2 \\ 0 & 11 & 6 \\ 0 & 0 & -\frac{6}{11} \end{bmatrix}$

5. $L = \begin{bmatrix} 1 & 0 & 0 & 0 \\ 0 & 1 & 0 & 0 \\ -2 & -4 & 1 & 0 \\ 0 & 0 & 0 & 1 \end{bmatrix}, U = \begin{bmatrix} -1 & 2 & -3 & 0 \\ 0 & -2 & 1 & -1 \\ 0 & 0 & -2 & -1 \\ 0 & 0 & 0 & 5 \end{bmatrix}$

7. $L = \begin{bmatrix} 1 & 0 & 0 \\ 2 & 1 & 0 \\ 0 & \frac{1}{2} & 1 \end{bmatrix}, U = \begin{bmatrix} 1 & -1 & -2 & 0 \\ 0 & 2 & 7 & -2 \\ 0 & 0 & -\frac{5}{2} & 0 \end{bmatrix}$

9. $L = \begin{bmatrix} 1 & 0 & 0 & 0 \\ 1 & 1 & 0 & 0 \\ -2 & -2 & 1 & 0 \\ 0 & -3 & -1 & 1 \end{bmatrix}, U = \begin{bmatrix} 1 & 0 & 0 & 4 & 2 \\ 0 & 0 & -2 & -2 & -3 \\ 0 & 0 & 0 & 6 & -2 \\ 0 & 0 & 0 & 0 & 0 \end{bmatrix}$

11. $L = \begin{bmatrix} 1 & 0 & 0 & 0 \\ 2 & 1 & 0 & 0 \\ 3 & 2 & 1 & 0 \\ 1 & -2 & 4 & 1 \end{bmatrix}, U = \begin{bmatrix} 1 & 0 & 1 \\ 0 & 2 & -2 \\ 0 & 0 & -1 \\ 0 & 0 & 0 \end{bmatrix}$

13. $P = \begin{bmatrix} 0 & 1 \\ 1 & 0 \end{bmatrix}, L = I, U = \begin{bmatrix} 2 & -3 \\ 0 & 2 \end{bmatrix}$

15. $P = \begin{bmatrix} 1 & 0 & 0 \\ 0 & 0 & 1 \\ 0 & 1 & 0 \end{bmatrix}, L = \begin{bmatrix} 1 & 0 & 0 \\ 2 & 1 & 0 \\ -1 & 0 & 1 \end{bmatrix}, U = \begin{bmatrix} 1 & -2 & -3 \\ 0 & 4 & 7 \\ 0 & 0 & 2 \end{bmatrix}$

17. $P = \begin{bmatrix} 0 & 1 & 0 & 0 \\ 0 & 0 & 1 & 0 \\ 1 & 0 & 0 & 0 \\ 0 & 0 & 0 & 1 \end{bmatrix}, L = \begin{bmatrix} 1 & 0 & 0 & 0 \\ 1 & 1 & 0 & 0 \\ 0 & 0 & 1 & 0 \\ 0 & -1 & 5 & 1 \end{bmatrix}, U = \begin{bmatrix} 1 & 2 & -1 & 0 \\ 0 & -3 & 4 & 2 \\ 0 & 0 & 1 & 1 \\ 0 & 0 & 0 & -5 \end{bmatrix}$

19. $L' = \begin{bmatrix} 3 & 0 \\ -6 & 5 \end{bmatrix}, U' = \begin{bmatrix} 1 & \frac{1}{3} \\ 0 & 1 \end{bmatrix}$

21. $L' = \begin{bmatrix} -2 & 0 & 0 \\ 1 & -2 & 0 \\ 1 & 2 & 7 \end{bmatrix}, U' = \begin{bmatrix} 1 & -2 & -3 \\ 0 & 1 & -\frac{5}{2} \\ 0 & 0 & 1 \end{bmatrix}$

Exercises 4.1, page 148

3. mn **5.** Conditions 3, 4, and 9 fail to hold.

7. **V** is a vector space with respect to the given operations.

9. Conditions 8 and 9 fail to hold.   **11.** Condition 10 fails to hold.

13. Condition 10 fails to hold.

15. **V** is a vector space with respect to the given operations.

17. Conditions 6, 8 and 9 fail to hold.   **19.** Conditions 8 and 9 fail to hold.

21. Conditions 5, 6, 8 and 9 fail to hold.

33. $\mathcal{F}$ is not a field since it is not closed with respect to addition.

Exercises 4.2, page 155

1. **W** is not a subspace of $\mathbf{R}^3$. The vectors $\mathbf{u} = (1, 0, 0)$ and $\mathbf{v} = (0, 1, 0)$ are in **W**, but $\mathbf{u} + \mathbf{v} = (1, 1, 0)$ is not in **W** since $1 + 1 \neq 1$.

3. **W** is a subspace of $\mathbf{R}^3$.   **5.** **W** is a subspace of $\mathcal{P}_2$ over **R**.

7. **W** is a subspace of $\mathcal{P}_n$ over **R**.

9. $\mathbf{W}$ is not a subspace of $\mathbf{R}_{2\times2}$. The vectors

$$\mathbf{u} = \begin{bmatrix} 1 & 2 \\ 3 & 4 \end{bmatrix} \quad \text{and} \quad \mathbf{v} = \begin{bmatrix} 1 & 5 \\ 6 & 7 \end{bmatrix}$$

are in $\mathbf{W}$, but

$$\mathbf{u} + \mathbf{v} = \begin{bmatrix} 2 & 7 \\ 9 & 11 \end{bmatrix}$$

is not in $\mathbf{W}$ because its first row, first column element is not equal to 1.

11. $\mathbf{W}$ is not a subspace of $\mathbf{R}_{2\times2}$. The vectors

$$\mathbf{u} = \begin{bmatrix} 1 & 0 \\ 0 & 2 \end{bmatrix} \quad \text{and} \quad \mathbf{v} = \begin{bmatrix} 1 & 0 \\ 0 & -2 \end{bmatrix}$$

are in $\mathbf{W}$, but

$$\mathbf{u} + \mathbf{v} = \begin{bmatrix} 2 & 0 \\ 0 & 0 \end{bmatrix}$$

is not in $\mathbf{W}$ since it is nonzero and not invertible.

13. $\mathbf{W}$ is not a subspace of $\mathbf{R}_{2\times2}$. The vectors

$$\mathbf{u} = \begin{bmatrix} 2 & 3 \\ 4 & -2 \end{bmatrix} \quad \text{and} \quad \mathbf{v} = \begin{bmatrix} 2 & 5 \\ 6 & 2 \end{bmatrix}$$

are in $\mathbf{W}$, but

$$\mathbf{u} + \mathbf{v} = \begin{bmatrix} 4 & 8 \\ 10 & 0 \end{bmatrix}$$

is not in $\mathbf{W}$ since $4^2 \neq 0^2$.

15. $\mathbf{W}$ is not a subspace of $\mathbf{R}_{2\times2}$. The vectors

$$\mathbf{u} = \begin{bmatrix} 1 & 1 \\ 0 & 0 \end{bmatrix} \quad \text{and} \quad \mathbf{v} = \begin{bmatrix} 0 & 0 \\ 1 & 2 \end{bmatrix}$$

are in $\mathbf{W}$, but

$$\mathbf{u} + \mathbf{v} = \begin{bmatrix} 1 & 1 \\ 1 & 2 \end{bmatrix}$$

is not in $\mathbf{W}$ since $1(2) \neq 1(1)$.

17. $\mathbf{W}$ is not a subspace of $\mathbf{R}_{3\times3}$. The vectors $\mathbf{u} = \mathbf{v} = I_3$ are in $\mathbf{W}$, but $\mathbf{u} + \mathbf{v} = 2I_3$ is not in $\mathbf{W}$.

19. $\mathbf{W}$ is a subspace of $\mathbf{R}_{n\times n}$.

21. $\mathbf{W}$ is a subspace of $\mathbf{R}_{2\times 2}$.

23. $\mathbf{W}$ is a subspace of $\mathbf{R}^3$.

25. $\mathbf{W}$ is a subspace of the set of all real-valued functions f with domain $\mathbf{R}$.

27. $\mathbf{W}$ is a subspace of the set of all real-valued functions f with domain $\mathbf{R}$.

29. $\mathbf{W}$ is a subspace of the set of all real-valued functions f with domain $\mathbf{R}$.

31. (a) Linearly dependent   **(c)** Linearly independent   **(e)** Linearly dependent

32. (a) Linearly dependent   **(c)** Linearly dependent

33. (a) The set spans P_2.   **(c)** The set does not span P_2.

34. (a) The set is not a basis for P_2.   **(c)** The set is a basis for P_2.

35. (a) $\{p_1(x), p_2(x)\}$ or any subset consisting of two distinct vectors.

(c) $\left\{ \begin{bmatrix} 1 & 0 \\ 1 & 1 \end{bmatrix}, \begin{bmatrix} 0 & 1 \\ -1 & 1 \end{bmatrix} \right\}$ or any subset consisting of two distinct vectors.

(e) $\{p_1(x), p_2(x), p_3(x), p_4(x)\}$

36. (a) $\{(1, 0, 0, 0), (0, 0, 1, 0)\}$ is a basis for $\mathbf{W}$.

37. (a) $[\mathbf{v}]_\mathcal{B} = \begin{bmatrix} 1 \\ 1 \\ 0 \end{bmatrix}$ **38. (a)** $[\mathbf{v}]_\mathcal{B} = \begin{bmatrix} 1 \\ 1 \\ -1 \\ 1 \end{bmatrix}$ **39. (a)** $[\mathbf{v}]_\mathcal{B} = \begin{bmatrix} 1 \\ 0 \\ 1 \end{bmatrix}$

40. (a) $[\mathbf{v}]_\mathcal{B} = \begin{bmatrix} 3 \\ -2 \end{bmatrix}$ **41. (a)** $\begin{bmatrix} 1 & 0 & 1 \\ 0 & 1 & 1 \\ 1 & 0 & 0 \end{bmatrix}$ **42. (a)** $\begin{bmatrix} 0 & 0 & 1 \\ -1 & 1 & 1 \\ 1 & 0 & -1 \end{bmatrix}$

43. (a) $\begin{bmatrix} 1 & 0 & 0 & 0 \\ 1 & 1 & 0 & 1 \\ 0 & 1 & 1 & 0 \\ 0 & 0 & 1 & 1 \end{bmatrix}$ **44. (a)** $\begin{bmatrix} 1 & 0 & 0 & 0 \\ -\frac{1}{2} & \frac{1}{2} & \frac{1}{2} & -\frac{1}{2} \\ \frac{1}{2} & -\frac{1}{2} & \frac{1}{2} & \frac{1}{2} \\ -\frac{1}{2} & \frac{1}{2} & -\frac{1}{2} & \frac{1}{2} \end{bmatrix}$

Exercises 4.3, page 163

1. (a) f is onto and one-to-one. f is an isomorphism.

(c) f is not onto, not one-to-one, and not an isomorphism.

(e) f is onto, not one-to-one, and not an isomorphism.

(g) f is onto, not one-to-one, and not an isomorphism.

(i) f is onto and one-to-one. f is not an isomorphism.

(k) f is not onto, is one-to-one, and is not an isomorphism.

3. Define f from $\mathcal{P}_n$ over $\mathcal{F}$ to $\mathcal{F}^{n+1}$ by $f\left(\sum_{i=0}^{n} a_i x^i \right) = (a_0, a_1, \ldots, a_n)$.

5. Define f from $\mathbf{W}$ to $\mathbf{R}^2$ by $f(a_0 + a_1 x - (a_0 + a_1)x^2) = (a_0, a_1)$.

7. Define f from $\mathbf{W}$ to $\mathbf{R}^3$ by $f\left(\begin{bmatrix} a & b \\ b & c \end{bmatrix}\right) = (a, b, c)$.

9. Define f from $\mathbf{W}$ to $\mathbf{R}^2$ by $f((x, y, 2y - 3x)) = (x, y)$.

11. Define f from $\mathbf{W}_1$ to $\mathbf{W}_2$ by $f\left(\begin{bmatrix} a & b & c \\ b & d & e \\ c & e & f \end{bmatrix}\right) = \begin{bmatrix} a & c & e \\ b & d & f \end{bmatrix}$.

12. (a) $r = 2$, $f(a_1(1, -1, 1) + a_2(0, 1, 0)) = (a_1, a_2)$

 (c) $r = 3$, $f(a_1(2, 0, 4, -1) + a_2(5, -1, 11, 8) + a_3(0, 1, -7, 9)) = (a_1, a_2, a_3)$

 (e) $r = 2$, $f(a_1 p_1(x) + a_2 p_2(x)) = (a_1, a_2)$

 (g) $r = 2$, $f\left(a_1 \begin{bmatrix} 1 & 0 \\ -1 & 1 \end{bmatrix} + a_2 \begin{bmatrix} 1 & 1 \\ 1 & 1 \end{bmatrix}\right) = (a_1, a_2)$

13. (a) $f(a_1 p_1(x) + a_2 p_2(x) + a_3 p_3(x) + a_4 p_4(x)) = (a_1, a_2, a_3, a_4)$ defines an isomorphism from $\mathbf{V}$ to $\mathbf{R}^4$.

 (b) $f(a_1 p_1(x) + a_2 p_2(x) + a_3 p_3(x)) = (a_1, a_2, a_3, 0)$ defines an isomorphism from $\mathbf{W}$ to a subspace of $\mathbf{R}^4$.

Exercises 4.4, page 169

1. (a) $\{1 + 2x^2, x + 3x^2, x^3\}$ **2. (a)** $\left\{\begin{bmatrix} 1 & 0 \\ -1 & 1 \end{bmatrix}, \begin{bmatrix} 0 & 1 \\ -\frac{1}{3} & -\frac{1}{3} \end{bmatrix}\right\}$

3. (a) $\{p_1(x), p_2(x), p_3(x)\}$

4. (a) $\{(3, 3, 2, 0), (1, 1, 1, 1)\}$ **(c)** $\{(3, 2, 2, 0), (1, 1, 0, 0), (1, 1, 1, 1)\}$

5. (a) $\left\{\begin{bmatrix} -1 & -2 \\ -2 & -1 \end{bmatrix}, \begin{bmatrix} 3 & 3 \\ 2 & 2 \end{bmatrix}\right\}$ **6. (a)** $\{1 + x + 4x^2, x\}$

7. (a) $\langle \mathcal{A} \rangle = \langle \mathcal{B} \rangle$ **8. (a)** $\langle \mathcal{A} \rangle \neq \langle \mathcal{B} \rangle$ **9. (a)** $\begin{bmatrix} 2 & 1 & 1 \\ 0 & 1 & -3 \\ 2 & 0 & 2 \end{bmatrix}$

Exercises 4.6, page 178

1. (a) $x_1 = -27$, $x_2 = 19$, $x_3 = -6$ **(b)** $\begin{bmatrix} 5 \\ 2 \\ 1 \end{bmatrix} = -27 \begin{bmatrix} 1 \\ -1 \\ 0 \end{bmatrix} + 19 \begin{bmatrix} 2 \\ -1 \\ 1 \end{bmatrix} - 6 \begin{bmatrix} 1 \\ 1 \\ 3 \end{bmatrix}$

3. (a) No solution   **(b)** Not possible   **5. (a)** No solution   **(b)** Not possible

7. (a) $x_1 = 1 + 2x_2 + x_4$, $x_3 = -x_4$; x_2, x_4 arbitrary

 (b) $\begin{bmatrix} 1 \\ 2 \\ 3 \\ 0 \end{bmatrix} = 1 \begin{bmatrix} 1 \\ 2 \\ 3 \\ 0 \end{bmatrix} + 0 \begin{bmatrix} -2 \\ -4 \\ -6 \\ 0 \end{bmatrix} + 0 \begin{bmatrix} -2 \\ 2 \\ 1 \\ 1 \end{bmatrix} + 0 \begin{bmatrix} -3 \\ 0 \\ -2 \\ 1 \end{bmatrix}$

9. (a) $x_1 = 5 - 3x_2 - 2x_4 - 2x_5$, $x_3 = 3 - 2x_4$; x_2, x_4, x_5 arbitrary

(b) $\begin{bmatrix} 2 \\ 13 \\ -5 \end{bmatrix} = 5 \begin{bmatrix} 1 \\ 2 \\ -1 \end{bmatrix} + 0 \begin{bmatrix} 3 \\ 6 \\ -3 \end{bmatrix} + 3 \begin{bmatrix} -1 \\ 1 \\ 0 \end{bmatrix} + 0 \begin{bmatrix} 0 \\ 6 \\ -2 \end{bmatrix} + 0 \begin{bmatrix} 2 \\ 4 \\ -2 \end{bmatrix}$

11. (a) $x_1 = 1 - 2x_2, x_3 = 1, x_4 = 0, x_5 = 1$; x_2 arbitrary

(b) $\begin{bmatrix} 1 \\ 6 \\ 1 \\ 6 \end{bmatrix} = 1 \begin{bmatrix} 1 \\ 2 \\ 1 \\ -1 \end{bmatrix} + 0 \begin{bmatrix} 2 \\ 4 \\ 2 \\ -2 \end{bmatrix} + 1 \begin{bmatrix} 0 \\ 1 \\ 2 \\ 3 \end{bmatrix} + 0 \begin{bmatrix} 1 \\ 4 \\ 4 \\ 5 \end{bmatrix} + 1 \begin{bmatrix} 0 \\ 3 \\ -2 \\ 4 \end{bmatrix}$

13. (a) 2   **(b)** 3   **(c)** The system is inconsistent since $2 \neq 3$.

15. (a) 2   **(b)** 2   **(c)** The system is consistent.

17. (a) 4   **(b)** 4   **(c)** The system is consistent.

19. All real numbers a and b such that $4a + b = 11$ except $a = \frac{1}{4}, b = 10$.

21. (a) $a = -3$ **(b)** All real numbers except 3 and -3 **(c)** $a = 3$

Exercises 4.7, page 184

1. (a) $x_1 = \frac{1}{2} x_3, x_2 = \frac{3}{2} x_3$; x_3 arbitrary   **(b)** $\{(1, 3, 2)\}$ **(c)** 1

3. (a) $x_1 = \frac{1}{3} x_4, x_2 = x_4, x_3 = \frac{1}{3} x_4$; x_4 arbitrary   **(b)** $\{(1, 3, 1, 3)\}$ **(c)** 1

5. (a) $x_1 = \frac{2}{5} x_3 + \frac{1}{5} x_4, x_2 = -\frac{1}{5} x_3 + \frac{2}{5} x_4$; x_3, x_4 arbitrary
(b) $\{(2, -1, 5, 0), (1, 2, 0, 5)\}$ **(c)** 2

7. (a) $x_1 = -\frac{9}{2} x_3, x_2 = -\frac{1}{4} x_3, x_4 = 0$; x_3 arbitrary
(b) $\{(-18, -1, 4, 0)\}$ **(c)** 1

9. (a) $x_1 = x_2 - x_5, x_3 = -x_5, x_4 = -2x_5$; x_2, x_5 arbitrary
(b) $\{(1, 1, 0, 0, 0), (-1, 0, -1, -2, 1)\}$ **(c)** 2

11. (a) $x_1 = x_2 = x_3 = 0$ **(b)** $\varnothing$ **(c)** 0

13. For $x = 1$, $\{(1, 2, 1)\}$; For $x = -1$, $\{(1, 0, 1)\}$

15. $\mathbf{v} = \mathbf{c} + \mathbf{w}$ where $\mathbf{c} = (4, \frac{1}{2}, 0, 0)$ and $\mathbf{w} = x_3(-4, -3, 2, 0) + x_4(2, 3, 0, 2)$

17. $\mathbf{v} = \mathbf{c} + \mathbf{w}$ where $\mathbf{c} = (3, -1, 0, -1, -1)$ and $\mathbf{w} = x_3(0, 3, 1, 0, 0)$

19. $\mathbf{v} = \mathbf{c} + \mathbf{w}$ where $\mathbf{c} = (0, -2, -1)$ and $\mathbf{w} = \mathbf{0}$

21. $\mathbf{v} = \mathbf{c} + \mathbf{w}$ where $\mathbf{c} = (3, 1, 0, 1)$ and $\mathbf{w} = x_3(2, 0, 1, 0)$

23. For $i = 1$, $X = \begin{bmatrix} 3 \\ -1 \\ 0 \end{bmatrix} + x_3 \begin{bmatrix} -1 \\ 0 \\ 2 \end{bmatrix}$; For $i = 2$, $X = \begin{bmatrix} -\frac{3}{2} \\ 5 \\ 0 \end{bmatrix} + x_3 \begin{bmatrix} -1 \\ 0 \\ 2 \end{bmatrix}$;

For $i = 3$, $X = \begin{bmatrix} -\frac{5}{2} \\ 4 \\ 0 \end{bmatrix} + x_3 \begin{bmatrix} -1 \\ 0 \\ 2 \end{bmatrix}$

25. For $i = 1$, $X = \begin{bmatrix} -4 \\ -1 \\ 10 \\ 0 \end{bmatrix} + x_4 \begin{bmatrix} -5 \\ -1 \\ 6 \\ 2 \end{bmatrix}$; For $i = 2$, $X = \begin{bmatrix} -\frac{9}{2} \\ \frac{5}{2} \\ 4 \\ 0 \end{bmatrix} + x_4 \begin{bmatrix} -5 \\ -1 \\ 6 \\ 2 \end{bmatrix}$;

For $i = 3$, $X = \begin{bmatrix} -\frac{9}{2} \\ \frac{1}{2} \\ 2 \\ 0 \end{bmatrix} + x_4 \begin{bmatrix} -5 \\ -1 \\ 6 \\ 2 \end{bmatrix}$

27. For $i = 1$, $X = \begin{bmatrix} \frac{3}{4} \\ 2 \\ 0 \end{bmatrix} + x_3 \begin{bmatrix} -1 \\ 4 \\ 4 \end{bmatrix}$; For $i = 2$, $X = \begin{bmatrix} -4 \\ 4 \\ 0 \end{bmatrix} + x_3 \begin{bmatrix} -1 \\ 4 \\ 4 \end{bmatrix}$;

For $i = 3$, $X = \begin{bmatrix} 2 \\ -12 \\ 0 \end{bmatrix} + x_3 \begin{bmatrix} -1 \\ 4 \\ 4 \end{bmatrix}$

Exercises 5.1, page 199

1. (b) $(S + T)(x_1, x_2) = (x_1, -x_2)$, $(2S - 3T)(x_1, x_2) = (2x_1 + 5x_2, 5x_1 - 2x_2)$

2. (a) T is a linear operator. **(c)** T is not a linear operator.
(e) T is a linear operator.

3. (a) T is a linear transformation. **(c)** T is not a linear transformation.

4. (a) T is a linear operator. **(c)** T is not a linear operator.

5. (a) T is not a linear operator. **(c)** T is a linear operator.

6. (a) T is a linear transformation. **(c)** T is a linear transformation.
(e) T is a linear transformation. **(g)** T is not a linear transformation.

7. (b) $\{0\}$ **9. (a)** $\{(3, 1, 2), (-2, 1, -3)\}$ **(b)** $\{(1, 1, 1, 0), (2, 1, 0, 1)\}$

11. (b) $\{1, x\}$ **(c)** $\text{rank}(T) = 2, \text{nullity}(T) = 4$

12. (a) $\text{rank}(T) = 3, \text{nullity}(T) = 1$ **(c)** $\text{rank}(T) = 2, \text{nullity}(T) = 1$

13. (a) $\{(1, 0, -\frac{1}{3}, \frac{2}{3})\}$ **(c)** $\{(1, -1, 1)\}$

15. (a) Define $T : \mathbf{R}^2 \to \mathbf{R}$ by $T(x, y) = x$.

19. (a) $\{(1, 0, 0), (0, 1, 0), (0, 0, 1)\}$ **20. (a)** $\{(1, 2, 3), (1, 1, 1)\}$

Exercises 5.2, page 212

1. $T(x, y) = (-y, x + 2y, x + y)$ **3.** $T(a_0 + a_1x + a_2x^2) = (a_1, a_2 - a_1)$

5. $\begin{bmatrix} 1 & 1 & -1 \\ 0 & 5 & -2 \\ 4 & 0 & 1 \\ 2 & 3 & 1 \end{bmatrix}$ **7.** $\begin{bmatrix} 2 & 4 & -2 \\ -1 & -2 & 0 \\ 0 & -1 & 2 \\ 0 & 0 & 1 \end{bmatrix}$ **9.** $(3, 0, 2)$ **11.** $(26, 16)$

13. $-14 - 15x$

15. (a) $\{(1, 0, -1, -1), (0, 1, 0, 2)\}$ **(c)** $\{(1, 0, -\frac{1}{2}, -\frac{1}{2}), (0, 1, -\frac{1}{2}, \frac{1}{2})\}$

16. (a) $\{(1, 0, -3), (0, 1, 2)\}$ **(c)** $\{(1, 0, 1, 0), (0, 1, 1, 0), (0, 0, 0, 1)\}$

17. (a) $\{(1, -1, 1, 0), (3, -2, 0, 1)\}$; $\{(1, 0, -2, 1), (0, 1, -3, 1)\}$;
   $\mathrm{rank}(T) = 2$, $\mathrm{nullity}(T) = 2$

   (c) $\{(6, -\frac{5}{2}, 1, 0, 0), (-7, \frac{5}{2}, 0, 1, 0)\}$; $\{(1, 0, -1, -1, 0), (0, 1, -\frac{14}{5}, -\frac{12}{5}, 0)\}$;
   $\mathrm{rank}(T) = 3$, $\mathrm{nullity}(T) = 2$

19. (a) $\{(\frac{4}{3}, 1, 1), (\frac{1}{3}, 1, 0)\}$

20. (a) $\{(-2, -2, -2, -2, -3), (-1, -1, -2, -2, -2), (0, 1, 1, 1, 2)\}$;
   $\mathrm{rank}(T) = 2$, $\mathrm{nullity}(T) = 3$

21. (a) $\begin{bmatrix} -1 & -2 & 0 \\ 2 & 1 & -1 \\ -1 & 0 & 1 \\ 0 & 1 & 0 \end{bmatrix}$ **(b)** $\{x - x^2, x - x^3, x^2\}$ **(c)** $\varnothing$

   (d) $\mathrm{rank}(T) = 3$, $\mathrm{nullity}(T) = 0$

23. (a) $\begin{bmatrix} 0 & 0 & 1 \\ 0 & 1 & 3 \\ 0 & 2 & 3 \\ 0 & -1 & 3 \end{bmatrix}$ **(b)** $\left\{ \begin{bmatrix} 1 & 5 \\ -3 & -9 \end{bmatrix}, \begin{bmatrix} 0 & 0 \\ 1 & 3 \end{bmatrix} \right\}$ **(c)** $\{ [0 \quad 1 \quad 0] \}$

   (d) $\mathrm{rank}(T) = 2$, $\mathrm{nullity}(T) = 1$

25. $\begin{bmatrix} 4 & \frac{3}{2} \\ -1 & 0 \\ 3 & 1 \end{bmatrix}$ **27.** $\begin{bmatrix} 4 & -\frac{5}{3} \\ 0 & 3 \end{bmatrix}$

Exercises 5.3, page 224

1. (a) $\begin{bmatrix} -1 & 1 & -1 \\ 0 & -1 & 1 \\ 1 & 1 & 0 \end{bmatrix}$ **(b)** $[\mathbf{v}]_\mathcal{B} = \begin{bmatrix} -1 \\ -1 \\ 3 \end{bmatrix}$, $\mathbf{v} = 1 + 2x + 3x^2$

3. (a) $\begin{bmatrix} -1 & 0 & -1 \\ 1 & 1 & 0 \\ 0 & -1 & 1 \end{bmatrix}$ **(b)** $[\mathbf{v}]_\mathcal{B} = \begin{bmatrix} -1 \\ 4 \\ -3 \end{bmatrix}$, $\mathbf{v} = \begin{bmatrix} 0 & -3 \\ 4 & -1 \end{bmatrix}$

4. (a) $\begin{bmatrix} 4 \\ -4 \end{bmatrix}$ **5.** $3x^2 + x - 1$ **7.** $\begin{bmatrix} 2 & 0 \\ 0 & 1 \end{bmatrix}$ **9.** $\begin{bmatrix} 2 & 0 & 0 \\ 1 & 2 & 0 \\ 0 & 0 & 1 \end{bmatrix}$

11. $\begin{bmatrix} -4 & -2 & 4 \\ 3 & \frac{3}{2} & -3 \end{bmatrix}$ **13.** $\begin{bmatrix} 2 & -1 & 0 \\ -1 & 3 & 3 \end{bmatrix}$ **15.** $\begin{bmatrix} 4 & \frac{3}{2} \\ -1 & 0 \\ 3 & 1 \end{bmatrix}$

17. (a) $\mathcal{A}' = \{\frac{1}{2}x - \frac{1}{2}x^3, x^2 - x^3, -x^2 + 2x^3, 1 - \frac{1}{2}x + x^2 - \frac{3}{2}x^3\}$, $\mathcal{B}' = \{1, x, x^2\}$

(b) $\mathcal{A}' = \{1, x, x^2, x^3\}$, $\mathcal{B}' = \{1 + x^2, 2 + x + x^2, 2x + x^2\}$

19. (a) $\mathcal{A}' = \{(-2, 0, 1, 0, -1), (-2, 0, 1, 0, 0), (7, 0, -3, 0, 1), (-2, 0, 0, 1, -2),$
$\qquad (-3, 1, 0, 0, 0)\}$
$\qquad \mathcal{B}' = \mathcal{E}_3$

21. $\mathcal{A}' = \{1, x, x^2\}$, $\mathcal{B}' = \{x, x^2, x^3, 1\}$

23. $\mathcal{A}' = \left\{ \begin{bmatrix} 1 & 0 \\ 0 & 1 \end{bmatrix}, \begin{bmatrix} 1 & 0 \\ 0 & 0 \end{bmatrix}, \begin{bmatrix} 0 & 1 \\ 0 & 0 \end{bmatrix}, \begin{bmatrix} 0 & 1 \\ 1 & 0 \end{bmatrix} \right\}$, $\mathcal{B}' = \{1, x, x^2\}$

Exercises 5.4, page 235

1. (a) Matrix of S: $\begin{bmatrix} 5 & 0 & -3 & 0 \\ 0 & 1 & 6 & -1 \\ 2 & -9 & 5 & 2 \end{bmatrix}$, Matrix of T: $\begin{bmatrix} 4 & 10 & 0 & 0 \\ 6 & -1 & 0 & 7 \\ -3 & 8 & 0 & -5 \end{bmatrix}$

Matrix of $S + T$: $\begin{bmatrix} 9 & 10 & -3 & 0 \\ 6 & 0 & 6 & 6 \\ -1 & -1 & 5 & -3 \end{bmatrix}$

(b) Matrix of $2S - 3T$: $\begin{bmatrix} -2 & -30 & -6 & 0 \\ -18 & 5 & 12 & -23 \\ 13 & -42 & 10 & 19 \end{bmatrix}$

3. (a) $ST(x, y) = (6x, -x + y)$, $TS(x, y) = (x + 2y, 6y)$

(b) Matrix of T relative to $\mathcal{E}_2$: $\begin{bmatrix} 1 & 1 \\ 2 & 0 \end{bmatrix}$, Matrix of S relative to $\mathcal{E}_2$: $\begin{bmatrix} 0 & 3 \\ 1 & -1 \end{bmatrix}$,

Matrix of ST relative to $\mathcal{E}_2$: $\begin{bmatrix} 6 & 0 \\ -1 & 1 \end{bmatrix}$, Matrix of TS relative to $\mathcal{E}_2$: $\begin{bmatrix} 1 & 2 \\ 0 & 6 \end{bmatrix}$.

5. (a) TS has matrix $\begin{bmatrix} 1 & -7 \\ 2 & 1 \\ 0 & -5 \end{bmatrix}$ relative to the bases $\{(1, 2), (0, 1)\}$ of $\mathbf{R}^2$ and $\mathcal{E}_3$ of $\mathbf{R}^3$.

(b) $\{(1, 2, 0), (-7, 1, -5)\}$ **(c)** $\varnothing$

7. (a) $\begin{matrix} \mathbf{u}_1 = & -2\mathbf{v}_1 - & \mathbf{v}_2 + \mathbf{v}_3 \\ \mathbf{u}_2 = & \mathbf{v}_1 - 2\mathbf{v}_2 \\ \mathbf{u}_3 = & & \mathbf{v}_2 \end{matrix}$ **(b)** $\begin{bmatrix} 0 & 0 & 1 \\ 1 & 0 & 2 \\ 2 & 1 & 5 \end{bmatrix}$

9. (a) $TS(a_0 + a_1 x + a_2 x^2) = \begin{bmatrix} 0 & a_0 \\ a_1 & a_2 \end{bmatrix}$ **(b)** $\left\{ \begin{bmatrix} 0 & 1 \\ 0 & 0 \end{bmatrix}, \begin{bmatrix} 0 & 0 \\ 1 & 0 \end{bmatrix}, \begin{bmatrix} 0 & 0 \\ 0 & 1 \end{bmatrix} \right\}$

(c) $\varnothing$ **(d)** $\begin{bmatrix} -\frac{1}{2} & 0 & -\frac{1}{2} \\ \frac{1}{2} & 0 & -\frac{1}{2} \\ -\frac{1}{2} & -1 & -\frac{1}{2} \\ \frac{1}{2} & 1 & \frac{3}{2} \end{bmatrix}$

11. (a) $TS\left(\begin{bmatrix} a_{11} & a_{12} \\ a_{21} & a_{22} \end{bmatrix} \right) = -a_{11} - a_{22}$ **(b)** $\{1\}$

(c) $\left\{ \begin{bmatrix} 1 & 0 \\ 0 & -1 \end{bmatrix}, \begin{bmatrix} 0 & 1 \\ 0 & 0 \end{bmatrix}, \begin{bmatrix} 0 & 0 \\ 1 & 0 \end{bmatrix} \right\}$ **(d)** $\begin{bmatrix} -1 & 0 & -1 & -1 \end{bmatrix}$

13. (a) $\begin{bmatrix} 2 & 4 \\ 4 & -4 \end{bmatrix}$ **(b)** $T^{-1} = \frac{1}{6}T + \frac{1}{6}$ **(c)** $T^3 = 7T - 6$

15. (b) $A^{-1} = -\frac{1}{12}A + \frac{5}{12}I = \begin{bmatrix} \frac{1}{3} & -\frac{1}{6} \\ \frac{1}{3} & \frac{1}{12} \end{bmatrix}$

17. Let S, T be the linear operators on $\mathbf{R}^2$ defined by $S(x, y) = (2x, y)$ and $T(x, y) = (y, x)$. Then $ST(1, 3) = S(3, 1) = (6, 1)$ and $TS(1, 3) = T(2, 3) = (3, 2)$, so $ST \neq TS$.

Exercises 6.1, page 243

1. (a) 7   **(c)** 11   **(e)** 8

2. (a) Even, $I = 6$ **(c)** Odd, $I = 11$ **(e)** Even, $I = 10$

3. (a) Apply the following interchanges in the given order: 5 and 3, 4 and 3, 2 and 3, 2 and 4, 2 and 5.

   (c) Apply the following interchanges in the given order: 3 and 1, 5 and 1, 4 and 1, 2 and 1, 2 and 4, 2 and 5, 2 and 3.

5. $\mathcal{I} = \sum_{k=1}^{n} \mathcal{I}(k) = (n-1) + (n-2) + \cdots + 2 + 1 = \dfrac{n(n-1)}{2}$ **6. (a)** $\dfrac{n!}{2}$

Exercises 6.2, page 247

1. (a) Odd   **(c)** Even   **(e)** Odd   **(g)** Odd   **(i)** Odd   **(k)** Odd

3. $a_{11}a_{22} \cdots a_{nn}$ **5.** $a_{11}a_{22} \cdots a_{nn}$ **7.** $8\det(A)$ **9.** $c^n \det(A)$

10. (a) -72 **(c)** 0   **(e)** $a_{11}a_{22}a_{33}$ **(g)** 1

11. (a) -6 **(c)** 0   **(e)** $-a_{13}a_{22}a_{31}$ **(g)** 1

12. (a) $|A| = -4$, $|L||U| = 1(-4)$

Exercises 6.3, page 255

1. (a) 0   **(c)** 32   **(e)** -32 **3. (a)** 24   **(c)** 0

4. (a) $-x^3 - x^2 + 5x - 3$ **(c)** $x^4 - 2x^3 - 7x^2 - 4x$

5. (a) $6, -3$ **(c)** $-1, 2$ **(e)** $3, 4, 2, 5$

7. (a) $B = \begin{bmatrix} 1 & 2 & 1 \\ -1 & -1 & 1 \\ 0 & 1 & 3 \end{bmatrix}$ **(b)** $AB = I_3$ **(c)** $A^{-1} = B$

Exercises 6.4, page 261

1. (a) -9 **(c)** $a^2 + b^2$ **(e)** -177 **(g)** 60

2. (a) The second determinant is obtained from the first by multiplying the first column by 2 and the second column by 3. Therefore the second determinant has the value $(2)(3)(139) = 834$.

(c) The second determinant is obtained from the first by multiplying the first row by $\frac{1}{2}$, the second row by $\frac{1}{7}$, and the third row by -1. Therefore the second determinant has the value $(\frac{1}{2})(\frac{1}{7})(-1)(-266) = 19$.

(e) The second determinant is obtained from the first by adding to row 3 the product of -1 and row 2. Therefore the second determinant has the same value as the first determinant.

3. $x = -1, x = -2,$ or $x = 4$ **5.** $2abc(a + b + c)^3$

12. (a) 1   **(c)** 0   **(e)** 0

13. (a) $x_1 = -3, x_2 = 8, x_3 = 4$ **(c)** $x_1 = -1, x_2 = 1, x_3 = 1$
    (e) $x_1 = 2, x_2 = -1, x_3 = 4$

Exercises 6.5, page 268

1. (a) $\frac{1}{8}\begin{bmatrix} -2 & -4 \\ 3 & 2 \end{bmatrix}$ **(c)** $\frac{1}{6}\begin{bmatrix} 6 & 0 \\ -5 & 1 \end{bmatrix}$

2. (a) $\frac{1}{2}\begin{bmatrix} -1 & 3 & 2 \\ 1 & -1 & 0 \\ 1 & -1 & -2 \end{bmatrix}$ **(c)** $-\frac{1}{9}\begin{bmatrix} -10 & 4 & -1 \\ 8 & -5 & -1 \\ 1 & -4 & 1 \end{bmatrix}$

(e) $\frac{1}{20}\begin{bmatrix} -34 & 8 & 36 \\ 12 & -4 & -8 \\ 33 & -6 & -32 \end{bmatrix}$ **(g)** $\frac{1}{24}\begin{bmatrix} -9 & 6 & 0 & -9 \\ 3 & 6 & 0 & 3 \\ -6 & 4 & 8 & 2 \\ -21 & 6 & 0 & 3 \end{bmatrix}$

5. 3

17. (a) 5   **(b)** 625   **(c)** $\frac{81}{5}$ **(d)** 125   **(e)** 25   **(f)** 5   **(g)** $\frac{1}{25}$

(h) $\frac{1}{5}\begin{bmatrix} 5 & 1 & -1 & 6 \\ 5 & -9 & -1 & -9 \\ 0 & 5 & 0 & 5 \\ 0 & 1 & -1 & 1 \end{bmatrix}$

19. The rank of A is either 1 or 2.

Exercises 7.1, page 279

1. $1, 1$ **3.** $\dfrac{5 + \sqrt{33}}{2}, \dfrac{5 - \sqrt{33}}{2}$ **5.** $1, -2, 3$ **7.** $1, 1, 2$ **9.** $1, 2, 2$

11. $-1, -1, 3, 3$ **13.** $\{2, -3\}$ **15.** $\{1, 2\}$ **17.** $8, (-5, -6); -1, (1, 3)$

19. 1 with eigenvector $x + 1$; 2 with eigenvector $x^2 + 2x + 2$

21. $\dfrac{1 + \sqrt{5}}{2}, \begin{bmatrix} \dfrac{1 + \sqrt{5}}{2} \\ 1 \end{bmatrix}; \dfrac{1 - \sqrt{5}}{2}, \begin{bmatrix} \dfrac{1 - \sqrt{5}}{2} \\ 1 \end{bmatrix}$

23. The spectrum is $\{2, 4\}$. $-1 + x^2$ and $2 + x$ are eigenvectors corresponding to 2, and $1 + x^2$ is an eigenvector corresponding to 4.

25. The spectrum is $\{1, -1, 2\}$. $\begin{bmatrix} 0 & 0 \\ 1 & 0 \end{bmatrix}$ is an eigenvector corresponding to 1, $\begin{bmatrix} -1 & 1 \\ 0 & 0 \end{bmatrix}$ is an eigenvector corresponding to -1, and $\begin{bmatrix} 0 & 0 \\ 0 & 1 \end{bmatrix}$ is an eigenvector corresponding to 2.

27. $1, (0, -1, -1)$ **39.** 1

Exercises 7.2, page 288

For Problems 1, 3, and 5, each eigenvalue is followed by its algebraic multiplicity and geometric multiplicity, in that order.

1. $1, 2, 1$ **3.** $2, 1, 1; 1, 2, 1$ **5.** $-1, 2, 2; 3, 2, 2$

7. (a) $\{(1, -2), (2, 1)\}$ **(b)** $\begin{bmatrix} -3 & 0 \\ 0 & 2 \end{bmatrix}$

9. $\begin{bmatrix} \lambda_1 & 0 & \cdots & 0 \\ 0 & \lambda_2 & \cdots & 0 \\ \vdots & \vdots & \ddots & \vdots \\ 0 & 0 & \cdots & \lambda_n \end{bmatrix}$, where λ_i is the eigenvalue of T that corresponds to $\mathbf{v}_i$.

11. (a) $\lambda_1 = 3$ with algebraic multiplicity 2, $\mathbf{V}_3 = \langle (0, 1, 0), (2, 1, 1) \rangle$; $\lambda_2 = -2$ with algebraic multiplicity 1, $\mathbf{V}_{-2} = \langle (5, 4, 3) \rangle$

(c) $\lambda_1 = 1$ with algebraic multiplicity 1, $\mathbf{V}_1 = \langle (0, 1, 1) \rangle$; $\lambda_2 = 2$ with algebraic multiplicity 1, $\mathbf{V}_2 = \langle (-1, 1, 0) \rangle$; $\lambda_3 = 3$ with algebraic multiplicity 1, $\mathbf{V}_3 = \langle (0, 0, 1) \rangle$

13. (a) $\lambda_1 = 0$ with algebraic and geometric multiplicity 1; $\lambda_2 = 2$ with algebraic and geometric multiplicity 2; $\lambda_3 = 5$ with algebraic and geometric multiplicities 2 and 1, respectively.

(b) $\mathbf{V}_0 = \left\langle \begin{bmatrix} 2 & 0 \\ 1 & 0 \end{bmatrix} \right\rangle$, $\mathbf{V}_2 = \left\langle \begin{bmatrix} -2 & 1 \\ 0 & 1 \end{bmatrix}, \begin{bmatrix} -3 & 1 \\ -1 & 0 \end{bmatrix} \right\rangle$, $\mathbf{V}_5 = \left\langle \begin{bmatrix} 1 & -1 \\ 0 & -1 \end{bmatrix} \right\rangle$

14. (a) $\lambda_1 = 1$ with algebraic multiplicity 1, $\mathbf{V}_1 = \langle (1 - i, -1) \rangle$; $\lambda_2 = 4$ with algebraic multiplicity 1, $\mathbf{V}_4 = \langle (1, 1 + i) \rangle$

(c) $\lambda_1 = 1$ with algebraic multiplicity 2, $\mathbf{V}_1 = \langle (1, 1, -3), (0, 1, -2) \rangle$; $\lambda_2 = 6$ with algebraic multiplicity 1, $\mathbf{V}_6 = \langle (2, 1, 1) \rangle$

(e) $\lambda_1 = 1 + i$ with algebraic multiplicity 2, $\mathbf{V}_{1+i} = \langle (1, 0, 1), (0, 1, 0) \rangle$; $\lambda_2 = 1$ with algebraic multiplicity 1, $\mathbf{V}_1 = \langle (0, 2, -1) \rangle$

15. For $\lambda = 3$:   **(a)** 1,   **(b)** 1,   **(c)** $\{1\}$;

For $\lambda = 2$:   **(a)** 2,   **(b)** 2,   **(c)** $\{1 - x, 1 - x^2\}$

17. For $\lambda = 1$:   **(a)** 1,   **(b)** 1,   **(c)** $\{1\}$;

For $\lambda = 2$:   **(a)** 2,   **(b)** 1,   **(c)** $\{2 + x^2\}$

Exercises 7.3, page 296

1. (a) A is not similar over **R** to a diagonal matrix.   **(b)** Not possible.

3. (a) A is not similar over **R** to a diagonal matrix.   **(b)** Not possible.

5. (a) A is similar over **R** to a diag$\{-1, -1, 3, 3\}$.   **(b)** $P = \begin{bmatrix} 1 & 1 & 1 & 0 \\ 0 & 1 & 0 & 0 \\ -1 & 0 & 0 & 1 \\ 0 & -1 & 0 & -1 \end{bmatrix}$

7. (a) T can be represented by a diagonal matrix.
   (b) diag$\{2, -3\}$, $\{(2, 1), (1, -2)\}$

9. (a) T cannot be represented by a diagonal matrix.   **(b)** Not possible.

11. (a) T can be represented by a diagonal matrix.
   (b) diag$\{3, 3, -2\}$, $\{(0, 1, 0), (2, 1, 1), (5, 4, 3)\}$

13. (a) T can be represented by a diagonal matrix.
   (b) diag$\{1, 2, 3\}$, $\{(0, 1, 1), (-1, 1, 0), (0, 0, 1)\}$

15. (a) A is similar over C to diag$\{1, 4\}$.   **(b)** $P = \begin{bmatrix} 1-i & 1 \\ -1 & 1+i \end{bmatrix}$

17. (a) A is similar over C to diag$\{1, 1, 6\}$.   **(b)** $P = \begin{bmatrix} 1 & 0 & 2 \\ 1 & 1 & 1 \\ -3 & -2 & 1 \end{bmatrix}$

19. (a) A is similar over C to diag$\{1+i, 1+i, 1\}$.   **(b)** $P = \begin{bmatrix} 1 & 0 & 0 \\ 0 & 1 & 2 \\ 1 & 0 & -1 \end{bmatrix}$

23. $\begin{bmatrix} 1 & 1 \\ 0 & 1 \end{bmatrix}$ **26. (a)** $\begin{bmatrix} 0 & 1 & 0 \\ 0 & 0 & 1 \\ -2 & 0 & 5 \end{bmatrix}$ **(c)** $\begin{bmatrix} 0 & 1 & 0 & 0 \\ 0 & 0 & 1 & 0 \\ 0 & 0 & 0 & 1 \\ -4 & 0 & -5 & 0 \end{bmatrix}$

27. (a) $P = \begin{bmatrix} 1 & 0 \\ 2 & 1 \end{bmatrix}$, $A^5 = A$ **(c)** $P = \begin{bmatrix} 1 & 0 & -1 \\ 1 & 1 & -1 \\ 0 & 0 & 1 \end{bmatrix}$, $A^{10} = \begin{bmatrix} 1 & 0 & -1023 \\ 0 & 1 & -1023 \\ 0 & 0 & 1024 \end{bmatrix}$

28. (a) $P = \begin{bmatrix} 1 & 0 & -2 \\ 1 & 0 & 1 \\ 1 & 1 & 1 \end{bmatrix}$, $A^{-1} = \frac{1}{4}\begin{bmatrix} 2 & 2 & 0 \\ 1 & 3 & 0 \\ 1 & 1 & 2 \end{bmatrix}$

Exercises 8.1, page 307

1. f is a linear functional.

3. f is not a linear functional since

$$f\left(\begin{bmatrix} 1 & 0 \\ 0 & 1 \end{bmatrix}\right) = 1, \quad f\left(\begin{bmatrix} 1 & 1 \\ 1 & 1 \end{bmatrix}\right) = 0,$$

and

$$f\left(\begin{bmatrix} 1 & 0 \\ 0 & 1 \end{bmatrix} + \begin{bmatrix} 1 & 1 \\ 1 & 1 \end{bmatrix}\right) = f\left(\begin{bmatrix} 2 & 1 \\ 1 & 2 \end{bmatrix}\right) = 3.$$

5. f is not a linear functional since

$$f\left(\begin{bmatrix} 1 & 0 \\ 0 & 1 \end{bmatrix}\right) = 2(1) = 2, \quad f\left(\begin{bmatrix} 2 & 0 \\ 0 & 2 \end{bmatrix}\right) = 2(4) = 8,$$

and

$$f\left(\begin{bmatrix} 1 & 0 \\ 0 & 1 \end{bmatrix} + \begin{bmatrix} 2 & 0 \\ 0 & 2 \end{bmatrix}\right) = f\left(\begin{bmatrix} 3 & 0 \\ 0 & 3 \end{bmatrix}\right) = 2(9) = 18.$$

7. f is a linear functional.   **9.** f is a linear functional.

11. f is a linear functional.   **13.** f is a linear functional.

15. f is a linear functional.   **17.** $\mathbf{c} = (2, -1, 4)$

18. (a) $\begin{bmatrix} 10 & 1 & 5 \end{bmatrix}$ **(c)** $\begin{bmatrix} 2 & 2 & 6 & 18 \end{bmatrix}$

19. (a) $f(\mathbf{e}_1) = 3, f(\mathbf{e}_2) = -2, f(\mathbf{e}_3) = 7$
   (c) $f(\mathbf{e}_1) = 2, f(\mathbf{e}_2) = 0, f(\mathbf{e}_3) = 4, f(\mathbf{e}_4) = 12$

20. (a) $A = \begin{bmatrix} 1 & 0 & 0 \end{bmatrix}, f(1 - x^2) = \begin{bmatrix} 1 & 0 & 0 \end{bmatrix}\begin{bmatrix} 1 \\ 1 \\ 1 \end{bmatrix} = 1$

21. $\mathbf{p}_1(x_1, x_2, x_3) = \frac{1}{6}(2x_1 + 2x_2 - x_3), \mathbf{p}_2(x_1, x_2, x_3) = \frac{1}{6}(4x_1 - 2x_2 + x_3),$

   $\mathbf{p}_3(x_1, x_2, x_3) = \frac{1}{3}(-2x_1 + x_2 + x_3)$

22. (a) $\mathcal{A}^* = \{\mathbf{p}_1, \mathbf{p}_2, \mathbf{p}_3\}$, where $\mathbf{p}_1(\mathbf{u}) = x_3 - x_2, \mathbf{p}_2(\mathbf{u}) = x_2 - x_1, \mathbf{p}_3(\mathbf{u}) = x_1$ for
   $\mathbf{u} = (x_1, x_2, x_3)$.
   (c) $\mathcal{A}^* = \{\mathbf{p}_1, \mathbf{p}_2, \mathbf{p}_3\}$, where $\mathbf{p}_1(\mathbf{u}) = \frac{1}{2}(2x_1 + x_2 - x_3),$
   $\mathbf{p}_2(\mathbf{u}) = \frac{1}{2}(-3x_1 - x_2 + 3x_3), \mathbf{p}_3(\mathbf{u}) = \frac{1}{2}(-x_1 - x_2 + x_3)$ for $\mathbf{u} = (x_1, x_2, x_3)$.

23. (a) $\begin{bmatrix} 0 & -1 & 1 \end{bmatrix}^T, \begin{bmatrix} -1 & 1 & 0 \end{bmatrix}^T, \begin{bmatrix} 1 & 0 & 0 \end{bmatrix}^T$
   (c) $\frac{1}{2}\begin{bmatrix} 2 & 1 & -1 \end{bmatrix}^T, \frac{1}{2}\begin{bmatrix} -3 & -1 & 3 \end{bmatrix}^T, \frac{1}{2}\begin{bmatrix} -1 & -1 & 1 \end{bmatrix}^T$

24. (a) $\mathcal{A}^* = \{\mathbf{p}_1, \mathbf{p}_2, \mathbf{p}_3\}$, where $\mathbf{p}_1(\mathbf{u}) = a_0 + a_1 + 2a_2, \mathbf{p}_2(\mathbf{u}) = -a_1 - a_2, \mathbf{p}_3(\mathbf{u})$
   $= -a_2$ for $\mathbf{u} = a_0 + a_1x + a_2x^2$.
   (c) $\mathcal{A}^* = \{\mathbf{p}_1, \mathbf{p}_2, \mathbf{p}_3, \mathbf{p}_4\}$, where $\mathbf{p}_1(\mathbf{u}) = a_0 + a_1 - a_2, \mathbf{p}_2(\mathbf{u}) = a_2, \mathbf{p}_3(\mathbf{u}) = a_0,$
   $\mathbf{p}_4(\mathbf{u}) = -a_0 + a_3$ for $\mathbf{u} = a_0 + a_1x + a_2x^2 + a_3x^3$.

27. $f(1 + x - x^2) = 3, f(a_0 + a_1x + a_2x^2) = a_0 + 2a_1$

29. (a) $\mathbf{p}_1 = 6\mathbf{g}_1 + \mathbf{g}_2 - \mathbf{g}_3, \mathbf{p}_2 = 53\mathbf{g}_1 + 11\mathbf{g}_2 - 10\mathbf{g}_3, \mathbf{p}_3 = -5\mathbf{g}_1 - \mathbf{g}_2 + \mathbf{g}_3$
   (c) $\mathbf{p}_1 = \mathbf{g}_1 + \mathbf{g}_2 + 3\mathbf{g}_3 - \mathbf{g}_4, \mathbf{p}_2 = \mathbf{g}_1 - 2\mathbf{g}_3, \mathbf{p}_3 = \mathbf{g}_1 + \mathbf{g}_4, \mathbf{p}_4 = \mathbf{g}_3$

32. (a) $\{4\mathbf{g}_1 - 3\mathbf{g}_2 - \mathbf{g}_3, 8\mathbf{g}_1 - 2\mathbf{g}_2 - \mathbf{g}_4\}$ where $\mathbf{g}_1, \mathbf{g}_2, \mathbf{g}_3$ and $\mathbf{g}_4$ are the elements of $\mathcal{E}_4^*$
   (c) $\{\mathbf{g}_1 + \mathbf{g}_2 - 4\mathbf{g}_3, \mathbf{g}_4\}$ where $\mathbf{g}_1, \mathbf{g}_2, \mathbf{g}_3$ and $\mathbf{g}_4$ are the elements of $\mathcal{E}_4^*$

Exercises 8.2, page 315

1. (a) $A = \begin{bmatrix} 4 & 6 \\ 0 & -2 \end{bmatrix}$ or $A = \begin{bmatrix} 4 & 3 \\ 3 & -2 \end{bmatrix}$

(c) $A = \begin{bmatrix} -2 & 0 & -3 \\ -10 & 0 & 8 \\ 0 & 0 & -7 \end{bmatrix}$ or $A = \begin{bmatrix} -2 & -5 & -\frac{3}{2} \\ -5 & 0 & 4 \\ -\frac{3}{2} & 4 & -7 \end{bmatrix}$

(e) $A = \begin{bmatrix} -7 & -9 & 0 \\ 0 & 6 & 0 \\ 0 & 0 & 0 \end{bmatrix}$ or $A = \begin{bmatrix} -7 & -\frac{9}{2} & 0 \\ -\frac{9}{2} & 6 & 0 \\ 0 & 0 & 0 \end{bmatrix}$

(g) $A = \begin{bmatrix} 2 & 6 & 8 \\ 0 & 4 & 10 \\ 0 & 0 & 6 \end{bmatrix}$ or $A = \begin{bmatrix} 2 & 3 & 4 \\ 3 & 4 & 5 \\ 4 & 5 & 6 \end{bmatrix}$

2. (a) $q(\mathbf{v}) = -2y_1^2 + 18y_2^2$ **(c)** $q(\mathbf{v}) = y_1^2 - 2y_1y_3 - 4y_2y_3 - 3y_3^2$

3. (a) $\begin{bmatrix} -1 & 0 & 0 \\ 0 & -3 & 0 \\ 0 & 0 & 4 \end{bmatrix}$ **(c)** $\begin{bmatrix} 144 & -12 & 0 \\ -12 & -12 & 12 \\ 0 & 12 & -16 \end{bmatrix}$

Exercises 8.3, page 322

1. (c) $\mathbf{v}_1 = \left(\dfrac{1 - \sqrt{3}}{3}, \dfrac{2 + \sqrt{3}}{6}, \dfrac{1 + 2\sqrt{3}}{6} \right)$, $\mathbf{v}_2 = \left(\dfrac{-1 - \sqrt{3}}{3}, \dfrac{1 - 2\sqrt{3}}{6}, \dfrac{2 - \sqrt{3}}{6} \right)$

3. $\mathbf{v}_3 = \frac{-1}{2\sqrt{15}}(2\mathbf{u}_1 + \mathbf{u}_2 - \mathbf{u}_3)$

4. (a) $\left\{ \frac{1}{5}(1, 2, -2, 4), \frac{1}{\sqrt{14}}(0, -1, 3, 2), \frac{1}{\sqrt{19}}(4, 1, 1, -1) \right\}$

(c) $\left\{ \frac{1}{3}(2, 2, 1), \frac{1}{\sqrt{2}}(-1, 1, 0), \frac{1}{3\sqrt{2}}(-1, -1, 4) \right\}$

5. (a) $\left\{ \frac{1}{2}(1, \sqrt{3}, 0), \frac{1}{2}(-\sqrt{3}, 1, 0), (0, 0, 1) \right\}$

(c) $\left\{ (1, 0, 0, 0), \frac{1}{\sqrt{2}}(0, 0, 1, -1), \frac{1}{\sqrt{3}}(0, 1, 1, 1), \frac{1}{\sqrt{6}}(0, 2, -1, -1) \right\}$

6. (a) $\left\{ \frac{1}{\sqrt{3}}(-1, 1, 1, 0), \frac{1}{\sqrt{3}}(0, -1, 1, 1), \frac{1}{\sqrt{6}}(2, 1, 1, 0), \frac{1}{\sqrt{6}}(0, 1, -1, 2) \right\}$

(c) $\left\{ \frac{1}{\sqrt{6}}(1, 2, 0, -1), \frac{1}{\sqrt{3}}(1, -1, 0, -1), (0, 0, 1, 0), \frac{1}{\sqrt{2}}(1, 0, 0, 1) \right\}$

14. (a) Yes   **(c)** No   **(e)** No

17. (a) $T^{-1} : \mathbf{R}^3 \to \mathbf{R}^3$ defined by $T^{-1}(x_1, x_2, x_3) = (x_3, x_2, x_1)$

Exercises 8.4, page 329

1. (a) $P = \frac{1}{\sqrt{6}} \begin{bmatrix} \sqrt{2} & 0 & 2 \\ \sqrt{2} & \sqrt{3} & -1 \\ \sqrt{2} & -\sqrt{3} & -1 \end{bmatrix}$ **(c)** $P = \frac{1}{3\sqrt{2}} \begin{bmatrix} \sqrt{2} & 0 & 4 \\ -2\sqrt{2} & 3 & 1 \\ 2\sqrt{2} & 3 & -1 \end{bmatrix}$

2. (a) Hyperbola, $P = \frac{1}{\sqrt{2}} \begin{bmatrix} -1 & 1 \\ 1 & 1 \end{bmatrix}$, $-7(x')^2 + 11(y')^2 = 8$

(c) Ellipse, $P = \frac{1}{\sqrt{5}} \begin{bmatrix} 1 & 2 \\ -2 & 1 \end{bmatrix}$, $(x')^2 + 6(y')^2 = 6$

3. (a) $q(x_1', x_2', x_3') = (x_1')^2 + 3(x_2')^2 + 3(x_3')^2$, $P = \frac{1}{\sqrt{2}} \begin{bmatrix} -1 & 1 & 0 \\ 1 & 1 & 0 \\ 0 & 0 & \sqrt{2} \end{bmatrix}$

(c) $q(x_1', x_2', x_3') = 9(x_2')^2 + 9(x_3')^2$, $P = \frac{1}{3\sqrt{2}} \begin{bmatrix} 1 & -3 & 2\sqrt{2} \\ 1 & 3 & 2\sqrt{2} \\ -4 & 0 & \sqrt{2} \end{bmatrix}$

(e) $q(x_1', x_2', x_3', x_4') = 2(x_1')^2 + 2(x_2')^2 - 4(x_3')^2 - 4(x_4')^2$,

$$P = \frac{1}{\sqrt{6}} \begin{bmatrix} -1 & \sqrt{2} & 1 & -\sqrt{2} \\ 0 & \sqrt{2} & -2 & 0 \\ 1 & \sqrt{2} & 1 & \sqrt{2} \\ -2 & 0 & 0 & \sqrt{2} \end{bmatrix}$$

Exercises 8.5, page 337

1. (a) $r = 2, p = 1, s = 0$ **(c)** $r = 1, p = 1, s = 0$

2. (a) $q(\mathbf{v}) = z_1^2 - z_2^2$ **(c)** $q(\mathbf{v}) = z_1^2$

3. (a) $P = \frac{1}{2\sqrt{3}} \begin{bmatrix} 1 & 0 & -2 \\ 1 & -\sqrt{3} & 1 \\ 1 & \sqrt{3} & 1 \end{bmatrix}$ **(c)** $P = \frac{1}{18} \begin{bmatrix} 2 & 0 & 4 \\ -4 & 3 & 1 \\ 4 & 3 & -1 \end{bmatrix}$

4. (a) $\mathcal{B} = \left\{ \frac{1}{2\sqrt{3}}(1, 1, 1), \frac{1}{2}(0, -1, 1), \frac{1}{2\sqrt{3}}(-2, 1, 1) \right\}$

(c) $\mathcal{B} = \left\{ \frac{1}{9}(1, -2, 2), \frac{1}{6}(0, 1, 1), \frac{1}{18}(4, 1, -1) \right\}$

5. (a) Negative semidefinite, $r = 1, p = 0, s = -1$

(c) Negative definite, $r = 3, p = 0, s = -3$

(e) Indefinite, $r = 3, p = 1, s = -1$

(g) Positive semidefinite, $r = p = s = 1$

(i) Negative semidefinite, $r = 1, p = 0, s = -1$

12. (a) $\frac{1}{3} \begin{bmatrix} 4 & 1 & 1 \\ 1 & 4 & 1 \\ 1 & 1 & 4 \end{bmatrix}$

Exercises 8.6, page 344

1. (a) $\begin{bmatrix} 0 & 2 & -1 \\ 2 & 1 & 1 \end{bmatrix}$ **(c)** $\begin{bmatrix} 3 & 5 & 1 \\ 5 & 5 & 2 \\ 1 & 2 & 0 \end{bmatrix}$ **(e)** $\begin{bmatrix} 5 + 4i & -5 + i \\ 3 - 6i & 3 + 6i \\ 6 + 4i & -6 \end{bmatrix}$

2. (a) 24   **(c)** -7 **(e)** $14i$

3. (a) $\begin{bmatrix} 32 & -4 & -11 \\ -8 & 1 & -40 \end{bmatrix}$ **(c)** $\begin{bmatrix} 84 & 0 & -15 \\ 0 & -3 & 3 \\ -15 & 3 & -9 \end{bmatrix}$ **(e)** $\begin{bmatrix} -9-i & -5+i \\ -9-3i & -6+3i \\ 3-i & 3 \end{bmatrix}$

4. (a) $\mathcal{A}' = \{(1, 0, 0, 0), (-4, 1, 0, 0), (1, -1, 1, 0), (2, -1, 0, 1)\}$,
$\mathcal{B}' = \{(1, 0, 0), (0, 1, 0), (-2, 5, 1)\}$

  (c) $\mathcal{A}' = \{(0, 0, 0, 1), (0, 1, 0, -2), (0, -2, 1, 0), (1, 0, 0, 0)\}$,
$\mathcal{B}' = \{(1, 0, 0), (0, 1, 0), (-1, -1, 1)\}$

Exercises 8.7, page 351

1. (a) $P = \begin{bmatrix} 1 & 0 & -6 \\ 1 & 0 & 4 \\ 0 & 1 & 5 \end{bmatrix}$ **(c)** $P = \begin{bmatrix} 1 & 1 & -2 \\ 1 & -1 & 3 \\ 0 & 0 & -2 \end{bmatrix}$ **(e)** $P = \begin{bmatrix} 1 & 1 & -3 \\ 1 & -1 & 2 \\ 0 & 0 & 1 \end{bmatrix}$

2. (a) $\{(2, 1, 2), (0, 1, -1), (-2, -1, -7)\}$ **(c)** $\{(2, 1, 2), (0, 1, 0), (1, -4, 3)\}$
  (e) $\{(2, 1, 2), (0, 1, 0), (-1, -2, -2)\}$

12. (a) $f((x_1, x_2), (y_1, y_2)) = g((x_1, x_2), (y_1, y_2)) + h((x_1, x_2), (y_1, y_2))$,
where $g((x_1, x_2), (y_1, y_2)) = 2x_1 y_1 + \frac{1}{2} x_1 y_2 + \frac{1}{2} x_2 y_1 - 4x_2 y_2$ and
$h((x_1, x_2), (y_1, y_2)) = \frac{1}{2} x_1 y_2 - \frac{1}{2} x_2 y_1$

Exercises 8.8, page 361

1. Those in parts (a), (b), (d), (e), (g).

2. (a) For part (a), $P = \begin{bmatrix} 1 & 1 \\ 0 & 5i \end{bmatrix}$;

  for part (c), the matrix is not Hermitian;

  for part (e), $P = \begin{bmatrix} 1 & 1 & -4-i \\ 0 & 5i & 5-2i \\ 0 & 0 & 9 \end{bmatrix}$;

  for part (g), $P = \begin{bmatrix} 1 & 1 & -1 \\ 0 & 0 & 2i \\ 0 & 11 & 1 \end{bmatrix}$

  (b) In part (a), $r = p = 2$; in part (c), the matrix is not Hermitian; in part (e), $r = p = 3$; in part (g), $r = p = 3$

  (c) Those in parts (a) and (b) are conjunctive, and those in parts (e) and (g) are conjunctive.

3. (a) $\frac{1}{3\sqrt{5}}\begin{bmatrix} 3 & 0 \\ 0 & 5i \end{bmatrix}$ **(c)** The matrix in 1(c) is not Hermitian.

  (e) $\frac{1}{3\sqrt{5}}\begin{bmatrix} 3 & 1 & -4-i \\ 0 & 5i & 5-2i \\ 0 & 0 & 9 \end{bmatrix}$ **(g)** $\frac{1}{\sqrt{66}}\begin{bmatrix} 3\sqrt{6} & 1 & -\sqrt{11} \\ 0 & 0 & 2i\sqrt{11} \\ 0 & 11 & \sqrt{11} \end{bmatrix}$

4. **(a)** Those for the matrices in parts (a), (b), (d), (e), (g).

    (b) For f in part (a), $\mathcal{A}' = \{(1, 0), (1, 5i)\}$;

       for f in part (e), $\mathcal{A}' = \{(1, 0, 0), (1, 5i, 0), (4 + i, -5 + 2i, -9)\}$;

       for f in part (g), $\mathcal{A}' = \{(1, 0, 0), (1, 0, 11), (1, -2i, -1)\}$

23. The diagonal elements must be pure imaginary, i.e., if $A = [a_{ij}]_n$ then $a_{kk} = bi$, $b \in \mathbf{R}$.

Exercises 9.1, page 368

1. **(a)** $(\mathbf{u}, \mathbf{v}) = (\mathbf{v}, \mathbf{u}) = -3$ 2. **(a)** $(\mathbf{u}, \mathbf{v}) = 15 + 3i$, $(\mathbf{v}, \mathbf{u}) = 15 - 3i$

4. **(a)** The rule defines an inner product on $\mathbf{R}^2$.

    (c) The rule does not define an inner product on $\mathbf{R}^2$. If $\mathbf{v} = (1, -1)$, then $(\mathbf{v}, \mathbf{v}) = 0$, and this contradicts the requirement that $(\mathbf{v}, \mathbf{v}) > 0$ if $\mathbf{v} \neq \mathbf{0}$.

    (e) The rule does not define an inner product on $\mathbf{R}^3$. If $\mathbf{v} = (0, 1, 0)$, then $(\mathbf{v}, \mathbf{v}) = 0$, contradicting the property that requires $(\mathbf{v}, \mathbf{v}) > 0$ if $\mathbf{v} \neq \mathbf{0}$.

5. Those in parts (a), (b), (e), (g).

7. $(\mathbf{u}, \mathbf{v}) = 5\bar{u}_1 v_1 + (-5 + i)\bar{u}_1 v_2 + (2 - i)\bar{u}_1 v_3 + (-5 - i)\bar{u}_2 v_1 + 7\bar{u}_2 v_2$
        $+ (-5 + i)\bar{u}_2 v_3 + (2 + i)\bar{u}_3 v_1 + (-5 - i)\bar{u}_3 v_2 + 6\bar{u}_3 v_3$

Exercises 9.2, page 373

1. **(a)** $\sqrt{11}$ **(c)** $\sqrt{5}$ 2. **(a)** $\sqrt{74}$ **(c)** $\sqrt{7}$

3. **(a)** We have $|(\mathbf{u}, \mathbf{v})| = |-13| = 13$ and $\|\mathbf{u}\| \cdot \|\mathbf{v}\| = \sqrt{14}\sqrt{38} = \sqrt{532} = 2\sqrt{133}$. Since $13 < 2\sqrt{133}$, the inequality is verified.

    (b) $\sqrt{14}$ **(c)** $\sqrt{78}$

5. **(a)** 3 **(b)** $2\sqrt{6}$

7. $d(\mathbf{u}, \mathbf{v}) = \sqrt{|u_1 - v_1|^2 + |u_2 - v_2|^2}$, where $\mathbf{u} = (u_1, u_2)$ and $\mathbf{v} = (v_1, v_2)$

13. $\cos \theta = \frac{17}{2\sqrt{91}}$

Exercises 9.3, page 379

2. **(a)** $\left\{ \frac{\sqrt{2}}{6}(3, 2\sqrt{2}, -1), \frac{1}{2}(1, -\sqrt{2}, -1), \frac{1}{6}(-3, \sqrt{2}, -5) \right\}$

5. **(a)** $\left\{ \frac{1}{\sqrt{3}}(1, -i, 1), \frac{1}{2\sqrt{6}}(3 + i, 1 + 3i, -2i) \right\}$

7. $\left\{ \left(\frac{1}{3}, -\frac{2}{3}, \frac{2}{3}\right), \frac{1}{\sqrt{2}}(0, 1, 1), \frac{1}{3\sqrt{2}}(4, 1, -1) \right\}$

8. **(a)** $\left\{ \left(\frac{i}{2}, \frac{1}{\sqrt{2}}, \frac{i}{2}\right), \left(-\frac{i}{2}, \frac{1}{\sqrt{2}}, -\frac{i}{2}\right), \left(\frac{1+i}{2}, 0, \frac{-1-i}{2}\right) \right\}$ 11. $\sqrt{3}$

17. **(a)** $\left\{ 1, x, x^2 - \frac{1}{3}, x^3 - \frac{3}{5}x \right\}$ **(b)** $\left\{ \frac{1}{\sqrt{2}}, \frac{\sqrt{6}}{2}x, \frac{\sqrt{10}}{4}(3x^2 - 1), \frac{\sqrt{14}}{4}(5x^3 - 3x) \right\}$

Exercises 9.4, page 383

1. (a) $\left\{\frac{1}{\sqrt{5}}(2, 1, 0, 0), \frac{1}{\sqrt{170}}(3, -6, 10, 5)\right\}$

   (b) Let $\mathbf{u}_1 = \frac{1}{\sqrt{5}}(2, 1, 0, 0)$ and $\mathbf{u}_2 = \frac{1}{\sqrt{170}}(3, -6, 10, 5)$. Then $\mathbf{v} = \mathbf{v}_1 + \mathbf{v}_2$ where $\mathbf{v}_2 = (\mathbf{v}, \mathbf{u}_1)\mathbf{u}_1 + (\mathbf{v}, \mathbf{u}_2)\mathbf{u}_2$ and $\mathbf{v}_1 = \mathbf{v} - \mathbf{v}_2$.

3. (a) $\langle 6x^2 - 6x + 1, 30x^3 - 30x^2 + 3x + 1 \rangle$

   (b) $a_0 + a_1 x + a_2 x^2 + a_3 x^3 = \frac{1}{30}(30a_0 - 5a_2 - 6a_3)(1) + \frac{1}{10}(10a_1 + 10a_2 + 9a_3)x$
   $$+ \frac{1}{6}(a_2 + a_3)(6x^2 - 6x + 1)$$
   $$+ \frac{1}{30}a_3(30x^3 - 30x^2 + 3x + 1)$$

Exercises 9.5, page 387

1. (a) No   **(c)** Yes   **2. (a)** Yes   **(c)** No

3. (a) $T(x, y) = \frac{1}{\sqrt{2}}(x - y, -x - y)$

15. Using $\mathcal{E}_3$ and $\left\{\frac{1}{\sqrt{2}}, \frac{\sqrt{6}}{2}x, \frac{\sqrt{10}}{4}(3x^2 - 1)\right\}$ as bases of $\mathbf{R}^3$ and $\mathcal{P}_2$, respectively, define $T : \mathbf{R}^3 \to \mathcal{P}_2$ by $T(a, b, c) = \left(\frac{1}{\sqrt{2}}a - \frac{\sqrt{10}}{4}c\right) + \frac{\sqrt{6}}{2}bx + \frac{3\sqrt{10}}{4}cx^2$.

Exercises 9.6, page 394

1. Those in parts (a), (c), (d).   **3.** Only $x = y = z = 0$

Exercises 9.7, page 400

1. (a) $T^*(a_1, a_2) = (a_1 + (1 - i)a_2, (1 + i)a_1 + 2a_2)$
   (c) $T^*(a_1, a_2) = (-ia_1 + a_2, ia_1)$

2. (a) $\left\{\frac{1}{\sqrt{3}}(-1 + i, 1), \frac{1}{\sqrt{3}}(1, 1 + i)\right\}$ **(c)** No such basis exists.

3. All of the matrices A are normal. A matrix U of the required type is given for parts (a), (c), and (e).

   (a) $U = \frac{1}{2}\begin{bmatrix} 0 & \sqrt{2} & -\sqrt{2} \\ \sqrt{2} & -1 & -1 \\ \sqrt{2} & 1 & 1 \end{bmatrix}$ **(c)** $U = \frac{1}{\sqrt{3}}\begin{bmatrix} i - 1 & -i \\ i & 1 + i \end{bmatrix}$

   (e) $U = \frac{1}{2\sqrt{2}}\begin{bmatrix} \sqrt{2} & 1 - i & 2 \\ \sqrt{2}i & 1 + i & -2i \\ -\sqrt{2} - \sqrt{2}i & 2 & 0 \end{bmatrix}$

Exercises 10.1, page 407

2. (a) $P(\mathbf{V}) = \langle (1, 1, 0), (-1, 0, 1) \rangle$, $P^{-1}(\mathbf{0}) = \langle (1, -1, 1) \rangle$, $D = \text{diag}\{1, 1, 0\}$
   (c) $P(\mathbf{V}) = \langle (-1, 0, 1, 0), (0, -1, 0, 1) \rangle$, $P^{-1}(\mathbf{0}) = \langle (1, 0, 1, 0), (0, 1, 0, 1) \rangle$,
   $D = \text{diag}\{1, 1, 0, 0\}$

7. Let P_1, P_2 be defined on $\mathbf{R}^2$ by $P_1(x_1, x_2) = (x_1, x_1)$, and $P_2(x_1, x_2) = (0, x_1 + x_2)$

Exercises 10.2, page 414

3. (a) T is diagonalizable.

(b) $\lambda_1 = 5$, $\mathbf{V}_5 = \langle (1, 0, 1), (0, 1, 0) \rangle$, $\lambda_2 = -2$, $\mathbf{V}_{-2} = \langle (-1, 0, 1) \rangle$

Exercises 10.3, page 427

1. In each part, direct computations show that A satisfies its characteristic equation.

(a) $A^3 - A^2 - 8A + 12I = 0$ **(c)** $A^4 - 8A^2 + 16I = 0$

2. (a) $(x + 3)(x - 2)^2$ **(c)** $(x + 2)(x - 2)^2$

3. $A = \begin{bmatrix} \pm\sqrt{-1 - bc} & b \\ c & \mp\sqrt{-1 - bc} \end{bmatrix}$ with $-bc \geq 1$.

4. (a) $T = 3P_1 - 2P_2$, where P_1, P_2 have matrices E_1, E_2 relative to $\mathcal{E}_3$ given by

$$E_1 = \begin{bmatrix} 2 & 1 & -1 \\ 1 & 2 & -1 \\ 3 & 3 & -2 \end{bmatrix}, E_2 = \begin{bmatrix} -1 & -1 & 1 \\ -1 & -1 & 1 \\ -3 & -3 & 3 \end{bmatrix}.$$

(c) $T = P_1 - P_2 + 3P_3$, where P_1, P_2, P_3 have matrices E_1, E_2, E_3 relative to $\mathcal{E}_4$

given by $E_1 = -\frac{1}{2} \begin{bmatrix} 6 & 4 & 5 & 2 \\ -12 & -8 & -10 & -4 \\ 0 & 0 & 0 & 0 \\ 0 & 0 & 0 & 0 \end{bmatrix}$, $E_2 = \frac{1}{4} \begin{bmatrix} 2 & 1 & 2 & 0 \\ -12 & -6 & -12 & 0 \\ 8 & 4 & 8 & 0 \\ -2 & -1 & -2 & 0 \end{bmatrix}$,

$E_3 = \frac{1}{4} \begin{bmatrix} 14 & 7 & 8 & 4 \\ -12 & -6 & -8 & -8 \\ -8 & -4 & -4 & 0 \\ 2 & 1 & 2 & 4 \end{bmatrix}$

5. (a) $A = D + N$, where the diagonalizable matrix D and the nilpotent matrix N are

given by $D = \begin{bmatrix} 5 & 2 & 2 & 2 \\ 4 & 3 & 4 & -4 \\ -8 & -4 & -5 & 0 \\ 0 & 0 & 0 & 3 \end{bmatrix}$, and $N = \begin{bmatrix} 2 & 1 & 1 & 0 \\ -4 & -2 & -2 & 0 \\ 0 & 0 & 0 & 0 \\ 2 & 1 & 1 & 0 \end{bmatrix}$.

(c) $A = D + N$, where the diagonalizable matrix D and the nilpotent matrix N are

given by $D = \frac{1}{2} \begin{bmatrix} 2 & 3 & 0 & -1 \\ 1 & 2 & 1 & 0 \\ 0 & 1 & 2 & 1 \\ -1 & 0 & 3 & 2 \end{bmatrix}$, and $N = \frac{1}{2} \begin{bmatrix} 0 & -1 & 0 & 1 \\ -1 & 0 & 1 & 0 \\ 0 & -1 & 0 & 1 \\ -1 & 0 & 1 & 0 \end{bmatrix}$.

Exercises 10.4, page 436

1. (a) $\mathcal{B} = \{(-4, 3, -1, 0), (-5, 3, -1, 0), (3, -2, 1, 0), (0, 0, 0, 1)\}$,

$$A = \begin{bmatrix} 0 & 1 & 0 & 0 \\ 0 & 0 & 1 & 0 \\ 0 & 0 & 0 & 1 \\ 0 & 0 & 0 & 0 \end{bmatrix}$$

(c) $\mathcal{B} = \{(-2, 4, 0, 0, 2), (-1, 0, 2, 1, 1), (1, 0, 0, 0, 0);$
$(1, -2, 0, 1, -1), (0, 0, 0, 0, 1)\},$

$$A = \begin{bmatrix} A_1 & \mathbf{0} \\ \mathbf{0} & A_2 \end{bmatrix} \text{ with } A_1 = \begin{bmatrix} 0 & 1 & 0 \\ 0 & 0 & 1 \\ 0 & 0 & 0 \end{bmatrix}, \text{ and } A_2 = \begin{bmatrix} 0 & 1 \\ 0 & 0 \end{bmatrix}$$

2. (a) $\mathcal{B}(\mathbf{v}, T) = \{(4, -3, 1, 0), (1, 0, 0, 0), (0, -1, 0, 0), (1, 1, 0, -1)\},$ $\begin{bmatrix} 0 & 1 & 0 & 0 \\ 0 & 0 & 1 & 0 \\ 0 & 0 & 0 & 1 \\ 0 & 0 & 0 & 0 \end{bmatrix}$

(c) $\mathcal{B}(\mathbf{v}, T) = \{(2, -4, 0, 0, -2), (1, 0, -2, 0, -1), (-1, -1, 1, 0, 1)\},$ $\begin{bmatrix} 0 & 1 & 0 \\ 0 & 0 & 1 \\ 0 & 0 & 0 \end{bmatrix}$

Exercises 10.5, page 442

1. D has matrix K and N has matrix L relative to $\mathcal{E}_6$, where

$$K = \begin{bmatrix} -1 & 0 & 0 & 0 & 0 & 0 \\ 0 & -1 & 0 & 0 & 0 & 0 \\ 0 & 0 & -1 & 0 & 0 & 0 \\ 0 & 1 & 0 & -2 & 0 & 0 \\ 1 & -1 & 0 & 0 & -2 & 0 \\ 0 & 1 & 0 & 0 & 0 & -2 \end{bmatrix} \text{ and } L = \begin{bmatrix} 0 & 1 & 0 & 0 & 0 & 0 \\ -1 & 0 & 1 & 0 & 0 & 0 \\ 0 & 1 & 0 & 0 & 0 & 0 \\ -2 & 0 & 1 & 1 & 1 & 0 \\ 2 & 1 & -1 & -1 & -1 & 0 \\ -1 & 0 & 1 & 0 & 0 & 0 \end{bmatrix}.$$

2. (a) $J = \begin{bmatrix} J_1 & \mathbf{0} \\ \mathbf{0} & J_2 \end{bmatrix}$, where $J_1 = \begin{bmatrix} 2 & 1 & 0 & 0 \\ 0 & 2 & 0 & 0 \\ 0 & 0 & 2 & 0 \\ 0 & 0 & 0 & 2 \end{bmatrix}$ or $\begin{bmatrix} 2 & 1 & 0 & 0 \\ 0 & 2 & 0 & 0 \\ 0 & 0 & 2 & 1 \\ 0 & 0 & 0 & 2 \end{bmatrix}$ and

$$J_2 = \begin{bmatrix} 7 & 1 & 0 \\ 0 & 7 & 0 \\ 0 & 0 & 7 \end{bmatrix}.$$

(c) $J = \begin{bmatrix} J_1 & \mathbf{0} \\ \mathbf{0} & J_2 \end{bmatrix}$, where $J_2 = [4]$ and $J_1 = \begin{bmatrix} -3 & 1 & 0 & 0 & 0 \\ 0 & -3 & 1 & 0 & 0 \\ 0 & 0 & -3 & 0 & 0 \\ 0 & 0 & 0 & -3 & 0 \\ 0 & 0 & 0 & 0 & -3 \end{bmatrix}$ or

$$\begin{bmatrix} -3 & 1 & 0 & 0 & 0 \\ 0 & -3 & 1 & 0 & 0 \\ 0 & 0 & -3 & 0 & 0 \\ 0 & 0 & 0 & -3 & 1 \\ 0 & 0 & 0 & 0 & -3 \end{bmatrix}.$$

(e) $J = \begin{bmatrix} 8 & 1 & 0 & 0 & 0 & 0 \\ 0 & 8 & 1 & 0 & 0 & 0 \\ 0 & 0 & 8 & 0 & 0 & 0 \\ 0 & 0 & 0 & 8 & 1 & 0 \\ 0 & 0 & 0 & 0 & 8 & 1 \\ 0 & 0 & 0 & 0 & 0 & 8 \end{bmatrix}$ or $\begin{bmatrix} 8 & 1 & 0 & 0 & 0 & 0 \\ 0 & 8 & 1 & 0 & 0 & 0 \\ 0 & 0 & 8 & 0 & 0 & 0 \\ 0 & 0 & 0 & 8 & 1 & 0 \\ 0 & 0 & 0 & 0 & 8 & 0 \\ 0 & 0 & 0 & 0 & 0 & 8 \end{bmatrix}$ or

$\begin{bmatrix} 8 & 1 & 0 & 0 & 0 & 0 \\ 0 & 8 & 1 & 0 & 0 & 0 \\ 0 & 0 & 8 & 0 & 0 & 0 \\ 0 & 0 & 0 & 8 & 0 & 0 \\ 0 & 0 & 0 & 0 & 8 & 0 \\ 0 & 0 & 0 & 0 & 0 & 8 \end{bmatrix}$.

3. (a) $P = \begin{bmatrix} 2 & -1 & 0 \\ 0 & 2 & 0 \\ 4 & 0 & 1 \end{bmatrix}$

4. (a) $\begin{bmatrix} 3 & 1 & 0 & 0 \\ 0 & 3 & 0 & 0 \\ 0 & 0 & 1 & 0 \\ 0 & 0 & 0 & -1 \end{bmatrix}$ **(c)** $\begin{bmatrix} 0 & 1 & 0 & 0 \\ 0 & 0 & 0 & 0 \\ 0 & 0 & 2 & 1 \\ 0 & 0 & 0 & 2 \end{bmatrix}$

5. (a) $\mathcal{B}' = \{(-1, 2, 0, -1), (-1, 0, 1, 0), (-1, 2, 0, 0), (0, -1, 1, 0)\}$,

$P = \begin{bmatrix} -1 & -1 & -1 & 0 \\ 2 & 0 & 2 & -1 \\ 0 & 1 & 0 & 1 \\ -1 & 0 & 0 & 0 \end{bmatrix}$

(c) $\mathcal{B}' = \{(1, -1, 1, -1), (2, -1, 0, 1), (-1, -1, -1, -1), (3, 2, 1, 0)\}$,

$P = \begin{bmatrix} 1 & 2 & -1 & 3 \\ -1 & -1 & -1 & 2 \\ 1 & 0 & -1 & 1 \\ -1 & 1 & -1 & 0 \end{bmatrix}$

6. (a) $A = \begin{bmatrix} 5 & 2 & 2 & 2 \\ 4 & 3 & 4 & -4 \\ -8 & -4 & -5 & 0 \\ 0 & 0 & 0 & 3 \end{bmatrix} + \begin{bmatrix} 2 & 1 & 1 & 0 \\ -4 & -2 & -2 & 0 \\ 0 & 0 & 0 & 0 \\ 2 & 1 & 1 & 0 \end{bmatrix}$

(c) $A = \begin{bmatrix} 1 & \frac{3}{2} & 0 & -\frac{1}{2} \\ \frac{1}{2} & 1 & \frac{1}{2} & 0 \\ 0 & \frac{1}{2} & 1 & \frac{1}{2} \\ -\frac{1}{2} & 0 & \frac{3}{2} & 1 \end{bmatrix} + \begin{bmatrix} 0 & -\frac{1}{2} & 0 & \frac{1}{2} \\ -\frac{1}{2} & 0 & \frac{1}{2} & 0 \\ 0 & -\frac{1}{2} & 0 & \frac{1}{2} \\ -\frac{1}{2} & 0 & \frac{1}{2} & 0 \end{bmatrix}$.

Exercises 11.1, page 451

1. (a) $\sqrt{21}$ **(b)** 4   **(c)** 7   **3. (a)** $(0, 0)$ **(c)** $(0, \frac{1}{4}, 1)$ **(e)** Diverges

Exercises 11.2, page 462

1. (a) 8 **(b)** 7 **(c)** $5\sqrt{2}$ **3. (a)** 7 **(b)** 7 **(c)** 7

4. (a) $\begin{bmatrix} 0 & 1 \\ \frac{\pi}{2} & 0 \end{bmatrix}$ **(c)** Diverges **(e)** $\begin{bmatrix} 0 & 0 \\ 1 & 1 \end{bmatrix}$

Exercises 11.3, page 465

3. (a) $A = \begin{bmatrix} -0.01 & 0.51 & -0.35 \\ -0.21 & -0.19 & 0.13 \\ 0.33 & 0.33 & -0.25 \end{bmatrix}, B = \begin{bmatrix} 0.11 \\ -0.02 \\ 0.10 \end{bmatrix}; \|A\|_r = 0.91 < 1$

(c) $A = \begin{bmatrix} -0.10 & 0.36 & -0.06 \\ 0.32 & -0.25 & -0.21 \\ -0.12 & 0.13 & 0.05 \end{bmatrix}, B = \begin{bmatrix} 0.17 \\ 0.25 \\ -0.24 \end{bmatrix}; \|A\|_r = 0.78 < 1$

4. (a) $x_1 = 0.07, x_2 = -0.02, x_3 = 0.09$
(c) $x_1 = 0.27, x_2 = 0.31, x_3 = -0.24$

6. (a) $\|S - X_4\| \le 0.84$ **(c)** $\|S - X_4\| \le 0.42$

Exercises 11.4, page 472

5. (a) $x_1 = 2.13, x_2 = 0.71$ **(c)** $x_1 = 1.36, x_2 = -0.63, x_3 = 1.63$

Exercises 11.5, page 480

1. (a) 10 **(c)** 20 **2. (a)** 1.0 **(c)** -2.0

3. (a) $\begin{bmatrix} 1.0 \\ 0.66 \end{bmatrix}$ **(c)** $\begin{bmatrix} 1.0 \\ 1.0 \\ -1.0 \end{bmatrix}$ **4. (a)** 10 **(c)** 20

Index

Addition
  of equations, 171
  of linear transformations, 192, 229
  of matrices, 145
  of subsets, 15
  of vectors, 2
Adjoint
  of a linear operator, 395
  of a matrix, 267
Algebraic multiplicity, 285
Annihilator, 309
Associated
  direct sum decomposition, 405
  eigenvector, 271, 274
  norm, 452
  norm of a matrix, 453
Augmented matrix, 174

Basis, 33
  cyclic, 430
  dual, 304
  orthogonal, 318
  standard, 38, 62, 166
Bijective mapping, 160
Bilinear form, 339
  complex, 353
  Hermitian complex, 356
  matrix of, 340
  rank of, 342
  skew-symmetric, 352
  symmetric, 346
Binary relation, 115
Bounded subset of R^n, 448

Canonical form
  for quadratic form, 334
  Jordan, 439
Cartesian product, 339
Cauchy–Schwarz inequality, 371
Change of variable
  linear, 315
  orthogonal, 319
Characteristic
  equation, 273
  matrix, 272
  polynomial, 273

root, 272
  value, 272
  vector, 272
Cimmino's method, 466
Closed subset of R^n, 448
Coefficient matrix, 174
Cofactor, 250
Column
  equivalent matrices, 116
  matrix, 67
  norm, 454
  operation, 95
  space, 125
  vectors, 125
Companion matrix, 257, 298
Complete set of projections, 405
Complex
  bilinear form, 353
  inner product space, 365
Components, 1
Conformable, 76
Congruence of matrices, 315
Conjugate
  of a matrix, 324
  transpose, 324
Conjunctive matrices, 356
Consistent system, 177
Continuous mapping, 448
Convergent
  sequence of vectors, 447
  series, 461
Coordinate, 38, 68
  matrix, 68
  projections, 304
Corresponding
  direct sum decomposition, 405
  eigenvector, 271, 274
  norm, 452
  norm of a matrix, 453
Cramer's rule, 259
Cross product, 32
Cyclic
  basis, 430
  subspace, 430

Determinant, 244
  Vandermonde, 257

Diagonal
  block matrix, 429
  elements, 67
  matrix, 67
Dimension, 38, 67, 143
Direct sum, 46
  associated, 405
Directed line segment, 23
Distance, 373
  between vectors, 32
Divergent sequence of vectors, 447
Division algorithm, 415
Divisor, 415
Dot product, 27
Dual
  basis, 304
  space, 304

Eigenspace, 282
Eigenvalue, 271, 273
  problem, 276
Eigenvector, 271, 274
Elementary
  column operation, 95
  matrix, 88, 104
  operation on vectors, 47
  row operation, 105
Equality
  of equations, 171
  of functions, 145
  of indexed sets, 34
  of mappings, 159
  of matrices, 67
  of n-tuples, 143
  of polynomials, 143
  of sequences, 144
  of vectors, 1
Equivalence, 124
  column, 116
  relation, 115
  row, 120
Equivalent
  matrices, 124
  systems of equations, 172
Euclidean space, 365

Fibonacci sequence, 280
Field, 141
Finite-dimensional vector space, 142
Form
    bilinear, 339
    complex bilinear, 353
    Hermitian, 358
    quadratic, 310, 350
    reduced column-echelon, 97
Frobenius inner product, 369
Function, 159

Gauss-Jordan elimination, 177
Geometric multiplicity, 282
Gram–Schmidt process, 319, 376
Greatest common divisor, 415

Hamilton–Cayley Theorem, 418
Hermitian
    complex bilinear form, 356
    congruent matrices, 356
    form, 358
    matrix, 356
    operator, 397
Homogeneous system, 179
Hyperplane, 466

Idempotent
    linear operator, 403
    matrix, 269
    principal, 425
Identity
    matrix, 67, 85
    operation, 48
Image, 159, 189
    inverse, 159, 189
Inconsistent system, 177
Indefinite quadratic form, 336
Index
    of Hermitian form, 358
    of an integer, 239
    of nilpotency of a matrix, 436
    of nilpotency of an operator, 424
    of a permutation, 239
    of quadratic form, 333
    of real symmetric matrix, 334
    set, 14
Inequality
    Cauchy–Schwarz, 371
    triangle, 372, 445
Infinite-dimensional vector space, 142
Injective mapping, 160

Inner product, 27, 365
    Frobenius, 369
    generated by a matrix, 367
    space, 365
    standard, 366
Invariant subspace, 421
Inverse
    image, 159, 189
    of elementary operation, 50
    of linear transformation, 233
    of matrix, 85
Invertible linear transformation, 233
Invertible matrix, 85
Isometric inner product spaces, 388
Isometry, 384, 388
Isomorphic vector spaces, 161
Isomorphism, 161

Jordan
    canonical form, 439
    canonical matrix, 437
    matrix, 437

Kernel, 197
Kronecker delta, 8, 47

Leading
    one, 59, 62, 97, 108
    variables, 176
Legendre polynomials, 378
    normalized, 380
Length, 28, 370
Limit
    of a sequence of vectors, 447
Linear
    change of variable, 315
    combination, 3
    dependence, 4
    functional, 301
    operator, 189
    transformation, 189
Linearly
    dependent, 3, 5, 8
    independent, 8
Lower triangular matrix, 133, 390
LU decomposition, 133, 134

Mapping, 159
    bijective, 160
    continuous, 448
    injective, 160
    one-to-one, 160
    onto, 160
    surjective, 160

Matrix, 67
    augmented, 174
    of bilinear form, 340
    characteristic, 272
    coefficient, 174
    column, 67
    companion, 257, 298
    of complex bilinear form, 354
    congruence, 315
    conjugate, 324
    conjugate transpose, 324
    conjunctive, 356
    of constants, 174
    coordinate, 68
    diagonal, 67
    diagonal block, 429
    elementary, 88
    Hermitian, 356
    Hermitian congruent, 356
    idempotent, 269
    identity, 67, 85
    inverse, 85
    invertible, 85
    Jordan, 437
    of linear transformation, 203
    lower triangular, 133, 390
    nilpotent, 269, 425
    nonsingular, 83
    normal, 393
    orthogonal, 269, 318
    polynomial, 416
    positive definite, 338
    positive semidefinite, 338
    probability, 281
    projection, 425
    of quadratic form, 312, 351
    reduced column-echelon, 97, 100
    reduced row-echelon, 108, 110
    row, 67
    scalar, 94
    sequence, 457
    series, 461
    similarity, 284
    singular, 83
    skew-Hermitian, 363
    skew-symmetric, 107, 269, 317
    square, 67
    symmetric, 107
    transformation, 191
    transition, 69
    triangular, 133
    unit lower triangular, 133
    unit upper triangular, 138
    unitary, 376
    of unknowns, 174

upper triangular, 83, 133, 389
zero, 67, 145
Metric, 373
space, 373
Minimal polynomial, 417
Minor, 250
Monic polynomial, 415
Multiplication
of linear transformations, 230
of matrices, 75
of a scalar and an equation, 171
of a scalar and a matrix, 94, 145
of a scalar and a vector, 2, 142
Multiplicity
algebraic, 285
geometric, 282

Natural ordering, 239
Negative
definite Hermitian form, 358
definite quadratic form, 336
semidefinite Hermitian form, 358
semidefinite quadratic form, 336
Nilpotent
linear operator, 424
matrix, 269, 425
Nonnegative linear operator, 415
Nonsingular
linear transformation, 233
matrix, 83
Norm, 28, 370, 373, 445, 452
associated, 452, 453
column, 454
corresponding, 452
row, 454
standard, 446
Normal
linear operator, 397
matrix, 393
Normalized
Legendre polynomials, 380
set of vectors, 318
Nullity, 179, 197
Nullspace, 179
Number field, 142

One-to-one mapping, 160
Order
of a determinant, 244
of a matrix, 67
Orthogonal
basis, 318
change of basis, 319
change of variable, 319
complement, 382

linear operator, 323, 384
matrix, 269, 318
operator, 384
projection, 382, 403
set, 30, 376
set of projections, 405
similarity, 389
vectors, 29, 30, 376
Orthogonally similar matrices, 389
Orthonormal set of vectors, 318, 376

Parallelogram rule, 24, 452
Parameters, 176
Partial sums of a series, 451, 461
Permutation, 239
Perp, 381
Polynomial
characteristic, 273
Legendre, 378
matrix, 416
minimal, 417
monic, 415
Positive
definite Hermitian form, 358
definite matrix, 338
definite quadratic form, 336
semidefinite Hermitian form, 358
semidefinite matrix, 338
semidefinite quadratic form, 336
Power series of matrices, 461
Principal idempotents, 425
Probability matrix, 281
Product
Cartesian, 339
cross, 32
dot, 27
inner, 27, 365
of linear transformations, 230
of matrices, 75
of a scalar and a linear
transformation, 192, 229
of a scalar and a matrix, 94, 145
of a scalar and a vector, 2, 142
scalar, 27
Projection, 30, 304, 403, 404
matrices, 425
orthogonal, 382, 403
scalar, 30
Proper
number, 272
value, 272
vector, 272

Quadratic form, 310, 350
indefinite, 336

index of, 333
matrix of, 351
negative definite, 336
negative semidefinite, 336
positive definite, 336
positive semidefinite, 336
rank of, 331
real, 310
signature of, 333

Range, 195
Rank
of bilinear form, 342
of linear transformation, 195
of matrix, 124
of real quadratic form, 331
Real
coordinate space, 1
coordinate vectors, 1
inner product space, 365
quadratic form, 310
Reduced
column-echelon form, 97, 100
row-echelon form, 108, 110
Reflexive property, 115
Relation
binary, 115
equivalence, 115
Restriction of a linear operator,
421
Row
equivalent matrices, 120
matrix, 67
norm, 454
operation, 105

Scalar, 1, 141
component, 30
matrix, 94
multiple, 2, 94
multiplication, 2, 142, 171
product, 27
projection, 30
Self-adjoint linear operator, 396
Sequence
convergent, 447
divergent, 447
limit of, 447
of matrices, 457
of vectors, 446
Series
convergent, 461
of matrices, 461
partial sums of a, 451, 461
power, 461

Series (*continued*)
    sum of, 451, 461
    of vectors, 451
Signature
    of Hermitian form, 358
    of quadratic form, 333
Similar matrices, 284
Singular matrix, 83
Skew-adjoint linear operator, 401
Skew-Hermitian
    linear operator, 401
    matrix, 363
Skew-symmetric
    bilinear form, 352
    linear operator, 401
    matrix, 107, 269, 317
Solution
    set, 171
    of a system of equations, 171
    trivial, 179
Space
    column, 125
    complex inner product, 365
    dual, 304
    Euclidean, 365
    inner product, 365
    metric, 373
    real coordinate, 1
    real inner product, 365
    unitary, 365
    vector, 142
Span, 18
Spectral decomposition
    of a linear operator, 411
    of a matrix, 425
Spectrum, 272, 273
Square matrix, 67

Square root
    of a linear operator, 415
    of a matrix, 338
Standard
    basis, 38
    basis of $\mathcal{P}_n$, 144
    basis of $\mathbf{R}_{m \times n}$, 146
    basis of subspace, 62, 166
    inner products, 366
    method of iteration, 463, 472
    norm, 446
Submatrix, 270
Subspace, 12, 152
    cyclic, 430
    invariant, 421
    spanned by a set, 19, 152
    zero, 13, 18, 33
Sum
    direct, 46
    of linear transformations, 192
    of matrices, 145
    of a series, 451, 461
    of subsets, 15
    of vectors, 2
Superdiagonal elements, 432
Surjective mapping, 160
Symmetric
    bilinear form, 346
    matrix, 107
    operator, 397
    points, 466
    property, 115
System
    consistent, 177
    homogeneous, 179
    inconsistent, 177

Trace, 83, 291, 302, 369
Transformation, 159

linear, 189
    matrix, 191
Transition matrix, 69
Transitive property, 115
Transpose, 107
Triangle inequality, 372, 445
Triangular matrix, 133
Trivial solution, 179

Unit
    lower triangular matrix, 133
    upper triangular matrix, 138
    vector, 31
Unitarily similar matrices, 389
Unitary
    matrix, 376
    operator, 384
    similarity, 389
    space, 365
Upper triangular matrix, 83, 133, 389

Vandermonde determinant, 257
Vector, 1, 142
    characteristic, 272
    column, 125
    component, 1, 30
    geometric interpretation, 23
    length, 370
    norm, 28, 370
    projection, 30
    space, 142
    sum, 2, 24
    unit, 31

Zero
    linear transformation, 190
    matrix, 67, 145
    subspace, 13, 18, 33